Catoen / Rees

Injection Mold Design Handbook

Bruce Catoen
Herbert Rees

Injection Mold Design Handbook

HANSER
Hanser Publishers, Munich

The Authors:

Bruce Catoen, Georgetown, ON, Canada; e-mail: catb0789@gmail.com

Herbert Rees, (deceased), formerly of Mono, ON, Canada

Carl Hanser Verlag
Postfach 86 04 20, 81631 Munich, Germany
eMail: info@hanser.de
www.hanserpublications.com
www.hanser-fachbuch.de

Editor: Dr. Mark Smith
Production Management: Der Buchmacher, Arthur Lenner, Windach
Cover picture: Courtesy of Husky Injection Molding Systems Ltd
Coverconcept: Marc Müller-Bremer, www.rebranding.de, Munich
Coverdesign: Max Kostopoulos
Typesetting: Eberl & Koesel Studio, Altusried-Krugzell, Germany

ISBN: 978-1-56990-815-0
E-Book ISBN: 978-1-56990-816-7

Preface

From 1990 to 2005 I worked together with Herbert Rees on editing and writing books and manuals for both Husky Injection Molding Systems, where I worked in various engineering capacities, and for Hanser Publishers. Herbert was passionate about molds, design and engineering. As the VP of engineering at Husky during some of its most formative years, Herbert worked closely with Husky's founder, Robert Schad, and together they developed many machine and mold technologies.

Herbert passed away peacefully on Saturday, September 18, 2010, at the age of 95.

Over the years Herbert repeatedly told me that "An injection mold is the heart of any plastics molding work cell. Since the objective of every molder is to produce as many good parts as possible, each and every day they MUST understand the process and details of designing an injection mold". Throughout my career I have seen this proven true again and again. Understanding the principles of an injection mold design is fundamental to the success of the molded product and the molding operation.

Mold design encompasses every aspect of mechanical engineering including dynamics, statics, thermodynamics, materials, heat transfer and stress. As a result of its broad application of engineering principles it is a difficult subject to master and it results in a long learning curve for engineers. Much of the learning I received during my career was tribal knowledge based on the application of engineering principles. Gaining this knowledge took being in the right place at the right time or learning by trial and error. My hope is that this book helps short-circuit the process of learning good mold design practices.

This book is designed to be a reference handbook for the mold designer, engineer, project manager and production manager. Since designing an injection mold all starts with the plastic part, the book will first focus on key features and details of plastics and the plastic part which are necessary for good mold design. The design of the main components of an injection mold will be discussed and good design practices, rules of thumb, and key calculations will be shared. More than 600 figures, images and tables are provided in the book to illustrate how a mold should be designed. Chapter 18 contains more than 40 reference mold designs graciously provided by mold-makers around the world. These references reinforce the previous chapters and illustrate how to apply the guidelines and principles from the book into a completed mold design. Finally, the process of testing and gaining customer acceptance of the mold for production will be detailed.

By using this book as a reference guide, the reader will be able to refer to it as needed to understand:

- Critical mold design features and design practices that will ensure a successful plastic part is molded
- Detailed steps, calculations and rules of thumb for mold design
- Critical aspects of mold design such as mold layout, mold shoe design, stack contruction, cooling ejection, runner systems and materials selection
- Plastic part design requirements for a good mold design
- Processes for testing and gaining acceptance of the mold for production.

Bruce Catoen, August 2021

Acknowledgments

There are many people who helped put this book together, and I am so grateful for the time and materials that were provided to me. I would like to thank my editors and reviewers: Mark Smith from Hanser; Glenn Anderson for reviewing Chapter 5 and his contacts throughout the industry; Dr. Davide Masato, Assistant Professor at University of Massachusetts Lowell, for reviewing Chapter 11 on ejection and Chapter 12 on cooling, as well as his additional contributions on surface finishes; Fabrice Fairy for reviewing Chapter 14 on melt distribution and his boundless enthusiasm for hot runners; Vince Travaglini for reviewing Chapter 2 on part design and the tremendous help from StackTeck Systems; Vince Lomax for reviewing Chapters 9, 10, and 11, his wisdom, and the additional materials he provided on mold testing and alignment; Jordan Robertson for reviewing Chapter 5; Randy Yakimishyn for reviewing Chapters 9 and 10; Rocky Huber for reviewing Chapter 11 on ejection; Don Smith for reviewing Chapters 11 and 12 and the reference mold designs in Chapter 18; Christian Krammer for reviewing Chapter 19 and the checklists; and Dr. David Barden from Clearly Scientific Limited for his excellent edits and suggestions. Thanks to Christina Fuges for her help in connecting me with mold-makers and other industry experts to review the materials in the book.

The book would not be what it is without the generous support of the many mold-makers, hot runner suppliers, software providers, and machine manufacturers for the use of their mold designs, figures and pictures. I would like to thank: Gene Altonen CTO of IMFLUX; Paulo Silva, Managing Director of Plasdan; Mario Haidlmair, CEO of Haidlmair; Adam Chuickshank, CEO of Fourmark; Peter Smith, CEO of DME; Vince Travaglini, President of StackTeck Systems; Rui Tocha, the Director General of Centimfe; Oliver Lindenberg, VP of Global Sales Moldmasters; Andy Stirn, VP of Injection Machinery and Aftermarket NA Milacron; Mike Ellis, Global Business Manager for Hot Runners at Husky Injection Molding Systems; Miki Bogar, Senior Manager of Global Marketing at Mold-Masters; Andreas Kliber, GM of FDU Hotrunner; Stefan Von Buren, GM of MHS; Angela Vitz-Schiergens, VP Hotset; and Don Smith.

I am very grateful for all the time and effort it took to provide high-resolution drawings and figures with explanations. In particular, I want to thank: Alberto Silva at Plasdan; Fabio Och at Fourmark; Beth Thompson at DME; Jordan Robertson at StackTeck; Peter Peschl at Haidlmair; Sylvia Schmidt from Hotset; Rob Irwin at Nypro Mold; and Brenton Huxel at iMFLUX. I would also like to thank Anthony Yang and Srikar Vallury at Moldex3D for the use of figures from the *Molding Simulation* book.

Very special thanks to Jigish Shah, who provided much needed assistance with the creation of many figures for the book.

I am so grateful to Jim Sykes for his contribution of Chapter 17 on dimensioning and tolerancing.

Many thanks to Willi Miller, John DiSimone, Gord Mackay, Denis Babin, Manfred Lausenhammer, Randy Yakimishyn, and all the design and mold assembly people who took the time to teach me about mold design over the years.

Special thanks to the family of Herbert Rees for the use of materials from his books.

Finally, I want to say a very special thanks to my wife Shelley for her support, encouragement, and patience during the time it took to write this book.

The Author

Bruce Catoen, B. A. Sc., Mechanical Engineering, P. Eng.

Bruce Catoen has more than 30 years of experience in the plastics industry and served as the Chief Technology Officer for Milacron and Mold Masters and as a senior executive at Husky Injection Molding Systems. Bruce is the named inventor on more than 50 patents and is author of the book *Selecting Injection Molds*. Through his consulting business, OASIC Consulting, Bruce advises senior executives on technology developments, business strategy, leadership and acquisitions. In addition to consulting, Bruce serves on two not-for-profit boards and mentors new engineers to Canada.

Contents

1 Introduction

Injection molding is a relatively new process compared to other manufacturing processes. However, in a very short period of a few decades injection molding has become one of the world's most productive and cost-effective means of producing a high-quality product.

Injection molding is a process by which plastic pellets are melted using heat and shear in an extruder and injected, at high pressure and flow, into an injection mold to form the part. While this process seems straightforward, it is full of engineering challenges and complexities.

An injection molding work cell can consist contain 4 to 16 separate elements (dryer, hopper loader, machine, hot runner, mold, robot, conveyer, etc.). Many of these elements will be standard catalog items. However, there will ALWAYS be one unique element in the work cell, and that is the mold. The mold is the **heart of the system** and all the other elements of the work cell must work together to make the unique plastic part. It is therefore critical and fundamental that the mold be designed with the utmost care and attention to detail, for if the mold does not operate as intended, then the entire work cell will operate in a subpar condition.

An old saying goes that "injection molders make money on weekends". The intended meaning is that a molder must run the first five days of the week to cover their costs, and they make profit when running throughout the weekend.

What is also written between the lines here is that injection molding is a 7-days-a-week, 24-hours-a-day business. Molders only make money when the molding work cell is producing parts. If the work cell cannot make a good part then efficiency is zero. *Molders only make money by putting good parts in the box. As a result, mold design is a critical aspect to every molded part.* Since a work cell could be in operation for many years, the mold must not only perform well on day one, but also until the last day that the production is needed.

Due to the importance of the mold design, it is critical that all levels of personnel in the molding plant and the mold-making facility understand the basics of good mold design, and the techniques used to create a mold that will allow it meet and exceed its intended purpose.

This book is therefore intended to be used, not just by the mold designer, but also by every person who comes in contact with the injection mold, so that they too can understand what makes a good mold and contribute in a meaningful way to building it.

With new, possibly difficult shapes, decisions on how to design the mold are usually left to the ingenuity of a mold designer. More frequently, *precedents* from earlier molds are used and re-applied. However, the mold designer and every person who will be involved in the molding operation must be aware of (and evaluate) new ideas, new methods, and developments, which when applied, would lead to better-quality, higher-productivity, simpler molds, and savings in the cost of the molded products.

Before proceeding with any mold design, the mold designer must understand what kind of mold should be selected. In other words, which features will be most suitable for the application to achieve the most economic *overall* manufacturing method for the product. This means not just specifying the number of cavities that will be required for the expected output, but also the *selection of mold materials and the degree of sophistication* of the mold. Any planned automation, especially in product handling *after molding*, can affect the mold layout, particularly spacing and orientation of the stacks. The mold designer must never lose sight of the ultimate goal: **to produce a part that meets or exceeds all specified requirements, at the lowest possible cost.**

The most important piece of information to know before deciding on the mold design is the quantity of parts to be molded. However, this is a piece of information, particularly with *new* products, that is often very difficult to obtain.

When looking at the overall cost of a plastic part, the per-piece cost of the mold is generally a few percent of the overall part cost (Figure 1.1 and Figure 1.2). However, the upfront cost of the mold may seem quite high. But due to the fact that it is a unique, one-off, engineered product made with very high-precision equipment to very tight tolerances by highly skilled tradespeople, the cost is realistic. On the other hand, a poorly engineered and manufactured mold is worthless, as it cannot produce a single good part.

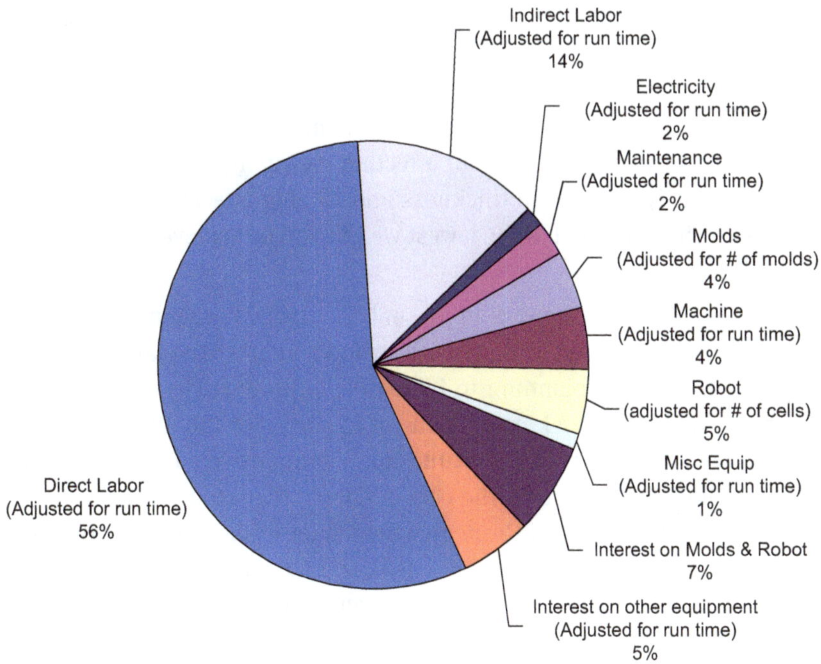

Figure 1.1 Conversion cost of an injection-molded medical part (resin excluded)

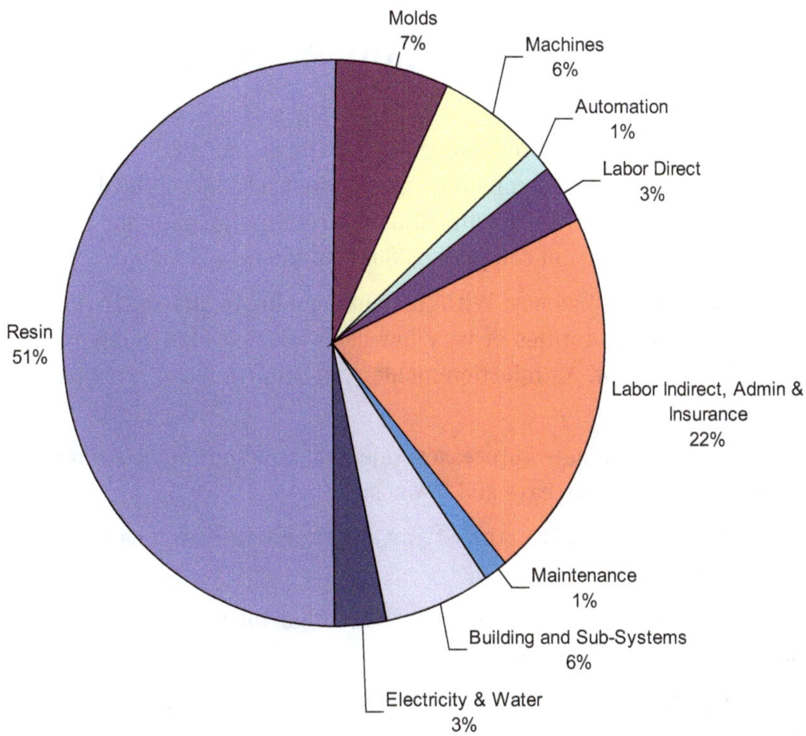

Figure 1.2 Part cost of an injection-molded pail

It should also be pointed out that of the total cost of almost all plastic products, the cost of the plastic material alone constitutes the greatest component. The most sophisticated, best-designed mold will not lower the cost of the product by as much as the reduction of just a few percent of the amount of plastic material, if it could be removed from the product without affecting its quality or serviceability. Most often, unnecessarily heavy wall thickness and ribbing affects the cost more than anything else. Chances are that the lowest weight will be achieved with the highest quality molds.

The foremost intent of this book is to present, in a logical sequence, the steps and choices available to the mold designer or decision maker when planning a mold for a new product, or when planning to increase the productivity for a product for which a mold exists. The book poses many of the questions that must be asked by anybody who needs a mold built. Any question left unanswered could significantly affect the productivity as well as the cost of a mold. For an experienced mold designer, the answers to many of these questions often come automatically, without being aware of the fact that a decision has been made. But even the most experienced mold designer can gain important information by systematically investigating all areas that can affect the design and the complexity of the mold, and checking to ensure that no obvious facts have been overlooked.

1.1 Benefits of Injection Molding

Today, injection molding is probably the most important method of processing plastics in the production of consumer and industrial goods, and is performed everywhere in the world. The benefits of using injection molding for a product or part of a product are vast and compelling. Some of the benefits are as follows:

- Low cost and high efficiency: With injection molding, a processor can produce parts in massive quantities at very low costs without high complexity or expensive skilled labor. An injection molder can set up a factory in a basic warehouse.
- Easily adapts to automated processes: Injection molding can be almost entirely automated with relative ease and low cost.
- Very high shape flexibility: Almost any shape and detail you can imagine can be injection-molded.
- Injection molding can produce parts with high tolerances and very highly detailed features or finishes.
- Excellent part properties such as light weight, high strength-to-weight ratio, excellent impact resistance, and low corrosion.

- Previously produced assemblies of multiple parts using other materials such as metal can be consolidated into one plastic part.
- Almost infinite color possibilities.
- Injection molding creates a net shape part without needing subsequent finishing.
- Plastics are easy, safe, and efficient to transport.
- Injection molding is a widely used and accepted process, so it is easy to find molders to make any part.

Before proceeding to use injection molding, the designer should always consider whether injection molding is the best solution to mold the part. Have alternative methods or product designs been considered or investigated, employing other manufacturing processes using the same or a similar materials, or using other materials which may permit a similar end product, possibly even with better quality, and/or at lower cost? A few typical examples of possible manufacturing alternatives to injection molding are:

- Thermoforming, foam molding, or blow-molding
- Coining and die-stamping (blanking)
- Machining.

The designer should also consider if other materials would be better suited to meeting the project objectives, such as:

- Paper (cardboard), wood, or cloth
- Metals (steel, aluminum, etc.)
- Glass or ceramic.

Once the decision has been made to use *injection molding* for a new product, a number of critical steps lie ahead, which will be addressed in this book:

- Plastic part design
- Factors affecting the design of an injection mold
- Mold design
- Testing and acceptance.

1.2 The Injection Mold

The heart of every injection molding work cell is the mold. It contains the form of the part that will ultimately be filled with plastic. It plays the most critical and fundamental role in the entire process - forming the part. The mold forms the desired end product. All of the other pieces of equipment in the molding work cell work to help produce a high-quality part. However, most of the other pieces of equipment in the work cell act in support of the mold.

1.2.1 The Role of the Injection Mold

Today, an injection molding work cell can contain up to 16 discrete pieces of equipment (see Figure 1.3). These devices all serve to help make a good-quality plastic part. In essence, the mold is the heart of the system, as all the other generic devices in the work cell allow the mold to make a good part. While the generic components of the work cell do not change, they must adapt every time a new mold is installed in the injection molding machine. It is therefore critical to understand the basics of an injection mold and what makes a good plastic part.

Figure 1.3 Fully automated injection molding work cell for DVD cases, containing 16 components in the work cell (Courtesy of Husky Injection Molding Systems Ltd.)

All of the supporting devices to the mold need to be correctly sized and functional in order to make the mold work well. *An injection molding work cell can be considered to be like a symphony orchestra, with the mold being the conductor. If the devices are not in tune with each other, then the whole work cell sounds like a bad high-school band, and they each contribute to making poor-quality plastic parts.* If, on the other hand, all of the supporting equipment is good working order, is correctly sized and maintained, then the resulting plastic parts will be better quality, and the work cell will run for longer without issues.

> The old saying "the chain is only as strong as the weakest link" holds true in injection molding.

1.3 What Is an Injection Mold?

An injection mold is a permanent tool, i.e., a tool that, if properly designed, constructed, and maintained will have a life expectancy (useful life) *well beyond* the time where the product itself becomes obsolete. This differentiates it from a "one-time use" mold such as a sand-casting mold, as used in metal foundries. A mold can be used to make products in a virtually infinite variety of shapes, made from injectable plastics. Common to all molds is the condition that it must be possible to remove the product after molding, without the need to destroy the mold (as is the case in sand-castings).

> There is an exception to this, the so-called "lost-core molding": There are injection molds for intricate products, such as intake manifolds for internal combustion engines, previously made from cast iron, which have an outside shape that *can* be molded with conventional (permanent "open and close") molds, but where the intricate *inside shape* is made from a molded, low-melting-point metal composite, which is inserted into the mold before injection, and then ejected together with the molded product. The metal is then removed by heat at a temperature above the melting point of the insert, but of course below the melting point of the plastic used for this product. The molded metal insert is thereby destroyed, but the metal will be reused.

A basic mold consists of two mold halves, with at least one cavity in one mold half, and a matching core in the other mold half. These two halves meet at a *parting plane* (parting line). Once the injected plastic is sufficiently cooled, the mold opens and the product can be removed by hand or be automatically ejected.

Because injection molding machines are mostly built with the injection on the stationary platen side, there is typically no built-in ejection mechanism on this side. If ejection from the injection side should be required - which is always the case in *stack molds*, and occasionally so in single-level molds - any required mechanism must be added to the mold, and occasionally to the machine; in either case, this adds complexity and increases costs. Only molds designed for using only air ejection do not require any external ejection mechanism.

Most products are removed (ejected) from the core. There are also many molds that need special provisions to allow the products to be removed from either the cavity or the core. This is the case for products with severe undercuts or recesses on the inside and/or the outside of the product, such as screw threads, holes, ribs or openings in the sides of the product, etc., or molds for insert molding.

Some of these design features of the product may require moving side cores, which are either inserts or whole sections of the cavity that move at an angle which is 90° to the "natural opening path" of the mold. Others may require special unscrewing mechanisms, either in the core or in the cavity side. The mold may require split cavities (or "splits"), i. e., the cavity consists of two or more sections, which are mechanically or hydraulically moved in and out of position, and then clamped together during injection. In some cases, the mold may require collapsible cores, or retractable inserts, which are all quite complicated (and expensive) methods.

Any of the above special features can add considerably to the mold cost when compared to a simple "up and down" mold where the products can be readily ejected with the machine ejectors during the mold opening stroke or after the mold is open, without the need for any of these complicated mold features.

Note that in this book, the term (simple) "up and down" molding is used, which comes from the earlier vertical molding machines, even though, today, most general-purpose injection molding machines are horizontal, and the mold opens and closes in a horizontal motion.

Example:

To illustrate how different mold features affect the mold cost, let a single-face mold with air ejection of the products cost *X* dollars. A similar mold, but with mechanical ejection, costs about 1.2*X*. A similar, air-ejected two-level stack mold will be about 1.8*X*. An unscrewing mold for a similar-size mold and product will cost about 2*X*.

1.3.1 Elements of an Injection Mold

There are books that show designs of numerous specific molds, but it is virtually impossible to show every possible configuration that may be required. It is more important for the designer, and any person requesting a new mold, to understand that a mold consists essentially of a number of elements, from which the most appropriate for the purpose is chosen.

The reader is encouraged to read one of the following books for more a more detailed understanding of the engineering of an injection mold:

- *Gastrow Injection Molds* [1]
- *Injection Mold Design Engineering* [2].

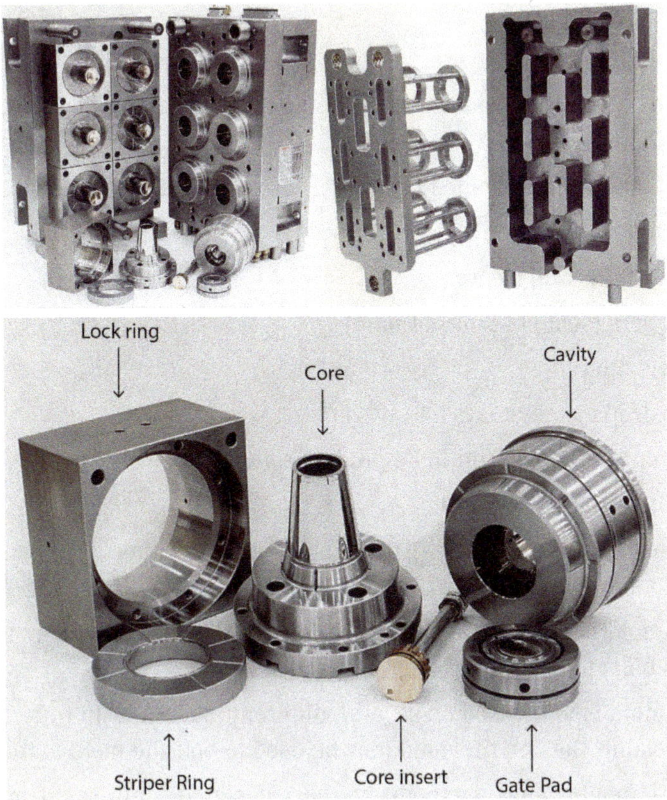

Figure 1.4 Parts of an injection mold (Courtesy of Husky Injection Molding Systems Ltd.)

[1] P. Unger (ed.), *Gastrow Injection Molds* (4th edn), Hanser Publishing, 2006.

[2] D. Kazmer, *Injection Mold Design Engineering,* (2nd edn) Hanser Publishing, 2016.

Every injection mold consists of the following basic elements (see Figure 1.4 and Figure 1.5):

1. One or more matching cavities and cores, defining the cavity space(s) (today, there are molds with anywhere between one and 256 cavities).
2. A method, or element, to duct the (hot) plastic from the machine nozzle to the cavity spaces. There is a choice between:
   - Cold runners (two-plate or three-plate systems)
   - Hot runners (various systems)
   - Insulated runners
   - Sprue gating (cold or hot).
3. Provision to evacuate air from the mold (venting). There is a choice between:
   - Natural venting
   - Vacuum venting.
4. Provision to cool the injected hot plastic sufficiently to allow ejection of the molded product.
5. Provision to eject the molded product. There is a choice between:
   - Manual product removal
   - Ejector pins and sleeves
   - Strippers (stripper rings or bars)
   - Air ejection
   - Free-drop ejection onto a conveyer
   - Various methods for in-mold product removal
   - Robotic product removal.
6. Provision to attach (interface) the mold to the molding machine. There are several methods to consider:
   - The mold is for one machine only. In this case the mold may be mounted with bolts to the platen
   - The mold is to be used on several, different machines. In this case, clamps and clamp slots on the mold may be used to bolt the mold to the platen
   - Quick mold-change methods (various designs). This could involve magnetic mounting.
7. Method of alignments of cavities and cores. There are several methods to consider:
   - No alignment feature provided in the mold. Also called flat parting line
   - Leader pins and bushings (2, 3, or 4)

- Leader pins and bushings between individual cavities and cores
- Taper fits between individual cavities and cores
- Taper fits between plates. These are also called side locks
- Any combination of the above.

8. Any number of (mold) plates to provide the necessary means for carrying and providing rigid back-up for the above elements.

In addition to the above parts, molds can have additional features, which will also be discussed in the following chapters. Each of these features can add (often considerable) costs to the mold, but in many cases they increase the productivity of the mold and reduce the cost of the product. Not all may be necessary, and each must be carefully considered when deciding on the type of mold that is most suitable (and most economical) for the job on hand.

Easy serviceability of the mold is important but often overlooked. It adds some mold cost, but saves much more in future servicing costs and downtime.

Ease of serviceability of the mold may affect the mold cost up front, but will ultimately reduce the lifecycle cost of the plastic part by reducing the need to remove the mold for service or repair. One example is the access to the hot runner for cleaning plugged gates or making minor repairs, such as changing a nozzle, a burned-out heater, or a faulty thermocouple at a hot runner drop. Building in functionality to conduct these repairs in the molding machine will cost more in the initial mold, but this will be easily recouped by reducing the downtime necessary to accomplish such repairs. By designing easy access to these components in the machine (without the need to remove the whole mold, or part of it, to the bench), such repairs can be made in less than an hour, instead of taking several hours. This work can also be done by the mold setup staff rather than getting the (expensive) mold makers involved.

Another area where valuable maintenance time can be saved is to design and provide easy access from the parting line to screws holding modular molding surface parts to their mounting plates, while the mold is in the machine. Since damage to the molding surfaces or parting lines can occur, it is advantageous to have the molding surfaces serviceable in the press. This is particularly valuable in high-cavitation molds.

Even minor changes to the part can dramatically lower or increase mold costs.

Defining what is *really* required considering the shape and complexity of the product and the required production quantities will enhance mold productivity. Alternatives and options should ALWAYS be considered and reviewed with all personnel before the mold design is finalized. Each department that interfaces with the mold will look at the design with a different set of eyes for their needs. It is of the utmost importance to include them in design reviews to ensure that the mold will have all the necessary features and functions.

Figure 1.5 shows a schematic of a basic injection mold with the key elements of the mold labelled with conventional terminology. It should be noted that the terminology used in the figure is used by the author, but there could be other names used for these components as well.

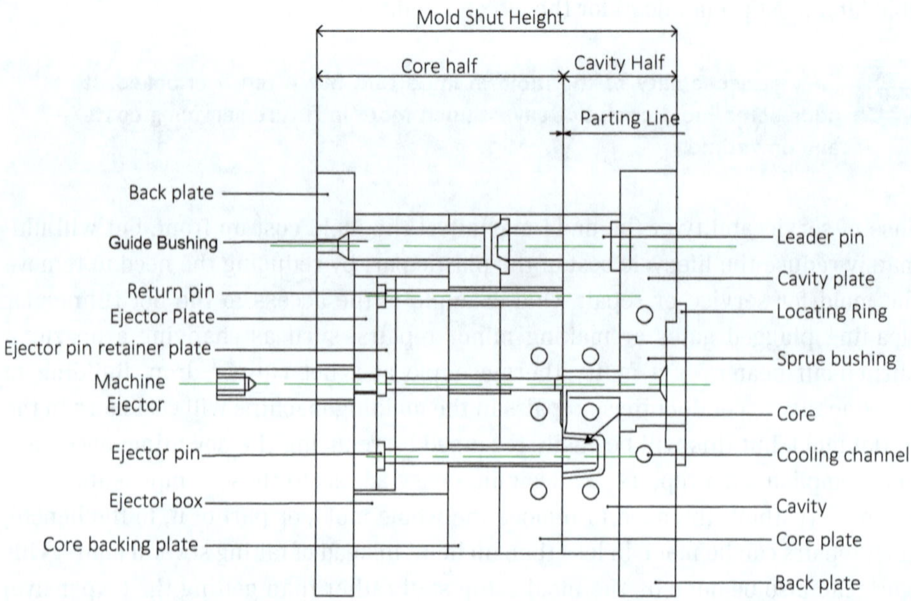

Figure 1.5 Mold terminology

■ 1.4 Classification of Molds

SPI has developed a set of standards to classify molds by their design and intended usage. Table 1.1 explains the different standards.

Table 1.1 SPI Mold Specifications [a]

Class	Cycles	Description	Mold base	Inserts	Other
101	> 1,000,000	Built for extremely high production. This is the highest-priced mold and is made with only the highest-quality materials.	Pre-hardened 28Rc steel Stainless steel plates	All hardened > 48Rc steel Cooled inserts	Guided ejection Wear plates on slides
102	< 1,000,000	Medium to high production mold, good for abrasive materials and/or parts requiring close tolerances.	28Rc steel	Hardened steels Cooled inserts	Some guided components Some corrosion protection
103	< 500,000	Medium production mold. This is a very popular mold for low to medium production needs.	8Rc steel	> 28Rc steel	Guiding optional
104	< 100,000	Low production mold. Used only for limited production preferably with non-abrasive materials.	Mild steel or Al	Mild steel or Al	None
105	< 500	Prototype only. This mold will be constructed in the least expensive manner possible to produce a very limited quantity of prototype parts. It may be constructed from cast metal or epoxy or any other material offering sufficient strength to produce minimum prototype pieces.	Mild steel or Al	Mild steel or Al	None

[a] For more details on mold materials and the use of the Rockwell hardness scale (Rc), please refer to Section 15.2.

■ 1.5 Continued Innovation in Molds and Hot Runners

While the use of molds dates back thousands of years, innovation continues in mold and hot runner design. There are thousands of patents on injection molding, and thousands more just on hot runners. The industry continues to innovate to provide customers with ever better ways to mold plastic products. The reader is encouraged to keep aware of the emerging trends in injection molding and to learn about how these new ideas could help to create a better injection mold. Some of the most recent trends are:

- *Conformal cooling of inserts using metal 3D printing:* Allows mold designers a additional level of design freedom in creating the cooling circuit, compared to drilling and milling multiple complex inserts and materials.
- *Direct 3D printing of plastic parts versus injection molding:* A potential threat to injection molding itself, it allows for customized creation of parts for joint replacements, running shoes, and other items that require high levels of customization.
- *Electrification of molding functions using servo motors and drives:* Functions such as rotations and stroking of pistons are now being electrified.
- *Internet connection of devices to make them "smarter",* called the industrial internet of things (IIoT or Industry 4.0).
- *Multi-material molding:* The creation of a part with multiple materials in a single process, e.g. toothbrushes, parts with integrated sealing, or parts with multiple colors.
- *Co-injection molding:* The creation of a part with multiple layers for extending shelf life, using recycled materials, and creating new aesthetics.
- *Use of gasses and liquids in the process* to core-out thick parts or to add/embed the gasses in the part for light-weighting.
- Continued development of *new resins and fillers* to create better plastics.

■ 1.6 The Injection Molding Machine

The accuracy of molding, and especially when molding products that are difficult to produce, is very dependent on the quality of the molding machine, its mechanical rigidity, accuracy of alignment, parallelism of platens, the quality of its controls, and the state of maintenance. As mentioned previously, the equipment in a molding work cell works together in unison, and the system is only as good as its weakest link. So a high-quality mold installed in a molding machine that is poorly set up or engineered will not make a good part. The machine must be able to meet the requirements of the mold that is being installed. A good machine, poorly aligned and maintained, can destroy a new mold in a matter of months. It is imperative that the molder's machine is in good shape to ensure that the mold will perform as intended, for the lifetime intended. If this is not the case, the mold may suffer from continuous problems and issues that cannot be rectified.

There is no point in buying a premium-priced mold only to run it in an out-dated machine.

Every good injection molding machine consists of the following basic elements (see Figure 1.6):

1. A rigid base that is welded (not bolted) together using stiff box steel members.
2. A rigid clamping unit, consisting of two cast or machined platens, for the mounting of the mold halves and provisions for guiding the platens (tie bars or linear ways). The thickness of the platens is a good indicator of the rigidity and quality of the machine.
3. Provision for moving the platens, preferably fast relative to each other, for opening and closing the mold in an adjustable fashion using a fast microprocessor. Toggle-style machines tend to be the quickest machines, and fully hydraulic machines the slowest.
4. Provision for clamping, i.e., holding the mold shut against the force of the injection pressures within the mold (in some machines, provisions 3 and 4 are combined). Clamp force can be built up using hydraulic fluids or electric motors.
5. Provision for ejecting the molded product(s) from the mold. The provision can be within the mold or, more commonly, using an ejection means on the machine such as an ejector plate behind the moving platen.
6. Provision to transform the raw plastic pellets into an injectable melt. This part of the machine is called the plasticizing unit or extruder. This is almost always done using a barrel and rotating screw. The melting of the plastic is done using a combination of shear from the rotating screw and heat from the barrel heaters.
7. Provision for injecting the melt into the mold (in most machines, provisions 6 and 7 are combined in one unit). Sometimes the creation of the melt and injection of the melt are split into two separate elements. In this case the injection unit is normally called a two-stage injection unit. The injection stage is then

normally an injection or shooting pot. The injectable melt is transferred to the pot by the extruder, and a separate means then injects the melt into the mold for the shooting pot. The advantage of this more expensive approach is that the injection is more accurate and the extruder can be creating more melt while the shooting pot is injecting plastic. This can result in significantly lower cycle times.

8. Cycle controls (sequencing logic, timers, etc.) and an interface for the operator to make adjustments to the process and to operate the machine in manual or semi-automatic modes.
9. Heat controls for all heaters in machines and molds. Some machines have a limited number of heat controls, and additional controls could be required for the molds, especially with larger hot runner systems. This point must be considered when estimating the mold cost.
10. Safety gates to protect operators and bystanders from all hazards when operating the machine.
11. Mechanical safety elements to prevent closing the machine when gates are open, in case of failures of the other (electric and hydraulic) safety measures.
12. Provision for cooling water distribution to the mold.
13. Provision for compressed air, for auxiliary actions required in the mold.

There are other features available, for example, for the convenience of quick mold installation, automation, set-up and operation of the mold and machine. These features are often offered as options that can be bought with the machine or added on later.

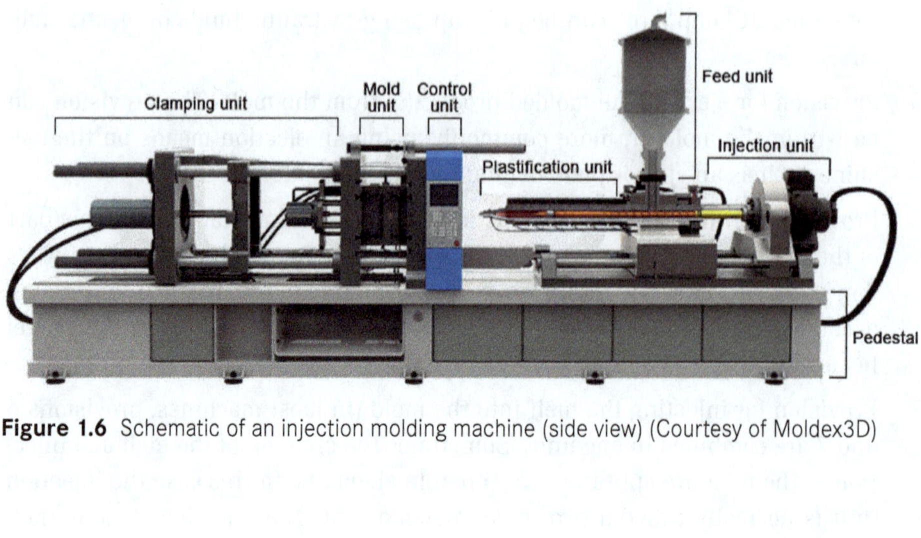

Figure 1.6 Schematic of an injection molding machine (side view) (Courtesy of Moldex3D)

2 Overview of Plastics for Mold Design

2.1 What Is Plastic?

Before considering the details of the product design, the mold designer must understand some details about the chemistry and characteristics of plastics used in injection molding.

A plastic is one of a large and varied group of materials consisting wholly or in part of carbon, hydrogen, oxygen, and other organic and inorganic elements. While solid in the finished state, at some stage in their manufacture plastics are melted (usually using heat and pressure), and thus capable of being formed into various shapes.

The process by which plastics are manufactured is called polymerization. Plastics are derived mostly from fractions of gas or petroleum recovered in the refining chemical reaction process. From these basic sources comes the monomer (a single "mer"), which is the basic hydrocarbon unit from which polymers are built. The monomer is then subjected to polymerization, by which these small molecules link together into huge molecules. Thus, the chemical reaction turns the monomer into a polymer.

A polymer may be defined as a high molecular weight distribution compound that contains comparatively simple re-occurring units. The way in which a monomer polymerizes (e.g. with chains packed close, or with chains separated by side branches) can affect its properties. Also, the size of the polymer chains and the type of molecule polymerized (e.g. substituting methyl groups for benzene rings) can affect the properties of the plastic. Polymerizing two or more different monomers together (called co-polymerization) or adding various chemicals to the basic resin can make drastic changes to the polymer's properties.

In general then, feedstocks (known as monomers) are polymerized by a chemical reaction into polymers, and the polymer is then made into useful forms for processing (e.g. granules, pellets, powder, etc.). Most plastics fall into one of two categories – thermosets or thermoplastics.

Thermoplastic resins have long molecules that are branched to one another by weak forces called van der Waals forces. Thermoplastics can be repeatedly softened and hardened by heating and cooling. An analogy would be a block of ice which can be softened, poured into a cavity of any shape, and cooled to become a solid. Examples are polypropylene, nylon, and polystyrene.

Thermoset resins consist of molecules which form cross-links with other molecules during polymerization. The chains are thus bonded together to form three-dimensional network. Once polymerized (or hardened), the material cannot be softened by heating without degrading the plastic. An analogy would be a hard-boiled egg, which has turned from a liquid to a solid during cooking, but cannot be converted back. Examples are epoxy, silicones, and phenolics. Thermoplastic resins are used almost exclusively in the injection molding industry.

There are literally thousands of thermoplastic resins commercially available today. The physical properties as well as the processing parameters used to mold these resins can vary dramatically. Appendices 8–11 list the general properties and processing conditions necessary for most of the resins used in injection molding today. These guidelines should be used for reference only: in all cases, the supplier should be contacted to obtain detailed information about a given resin.

2.2 Plastics Terminology

The following are some of the main terms used to describe the characteristics of a particular grade of injection molding resin.

Molecular Weight (Distribution)

Not all molecules are the same size. A molecular weight distribution gives the engineer a normally distributed average molecular weight. A material with a narrow molecular weight distribution (NMWD) has a molecule size which is well-controlled (i.e., a small standard deviation). On the other hand, a material with a broad molecular weight distribution (BMWD) is one which could have molecular chains of radically different sizes. Therefore, a NMWD resin will have a more consistent and usually enhanced performance compared to a BMWD resin.

Linear versus Branched Polymer

A linear polymer has no side-branching, while a branched polymer has a three-dimensional order. Note that branches of a polymer do not link with another molecule. Linking is the reaction which takes place during the molding of a thermoset resin. Linking does not occur in thermoplastic resins without degradation – a cross-linked thermoplastic resin would be basically broken down into carbon.

Homopolymer versus Copolymer

Homopolymers contain only one repeating unit, while copolymers have two or more monomer units incorporated into a single polymer. There are several types of copolymers. The second monomer and the distribution of it within the polymer can vary the material properties dramatically. ABS is a good example of a copolymer. Random copolymers show characteristics that are intermediate between the two. See Figure 2.1.

Homopolymer	-AAAAAAAAAAAAAAAAAA-
Linear copolymer	-ABABABABABABABA-
Random copolymer	-ABBAABABBABA-
Block copolymer	-AAA-BBB-AAA-BBB-
Graft copolymer	-AAAAAAAAAAAAAAAAAA- B B B

Figure 2.1 Schematic of some common types of polymers

Nucleated Polymers

Nucleating agents accelerate the formation of crystallized regions, by serving as a nucleus from which the crystallization can begin. A nucleated polymer will have many small spherulites instead of fewer larger ones. The most common nucleating agents are talc, silica, kaolin, and other polymers.

Crystalline versus Amorphous Polymers

Crystalline polymers have quite different characteristics from amorphous polymers. Table 2.1 compares the characteristics of the two materials. The materials also behave differently on heating and cooling compared to amorphous plastics.

Table 2.1 Comparison of Crystalline and Amorphous Polymers

Property	Crystalline	Amorphous
Transparent?	No	Yes
Resistance to chemicals?	Excellent	Poor
Prone to stress-crazing?	No	Yes
Propensity to shrinkage?	High	Low
Strength	High	Low
Viscosity	Low	High
Melting point	Sharply defined	Range
Volume change	Greater	Lower
Heat content	Greater	Lower

2.3 Polymer Orientation

When resin is injected into a mold, the coiled chains must untangle and orient themselves parallel to the flow axis in order to slide past each other easily. The smaller the orifice the resin has to flow through, the higher the orientation. Therefore, the gate area is usually the area with the highest orientation.

Orientation varies greatly in the cross-section of the molded part. When the material enters the cavity, the resin that hits the cold wall and freezes retains high levels of orientation. The resin at the center of the wall section can relax and shrink during the cooling stages of the cycle. Orientation increases properties in the flow direction, but decreases the properties in the transverse direction. Orientation due to differential cooling can be the major cause of plastic part warpage.

2.3.1 Shrinkage

One of the most misunderstood areas of mold design is shrinkage. Every material (metals, plastics, gases, liquids) expands as its temperature increases (heat expansion) and returns to its original volume, if cooled down to the original temperature. The problem with all plastics is the additional characteristic of compressibility. All solid materials compress under load, but most not as much as plastics. When pressure is applied to plastics (or to hydraulic oil, but not to water), plastics will compress significantly (i. e., reduce in volume) in proportion to the amount of pressure applied. This may be (within the range of molding operations) as high as 2% of the original volume.

Thus, we now have two conditions that work against each other: heat expansion and compressibility. As the plastic is injected, it is both hot and therefore expanded, but also under significant pressure, which reduces its volume. This makes it very difficult to arrive at a true shrinkage factor, because the actual change in volume depends on the type of plastic, the melt temperature, the injection pressure required to fill the cavity space, and the temperature at which it will be ejected from the mold.

Shrinkage is affected by many variables in the molding process. Some of the more important factors are:

- Type of plastic being injected
- Any additives or fillers added to the plastic material
- Heating and cooling
- Injection pressure
- Ejection temperature
- Variation of temperature across the ejected part.

For practical purposes, and for many products and molds, the shrinkage factors supplied by materials suppliers can be used. However, these figures indicate only a range within which to choose, usually between 0 and 5%. In some cases, where the volume or size of a product is important, this is not accurate enough. With crystalline plastics, such as polyethylene (PE), polypropylene (PP), and polyamide (nylon), the shrinkage factor is much higher than with amorphous plastics, such as polystyrene (PS) and polycarbonate (PC). Plastics filled with inert substances, such as glass, carbon fibers, or talcum, have a much lower shrinkage than that for the same but unfilled material. Shrinkage figures should be obtained from material suppliers, for guiding purposes.

In order to mold parts at an economically viable rate, they should be ejected as soon as possible without damaging the product. At this stage the product is still warm, and will shrink outside the mold, sometimes fast, sometimes slowly, depending on the ambient temperature and the plastic's characteristics.

There are also other considerations, such as the shape of the product. For example, if there are large differences in wall thickness or the shape is non-symmetrical, the product will cool rapidly in some areas, while remaining hot in others. This may result in warping, sink marks, or voids due to uneven shrinking once the product is ejected, because there will be still pockets of hot plastic under the cooler skin, which continue to contract, and pull on the already cooled areas. Shrinkage will also be greater in the direction of the plastic flow than across it. Due to all these factors, it is very difficult to predict shrinkages accurately in injection-molded parts. As a result, the kind of tight tolerances used in the metal industry cannot be maintained.

Mold makers generally do not guarantee the shrinkage values for the parts due to the unpredictable nature of plastic shrinkage. However, historical data can be used to guide the customer, or if necessary a prototype mold can be built. Shrinkage of height, diameter, and in the lip area may all be different. The following shrinkage sizes are commonly used but are guidelines only. The designer or project engineer should see material suppliers for more detailed and specific information for their application.

Table 2.2 General Shrinkage Guidelines (see material suppliers for detailed numbers)

Polymer	Shrinkage
PS	0.5%
PP	1.8–2.0%
PE	1.6–1.8%
LLDPE	1.5%

CAE analysis and simulation tools can be very helpful in determining where areas of high potential shrinkage may occur. The designer can then focus on this area to improve the cooling or part design to eliminate the issue. Figure 2.2 shows a simulation of a computer screen. The areas in red indicate areas of high shrinkage. The designer can then focus on these areas for improvement in the mold design before the mold is built, or they can apply a different localized shrinkage to compensate for the differences.

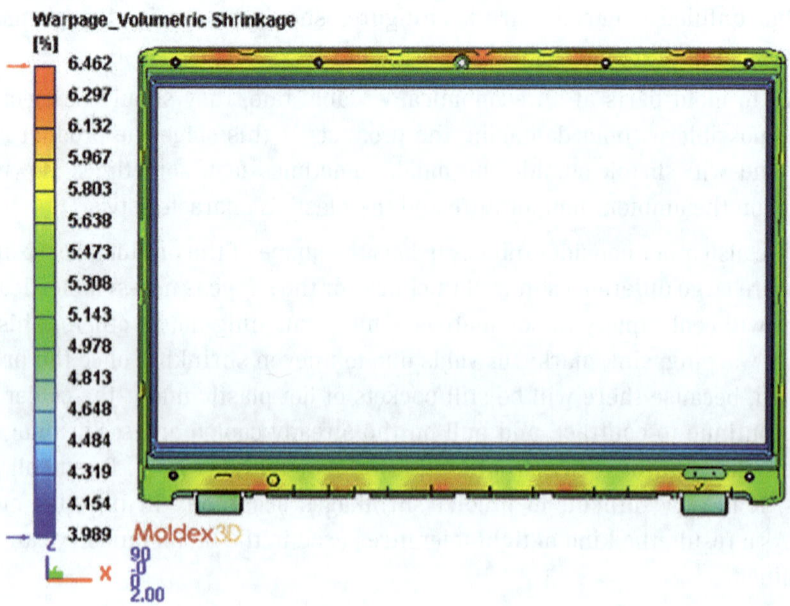

Figure 2.2 CAE prediction of shrinkage [1] (Courtesy of Moldex3D)

2.3.2 Multiple Shrinkages

The areas within the cavity spaces close to the gate see higher pressures, so the shrinkage there will be less and will require a smaller shrinkage factor. Conversely, near the end of the flow through the narrow cavity space, the pressure in the plastic is much lower than near the gate, and a higher shrinkage factor will apply. In some applications, more than two shrinkage factors may have to be selected within one cavity.

It is also important to establish at what temperature the product will be ejected. If it is ejected while still hot, it will shrink more outside of the cavity space as it cools

[1] Source: M. Wang, R. Chang, C. Hsu, *Molding Simulation: Theory and Practice,* Hanser Publishing, 2018 (Figure 3.22).

to room temperature. If ejected later, when it is cooler, it will shrink less, as measured in comparison with the steel sizes of the cavity and core. This is sometimes, but uneconomically, used to arrive at the proper size of a product such as a container or lid. If a molded product is too small because an insufficient shrinkage value was added to the product dimensions when specifying the mold steel dimensions, the proper product size can be achieved by ejecting it later, when it is cooler, but this means a loss in productivity. With high-production setups, the proper procedure is to resize the mold.

2.4 Additives

Below are some additives that are commonly added to plastics to improve some aspect of their function. As these additives can have significant effects on how the plastic behaves, it is important to know exactly what additives will be in the resin to be molded.

Plasticizers are monomeric liquids or low-melting solids of low volatility that can be used to improve the flow of the resin in the molten state. Plasticizers make the polymer more flexible, increase elongation, and lower the glass transition temperature. As a result, thinner-wall sections can be filled when these additives are used.

Fillers such as iron, copper or talc are used to increase the frictional heating during injection and improve the thermal conductivity during cooling. This results in a resin which can be molded at shorter cycles.

Lubricants are used to facilitate removal of the part from the mold, and to increase the smoothness and gloss of the molded part.

Antistats: Since plastics are excellent electrical insulators, they hold a static change very well. The surface conductivity of plastics is 1020 times greater than metals. As a result, any surface charge will dissipate slowly and attract dust and contamination. Antistats such as glycols and amines are used to combat these phenomena.

Reinforcements are used to strengthen or reinforce the properties of the virgin resin. A reinforced resin is basically a composite in which the fiber is the reinforcement and the resin is the matrix. Reinforcements used are usually very stiff and strong but brittle (e.g. glass and carbon fiber).

Colorants are used to modify the color of the molded part. Colorants are classified as either dyes or pigments. Pigments are used most often in injection molding. The color can affect the viscosity of the material, since the carrier for the pigment is usually a low-viscosity material that melts easily to allow for better dispersion of the color.

Flame retardants such as antimony trioxide are used to limit the onset of ignition and smoke generation of the resin.

Heat stabilizers are used to resist melting of the plastics. Since plastics soften and melt at a relatively low temperature they cannot generally be used in high temperature applications for extended periods.

UV stabilizers such as hydroxybenzophenes and dialkyldithiocarbonates prevent the breakdown of polymers when exposed to ultra-violet light. Ultra-violet light is absorbed by the polymer when it is exposed to sunlight. If the plastic absorbs enough energy to break the chemical bonds the resin can degrade. For outdoor applications, a UV stabilizer is required to avoid degradation.

Impact modifiers are rubbery, high-MW polymers that are added to the virgin polymer in large quantities. The result is a polymer which can resist far more shock than the virgin resin. Examples of such resins are impact-modified EVA and ABS.

Nucleating agents improve the onset of crystallinity by supplying the molten resin with a nucleus (spherulites) from which the crystal can grow.

■ 2.5 Mechanical Properties of Plastics

A detailed list of plastic properties and processing conditions can be found in Appendices 8–11.

Tensile strength (also called the ultimate strength) is the maximum load the resin can take per unit of cross-section. In tensile testing, it is the pulling stress necessary to break the specimen.

Yield strength is the lowest stress the material can handle without experiencing plastic (permanent) deformation. Plastics in general do not show a well defined yield point because of the high elongation that the polymer can withstand.

Flexural strength is the strength of the tensile bar specimen in bending. It is the maximum tensile strength the surface fibers of the resin can withstand prior to failure.

Izod impact strength is a measurement of the resistance to shock loading. It is determined by holding a notched specimen at one end, striking it, and determining the energy absorbed.

Heat deflection temperature (HDT) is the temperature at which a standard test bar of the resin deflects by a specified amount under a static load.

All of the details of these tests can be found in the ASTM standard test procedures.

NOTE: When looking at the properties of plastics, keep an open mind to accept apparent contradictions that make it difficult to pin common labels on different families of plastics. Even within various types of a single family there can be large differences. For example, low-density polyethylene (LDPE) is flexible while high-density polyethylene (HDPE) is rigid and tough (but it can also be flexible!).

2.6 How Molten Plastics Behave

The study of flow and deformation of viscous fluids (called rheology) is used to determine the state of the material in the hot runner, cold runner or in the cavity. In order to understand how plastics behave when they flow (the rheology of the plastic), one must understand the principles of viscous flow relationships.

2.6.1 How Plastics Flow

Consider two parallel plates, one of which is moving with respect to the other (see Figure 2.3). This relative movement will cause a flow to be set up within the fluid. The fluid which is next to the moving plate will assume the velocity of the moving plate. The velocities of the intermediate layers will be in proportion to their respective distances from the plate at rest.

$$u = V \times (y/d) \quad \text{(see Figure 2.3)} \tag{2.1}$$

where
V = average velocity

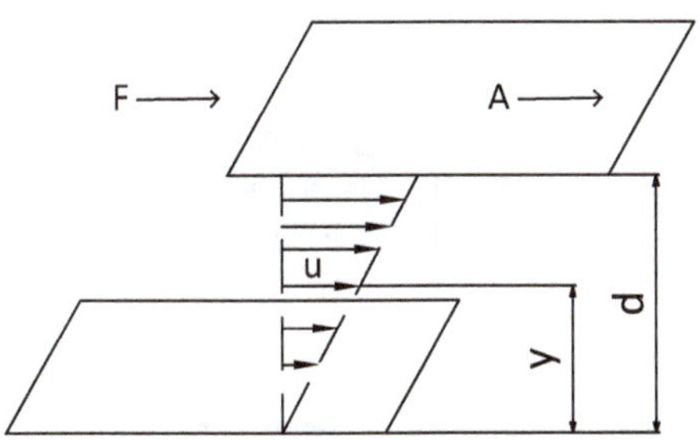

Figure 2.3
Viscous flow relationships

Isaac Newton observed that the shearing force, F, applied over the surface area A is proportional to the velocity gradient (du/dy) within the fluid. He formulated this observation as the following equation, which shows the relationship between shear force and velocity:

$$F / A = \tau = \eta \times du/dy \tag{2.2}$$

where
η = viscosity
τ = shear stress

Viscosity is a measure of the resistance to flow. A highly viscous fluid will have a high resistance or viscosity. For example, plastics have a viscosity magnitude of 10^3, glass is 10^{20} and water is 1. Equation 2.3 shows the relationship between stress, shear and viscosity.

$$\tau = \eta \times \text{Shear rate} \tag{2.3}$$

In a round channel, *shear rate is defined as the rate of change of velocity of the moving plastic with respect to the change of radius of the round channel, and is a maximum at the wall of the runner or cavity.* The shear rate increases with the volume flow rate, but more significantly with a reduction of the melt channel diameter, as shown in Equation 2.4.

$$\text{Shear rate} = 4 \times Q / \pi r^3 \ \left(\text{in s}^{-1}\right) \tag{2.4}$$

2.6.1.1 Pseudo-Plastic Behavior

Fluids can be classified according to their flow behavior. Many common materials such as water, gas, oil, and mercury are known as Newtonian fluids (NFs). A Newtonian fluid's viscosity does not change as a function of increased shear rate. Therefore, the shear stress is proportional to the shear rate (see Figure 2.4).

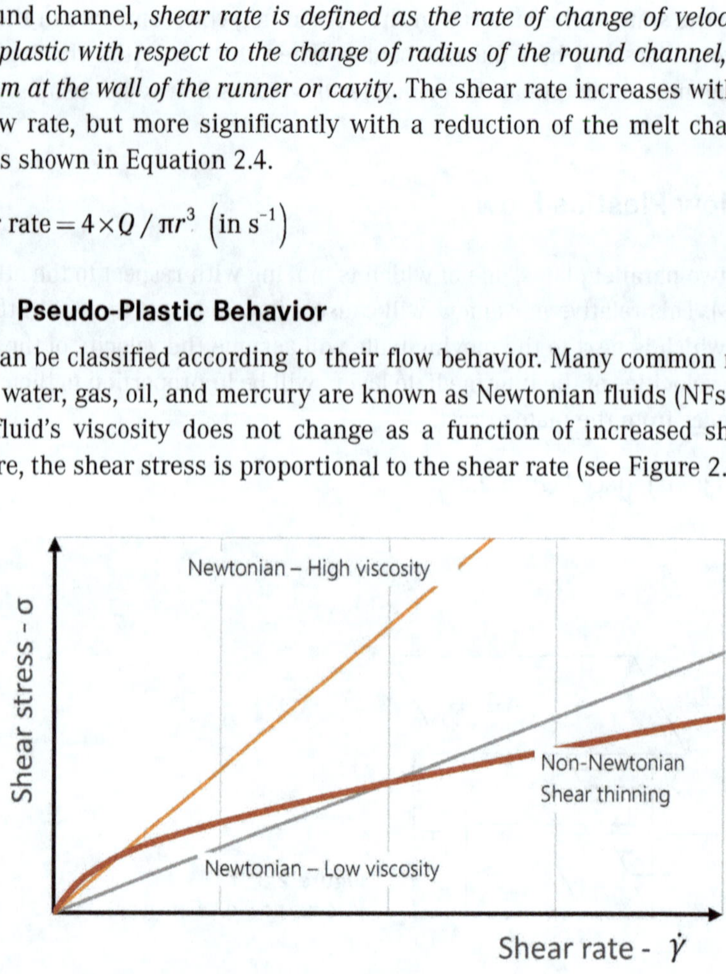

Figure 2.4 Shear stress versus shear rate [2]

[2] A. Ostergard, *Rheology Basics, www.fluidan.com/rheology-basics/*

Plastics, on the other hand, are known as non-Newtonian fluids (N-NFs). For N-NFs the viscosity changes as a function of temperature and shear rate, as shown in Equation 2.5.

$$\eta_{\text{plastics}} = f\left(\text{Temperature, Shear rate}\right) \tag{2.5}$$

In the case of plastics, as temperature and shear rate increase, viscosity decreases, but not linearly. This behavior is type of non-Newtonian fluid called a pseudo-plastic. As the shear rate increases in N-NFs, the viscosity decreases and the shear stress applied to the plastic does not increase proportional to the shear rate (see Figure 2.4). Most plastics exhibit a form of non-Newtonian behavior called pseudo-plastic behavior. For a pseudo-plastic polymer, the viscosity decreases as the shear rate increases (see Figure 2.4). This affect is called shear-thinning.

The shear-viscosity curve (Figure 2.5) is like a fingerprint of the resin because it indicates how the resin will flow at a given shear rate and a given temperature. A device called a rheometer is used to create these curves and points. However, in most cases the shear-viscosity curves are not published. Instead, an ASTM standard test (called the melt index) is conducted on the resin to determine how it flows under a given set of conditions. The melt index is an ASTM measure of the amount of plastic, in grams, that can be pushed through a defined orifice (a), at a certain temperature (b) and under a certain load (c) in 10 min. The higher the melt index the easier the material will flow. This is basically a single point on a shear-viscosity curve.

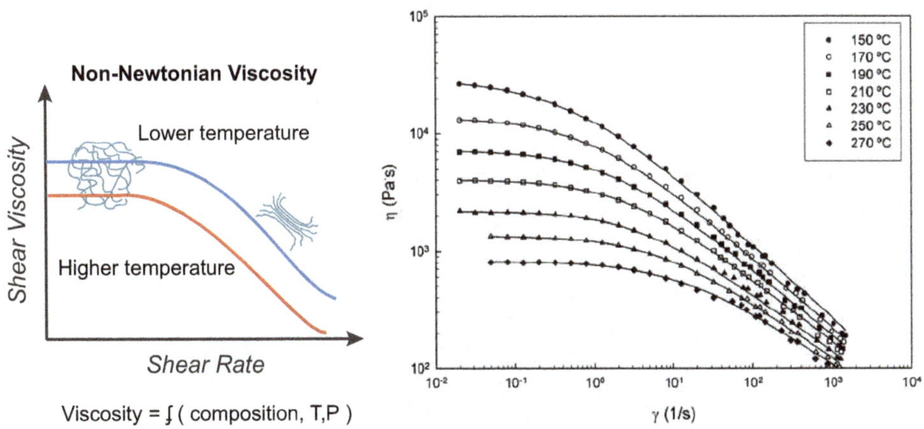

Figure 2.5 Example shear-viscosity curve for plastic

2.6.2 Plastic Flow in Runners and Cavities

Due to the behavior of plastic described in the above sections, plastic flows in a characteristic fashion called a "fountain-type flow". Figure 2.6 and Figure 2.7 show how this looks in a cavity or runner system. Fountain flow means that the material at the center of the melt has a higher velocity than the material at the wall. This has implications for part design, mold design and melt distribution (runner) layout, as the material closest to the walls experiences higher shear rate and shear stress (thus changing its viscosity). At the same time, this material is also cooled by the core, cavity, or cold runner system. Note that the cooling effect is not present in hot runner melt channels. The impact of this type of behavior on mold design and melt flow paths will be discussed throughout the book.

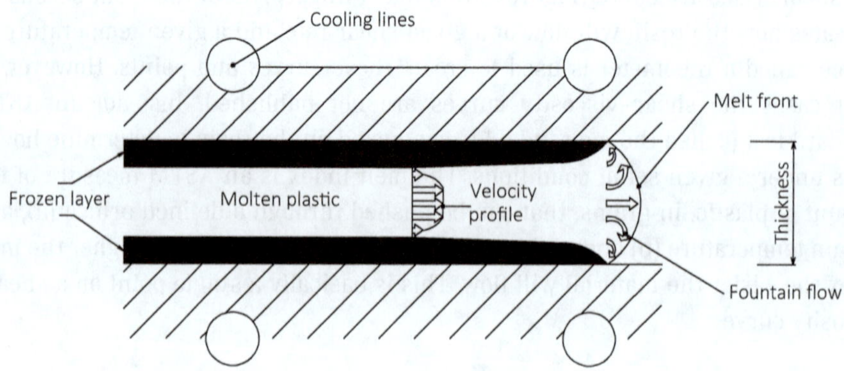

Figure 2.6 Fountain flow behavior of plastics

The impact of this flow behavior can create preferential flows in runners and in the part. In particular, when there are wall thickness changes in the part, the material will always flow along the path of least resistance. In the case of a split in the flow, like at an intersection in the runner system, the material in each sub-branch may not be identical.

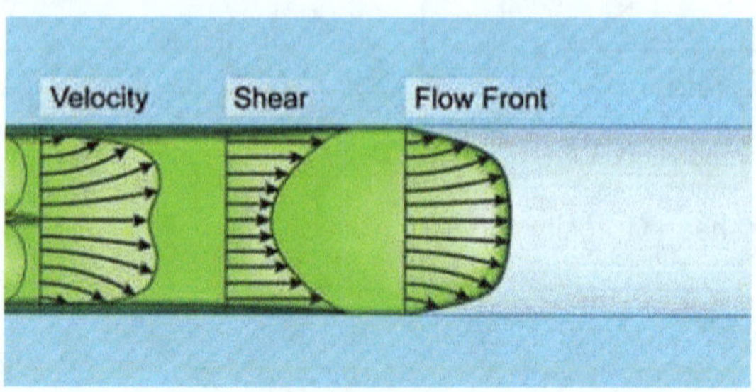

Figure 2.7 Flow in a hot runner melt channel

A good way to visualize this flow is to look at a part that has been made with two different color materials, as shown in Figure 2.8. It can be seen that the first material to enter the cavity (white) ends up on the cavity walls. The later material to enter the cavity follows the path of least resistance and ends up in the center of the part. Note that when flowing around the corner the colored material "cuts" the corner where the flow has the least resistance. This effect is also seen when looking at sections of insulated runner systems (see Figure 2.9 and also see Chapter 14 on melt distribution).

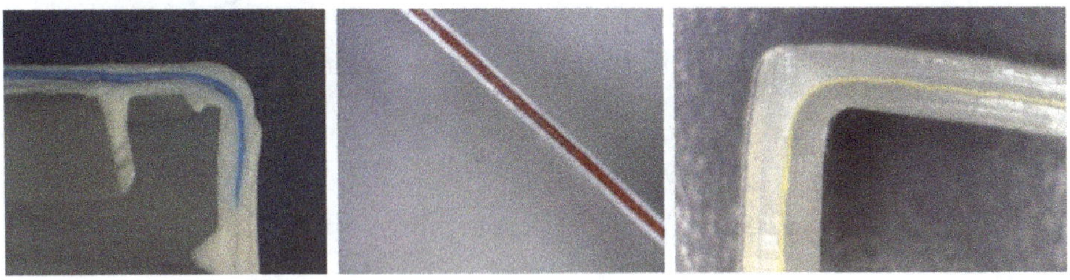

Figure 2.8 Two-color flow in co-injected plastic parts illustrates how plastic flows in the part

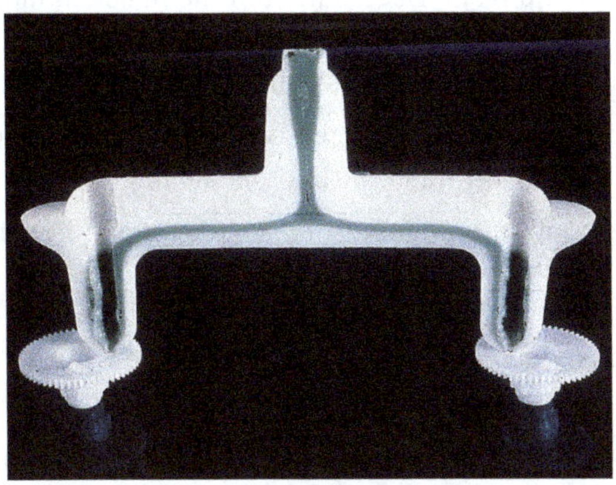

Figure 2.9 Flow in an insulated runner system showing the preferred flow path of the molten plastic

When the plastic flow splits into branches, this fountain flow behavior can cause an imbalance in filling in the runners or in the cavity itself. As the number of subbranches or splits in the flow is increased, this imbalance in the flow is exaggerated. Eventually, if too many branches are created the flow differences may cause problems in filling all the parts equally (Figure 2.10). The impact of this behavior on the design of runner systems will be discussed further in Chapter 14.

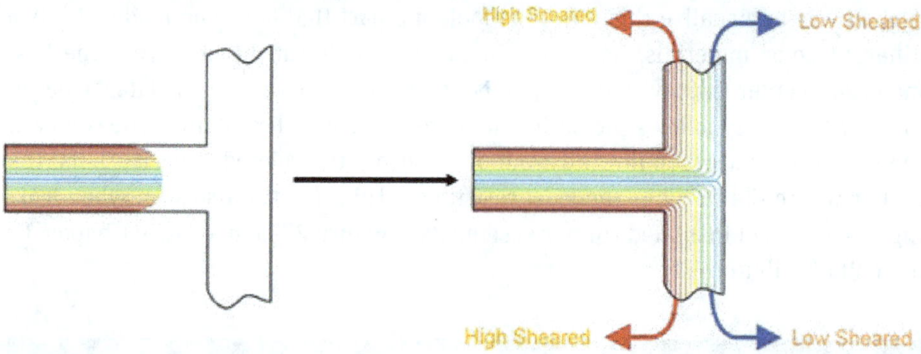

Figure 2.10 Flow of plastic at an intersection in the runner system

■ 2.7 Degradation

Degradation is defined as any process which reduces the physical, chemical, or thermal properties of the resin. Degradation in the injection molding cycle generally occurs as the result of moisture, heat, shearing, or time. In terms of severity, heat is the most prominent, then time, shear, and moisture.

Hydrolysis or moisture degradation occurs when a hygroscopic (water-absorbing) resin comes into contact with water during the molding cycle.

Shear-sensitive materials are profoundly affected by shear. They become very easy to flow when they are sheared at a relatively low rate. This would be observed on the shear-viscosity curve (Figure 2.5) as a steeper curve. Shear-rate sensitivity can be advantageous when processing, because a faster injection rate will make the resin much less viscous and thus easier to inject.

If left above their melting point for extended periods of time, most plastics will degrade. The higher the temperature and the longer the resin is exposed to a high temperature, the more the resin will degrade or break down into its carbon base. Some materials degrade faster than others, and so must be kept in the molten state for as short a period as possible. Figure 2.11 indicates the effect of residence time and melt temperature on the time for a resin to degrade.

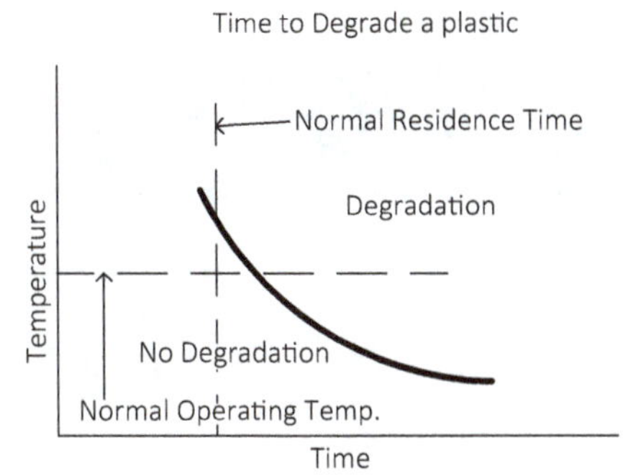

Figure 2.11 Time to degradation for a plastic

2.8 Selection and Requirements for Plastic Materials

Once injection molding has been selected as the preferred method for producing the part, the next major step in the design process will be to select the type of plastic to be used for the product. There are many criteria to consider when selecting the type of plastic to be used, and there are literally thousands of grades of plastic to choose from. The starting point for selection of the right plastic material is the requirements for the product. Table 2.3 can be used as a partial checklist for material selection. The mold designer should fully understand this before designing the mold.

Table 2.3 Checklist for Material Selection

Criteria	Range of values and/or relative importance	Options available
Color		
Weight		
Wall thickness		
Temperatures of operation		
Tensile strength required		
Compressive strength required		
Flexibility (modulus of elasticity)		
Hardness (soft or hard)		
Toughness or impact resistance		
HDT (heat deflection temperature) under load		

Table 2.3 Checklist for Material Selection *(continued)*

Criteria	Range of values and/or relative importance	Options available
Thermal conductivity		
UV resistance required		
Flame resistance		
Contacting solvents, fluids or gases		
Expected product life and reuse		
Expected material cost by weight		
Material being replaced		
Ergonomic considerations		
Surface finish or engraving?		
Product quality and appearance		
Recycling requirements		
Fillers or additives allowed		
Other		

3 Plastic Part Design for Mold Designers

Today, plastics are used in almost every area, from small bottle caps, disposable cutlery, and packages for dairy products, to large containers, such as laundry baskets and garbage pails. Plastics have transitioned from a "cheap" substitute for metal and glass to the material of choice providing almost unlimited design freedom, unique properties, and significant cost savings. Figure 3.1 and Figure 3.2 shows various durable and disposable injection-molded parts.

Figure 3.1
Molded products of various sizes (Courtesy of Husky Injection Molding Systems Ltd.)

Figure 3.2 Small and large technical (engineering) products, heavy-walled jars for cosmetics, and tubular containers with integral, hinged lids (Courtesy of Husky Injection Molding Systems Ltd.)

This chapter contains suggestions and guidelines for the *plastic product design and how it may impact the mold design and the productivity of the mold.* It is not intended to provide an extensive guideline on plastic part design, unless it materially affects how the subsequent mold may be designed. The intent is to provide suggestions on how the plastic part may be modified to ease mold design.

A new mold is usually required:

- For a new product
- If an existing product is being re-designed
- To increase the productivity and the output of the production facilities already in place. This usually provides a good opportunity to re-evaluate and improve the product, and to reduce manufacturing costs, particularly through the reduction of the plastic mass of the product. One particularly frequent productivity improvement is light-weighting of the product.

During the lifecycle of a product consideration is given to light-weighting, as it is the single biggest contributor to reducing cost. Examples of the extent of light-weighting achieved in common parts can be seen in Figure 3.3. In fact, many products are initially launched relatively heavy and then light-weighted as more is known about how the product performs.

The mass of the plastic accounts for a significant portion of the cost of every product. Reducing wall thickness and reduction of unnecessarily heavy cross-sections will not only reduce the cost of plastic material for the product, but will also result in - sometimes significantly - faster molding cycles. The result is that more of the products can be made per hour at lower cost than was possible with the preceding design. In such a case, important considerations are:

- The output of the plasticizing unit and the dry cycle of the machine manufacturing the product before the planned changes.
- If there was special handling equipment (product removal, stacking, printing, etc.) with the old mold, will it be able to handle the greater output, or will it need improvements as well?

For more in depth reading on plastic part design, the reader is encouraged to consult the following books and articles:

- *Plastic Part Design for Injection Molding: An introduction* [1]
- *Designing Plastic Parts for Assembly* [2].

[1] R. A. Malloy, *Plastic Part Design for Injection Molding: An Introduction,* Hanser Publishing, 2010.

[2] P. A. Tres, *Designing Plastic Parts for Assembly* (8th edn), Hanser Publishing, 2017.

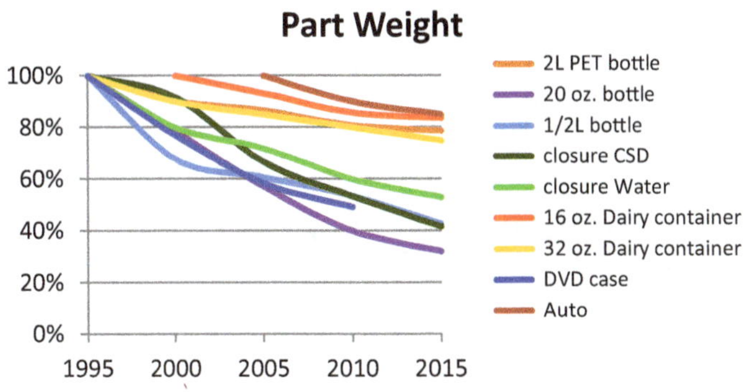

Figure 3.3 Trends of light-weighting over time in some commonly molded parts

■ 3.1 Plastic Part Drawing

Occasionally, only samples or CAD models of a new product are available. 3D printed models are becoming more common to visualize the true part before steel is cut. This may be of some advantage to better visualize the product, but it is absolutely necessary, to minimize risk for all parties involved in the final decision, to have a complete detailed drawing of the product, showing all features, tolerances, and specifications.

It is critical that complete product drawings are available for the mold designer before any mold design (and certainly manufacturing) is started.

It is also essential that the part drawing is signed off by the end customer before the start of mold design, so that any changes in the design are noted and the corresponding change costs accounted for and billed accordingly to the customer. In today's fast-paced environment the mold designer must sometimes work in parallel with the customer by completing some of the mold design while the part is still being finalized.

An example of a complete part drawing is shown in Figure 3.4.

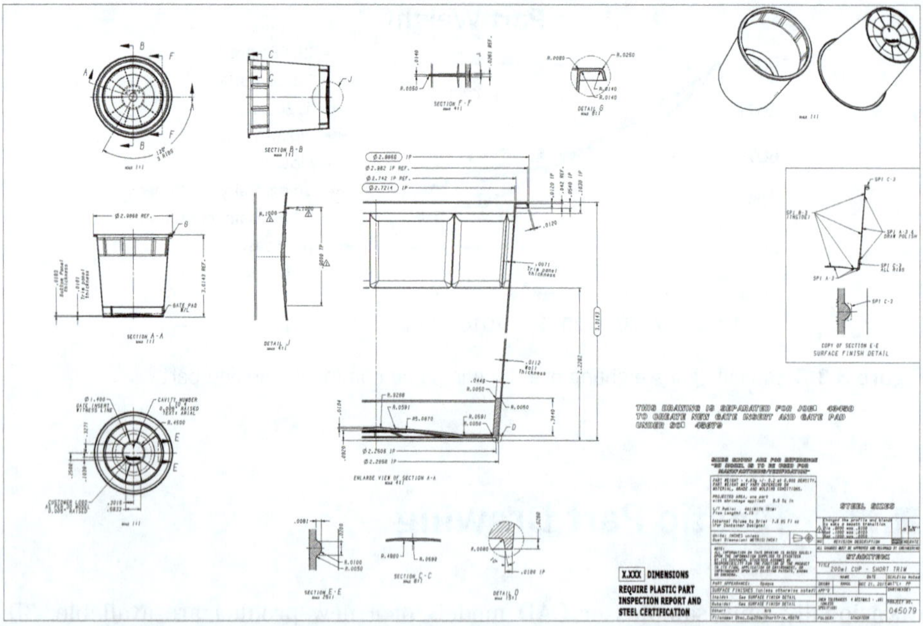

Figure 3.4 Example part drawing (Courtesy of StackTeck Systems)

A complete plastic part drawing should contain the following information at a minimum:

- A 3D solid model of the part or (at minimum) a completely dimensioned plan and section view
- Any small or intricate details blown up into additional sections or views, for example ribs or bosses
- A detail showing any fits to other components
- Engraving, artwork and cavity numbering
- Mold label information (if in-mold labeled)
- Surface finishes
- Any stacking details (if appropriate)
- Part weight (assumed plastic density)
- Part volume (if filled and weighed)
- Center of gravity (if appropriate)
- Identification of plastic or steel sizing. If plastic size dimensions are used on the drawing then shrinkage rates must be added as a note
- Identify all split lines and parting lines, including any intentional mismatches.

The checklist in Table 3.1 can be used as a part drawing critique during the part design review meeting.

Table 3.1 Review Checklist for Plastic Part Design

Checklist item	Completed (Y/N)
1. Is the drawing a plastic part drawing or "steel part drawing"? A steel part drawing is the plastic part with shrinkage dimensions applied, so that the mold designer does not need to add shrinkage. This is often used when the shrinkages are not uniform around the part.	
2. Is the shrinkage defined?	
3. Are part weight and tolerances defined?	
4. Are all drafts defined, including ribs, bosses and sidewalls?	
5. Are there no sharp corners on the drawing? Are minimum radii of at least 0.25 mm shown?	
6. Are all intentional mismatches between core and cavity shown and defined?	
7. Will the part fill? Review the L/t ratio (length of flow/thickness)	
8. Are the parting line and all split lines defined?	
9. Are locations where sinks may occur (like at the end of a rib) called out?	
10. Is the gate position defined and an acceptable gate vestige called out? Normally the acceptable vestige for a valve gate is flush with the molding surface or slightly into the molding surface to prevent interference. If the gate is a hot tip then normal acceptable vestiges are around 50–75% of the gate diameter.	
11. Is allowable warpage called out?	
12. Do the parts need to stack? If so, is the stacking height shown, and is there a diagram showing the stacking of the parts?	
13. For critical dimensions is the dimension going to be left "steel-safe" for the first run so that the sizing can be adjusted?	
14. Are all ribs shown in plan view, top view and side view?	
15. For multi-cavity molds is the cavity numbering identified?	

All sharp corners on the part drawing should be investigated and eliminated if possible. A minimum radius of 0.25 mm (0.010") should be used on plastic parts if possible. For radii bigger than 0.8 mm (0.030") the stress concentration is mostly eliminated.

While creating the plastic part drawing, the designer has the greatest opportunity to decide on the most suitable design for the mold, and/or to make suggestions on how the product design might be modified to improve the productivity, to simplify the mold design, and to reduce mold costs. This is also the time to consider any ancillary equipment required for this production. An opportunity graph (Figure 3.5) shows symbolically the value of planning a project. *Making improvements, revisions, and selections at the outset of a project offers the greatest potential to affect the final outcome of the project, including the part cost.* After concept analysis, once the elements of the project have been agreed upon and as engineering of the mold progresses, the opportunity to make conceptual changes or improvements diminishes, and any costs associated with it will increase. By the time the project reaches completion and gets into testing and production, the opportunity to make changes is low, and any costs could be very high.

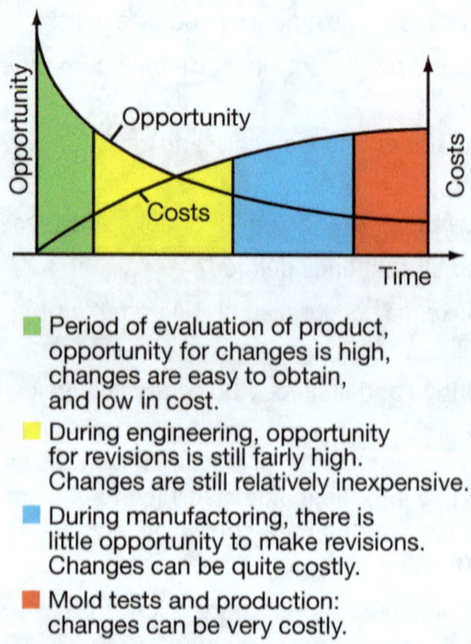

Figure 3.5 Opportunity versus cost versus time graph

■ 3.2 Product Shape: How Can the Product Best Be Molded?

A product designed for injection molding can be generally defined as flat or a box shape with some sidewalls on it. Although this is a significant oversimplification, most injection molded parts fall into one of these two categories.

The designer must consider three things right from the start (for ease of mold design):

- How will the cavity space be filled with molten plastic?
- How will the mold open?
- How will the part be ejected?

Perhaps it is the case that the answers to these three questions are obvious. If not, a designer can search the web or reference books for how a particular type of part is molded and ejected - injection molding has been around for more than 50 years, and almost every type of part has been molded. Chapter 18 of the current book provides a number of reference mold designs that the mold designer can review. A particularly good reference is the textbook *Gastrow Injection Molds* [3].

Another good way to understand how the part is made is to look at other similar parts on the market and identify the parting lines, split witness lines and ejector pin witness lines, etc. It is also advisable to consult with another knowledgeable colleague, and/or with anyone else who is familiar with the type of product for which the mold is to be built, and discuss the problems with making and of operating such a mold, to get their input regarding the proposed product design.

One of the most useful tools available for the mold designer is the use of computer-aided engineering (CAE) molding simulation software. The cost of this software is quite reasonable given the value it can provide on filling, gating, cooling, warpage, shrinkage, and more before steel is cut. The tools available from these suppliers allow the mold designer to simulate their mold concept in detail and apply molding conditions to them. They can then get a multitude of different outputs from the software that allow the mold designer to evaluate the resulting mold design, make modifications, and continue to iterate until the optimal design is found.

Figure 3.6 shows how a problem with a cellphone housing was resolved using CAE flow analysis. The existing mold was simulated on the software and compared to the actual flow-front progression of the tool. The resulting stress mark was duplicated in the software, and then modifications were made to the part until the stress concentration was removed. The tool was then modified to the recommendations made from the software, and the stress mark was eliminated.

[3] P. Unger (ed.), *Gastrow Injection Molds* (4th edn), Hanser Publishing, 2006.

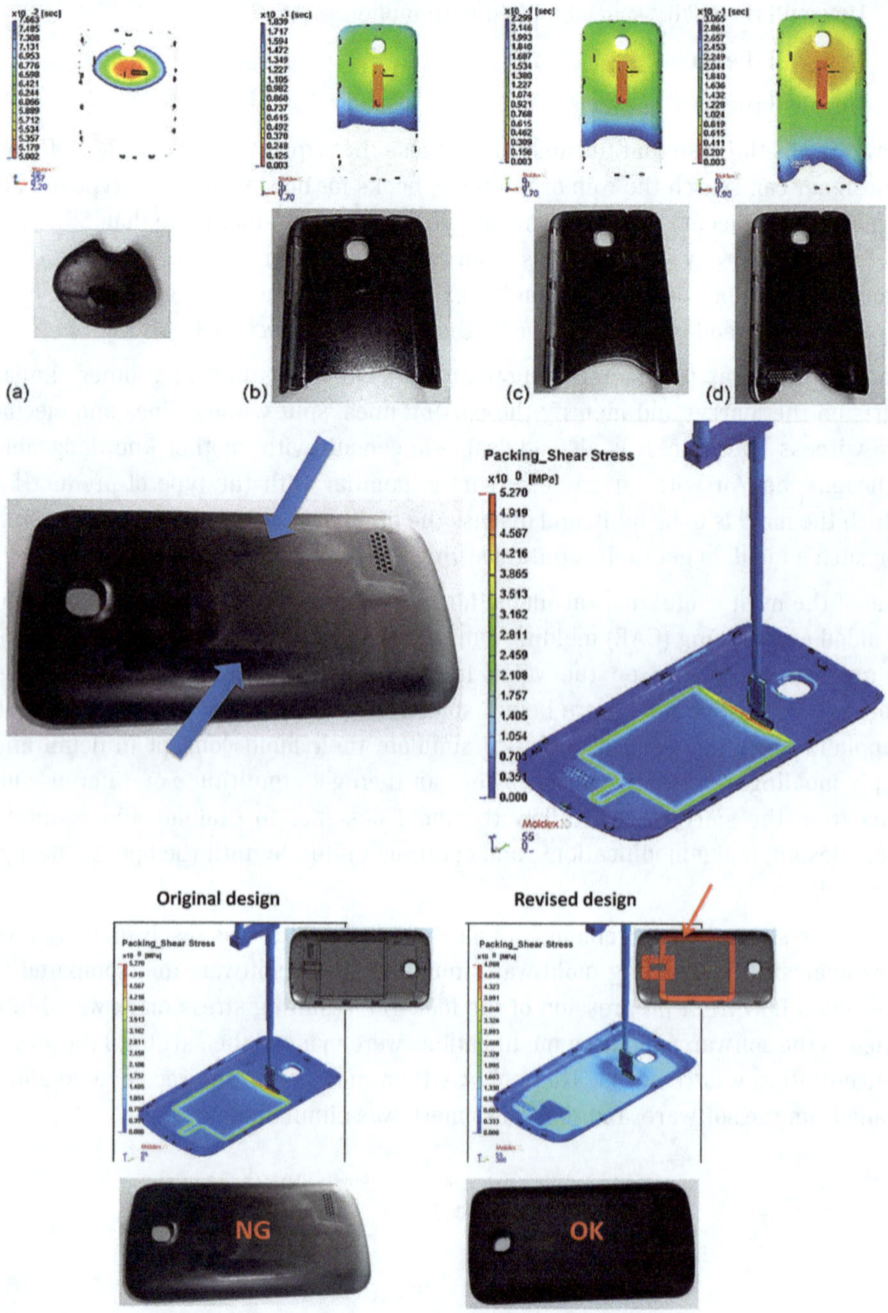

Figure 3.6 Use of computer-aided simulation tools (filling, cooling, etc.) to evaluate how to best mold a part and optimize the mold design. In this example, a stress mark on a cellphone housing was improved by comparing the progression of a mold flow front against a simulation [4]

[4] Source: M. Wang, R. Chang, C. Hsu, *Molding Simulation: Theory and Practice,* Hanser Publishing, 2018 (Figures 6.42–6.44).

3.3 Parting Line (P/L)

In many products, the location of the parting plane (usually called the parting line, P/L) is obvious. It is along the largest cross-sectional dimension of the product, at right-angles to the motion of the opening and closing of the mold, and should preferably be in *one* plane. This is the least expensive, and fortunately, the most frequent case. However, there are many cases where the P/L cannot be located there, and requires special consideration. A few examples are listed below:

- Simple parting lines (Figure 3.7).

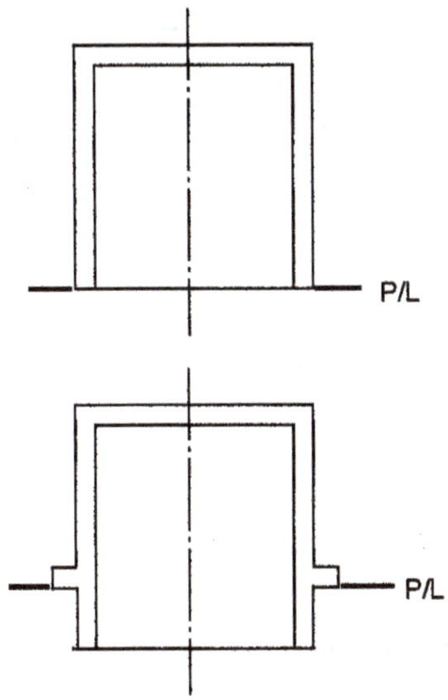

Figure 3.7 Examples of straight, simple parting lines: (Top) At the opening; (Bottom) At the largest diameter

- Sometimes, the P/L *must* be stepped or angled because of the shape of the product (Figure 3.8).

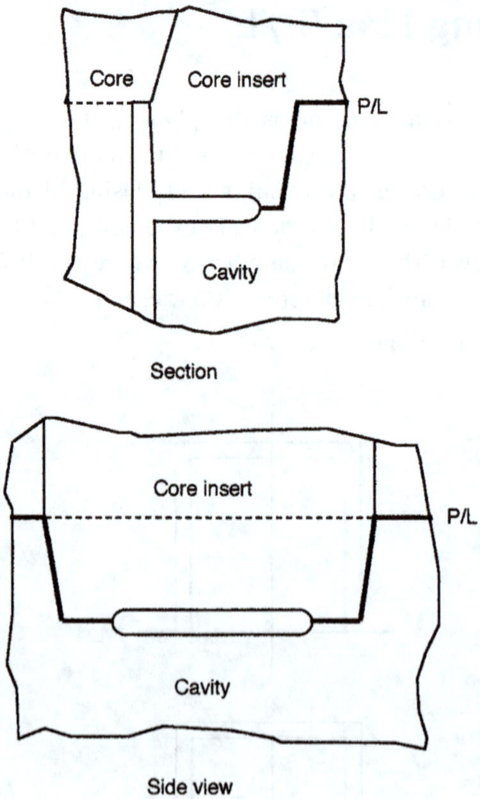

Figure 3.8 Example of simple mug handle, using an offset P/L

- It may be of advantage to place the P/L at a level that is not at the largest cross-section, to force the product to stay on the side from where it will be ejected, as can be the case with flat products. This would not affect the mold cost. However, flat products often cause trouble at ejection, because they do not always stay reliably with the side from where they are ejected. Additional mold features, such as sucker pins, or undercuts in the side of the product ("pull rings") may be required to hold the product on the ejection side, to make sure that the mold can operate automatically, without interruptions (Figure 3.9).

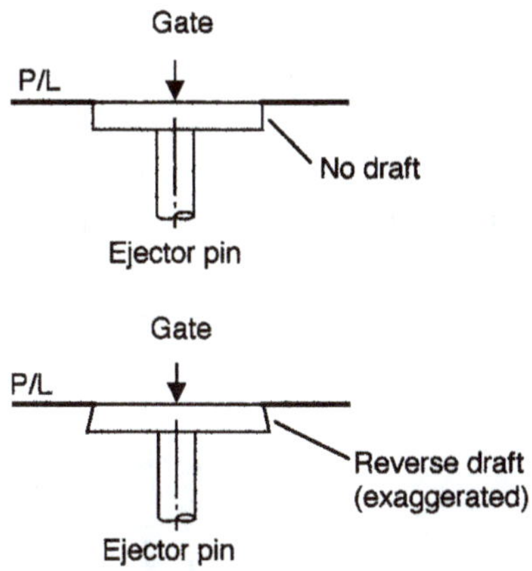

Figure 3.9 Typical flat piece with undercut below parting line to retain it to one side. Note that sometimes a change in the surface finish can cause the part to stay on the core side as well

- The P/L is curved. This is sometimes unavoidable because the product shape will not permit a straight P/L, for example in some toys, but occasionally also in technical products. A typical example is the P/L for plastic forks or spoons. In all these cases, the matching of the P/L is difficult and expensive. It may need special, costly grinding equipment or expensive fitting by hand (called "bluing") (Figure 3.10).

Figure 3.10 Typical mold profile for cutlery

■ 3.4 Uniform Wall Thickness

An injection-molded product should, as much as possible, have a uniform wall thickness for even filling and cooling. It is better to modify the geometry of the part to eliminate thick sections such as shown in Figure 3.11A, rather than have to work out how to cool the thick sections and prevent unsightly sinks, voids and warpage. The poor design shown in Figure 3.11A will never end up looking like this in the molded part due to the changes in wall section. The better design shown in Figure 3.11B will be easier to mold and result in a part that is more consistent and has a higher-quality appearance. When considering a change from another material such as metal, wood, ceramic, or glass, the mold designer cannot just assume that this part can be replicated in plastic. The process for making the part in plastic must be considered and the part altered to take advantage of the molding process and account for its disadvantages. Decades ago, when plastics were being substituted for other materials, the biggest mistake designers made was to just try to make a duplicate of the existing non-plastic part, rather than considering how to best make the part in plastic. Once designers understood the advantages of using plastic for part consolidation, light-weighting and other features helped plastic evolve from being "cheap" to being the material of choice.

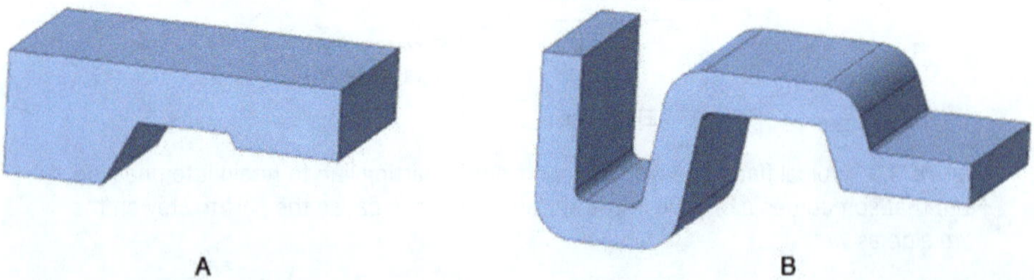

Figure 3.11 Maintain uniform wall thickness throughout the part: (A) Poor design; (B) better design

Plastic will always flow along the path of least resistance, and this must be taken into consideration when designing the part for moldability. There are a few good reasons for keeping the wall thickness as uniform as possible:

- The thicker the plastic wall section, the longer the cycle will be. The thickest portion of the molded part will control the cycle time if the part is evenly cooled in the mold.
- The thicker sections will shrink more than a thinner section, and this can cause sinks, voids and unexpected warpage in the cooled part. There are some areas where sinks are unavoidable, such as at the base of ribs and in some corners.

If differing wall sections are required, it is advisable to make the transition between thicknesses as smooth as possible and to avoid any sudden changes in wall sections, which can cause high stresses in the cooled part and very noticeable sinks or voids (see Figure 3.12 for best practice on transitions).

As a general rule, a part should be thickest near the gate and thinnest at its furthest point from the gate.

> Flowing thick to thin using the gentlest possible transitions is a general rule for plastic part design.

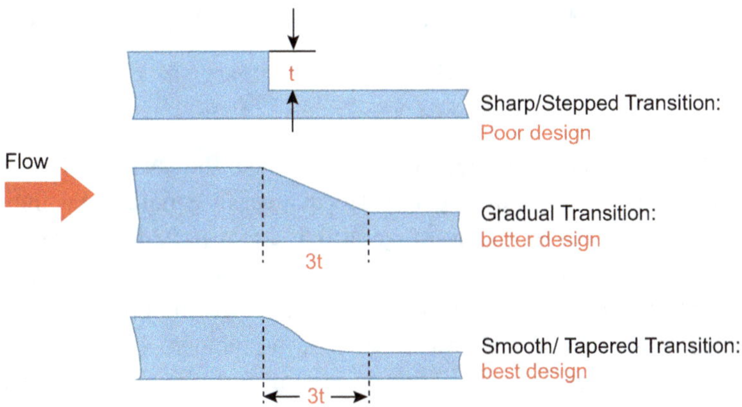

Figure 3.12 Wall thickness transitions - bad versus preferred [5]

As mentioned above, plastic will always flow along the path of least resistance, and Figure 3.13 shows how parts with uneven wall thicknesses fill. As can be seen, the plastic in the thickest areas is flowing more easily and progressing further than in the thinner areas. In some cases this can be detrimental when it causes weld lines, sinks, voids, and other molding defects. However, thicker sections can also be helpful by filling parts of the design where the plastic has further to travel.

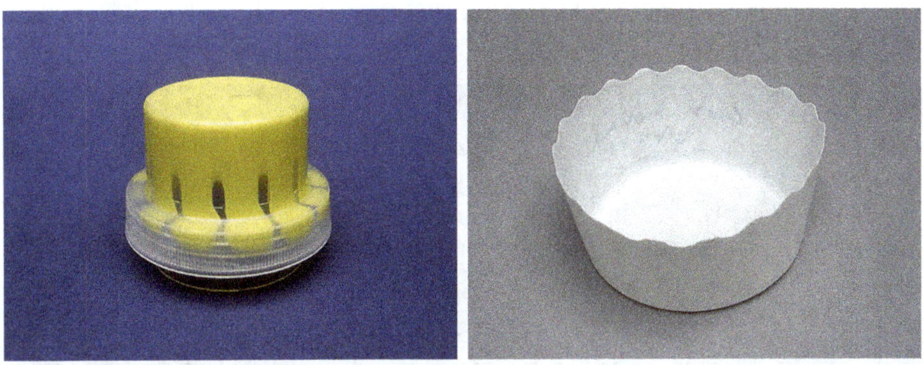

Figure 3.13 Plastic always flows along the path of least resistance (or thicker sections rather than thinner ones)

[5] Source: M. Wang, R. Chang, C. Hsu, *Molding Simulation: Theory and Practice*, Hanser Publishing, 2018 (Figure 3.3).

3.5 *L/t* Ratio (Length of Flow Divided by Wall Thickness)

To keep costs down there is always the desire to make the part as thin as possible while still achieving the required part performance. In general, injection-molded parts have wall thicknesses in the range of 0.5–2 mm (0.020–0.080"). There are dramatic cases of wall thicknesses of up to 10 mm (0.4") or as low as 0.3 mm (0.010"). However, in the quest to reduce plastic consumption, the industry is pushing the boundaries of how thin a part can be molded. The most aggressive market segment in this respect is the packaging industry. Due to advances in plastic flow characteristics, as well as machinery and mold design improvements, it is now possible to mold very thin sections down to between 0.3–0.5 mm (0.010–0.020"). A good measure of how aggressively a plastic part can be thin-walled is the *L/t* ratio (length of flow from the gate to the longest end of flow, divided by the wall section. *L/t* ratios of up to 300 can be achieved in smaller parts like margarine containers using very low-viscosity plastics and up to 400 in larger parts like pails and buckets (Table 3.2).

$$L/t \text{ ratio} = \text{Flow length from gate to furthest point} / \text{Average wall thickness} \tag{3.1}$$

Table 3.2 Range of Achievable *L/t* Ratios

Part type	Resin	*L/t*
Heavy-walled parts	Low MFI [a] (PET, engineering resins)	50–100
Most parts easy to mold, high cavitation molds	Engineering resins	100–200
Higher cavitations	Commodity resins with high MFI [a]	200–300
Ultra-thin wall parts (< 0.5 mm) and very long flow-length parts with thick walls (pails)	Very high MFI resins	300+

[a] MFI is a measure of the ease of flow of a resin. It is defined as the mass of polymer, in grams, flowing in 10 min through a capillary rheometer of a specific diameter and length by a pressure applied. The higher the number the easier it will flow. ASTM 1238 and ISO 1133 standard testing methods are used to describe the required testing protocols and generate the results.

The size of the part (length of flow and wall section) and of course the viscosity of the plastic to be molded affects the achievable *L/t*. Very small parts with wall sections of less than 0.5 mm (0.020") have maximum *L/t* ratios of about 200. Much larger parts like pails, white goods, and automotive parts can have *L/t* ratios of about 400.

The number of cavities to be molded can also affect the achievable L/t ratio. Up to a 10% reduction in achievable L/t ratios can occur for large cavitations (>16).

Figure 3.14 shows sections through a molded tumbler. On the left is the original part. Note that the plastic flows from thin to thick to thin to thick, and finally to thin. This makes the part very difficult to mold, On the right is the redesigned part. Modifying the core to flow from thick to thin reduced the part weight by 20% and at the same time, decreased the cycle time by about 20%. Finally, it eliminated many molded defects caused by the thick-to-thin transitions.

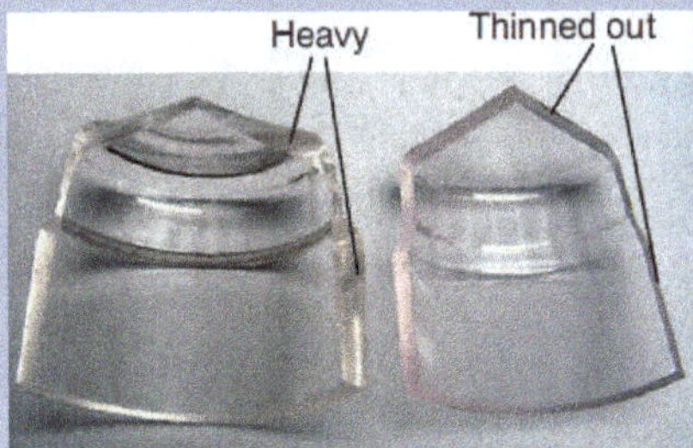

Figure 3.14 Sections through an injection-molded tumbler

3.6 Drafts

In order for a mold to open, the core must be able to open away from the cavity to eject the part. If there is no draft between the core and cavity, then the mold may "lock" during the injection process. If the design of the plastic part causes this to happen, then the opening force of the machine may be unable to separate the core from the cavity.

The following general rules for drafts should be adhered to:

- There should be drafts on all surfaces parallel to the direction of mold opening.
- Both the core and cavity should have corresponding drafts to keep wall thicknesses uniform.
- If the surface finish has significant texture, 1° of draft should be added for every 0.025 mm (0.001") of texture depth.
- Plastic part drafts must be a minimum of 0.5° for air ejection and 1° for rib drafts (0.5° per side).
- Mechanical ejection should be considered when draft angles are less than 1°.

In the event that these rules cannot be respected, then consideration must be given to more elaborate means to facilitate mold opening, such as split cavities and cores. These types of mold designs are significantly more costly and should be avoided if at all possible.

Any negative draft on the part MUST be questioned and examined. There are some cases where a negative draft might be acceptable – for instance, where a stripper ring or stripper plate will mechanically eject the part. In this case the mechanical force of ejection will cause the still-warm molded part to temporarily deform to allow the part to be ejected. Good examples of this are the undercut on a lid, stacking shoulders, and small ribs.

> Surface finishes can have a dramatic impact on whether a draft angle is sufficient for ejection. A deep etching or engraving may act as an undercut and prevent ejection without plastic part damage.

3.7 Corners, Fillets, and Chamfers

Sharp corners or fillets must be avoided in plastic part design if at all possible. Sharp corners can cause stress fractures of the part, are difficult to fill completely, increase pressures required to fill the part, and make mold manufacture more complex. As the radius or fillet gets larger, the stress concentration produced decreases rapidly. CAE analysis can help predict any stress concentrations in the proposed plastic part. These can then be addressed before the mold is designed.

The corner radii should match to produce an even wall section throughout the corner, as shown in Figure 3.15. If the outside radius is too large, then the filling can be restricted and the part weakened. If the outside radius is too small, then the part becomes too thick in this region, causing localized racetrack filling of the part and wasted plastic in the part.

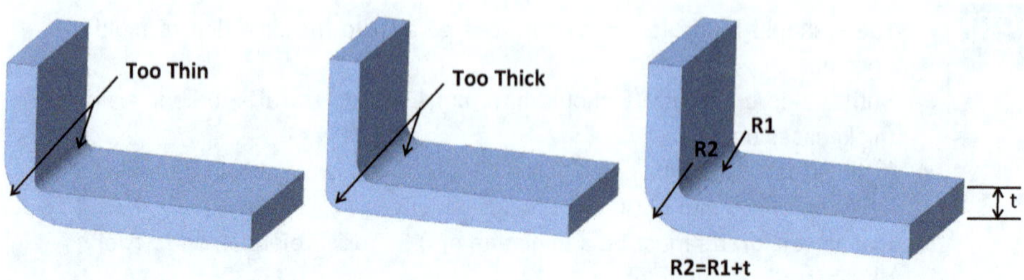

Figure 3.15 Corner design

As a rule of thumb, any external corner or radius should be a minimum of 150% of the wall thickness, and the corresponding internal radius should be at least 50% of the wall thickness.

3.8 Ribs and Bosses

Ribs and bosses are commonly used in plastic parts for reinforcement and for assembly. While needed for the physical strength of the molded part, they can cause molding defects (such as sinks voids, shorts, and trapped air) if incorrectly specified. To prevent defects in the molded part, the following guidelines should be followed for ribs and bosses:

- Rib height should be less than 3 × the part wall thickness at the rib.
- The draft on all ribs and bosses should be 0.5° or more per side.
- The span between two ribs should be > 2 × the wall thickness at the rib.

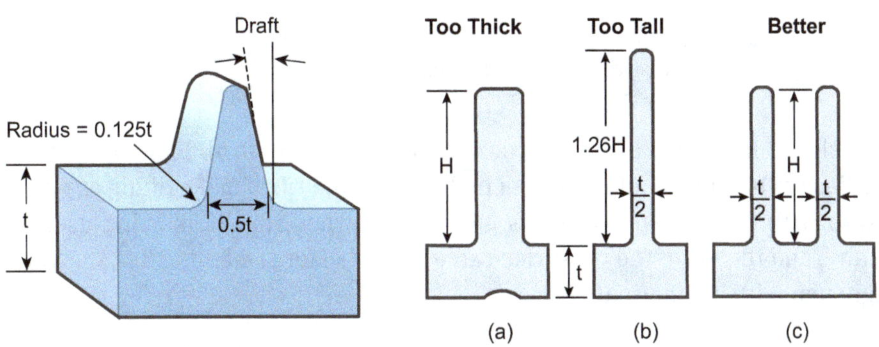

Figure 3.16 Recommended dimensions for a rib [6]

[6] Source: M. Wang, R. Chang, C. Hsu, *Molding Simulation: Theory and Practice,* Hanser Publishing, 2018 (Figures 3.11 and 3.14).

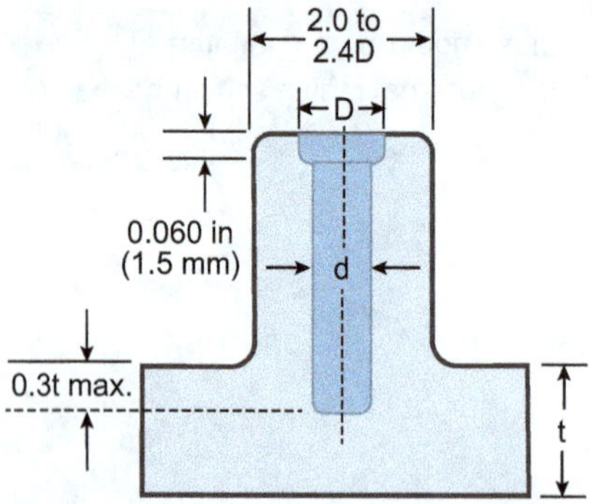

Figure 3.17
Recommended dimensions for a boss [7]

A good way to hide the location of a rib or a boss is to put a transition in this area so that the sink is not noticeable (Figure 3.18).

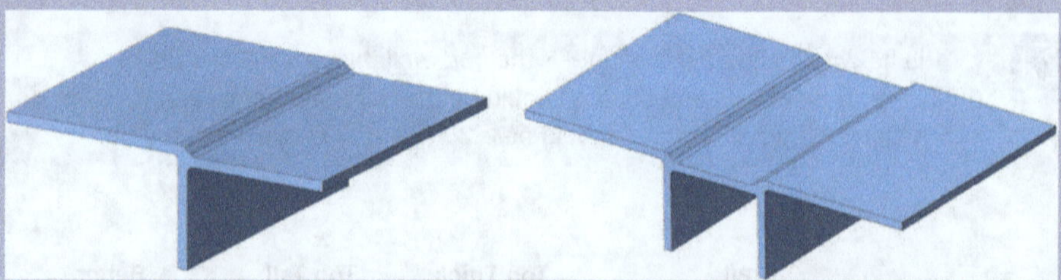

Figure 3.18 Hiding the location of a rib or boss

A rib should not end in a sharp corner, as this makes the rib difficult to fill. Instead a radius or chamfer should be added to the end of the rib to make filling easier (see Figure 3.19). The exception to this guideline is when the rib is being used as a feature, and the top of the rib surface area is needed for something like stacking or alignment with another part.

Figure 3.19 Avoid sharp corners at the end of ribs

[7] Source: M. Wang, R. Chang, C. Hsu, *Molding Simulation: Theory and Practice,* Hanser Publishing, 2018 (Figure 3.16).

■ 3.9 Rim Designs

Figure 3.20, designs A–C, shows round rim designs frequently specified on plastic parts. The large bead rim shape (Figure 3.20A) is not recommended. As the rim is at the end of the flow path, the injection pressure is greatly reduced, and such a rim will be difficult to fill and pack. The result is either a void, usually with amorphous plastics such as PS which freeze rapidly on the surface (Figure 3.20A1), or a sink, usually with crystalline plastics which freeze slower and pull the still-soft plastic skin toward the center, as shown, exaggerated, in Figure 3.20A2 and shown in Figure 3.21. The bead rim dimension will be difficult to control dimensionally and will cause sealing issues or fit issues with the mating component. A better plastic part design would avoid such large cross-sections at the end of any flow path and the mold maker should explain it to the product designer, and request a change as suggested in Figure 3.20B or C which will cycle much faster and not have the risk of sinks or voids. An additional benefit of designs B and C is that these options use less plastic and permit faster cycles, resulting in lower product costs.

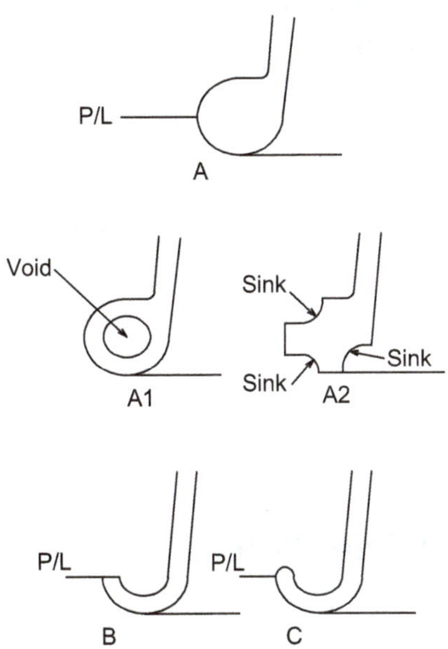

Figure 3.20
Beaded rim, with void and sinks

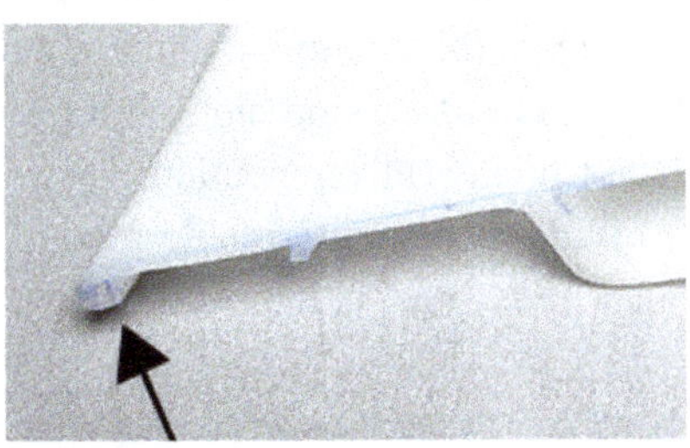

Figure 3.21
Cut section of the rim of a pail with a large void

3.10 Stripped Undercuts

To strip an undercut, the lead angle (see Figure 3.22) needs to be between 30° and 45° in order that the plastic can travel up the incline and out of the undercut. The surface finish of the part in this area needs to be smooth and highly polished to avoid any hang-up points. The size of the undercut to be stripped depends on the flexibility of the resin and the total surface area of the undercut. The undercut itself can usually be in the range 0.35–0.75 mm in depth per side (that is $D - d = 0.7$–1.5 mm). More detail on stripper ejection of undercuts can be found in Chapter 11.

Certain resins that are stiff, such as PS, PC, nylons and other reinforced resins should not be stripped. Instead these parts should have split cores or cavities or other motions that free the undercut before ejecting the part.

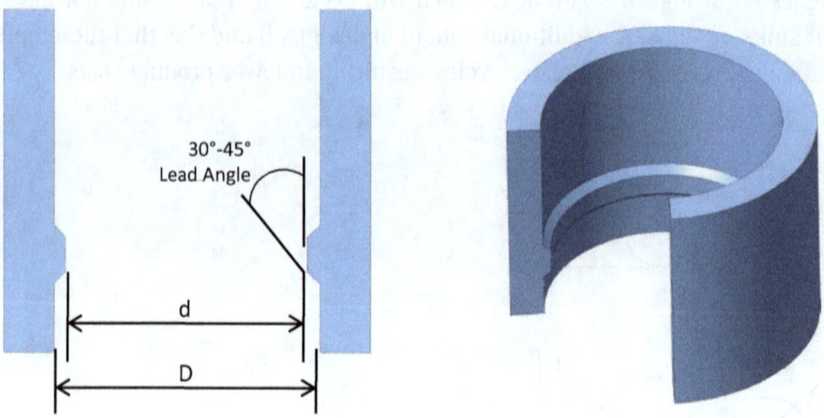

Figure 3.22 Stripped undercut should have a generous lead-in, smooth radii and polished molding surfaces

3.11 Sidewall Windows in the Part

Sometimes the plastic part requires openings on the sidewall of the part. This presents a challenge in mold design, as these openings need to be filled with steel during injection, and then the steel needs to retract from the opening during mold opening and ejection. One way to create these holes is to use a side motion in the mold that occurs during opening and ejection (see Chapter 11 on ejection). However, if the plastic part is designed with this in mind, the expensive side motion may be avoided. Figure 3.23 shows an example of where the shape of the window and the core and cavity design can avoid this added complexity. Also see Chapter 18 for examples of mold designs with sidewall windows such as crates and boxes.

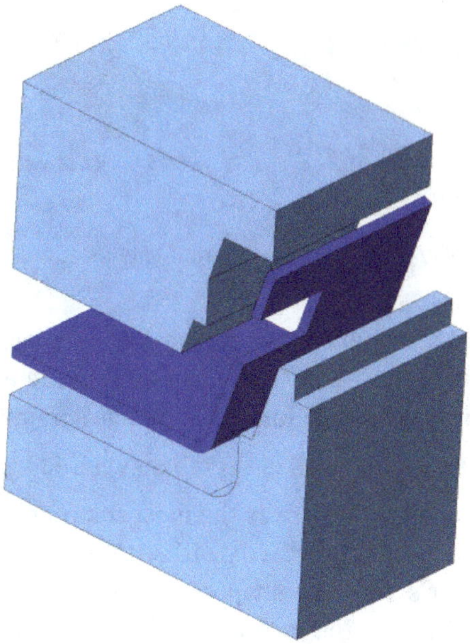

Figure 3.23 Creating a window in the side of the part without needing side opening in the mold design

■ 3.12 Gate Location and Number of Gates

A plastic part should be gated in the thickest section of the product, and preferably with one gate. *Flow should always go from thick to thin to prevent molding defects such as sinks, voids and warpage.* Keep in mind again that plastic always flows along the path of least resistance. Plastic will therefore preferentially flow down ribs, beads or gussets, which could cause issues with weld lines or part strength.

The right gate location on the part is sometimes obvious (as in the middle of a round lid), and sometimes complex (as in an instrument panel for a car dashboard). In complex cases, CAE flow analysis is an excellent tool to find the right location(s) for gates on a molded part. It allows the mold designer to see exactly how molten plastic will fill the part, and it can be used to assess different gate locations through very accurate simulations. Figure 3.24 is a good example of the power of CAE simulation, as it shows quickly and vividly the differences between two and three gates per part and the resulting weld lines.

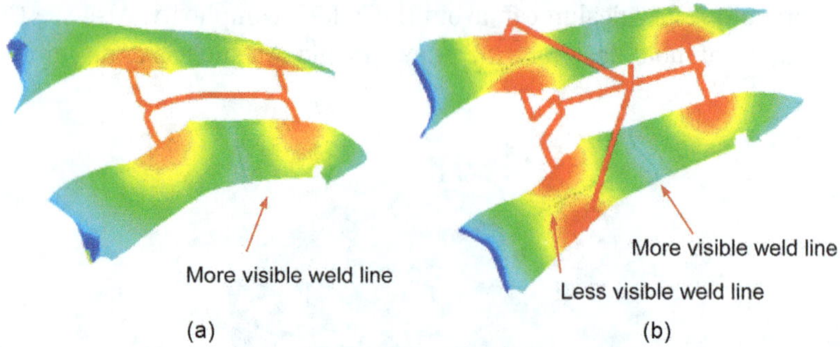

Figure 3.24 Weld line quality estimated with different gate locations by CAE analysis [8]

The use of flow analysis for determining the optimal gate location is highly recommended if there is any doubt, because the cost to relocate a gate position can be very high, and sometimes result in almost a complete scrapping of the inserts.

The following are some general rules of thumb on where to gate for *hot runners*:

- **Outside center gate (OSCG).** This is the most common gate location when hot runner gating. The gate is in the cavity, either in the true center of the product, e.g. in most round or symmetric shapes like containers, or in a place close to the center (Figure 3.25). If the gate cannot be placed in the center then flow leaders can be used to modify the flow path.

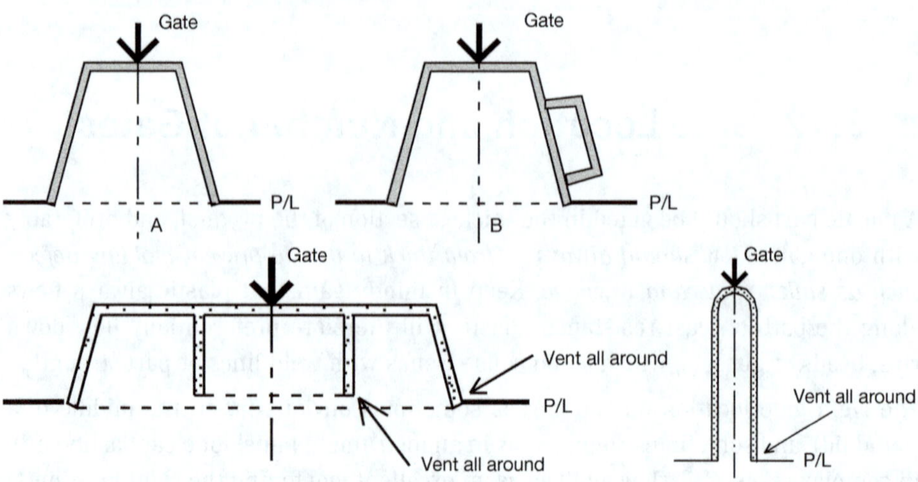

Figure 3.25 Outside center gating locations

[8] Source: M. Wang, R. Chang, C. Hsu, *Molding Simulation: Theory and Practice,* Hanser Publishing, 2018 (Figure 3.34).

- **Inject against the core.** It is always best to inject against a solid part of the core, to create an evenly distributed flow of plastic as it flows away from the gate. It is not generally recommended to shoot into an open space, or into a slot (forming a rib), because it may cause a portion of the plastic to flow away rapidly from the gate and cause serious backflow conditions. Note there are certain types of hot runner gates (such as the FDU gate from FDU hot runners) that have been designed to be used on a rib or boss.
- **Inside center gate (ISCG).** The nozzle and gate is located in the core, usually in the center, and the plastic shoots at the cavity wall (Figure 3.26). This gate is sometimes selected if gate marks are not permitted on the visible (outside) of the product. It is often demanded with high-end cosmetic packaging or in-mold labeling. Note that a much longer hot runner nozzle must go through the core, meaning that there is little (if any) space for good core cooling. Typically, the cycle times are several times longer than the same shape would be with OSCG due to the limited core cooling. Also, because it will also be necessary to eject from the injection side of the mold, the cost of the mold will be much higher.

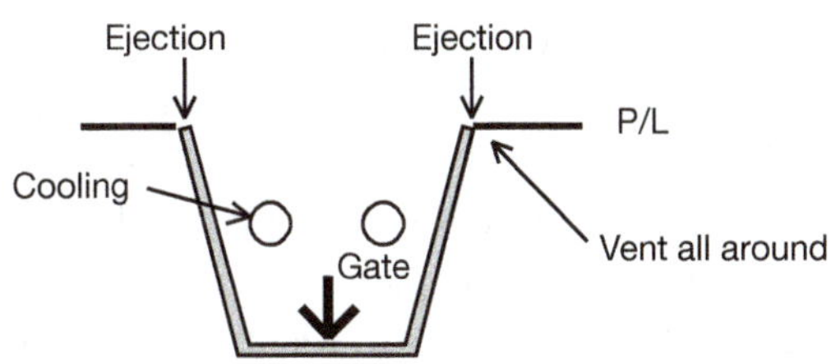

Figure 3.26 Inside center gating locations

- **Gate near the end of handled products.** Due to the high flow rate and shear in the gate region, this area has a concentration of stress and can be more brittle. As a result, for handled items like a toothbrush or razor the gate should be placed at one end of the handle to avoid breakage or warpage of the part (Figure 3.27).

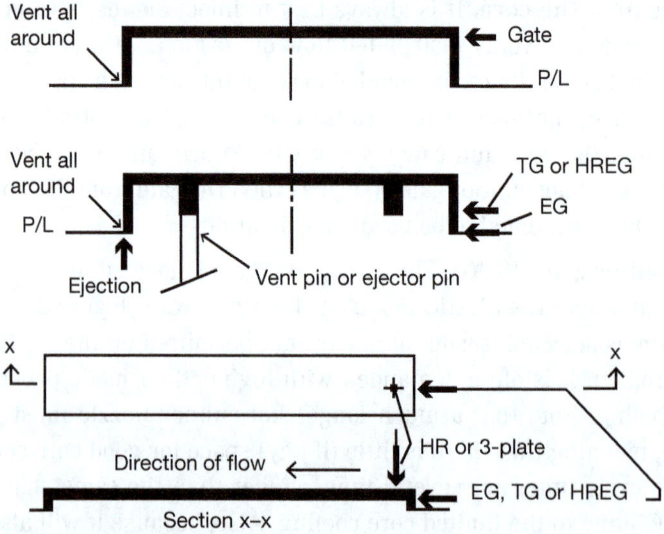

Figure 3.27 Gating locations for long parts; EG = edge gate, TG = tab gate, HREG = hot runner edge gate

- **Gate into the heaviest area.** As the plastic flows away from the gate, it comes into contact with the cold mold walls, and stays there, thereby making the passages for the following plastic narrower. As the passages get narrower, more pressure will be required to keep the plastic flowing until the mold is filled. It is therefore always better to gate into a heavier area, which may then gradually reduce in thickness. It is very difficult to fill a heavy rim of a product if the plastic has to squeeze through narrow walls to get there.
- **Gate at or near the edge of large panels.** It would be relatively easy to gate a large panel at or near the center because the *L*/*t* ratio would be about the same in all directions. However, this would create a weak spot at the gate area, and the panel is more likely to break there if stressed. A gate mark in the center may also be objectionable. The right way for such large panels (automotive, etc.) is to gate at or near the edge. Considering the same product as before, because the *L*/*t* ratio is now much greater for at least half of the panel, it may be necessary to increase the number of gates to bring the *L*/*t* ratio to a reasonable value. However, there may now be another problem: several streams from the gate will join somewhere between them and must properly weld, to avoid weld lines and weak spots in the panel. Higher molding temperatures will then be necessary.
- **Gating products with live hinges.** Live hinges are used to connect two products such as a box and lid with one or more thin connecting strips that will act as hinges (Figure 3.28). With some materials (usually PP) the strips can be

bent millions of times without breaking. The secret is to locate the gates so that the molecules in the very thin strip(s) are oriented leading from one part of the product to the other. Usually, one gate into the larger part is enough. When gating both parts, it is very important that the distances from the gate to the strip(s) are different by a large margin, so that any weld lines do not occur in or near the strip(s).

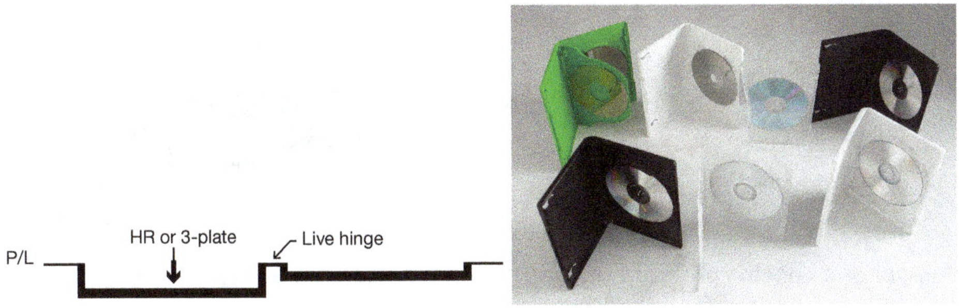

Figure 3.28 Gating of a live-hinged product such as these DVD cases using hot runner (HR) or 3-plate cold runners

3.12.1 CAE Filling Analysis for Gate Location Optimization

If there is any doubt about where the gate should be placed, it is recommended that the designer undertake a CAE analysis of the gate to understand the implications of different gate locations. Figure 3.29 shows a good illustration of how different gate locations can impact the resulting warpage on a molded part. In this case, using a hot runner with six gates results in a much more balanced filling of the cavity. The end result is a much flatter electrical housing than a two-gate cold runner system.

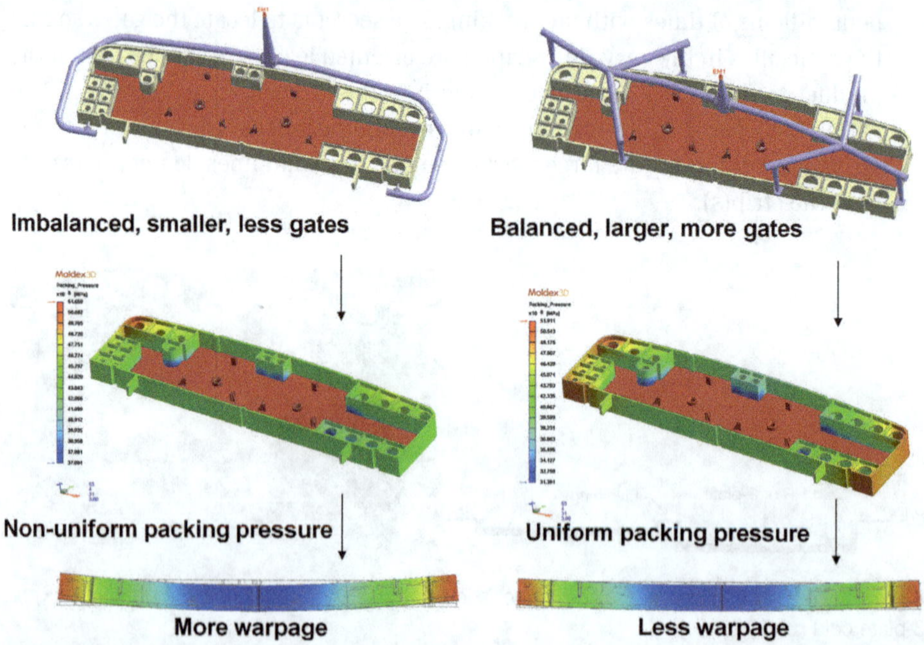

Figure 3.29 CAE molding simulation of different gate configurations [9] (Courtesy of Moldex3D)

Figure 3.30 shows how the flow simulation software can identify possible weld line issues on a scanner part before the mold is built. The mold designer was able to modify the gating locations to prevent the weld lines from being located in critical areas.

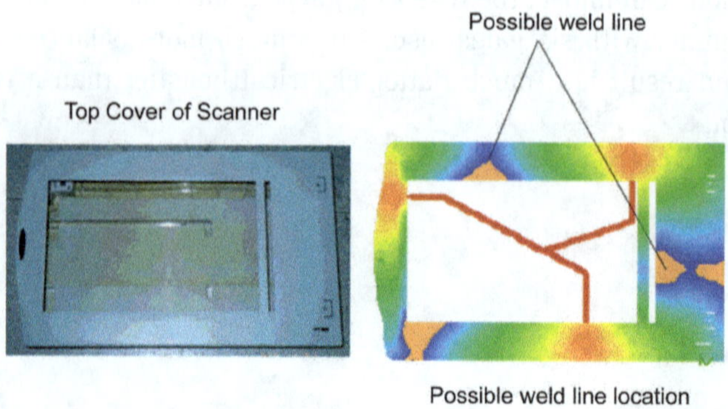

Figure 3.30 CAE prediction of flow front progression, weld lines, weld strength, shorts and voids [10]

[9] Source: M. Wang, R. Chang, C. Hsu, *Molding Simulation: Theory and Practice,* Hanser Publishing, 2018 (Figure 9.31).

[10] Source: M. Wang, R. Chang, C. Hsu, *Molding Simulation: Theory and Practice,* Hanser Publishing, 2018 (Figure 3.32).

In the case of Figure 3.31, the mold was built and the molder had re-occurring problems with voids on the ring of the part, which was gated with three hot runner gates. The location of the voids can be seen in the center panel. The flow analysis software was able to accurately replicate the location of the void and allow the designer to iterate new designs to resolve the air entrapment problem. The solution to the problem was to reduce the thickness in areas where the flow front progressed faster and increase the wall thickness where it lagged. In the end the void was avoided, and the end of flow occurred at the parting line so proper evacuation of air could occur.

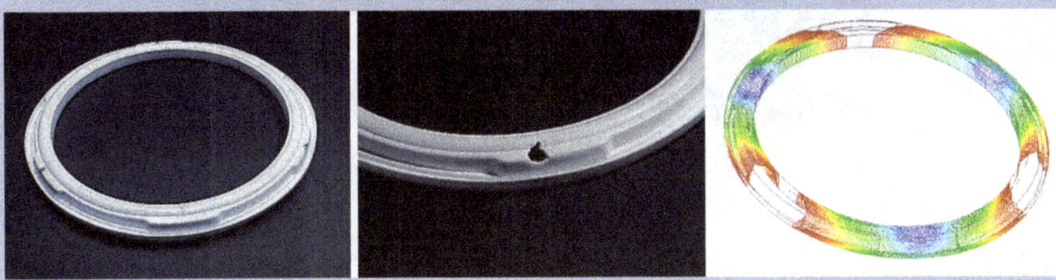

Figure 3.31 Using CAE flow simulation software to determine flow fronts

3.12.2 Two or More Gates per Cavity - Large Products

Two or more gates per cavity are sometimes required for very large products or parts that cannot be easily filled with a single gate due to some geometric constraint (e.g., automotive products, bottle crates, thin parts, many holes, highly viscous resin, etc.). In these cases, the flow distances from a single gate would be too long and/or one gate could not supply all the plastic required to fill the cavity in time before freezing off (see the example in Figure 3.32). It is also used in molds for large containers where the flow through two or more gates speeds up the filling compared with a single gate.

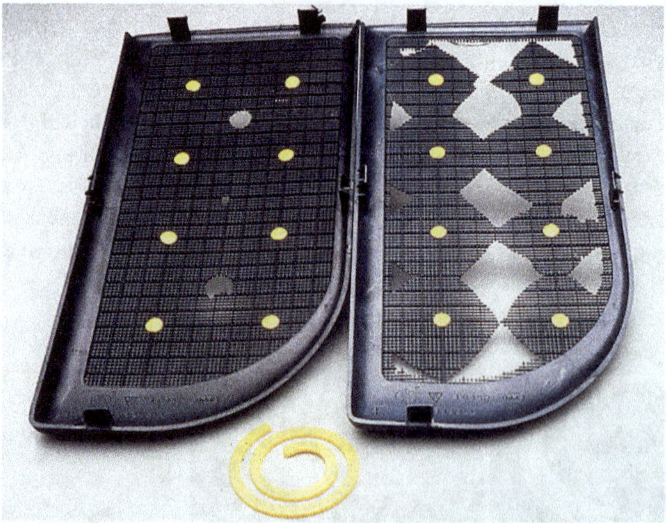

Figure 3.32 Multiple gates on a single part are sometimes needed for large or complicated parts

There are potential problems connected with multiple gating of a cavity:

- *Freeze-off of gates:* With more than one gate per cavity, freeze-up is no problem for cold runner (two-plate or three-plate) gates, because the plastic in the gates is ejected with the product and the gates are always open for the next cycle.

  In hot runner molds with open gates, the plastic pressure opens the gate, which has partially frozen at the end of the previous cycle. If the pressure at the gates is uneven, or if one gate is cooler than the other, the cavity will be filled from the gate that is easier to break through, and no plastic will enter from the other gate(s). For valved gates, this is not a problem, since all gates are opened mechanically or with special heaters.
- *Minimum distance between gates:* Hot runner gates require a necessary minimum distance between gates because of the physical dimensions of the hot runner components as specified by the hot runner supplier.
- *Weld lines and venting:* At the meeting front(s) of two or more streams of plastic, there will be a weld line, which can cause a weakness compared to the strength in other cross-sections of the product. It may also create an unsightly line or even voids on the surface of the product.
- Strength in the area of weld lines can be improved if the air trapped between the inrushing fronts of plastic is permitted to escape, thereby preventing weakening of the weld because of air entrapment and/or burning because of the rapid rise in temperature as the trapped air is compressed. It is important that the plastic arrives at the junction points (welds) hot enough to form an acceptable joint.

3.12.3 Gate Dimple

A dimple is a slight enlargement in the thickness of the product opposite the gate (see Figure 3.33). The purpose of a dimple is to facilitate the plastic flow away from the gate. A dimple opposite the gate creates a small, hot pool of plastic in this well-cooled area of the mold, so that by the time the cavity space is filled, this pool is still warm enough to suck (by shrinkage) any remnant of the vestige into it, and make it (almost) disappear. The dimple can in most cases hardly be seen, but can make a big difference in the quality of the product, and can also improve the cycle time.

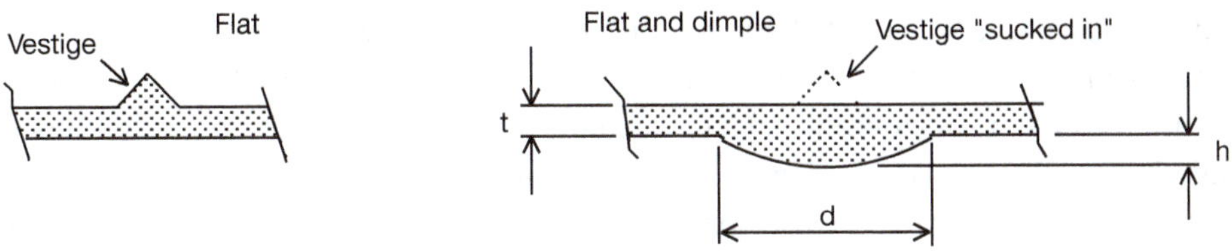

Figure 3.33 Suggested dimple geometry

Dimples should always be used, unless the customer has serious objections. Without a dimple, particularly in thin-wall molding, the product quality may be seriously affected, and the molding cycle may increase. The following dimple dimensions are suggested:

- Diameter: d = 4–10 mm
- Height: $h = 0.75\,t$, for $t < 0.75$ mm, or $h = 1.0\,t$, for 0.75 mm $< t <$ 1.5 mm.

For heavier walls ($t >$ 1.5 mm), there is little need for a dimple: the plastic finds enough area to flow away unrestricted from the gate.

3.12.4 Recessed Gate

A small area can be recessed below the level of the product (Figure 3.34). Even if a protrusion is left standing, it should not go beyond the level, which might later be covered with a label that could be damaged by a gate protrusion or gate vestige. A protrusion may also be undesirable at the underside of a flat-bottomed part, as it can cause a stress riser at the gate and "rocking" of the plastic part.

If the gate can be recessed, the depth of the recess h_r should be at least equal to t. With a wall thickness $t <$ 1.5 mm, the recess should be combined with a dimple (Figure 3.34B), similar to a level surface. Note that the recess shown is spherical, but it could also be any shape that suits the product.

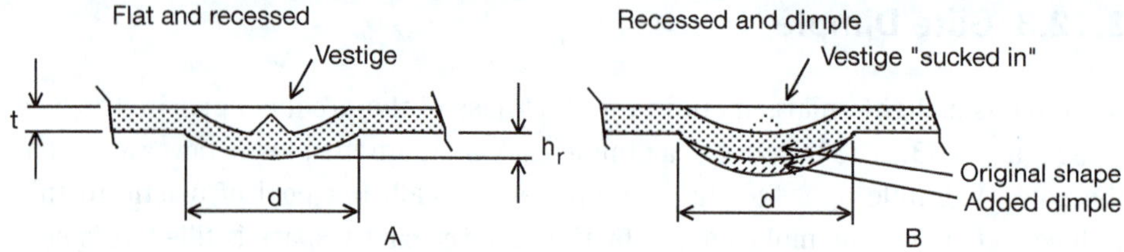

Figure 3.34 Suggested recessed gate geometry

3.12.5 Deep Undercuts and Mold Design Complexity

Take another closer look at the part to be molded. Is there a need for side cores, cavity splits or other methods to release severe undercuts or threads? Any of these features will add considerable cost to the mold (and to the cost of the product), not only because of the added complexity of the molding surfaces, but also because each stack requires much more space than a simple stack without side cores (see Figure 3.35 for an example of a complex mold with side cores). For the same number of cavities, a much larger mold and therefore often also a larger machine size may be required just to accommodate the mold in the available platen area, even though the clamping forces required would be little more than for the mold without side cores or splits.

Side cores and splits, etc. will lengthen the cycle time and reduce productivity compared to molds that do not have such features. These types of molds will also have significantly more maintenance costs due to the complexities of the mold.

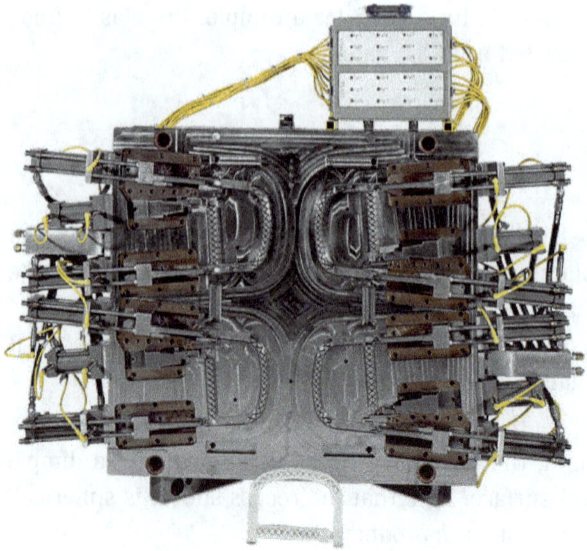

Figure 3.35
Four-cavity handle mold with three side actions per cavity (Courtesy of Top Grade Molds)

3.13 Re-Design of the Product to Avoid the Need for Side Cores

In some cases, round holes or odd-shaped openings generated by using side cores or split cavities could be redesigned without sacrificing the usefulness of the product, and possibly without significantly changing the appearance, by creating such holes or openings in the side walls (or even in ribs inside the product) with a design method where core and cavity meet on a shutoff. This may require the use of special inserts in either or both cavity and core, which may necessitate a change in the shape (or in the draft angle) of the side wall of the product, or require an opening in the bottom of it. In many cases, this could be acceptable for the end use of the product and allow a much simpler, less costly mold. By just giving a bit more thought to the product design before planning and designing a mold, and by understanding the application for which the product is used, a little redesign can often result in spectacular savings in mold and product costs.

3.13.1 Selecting Other than the Conventional Parting Line

Occasionally, the obvious choice of the parting line location would require a side core. However, by slanting the P/L, the product could be molded with a simple open-and-close mold. A good example is a louver (Figure 3.36), but the principle applies to any similar case. In the case of a louver, if the mold is designed with a parting line perpendicular to the mold opening it would require a core pull or side core due to the creation of a deep undercut. However, by tilting the part in the mold slightly, the undercut is eliminated. The cost of a mold with a "slanted" parting line (P/L) is slightly higher than that of a mold with an ordinary P/L, but much lower than a mold with a side core.

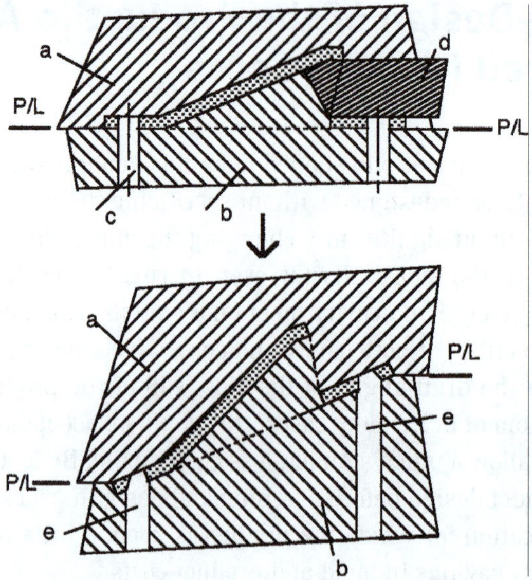

Figure 3.36 Example of a louver: (Top) A side core is needed; (Bottom) A tilted orientation becomes an "up-and down" mold without any side cores

■ 3.14 Shape of Threads and Undercuts

Often, a design specifies threads or undercuts on the inside of the product (Figure 3.37). Is the specified shape of thread or undercut designed with molding in mind? Many such threads or undercuts could be molded without unscrewing, or the need for collapsible cores, by changing the shape of the undercut so that the product can be stripped off the core. If these changes could be made, the undercuts could easily be forced out of the grooves that created them when pushed by ejectors or a stripper. Figure 3.41 shows examples of closure molds for stripped and unscrewed closures. Note the large superstructure above the unscrewing mold.

To create a thread profile for stripped ejection, the thread should have the following features:

- Maximum available radius at the crest and root
- Thread runouts should be short and cut off to avoid cross-threading
- Limit thread pitch to less than 32 threads per inch
- Avoid tapered threads.

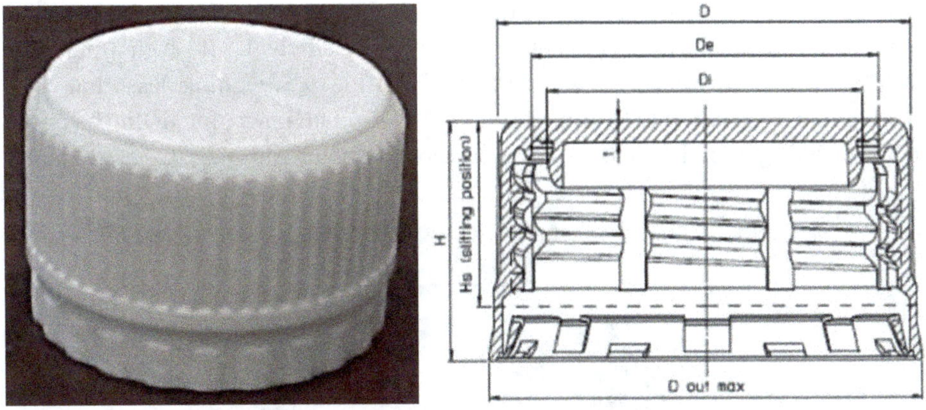

Figure 3.37 Typical bottle cap with tamper-proof ring and stripped thread for simpler ejection (no unscrewing mold required). This product is outside-gated, using a hot runner hot tip gate

Figure 3.38 shows the difficulties of a typical unscrewing mold (see Chapter 11 on ejection for more details and Chapter 18 for example mold designs). The core must rotate out of the cap before it can be ejected. This makes core cooling more difficult, as the coolant must pass through a rotary joint, resulting in 30% longer cycle times than a stationary core. Unscrewing molds are much more complicated than bump-off (or stripped) closure molds (see Section 18.22 for an example), and have the potential of additional issues such as seizing of components and scoring. A 48-cavity unscrewing closure mold would probably require a 500 t (5500 kN) machine to have adequate tie bar spacing. Due to the limited core cooling and larger machine, the mold cycle time would likely be > 10 s, meaning a productivity of ~17,000 caps per hour.

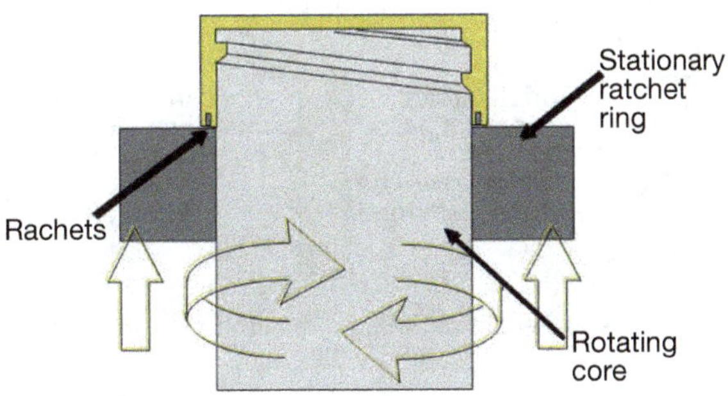

Figure 3.38 Schematic showing the difficulties of a typical unscrewing mold

Figure 3.39 shows a schematic of a much simpler mold, where the thread (and the cap) can be stripped. Here, core cooling can be very efficient. The cycle time for a typical (28 mm) bottle cap made from HDPE MFI 19, weighing less than 3 g, molded in a 48-cavity mold running in a 225 t (2500 kN) machine, is in the order of 3 s, meaning a productivity of ~57,000 caps per hour.

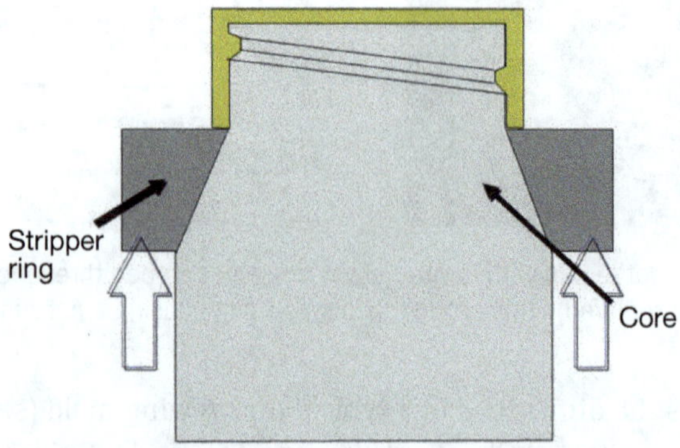

Figure 3.39 Mold where thread can be stripped

Figure 3.40 exemplifies how a small change in the angle of the flank of the thread can allow a thread to be stripped from the core, rather than requiring an unscrewing mold. Small changes like this can have a major impact on product cost, because mold cycle, cost, and maintenance will be significantly improved with a stripped product.

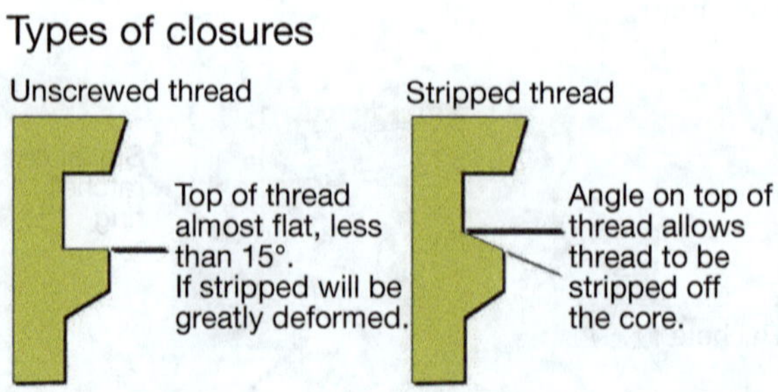

Figure 3.40 Change in flank angle allows thread to be stripped much more easily than if unscrewed from the core

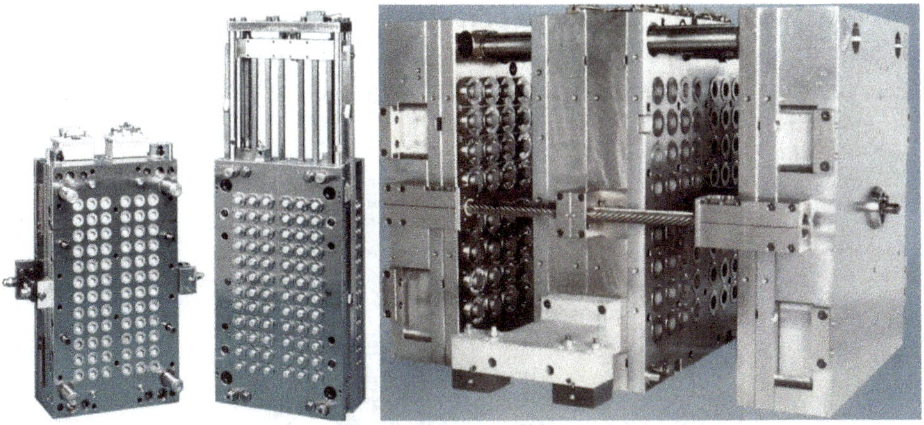

Figure 3.41 (Left) 72-cavity unscrewing mold (Courtesy of StackTeck Systems). (Right) 2×64-cavity stripped closure mold

3.15 Need for Multi-Stage Ejection

Sometimes the part design may require that the part is ejected in multiple stages to free it from the steel in the mold. There are several examples in Chapter 18 that show ejection of this type of part. This applies to a shape or design feature of a product consisting of:

- Deep ribs, bosses and undercuts on the core side, such as the example shown in Figure 3.42 (even without the thickening at the end of the rib as shown in this illustration).

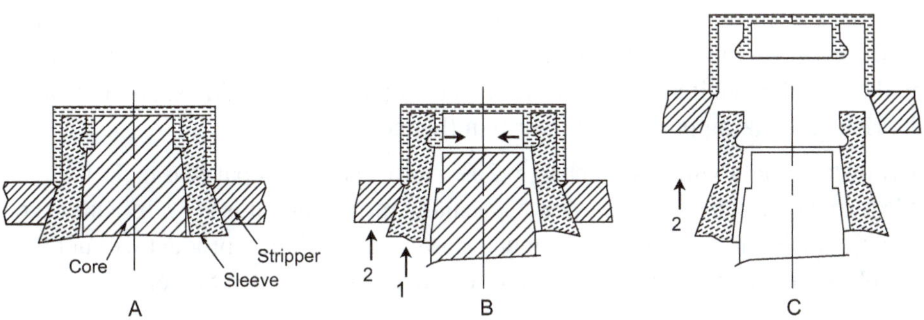

Figure 3.42 Schematic of a two-stage core ejection sequence: (A) mold is closed; (B) mold opens and the sleeve and stripper follows core for a limited distance to ensure that the rib becomes free; (C) once the mold is open past the undercut, then the second stage or stripper ring can advance

- Deep ribs or undercuts on the cavity side. An example is shown for a false-bottom cup in Figure 3.43. The depth of the rib *F* and the ratio of the thickness of the rib *t*, as well as the draft angles of the rib, are critical considerations to determine if multi-stage ejection is needed.

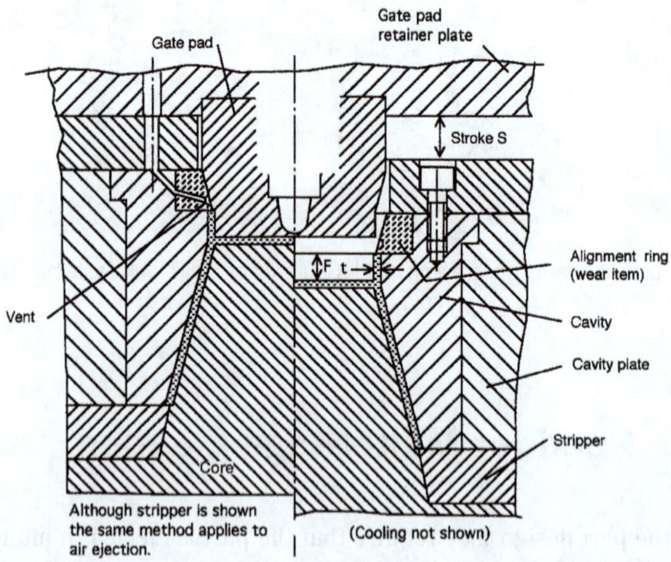

Figure 3.43 A product with deep ribs in the cavity is ejected in two stages. On opening, the part is kept on the core by moving the cavity with the core for a distance referred to as the "stroke" (*S*) to free the rib. At this point, the cavity plate stops and the mold continues to open to the "mold-open" position, whereupon the part is ejected from the core by the stripper ring

In both cases, if the ratio $F/t > 2$, or if there is any thickening at the end of the rib, either a "two-stage ejection" or a "moving cavity" are necessary, which will *increase the mold cost* by about 15–20%. In both cases, it is important to provide especially good *venting* at the end of the ribs to ensure proper filling. Failure to use these methods will make it very difficult to withdraw (eject) the products, and increases the risk of breaking portions of the rib in the mold.

A two-stage mold will cost about 15–20% more than a comparable mold without this feature. Also, because the sleeve is usually rather thin, it is very difficult to get cooling into it. The mold will cycle much slower than a similar product without this complication, and the maintenance cost of such molds is much higher.

Moving cavities are more complicated and cost about 10% more than a mold without this feature. Some molders use it (despite its higher cost) for products even without a false bottom, because they can cycle even faster than a mold with a conventional cavity.

3.16 Post-Molding Operations versus Mold Complexity

While it is possible to mold almost any shape or feature, the cost and complexity should always be weighed against alternatives. Sometimes, molds can be much simplified by doing additional work to the product after molding. Post-molding operations are of particular importance whenever relatively small quantities are to be made. For example, one or a few simple holes or openings in the side wall of a product would require a side core in the mold, but such holes or openings could also be drilled or die-stamped after molding. Such additional operations may require a drilling fixture or a stamping die. The actual time (direct labor) for such post-molding operations and any costs for tools or fixtures would have to be added to the total cost of the product. But such post-molding operations could also take place later at the assembly line, where the product is assembled or packed, without any additional labor cost, if properly integrated into the process. Again, it is the overall cost of the end product that is important, not just the cost of the mold or the molded piece itself. In many cases, the savings in the mold cost achieved by eliminating a side core (or some other complications of the mold) can be substantially greater than the combined additional cost for fixtures or tools, plus the cost of the additional direct labor to finish the product.

A typical example for this would be the need for small holes for a hinge pin (for a hinged lid), located in two lugs projecting from the bottom of a product (see Figure 3.44). The plastic melt is injected into the bottom of the product, near the lugs. It is of course feasible to mold these holes, but it could be quite difficult to arrange the side cores required for such holes as well as the actuation for such side cores, without interfering with the gating and the cooling layout in this area. It would be, however, quite easy to just mold the lugs as projections from the container bottom, and then drill the holes, using a simple drilling fixture.

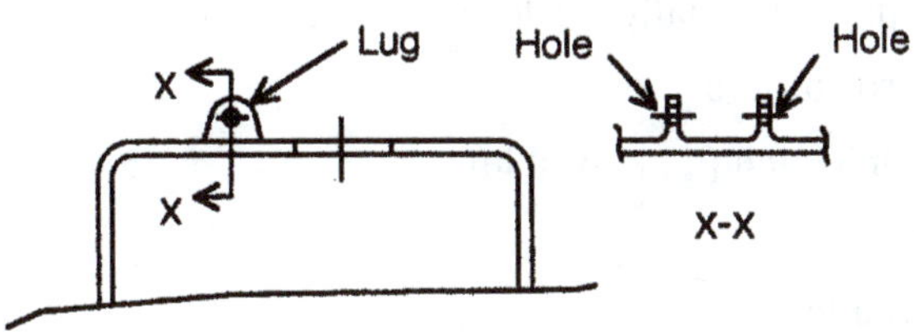

Figure 3.44 Lugs with holes on plastic part

Another common example of this is with tamper-evident bands on closures. In some cases the tamper-evident band is slit after molding to eliminate the need for slides to create shut-offs in the mold.

3.17 Plastic Part Tolerances and Effect on Mold Design

Next, the mold designer should look at the specifications relating to accuracy and tolerances.

Unfortunately, it is often the case that after a product has been conceived, the design has been either just sketched by the inventor or an artist, or a model has been created. This information has then been passed on to a designer to be put into CAD. This may result in a good visual description of the new product, but to be practical for manufacturing, any drawing must be fully dimensioned, and *intelligently* toleranced. To design a product for injection molding requires certain knowledge of this technology. A design that may be suitable for one method of processing plastics (or other materials) may be unsuitable or impractical for another process, even though the end use is the same.

For example, a disposable drinking cup of a specified capacity could be made from paper or Styrofoam, and be thermoformed from sheets, injection-molded, or made by another, entirely different, new method or material. The final product design for each of the above-cited materials and methods would most likely look different, to suit the method of manufacturing and the selected material.

Also, while the *dimensional accuracy* of the product for its final use (i.e., as a drinking cup) may be of little importance, its actual dimensions will require high accuracy because of demands *not related to its use,* such as stacking height (e.g., for packaging), ease of releasing the individual cups from the stack (as required in automatic vending machines), and printing – but mainly because even small variations in wall thickness may have a great effect on the mass of plastic used for each unit and on the molding cycle.

A design for a metal product is different from the design for a similar product made by injection molding, even though the products may be fully interchangeable in their use. This applies especially for design features such as:

- Radii and sharp corners
- Flow path for injection (if applicable)
- Wall thickness
- Ribbing and reinforcements
- Openings (round or shaped).

These features, by their presence or absence, not only affect the making of the mold (and its cost), but also affect the speed of the molding operation itself. It is very important to understand that it is relatively easy to achieve close tolerances for the mold parts, which are usually made from metal. However, the plastic products made by the mold do not solely depend on the mold dimensions. The designer must be aware that the final size of the product is greatly affected by variations in the shrinkage of the plastic (see Appendices 8–11), which in turn are caused by variations in molding conditions (pressures, temperatures, and timing) and by variations in the composition of the plastic not only from batch to batch, but also from manufacturer to manufacturer. All this makes it very difficult to mold products dimensioned within close tolerances.

Using unsuitable, old, and/or poorly maintained machine tools makes it even more difficult to make mold components to close tolerances. The accuracy of the work depends much on the skill of the machinists, and even with good checking equipment can become time-consuming, because it requires frequent measuring of the closely toleranced dimensions. The alternative is to use good machine tools, or even machines specially designed or adapted for certain steps in the manufacture of the mold parts, requiring much higher investments by the mold maker. Either one of these conditions affects the cost of machining and explains why close tolerances can be expensive to achieve.

Note also that dimensions are affected by the ambient temperature of the machine shop and that even when cooled by cutting fluids, the work pieces heat up during machining. They will measure larger when warm immediately after cutting than after cooling to room temperature. Of course, the larger the dimension, the larger the dimensional differences caused by heat expansion.

Always remember that tighter tolerances mean higher mold costs, maintenance, and inspection. As can be seen in Figure 3.45, the mold cost increases exponentially with the tightness of the tolerance.

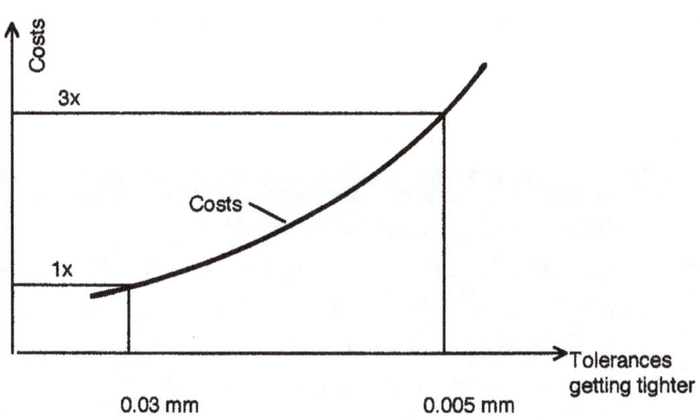

Figure 3.45 Relationship between tolerances and mold cost

Without giving actual cost figures, the curve just shows how costs can increase, as the tolerances get tighter. *The cost to achieve a 0.005 mm (0.0002") tolerance can be three times the cost of a 0.03 mm (0.0012") tolerance.*

Other points that should be clarified when looking at product dimensions with close tolerances are:

- How will these dimensions (or the entire product) be checked (measured) on the finished product? With vernier calipers, micrometers, gages, measuring machines, fits with other products?
- When will the parts be checked? Immediately after ejection, one hour later or 24 hours later?
- Will there be 100% inspection or statistical (random) inspection?

Clarifying all this ahead of time can avoid much future unpleasantness or arguments.

3.17.1 General and Specific Plastic Part Tolerances

All tolerances must be specified on the product drawing and must be looked at by the mold estimator or designer when starting the project to see if they are reasonable. The Plastics Industry Association (formerly the Society of the Plastics Industry (SPI)) has a suggested list of practical general tolerances for injection-molded products. For more information, go to their website, *www.plasticsindustry.org*.

In most cases, these tolerances are satisfactory and achievable. Specific, closer tolerances may require experiments to be made with cavity and core sizes, and under various molding conditions, to achieve the required sizes. This can mean considerable added costs for the mold maker, a higher initial mold cost, higher maintenance costs and lower yields.

Table 3.3 suggests tolerances to be used on plastic product drawings (radii are not toleranced):

Table 3.3 Suggested Plastic Part Tolerances

Measurement	Tolerance
Product weight	±10% on projected weight (range ±2%)
Wall thickness	±0.03 mm (0.013 mm in special cases like thin-wall packaging)
Fit diameter	Up to 75 mm ∅ → ±0.20 mm
	Up to 106 mm ∅ → ±0.25 mm
	Up to 160 mm ∅ → ±0.30 mm
	Up to 300 mm ∅ → ±0.64 mm
Overall height	±0.5% or 0.13 mm minimum
Stacking height	±0.5% or 0.13 mm minimum

Note that the steel size requirements, and thus the difficulty of manufacture, are dependent on the plastic tolerances on the product drawing.

3.17.2 Are Special Fits with Matching Products Required?

Often, certain dimensions of a product are specified with unnecessarily close tolerances, when the designer wanted to convey that the product should fit suitably on another product (tightly or loosely), typically, a container and a matching lid. This requirement must be clear. Especially, when molding plastics with high shrinkage factors (e.g., PP or PE), it can be difficult to arrive at the proper "steel" dimensions, and some experimenting may be required to achieve the required fit. Specifying the matching diameters with standard, loose tolerances may yield pieces correct in size, but wrong because the fit is not as desired. The alternative, providing closer tolerances, could be unreasonable, because the dimensions of the molded product depend not solely on the steel dimensions of the stack parts but also on the molding parameters.

In such cases, it is advantageous to complete the more complicated mold first and test it in actual molding conditions until the best cycle time is established. The critical mold parts of the matching product (e.g., the lid) should be finish-machined only after having established what the actual molded container dimensions are. This could require completing the lid mold with only one cavity, using assumed suitable dimensions, testing the unfinished mold until the best cycle is achieved, and then adjusting the assumed dimensions so that the proper fit can be achieved. All lid mold parts can then be finished.

Some mold makers will make an entire mold at a "steel-safe" condition in the areas requiring fitting. Once the mold has been run and parts measured, these steel-safe dimensions are re-machined to a final dimension.

An alternative to this approach is to build a single-cavity prototype of the two parts to be matched. Once these parts have been molded then re-cuts can be made to determine the final sizes for the desired fit. Note, however, that this has limited validity, because the shrinkage on a single cavity mold can differ from the production tool due to fill, packing, and cooling differences.

3.17.3 Tolerances for the Filling Volume

This applies mostly to parts into which a more or less viscous product will be filled by volume to within closely specified limits (typically, containers for margarine, paint, etc.). In their end use, it is important for the seller that a minimum amount must be filled into the package without shortchanging the buyer, but also they should not be overfilled, which would mean a loss for the seller. There should be clearly defined fill lines (usually inside the container) to mark the minimum and maximum volumes. This can be a problem with plastics with large shrinkage factors such as PE and PP. It requires special consideration when dimensioning the cavity and core because of the unavoidable variations in shrinkage values, as the plastic flows away from the gate and slowly cools, and as the injection pressure within the mold decreases. The same considerations apply to measuring cups or vials which have the various levels (or volumes) indicated by lines on the sides of the product. It may be necessary to first test the mold to find the best cycle times, and then establish the location of the measuring lines.

3.18 Stacking of Products and Free Dispensing

Product design features are often needed for efficient stacking, shipping, and storing of empty products in bulk, but have nothing to do with product use (i.e., ribs, lugs, shoulders, draft, and undercuts) (see Figure 3.46, Figure 3.47, and Figure 3.48). These features often compromise the product function. For example, recycling bins lose valuable internal volume due to need for a large draft on all sidewalls for stacking empty bins. The walls ideally should be vertical (like cardboard boxes) but this makes the products difficult to transport and injection-mold. In the case of food containers, these stacking features become useless as soon as the product is added, and then different features are used on the product design to permit proper stacking and shipping of filled containers.

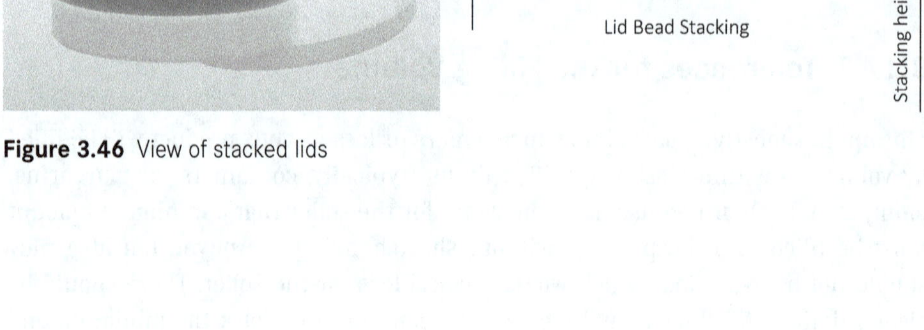

Figure 3.46 View of stacked lids

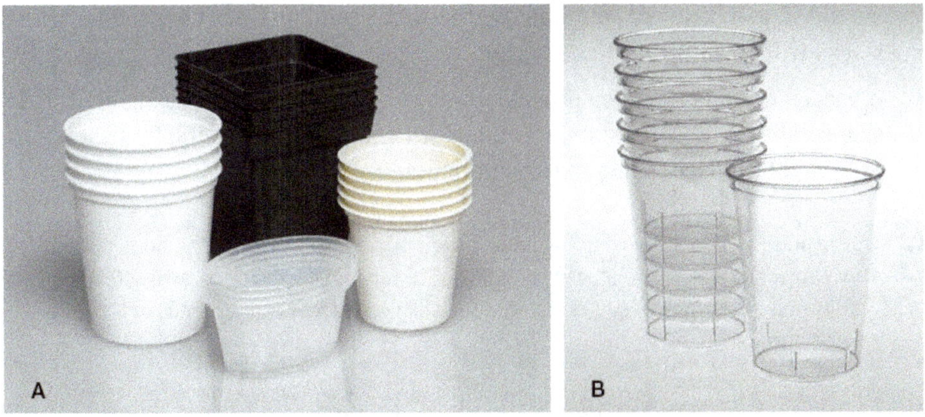

Figure 3.47 Views of stacked products on (A) stacking shoulders, and (B) stacking lugs

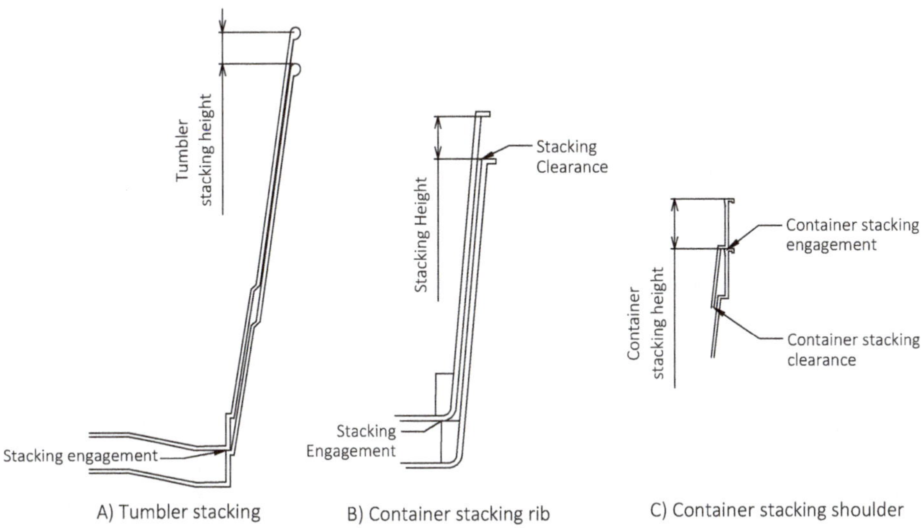

Figure 3.48 Container stacking details

Any product stacked for shipping must have a clearly defined stacking height (Figure 3.46 and Figure 3.48), which is usually created by resting the outside or the bottom of one piece on the inside stacking provision of the following piece. These provisions for stacking can be "stacking lugs", or clearly defined steps in the product. The purpose of these stacking lugs (or steps) is:

- The products must not jam when pushed together, which would make it difficult to separate them when required by the user.
- They will ensure a total stack height of a certain, specified number (e.g., 20, 25, 40, etc.) of the products when stacked. The stack height should be suitable

for the size of boxes or containers (preferably, standard size cartons) in which stacks will be shipped.

- If special cartons are to be provided, it may be necessary to investigate if their size will be suitable for standard rail or sea shipping containers, for best use of the available space inside these containers.

> Ensuring a proper stacking height is essential, as a small difference adds up when many parts are stacked together. For instance, a 0.05 mm (0.002") difference in stack height is 2.5 mm (0.100") when 50 parts are stacked on one another.

Stacking is more difficult if the angle of the sidewalls is small. Obviously, a cylindrical container (0° draft) cannot be stacked at all. A typical disposable drinking cup has an angle of approx. 7°. With certain considerations such as gaps and clearances, parts can be stacked with draft angles down to 3–4°. Larger angles stack and de-nest more easily.

Problems can also arise when parts are used in an assembly line or in a dispensing mechanism (e.g., vending machine), where it is important that the parts will release easily, without fail, from the stack, i.e., they must not get "hung up" by vacuum or by friction because the gap between two stacked containers is too small, even though they are properly stacked as designed. Even the surface finish is a variable in de-nesting of parts. A polished finish is harder to de-nest than a rougher one. When the gap between two sidewalls is very close, static electric charges may also prevent the lowest part from falling from the stack when desired. Some dispensers have mechanical separators and don't depend on gravity, but it is preferable not to depend on having such separators (added costs). It is highly recommended to make sure that any stacking height dimensions are carefully checked *before* beginning to build a mold. If they are wrong, the mold must be changed after finishing, or the packaging (carton size) has to be redesigned after the height of the stack was not as originally planned. Occasionally, a mold maker may decide to make slots in the mold cavity for the stacking lugs by electric discharge machining (EDM) into the core only *after* the mold is finished, rather than do it before and then have to increase the height later. The disadvantage of this method is that the mold has to be dismantled to be able to machine the cores (costs!). The advantage is that a minimum stacking height can be achieved.

A part drawing must show the stacking arrangement of the parts and be signed off by the customer. For reference, below are some guidelines for stacking clearances and stacking engagement used in the thin-wall container and lid industry.

Recommended stacking clearances are as follows:

Plastic lids (Figure 3.46):

Stacking shoulders and beads: 0.3 mm < clearance < 0.5 mm

Stacking ribs: 0.20 mm < clearance < 0.45 mm

Plastic containers (Figure 3.48):

There is a definite relationship between the required stacking clearance, the container height and its sidewall draft (Figure 3.49).

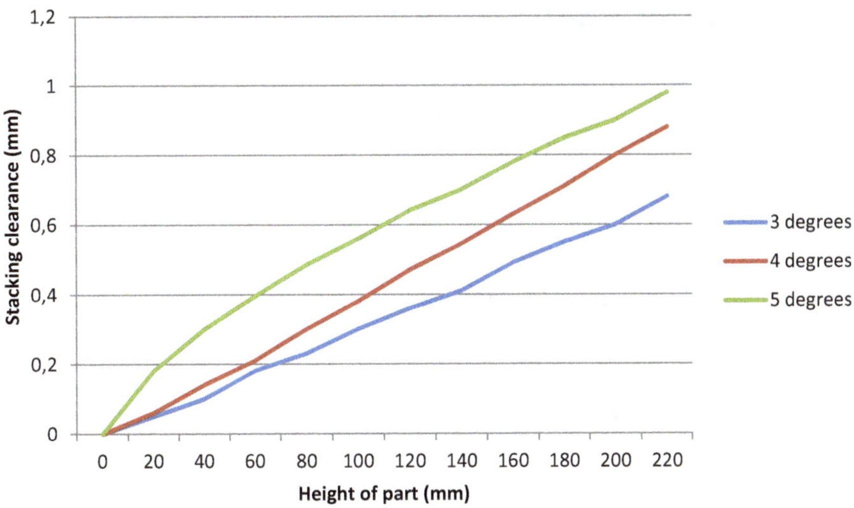

Figure 3.49 Stacking clearances used for molded containers

Stacking clearances allow the parts to "de-nest" easily for printing, labeling, and dispensing. However, there must be good engagement between nested parts so that they do not compress and lock together. The following tables provide recommendations on stacking engagement for common thin-wall molded parts as a reference for other parts.

Table 3.4 Recommended Stacking Engagements for Containers

Stacking type	Nominal engagement (mm)	Minimum engagement if part shifted to one side (mm)
Shoulder	0.8	0.6
Stacking ribs	1.0	0.8

Table 3.5 Recommended Stacking Engagements for Lids

Stacking type	Nominal engagement (mm)	Minimum engagement if part shifted to one side (mm)
Shoulder	0.5	0.2
Stacking ribs	1.3	1.3
Bead	0.6	0.25

3.19 Deliberate Mismatches for Easy Mold Design

Trying to produce a "perfect match" between two surfaces is not only difficult to achieve but also very costly. Designers therefore create deliberate mismatches for ease of manufacturing and improved part aesthetics.

There are two areas of "deliberate mismatch" to consider, and two typical examples are shown. There are many variations of matching parting lines, or between lids and covers, but the basic principle applies to all of them.

3.19.1 Mismatch at the Parting Line

Always consider:

- Is the rounded edge really necessary?
- Is the sharp edge really necessary?

First, it must be clarified, whether a rounded edge is *really necessary* for the product. In many cases, the product designer may not be aware of the possible additional cost involved to produce a round edge as in Figure 3.50a or c, and will often agree that a simple, "sharp" edge, as in Figure 3.50b or d, would be just as acceptable for the application.

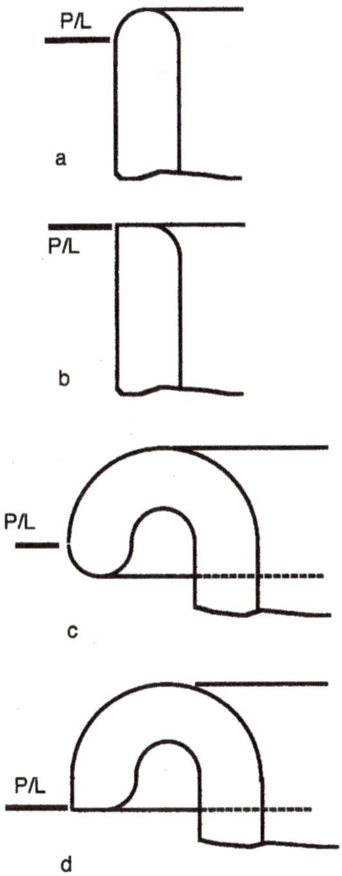

Figure 3.50 Typical round edges where a "sharp" edge could be considered

Figure 3.51 shows just one of several designs of a round edge, with the ideal case (1) having a perfect match at the parting line. However, due to the buildup of manufacturing tolerances of the mold parts, such an ideal case is not practical. In reality, the nominal diameter *D* of the cavity, or of the core, will be either larger or smaller than the matching one, and create either a hook (2), which is generally not tolerable, or a small step (3), which in most cases is perfectly acceptable. Note that the actual differences caused by the tolerances of the diameters are small, usually less than 0.1 mm (0.004"), so that a step would not be more than about half this amount. However, a step is much less noticeable than a hook.

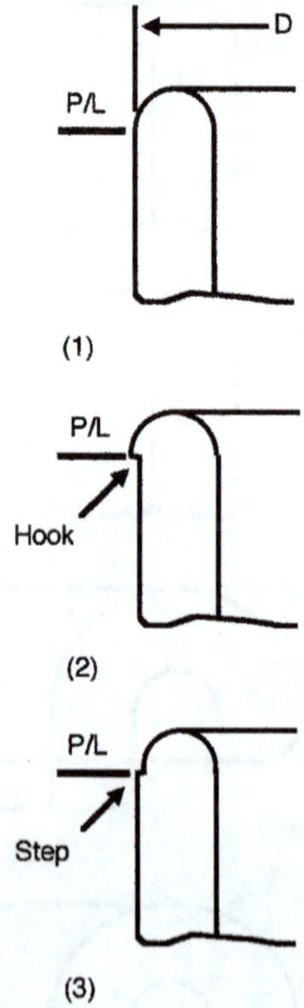

Figure 3.51 Round edge: (1) Ideal; (2) With "hook"; (3) With "step"

In fact, a mismatch can be corrected by very time-consuming handwork, by grinding or stoning (polishing), but this should be avoided because of the high cost.

The suggested proper (and most economical) approach is to dimension the matching diameters so that there is always a step, as shown in item (3) of Figure 3.51, of a magnitude between 0 and 0.1 mm (0–0.004").

If the mismatch is to be in a location where the consumer uses their hand or lips, then the location of the mismatch must be verified with the molder or OEM to ensure this will not be a problem once molded. It may be preferable to offset the mismatch in one direction or another to hide the mismatch.

3.19.2 Mismatch between Two Matching Pieces, such as Box and Lid

The conditions are similar when designing and building molds for "matching" boxes and lids. Here, a deliberate mismatch is even more important, because the products may come from different cavities and even molds, made under varying molding conditions, and the mismatch due to build up of many tolerances (in cavities for both products) could be much larger.

Figure 3.52 shows the ideal condition (1), which is difficult to achieve, and a way to minimize the effect of a mismatch between matching parts (2). There is also another way shown by adding a "decorative" band to the larger part (3).

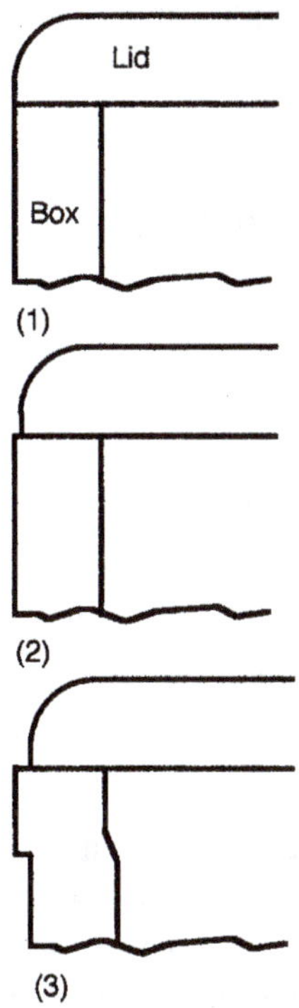

Figure 3.52 Mismatch avoidance options between box and lid

■ 3.20 Surface Finish

The finish of the molding surfaces, and the fitting surfaces where mold parts meet, are important cost factors. The finer the machining finish, and the more hand-finishing is required, the higher is the mold cost. This appears to be obvious but is often overlooked or neglected.

The relationship between surface finish and costs and the relationship between tolerances and costs (as shown in Figure 3.45) are very similar and apply here too.

All agreed-upon finish specifications must be shown on the finally approved *product drawing*. SPE (Society of Plastics Engineers, *www.4spe.org*) have produced a series of standard finishing specifications, which can also be translated into the finish in *microns* (thousands of a millimeter). They are a good method of specifying finishes, but additional information may be required on the drawing to *clearly specify to which areas* these specifications apply. The mold designer should never accept a *general finish* unless it is easy to produce, or the cost of it will be factored in the mold cost.

Appendix 13 contains a table of typical molding surface finishes for molded products.

3.20.1 Finish of Molding Surfaces

Molding surfaces (the areas in contact with the plastic product) are finished for the following reasons:

1. To provide the required appearance or function of the product.
2. To ensure that the product can be easily ejected from the mold. However:
   - Occasionally, a relatively rough surface in specific areas may be beneficial to keep the product on the side of the mold from where it will be ejected
   - On the other hand, sometimes, a high polish is needed for appearance or function, but could be detrimental to easy ejection, depending on the design of the product.

Especially with very thin-walled products, the surface finish of the cavity space affects the plastic flow into the cavity space. A finer finish results in faster filling and shorter cycle times. In some cases (notably with PS), flash chrome plating over a highly polished area can increase the productivity of the mold by up to 10% (see Figure 3.53 and Figure 3.54).

Figure 3.53 Typical PS tumblers

Figure 3.54 PS tumblers and core/cavity show the highly polished finishes required to achieve the glass-like look of the molded cup (Courtesy of Husky Injection Molding Systems Ltd.)

Finishing (polishing, etc.) the mold parts is generally an expensive activity in the mold-making process because much handwork is required, and should be limited to those areas that *really* require it. Most mold makers today use hand-operated mechanical and some fully automatic methods to finish a surface, but there is still much need for hand-finishing wherever the shape of the product does not allow easy access for mechanical or automatic equipment (a typical polishing department is shown in Figure 3.55).

Figure 3.55 Polishing area in a shop (Courtesy of Husky Injection Molding Systems Ltd.)

The purpose of finishing, in general, is to remove the tool marks remaining on the surface of a work piece. In many cases, the rough, "as-machined" finish after metal removing operations (turning, grinding, milling, etc.) could be quite satisfactory for the appearance of the product, for example on the inside surface of a technical product (enclosures, boxes, television cabinets, etc.), but this may not always be satisfactory for the ejection of the product, because the plastic will not easily (or not at all) slide over a surface that is too rough. It is also important to consider in which direction the rough machining grooves lie: to be in line with the ejection could be satisfactory, but across it is usually not acceptable.

Also, the draft angle of a wall (or of the sides of a rib) is important. With little draft (a small draft angle), the surface finish must be much better, whereas with a large angle (approx. 5° or more), a much rougher finish, such as "as-machined", could be permissible. With the need to design for less and less mass, the draft angles, especially of ribs, must be kept small, and these walls therefore need a good finish, but not necessarily a polish. A good finish *in line* with the ejection motion ("draw-stoning") will usually be good enough. If ejectors can be placed under such ribs, the finish becomes even less of a problem. In addition, we must always consider what would happen if a piece of plastic breaks off inside a rib. It may save time in the making of the mold but can become expensive later, if the service personnel are frequently required to remove some broken-off bits of plastic from the mold, causing severe delays in production.

Grinding and electric discharge machining (EDM) leave smaller tool marks on the worked surface. Such surfaces may not need any further finish, except polishing where required for appearance. An EDM finish can vary from rough to very fine, and in the latter case may not require any polishing at all. A rough finish is the result of high currents and faster cutting speed, and therefore requires less time.

In addition, with today's new methods of finish turning and milling hardened surfaces, the finish achieved is often as good as a ground finish, and no further polishing is required.

3.20.1.1 Molding Surface Finish for Ease of Ejection

There is a relation between molding surface finish and ejection. The designer must be aware that the finer the finish of the plastic part, the more attention must be paid to parts sticking to the core and cavity. General guidelines regarding surface finish are listed below:

- Specify finishes that will permit easy ejection, such as grinding or draw-stoning, in the direction of the withdrawal of the core, core pins, or side cores.

- Where it is expected that the product will not naturally stay with the core when the mold opens, a rougher finish should be specified, and milling, turning, or grinding across the direction of withdrawal may be required to keep the product on the desired side. In extreme cases, undercuts that are not specified on the product may have to be added. The size, number, and location of such undercuts must be discussed with the customer.
- Use as rough an EDM finish as possible to permit faster EDM speeds and less machining time.
- Certain plastics (e.g., LDPE, PP) require a rough finish for easier release. Typically, the cores of LDPE overcaps are vapor-honed (sand-blasted) to provide these rough surfaces. Often, after a time in operation, the mold surface may have to be roughed-up by hand, using fine sandpaper, to provide the necessary rough surface condition.
- Cavities may have to be chrome-plated and/or treated with Poly-Ond™, Dicronite™, or other special finishes, to ensure that the products stay with the cores.

3.20.2 Texturing of Surfaces

There are also other surface finishes for appearance, such as *texturing,* to create leather, basket weave, or other patterns (see the tumbler in Figure 3.56 as an example). If it is a deep pattern, it should be clear if any related dimensions apply to the highest point of the pattern or to the base where it is applied to. A rough EDM finish is a good and inexpensive solution for a good-looking *matte* surface.

Figure 3.56 The etched cavity wall gives this tumbler a frosted look

3.21 Fitting Surfaces of Mold Parts

This applies to all surfaces of mold parts that abut on other mold parts, but are *not in touch with plastic*. Usually, grinding or fine machining surfaces are required where the dimensions stack up and their sum must be held to close tolerances. Otherwise, ordinary turning and milling surfaces are sufficient.

We have dwelled on the finishing of mold parts to highlight the importance of properly specifying *how* and *where* a mold (mold part) needs to be finished (polished) because of the cost. The mold designer should analyze whether the finishing specifications shown on the product design are realistic and really necessary for the functioning or use of the product, and discuss it with the product designer. This can result in great savings, reduced delivery time, and improved productivity (output of the mold).

3.22 Engravings

The term engravings covers lettering, lines, ornaments, logos, and other types of identification or other marking of the plastic part.

3.22.1 Engravings versus Applied Labels

Engravings in the mold represent a one-time cost. Therefore, in the long run the cost of the finished product is less than the cost of applying labels made from paper or plastic film to the molded product. However, an engraving on the molding surface will have limitations such as color, depth, size, and location (see Figure 3.57).

If labels are applied in a separate operation, this cost must be added to the cost of the product. In recent years the use of in-mold labeling has become quite popular, and the technology has advanced significantly to allow for full wrap of the parts. In-mold labels are vibrant and rich in color but are slightly more costly than standard multi-color dry offset printing. In either case, the cost of the labels, or ink and additional equipment, must be added to the part cost.

Figure 3.57 Engraved products

We must not forget that the same product could be used for different end-user applications (for example, different chemicals are sold in containers of the same size) and/or for different end users (manufacturers). In either case, labels applied after molding would make more sense than changing mold components for a different engraving. Whether to use engraving throughout, labels alone, or part engraving and part labeling, must be decided in view of the quantities of pieces to be produced and the flexibility needed in each case. There are other methods of applying information to a plastic product such as printing, hot-stamping, and others.

As a general guideline, we can assume that the cost of the molded product increases approximately as follows:

- Engraving: By a small amount
- Printing: By approx. 10%
- Labeling: By approx. 20–50%.

3.22.2 Two-Color and Two-Material Engraving

Buttons (typewriter keys, pushbuttons, automotive knobs and buttons, etc.) with two materials or colors molded in one molding setup, are another (quite complicated) method of marking molded surfaces (Figure 3.58 and Figure 3.59). Originally, these buttons or keys were molded with (depressed) engraved "text" (alphabet letters, symbols) and the molded recesses thus created were then filled with paint. This was expensive hand work. In addition, raised engraving is very expensive to make in the mold (see below). Today, most mass-produced keyboard keys for computers, etc., are printed by various methods.

Figure 3.58 Printed keyboard keys

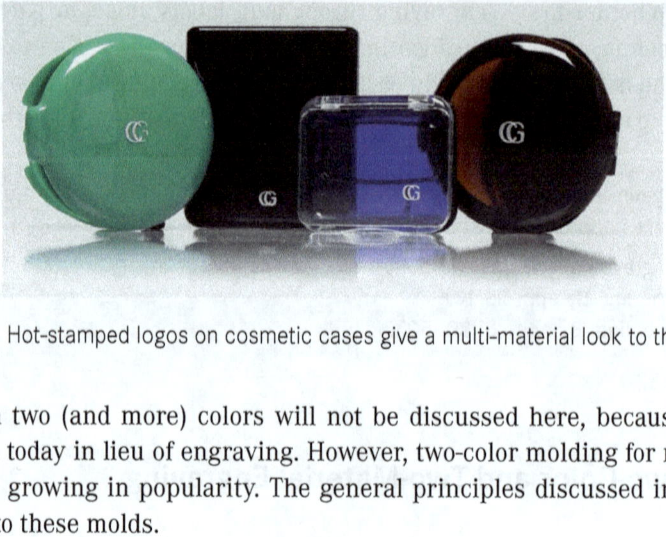

Figure 3.59 Hot-stamped logos on cosmetic cases give a multi-material look to the products

Molds with two (and more) colors will not be discussed here, because they are rarely used today in lieu of engraving. However, two-color molding for many other products is growing in popularity. The general principles discussed in this book also apply to these molds.

3.22.3 Depth of Engravings

It is important to understand that engravings that are to appear *depressed* (appearing engraved) in the surface of the product are created by *raised* features in the mold. Conversely, *raised* features in the product are created by *depressed* (engraved) features in the mold. It is amazing how many product designers do not realize that it is fairly easy to engrave *into* a steel surface, but very time-consuming (and costly) to create engravings *projecting from* a surface.

Many designers, when confronted with these facts, confessed they did not know that it makes such a difference, and readily changed their design to "raised in the product". The only time when it may be *really* necessary to have the engraving depressed in the *product* is when the lettering will be filled with paint, after molding, for better readability, special effects, or for special, artistic designs, usually associated with high-quality products, such as technical enclosures for hand-held devices (cellphones, etc.) containers for cosmetics (compacts) and so forth.

Occasionally, when the raised lettering in the product is objectionable, there is always the possibility of depressing a panel and have the engraving on this panel, so that the top of the engraving is level or slightly below the main surface (see Figure 3.60).

From Figure 3.60 engravings should be no deeper than H = 0.25 mm (0.01"). The width of the engraving should be 1.5–2 × the depth, and the engraving should have an angle of 30° on all walls for ease of ejection and visibility.

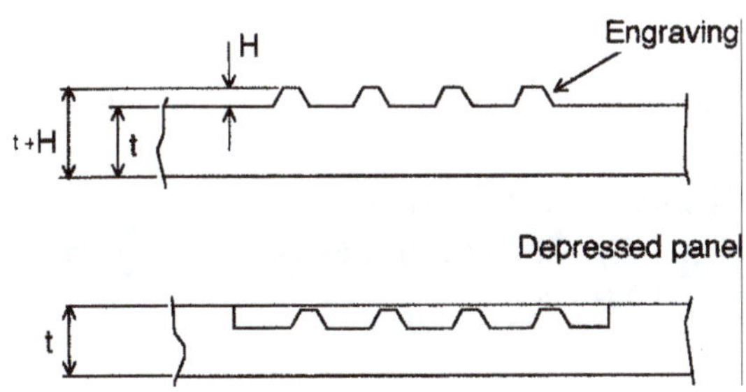

Figure 3.60 (Top) Raised engraving on top of product. (Bottom) Raised engraving in depressed panel. t = wall thickness of the product; H = height (depth) of engraving

3.22.4 Font Style and Size of Artwork

For general applications, such as cavity marking or manufacturer's identification, the style (font) or size of the lettering may not be very important. The lettering should be (pleasantly) proportional to the size of the product and be easily readable. The mold maker may have only a certain range of styles and sizes available, so using these will be less expensive. If the engraving has special requirements (as in the case shown in Figure 3.61), the product designer must supply the artwork from which the necessary templates or models are made for machining. The mold designer and product designer must agree on the form of artwork best suitable for the mold maker, as there can be costs involved in preparing such artwork, for example in the size (photo-enlargement) and material (Mylar film, etc.) required.

Figure 3.61 Picture of artwork on a molded part

The smallest acceptable size of engraving should be considered.

> A suggested minimum size of molded in lettering is 8 pt, to be readily legible, but 6 pt could be used in exceptional cases. The recommended depth is 0.08–0.13 mm (0.003–0.005").

In all cases of engravings, it is also important to consider the cost of removing the burrs (by hand or mechanically) after cutting the steel, to prevent unsightly, fuzzy outlines of the engravings on the molded products.

3.22.5 Polarity of Engraving

We shall define *positive* engraving as any engraving such that will appear "readable" to the user. *Negative* engraving is the inverted image, e.g., as ordinary lettering would appear in a mirror. This may seem obvious but it still does require some comments. Most engravings are viewed from the *outside* of the product (top, side, or bottom), regardless of whether the plastic is opaque, transparent, or translucent. In all these cases, the engraving must be *negative* to appear in the molded piece as "readable" (positive). This is also important where the features of the engraving may not be so obvious, such as in the case of logos or trademarks, which may appear to the casual observer to be symmetrical, but may have some asymmetrical features, which must be seen by the user in the proper orientation (polarity).

In some products molded from transparent or translucent plastics, the required lettering or marking could be molded on the *inside* of the product, so as to be read by the user *through* the plastic. In these cases, the engraving must be *positive* in the mold steel. This is the case in measuring cups, if the engraving is on the core (see Figure 3.62).

Figure 3.62 Gate pad engraving in steel and in the bottom of the molded container

3.22.6 Are the Locations Selected for Engraving Practical?

The product designer usually places the lettering, lines, or symbols at locations where they are best suited for the end user, but occasionally such engravings could be difficult to produce by the mold maker in the location specified. This could be the case where engraving inside a pocket in the mold would be difficult or even impossible, and would require inserts, or EDM with special electrodes. In some cases, the engraving could be *too close to the edge* of the mold steel, thereby increasing the risk of early failure of the mold steel due to stress cracks. A minimum of 2 mm between any engraving and the edge of the mold steel is suggested.

Here again, the mold designer and the product designer must work together to find the most suitable compromise between the product requirement and mold cost.

3.22.7 Engravings in the Walls and Bottoms of Products

Engravings can be either on the cavity wall or on the core (they could also be on inserts in either cavity or core). In either case the engraving must be designed such that the part will easily release from the cavity during mold opening and eject from the core.

3.22.7.1 Engravings on the Outside of the Product (Engraved Cavities)

Containers usually require markings on the outside of the sidewalls or in the bottom. Markings in the bottom are often required to show trademarks, patents, product identifiers, batch identifiers, dates of manufacture, or others. Engravings in the sides are occasionally required (usually with transparent or translucent plastics) to indicate liquid levels inside a container.

Engraving into the bottom of a cavity is usually not difficult to manufacture, especially if most of the bottom of the cavity is an insert in the cavity block (such as shown in Figure 3.62). Alternatively, it is not too difficult or costly if inserts with the required engravings are placed either in the solid cavity bottom, or within a large cavity bottom insert (creating "inserts within an insert"). However, serious challenges can arise when laying out the cooling circuits in such complex cavity bottoms. Good cooling in the gate area is very important for fast molding cycles, and inserts make it more difficult to lay out efficient cooling channels. A poorly cooled cavity bottom, especially near the gate, will result in a longer molding cycle. In this case, the preferred method is to have a solid insert for much of the cavity bottom. If there are changes required in the engraving, it is not too difficult or expensive to change the bottom. This may result in having and storing a number of different bottoms for the cavity for the various applications or end users of the product, which are also costs to be considered.

Mechanical engraving in the bottom of deep cavities is difficult, because long, unsupported engraving cutters will by necessity operate at a slower speed for the required accuracy and cleanliness of cut. Long EDM electrodes can be used, although they are slow and expensive. However, this method has the advantage that it can be done even after the cavity is finished.

A method not much used today is the "hobbing" of the engraving into the bottom of a cavity. This method was used extensively in molds built about the middle of the last century (both for small compression and injection mold cavities). This method can be used only in soft steels, and requires special heat treatment (carburizing and hardening) of the steel after hobbing. It is still occasionally used today.

The injected plastic, as it cools inside the mold, shrinks away from the cavity wall. Provided the depth of engravings into the cavity walls is not too deep, there is usually no problem with ejection. As the product shrinks toward the core, it will not "hang up" in the cavity as the mold opens. However, the clean withdrawal of the molded piece from the cavity depends also very much on the draft angle of the sidewall, on the wall thickness of the product in this area, and on the type of plastic used.

There is no easy formula to indicate what is possible and what is not, but as a general rule it can be stated that:

- Any engraving (by chip removing or EDM) in the sidewall inside a cavity, especially in a small one, is very difficult and can be very expensive. Shallow engravings "burnt" with EDM are easier to achieve, but there is the problem of matching the engraved electrodes to the shape (curvature) of the cavity wall so that the depressions created with EDM are uniform both in depth and appearance and do not exceed the critical depth beyond which the product cannot pull out of the cavity. The suggested maximum depth is in the order of 0.1 mm (0.004") or even less for difficult cases, such as those explained in the following points.
- Walls with heavier thickness allow deeper engravings because they shrink more and let the product withdraw more away from the cavity.
- The greater the shrinkage factor, the easier the engraved portion pulls away from the cavity.
- The greater the taper of the sidewalls, the easier the product will pull out of the cavity. Engravings in sidewalls with tapers of less than approx. 5° are more difficult to withdraw than from walls with larger tapers.
- Hard plastics such as PS will offer more resistance than more flexible plastics such as PP if they are "caught" by the edge of too deep a depression. This will show up as drag marks on the part. However, there are many molds successfully producing even thin-walled PS products with decorations on their outside walls.
- The angle and shape of the sides of the engraving within the sidewall of the cavity must be so that it offers little resistance as the mold opens, and the edge of the engraved projection in the product slides past the engraved depression in the sidewall.

Any deeper engravings in the side walls, or where there is not enough draft angle, will require the designer to place the engravings either on moving side cores in the cavity or on split cavities. Both methods would require more space, much larger molds, and add considerably to the mold cost. Such molds will usually also potentially produce more scrap, require longer molding cycles, and thereby increase the cost of the product even more.

Figure 3.63 shows heavy-walled tumblers engraved with an artistic pattern on the outside, produced by engraving (texturizing) the inside of the cavity. This engraving is not deep enough to require a split cavity. Note the stacking lugs visible through the plastic. They are used to stack the parts in a dense pattern.

Figure 3.63 Tumblers engraved on the outside

3.22.7.2 Engravings on the Inside of the Product (Engraved Cores)

The following comments apply to engravings into the top or the sides of the core.

Engravings in the sides are often required with transparent or translucent plastics, e.g., to indicate liquid levels inside a container (measuring cups, vials, etc.). The markings are usually lines indicating the proper height and lettering to identify the values. Such products are made mostly from clear PS, SAN, acrylic, or PC that has low shrinkage factors. This makes it relatively easy to calculate the dimensions where the measuring lines should be located. If such products are made from high-shrinkage materials, such as PE or PP, the high shrinkage factor makes it more difficult to predetermine the proper location for the level markers. In such cases, especially if the accuracy of the measuring lines is important, it may be necessary to finish the mold first, complete with the lettering, but to engrave the measuring lines only after the mold has been tested and run on an optimal cycle, because the volume of the container can vary substantially when operating at different operating conditions of the mold.

Except for very stiff plastics (such as PS, SAN, or PC) or sometimes for air ejection of softer plastics, lines and lettering on the core present fewer problems, because the plastic will stretch during ejection and let the plastic slide out of the engravings. This is possible because at the time of ejection, the cavity has already moved away from the product and there is ample room for the plastic to stretch during ejection. However, the deeper the engraving, the more important it is to make sure that the sides of the engravings are tapered and/or rounded sufficiently to allow easy sliding out of the engravings. The draft of container sidewalls can be quite small. A 1° taper could be acceptable, as long as the engraving is not too deep and the side of the engraving in the direction of the ejection is smooth and chamfered or rounded.

> Engraving into the side of a core is usually not difficult to achieve. The depth should be in the order of 0.1 mm, but less is recommended for small draft angles (less than 1°) of the core. While it is feasible to produce raised "engraving" on the core, it is extremely difficult to machine and then to finish the molding surface of the core, and the mold would therefore be very expensive.

The top of the core can be a good location to engrave the cavity number - this is easy to produce and is frequently done in technical products and enclosures. The designer must be sure that it can be easily read. If it is to be read from the inside, the engraving must be negative, whereas if it is to be read from the outside (through the plastic), the engraving must be positive.

3.23 General Appearance of the Product

3.23.1 Flatness

Where very high polish is required, common polishing practices can result in waviness of the surface, which may not be acceptable for products requiring near-perfect flat areas with optical clarity. In such cases it may be necessary to provide the mold with inserts for the areas requiring the optical finish. They can then be polished separately, on appropriate lapping equipment, which can guarantee flatness. A typical example is the top surface - both on the core and the cavity side - of Petri dish bottoms and lids made from crystal PS (see Figure 3.64).

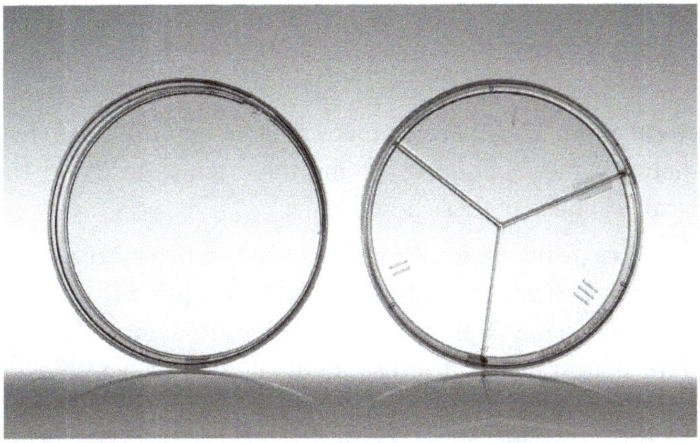

Figure 3.64 Petri dishes require optical clarity and flatness

Flat surfaces may be easy to machine but molding them can be a problem, particularly when materials with high heat content and low thermal conductivity are used, such as PE or PP (see Figure 3.65). Taking this into consideration is especially important when the products are to be ejected as early as possible to achieve fast molding cycles, i.e., while the products are still warm but rigid enough to allow ejection without damage. A flat, relatively large area in the mold is usually easier to cool than corner areas, or heavy rims or intricate sections in the product. However, the surrounding areas, which are often thicker and almost always poorer-cooled, stay hot longer and will continue to shrink after ejection and thereby tend to deform the already cold, flat areas. Typical examples are rectangular flat trays or other flat shapes surrounded by heavier rims. Such rims stay warm longer and distort the flat areas while they cool down to room temperature.

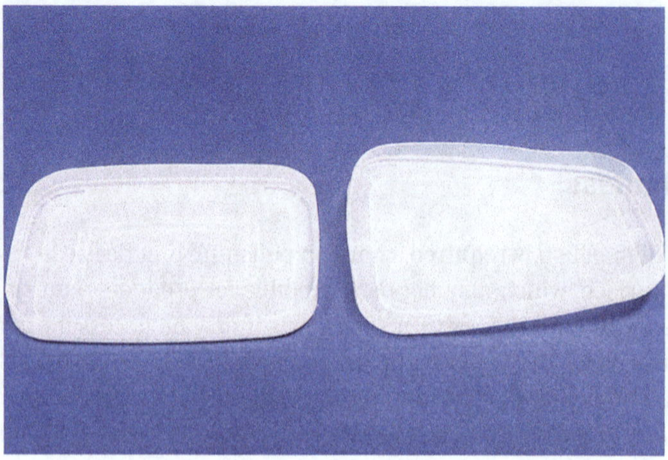

Figure 3.65 Flat parts can look like potato chips if the mold and part are not designed properly. A stepped ring was added to the part to add rigidity and eliminate warpage. Using flow leaders or multiple gates could be other options to correct the warpage. Alternatively changes in the core to cavity temperature may also help

There are several approaches to solve this problem, but as always, they need full cooperation between the mold designer and the product designer. The following are some typical examples of these approaches:

- The flat surfaces at the bottom of a container can be designed in the mold as curved (or arched) so that the plastic, as it cools outside the mold, will shrink to a less arched shape or even become flat (see Figure 3.66). If it does not matter to the appearance and/or usefulness of the product, this is a preferred solution. The curvature of the arch must be selected to suit the anticipated cycle time. It is suggested to ask the product designer for a wide tolerance on the curvature of the arch so that it will be still acceptable for the purpose of the product, regardless of the actual shrinking experienced, which may change with changes in the molding conditions and with the plastic batches.

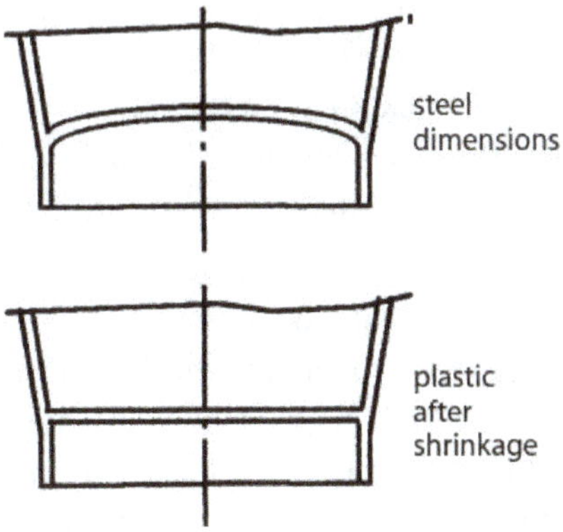

Figure 3.66
Schematic of cup with arched bottom

Note that with typical small containers, such as drinking cups or cottage cheese containers, the use of a flat-bottom mold will cause the bottom to pull towards the center and deform (pull) the sidewalls inwards as the product continues to cool outside the mold, even with good cooling and walls of equal thickness. This deformation may be objectionable. In such cases, the arching of the bottom of the mold is absolutely necessary.

- The flat surface of a lid can be modified by adding some steps or expansion loops so that, as the top of the lid shrinks, the steps or loops will bend due to the pull from the shrinking, and so prevent warping of the lid. This is particularly advantageous with large lids, as for pails, etc. (see Figure 3.67).

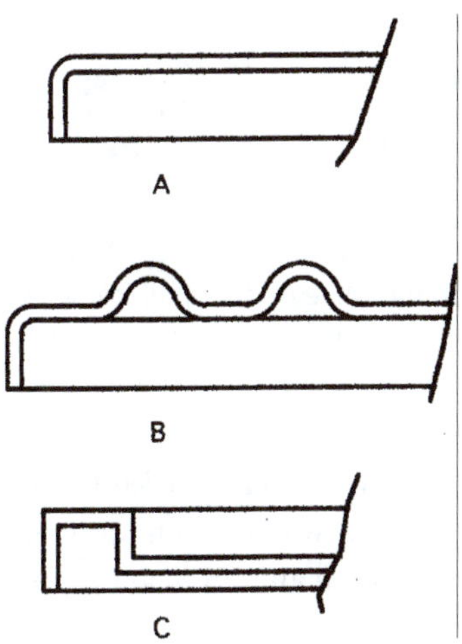

Figure 3.67
Lids with added steps or loops to hide the difficulty in achieving flatness. These ridges also add strength to the part

- Large trays or lids, especially those that are rectangular and must be flat, are always difficult to keep from warping (potato-chipping). It is very important that an equal wall thickness throughout the tray is maintained so that there are no warmer pockets of plastic, which will take longer to cool and shrink after the rest of the molded piece is cooled. If the rim must be thicker, more emphasis must be given to the cooling of the thick areas so that all the plastic in the mold is cooled evenly. If this is not the case, longer cycle times will be required to achieve flatness, or costly shrinking fixtures may have to be planned.
- It is also important that the flow lengths from the gate(s) to the rim are as symmetrical as possible to permit the plastic to arrive to all parts of the rim at the same time. This depends also on the thickness of the product, where heavier sections permit easier and faster flow. This can affect the selection of the hot runner system (e.g., using more than one drop) and adds costs to the mold.

  Figure 3.68 shows how the flow in a tray can be improved by machining so-called flow leaders into the cavity or core, which are slight thickenings in the wall thickness in those areas, to allow a faster flow and so help to equalize the filling pattern in a mold. Such thickenings will add a very small, scarcely visible, amount of plastic, but will ensure better trays that suffer from less warping. Flow analysis of such a product will show where such flow leaders are required. In Figure 3.68, T2 may be 10% greater than the wall thickness T1, and the width of the flow leader would be approx. 10–20 mm (0.38–0.75").

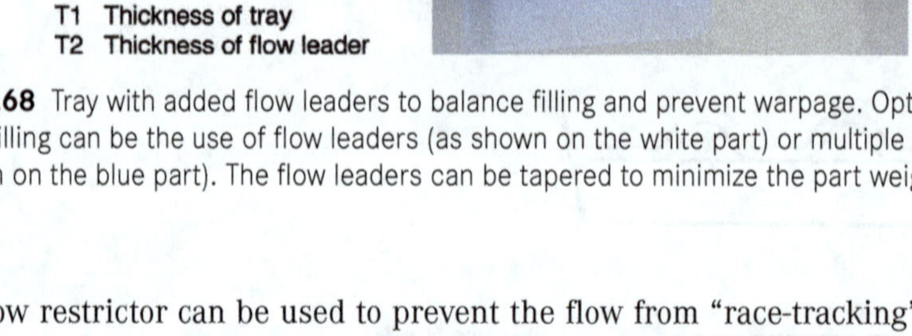

Figure 3.68 Tray with added flow leaders to balance filling and prevent warpage. Options for even filling can be the use of flow leaders (as shown on the white part) or multiple gates (as shown on the blue part). The flow leaders can be tapered to minimize the part weight increase

- A flow restrictor can be used to prevent the flow from "race-tracking" around one feature and creating a void. By adding a reduced wall section strategically in the part, the flow of the plastic during filling can be altered to assist in solving a flow defect.

3.23.2 Sinks and Voids

Sinks (or sink marks) are surface flaws (imperfections) of the product resulting from either incomplete filling during injection, excessive local shrinkage, or a combination of both. During injection, the hot plastic flows through the cavity space in contact with the cooled mold walls. This causes the plastic layer near the walls to solidify, thereby reducing the passage for the flow. It requires more effort (higher pressure, higher melt, and/or higher mold temperature) to completely fill the subsequent portions of the cavity space.

Figure 3.69 shows the plastic flow through the cavity space. The frozen plastic layers close to the cold walls reduce the passage through which the plastic has to flow on its way to fill the cavity.

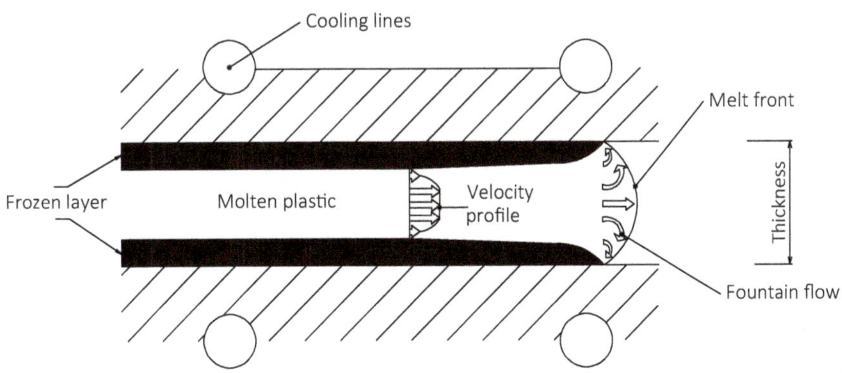

Figure 3.69 Plastic flow through the cavity space

The shrinkage factor must also be considered. To avoid poor-quality products (voids) and/or unsightly shrink marks caused by the shrinkage as the plastic cools, pressure must be kept on the plastic already in the cavity space by the so-called injection hold pressure, to add more plastic into the cavity space and make up the lost volume due to cooling. This is useful only as long as the gate is not frozen, i.e., as long as plastic can still pass through the gate. The hold time adds to the cycle time and adds cost to the products.

Ideally, for best flow, the cross-section through which the plastic flows away from the gate should be largest near the gate and from there gradually diminish toward the end of the flow. However, this is not practical because a lot of plastic would be wasted. The next best thing is to make sure that, at least, the same cross-section is maintained throughout the mold. This is not always possible because of the requirements of the product, but it should be attempted.

Possibly the worst condition is if a heavy area must be filled after the plastic has passed through a long, narrow path and has suffered a large pressure drop. Such remote heavy sections (typically, the rim of the product), even when they are completely filled, see much lower injection pressures and because the amount of shrinkage is greatest where the pressure is the lowest, these areas will experience much shrinkage and result in sink marks or voids (more about rim shapes in Chapter 11 on ejection).

Sinks and voids appear often at the intersection of ribs and walls, or in general at any localized thickening of the plastic required for functional reasons, such as hubs, and so forth (for design suggestions see Section 3.8). Because the thick section of the plastic remains hot longer than the thinner sections, the plastic will continue to shrink there. While the plastic is still relatively soft, it will pull the already more-or-less cooled surface towards the center of the heavy, hot section, thereby creating dips in the nearby surfaces (Figure 3.70). In many applications, a sink on a surface visible to the user may be acceptable, but how much of a sink is acceptable, as well as its probable location, should be agreed upon before designing the mold. *A solution to hide the sink may be to use a rougher surface finish in the area of the sink.*

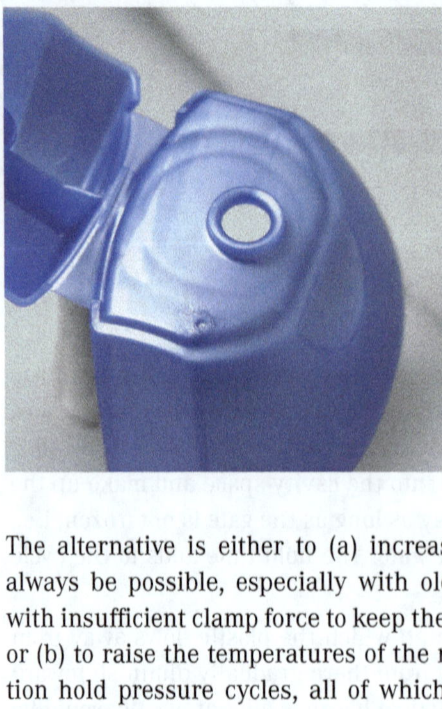

Figure 3.70
Intersections of ribs and thick sections can cause sinks or voids

The alternative is either to (a) increase the injection pressure, which may not always be possible, especially with older injection molding machines (or those with insufficient clamp force to keep the mold closed against the higher pressures), or (b) to raise the temperatures of the melt or/and the mold and use longer injection hold pressure cycles, all of which will result in longer molding cycles and higher product costs.

If the plastic surface is already so stiff that it cannot be pulled in (or sink), the still-hot plastic will shrink away from the center toward this stiffer outer skin, and will create a "void". A void is a hollow space inside the plastic and contains a vacuum. In opaque plastics a void cannot be seen, but it can be undesirable because it weakens the plastic, in a similar manner to porosity. Such a weak spot, e.g., in a hub designed to receive a screw, would not be as strong as expected.

If a transparent or translucent plastic contains a void, it is visible and can look like a chain of round or elongated bubbles near the center of the heavy section. To eliminate this defect, the molding conditions must be changed to ensure that injection pressure is maintained in this critical area, which often requires higher temperatures and results in longer cycle times. To remedy this problem, the product design should be modified to eliminate any thick spot(s), or add venting to the area.

Voids can be easily seen by cutting the suspect section with a saw or by drilling a small hole into it from the nearest surface while holding the product under the surface of a pail of colored water. As the drill breaks into the void, the water is sucked past the drill into the void, and can be seen as the colored fluid fills it.

The possibility of voids or any potential defects caused by heavy sections in a product must be discussed at the time a job is started, and not after the mold is completed.

3.23.3 Witness Lines

Witness lines appear on the product wherever mold parts or inserts join on the molding surface (Figure 3.71). No matter how good the fit of the mold parts and how well polished the surface is at this spot, there will always be a more or less fine line visible on the product. When the gap between the mold parts is too large, it will flash, i.e., the plastic will enter the gap during injection, and if it can pull out during ejection, it will be an unsightly thin projection from the surface of the molded piece. At best, it may not affect the overall appearance or serviceability of the product, but it is still the sign of poor workmanship.

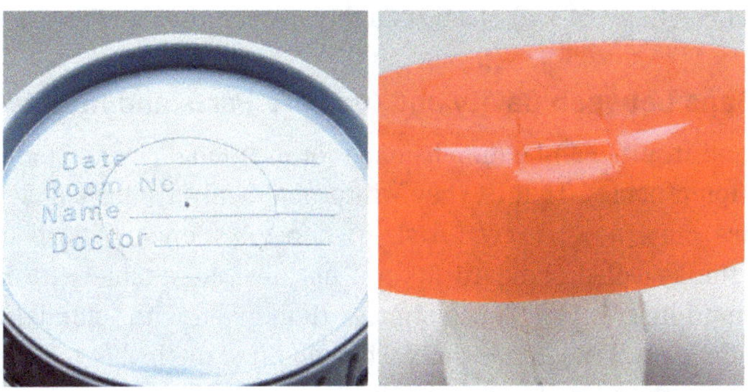

Figure 3.71 The gate insert witness line can clearly be seen on this worn insert for a specimen cup lid

> In general, gaps in the mold of less than 0.01-0.03 mm (0.0004-0.0012") will not flash, depending on the type of plastic, the melt temperature, and the injection pressure.

3.23.3.1 Parting Line Flash

A common misconception is that if a mold fits a machine it should be able to run on it. There are many considerations for ensuring that the mold will run acceptable parts in the machine. Chapter 4 covers in detail the requirements fo the machine to ensure that the mold will run acceptable parts. In the case of parting line flash, the mold designer must ensure that the machine will have sufficient clamping force to keep the mold firmly closed during the filling and hold stages of injection. If there is insufficient tonnage the mold will flash, no matter how well the mold is designed and manufactured.

Good molds have venting around most of the length of the parting line, if not all of it. As long as the mold inserts and the vents are designed properly, there should never be a flash on the parting line. If flashing does occur despite a properly designed and built mold, there could be a number of other causes:

- The selected molding machine has not enough clamp force (tonnage) for the projected area of the cavity and the required injection pressure, or was not adjusted properly. This is a rather common occurrence, especially when the mold is installed in a machine with low-tonnage (clamping force) because a suitable machine was not available.
- Flashing at the parting line could be caused by a poorly maintained and out-of-alignment molding machine, which could be the result of poorly adjusted tie bars, worn tie bar bushings, worn or bent machine platen surfaces, a poor hydraulic system, and so forth. Too often, a mold is blamed for flashing when the real reason is a poorly maintained machine or one with an insufficient tonnage.
- There may be (hard-to-see) plastic caught between the mold faces.

3.23.3.2 Gaps between Cavity and/or Core Parts and Inserts

Mold parts and inserts can be dimensioned for locational clearance fits (standard geometrical tolerancing). That is, they will probably always have a clearance, but this clearance must never exceed a size where plastic could flash into it. This means that the so-called standard fits are not always suitable with larger mold components and inserts. Smaller tolerances than given with “standard fits” must be specified for some mold dimensions. This modified method of tolerancing fits is quite suitable for the assembling of molds.

Because gates wear out in time and get easily damaged, often during startup of the mold, molders prefer that the gate is in an insert (see Figure 3.72), which can be replaced when necessary, so avoiding having to rebuild the entire cavity. Welding a worn gate is sometimes successful, if done by experienced welders, but in most cases, even the best weldment will be visible on the product, especially on a highly polished surface. This kind of repair is generally not recommended.

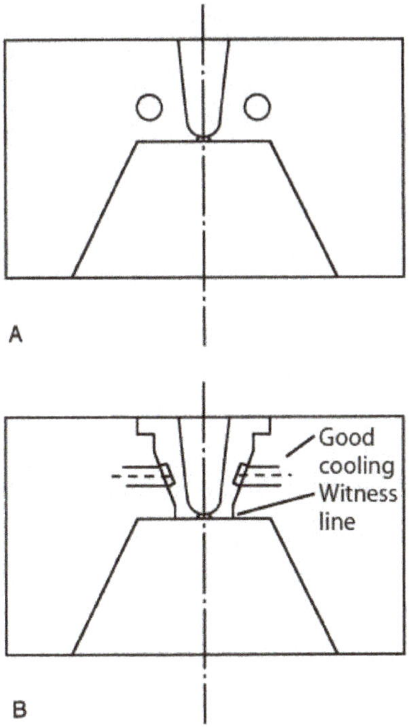

Figure 3.72 Typical cavity with and without gate inserts

Any gate insert will leave a circular witness line (Figure 3.71). It must be pointed out to the customer at product sign-off and before beginning the mold design that the mold would be less expensive to build without an insert, but the costs of upkeep will be much higher, especially with high production requirements, which in the long run will affect the cost of the product. Figure 3.73 shows a modular cavity, the matching cavity bottom, and the beryllium-copper alloy gate insert for the cavity bottom. The witness line will be a small circle around the gate.

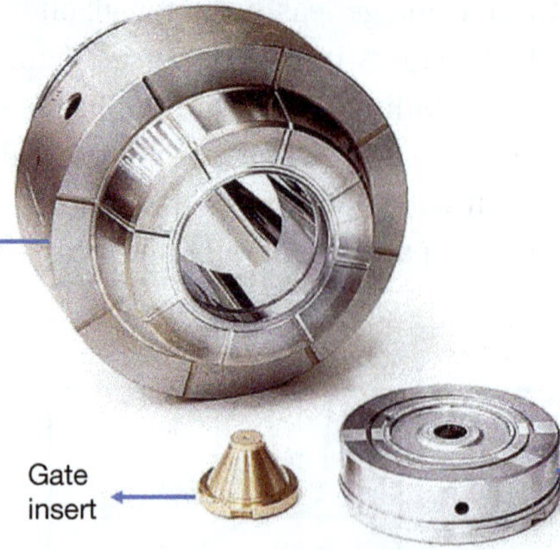

Figure 3.73 Modular cavity (Courtesy of Husky Injection Molding Systems Ltd.)

It is also important to realize that a properly designed gate insert permits better cooling around the gate, which is very important for achieving a faster molding cycle. The added cost of the insert can be easily justified with the increased productivity of the mold. The cost of a gate insert may be between \$200 and \$1000.

3.23.3.3 Clearance of Ejector Pins and Ejector Sleeves

It is important to consider the locations and sizes of the ejector pins carefully. In many molds, fewer pins of appropriate size in more suitably placed locations could save significant mold cost and improve operation.

Ejector pins are almost always produced by mold part supply manufacturers with specified, standard, close tolerances so that when the mold is built with the bore dimensions for these pins, as suggested by the pin manufacturer, they will not flash and they will always leave a witness line. The clearance between ejector pin and bore is actually desirable, because it creates a good, self-cleaning vent.

Ejector pins (and sleeves) and the bore in which they slide will wear over time, especially when abrasive plastics are molded, but also, frequently, when the mold is poorly designed or built, and the pins are allowed to rub on the side of the bores. If the ejector plates are properly aligned, with their mass suitably supported by guide pins (see Chapters 9 and 11) and the ejector pins (or ejector sleeves) are allowed to float freely in their retainer plates, little wear should be expected. But even under the best conditions, after the mold is in operation for a long time, the clearance in the bores will increase, and the plastic will begin to flash around the pin. Such flash may also create points of hang-up and prevent fully automatic ejection.

The proper way to repair the mold is to increase the bore size and replace the ejector pins with a larger size. Standard oversize pins are commercially available, usually 0.13 mm (0.005") larger in diameter than the original, nominal standard sizes. Allowing space for oversize bores should be considered when designing the mold. This is sometimes difficult with small technical products, where small pins barely clear the space between walls or under ribs.

Note that some molders use double or even multiple ejection strokes to overcome the problem of hanging-up due to flashing. However, this adds considerable time to the cycle, adds wear to the machine and mold, and is not recommended.

3.23.3.4 Hiding the Gate Mark

Occasionally, to improve the appearance of the product, it may become necessary to use certain "tricks" to hide the gate marks. A pinpoint gate mark (vestige) from a three-plate or a hot runner mold can be hidden inside an ornament or inside the closed portion of a letter appearing on the same surface (typically, inside a letter O, P, D, etc. - see Figure 3.74). In other cases, an ornament could be created around a gate, such as several concentric circles, with the gate as the center. In some cases, where the vestige of a valve gate is on the visible top surface near one end of the product, a similar "fake vestige" near the opposite end can be created with an inserted round pin, which serves no other purpose than to make the product appear symmetrical, which is usually more pleasing to the eye.

Figure 3.74 Gate mark hidden in the letter Y of a Playdoh lid

3.23.4 Weld Lines

Weld lines are created when two semi-molten streams of plastic within the cavity space intersect. They appear on the surface as more or less pronounced lines. These streams may originate from two separate nozzles in the same cavity, or when a projecting pin or other interruptions in plastic flow breaks the flow into two separate streams. In some products, these weld lines can be objectionable for aesthetic reasons. In others, they may be acceptable. When the plastic surface is to be printed after molding, such lines could present an unacceptable visible flaw. Note that a weld line also weakens the product, because when the plastic streams rejoin they are already colder and may not fuse sufficiently. Higher mold and melt temperatures can sometimes improve or even eliminate a weld line, but this means longer cycle times (lower productivity). When molding pigmented (colored, particularly metallic) plastics, weld lines can show up very distinctly.

Plastic always finds the easiest path (path of least resistance) within the mold. Typically, a heavier rim or ribs allows more and faster flow than the adjoining thinner walls, so that the plastic, after the rim or ribs are filled, will fill the remainder of the cavity space from there (back-fill), and create weld lines where these flows meet. This is an important consideration for locating the gate(s). If weld lines are created by such conflux of plastic, it is important to provide vents at these spots to let any trapped air escape. These features add cost to the mold.

Figure 3.75 Weld lines on a cosmetic case. The plastic flows from the gate through the grille and rejoins afterwards, forming the weld line

3.23.4.1 Location of Weld Lines

The location of weld lines (see Figure 3.75 and Figure 3.76) can be very accurately predicted with the help of a flow modeling computer program such as Moldex3D or other software. If the area where the weld line is expected to form is critical for strength, a change in the gate location may be a way to change the location of a weld line. In addition, it may be possible to facilitate the plastic flow to a certain path (e.g., by slightly increasing a portion of the cross-section of the flow path) so that the plastic reaches the critical point and flows around it before it meets the other stream, thereby relocating the weld line to an area where it will do less harm to the strength of the product. Such slight increase in the thickness may be hardly noticeable and, while increasing the amount of plastic insignificantly, it ensures a stronger product. Since this is a change to the product, the approval of the product designer may be necessary.

Figure 3.76 This lid for an ice cream container has a cardboard insert, and the plastic is molded in a ring around it. The weld line is formed opposite the gate location

In many cases, a weld line is unavoidable, and the only option is to influence its location by the positions of the gate, flow leaders/ribs, and flow restrictors, and changes in thickness of the part. To improve the appearance and strength of a weld line, a molder will typically increase the mold temperature, injection speed, melt temperature, or the degree of venting. A rougher surface finish will also reduce the visibility of a weld line. A filling simulation done at the beginning of mold design may avoid the problem of unacceptable weld lines (see Section 3.12.1).

3.23.5 Surface Defects (Flow Marks, Splay, Record Grooves, Haze, Jetting, Hooks, and Ripples)

Figure 3.77 shows four examples of surface defects that caused by improper settings or functionality of the molding machine.

Figure 3.77 Typical surface defects on a cosmetic jar created by inconsistent flow of plastic from the machine and thick to think transitions in the plastic part

The surface may show unsightly flow marks as the plastic flows away from the gate. This is often the case when injecting into the open surface of a product. A simple remedy is to ensure that the plastic coming from the gate is immediately broken up by "shooting" against the core if that is possible. This is always the case with hot runner molds and with cold runner, three-plate molds where the gate is opposite the core. With two-plate molds, if there is no sidewall in the product, it is best to shoot against a core pin located in the cavity space (a pin or projection either in the cavity or in the core).

Products with many thick and thin sections may also show such flow defects. Slower injection and adjusting the temperatures and pressures may be the only remedy. This will increase the cycle time and affect the cost of the product.

3.24 Identification of the Molded Piece

Every molded piece should have certain identifications molded into it. The product designer will sometimes specify the location of some of the engravings, but often leave it to the mold designer to decide their location, size, and style. Before starting, the mold designer must know which engravings will be required. Some of the markings could be created by special, often standard inserts - for example, the date (week, month, year) when the product was molded. They should be located so that the date can be changed easily without major work on the mold. These markings are often mandatory for food packaging, pharmaceutical, and other products.

Typical markings, in addition to any specific instructions, are:

- The seller of the product
- The molder, if not the same as above
- Resin type for recycling (including the standardized recycling logo)
- Cavity (and mold) numbering (position, location)
- Patent number, if applicable
- Made in ______ (the country of origin)
- Date of manufacture
- Special information: "Food grade", "Microwave safe", "Dishwasher safe", etc.
- Other information as required by law
- Logos
- Product name
- Bar code.

Figure 3.78 Identification of a molded piece (usually with laser engraving 0.08-0.013 mm (0.003-0.005") deep)

In some countries, some of these markings are mandatory, and it is up the product designer to specify what is obligatory and what is optional in the countries where the product will be used and/or sold.

Occasionally, there is resistance to show cavity numbers, but it is absolutely necessary for the molder to know from which cavity and from which mold a piece has originated, in order to be able to find the cause of any defects. If consecutive numbers are not desired, some code could be used for identification of the stack from which a faulty piece has been ejected.

It is always difficult to add engravings to a mold or to make revisions to existing engravings. If it is anticipated that changes will be needed in the future, the mold designer should design the mold with the necessary inserts so that changes can be made more easily, or put it on an inexpensive part of the stack. It may add some costs at the time of building of the mold, but will be much less expensive than making changes to a finished mold. This is often the case when the same mold could be used for products sold to different customers.

> Every mold insert (including spares) should be marked with the mold and cavity number to ensure that after disassembling the mold for maintenance or repairs, the various mold parts are assembled in the same relationship as before disassembly. This is typically done using laser engraving.

3.25 Product Strength Requirements

Plastic products must not only look aesthetically pleasing but they must also be strong enough and suitable for the application for which they are intended. The molding material (the plastic) is usually specified by the product designer, so as to create a product with all the desired properties. This material selection can affect the selection of features to be built into the mold. Not every mold can be used for every type of molding material. A mold that is designed for one type of plastic will likely not be suitable for another, or at least not perform as well. For example, air ejection could be used for a product made from PP, but the same product in PS would require a stripper ring ejection. Using a different plastic from what was originally intended may also present difficulties in filling and packing.

The addition of fillers to the plastic can increase the strength of the product but will also affect the shrinkage factors, usually reducing them. This means that the size of the product will be affected when molding a material different from the one for which the mold was built. In addition, fillers can be abrasive and wear parts of the mold sooner than unfilled plastics. This can affect the mold design and construction, and therefore mold parts subject to erosion should be easily replaceable when worn. It may also require the use of better-wearing steels. This is particularly important for gates in hot runner molds and all other areas where the hot, abrasive plastic flows at high speed over the molding surface, but this kind of abrasive action can also be the case in molds with cold runner gates. These provisions will increase the cost of the mold.

3.25.1 The Role of Gate Location in Increasing Product Strength

Gate location is important for the ability to fill the mold. Some large cavities may require two or more gates to properly fill, which of course affects the cost of the hot or cold runner system.

The gate location is also important because, to a large extent, it determines the direction in which the plastic molecules will be oriented within the molded product. In general, the strength of plastic in the direction of flow is greater than in the direction across the flow. Near the gate, when the plastic flows in all directions, the strength is relatively low and can be compared with the low strength at a weld line. Therefore, if maximum longitudinal strength is required in an elongated product, gating near the end of the product is very important (see Figure 3.79). Thin-walled products are especially vulnerable to this difference in strength, because the thinner plastic cools faster and does not have a chance to re-align the molecules randomly for better strength in all directions, as is the case with heavier-wall products where the plastic stays hot longer between the cooled outer layers and gives the inside core of the plastic time to re-align.

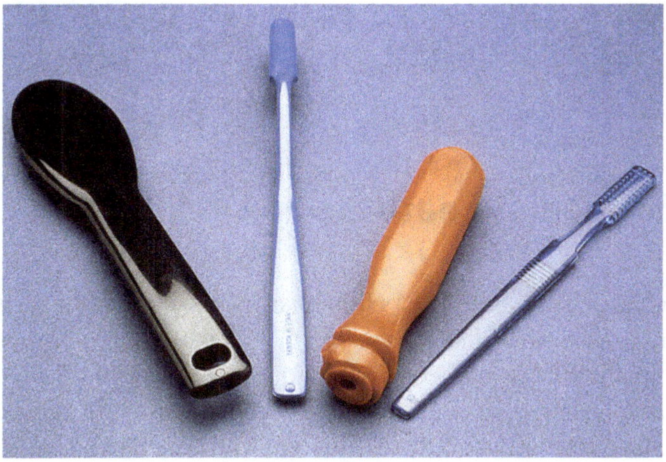

Figure 3.79 Typical elongated products, gated at or near the end for maximum strength

It is important to recognize the importance of the gate location. It may be easier (and cheaper) to locate the gates so as to employ the shortest runner length (cold or hot runner), but this could result in the product having lower strength, particularly with elongated products such as shown in Figure 3.79, which are subjected to bending forces in use. There are also other issues that affect the gate location:

- The gate needs to be hidden or recessed, so that any protruding gate mark cannot hurt (scratch) anyone using the product (this is very important in medical products).
- A protruding gate mark could also be in the way of a paper label that will be applied there.
- Some protruding gate marks are protected with gate rings for strength, typically at the bottom of large containers (see Figure 3.80).

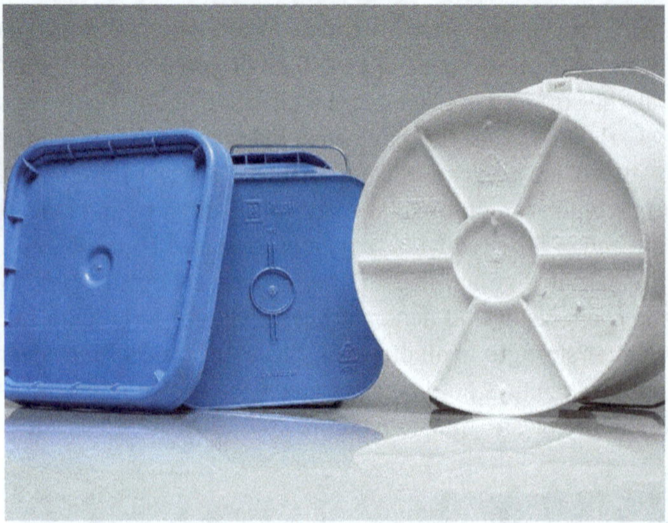

Figure 3.80 Gate protection rings are used to increase a pail's impact resistance

3.26 Special Features

A big advantage of molding products from plastic is that different parts of a product assembly can be consolidated into one molding, and that many features can be molded directly into the part versus added later in assembly. Even though this may result in a better appearance, better functioning, or lower assembly costs of the final product, the design of the cavities and cores can nevertheless sometimes be made more difficult. In addition, some designers, when redesigning a product which was originally made from metals, do not appreciate or understand the intricacies and problems of mold design and molding, especially if they are used to designing for metals or other materials. It is therefore always suggested to scrutinize the drawing for the proposed product to see if its design is appropriate for plastics.

3.26.1 Holes and Counter-Bores for Assembly Screws or Rivets

In general, it makes for a more pleasant appearance if mounting screws, used for example to fasten a cover to a base, are recessed, i. e., the screw head will be below the surface of the cover. If the counter-bore for the screw head is deep and its diameter is small, the core pin creating this counter-bore becomes very difficult to cool efficiently, and a redesign may be necessary in this area.

Plastics lend themselves well to the use of self-tapping screws. The design (dimensions) of screw holes must be to the specifications provided by the manufacturer of such screws, and the screws must be selected to suit the plastic material with which they are used. The holes for the (usually small size) screws are molded with very small core pins. These pins are very difficult to cool or cannot be cooled at all. Because the hubs or walls where the screw holes are located are much heavier than the adjoining product walls, the cooling time will be longer than expected for the rest of the product. This problem may result in grossly underestimating the molding cycle time and lead to a mold with less productivity than expected. Many molds have been built with excellent cooling in areas for which it was easy to provide cooling. However, because there was no effective cooling at all near critical points, such as some small core pins, they were a waste of effort and money. In other words, despite the expensive (and unnecessary) good cooling for most areas, such molds ended up having a very slow molding cycle because of poor cooling in one area.

As with many things, the old proverb of "the chain is only as strong as the weakest link" applies here. It is important for the mold designer to concentrate on the mold parts that are the most difficult to cool, as they will be the areas that determine the ultimate productivity (or cycle time) of the mold.

3.26.2 Hinges and Snaps

Snapping two molded pieces together (or snapping one molded piece over a metal piece) is an ideal application for plastics products.

Typical examples are containers for dairy products using a PE snap-on lid over a PP container, a PE lid on metal can (coffee can), cell phone assemblies, automotive applications, DVD cases, and many more. Some of these snaps have a long, successful history and the dimensioning is well documented. However, the designer is cautioned that such snaps depend on the amount of shrinkage experienced (a) in the mold during cooling and (b) after ejection, while cooling outside the machine.

Also, from experience, a mold can produce perfect (as expected) products in one machine, but not in another, which could result in wrong snapping forces ("opening forces"). More modern machines can use higher injection pressures and speeds and can therefore also cycle faster. In doing so, the shrinkage values will change. For example, when receiving a repeat order for a mold (e. g., a lid), it will be correct to assume that the steel dimensions of cavities and cores for the new mold will be the same provided that the mold will run on the same or a similar machine. However, if a more recent or better injection-molding machine will be used, it will most likely be necessary to establish new core and cavity dimensions. This can be costly and must be considered in estimating the mold cost. In the long run, faster cycles can save a lot of money.

Figure 3.81 DVD cases with hinges

When considering a mold design for a part with a hinge it is *critical to confirm that the gate location will not create a weld line at the hinge location.* If this occurs, the living hinge will break after a short number of flexes. Figure 3.82 shows several samples of small and large over-caps for containers with integral, hinged snap-on lids that snap close over the opening in the over-cap. Some of these over-caps are produced with simple open and close molds (A), some need intricate side cores (B) and much more complicated molds. Also shown is a very large box (C) with snap-on lid, with an integral, molded hinge connecting the two parts of the part.

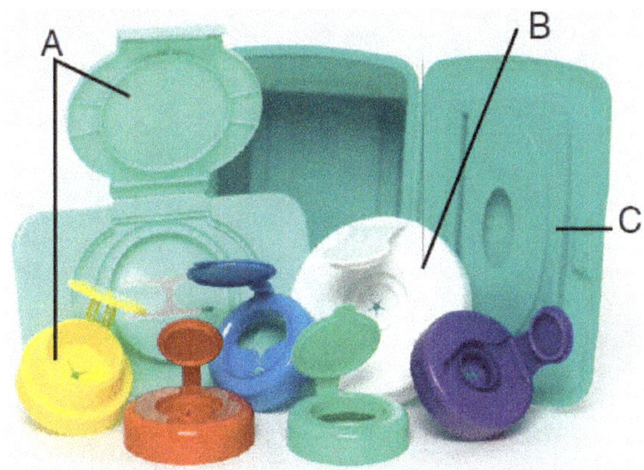

Figure 3.82 Examples of over-caps for containers

4 Specifying the Right Machine for the Mold

The designer must ensure that the mold design will function correctly in the machine it will be mounted in. In order to do this, the designer must work with the customer's project manager to ensure that the machine is specified correctly. This chapter will take the designer through the process of selecting the right injection molding machine.

Sometimes, the mold cost will exceed the cost of the machine. It does not matter how well a mold is built if the machine cannot meet the molding requirements to produce quality products.

> **While considering the purchase of an injection mold, it is always CRITICAL to make sure the machine can do the job.**

Machine Information

Below is a basic checklist along with general rules of thumb to ensure that the machine being considered for the job will be acceptable:

- **Tie bar spacing:** This is needed to determine if the mold will fit into the machine and if the parts can be ejected without interfering with the machine tie bars.
- **Stroke and shut height:** These are needed to ensure that the mold will fit in the machine and that it can open enough to eject the plastic part by free drop or with automation. See Section 4.2 for more details.

> *As a rule of thumb, to ensure that the part can be "free-dropped" manually from the machine, the required mold opening stroke must be 2.5 × the part height.*

- **Injection speed (average and peak) and pressure:** This is needed to ensure that the machine has the power to fill the molding space. This is especially critical when the part is very thin (under 0.5 mm).

- **Maximum injection shot size and plastic recovery rate capability:** These are needed to determine throughput. See Section 4.5 for more details.

> *The required shot size should ideally be between 20 and 70% of the machine capacity. Check with your machine supplier if it is outside this range.*

- **Platen rigidity:** Are the platens rigid/robust enough to carry the mold weight? Consult the machine supplier for the maximum mold weight allowed in the machine for the moving and stationary. If the mold is extremely thick or heavy then supports can often be added between the mold and the tie bars or linear ways of the machine.
- **Available clamp tonnage:** This is needed to ensure that the mold will not flash or breathe open during injection. There are rules of thumb to determine the required clamp tonnage, and 3D modeling can also be used to more accurately predict the required tonnage. First you must calculate the projected area that the mold will create when the clamp is forced open. To determine this, calculate the planar area of the mold on the platens and multiply by the number of cavities. If a stack mold is being considered, calculate the tonnage for one face and add 10% for every additional level of the mold. This topic will be discussed further in Section 4.1.
- **Platen parallelism:** Measured diagonally along the platens in multiple directions.

> *This number should be as small as possible, but generally 0.001-0.002" (0.025-0.05 mm)*

- **Clamp speed (or dry cycle) requirements:** This is the time for the mold to open to an industry-standard distance, eject the part, and close again.

> *For thin-wall fast-cycling applications, this number should be between 1 and 2.5 s. For most other applications it can be between 3 and 5 s.*

- **Shut-off nozzles:** The mold may need a shut-off nozzle for fast-cycling or high-pressure applications to separate the extruder pressure from the pressure in the mold during cooling and recovery. The shut-off nozzle also acts to prevent drooling of melted plastic from the machine nozzle if a carriage retract is used on each stroke. See Section 4.6 for more details.
- **Screw design:** Most machines come equipped with a general-purpose (GP) or barrier-type screw. However, there are literally hundreds of thousands of screw designs used to optimize melt creation. You should check to see what screw is

in the machine, and confirm that this screw will do a good job in creating a uniform, homogenous melt for the mold. On many occasions the problems with poor-quality moldings can be traced back to the screw design.

- **Accuracy and repeatability of controls:** Many older molding machines have control systems which are slower or more limited in terms of flexibility or output. When selecting a molding machine for your application, try and use a machine with a control system that is less than 10 years old to take advantage of the newer technology and additional computing power available.
- **Operator and service access:** Ensure that the machine will permit the operator and service personnel to maintain and service the mold in the machine for most issues. Removing the mold from the machine is a time-consuming and costly activity that should be avoided if at all possible.
- **Mold protection sensitivity:** A mold can be severely damaged if the machine applies clamp tonnage when a plastic part or piece of the mold is jammed in-between the mold halves. Clamp tonnage should never be applied if the mold cannot close properly. The mold protection function on the machine is designed to sense if there is something preventing the mold from closing normally. Once it senses this, the machine should open back up and provide an alarm to the operator. The easiest way to test the mold protection on a machine is to tape a thin piece of plastic or cardboard onto the parting line (about the thickness of a credit card or no more than 50% of the wall thickness of the plastic part you are molding). Once this is done, dry-cycle the machine. The machine should no build clamp tonnage. If it does, there is a problem with the mold protection setup or function. This MUST be addressed prior to starting volume production, or the machine may destroy the mold, and production will be halted for weeks until the mold is fixed.

Mold information

To determine the right machine, the following information on the mold is required:

- **Mold dimensions:** Mold width (horizontal), height (vertical), and shut height (thickness between platens).
- **Core and cavity locations:** The locations of the cores and cavities on the mold should be inside the opening between the tie bars for free-drop capability and for robot removal.
- **Mold opening stroke required:** This is usually 2.5 × the part height for free drop in manual.
- **Mold mounting locations:** These must align with the mounting on the machine (normally SPI or Euromap standards are used). If clamp slots are used in the mold design, then the mounting holes for the clamps must be available.

- **Ejector rod locations:** These must match the machine ejector locations (normally SPI or Euromap standards are used). If this is not the case an ejector box may be required on the mold to match with the machine locations.
- **Locating ring size:** The locating ring is used to align the mold to the machine and to ease installation. There are a few standards in the industry, but most molds have a 4" (101.6 mm) or a 120 mm locating ring.
- **Part dimensions:** Including wall thickness.
- **Flow length:** This is the length of the longest flow path from the gate.
- ***L/t* ratio:** This is the ratio of the length of flow from the gate to the furthest parting line divided by the average all thickness. This ratio is used to determine the difficulty of filling the part. *L/t* ratios above 200 may require a machine with enhanced injection speeds, pressure and more rigid clamps to resist the injection force.
- **Part weight.**
- **Runner weight** (if cold runner).
- **Cavitation.**
- **Nozzle radius and orifice:** The machine nozzle radius and orifice must match the mold sprue bushing radius and orifice so that plastic does not spray out of this interface during injection. Also a mismatch in this area can create flow problems in the mold and also the potential of burning or degradation of plastic before injection.
- **Plastic to be molded:** Including color, additives and shear/viscosity curves for modeling.

4.1 Required Clamp Size (Tonnage)

Molding machines build clamping force to keep the mold closed during injection of molten plastic from the extruder. This is accomplished using hydraulic or electrical power coupled with a piston or a multi-point toggle system and stretch in the tied bars. It is not the intention of this book to go into the design of the machine's clamping system, but the sizing of the machine is of great importance to the mold designer.

In order to calculate the required clamp tonnage for the mold, we must calculate the injection pressures needed to fill the part, and then calculate the projected opening area that these forces will act on to try and force the mold open (see Equation 4.1). Clamp tonnage required can be calculated using the formula below:

$$T_c = P_{inj} \times A_p \times \text{number of cavities} \times \text{SF} \tag{4.1}$$

where:
T_c = clamp tonnage (metric tonnes or US tons). Conversion: 1 metric tonne = 1.1 US tons
P_{inj} = injection pressure (psi or kPa)
A_p = projected area per stack (in^2 or cm^2) (area where pressure acts to open the clamp)
SF = safety factor = 1.2 at a minimum

The difficulty is how to determine the value of the injection pressure (P_{inj}). We can easily calculate the injection pressure inside the machine nozzle, which is directly related to the size of the injection cylinder of the machine and the hydraulic (oil) pressure supplying the injection cylinder. The injection pressure at the machine nozzle, in general, is adjustable between any low values, to a high of about 140 MPa (20,000 psi) in most molding machines, or to 200 MPa (29,000 psi) or even more in some machines. This pressure, however, is greatly reduced (by pressure losses due to friction) by the time the plastic passes through the machine nozzle orifice, the runners, and the gates, and as it flows through the narrow passages of the cavity space. The flow also depends largely on the viscosity (defining the ease of flow) of the plastic, which depends on its chemistry and on its temperature (the higher the temperature, the lower the viscosity and the easier the flow).

Computer-aided engineering (CAE) simulation software can estimate the injection pressures most accurately if the boundary conditions are well known and the plastic viscosity versus temperature and shear has been correctly modeled.

If the mold designer does not have access to this sophisticated modeling software, there are a couple of rules of thumb that can be used very effectively. In both cases the projected area (A_p) that the mold will create when the clamp is forced open must be calculated (see Equation 4.1). To determine this, calculate the planar area of the stack (see Figure 4.1) on the platens and multiply by the number of cavities. The following rules of thumb can now be applied.

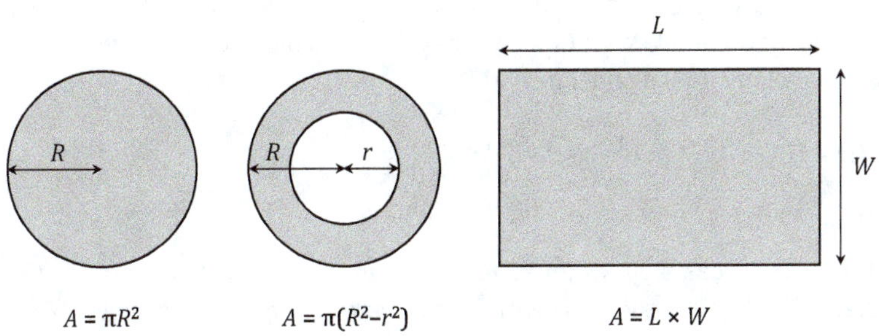

Figure 4.1 Examples of planar areas for calculation of clamp tonnage

Rule of thumb 1 for clamp size calculation:

Calculate the projected area of the product and then make an assumption of the *average* pressure the cavity space will see during injection. Multiply these together using Equation 4.1. From experience, we know that thick-walled products need lower pressure to fill the mold, and thin-walled products need higher pressures. Also, if good surface definition is required, maximum pressures will be needed. The following values can be used for a quick but conservative estimate:

Table 4.1 Estimated Injection Pressures for Use in Calculating Clamp Tonnage Requirements

Average wall section	Average injection pressure required
2-3 mm	40-60 MPa (3000-4000 psi)
1-2 mm	60-70 MPa (4000-5000 psi)
<1 mm	80-100 MPa (5500-7000 psi)

Note that in "Rule of thumb 1" we are using the *average injection pressure within the cavity*, not the *indicated* injection pressure at the machine nozzle, which could be in the order of 140,000 kPa (20,000 psi) or more. If the machine cannot provide these required high pressures, the product cannot be molded successfully on it, unless the plastic is injected with very low-viscosity (very hot) plastic, which, as we have already seen, is as a rule not desirable, mainly because it increases the cycle time.

For the next step, we take the above assumed pressure and multiply it by the projected area of the product to arrive at the clamp force required to hold the mold closed during injection.

Example calculation of required clamp tonnage using Rule of thumb 1:

Assuming a projected area of 64.5 cm^2 (10 in^2) and an assumed injection pressure of 35,000 kPa (5000 psi) necessary to fill the cavity, what will be the required clamp tonnage per cavity?

$$64.5\ \text{cm}^2 \times 35{,}000\ \text{kPa} \div 10{,}000 = 225\ \text{kN}$$

$$(\text{or } 10\ \text{in}^2 \times 5000\ \text{psi} = 50{,}000\ \text{lb} = 25\ \text{tons})$$

We require at least 225 kN (or 25 tons) per cavity, and the number of cavities selected will define the minimum clamp size of the machine selected. Conversely, we can determine the maximum number of cavities possible in a machine by dividing the available clamp force of the planned machine by the force required for each cavity. For this example, for a 100 ton clamp, we obtain

100 tons ÷ 25 tons = 4

This means a maximum of four cavities could be run in the press.

Example using Rule of thumb 1:

How much clamp tonnage is required to mold a disk of 100 mm diameter, with a thickness of 2 mm?

The mold has four cavities, so using Equation 4.1:

T_c per cavity = ($10^2 \times \pi \div 4$) cm^2 × 60 MPa = 470 kN (approx. 52 US tons) per cavity

T_c = 4 cavities × 52 tons/cavity × 1.2 safety factor = 250 US tons is needed to mold this disk.

A 300 ton machine will probably be selected, as this is probably the next available standard-size molding machine.

Rule of thumb 2 for clamp size calculation:

In this method an empirical factor, RCT, is applied to the projected area of the mold. Refer to Table 4.2 to find the average wall thickness of the part to be molded. Multiply this by the total projected area of the mold that is subjected to injection pressure, using Equation 4.2. The clamp tonnage required can now be calculated as follows:

$$T_c = A_p \times \text{\# of cavities} \times \text{RCT} \tag{4.2}$$

where:

T_c = clamp tonnage (metric tonnes or US tons). Conversion: 1 metric tonne = 1.1 US tons

A_p = projected area per stack (in^2 or cm^2) (area where pressure acts to open the clamp)

RCT = required clamp tonnage from Table 4.2

Table 4.2 Clamp tonnage requirements by wall thickness OR plastic flow characteristics

Average plastic part thickness (mm)	Range of melt index of resin	RCT (Clamp tonnage)	
		(tonnes/cm^2)	(tons/in^2)
>2 mm (> 0.080")	Easy flow, high melt index Commodity resins	0.17–0.25	1–1.5
0.6–1.5 mm (0.025–0.060")	Easy flowing engineering resins Medium melt index commodity resins	0.34–0.42	2–2.5
<0.6 mm (< 0.025")	Engineering resins Low melt index commodity resins	0.5–0.68	3–4

Example using Rule of thumb 2:

What size clamp will be required by an 8-cavity mold running an LLDPE lid 0.7 mm thick and 125 mm diameter?

Using the estimated tonnage requirements from Table 4.2 and Equation 4.2:

$T_c = A_p \times \#$ of cavities $\times$ RCT $= (12.5)^2 \times \pi/4 \times 8 \times 0.40 = 392$ tonnes (431 US tons).

The customer should choose a 400 metric tonne (or 500 US ton) machine.

In the above examples we have omitted the impact of the projected area of the *cold* runner system (hot runners do not add to the projected area, as they are fully contained within the hot runner manifold).

- With *two-plate* (cold runner) molds, the projected area of the runner system must be added to the sum of the projected areas of the cavities. For relatively large products, the area of the runners is usually small in comparison to the area of the cavities, but with small products, the area of the runners must be seriously considered, as it could be as large as (or even larger than) the area of the cavities.
- With *three-plate* molds, it is very rare that the projected area of the runner system is larger than the projected area of the sum of all cavities, and therefore only the sum of the projected area of the cavities needs to be considered. If the projected area of the runner system is larger than the sum of the cavity areas (which could be the case for very small products), the projected area of the runners must be used to calculate the clamping force.

If a stack mold (a mold consisting of two cold halves back to back) is being designed, calculate the tonnage for one face and add 10% for every additional level of the mold.

It is always best practice to use the minimum clamp tonnage required to mold a good part to reduce wear on the mold. However, if there is ample clamp tonnage capacity available, it is better to inject with lower melt temperatures and higher injection pressures, as this will reduce cooling, pack, and injection times. The result will be higher productivity and lower product cost.

■ 4.2 Required Opening Stroke and Shut-Height

The mold must be able to be physically fit into the machine and be stroked open and closed to eject the part. This means that the mold must be at least thick enough to fit into the machine at its smallest stroke and thin enough to fit in the machine at its maximum opening. In addition to this requirement, it must be possible to open the mold to the required opening stroke to eject the parts automatically. Figure 4.2 illustrates the different terminology relating to stroke and daylight in a molding machine.

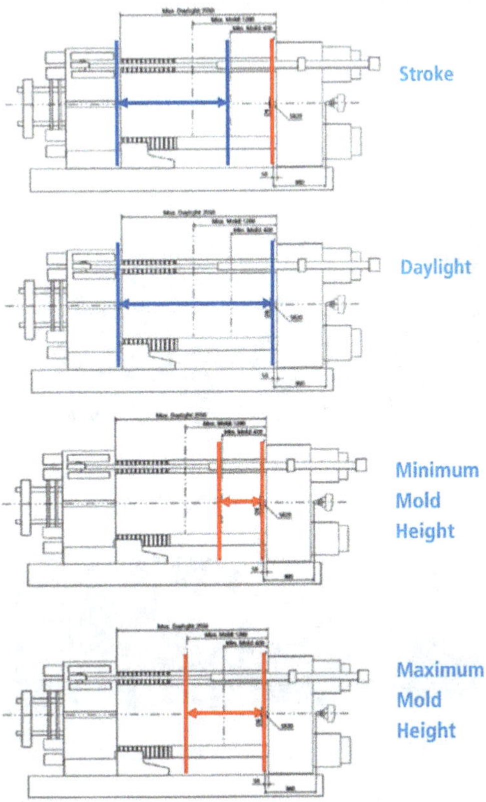

Figure 4.2
Machine stroke and shut height [1]

Maximum daylight in a molding machine is the largest opening from platen to platen that the machine is capable of with the clamp in its fully open position.

[1] *https://www.plasticsmachinerymanufacturing.com/injection-molding/article/21151087/a-guide-to-sizing-injection-molding-machines*

It is important to note that there are two basic types of machine opening stroke, one based on toggles (linkages) and other other based on pistons (hydraulics) (Figure 4.3). Daylight is calculated differently in these machines, and this can be confusing:

Toggle machine daylight calculation: The stroke available is independent of the daylight. Therefore the maximum stroke is always available at each shut height.

$$\text{Toggle maximum daylight} = \text{Maximum mold shut height} + \text{Stroke} \tag{4.3}$$

Piston machine daylight calculation: The shut height of the mold increases as the available stroke decreases. The maximum stroke of the machine is achieved at the minimum shut height.

$$\text{Piston maximum daylight} = \text{Minimum mold shut height} + \text{Maximum stroke} \tag{4.4}$$

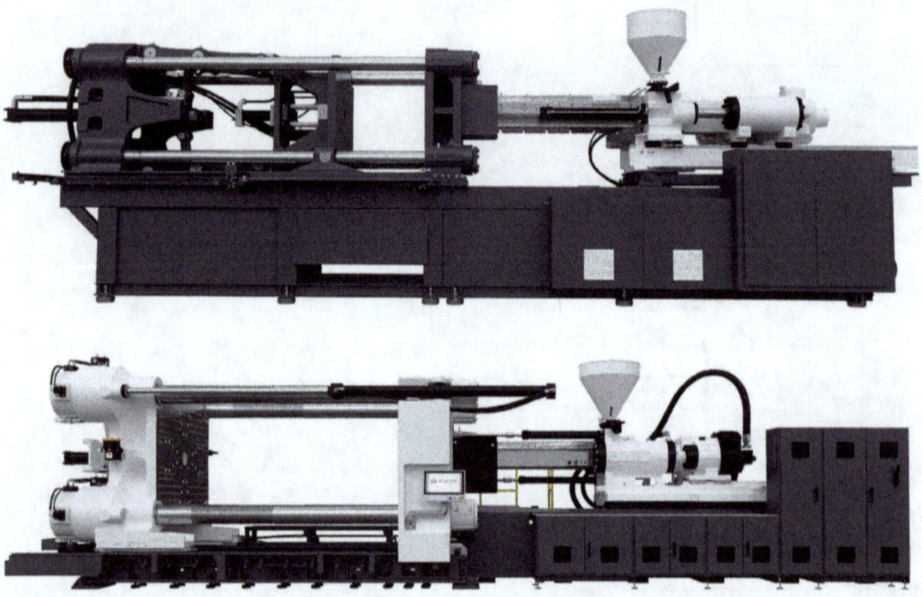

Figure 4.3 (Top) Toggle machine; (Bottom) Two-platen machine (Courtesy of Milacron Marketing LLC)

4.3 Platen Mounting Pattern and Ejector Positions

The locations of the mounting hole or clamp slots on the mold must follow the patterns and available holes on the machine. Before the start of designing the mold, the machine mounting pattern, ejector hole positions, and ejector rod sizes must be known.

There are industry-standard mounting and ejector patterns that almost all machines abide by. However, this should always be checked and verified. The major machine standards are SPI and EUROMAP.

4.4 Locating Ring Size and Machine Nozzle Size

An additional piece of information required about the machine is the size of the locating ring, so that the mold can be designed to match it. Again, there are a number of standard sizes used in the industry (101.6 mm or 4" for the US, and 120 mm for Europe). This should be confirmed by the customer.

The machine nozzle radius and melt channel size must match the sprue bushing (entrance of the melt in the mold). Many defects can occur if this is not the case.

4.5 Extruder Sizing and Specification

Today, practically all machines use an extruder to "plasticize" the material. The extruder consists of a plasticizing screw of appropriate design, rotating (for plasticizing) inside an externally heated barrel. In most machines, the screw is driven by a hydraulic drive, but today an increasing number of machines drive the screw by an electric motor, which is more efficient and saves energy costs. As the screw rotates, the raw plastic, which enters usually near the drive end, is pushed against the inside wall of the barrel. The friction generated between the plastic in the rotating screw flights and the barrel heats the plastic, and the "melt" gradually moves forward toward the end ("tip") of the screw, where it accumulates while the screw retracts, pushed back by the pressure exerted by the plastic. The screw stops when the desired shot volume is reached. At the "injection" signal, in a *ram screw*, the plastic accumulated in front of the screw is pushed out through the machine nozzle into the mold. In *two-stage* injection machines, the extruder is used to fill an injection cylinder (the "shooting pot").

Note that as a rule, the heaters surrounding the barrels contribute less than 20% of the melting power to the plasticizing process. The heaters are there mainly to allow start-up after a shut-down, when the screw and barrel are cold and filled with cold plastic. Heating the screw from the outside and melting the frozen plastic enables the drive to turn the screw again.

The amount of plastic an extruder can convert to injectable melt depends essentially on, and is limited by, the size (or power) of the drive motor (kW or horsepower), torque, screw speed, screw diameter, screw length, and on the design of the screw (screw design is a specialized area of engineering and not within the scope of this book). Many machines are equipped with so-called "general-purpose" (GP) screws, which do a *fairly good* job for most materials, but do not work as efficiently as a screw designed for a specific material. This is an important consideration when planning production. A mold in a machine equipped with the most suitable screw will perform better (i.e., deliver better quality melt faster and with less power) than when using a GP screw.

Today, most machines are equipped with easy screw-change features. Note also that the condition of the screw is very important. A worn screw (and/or barrel) will have greater clearances than when new, and therefore will produce less than the original specifications indicate.

4.5.1 Extruder Shot Size Calculation

Once the actual or estimated mass of the planned product is known, we must consider the proper sizing of the machine on which the mold will be run. Shot size is an important characteristic of the molding machine, which affects the molding cycle. In order to calculate the required extruder size, we must understand the difference between "shot size" and "rated shot size".

Important note on shot size calculation:

All injection molding machine specifications rate the extruder size using polystyrene (PS). The actual shot size and plasticizing capacity must be adjusted to the plastic being processed. Since the density of PS can be 10–40% different than some plastics, this is a critical calculation.

Rated shot size is the amount (in grams or ounces) of PS (melt density = 0.95 g/cm^3) that the injection system (ram screw or two-stage) can inject at every cycle, and is indicated in all machine specifications.

Actual shot size depends on the mass (W) of the molded product (in grams).

There are several common methods to determine the mass:

- If you are using a 3D CAD system, simply calculate the volume of the space between the core and cavity. Alternatively, if the part drawing is a 3D model, simply interrogate the model for volume or mass.
- Weigh a sample (or a handmade model). If the sample (or model) is of the same material as the desired product, this is the mass of the product. Otherwise, divide by the specific weight of the sample (model) and multiply with the specific weight of the desired plastic.
- Establish the volume by completely immersing the sample (or model) in a graduated container, filled partly with water. The difference in filling levels gives the volume.

Multiplying the established volume by the specific weight of the desired plastic gives the estimated mass (weight) of the product (see Appendix 8 for charts of average specific densities (weights) for various plastics).

Now we must again differentiate between hot runner molds and cold runner molds (two-plate or three-plate). With hot runners, only the mass of each product, multiplied by the number of cavities planned, needs to be injected and considered in the shot size calculation. With cold runners, in addition to the mass of the sum of products to be molded, we must inject the estimated mass of the runners.

4.5.1.1 Shot Size Calculation for Cold Runners

If a two- or three-plate cold runner system is selected, the mass of the cold runner must be added to the shot size. And finally, the number (N) of cavities should be determined. All these calculations will have to be repeated several times, with different assumptions, before settling on the final selections for the planned mold. The shot weight for a cold runner system is therefore:

$$\text{Shot weight}\left(\text{SW}\right)=\left(N\times W\right)+\text{Mass of the cold runner}\left(\text{MCR}\right) \tag{4.5}$$

where
N = number of cavities
W = mass of the molded product

The runner size (mass) is very important. The mass of a cold runner can be small and represent only a few percent of the shot weight. But it can also be quite large, as shown in Figure 4.4. In three-plate molds, and in some two-plate molds, MCR could be as much, or even more than $N \times W$, particularly if the products are very small and a large network of runners is required.

Figure 4.4 shows a runner system for a 32-cavity two-plate mold for caps. The mass of the runner system is about 25% of the mass of all the caps per shot. The runner system is fed in the center by a hot sprue (not shown) to avoid an otherwise large, cold sprue. The cavities are tunnel-gated, and the products and runners are separated after ejection.

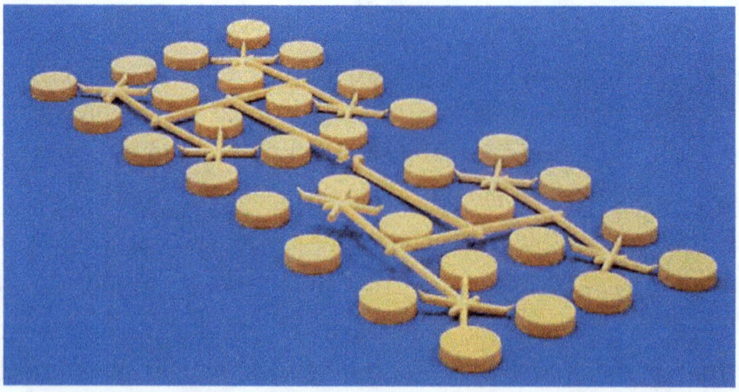

Figure 4.4 Runner system for a 32-cavity two-plate mold for caps

Figure 4.5 shows a runner for a 72-cavity three-plate mold for a small cap. The runner takes almost as much plastic as the products themselves. The cycle time is controlled by the cooling time of the (clearly visible) heavy distribution runners. A hot runner system for such a mold would reduce the injected mass by half and the mold would cycle twice as fast, but would be more expensive to build in the first place. This is a typical example where both methods must be considered, in view of the total requirements of the product.

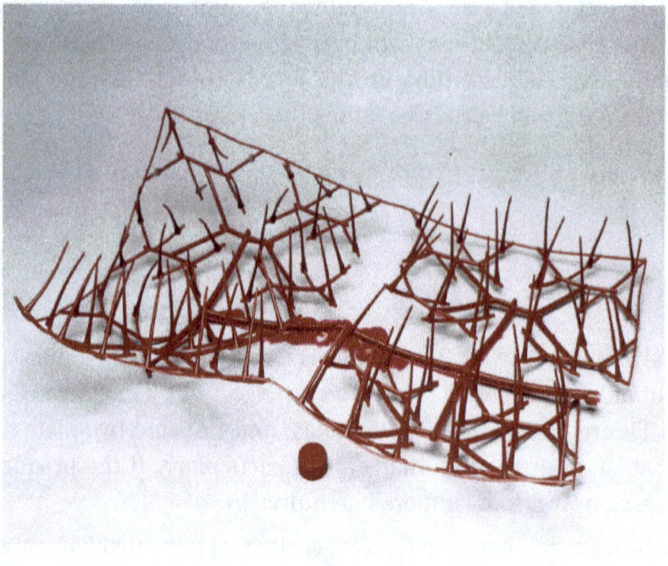

Figure 4.5 Runner for a 72-cavity three-plate mold for a small cap

4.5.1.2 Shot Size Calculation for Hot Runners

In the case of a hot runner mold, the shot size calculation is simply the mass of one cavity multiplied by the number of cavities. The shot weight (SW) for a hot runner systems is therefore:

$$\text{Shot weight } (\text{SW}) = N \times W \tag{4.6}$$

4.5.1.3 Extruder Sizing Calculation

Important note on extruder sizing:

The shot capacity of the machine is rated in polystyrene (PS melt density = 0.95 g/cm³). The shot capacity must be adjusted for the plastic to be used for the planned product, and its *melt* density must be used for this calculation, not its *solid* density.

With the part weight, the runner weight (if cold runner), and the melt density of the plastic to be molded, the required extruder shot size can be calculated as follows:

$$\text{SSR (g of PS)} = \frac{(\text{Part weight (g)} \times \text{\# of cavities}) + \text{Cold runner weight (g)}}{\text{Melt density (g/cm}^3)} \times 0.95 (\text{g/cm}^3 \text{ PS})$$

$$\text{Minimum extruder size} = \text{SSR} \div 0.7$$

$$\text{Maximum extruder size} = \text{SSR} \div 0.2 \tag{4.7}$$

where

SSR = shot size required

Note: Conversion factor between grams and ounces: 28.3 g = 1 oz.

Example:

Calculate the acceptable extruder sizes for a polypropylene printer tray weighing 45 g molded in a four-cavity hot runner mold.

SSR = (120 × 4) ÷ 0.70 × 0.95 = 244 g of PS = 8.6 oz

Min. extruder size = 350 g = 12 oz

Max. extruder size = 1.2 kg = 43 oz

Although this is a simple calculation, it is often overlooked (especially the use of melt density versus solid density). Failure to do this calculation could cause significant problems for the duration the mold is run. In some cases it could cause the mold to be completely incompatible with the machine!

It is suggested *not to use more than approximately 70% of the rated shot capacity* of the machine to ensure that:

- Enough shot capacity is available
- There is a cushion of material remaining in the extruder
- The solid plastic pellets have enough time to melt properly
- Enough plastic is available in the event that the check valve or ball check in the machine develops some leakage back to the extruder as a result of wear.

The injection capacity can also be *too large* for a required (small) shot size. The practical lower limit for the shot size (the distance the screw retracts for the shot) should not be smaller than 0.5 times the screw diameter or *less than 15–20% of the machine shot capacity*. If the shot size is smaller than this value, injection will be inconsistent because *some* stroke is required to reset the check ring or ball check. If this is the case, a smaller machine or a smaller injection unit should be selected. Most newer injection molding machines have multiple screw and barrel combinations available for shot size flexibility in a machine of a given tonnage.

4.5.2 Requirements for Machine Plasticizing Capacity

The plasticizing capacity is defined as the amount (mass) of plastic an injection unit can convert per hour from cold pellets into a homogeneous, thoroughly heated and mixed plastic melt, at the required temperature, ready for injection.

All machine specifications rate the capacity in kg/h, but that really means the amount the extruder can plasticize if it runs *continuously!* But no reciprocating screw or ram screw (RS) machine can run continuously. The screw cannot turn when pushed forward by the high injection forces. As we have described earlier, while the screw turns, the plastic in the barrel is melted and moves forward toward the screw tip, past the check valve and accumulates there until enough plastic is made up for the next shot. When the desired shot volume is ready, the screw stops and waits for the signal to inject. At this moment, the screw is pushed forward to inject the plastic into the mold. The screw *does not plasticize* (a) during the time when the screw is stopped (and waiting before injecting), (b) while injecting (injection cycle), and (c) while holding the screw forward (low-pressure hold cycle). The sum of these times must be subtracted from the total available time; therefore, less time is available for plasticizing. The concept of "plasticizing per hour" is really a guide for general sizing.

To ensure that the machine will do an adequate job of plasticizing for the planned mold, we must consider the hourly throughput *adjusted for the type of plastic used*, not the figure specified in the machine specifications, which is for PS. The plastic pellets enter the extruder as a solid and they are melted through a process of shearing and heating in three sections of the extruder: feeding, transition, and metering. This process is time-dependent and as a result each extruder on an injection molding machine is limited to the maximum throughput stated in the specifications (in kg/h or lb/h). It is critical to calculate the expected throughput of the mold to ensure that it is within the specification limits of the machine.

Note: Usually, the higher the back-pressure, the better is the quality of the melt, but at the same time, this means that more power (kW, horsepower) is drawn from the screw motor and less melt is pushed ahead of the screw. The amount plasticized is directly proportional to the speed of the screw (in rpm), its diameter, length and design. There is a limit to the available power of the motor and to the screw speed (as stated in the machine specifications), which limits the amount plasticized per unit of time.

Rule of thumb:

It is wise not to use the total plasticizing capacity of the machine, but rather use only about 80% of it, to allow for loss due to wear in the extruder, but also to have the possibility to increase the output if the mold can be cycled faster than anticipated. If the output of the extruder is less than required, it means that although the mold *can* be run on the machine, it will run slower and therefore with lower productivity than planned, because the mold will need to wait for the extruder to make up a new shot each cycle.

The available recovery time for the screw can be calculated as follows:

$$\text{RTA}\ (\text{s}) = t_{\text{cycle}} - t_{\text{inj}} - t_{\text{hold}} - t_{\text{pullback}} - t_{\text{decompression}} \tag{4.8}$$

where

RTA = recovery time available

Note that if a two-stage injection unit is to be used then the available recovery time is $t_{\text{cycle}} - t_{\text{transfer to shooting pot}}$.

Once the available recovery time (RTA) has been calculated, the required throughput of the extruder can be calculated as follows:

$$\begin{aligned}&\text{Required extruder throughput}\ (\text{kg/h}) = \\ &\text{SSR}\ (\text{g of PS}) \div \text{RTA}\ (\text{s}) \times 3600\ (\text{s/h}) \div 1000\ (\text{g/kg}) \div \text{SF}\end{aligned} \tag{4.9}$$

where

RTA = recovery time available (s)

SSR = shot size required (kg, g, or oz)

SF = safety factor = 0.8 (recommended)

As a side note, the throughput of the mold is calculated using the cycle time instead of the RTA, and as a result it will be lower than the required extruder throughput because the extruder can only run at certain times during the cycle. The mold throughput (not needed for extruder sizing but noted here for clarity) would be as follows:

$$\text{Mold throughput }(\text{kg/h}) = \text{SSR }(\text{g of PS}) \div \text{Cycle time }(\text{s}) \times 3600\ (\text{s/h}) \div 1000\ (\text{g/kg}) \tag{4.10}$$

> As a result of the fact that the extruder can only run during certain portions of the cycle (RTA), the required extruder plasticizing capacity will be higher than the throughput of the mold. The example below illustrates the point.

> *Example:*
>
> A 36 g HDPE blender casing is molded in a two-cavity mold with a 50 g cold runner system at 25 s (with 15 s available for recovery). What is the throughput of the mold? What is the throughput required of the extruder? The melt density of HDPE = 0.76 g/cm^3.
>
> SSR = (36 × 2 + 50) ÷ 0.76 × 0.95 = 152.5 g of PS (Equation 4.7)
>
> Throughput of the mold = 152.5 ÷ 25 × 3.6 ÷ 0.8 = 22 kg/h (Equation 4.10)
>
> Required extruder throughput = 152.5 ÷ 15 × 3.6 ÷ 0.8 = 45.75 kg/h (Equation 4.9)
>
> This value should be checked with the extruder capacity to ensure that the mold will not be limited by the machine.

Alternatively, the following equation can be used to accurately size the extruder if provided by the machine supplier:

$$\begin{matrix}\text{Required}\\ \text{extruder}\\ \text{throughput}\\ (\text{kg/h})\end{matrix} = \begin{matrix}\text{Instantaneous}\\ \text{recovery (g/s)}\\ \text{(adjusted for the right resin)}\end{matrix} \times \text{RTA (s)} \times \#\ \text{shots/h} \times 0.8 \tag{4.11}$$

What happens if there no such larger machine or extruder is available and there is insufficient plasticizing capacity? The mold will be able to run, but it will not run at the expected speed, because the system will have to wait for the shot to be made and will therefore have less output than planned. Normally if the machine is extruder-limited, the machine operator will extend the cool time of the machine to allow the extruder more time to rotate and to ensure that the cycle time stays consistent.

4.5.2.1 Impact of Light-Weighting the Product

The importance of understanding the plasticizing capacity of a machine becomes clear when planning to redesign a product for less mass ("light-weighting"). The reduction of mass not only saves plastic, but also decreases (in most cases) the cooling time, thus increasing productivity of the mold.

The preliminary questions to ask are: "Will the new mold require more plastic per hour?" and "Is the existing injection unit large enough?" But even if no plastic can be saved by light-weighting, better mold design, and (in particular) better cooling, better ejection methods, and other improvements can result in a substantial decrease in cycle time; in other words, more pieces per hour. Is the plasticizing capacity now large enough for this new mold, on the same machine as before?

Example of the effect of light-weighting on extruder throughput requirements:

A product has an original part weight of 40 g and is molded in PP (melt density 0.80 g/cm^3) a 12-cavity hot runner mold, running at four shots per minute (15 s cycles). The time available for recovery of the screw is 8 s.This product is later redesigned to have a weight of 35 g and will be able run at 6 shots per minute (10 s cycles) and the time available for extruder recovery will be 5 s. The existing extruder has a rating of 400 kg/h.

What will be the difference in extruder throughput required? Will the new mold be able to run in the existing machine?

The old design of the product required the following throughput:

SSR (old) = 40 × 12 × 0.95 ÷ 0.80 = 570 g of PS (Equation 4.7)

Required extruder throughput = 570 ÷ 8 × 3600 (s/h) ÷ 1000 (g/kg) ÷ 0.8 = 320 kg/h (Equation 4.9)

With the redesigned product, we now require:

SSR (new) = 35 × 12 × 0.95 ÷ 0.80 = 500 g of PS (Equation 4.7)

Required extruder throughput = 500 ÷ 5 × 3600 (s/h) ÷ 1000 (g/kg) ÷ 0.8 = 450 kg/h (Equation 4.9) The difference in extruder throughput is 450 - 320 kg/h = 130 kg/h

The new mold will be cycle-limited in the existing machine.

The example above highlights three important consequences:

1. The productivity can be increased by 50%:

   Before, 12 cavities × 4 shots/min × 60 min = 2880 pieces per hour were produced. After redesign, 12 cavities × 6 shots/min × 60 min = 4320 pieces per hour can be produced, requiring much fewer machine hours, and thus also reducing the product cost. The increase in productivity is significant. However, the cost savings could be somewhat less if a larger machine is required.

2. The product cost has been greatly reduced by using less plastic, the difference being 5 g per unit or 5 kg per 1000 pieces. At an estimated cost for a commodity plastic of approx. $1.00 per kg and an annual requirement of 10,000,000 pieces, this amounts to a saving of $50,000 per year. If it was an expensive engineering plastic, the savings would be spectacular.
3. If the injection unit is not capable of supplying the new required melt quantities, the whole effort of redesigning for better productivity would have been economically useless and the money wasted. The machine would have to continue to cycle at the old, lower speed.

All the above must be considered and the calculations must be done every time a redesign is contemplated.

4.5.3 Screw Selection

Some types of plastics, especially engineering plastics, require a more specialized screw design. These plastics may require a different *L/D* ratio (the ratio of the active length of screw over the screw diameter) and a different compression ratio (ratio of height of screw flights from the feed zone to the final "metering" zone, near the screw tip). If molds for plastics other than the most common ones are used, this should be discussed with suppliers of the plastic intended to be used, with the machine designer, or with a specialist in plasticizing screw design. Many aftermarket options are available to the molder for screw designs.

4.6 Machine Nozzle Selection

The machine nozzle is where the injection unit intersects with the mold at the mold sprue bushing. The type of machine nozzle and the size of the orifice used can limit the productivity of the injection molding work cell if the incorrect design is used. Given that the machine nozzle is only a small fraction of the cost of the machine, it is important to consider the right machine nozzle for the application being considered.

4.6.1 Open Nozzles

All molding machines come standard with "open" nozzles, i.e., the tip of the machine nozzle is open and will let plastic pass through freely from the end of the screw either into the open air or into the sprue bushing of the mold, while injecting. In order to achieve good plasticizing, we must provide some controlled low back-pressure in the injection cylinder, acting on the injection piston. This back-pressure is usually only in the order of about 5–10% of the injection pressure, but the pressure is high enough at the tip of the screw (while the screw is plasticizing) to push the plastic through the open nozzle.

With an *open* nozzle, the screw *must not turn* (plasticize) unless it is blocked, because:

- With cold runner molds, the nozzle is pressed against the mold sprue bushing. The plastic in the sprue acts as a stopper and the screw can start rotating and producing as soon as the injection (or injection hold) pressure ends. But as soon as the mold opens, the screw rotation and the back-pressure must stop, otherwise, the plastic will be pushed into the now empty sprue bushing and into the open mold.
- With hot runner molds, the nozzle is also pressed against the sprue, and the screw can start rotating as soon as the injection (hold) pressure ends. The mold *could* be opened safely even with the screw plasticizing, but *only if all gates* were frozen sufficiently to stop the plastic from drooling out of the gates; otherwise, plastic will drool into the open cavities, which is of course unacceptable. In these cases, the screw must also be stopped as soon as the mold opens. With most valve-gated molds, the gates are mechanically closed after injection, and a shut-off nozzle (see below) would not be required.

In both these cases, the time available for plasticizing is limited to the "cooling" cycle. While the mold is open, the screw is stopped. In molds with long cooling cycles, there is usually sufficient time for plasticizing the next shot volume.

Example 1:

Let us assume a mold and machine with a 4 s dry cycle, an injection and hold cycle of 2 s, and a cooling cycle of 6 s. The total cycle is therefore 12 s. When using an open nozzle, the maximum time the extruder can run is 6 s, which is the same length of time as the cooling cycle. If the amount of plastic needed for the shot to be injected can be plasticized in 6 s or less, there is no problem with an open nozzle. Note that in this example, the screw can run only 50% of the time; therefore, the extruder uses only 50% of its rated capacity.

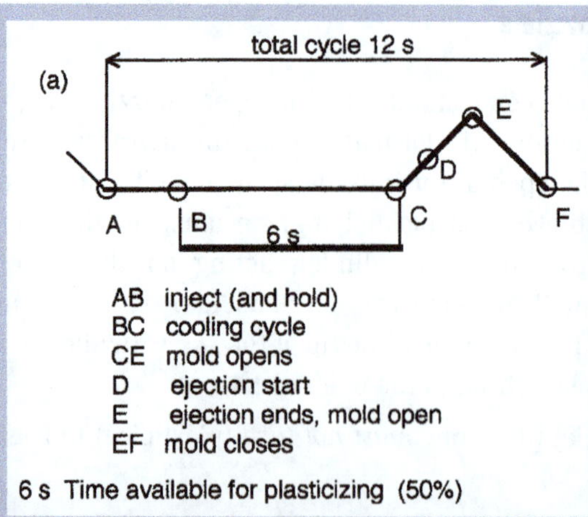

Figure 4.6 Graphic illustration of Example 1

Example 2:

Let us assume the same machine and mold conditions as in Example 1, but this time not enough plastic can be plasticized in the 6 s between the end of injection and the end of cooling. If we assume we need 9 s to generate the melt for the next shot, we must increase (unnecessarily) the "cooling" time by 3 s (from 6 s to 9 s), for a total cycle of 15 s. This represents a severe loss of productivity (4 versus 5 shots/min). This extruder uses 60% of its rated capacity

We should therefore look for an alternative - either the use of a shut-off nozzle, or a machine with a larger extruder. It should also be noted that the shut-off nozzle will only be effective if there is extruder capacity to run during clamp open/close time.

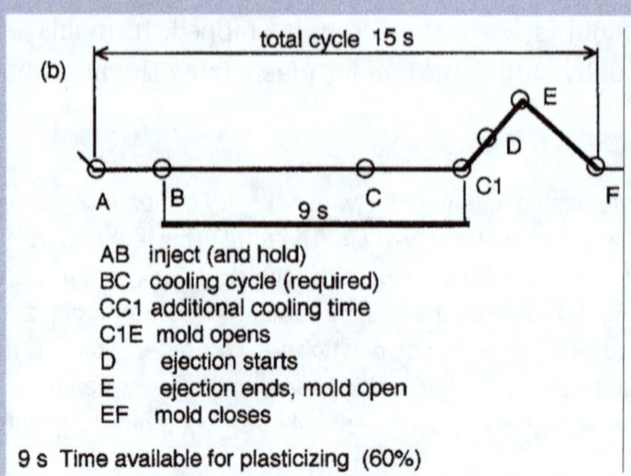

Figure 4.7 Graphic illustration of Example 2

It is also important to understand that by adding unnecessary cooling time, the products will eject cooler than necessary and shrink less (they will be larger than expected). This can be significant when molding plastics with high shrinkage factors. Also, products cooled too much inside the mold may become overstressed in some areas as they shrink onto the core and therefore fail early in use. Theyt may also stick to the molding surfaces and be hard to eject. To overcome both problems, the melt temperature should be higher to ensure that the product will not be "overcooled" in the mold. This adds to the product cost, because it requires not only more energy for heating the plastic higher than necessary, but it will also require more energy for cooling it. The use of a shut-off nozzle (see below) eliminates the extra cooling time.

Example 3:

Let us assume the same machine and mold conditions as in Example 1, but here the *necessary* cooling time is 9 s, for a total cycle of 15 s. In this case, the screw has sufficient time for plasticizing, up to 9 s out of a 15 s cycle. The extruder uses up to 60% of the cycle time.

This illustrates that with longer cooling cycles, a simple open nozzle is adequate.

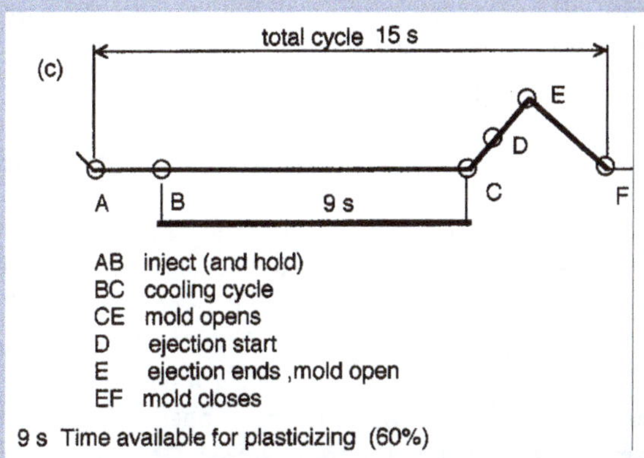

Figure 4.8 Graphic illustration of Example 3

4.6.2 Shut-off Nozzles

For short cycle times, a shot-off nozzle can greatly increase the productivity of mold and machine. The basic principle of the shut-off nozzle is to provide a mechanical stop *within the machine nozzle*, which closes and opens the flow path of the plastic from the extruder to the nozzle tip. Shut-off nozzles come in various executions, such as shuttles, rotary cocks, or pins, and are usually operated by compressed air or by hydraulic pressure (oil).

Figure 4.9 shows a photo of a complete shut-off nozzle. In this design, the lever (A) pushes a pin inside the nozzle to close the nozzle opening (B). When injecting, the plastic pressure pushes the pin to the right, thus opening the nozzle opening to let the plastic enter the mold. The lever is operated by link (C) connecting it with a hydraulic or air actuator (not shown).

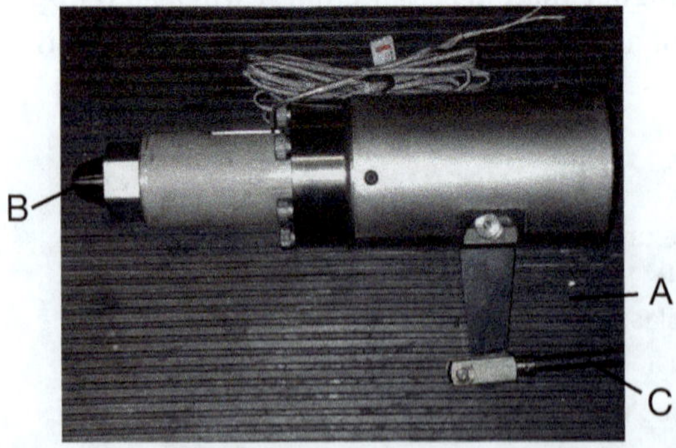

Figure 4.9
Shut-off nozzle

Even though the shut-off nozzle represents an added *one-time* expense, their use is very desirable, especially with short cycle times, because it allows the screw to plasticize while the mold is open.

> *Example 4:*
>
> Let us again assume a mold and machine with a 4 s dry cycle, an injection and hold cycle of 2 s, and a cooling cycle of 6 s. The total cycle is therefore 12 s. Using a shut-off nozzle, the time the extruder can run now is 6 s (the cooling cycle) plus 4 s (the dry cycle), equaling 10 s. The screw has now enough time to plasticize 10 out of the 12 s full cycle, or 83% of the rated capacity. This is a tremendous improvement over the use of an open nozzle. Even a smaller extruder (or machine) could be used for this job.

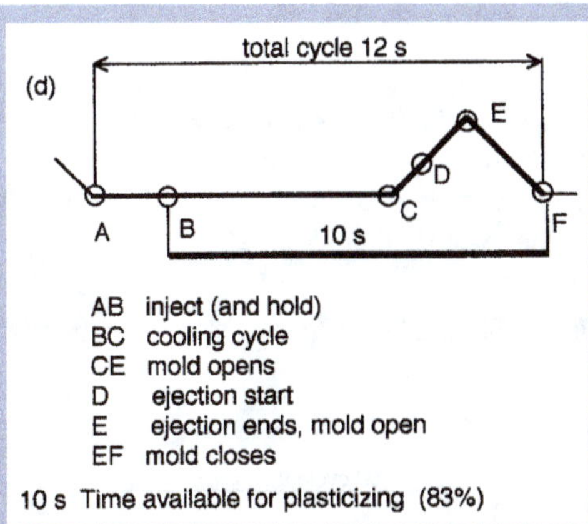

Figure 4.10 Graphic illustration of Example 4

Example 5:

This example illustrates an extreme case: A machine with a 2 s dry cycle runs a mold with a 1 s injection cycle (no hold time). The cooling cycle is 1 s, for a total cycle of 4 s (15 shots/min). With an open nozzle, the screw would have only 1 s time to make up for the next shot or can run at most 25% of the rated capacity. With a shut-off nozzle, the screw could plasticize for 3 s (adding the dry-cycle time and the cooling time). The screw could therefore plasticize during 3 s out of 4 s, or at 75% of the rated capacity.

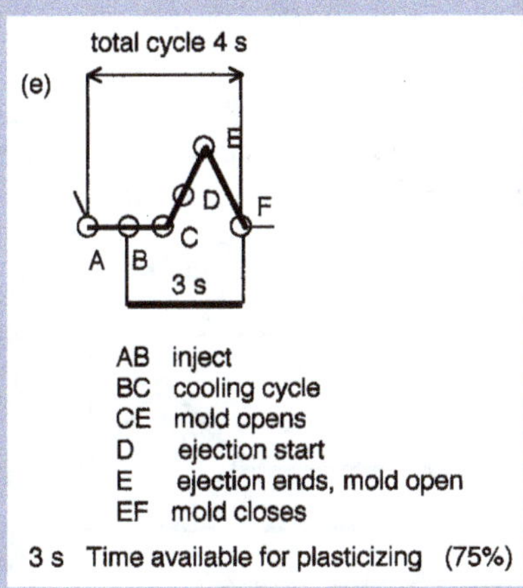

Figure 4.11 Graphic illustration of Example 5

Example 6:

A molder planned to operate a mold in a machine with a 4 s dry cycle, at a 9 s total cycle, or 400 shots/h. However, there was not enough time for the extruder to deliver the required shot size in time for the next shot. The cooling cycle had to be lengthened from 3 s to 6 s and the total cycle increased from 9 s to 12 s. In other words, there was about 25% less production than expected.

With the use of a shut-off nozzle, the time available to extrude could be raised to 7 s, plenty of time for this job.

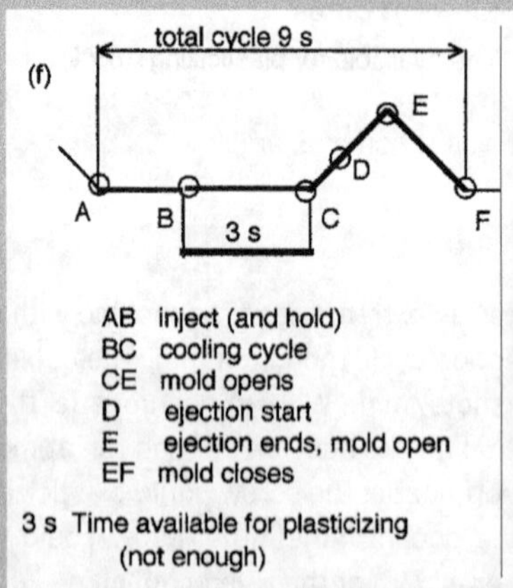

Figure 4.12 Graphic illustration of Example 6

Example 7:

Here the revised conditions suggested in Example 6 are demonstrated, showing that using a shut-off nozzle increases the time available for plasticizing.

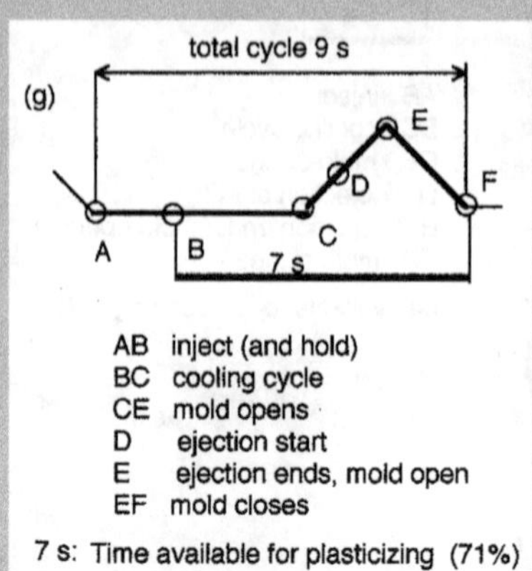

Figure 4.13 Graphic illustration of Example 7

Example 8:

If the mold was moved to another, same-tonnage machine, but with a shorter dry cycle time of only 3 s then as shown below, the overall cycle could be reduced by another second, to a total of 8 s, thus increasing the production to 450 shots/h, or 11% over the original estimate.

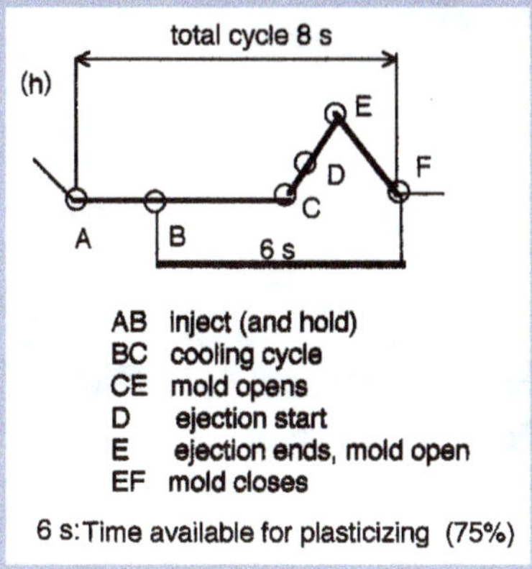

Figure 4.14 Graphic illustration of Example 8

4.7 Injection Unit Selection

The injection unit selected for the mold being designed is critical. The injection unit and the mold are intimately connected via the sprue and nozzle. If the machine does not produce a homogenous melt in the timeframe required by the mold, then many problems can occur. Ensuring that the injection unit is properly sized for the molding application is extremely important.

4.7.1 Reciprocating Screw versus Two-Stage Injection Units

Figure 4.15 shows the difference between an RS (reciprocating screw or ram screw) injection system and a two-stage injection system.

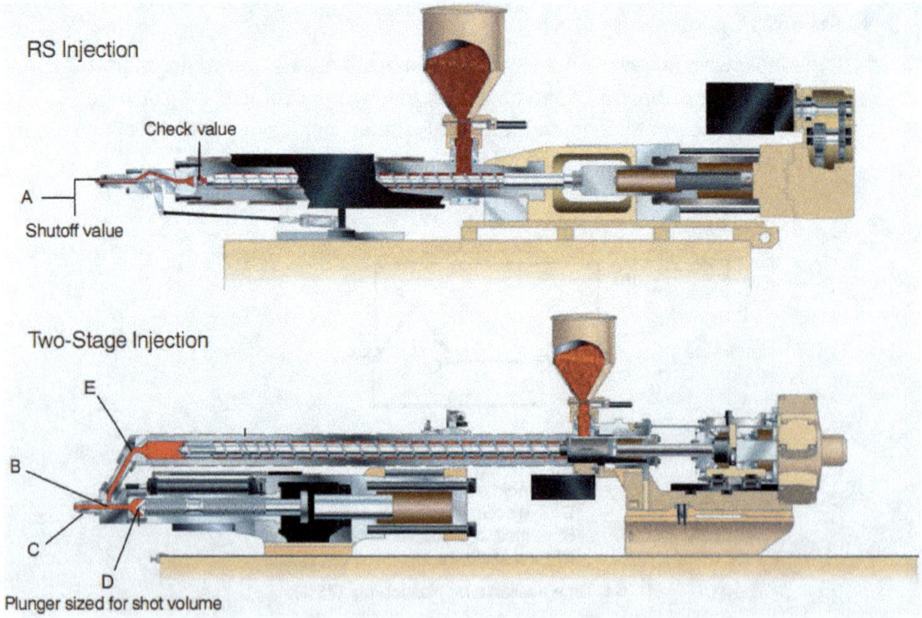

Figure 4.15 (Top) Reciprocating screw (or ram screw) injection system; (Bottom) Two-stage injection system (Courtesy of Husky Injection Molding Systems Ltd.)

In the *RS system,* the plasticizing and injection functions are combined into one unit. While the screw rotates, the plastic is transformed into a melt, which accumulates in front of the check valve. A shut-off nozzle (A) keeps the plastic locked, while the rotating screw retracts to prepare the shot volume for the next shot. The screw stops when the shot size or required volume has been created by the screw (when the injection unit reaches the shot size set point in mm). The check valve or ball check at the tip of the screw prevents the melt flowing back into the extruder during injection.

In the *two-stage system,* the two functions are separated. A rotating valve (B) between the nozzle (C), the shooting pot (D), and the duct (E) coming from the extruder direct the plastic flow as required. In one position, the valve lets the plastic melt coming from the extruder enter the shooting pot and drive the shooting piston back until the pot is filled for the next shot. In the alternate valve position, the plastic can be injected into the mold, without pushing it back into the extruder (this position of the valve is shown). The extruder is also reciprocating, i. e., during the injection cycle, the screw can continue to rotate and plasticize while retracting. There is no screw check-valve required.

Two-stage injection is a special method, in which the plasticizing uNit (with a reciprocating extruder) *can* run continuously if needed. The melt is fed into an injection cylinder ("shooting pot") where it is stored and from where it is injected into the mold with a piston. The molding machine system is more complicated and expensive than an RS system, but it can be easily justified economically whenever large shot sizes are injected because a much smaller, much lower cost, extruder can be used. In addition, the extruder can be running continuously, i.e., without stopping at every cycle. This can also be of advantage for the quality of the melt, because the melt temperature will be more constant.

Another advantage of two-stage injection is that it allows for highly consistent small or large shot sizes, in the same machine. This is due to the fact that no check-ring or ball-check system is needed to prevent melted plastic from flowing back up the screw. A small-diameter shooting pot will also permit extremely high injection pressures where needed, under exceptional conditions.

4.7.2 Injection Speed and Pressure Requirements

Injection speed, in this context, is the amount of plastic per second that can be pushed out of the machine nozzle into the mold as the extruder screw (or piston, in two-stage systems) moves forward to inject. Injection speed is usually specified in cm^3/s (or in^3/s). Assuming that there is no loss (due to leakage or poor design of the check valve in front of the screw), the entire plasticized volume of melt waiting to be injected will be pushed into the mold. This volume V can be easily calculated as the cross-sectional area A of the barrel (with a bore D), multiplied by the stroke S of the screw:

$$V(\text{mm}^3) = A(\text{mm}^2) \times S(\text{mm})$$

or

$$V(\text{mm}^3) = D(\text{mm}^2) \times \frac{\pi}{4} \times S(\text{mm})$$

For this discussion, we will ignore the fact that plastic expands as it is heated and therefore the volume of the melt is – in some plastics – actually up to 30% greater than the volume of the cold plastic. We also ignore the fact that the usable volume is actually somewhat smaller because of the "cushion" left at the tip of the screw, after every injection stroke or cycle.

A (the screw area) is a constant; therefore, the (shot) volume V is directly proportional to the stroke S. Each extruder screw (or shooting pot) has a specified, maximum stroke S_{max}. The stroke is the same as the stroke of the (hydraulic) injection cylinder, which pushes the screw (or piston) forward during injection. In most injection molding machines, for the "standard size" screw (or piston), the ratio of area of the screw (piston) and of the injection cylinder is 1 : 10, so that the oil pressure of the hydraulic oil in the injection cylinder can be simply multiplied by 10 to arrive at the melt pressure of the plastic at the machine nozzle. For example, for an indicated oil pressure of 14,000 kPa (2000 psi), we get a plastic injection pressure of 140,000 kPa (20,000 psi).

In most machines, the extruders can be easily changed (within limits) for smaller or larger screws and barrels; however, the hydraulic injection cylinder remains unchanged. Therefore, using the above formula, a 25% larger screw diameter will result the maximum shot volume increasing by about 56%, at the same time as reducing the maximum possible injection pressure by 36%. On the other hand, reducing the screw diameter by 20% will result in about 36% less volume, but with an increase of 56% in the injection pressure. This can (and should) always be calculated, and it may often be possible to run a mold that does not need high injection pressures on a smaller machine, just by changing the extruder screw and barrel to a larger diameter. Note that as the screw diameter decreases the volumetric injection rate (cm^3/s or in^3/s) goes down.

The impact on the extruder can be summarized in Table 4.3.

Table 4.3 Impact of Screw Diameter on Extruder Performance Characteristics

Screw diameter (mm or in)	Shot size capacity (g, kg, oz)	Injection rate (cm^3/s or in^3/s)	Plasticizing capacity (cm^3, in^3, oz)	Injection pressure (MPa or psi)
Decreasing	Smaller	Lower	Lower	Higher
Increasing	Larger	Higher	Higher	Lower

The *injection speed* is essentially dependent on four factors:

1. The viscosity of the injected plastic (which varies as a function of the injection rate and the temperature - see Figure 4.16)
2. The available injection pressure
3. The resistance against flow within the runner system, the gates, and the cavity space
4. The volume of hydraulic oil flowing into the injection cylinder.

Points (1) and (2) above are closely related: plastic at higher temperatures (with lower viscosity, see Figure 4.17) will require lower pressures to fill the cavity space, and plastic at lower temperatures (higher viscosity) will require higher pressures. This fact may appear obvious, but it has some implications on the selection of the mold, especially its construction. In thin-wall products, in order to push the more viscous (or colder) plastic through the narrow flow paths between cavity and core, higher pressure will be required than for similar, but thicker-walled products. This higher pressure means that more forces are generated inside the mold, which affects the stresses on some of the mold components, but more importantly, it will require higher clamping forces to ensure that the mold will not open up during injection and cause flashing. The parting line (surface) area must be large enough to support the higher clamp forces without being damaged. This could mean that a larger mold is required.

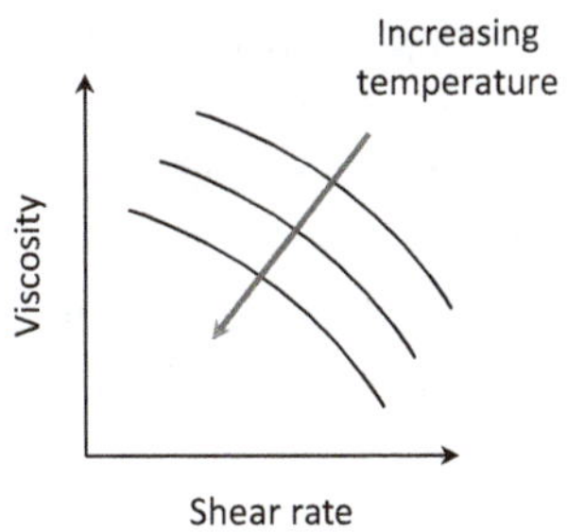

Figure 4.16
Typical shear rate-viscosity relationship for plastics

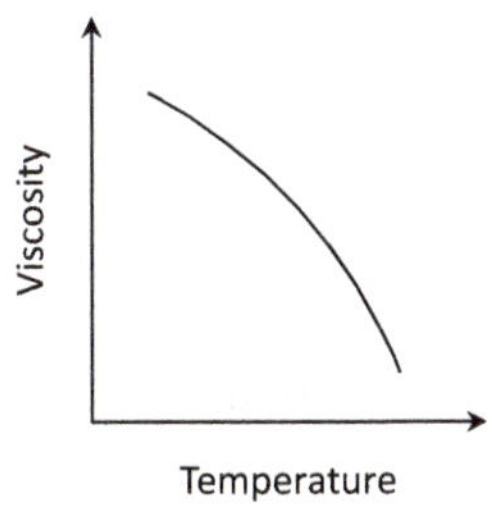

Figure 4.17
Typical temperature-viscosity relationship for plastics. The higher the temperature, the lower the viscosity

To prevent "hobbing" and damage to the mold when steel meets steel under very high clamp forces, it is recommended that the clamping pressure at the parting plane be *less than* 42,000 kPa (6000 psi).

The alternative, to make the plastic less viscous by raising its melt temperature, is not generally a good solution, although some operators who do not understand the implications use it frequently. Higher melt temperatures can make the mold flash more easily, but more importantly, raising the melt temperature means adding more energy *(costs!)* to the plastic, and the hotter injected plastic needs more cooling before ejection, requiring additional (cooling) energy *(more costs!)*. Since it will take more time to cool the hotter plastic, it also will require longer molding cycles, resulting in *lower productivity*.

It does take more energy to generate the higher pressures, but much less than the energy required for the added heating and cooling. Older machines were often limited by injection pressures to a maximum of about 140,000 kPa (20,000 psi); most modern machines can deliver injection pressures of 200,000 kPa (29,000 psi) and even more by providing a selection of extruder barrels and screws, or shooting pots, of smaller sizes.

All this explains why it is usually advantageous to use high injection pressures when molding thin-wall products of any size, whether disposable goods or not. While the increased strength may add costs to the mold, the savings in energy and the increased production readily pays this additional cost back.

As for the third limiting factor on injection speed, much can be done to improve the plastic flow within the mold, in the runners, and in the cavity space by optimizing the gate location and size, occasionally by thickening the walls in certain areas, and by making sure that the finish of the molding surfaces is best suited for the plastic injected. As stated earlier, flash chrome-plating a highly polished surface can greatly improve the flow and reduce the molding cycle. Also, hot runner systems, especially the channels inside the manifolds, must be properly sized for optimal conditions, and have all channels smoothly finished.

The injection force is generated by the injection cylinder. The piston within the injection cylinder is driven forward by the hydraulic pressure oil as soon as the hydraulic valve admits oil to the cylinder. The speed of the piston motion is therefore proportional to the amount (volume) of oil flowing into the injection cylinder. This volume depends on the source of the pressure oil, but also on the flow restrictions in piping and valving between the source and the cylinder. In many injection-molding machines, the source of the pressure oil is a hydraulic oil pump, either a single pump, supplying the whole machine (clamp and injection, and any accessories requiring pressure oil), a pump dedicated to the injection system, or a combination of pump(s) and hydraulic accumulators.

Hydraulic pumps are rated by their output, in liters or gallons per minute. The relation between the pump output and the volume of the injection cylinder limits the injection speed. If we have, for example, an injection cylinder with a diameter of 150 mm and a stroke of 150 mm, the volume is 2.65 L. A pump (in this example) may have an output of 40 L high-pressure oil per minute (60 s). It will therefore take 60 s × 2.65 L ÷ 40 L = 3.98 s to fill the cylinder. This translates into a maximum injection speed of 150 mm ÷ 3.98 s = 37.74 mm/s (or about 1.5 in/s). This would be the speed with which the plastic can be pushed into the mold. For some products this may be adequate, for others, especially thin-walled products, this may not be fast enough. If we want to increase the injection speed, we have one of two options:

- Equip the machine with a larger pump, also requiring a larger drive motor (and larger controls)
- Equip the machine with an *accumulator system*, i.e., adding one or more pressure vessels (accumulators) where hydraulic pressure-oil can be stored during the time when it is not needed during a molding cycle (Figure 4.18). During injection, the pressure oil coming from the accumulator is added to the pressure oil coming from the pump. Note that this method is only viable if there is

enough time available during each cycle to recharge the accumulator. If the molding cycle is very short, there may not be enough time for it, and a larger pump and motor will be required.

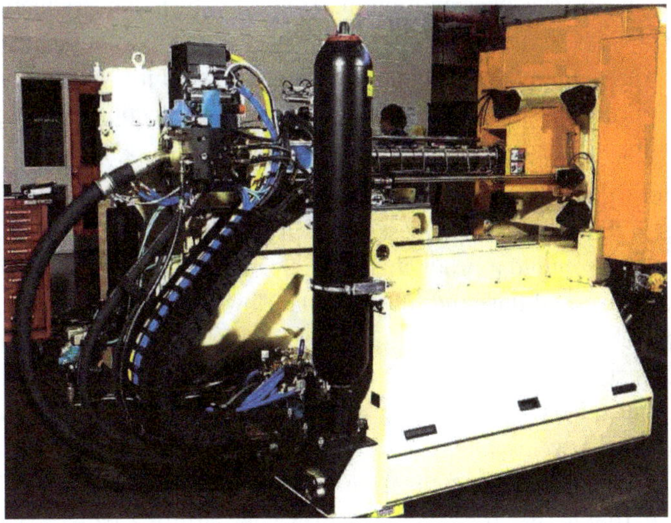

Figure 4.18 Injection unit equipped with an accumulator to increase injection speed (Courtesy of Husky Injection Molding Systems Ltd.)

It is therefore important for the decision maker to ascertain that the machine for which the mold is planned has sufficient injection speed to take advantage of the higher capability of the mold that is planned. There is no point in having a mold designed that could run, for example, 15 shots/min if the injection system of the planned machine is capable of providing only 10 shots.

Note: There is some difference in injection time when filling a mold with more cavities as compared to a mold for the same product with fewer cavities, because of the longer distances the plastic must flow (greater pressure drop) and the greater total amount of plastic that must be injected. However, at this point, a conservative estimate can ignore such differences, which may amount only to a fraction of a second. Once a mold layout is available, these differences in injection time could be calculated.

4.7.2.1 Need for High Injection Pressures

High injection pressures are needed to ensure that the mold is completely filled during the injection cycle, with the desired clear surface definition. There are several challenges to consider.

1. The thinner the wall thickness of the product, the more difficult it is to push the plastic through the gap between cavity and core, thus requiring higher pressures. Since material (the plastic) usually accounts for 50–80% of the total cost of a molded product, it is highly desirable to reduce the weight (mass) of plastic injected to a bare minimum. This usually means reducing the wall thickness as far as possible without affecting the usefulness of the product. Over the years, many products have been redesigned just to reduce the plastic mass of a product. This is also why many modern injection molding machines provide higher injection pressures than older ones.
2. The colder the injected plastic, the higher its viscosity, and the more difficult it becomes to fill the mold. The cost of the product depends directly on the cycle time required to mold a product. The higher the melt temperature of the plastic, the easier it will flow and fill the mold. However, higher melt temperatures also require increasing the cooling cycle time to bring the temperature of the injected plastic down to a level where the product can be safely ejected without distorting or otherwise damaging it. This means more power (for heating and cooling), longer cycles, and therefore higher costs. It is often better to inject at the lowest possible temperatures, even if more pressure is needed to fill the mold. Note that higher injection pressures will require greater clamping forces and a stronger, possibly larger, machine. Another solution to the problem might be to select a plastic that flows more easily. Such plastics, however, are usually more expensive and may not be as strong as desired.
3. High injection forces are needed for good surface definition. Typically, this is important when molding articles such as compact discs, where the clarity and precision of the surface definition is in direct relation to the quality of the sound reproduction of the recording.

5 Factors Affecting the Design of an Injection Mold

In this chapter we will examine the environment that the proposed mold will need to operate in. This means the machine into which it will be mounted, the services such as air and cooling, and the plant environment itself. All of these areas can have a significant impact on the design and productivity of the mold. Although the mold is the heart of every injection molding system, it cannot operate efficiently if it is not able to function in a satisfactory environment. We will examine these external factors and the impact they can have on the design of the injection mold. Once these items are understood, the designer can accurately estimate the cavitation of the mold, the materials of construction, and the sophistication required of the mold.

5.1 Total Equipment Productivity (TEP) Objectives

A mold's productivity is ultimately measured by how many good parts it can produce over time. A mold that runs very fast but is frequently down for maintenance or repair will not result in the lowest part cost and highest productivity. Conversely, a mold that produces parts consistently over time but runs too slow is also not optimal. The same can be said for each component in the injection molding system.

One way to quantify productivity is to measure the total equipment productivity (TEP). TEP is the product of the quality (or yield), the reliability (or uptime), and the cycle time (or speed of production) (Figure 5.1).

$$\text{TEP} = \text{Reliability}(\%) \times \text{Quality}(\%) \times \text{Cycle time (as \% of expected cycle time)}$$

$$= \frac{\begin{matrix}\text{Production hours}\\ \text{(auto-cycling)}\end{matrix}}{\text{Available hours}} \times \frac{\begin{pmatrix}\text{Parts} \\ \text{produced}\end{pmatrix} - \begin{pmatrix}\text{Scrap} \\ \text{parts}\end{pmatrix}}{\text{Parts produced}} \times \frac{\begin{pmatrix}\text{Standard} \\ \text{cycle}\end{pmatrix} - \begin{pmatrix}\text{Actual} \\ \text{cycle}\end{pmatrix}}{\text{Standard cycle}} \tag{5.1}$$

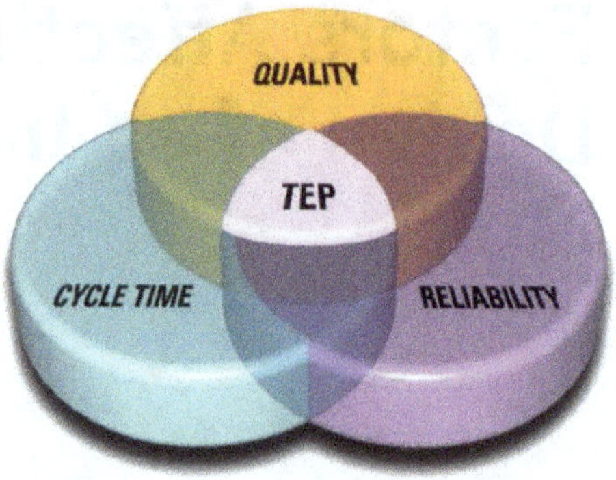

Figure 5.1 Total equipment productivity (TEP)

Reliability or update can be affected by the time to maintain the mold, the frequency of stoppage, the scheduled uptime, and the time to repair breakdowns. The cycle time can be affected by the machine performance, the cooling, and the plant environment. Finally, the yield of good parts is affected by molding defects, color change, gate quality, dimensions, loss of cavitation, contamination, and other molding problems.

The average molding operation achieves in the neighborhood of 70–75% TEP. A very good custom molder can achieve TEPs greater than 80%, and good dedicated systems achieve values higher than 90%. Understanding the objectives for the system will help the designer select the appropriate level of sophistication in the mold and costs.

Example TEP calculation:

A molder is running a 48-cavity closure mold at 4 s cycles. They have scrapped an average of 10,000 parts per day. They average 22.5 auto-cycling hours per 24 h of available time. The expected cycle time is 3.8 s for this part. What is the TEP of this system?

Expected parts per hour = 3600 ÷ 3.8 × 48 = 45,474 parts per hour

Actual parts per hour = 3600 ÷ 4.0 × 48 = 43,200 parts per hour

Uptime = Hours running ÷ Hours available = 22.5 ÷ 24 = 94% (0.94)

Scrap per hour = 5000 ÷ 24 = 417

Scrap percentage = (Actual parts/h – Scrap/h) ÷ Actual parts per hour = 1% (0.99)

Cycle efficiency = 3.8 ÷ 4.0 = 95% (0.95)

TEP = 0.94 × 0.99 × 0.95 = 0.88 (88%)

Since the TEP equation multiplies all the factors associated with processing, the impact of a reduction in any of the three factors has an impact on the number of good parts in the box at the end of the day.

For example, let's look at a 48-cavity 2.9 g closure mold running a standard cycle of 5.5 s. It runs at 90% uptime and 3% annualized scrap. The impact of a 1% change in cycle time, weight, scrap, and uptime results in important savings to the part cost.

Table 5.1 The Impact of 1% on the Part Cost Using a 48-Cavity Closure Mold

	Baseline	1% Decrease in cycle time	1% Decrease in weight	1% Decrease in scrap	1% Increase in uptime
Cavitation	48	48	48	48	48
Weight (g)	2.9	2.9	2.87	2.9	2.9
Cycle (s)	5.5	5.45	5.5	5.5	5.5
Uptime (%)	90	90	90	90	91
Scrap (%)	3	3	3	2	3
Impact	–	2M parts	13,000 kg	2.2M parts	3 days
Impact ($)	–	**5200**	**8900**	**5900**	**6300**

5.2 Cycle Time

The cycle time expectation of the mold is a key criterion in determining the type of mold to be designed. The cycle time affects almost every aspect of how the mold will be designed. Establishing and agreeing to a cycle time estimate is fundamental for the designer to know before starting the mold design. Chapter 6 covers cycle time estimation in detail, as this is such a key factor in the mold design.

5.3 Projected Annual Requirements

How many pieces of the product will be made from the planned mold on an annual basis? How many years will the product be in existence before it is phased out? These could be the most important questions to ask before deciding on the type of mold required for any job. But they are also often the most difficult questions to answer, particularly if the product is new on the market. It is nearly impossible to predict whether the product will find the hoped-for acceptance and increase in sales, or if the product will not be accepted as expected.

If 1,000,000 pieces of a new product are to be molded, the question is:

- Is this a limited production run, say within four months (usually as soon as possible)?
- Is this quantity needed every year, for an unspecified number of years?
- Is this quantity needed over the expected life of the product, e.g., 5 years, in which case the annual requirement is only 200,000 pieces?

The trend over the last 20 years or so has been to use larger and larger cavitations to meet the volume requirements on an annual basis. Figure 5.2 illustrates how cavitations have been increasing steadily over time as new mold and hot runner technologies emerge, and machine speed and capability has improved.

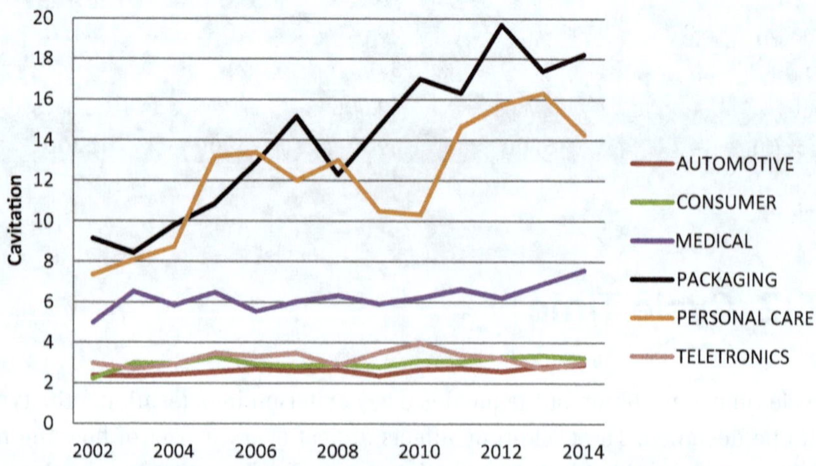

Figure 5.2 Cavitation trends over time by product type

■ 5.4 Purpose of the Mold

The designer must understand the reasons that the mold is being created. It could be for a short run, for testing, for experimentation, for a new product, or it could be to increase production of an existing part. Whatever the reasons are for buying the mold, they should be understood so that the correct decisions are made in the design process to meet the customers' intended use for the mold.

5.4.1 Prototype Mold

Prototype molds, such as shown in Figure 5.3 and Figure 5.4, are required to make samples of a new product for evaluation of a newly developed shape, to see how the product appeals to the eye and/or to the touch, how well it functions, and if it meets the strength and durability requirements. Molded samples can be subjected to the expected stresses and wear, and the results are better than testing a handmade (machined or assembled) model. The result also could be more accurate (and possibly cheaper) than a computer simulation.

Figure 5.3 Typical prototype mold for a lid, capable of 4–8" lid prototyping (Courtesy of Husky Injection Molding Systems Ltd.)

Figure 5.4 Single-cavity prototype mold for production 2 × 2 system. In this case, the prototype stack was used as a spare in the production mold (Courtesy of Husky Injection Molding Systems Ltd.)

Because it is only important to mold the overall shape of the product, without worrying about the productivity of the mold, cost savings can be made everywhere in the design. Mold materials such as mild steel, aluminum, and even plastics (epoxy, etc.) can be selected, as long as they are sufficiently strong and resistant to the heat and the pressure of the injected plastic. Working to close tolerances is usually not necessary. Generally, there will be no need to worry about surface appearance (polish, engraving, even flashing). There is no need for extensive cooling channels, as it will take just a little longer to cool the plastic before being able to remove the molded sample from the mold. In many cases there is also no need for an ejector mechanism. An air jet directed against the edge of the product at the parting line, or a few simple ejector pins that can be manually pushed to eject, may be all that is required.

Other features of the product, such as internal or external threads, can be produced by using loose inserts in the mold that can be ejected with the product and then unscrewed by hand. Loose inserts can also be used for odd shapes in the sides of the product, which would otherwise require side cores. Round holes or simple openings in the sidewalls could be machined after the molded piece is cold. These are just some of the mold features that can be omitted to simplify the stack and to reduce the cost of the prototype mold. If prototypes are frequently required, the stacks could be mounted in a common mold shoe, thus saving even more costs. The runner system would normally consist of a simple sprue gate directly into the product, or a sprue and short runner could be used for edge gating. The gate will then be cut manually.

5.4.2 Experimental Mold

This type of mold is different from the prototype mold. It will be used mostly to establish the behavior of the plastic in a newly developed product during injection. Some of the shortcuts used in a prototype mold to save costs can be used, but in general, the mold is closer to a simple, single-cavity production mold (see Figure 5.5). The gate should be located as planned for the production mold. The mold could also be used to establish the most suitable location of the gate and the method of gating for the product.

Such a mold would normally need to produce a product with a proper finished appearance. Note that especially in thin-walled products, the finish affects the flow of plastic through the cavity space. Cooling efficiency is not as important as in a production mold, but some cooling should be provided to maintain a stable mold temperature. Because the quality of a molded piece depends very much on the accurate repetition of the cycle time, an ejector mechanism should be provided rather than manual product removal, to eliminate any operator-created variations in ejection (and cycle) time.

Figure 5.5 Typical four-cavity experimental mold that will emulate the behavior of the production mold (48 up to 144 cavities) (Courtesy of Husky Injection Molding Systems Ltd.)

An important feature of an experimental mold is often the ease with which some stack parts can be changed. This adds costs but will make experimentation easier. Experiments with such molds can also determine the effect on molding cycles when areas of the mold are not cooled, little-cooled, or well-cooled. Such information can be valuable before an expensive, multi-cavity production mold is designed. The difference between ordinary and exceptional cooling could mean a lot in engineering the production mold. A reduction in cycle time achieved by exceptional cooling could be insignificant and not worth the additional costs and complications to the mold.

5.4.3 Combination of Prototype and Experimental Mold

This applies when an inexpensive mold is required to establish the shape of the product, but when at the same time it is planned to explore market acceptance of such a product by manufacturing a few hundred or even thousands of samples for field testing. Typically, such molds should run fully automatically, but there is no need to achieve maximum efficiency in molding (using better cooling, better runner system, etc.), and nor is there any need to apply special finishes or (most) engravings. Such molds can also be used to establish shrinkage conditions.

5.4.4 Production Mold

Production molds are any type of molds other than prototype and experimental molds. At this point in the planning for a new mold, it becomes necessary to have information on the following:

- How many pieces will be required?
- What will be the molding cycle?
- What environment will the mold run in (machine, plant, etc.)?

Since the mold type and number of cavities will depend primarily on the quantities required to be molded, we must first differentiate between the various possibilities as they present themselves, before deciding on the kind of mold that will be most appropriate.

5.4.5 New Products

The new, untried product is a common case, and can be part of a new invention or an existing product previously made from a different material. Will the market accept it in its new shape, made from injection-molded plastics? Will it require modifications after complaints or suggestions from the field after being launched on the market? Or will it be a disappointment for the seller, and soon disappear? Unfortunately, the customer (molder or OEM) takes all the risk when investing in the required mold.

Of course, it would be convenient to keep mold cost as low as possible, but we know that this may increase the cost of the products in the long run. The cost of a mold with a large number of cavitations may also affect the timing of the launching of the product. Should a large production be anticipated, which will require a multi-cavity mold of high quality? In this case, if the product is not accepted in the field, the loss could be substantial.

But there is also another, equally serious problem when launching a new product: the investor was overly cautious and is waiting for the acceptance in the field. If the product is a great success, the first mold will probably not have been designed for the unanticipated high demand. What would be the best strategy at this point? Make another mold (or even several molds) similar to the first one and run them side-by-side? This approach may have the advantage of lower additional investment while providing more flexibility. It is easier to find several smaller machines than one larger machine. However, a larger system, using a high-production mold, with more cavities, better runner systems, better cooling, better ejection, more automation, and therefore higher uptime, will result in the lowest cost of the products.

Often there is no clear answer to the above questions. It may depend on the expected life of the product, which is often just as difficult to estimate. Some products are seasonal and the demand finds an early saturation point. Some products increase in demand until some competitive, similar, or even better product comes along, in which case demand for the original product could sooner or later disappear. One possible approach is to build a mold for the initially estimated volume *plus* 25% for surge demands, unanticipated stoppage, and some growth.

5.4.6 Existing Product, Large Quantities

Some products are "timeless", meaning that their annual quantities are more or less constant and known, such as the stadium cup mold shown in Figure 5.6. Their use may vary within seasons and even with the economy in general, but they remain essentially unchanged. This applies to many technical articles, as well as many packaging products (such as food containers), and medical products. In these cases, it is not difficult to establish annual requirements and a projection for how long the product will be in demand. In addition, it is always important to consider the whole system, i.e., machine, mold, and any after-molding operation (automation, product handling, packaging, assembling, etc.) that will yield the lowest-cost product. With long and high production runs, even high mold cost is insignificant per unit produced, and helps lower the product cost, provided it runs faster, longer, and with higher-quality products.

Figure 5.6 High-volume production mold for a stadium cup (2 × 12, 24 cavities, air eject, modular construction) (Courtesy of Husky Injection Molding Systems Ltd.)

5.4.7 Limited Quantities, Limited Product Lifecycle

Sometimes, a product is required in limited quantities, or for a one-time occasion only. This may be the case where a molded piece is designed for a special occasion or application. The quantities are relatively small but usually known. Frequently, a molded piece will be required as a promotional item, such as giveaway items to retail customers. Such promotions are usually limited in time and the requirements are stipulated at the beginning of an advertising campaign. Usually, such promotion needs fast delivery of the molded pieces, and the total amount of pieces must be supplied in a very short time span.

A decision will have to be made: Should the order be produced on a large, multicavity mold? This will yield the best piece cost but will require a larger machine, which may not always be available at the time the mold is ready for production. The mold cost will be higher but the cost per molded piece is probably insignificant. The problem is that such larger molds will take longer to build and there may not be enough time. Also, it leaves the molder vulnerable, in case of machine or mold breakdowns, in which case there could be no production at all.

As an alternative, several smaller, identical molds could be built which are simpler and can be made faster by contracting out to more than one mold maker if necessary. These (smaller) molds can be built faster and then be run on smaller machines, which are also usually easier to locate; if necessary, even at different custom molders. This will probably increase the piece cost, (a) because the smaller and less expensive molds are not as productive as a larger mold, and (b) because of the added cost of dealing with more than one source. However, this approach will also ensure that any breakdown will be less serious to the customer. In all these cases we assume that the molds are complete, self-contained molds.

5.4.8 Short Runs, Small Production Requirements

For short runs or for very small total production, when products are relatively simple and small, and when it is known that the annual requirements are also small, there are two alternatives:

- Individual molds, with the smallest number of "high productivity" features (especially good cooling, hot runners, etc.).
- Making inserts for so-called "universal mold shoes" which can be purchased from mold supply-house catalogs.

Such mold inserts for universal mold shoes (see Figure 5.7) are available for quick deliveries and have all the necessary features of the mold built into a standardized format. The inserts can be mounted in the universal shoe and run singly, in pairs, in multiples, or even in combination with inserts for another product. Mold changing is usually simple and fast. These mold shoes can be used for a different product in the future and the cost of the mold shoe saved in future projects.

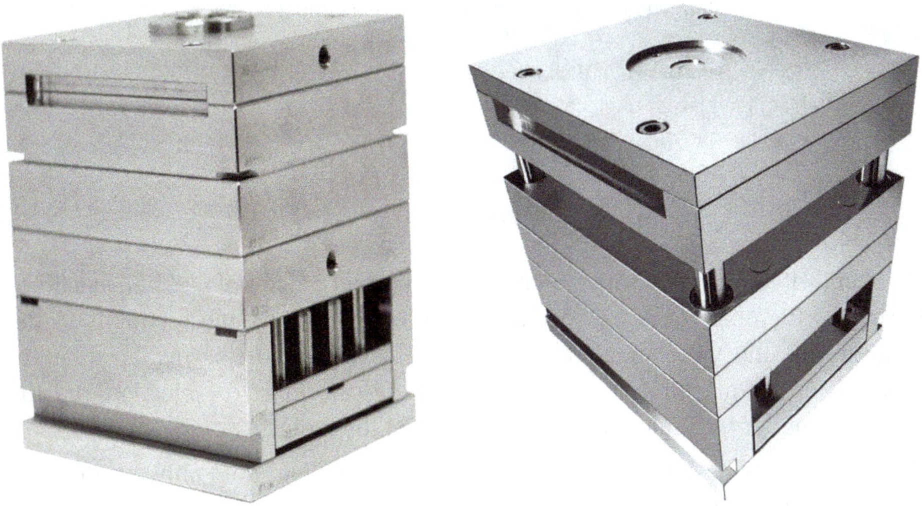

Figure 5.7 Standard mold bases with ejector boxes (Courtesy of DME)

Use standard mold shoes if possible, or use simple, dedicated molds.

5.5 The Injection Molding Machine

The designer must ensure that the mold design will function correctly in ALL the machines it will be mounted in. In order to do this, the designer must work with the customer's project manager to ensure that the machine is specified correctly. Chapter 4 guides the designer through the process of selecting the right machine for the mold.

The mold must be equipped with all features that are compatible with all machines it will be mounted in. This applies to several areas of the mold:

- Shut height
- Any downstream automation
- Mold mounting (including any systems for quick mold changing)
- Locating ring size
- Sprue bushing size and shape
- Machine ejector pin locations
- Cooling- and air-circuits
- Hydraulic functions
- Electrical connectors.

The mold will often be required to operate in different models of molding machines during its lifetime. This may result in some complications and additions in the mold layout, and will commonly increase the mold cost. In particular, different locations of the machine ejectors and mounting can affect the ejection *and* the cooling layout of the mold and the overall size of the mold.

If a mold is to be designed for one machine only, in one location only, it can result in a simpler mold. For example, there would be no need to provide for various sizes of locating rings and ejector mechanism, and the mold mounting provisions could be designed for the pattern of the selected machine only.

As part of the confirmation of order, the customer must specify all the machines that the mold is expected to run in and all the associated information to ensure that the mold will function in all required machines.

5.6 The Plant Environment

We tend to assume that the mold will be operated under ideal conditions, but this is typically not the case. The environment in the molding shop can vary from very cold to very hot, from dry to very humid, from clean to dusty and dirty. With sudden changes in any of these conditions, a molding operation can be affected significantly. High humidity will affect the mold itself (corrosion) and will affect the cycle time (productivity) of the mold, because the mold may sweat, causing difficulty in ejection or cooling. Rapid temperature changes may even affect the operation of a machine and mold, and lead to breakdowns and loss of production. The anticipated plant environment must be understood before undertaking the mold design, to ensure that the mold to be built will perform as expected. If there are conditions that the mold cannot tolerate, then the designer or project engineer must inform the customer that these conditions need to be improved, in order for the mold to function as expected for its expected lifetime.

5.6.1 Condition of Ambient (Shop) Air

In addition to impacts to production and quality, the condition of the ambient air can affect the mold surfaces and the exposed surfaces of the tool through corrosion. In cases where the ambient air will be humid or if the mold temperatures are below 10 °C, protection of the surfaces of the mold is recommended. Chapter 15 covers mold materials in more detail, but there are four ways to address this concern:

1. Use pre-hardened plate steel and paint the mold shoe
2. Use pre-hardened plate steel and add ENP (electroless nickel plating)
3. Use stainless steel (preferred method - all the surfaces, including cooling channels are protected)
4. Flash chrome plate the molding surfaces.

It is not just in the production environment that the mold can rust. It is frequently during storage that a mold rusts. A common approach in many shops is to protect the molding surfaces before the mold is put into storage by using silicon spray (frequently called "Mold Saver") or to just apply plain, clean machine oil. Many shops paint the outside of the mold shoe with a permanent oil paint, to protect the outside of the mold against corrosion.

A typical example:

A molding shop operated eight identical machines in each of two parallel rows; all were molding the same or very similar products with the same type of mold. They all worked fine, except the last machine in one row, which stopped frequently, without apparent reason. After checking for machine problems, such as possible power fluctuations, poor supply of cooling water, etc., it was noted that this last machine was close to an emergency exit door, which was supposed to be closed all the time; however, on some days, the workers kept the door jammed open to improve the shop ventilation. The draft from the cooler outside air was enough to affect operation. After ensuring that the door stayed closed at all times, there were no more problems reported with this machine.

> *Another example:*
>
> A mold and machine worked perfectly, but occasionally, for several hours, it produced pieces with surface blemishes that looked like blisters. Investigation showed that it happened only on very humid days. This particular operation required a rather long mold open cycle. On humid days, the water in the air condensed in tiny droplets on the cold mold cores during the few seconds the mold was open, and the cavities and cores were exposed to the shop air. The droplets appear as blisters on the surface of the product. After slightly increasing the cooling water temperature to bring it above the dew point, the problem disappeared. The penalty was a slightly longer cycle time, but it ensured continuous production of quality products.

The best choice is to make the entire mold shoe from stainless steel (SS). The basic cost of SS is higher than the cost of mild steels or pre-hardened machinery steels. However, when SS is bought in large quantities, the cost difference can be much less. When molds are expected to run for a long time, the advantage of SS over other steels can justify the higher cost. We must not forget that chrome plating or ENP (which is needed for rust protection on pre-hardened plate steel) also cost money. We must also consider the costs of transportation to and from the nickel or chrome plater, the additional time required for these operations, the lack of control over the transport, and the dependence on an outside supplier.

Another problem with chrome plating is that any change (requiring re-machining) of a chromed surface requires that the chrome must first be removed from the base steel part. This requires shipping the part to the plater for removing the coating by a process similar to plating. After re-machining, the changed part must again be shipped to be plated. This is an expensive and time-consuming procedure.

The use of full-hardened (or pre-hardened) SS for cavities, cores, and inserts is quite common today, even though the initial steel cost is higher. When comparing this to the cost of a high-carbon tool steel such as H13 and the costs and time required for plating, the total cost will be more than using SS.

Another solution for all these issues is to provide the molding plant and the mold storage facilities with air conditioning, or at least with controlled, low-humidity air. Occasionally, it can be useful to surround the machine with a shroud, and use a portable dehumidifier to keep the environment immediately around the mold and machine at a desired low humidity (see Figure 5.8).

Figure 5.8 Plan of molding plant in which the environment is well-controlled (Courtesy of Husky Injection Molding Systems Ltd.)

5.7 Coolant Supply

The available cooling water supply (quantity, quality, and pressure of the coolant) must be considered when designing an injection mold. Remember that for water cooling to be effective, the water must flow fast enough to establish turbulent flow. Turbulent flow removes significantly more heat per liter (or gallon).

There is no point designing a mold with an expensive, elaborate cooling system if there is not enough coolant flow and pressure available to take full advantage of it. Some mold plants develop from only a few to a high number of machines, but without a corresponding increase in the cooling water supply. This can result in the molds running much slower than they could if the cooling water supply had been sufficient.

Good cooling of a mold depends not only the coolant *temperature,* but also on the volume of coolant that flows through the mold, measured in liters or gallons per minute. This *volume* depends essentially on the pressure differential between the IN and OUT cooling channels in the mold and on the method of distribution through the mold.

Guidelines for designing the cooling circuits in the mold can be found in Chapters 8, 9, 10, and 12.

5.7.1 Is the Cooling Water Clean?

Cooling water must be free from contaminants and/or oxidizers, which corrode the inside of the cooling channels. This is where using stainless steel for mold bases and inserts is of great advantage. The coolant must also be free from lime and dirt, which will gradually settle in corners of the cooling system and plug the cooling channels, especially if the channels are small and in elaborate circuits, as is often required in high-production molds to cool small mold parts. Under such bad conditions, a mold will probably run satisfactorily and produce as planned for the first few months, but because of buildup of dirt in the cooling channels, the mold will gradually lose its cooling efficiency and run slower than it could with good, clean coolant. Dirt in the water will also require more mold maintenance, as the channels will have to be cleaned from time to time. Such unnecessary costs are often overlooked while worrying about the high initial mold cost.

> If a mold ran initially well after the start of production but performance deteriorated over time, the culprit can often be poor cooling water and plugged cooling channels. Always check the condition of the cooling water and the cooling channels before considering major changes to the mold design.

Rust inside the mold cooling circuits, as shown in Figure 5.9, is an insulator and will eventually slow the molding cycle as it builds up. Although stainless steel and other corrosion-resistant materials are higher cost, the impact of the higher initial cost spread over the life of the mold will be negligible.

Figure 5.9 Rusted mold components can significantly affect the mold performance over time

5.7.2 Chiller Sizing Guidelines

As stated previously, the mold must be fed with a sufficient volume of cooling water at the required temperature to meet the required cycle time and many quality criteria. Therefore, the mold designer must ensure that the chillers to be used with the mold are a sufficient size. In order to calculate the required chiller size for the mold, the designer must calculate the estimated heat loads for each factor associated with the process, and then add up the results to obtain the estimated tonnage of cooling capacity required. The calculation of required cooling in the mold and machine can be complex, but over the years values have been determined empirically to adequately size the chilling needs for the system. The following calculation and the values in Table 5.2 [1] can be used to size the chiller for the mold:

$$\text{Chilling required for mold}\left(\text{tons}\right)=\text{Resin consumed}\left(\text{lb/h}\right)\div K\left(\text{lb/h/ton}\right) \quad (5.2)$$

Table 5.2 Chilling Tonnage Requirement by Resin Type for Equation 5.2

Resin type	*K* (lb/h/ton)
HDPE	30
LDPE	35
PMMA	35
PP	35
PA (Nylon)	40
PPO	40
ABS	50
PS	50
Acetal	50
PET	45
PVC	75

To this value, we must remember to add the chilling requirements for the machinery in the work cell if the mold and machine are connected to the same cooling equipment. Sometimes the machine may run on plant or tower water. This is water supplied to the plant for all other equipment in the plant, and it is normally cooled by a cooling tower on the roof of the plant. If both the machine and mold are using the same cooling supply, then the following additional chilling capacity must be added to the workcell chiller requirements:

- 0.1 ton/horsepower of hydraulic or servo motors
- 0.5 ton for feed throat cooling (if needed on the machine)

[1] Conair group, *https://www.conairgroup.com/blog/chiller-sizing-part-1-the-easy-way/*

- 0.15 ton/kW of power drawn (not connected power) by all heat sources on the mold (primarily the hot runner)
- 0.2 ton/pump-hp of thermolators.

The total chilling requirement is therefore mold + the machine cooling requirement.

Example:

A molder wishes to inject PET at a rate of 600 lb/h in a machine with a 90 horsepower motor and with a hot runner using 30 kW of power. What size chiller is required?

Mold chiller requirement = 600 ÷ 45 = 13.3 tons

Machine chiller requirement = 0.1 × 90 = 9 tons

Hot runner chiller requirement = 0.15 × 30 = 4.5 tons

Total requirement = 26.8 tons (use a 30 ton chiller)

5.8 Power Supply

Electric power supply is not always as stable as desired, especially outside the larger industrial areas of North America and Europe. In many parts of the world, especially in developing countries, there are often considerable voltage fluctuations because of weak and overloaded power lines. As a result, molders experience occasional, and sometimes even daily, "brownouts" (periods of lower voltage) and are often plagued with complete power failures (blackouts) lasting anywhere from just minutes to many hours. To say the least, these stoppages are annoying, but they can also be very expensive if a mold stops frequently just because of failure of the machine controls.

Voltage fluctuations affect molding operations for two main reasons:

- *Logic controls* are sensitive to voltage fluctuations and may require voltage stabilizers. Although this is a machine requirement, it needs to be pointed out. Every time the machine stops, the mold also stops producing. In general, electronics are quite sensitive to high ambient temperature.
- *Melt temperature.* Virtually all heaters in molds and molding machines today are electric *resistance* heaters. The heat output of a resistance heater is proportional to the *square* of the voltage applied. A drop of just 10% in voltage will reduce the heat output by 20%. While the barrel heaters of the extruder are always thermostatically controlled, a transformer, without feedback, often con-

trols the machine nozzle heaters. With heat controls, any reduction in voltage (and temperature) will be automatically compensated by having the heaters ON for longer time periods. In hot runner molds, the hot runner manifold heaters are always equipped with thermocouples; however, because of the high initial costs (in the mold, and for the associated external controls required) many molds do not have heat controls on the nozzle tip heaters, and can therefore experience major temperature variations as the voltage varies. This will lead to trouble in the mold's performance. Even so, today, about 80% of the high-production hot runner molds are equipped with thermostatically controlled nozzles, as the added costs can be easily justified by the increased productivity.

Cold runner molds: With such molds, power interruptions, while annoying, are not serious. If an interruption is only of short duration - in the order of a few minutes - the plastic in the injection unit is probably still hot enough so that production can resume immediately, without causing problems. If the interruption takes longer, it will take again the time necessary to heat up the injection unit before resuming production after purging.

Hot runner molds: With these molds, power interruptions can be more serious if a good hot runner controller is not used with the mold. Short interruptions of a minute or two can be tolerated without problems, but any longer stoppage will cause the plastic in the manifold and the hot runner nozzle:

- to *degrade*, especially heat-sensitive plastics in the still-hot manifold, and
- to *freeze* sooner, because the masses of the manifold are much smaller than the masses of the extruder. It takes time to heat up the whole system to operating temperatures, and the plastic both within the injection unit and in the hot runner system must be first purged before resuming operation.

Modern hot runner controllers contain features such as standby temperature settings that decrease the hot runner temperature when they detect that the mold is not in operation for a period of time. These standby temperature features maintain the temperature at a low enough level that degradation will not occur.

A good hot runner system should be ready to resume production about 10-15 min after any interruption.

6 Cycle Time Estimation

Once the plastic part is finalized and optimized for injection molding (Chapter 3), the machine(s) in which the mold will be run is specified (Chapter 4), and all key external factors are understood about the molding environment (Chapter 5), the cycle time can be estimated. The output of the mold can be expressed the as cycle time (seconds), or in shots per minute or parts per hour (if the cycle and cavitation is known):

$$\text{Output}\left(\text{shots/h}\right)=3600\left(\text{s/h}\right)\div\text{Cycle time}\left(\text{s}\right) \tag{6.1}$$

$$\text{Output}\left(\text{parts/h}\right)=\text{Output}\left(\text{shots/h}\right)\times\#\text{ of cavities} \tag{6.2}$$

The cycle time of an injection mold can be broken down into eight parts (see Figure 6.1). A correctly sized machine for the application will have an injection unit designed such that it can recover the required shot size within the overall cycle time, minus the ejection and hold times. If this is not the case, then the machine may be extruder-limited and the cycle time will be affected by the additional time required to wait for the next shot to be created by the extruder. In these cases, the cool time must be unnecessarily extended to increase the cycle time.

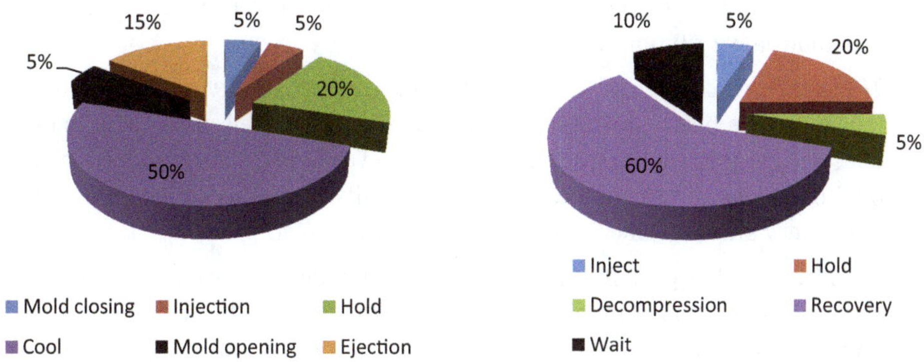

Figure 6.1 Breakdown of clamp cycle time (left) and extruder cycle time (right) by the stages of an injection molding cycle (varies by application)

At the highest level, the main elements of cycle time are shown in Equation 6.3.

Parts of the molding cycle:

$$\begin{array}{c}\text{Cycle}\\ \text{time}\\ \text{(s)}\end{array} = \begin{array}{c}\text{Machine}\\ \text{dry cycle}\end{array} + \text{Injection} + \text{Hold} + \text{Cooling} + \begin{array}{c}\text{Mold}\\ \text{open}\\ \text{time}\end{array} \tag{6.3}$$

where:

Machine dry cycle = the time for the machine to close, clamp up, unclamp, and open to the desired stroke

Mold open time = dwell time the mold is open to allow parts to be removed from the press

The cycle time of any molding system depends on a number of parameters, which will be discussed in detail in the following sections. It is important to understand these dependencies before settling on a reasonable figure for the cycle time.

6.1 Factors Impacting Cycle Time

6.1.1 Type of Plastic

The type of plastic to be molded has a significant impact on cycle time. Plastics have many material characteristics that can impact the way plastic is injected, cooled and ejected (Figure 6.2). For reference, the characteristics of injection-molded plastics are listed in Appendices 8–11. There are several parameters to consider:

- *Melt temperature* required to be able to inject and to fill the cavity space. Higher melt temperatures require longer cooling times before the pieces can be ejected. Note that higher temperatures reduce the material viscosity, which can allow faster fill times.
- *Thermal conductivity* of the plastic. When heat conductivity is lower, it takes longer for the heat within the melt to travel to the cooled mold walls than it does with higher conductivity. However, with very thin-walled products, without heavy sections, the difference in conductivity can be insignificant because of the very short distance the heat travels to the cooled walls.
- *Injection speed,* especially through the gate(s). Some plastics are sensitive to shear stresses caused by high injection speeds (especially in the gate area), and exhibit degradation of the plastic flowing into the cavity space. In this case, lower injection speed (and pressure) will be required, which will affect the molding cycle. Some plastics, especially commodity resins such as PS, PP,

and PE, are mostly unaffected by high shear stresses, but many heat-sensitive plastics can be damaged (degraded) by high injection speeds.

- *Crystallinity of the plastic.* To reach the required melt temperature, crystalline plastics (such as PE and PP) require more heat input than amorphous plastics (such as PS). The resulting higher heat content in the plastic means that it will then require more time to cool before ejection is possible. Occasionally, a mold may be planned to produce both types of plastics: in this case, both the shrinkage and the molding cycle time will be different.
- *Fillers.* Filling of the plastic with inert materials (fibers, talcum, etc.) can also adversely affect the cycle time. In addition, the shrinkage will be smaller than when using the corresponding unfilled plastic. We must also consider that many types of filler tend to erode the mold materials, especially the gate; when selecting the materials for the gates and the mold in general, this must be taken into account because of its effect on the mold cost. For example, a high-strength, wear-resistant steel for a gate insert is good for resisting abrasion, but because it is also a poor heat conductor, the gate area is more difficult to cool. This will result in a slower molding cycle. This is especially important with fast-cycling molds.

Figure 6.2 The resin to be molded can significantly affect the cycle time due to its mechanical and thermal properties

6.1.2 Wall Thickness of Product

The wall thickness plays the most significant role in the cooling process and thus influences the cycle time the most. Ideally, the walls should be uniform throughout the whole surface of the product; however, this is rarely achievable or practical, except in some lids (covers), some containers, and some flat products. Most products have thicker sections in design features such as hubs for fasteners, but even with otherwise uniform walls, there will be thicker sections at the junctions of walls and at the base of ribs (due to the need for positive draft angles to eject the parts). It is not the thickness of the uniform parts of the wall that governs the cycle time, but these thickest sections.

The situation is further complicated by the fact that these heavier sections are usually more remote from the areas that are easily provided with good cooling, and therefore depend on longer paths for the heat to travel to the cooling channels. Materials with better heat-conducting proprties are occasionally used to help to remove the heat faster from these "hot spots".

Charts and computer programs are available to relate the wall thickness to the cooling time. These charts are mostly based on the simplest (and hence often unlikely) cases, namely perfectly uniform wall thickness throughout, and equally well-cooled surfaces throughout the mold. Figure 6.3 shows the relationship between wall thickness and cooling time, and shows that thinner walls reduce the cooling time at a more than linear rate. A paper by BASF does an excellent job of explaining how to estimate the cooling times in a plastic part of a given geometry [1]. Molding simulation software (also known as CAE software) can be also used to estimate components of the cycle time such as mold cooling time (see Figure 6.4), and are very accurate if the inputs to the model are accurate.

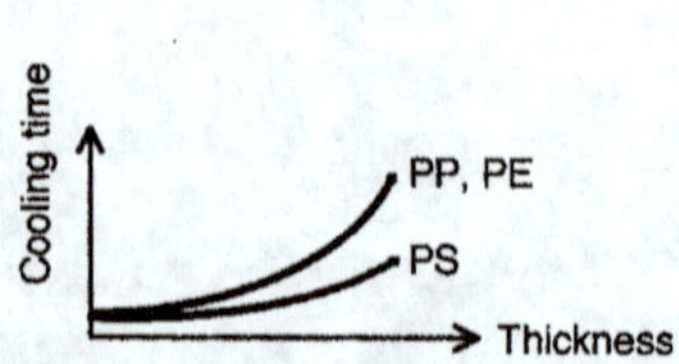

Figure 6.3 Schematic showing the relationship between material, uniform wall thickness, and cooling time

[1] *Displaying cooling times in injection molding*, BASF, 2007, *http://www2.basf.us/PLASTICSWEB/displayanyfile?id=0901a5e1801499d3*

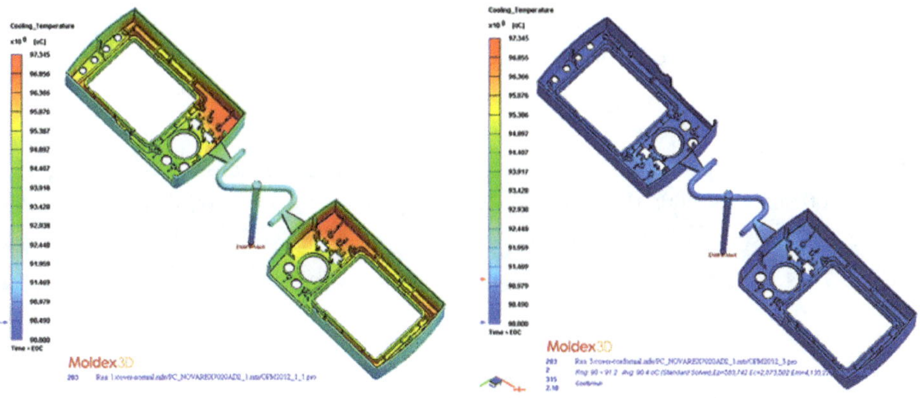

(a) (b)

Figure 6.4 Molding simulation software (also called CAE analysis) can be used for determining the cooling component of the cycle time: (a) Using poorly designed cooling channels; (b) Using optimized cooling channel layouts (to be discussed later) [2] (Courtesy of Moldex3D)

6.1.3 Mold Materials

The selection of mold materials for the stack parts has an important effect on the molding cycle. There is some, but relatively little, difference in heat conductivity between the various hardened alloy steels commonly used for mold making (e.g. H13 versus 420SS) (see Appendix 1 for more details on the thermal properties of mold steels). The conductivity of pre-hardened machinery steels that are often used for larger stack parts is somewhat better but still poor (e.g. 4140 versus P20). Mild steels have a still better conductivity, but are rarely used for stack parts because of their low physical strength, poor polishing quality, and poor consistency of grain structure.

Metals with much higher conductivity, such as aluminum and copper, are used cautiously because of their softness. Certain aluminum alloys are easily machinable and relatively inexpensive, and are used in *blow molds* where pressures are much lower, occasionally in injection molds in areas of low stresses, and even in prototype molds.

Beryllium-copper (BeCu) alloys have heat conductivity about 4 to 7 times better than steel. Their use in mold stacks (usually as inserts in cavities and cores) is often advantageous in areas that require the highest heat removal rate possible (see Figure 6.5). In fast-running molds, the difference between using steel versus BeCu can be a few seconds, or even just a fraction of a second. It may not appear to be

[2] Source: M. Wang, R. Chang, C. Hsu, *Molding Simulation: Theory and Practice,* Hanser Publishing, 2018 (Figure 8.39).

much, but a saving of 0.5 s can translate into a large increase in production for a mold that runs cycles less than 6 s.

> *Example:*
>
> A mold with steel cavities runs at a 4 s cycle, and so produces 3600 ÷ 4 = 900 shots per hour. By using BeCu for the cavities, it can run at a 3.5 s cycle, producing 3600 ÷ 3.5 = 1028 shots per hour, an increase of 14% in productivity!

Figure 6.5 This two-cavity lid mold uses BeCu inserts (copper color) on the cavity ring and the gate inserts to significantly cut cooling time (Courtesy of Top Grade Molds)

The reason that BeCu and other copper alloys (such as Moldmax) are not used more in molds is that it is much more expensive than steel and not as strong. Because BeCu is softer than hardened steels, it is usually inserted into steel cavities or cores, and it should never be used on the parting line, on alignment components, or on shut-off tapers. With larger pieces, there is the possibility of porosity in the BeCu. Smaller parts can usually be machined from forged or drawn rods and bars, but for larger parts, pressure-casting of the blanks for mold parts is preferred to ensure good consistency and no porosity.

6.1.4 Efficiency of Cooling

The purpose of mold cooling is to remove, in the shortest possible time, the heat energy that entered into the cavity space during the injection of the hot plastic melt (Figure 6.6). The higher the efficiency of removing the heat, the higher the productivity the mold will have.

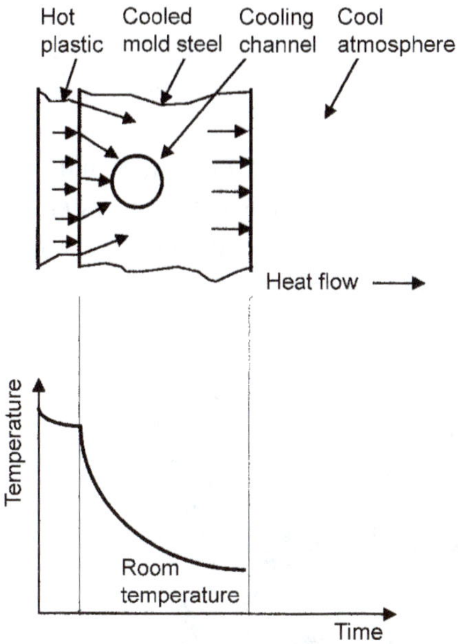

Figure 6.6
Mold cooling schematic from the plastic to atmosphere

The cooling of the plastic part is partially accomplished while in the mold, and the rest of the cooling happens after the part is ejected. Since excessive cooling reduces productivity, the parts should be ejected as soon as possible. Depending on the type of molding being conducted and the impact on part cost, different methods of post-mold cooling the following methods can be considered.

6.1.4.1 Molds for Small-Scale Production (Fewer than Approx. 1000 Pieces)

With any injection mold, over time, the heat will dissipate through the mold into the surrounding air and into the machine platens. This could be considered as sufficient in cases where only a few pieces are required and where the cycle time is not important. In this case, there would be no need to provide any mold cooling at all.

6.1.4.2 Molds for Large-Scale Production of Products 0.4–1.3 mm (0.015–0.050") Thick

On the other end of the scale for cooling efficiency is the cooling for a mold for fast-running, thin-walled products (Figure 6.7). The mass of each product may be relatively small, but because the cycle times are short, large amounts of heat (in the hot plastic) per unit of time enter the mold, which must be well-cooled to ensure that its temperature is kept stable and at an optimal (low) temperature. This will require the best possible cooling methods, which are more costly to design and to manufacture.

These molds use the most suitable (and sometimes expensive) mold materials to facilitate the rapid removal of heat. The higher costs incurred will usually be worthwhile, because they result in a mold with higher productivity and in lower costs per molded piece. Molds of this type are often coupled with automatic part removal systems to reduce the mold open time, the time to eject the parts and any post-mold stacking, assembly boxing, or labeling. Top- or side-entry robots enter the molding space and retrieve the parts while preserving orientation. This allows the parts to be handled more easily when removed from the machine.

Figure 6.7 Four-cavity mold for a 4 lb (PP) margarine tub (Courtesy of Husky Injection Molding Systems Ltd.). The mold exhibits modular construction, floating cores, and a cavity lock. Air ejection (absence of an ejector box) allows a very short shut height of the mold. Air jets (A) (four per cavity) and the two blow-down air jets (B) on top assist fast removal of products from the molding area. Note the intricate system of continuous vent gaps (C), venting grooves (D), and channels (E) to permit fast filling of the cavities. The 6.0 s cycle for this mold results in a productivity of 2400 tubs per hour

6.1.4.3 Molds for Large-Scale Production of Products >3 mm (0.120") Thick

As the plastic cools during molding, it shrinks onto the core, away from the cavity walls. After losing contact with the cavity walls, even the best cavity cooling will not do much in removing heat from the product. But the relatively short time the plastic is in contact with the cold cavity walls is enough to create a rather thin, rigid, but still warm surface, while the core cooling continues to remove heat from the inside of the plastic walls. While the rigid outer skin allows early opening of the mold and to pull the product (while still on the core) out of the cavity without risk of damage, it would not be possible to eject it at this time, because the outside of the product could be damaged despite the rigid skin.

It is usually fairly easy to provide adequate cooling to the cavity but often not easy to cool the core, because:

- The core is usually much smaller than the cavity
- There may be ejector pins or other ejection devices going through the cores
- There are sometimes air channels in the core.

The latter two conditions reduce the space to accommodate effective cooling lines.

But there are a number of possibilities to increase the productivity:

- The first method, going back to the earliest manufacture of injection-molded products, is simple and still being used occasionally for very heavy, simple, and not particularly closely dimensioned products. The mold opens and the externally cool (but internally still hot and soft, and therefore easily damaged) products drop into a container with cooled, circulating water, from which they are then removed and dried, either by hand or by a conveyor carrying them through an air cooling tunnel. This is a rather crude method, but can be quite efficient in some cases. For nylons and other materials that require high water content to reach their optimum physical properties, this water immersion may even be an advantage. For other materials, water absorption could be damaging, and therefore this method should not be used.
- Another typical method is the handling of flat products such as trays, but also other - usually larger - shapes that tend to warp after being removed early from the mold. The still hot (and easily damaged) product is placed into a cooling fixture (some are simple, others quite elaborate), where it is held by weights, clamps, or in a mechanically locked frame or any other suitable method, until it is cool enough and keeps its proper (as molded) shape without warping. There could be a small number of such fixtures beside the machine where pieces are successively placed as they are ejected, and then packed once they are cold. This method is also rather crude and labor-intensive, but can shave quite a few seconds from the cycle time. The alternative is to keep the mold closed until the molded product is cold enough and will not warp after ejection.
- Hot products can be held in actively cooled fixtures (post-mold cooling). Typical examples of post-mold cooling are the heavy-walled preforms used for the manufacture of blown PET bottles. These products are required daily by the billions worldwide, and every fraction of a second saved on the cycle time amounts to huge savings over a period of years. The biggest problem is the intense cooling and the long time required to cool the very thick walls which, in the next step of operation, will be reheated and blown up to the final bottle shape in special machines. The usual wall thickness of a preform is in the

order of 3–4 mm (0.120 to 0.160"). Cooling must be very intense and efficient to prevent crystallization of the plastic as it cools, requiring very cold water with large flow rates. But while it is relatively easy to cool both the cavity and the core, the cycle could be still in the range of 30–35 s or even more. As a result of much research, and many improvements to in-mold and post mold cooling, these preforms can now be produced at cycle times of 12 to 16 seconds. It is worth noting that since the early days of these molds being used, the number of cavities has increased from 4 to 192 in many production molds. Accounting for both cycle time and cavitation increases the productivity of these parts has increased more than 100 times in that last 30 years. The key to reducing the cycle time in thicker walled parts is to increase the time of the heat transfer from the plastic to the cooling media (water or air), either in special cooling fixtures or by removing the products earlier from the cavities but leaving them longer on the cores, as explained below:

The still very hot products can be ejected from the core into water-cooled, tightly fitting sleeves, advanced by a robot into the molding area. The whole array of still-hot preforms is then transported out of the molding area and a new injection cycle can begin. There are several executions of this method and the cycle times have been reduced by half, two-thirds, or even better. This significant increase in productivity is not inexpensive, but the cost of the necessary equipment can be recouped in a very short time by the savings achieved.

Figure 6.8 shows a 48-cavity mold for PET preforms from the rear of the machine. It shows (right) the array of cores (A) on the moving platen. The side-entry robot (B) carries 3 × 48 cooling receptacles (C), their position timed so that every time after the machine ejects an array of still-hot preforms into the cooling receptacles, the robot plate shifts so that by the next cycle, another (empty) set of cooling receptacles faces the cores. Before the third cycle of unloading the mold, the now-cold preforms of the first unloading cycle are ejected. For every ejection, the plate swings 90° so that the now-cold preforms drop onto a conveyor (D) for removal to a shipping crate. At a 12–14 s cycle time, this system yields between 14,400 and 12,340 preforms per hour.

Figure 6.8 48-cavity mold for PET performs with cooling receptacles (Courtesy of Husky Injection Molding Systems Ltd.)

6.1.4.4 Molds for Most Other Products

Molds for most other products are equipped with a cooling system somewhere between these extremes. Often, in molds for intricate shapes requiring many stack inserts (usually technical products), the cavities and cores are almost impossible to cool close to the cavity space. Cooling can only be achieved by conducting the heat through the stack walls and the inserts to the cooled mold plates located immediately behind them and/or surrounding them. This results in slower heat removal than cooling the cavity or core walls directly, but is often the only way to keep the mold at a stable temperature. Similarly, in some products, which have unavoidable hot spots (thick sections), there is not much point in providing excellent cooling for the areas that could be easily cooled just because there is enough room for such cooling, unless it is possible to provide better cooling to these hot spots.

> The slowest-cooling area of the mold always governs the molding cycle.

It is amazing how many mold designers and mold makers overlook this point and then wonder why "despite the massive cooling provided" in the mold they cannot achieve a better cycle time. As already pointed out, especially in molds for containers or other cup-shaped products, some mold parts, such as cavity blocks, have ample space for cooling circuits, while the core, which really should have *more* cooling, does not provide enough space, because it is much smaller than the cavity block and has to accommodate the ejector mechanism as well, which encroaches on the available space for cooling and air channels. Unfortunately, many mold designers don't understand this, and only see the large available space in the cavity. As a result, they provide more cooling than is necessary for the job in non-critical areas, and thereby waste money.

6.1.5 Venting

Venting is another feature that can affect the molding cycle. It is critical to evacuate the air in the cavity space in front of the inrushing plastic during injection. Venting is important for any mold to ensure that the plastic can enter freely into all areas of the cavity space, and must be properly specified. Venting is especially important with fast-running molds.

> Although some ribs and bosses may fill without venting, all points where the molten plastic flow front finishes must be vented.

> A typical example is an experiment with a four-cavity mold for a small disposable container. With standard, venting, the mold ran at 17 shots/min. It had continuous vent gaps along the rim. By simply providing more vent *channels* (eight instead of four) to allow the air to escape more easily into the open, the production could be increased to 20 shots per minute, an increase in productivity of more than 17%!

6.1.6 Ejection

As mentioned in Section 6.1.4, many molding systems use automation to remove the parts while barely cool, to reduce the high cost of cooling time in the molding machine. For other molding applications, automation cannot be justified and the parts are simply allowed to free-fall onto a conveyer belt. While ejection typically constitutes only a small portion of the overall cycle time (see Figure 6.1), methods to reduce the dwell time that the machine sits open waiting for the parts to be ejected can be cost-effective and efficient. Figure 6.7 shows a four-cavity mold for a 4 lb (PP) margarine tub with some simple blow-downs (from Exair) that are used to accelerate the parts out of the molding area. Cartesian coordinate robots (three-axis) or six-axis robots (Figure 6.9) can provide flexible and cost-effective methods to retrieve the part from the mold. While these methods may add some time to the cycle, they provide additional benefits in terms of quality and post-mold assembly, stacking, and packaging that outweigh the cycle time penalty.

Figure 6.9 Three-axis robot used for part removal (Courtesy of Milacron Marketing LLC)

6.1.7 Molding Machine

An injection molding machine has many discrete functions that add up to create the overall cycle time. The machine controller allows the operator to make adjustments to the machine to modify and fine-tune the cycle time elements to make acceptable-quality parts in the shortest possible time.

Several features of the molding machine affect the mold productivity and will be discussed in the following sections of this chapter. It is important to be familiar with the machine for which the mold is to be built, in order to arrive at a more accurate estimate of the probable cycle time. In case of similar or even identical products and molds, the cycle time can vary considerably when run on machines of different make and size. Machine factors affecting the mold productivity are dry cycle, injection speed/pressure, tonnage, and recovery time.

6.1.7.1 Machine Dry Cycle

The dry cycle is probably the most significant variable from machine to machine, and the feature that can influence the cycle time more than other factors. The dry cycle is defined as the time (in seconds) it takes the moving platen to complete the whole stroke – moving from the mold-open (MO) position to the mold-close (MC) position, clamping up, unclamping, and then returning to the MO position. Obviously, the larger the masses (platen and moving mold half) to be moved, the more power will be required to accelerate, to move, and to decelerate them for a soft stopping in both the MC position and in the MO position. Therefore, smaller machines can have faster dry cycles than larger machines, but also, premium machines have faster dry cycles than lower-performance machines. Machines in the 500 kN to 10,000 kN (50–1000 ton) range can have dry cycles from 1.0 s up to about 10 s. But there are also other, mainly older machines with dry cycles up to 20 s! It is obvious that for large required annual production volumes and a mold with a short molding cycle, the length of the dry cycle is much more important than with a mold that requires a long molding cycle.

Figure 6.10 depicts an approximate motion curve of the opening and closing of the moving platen. It is typically an S-curve. The platen initially moves rather slowly, but reaches a high speed for much of the closing stroke, until it slows down again and finally creeps the last few millimeters before reaching the closed position, when the stretching of the tie bars begins as the clamping force is created. Similarly, but in reversed order, the mold is first unclamped, it starts opening slowly, accelerates and moves fast until it slows down again for a gentle stopping in the MO position.

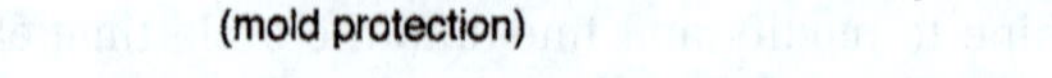

Figure 6.10 Schematic showing the closing motion of the clamp

Note that the following illustrations (Figure 6.11 and following) show the speed (velocity) of the moving platen as a *straight line* that really represents the average speed from start to stop (see Figure 6.10). The opening speed is not necessarily the same as the closing speed. In most machines, segments of the speeds in either direction can be adjusted to best suit the molding conditions, but molds are often run at the maximum available speed.

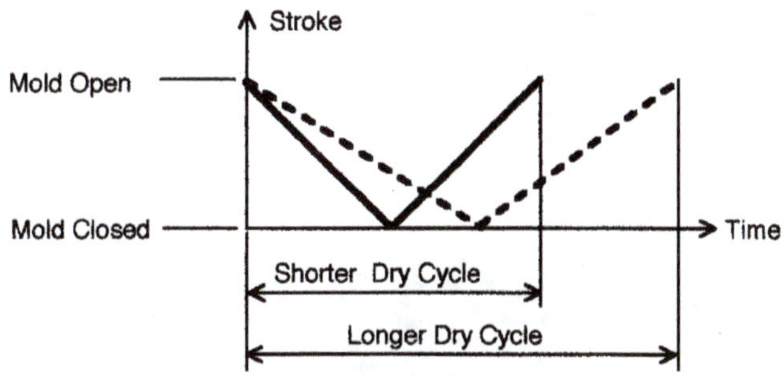

Figure 6.11 Schematic showing a shorter and a longer dry cycle

Note that the straight line representing the average speed terminates sooner than at the end of the tie bar stretch time, i.e., at the moment when the mold halves meet and the tie bars begin to stretch. This fact allows the injection to start just at the moment the mold halves touch and the tie bars begin to stretch. Any significant forces inside the mold will commence only when the cavity spaces are almost completely filled with plastic. The filling takes usually longer than the time required for the final clamp-up.

Starting the injection sooner, even by only a second or even a fraction of a second, will result in a significant gain of cycle time and productivity, especially with molds running at short cycles.

The difference between the time required for the shorter and the longer dry cycle is *wasted time.*

Example:

If the estimated combined injection and cooling time is 3 s and the dry cycle is 3.5 s, because MO = 0, the molding cycle is 6.5 s; this corresponds to 554 shots per hour. If the dry cycle were 5 s, the molding cycle would be 8 s, and the mold would yield only 450 shots per hour, a considerable loss of production. This is certainly enough to seriously consider the choice of a faster machine, especially when the expected production is high.

Mold-open time is also called ejection time. This is the time that the mold stays open waiting for the parts to be ejected. Production with zero mold-open time (MO = 0) has been achieved on many smaller molds, even with short dry cycles (in the order of 2 s), but can also be achieved with molds equipped with automatic unloading equipment where the action of the “take-out”, i.e., the mechanism that reaches into the mold to remove the products while the mold is opening and closing, is mechanically linked with the mold-open and mold-close motion. Figure 6.12 shows a 2 × 8-cavity stack mold for dairy tubs with modular construction, built for a machine equipped for stack molds. Following air ejection, cam (A) operated swing arms (B) with suction cups (C) pick the products from the cores and deliver them into the chutes (D) - shown in blue - on the side of the cavity plate. This mold operates with zero mold-open time (MO = 0) and has a productivity at a 4.5 s cycle of 12,800 tubs per hour.

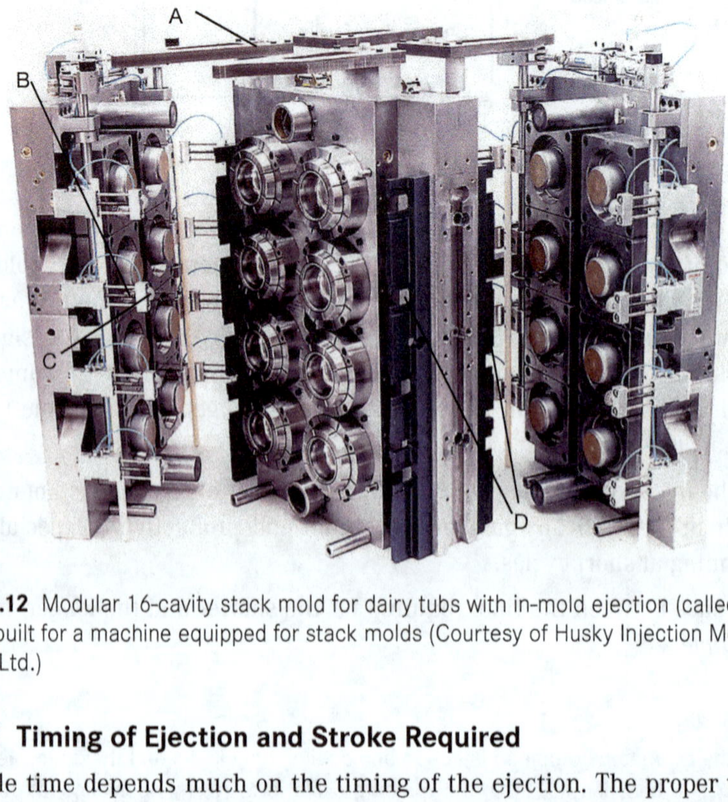

Figure 6.12 Modular 16-cavity stack mold for dairy tubs with in-mold ejection (called swing chutes), built for a machine equipped for stack molds (Courtesy of Husky Injection Molding Systems Ltd.)

6.1.7.2 Timing of Ejection and Stroke Required

The cycle time depends much on the timing of the ejection. The proper timing of the molding machine to suit the mold is often overlooked. Clear instructions specifying *when* ejection should start are most important to achieve the best cycle time. The mold must of course open up sufficiently to allow the products to eject, before closing again safely, without trapping pieces on their way out of the molding area. But any unnecessarily long stroke represents a waste of time and loss of production. There is, however, the need to service the mold, usually during startup of the mold. There must be sufficient space between the open mold halves, at least 150 mm (6") for small molds, and more for larger ones to permit access, as may be required for cleaning a frozen gate or removing a broken piece of the product that failed to eject. In the case of a multi-cavity mold for a flat product, such as a lid for containers, there could be an opening stroke of as little as about 2.5 × the height of the product, provided the products will clear the molding area fast enough so they will not be trapped by the closing mold half. In some machines the stroke can be adjusted very easily so that the mold can be set up for production with a very short stroke, but can be quickly opened up wider when servicing is required. The relation between stroke length and cycle time can be easily seen in Figure 6.13.

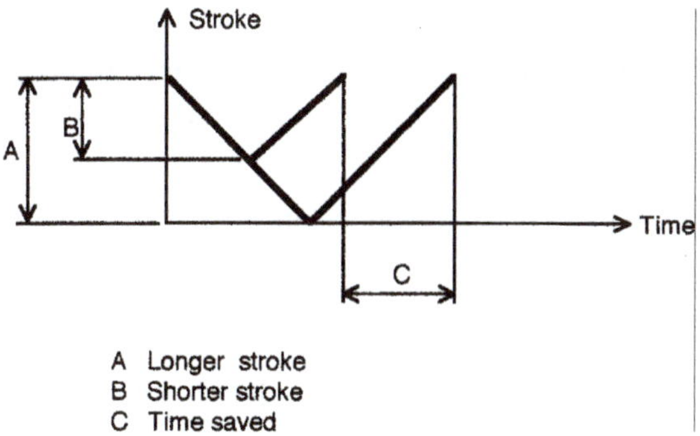

Figure 6.13 Schematic showing relation between stroke length and cycle time

Figure 6.14 shows a 2 × 4-cavity family stack mold for rectangular flat boxes and covers, equipped with guide rails. As the mold closes, the guide rails (A) are pushed out of the way of the closing cavities and cores by the action of the cams (B). As the mold opens, the guide rails close in as the products are pushed in between them. When the ejector pins retract, the molded pieces are free, but oriented as they come out of the cavities, to fall onto conveyors (not shown) that transfer them to assembly stations. There is no mold open time required.

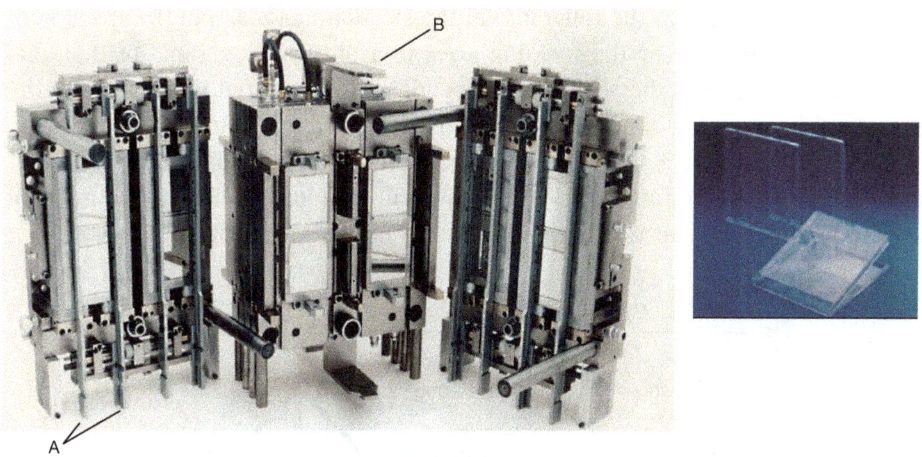

Figure 6.14 An 8-cavity family stack mold for rectangular flat boxes and covers (Courtesy of Husky Injection Molding Systems Ltd.)

Mold-Open (MO) Time Requirements (General Observation)

The time that the mold stays open (mold-open or MO) time is an important factor to consider, because it adds time to the cycle without contributing to the molding process; it should be kept to a minimum or better yet, not be used at all (MO = 0). The need for MO time depends very much on the product and the mold design, the size of the mold, the type of machine, and the selected ejection system. The ideal arrangement would be:

a) to use no MO time (to run the machine "on the fly"), or

b) to use as short a mold opening stroke as possible, or

c) both (a) and (b).

In many molds it is possible to start ejection shortly after the product has partly pulled out of the cavity; by the time the mold has reached the fully open position, the products have cleared the molding area making it safe to re-close the mold. In some cases, air assist (air curtain or air blasts, see Figure 3.17) will assist the product in going out of the way of the closing mold. Figure 3.34 shows a schematic of a mold cycle not requiring MO time.

In Figure 6.15, the point denoting the start of ejection has been selected arbitrarily. If ejection starts *too soon,* there is the danger that the product will be pushed back into the cavity and will fail to eject. If starting *too late,* time is wasted. The "time for ejection" is shown ending at the end of the molding cycle. This is not absolutely necessary; it could extend into the next cycle, as indicated by an (arbitrarily placed) vertical line L shown in the graph. When this additional time ends depends on the shape of the product, on the reliability of the ejection method, and the speed with which the products clear the molding area, to avoid trapping some of them between the closing mold halves. It also depends on the layout of the cavities in the mold, which affects the time taken for the piece ejected from the uppermost cavity to clear the bottom edge of the mold. For example, products falling from a four-cavity mold laid out with four cavities in a row will take longer to clear the molding area if the row is in a vertical rather than in a horizontal layout.

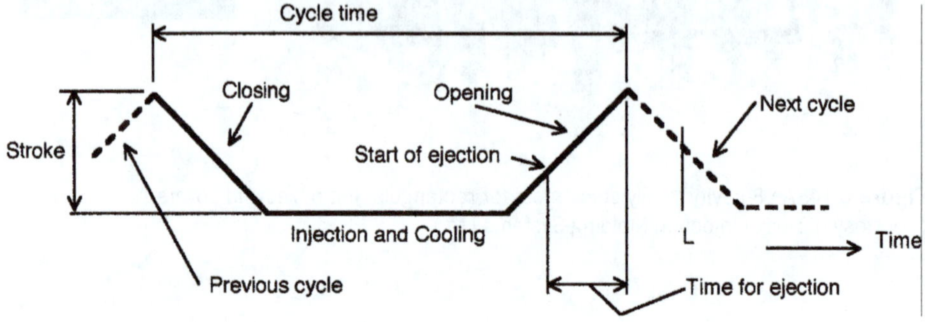

Figure 6.15 Schematic of ejection timing without MO time

MO Time in High Cavitation Molds

Especially with high cavitation molds (Figure 6.16), it may not be possible to clear all molded pieces out of the molding area in time, and some MO time will have to be added. The need for this MO time should be closely scrutinized. While it is sometimes not possible to avoid MO time, it is quite common that the setup persons add extra MO time "just to be sure" that the pieces have cleared, rather than to ensure that the ejection, including air assist, is properly adjusted and timed as intended. The length of the ejection stroke, when to start ejection, when to start and to stop blowing, etc., must be recorded for the next time the mold will be used in the instructions for the setup of this mold, but also in the job file for future reference.

Figure 6.17 shows the same cycle as in Figure 6.15 but with the addition of MO time, providing more time available for ejection but also lengthening the cycle time. Here again it would be possible to use a portion of the closing time of the next cycle to extend the time available for ejection (line L).

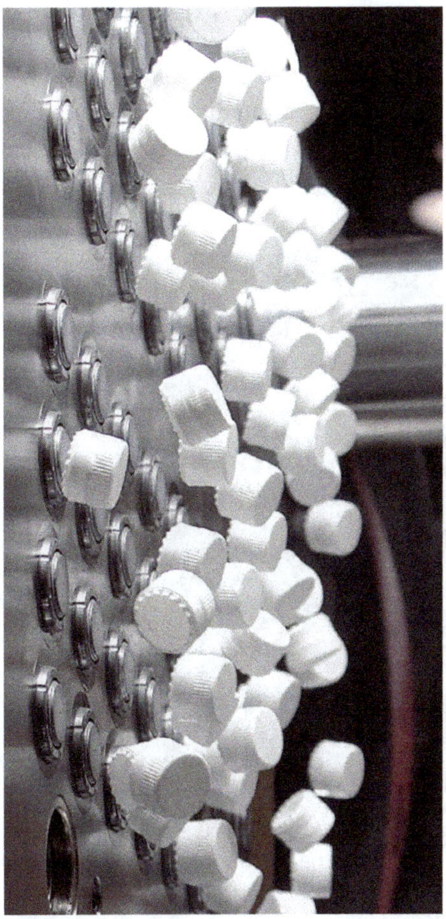

Figure 6.16
To reduce cycle time in this 96-cavity mold, closures are ejected before the mold is fully open. This reduces the amount of mold open (MO) time needed to ensure proper ejection, and is sometimes referred to as "eject on the fly" (Courtesy of Husky Injection Molding Systems Ltd.)

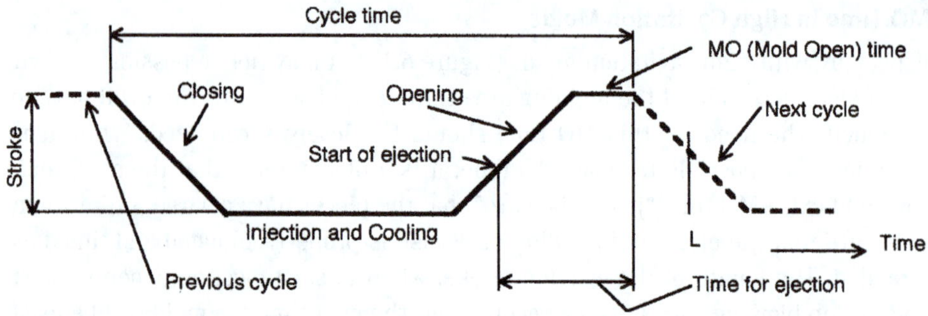

Figure 6.17 Schematic of ejection timing with added MO time

Following is a list of reasonably achievable MO times to be added for various ejection methods:

- Air ejection only: 0.1–0.5 s
- Machine ejector (hydraulic): 1.0–2.0 s
- Robot: 0.5–2.5 s.

The open stroke required for ejection must be such that the parts reliably clear all parts of the mold and machine (Figure 6.18). This is normally in the range of 1.5–2.5 × the product height. Much more detail on part removal is covered in the Chapter 11 on ejection.

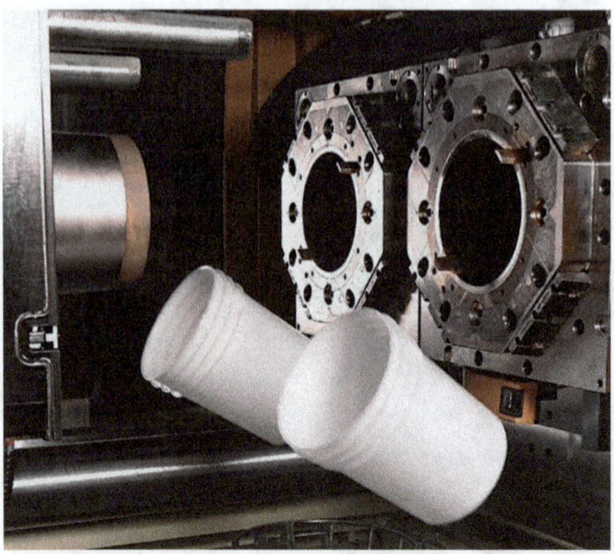

Figure 6.18 1000 tonne molding machine molding two large industrial pails "on the fly" (pail weight 800 g, cycle time 18 s). Note how the parts clear the cores and cavities and fall in a similar pattern, indicating consistent ejection onto a conveyer (or pail catcher) under the machine (Courtesy of Husky Injection Molding Systems Ltd.)

Multiple Ejection Strokes

> Don't use multiple ejection strokes just because they are available.

Most molding machines are equipped with the feature of double or multiple ejection strokes. The idea behind this system is that if the original (first) ejection stroke has failed to eject some of the molded pieces, additional strokes will shake these pieces loose from where they hang up so that they fall free, to clear the mold before the next cycle will start. A well-designed mold has no need to use this system: the molded pieces must eject on the first stroke. Additional strokes not only add to the cycle time but are also noisy and wear out machine and mold.

Automatic Product Removal

> Cavities in the mold can be laid out so that the products are in the proper attitude (and relative relation to each other) for subsequent operations. With suitable take-offs, they can be ejected and transferred to ancillary equipment for stacking, assembling, etc., and save much on unnecessary handling such as unscrambling and rearranging.

Molds with integrated mechanized product removal, which act in mechanical or timed synchronization with the mold opening and closing stroke, will remove the moldings as soon as the mold has opened far enough to allow the take-off member to enter the space opposite the core or cavity. The products are then grabbed (mechanically or with suction cups) and the take-off arm withdraws ahead of the re-closing mold (see Figures 6.12 and 6.14). This method has been used successfully even in multi-cavity molds, cycling as fast as 3 s (20 shots/min). Such special ejection methods should be considered, especially when it is required to *maintain the orientation* of the products outside the mold. These methods also allow shorter strokes, thus reducing the time for the products to clear the molding area (swing chutes, guide rails, and other systems). Typically, there is no MO time required.

While these integrated systems could add considerably to the mold cost, this is quickly recovered by saving substantial cycle time and by the increase in productivity. It also eliminates the need for subsequent orientation of the molded pieces (for stacking, assembling, decorating, etc.), which often requires expensive orienting devices further down the line.

Robots

Robots can often be used to unload large products, or for very large multi-cavity molds for smaller products, especially with relatively long molding cycles. The principle is similar to automatic product removal, except that for safety reasons the mold will have to stay open longer. As a rule, the robot's "end-of-arm tooling" will not enter the molding area until the mold has stopped in the MO position, and the ejection will not start before the end-of-arm tooling is in front of the cores (or cavities) from where the products will be removed. Similarly, the mold will not close before the products and all robot mechanisms are safely out of the molding area. Typically, MO time is 0.4–2 s.

Figure 6.19 shows a high-speed side entry robot for removal of DVD cases. This robot accelerates at more than 9 × gravity and adds 0.5 s of mold open (MO) time to the cycle of 6 s. In addition to part removal, the robot automatically flexes the hinge and then places the parts on a conveyer for post-mold assembly and packaging. Without a robot these parts could not be produced reliably.

Figure 6.19 High-speed side entry robot for DVD case removal (0.7 s part removal time) (Courtesy of Husky Injection Molding Systems Ltd.)

Figure 6.20 shows a typical robot (from the rear of the machine), removing very large industrial pails from a single-cavity mold. The end-of-arm tooling (A) grabs the pail with four suction cups (B) as it is ejected from the mold. The pail is then raised and moved out of the molding area, where the pail is rotated into a vertical attitude and stacked, opening down, onto the previously ejected pieces (C).

Figure 6.21 shows a machine with a 2 × 16-cavity stack mold for small containers. After the mold is open, two parallel arms (A) enter the mold between the two levels (B) to face the cores, from where the products are ejected into two arrays of 16 receptacles (C) on robot arms, held there by vacuum. The arms retract, and the mold can start the next cycle. Once outside the mold, the plates with the receptacles (C) rotate 90° to deposit the products onto a conveyor (D) that guides them, still oriented as they came out of the cavities, to the next station in the process. This high speed robot adds 0.4–0.7 s of MO time.

Figure 6.20 Typical robot (from the rear of the machine), removing very large industrial pails from a single-cavity mold (mass 2.0 kg (PP), 40 s cycle, productivity 90 pieces/h). (Courtesy of Husky Injection Molding Systems Ltd.)

Figure 6.21 Machine with a 2 × 16, 32-cavity stack mold for small containers (Courtesy of Husky Injection Molding Systems Ltd.)

Ejection Stroke, Air Assist

The ejection method used in the mold will also affect the time needed for ejection. The mold must be properly designed with sufficient stroke to ensure that the pieces will not hang up on the core (ejector side) or on the ejectors themselves. On the other hand, poor adjustment during setup of the mold will have the same effect as a poorly designed mold. The ejection stroke does not necessarily have to be long, if other methods are used, such as mold-integrated take-offs, robots, or if compressed air is used to eject from the mold (Figure 6.22). Combined with a short ejection stroke, which will push the molded pieces off the cores just far enough to break a vacuum or its hold on the cores (e.g., products with undercuts, or with very little draft), air assist will blow the product out of the mold and into the waiting robot (shown at the other end of the mold in the figure).

Figure 6.22 Air-ejected medicine cups (2 × 32) into a high-speed robot (Courtesy of StackTeck Systems)

Any "air only" method can be quite tricky and needs experience in designing and setting up to prevent hanging up caused by a vacuum created by the air blowing out of the mold, between the product walls, and the core. With some materials and some shapes, pure air ejection (without any mechanical ejectors) allows the construction of a better, and at the same time, lower-cost mold. This is often used for PP and PE containers. These molds can be smaller (no need for an ejector system), they run quieter, faster, and they last longer because there is no wear of moving parts, such as stripper rings, etc. However, the mold maker also needs experience in this design, and not all product shapes are suitable for this system. The details of how to design a mold with proper air ejection will be covered in Chapter 11 on ejection.

6.1.8 Impact of Cold Runners versus Hot Runners on Cycle Time

As will be discussed in Chapter 14, there are many advantages to a hot runner system. One of the key advantages is the *cycle time*. A hot runner system's cycle time is governed solely by the time it takes to mold the product. Cold runner molds are often governed by the time it takes to cool the cold runners versus the plastic part to be molded. This can result in a significant difference in the molding cycle. *Often a hot runner mold will be able to cycle twice as fast as a cold runner system and run in a smaller and faster machine.*

Figure 6.23 shows a runner system for a 32-cavity two-plate mold for caps. The mass of the runner system is about 25% of the mass of all the caps per shot. The runner system is fed in the center by a hot sprue (not shown) to avoid an otherwise large, cold sprue. The cavities are tunnel-gated, and the products and runners are separated after ejection. The cycle time for this system is governed by the time it takes to cool the runners. A hot runner system for this part would cycle at around 6 s or less. The cold runner system runs over 12 s.

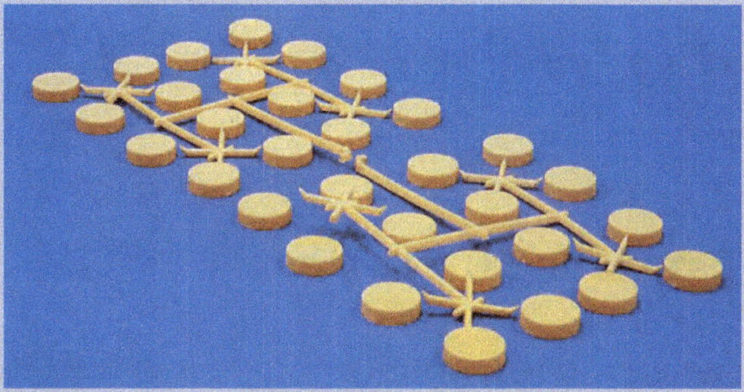

Figure 6.23 Runner system for a 32-cavity two-plate mold for caps

Figure 6.24 shows a runner for a 72-cavity three-plate mold for a small cap. The runner takes almost as much plastic as the products themselves. The cycle time is controlled by the cooling time of the (clearly visible) heavy distribution runners. A hot runner system for such a mold would reduce the injected mass by half and the mold would cycle twice as fast, but would be more expensive to build in the first place. This is a typical example where both methods must be considered, in view of the total requirements of the product.

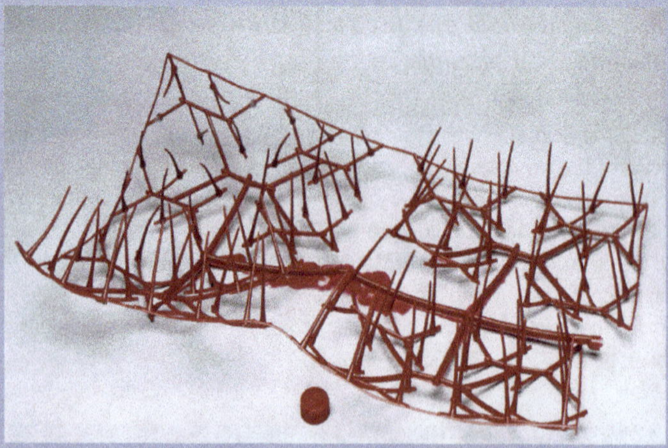

Figure 6.24 Runner for a 72-cavity three-plate mold for a small cap

6.2 Estimating Cycle Time

Now that all the factors to be considered for cycle time have been discussed, an estimate of the cycle time can be made. A reasonable estimate of the cycle time can be found by using the following figures and tables and adding the components of cycle time together to achieve an estimated cycle time, as in Equation 6.4. All the factors discussed previously in this chapter must be considered when looking at the results of this equation. Are there factors in the plant, cooling, machinery or other areas that may affect this number either positively or negatively? If this is the case then the resulting number should be multiplied by a factor K.

Equation for cycle time:

$$t_{\text{cycle}} = \left(t_{\text{cooling}} + \left(t_{\text{injection}} + t_{\text{hold}}\right) + t_{\text{dry cycle}} + t_{\text{ejection}}\right) \times K \tag{6.4}$$

where:
K is a factor to adjust the number based on plant, machinery, cavitation and molding conditions
$t_{\text{dry cycle}}$ (machine dry cycle) can be obtained from the machine supplier for the required stroke
t_{cooling} can be obtained by using the equation for the cooling of the plastic between two plates defined by the 1D steady-state conduction equation shown in Equation 6.5. Remember that the thickest wall section sets the minimum cooling time required.

1D cooling between plates:

$$K \times \frac{d^2 t}{dx^2} = 0 \tag{6.5}$$

where
K = thermal conductivity of the resin
t = temperature
x = dimension through the wall

For ease of use by the mold designer, this equation has been graphed in Figure 6.25 and Figure 6.26 for different materials.

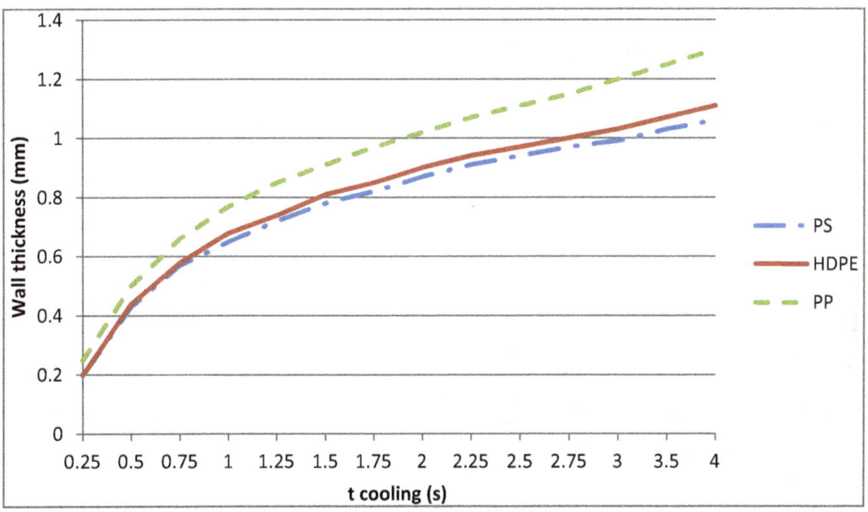

Figure 6.25 1D steady-state conduction through a plane (t_{cooling}) (empirical data)

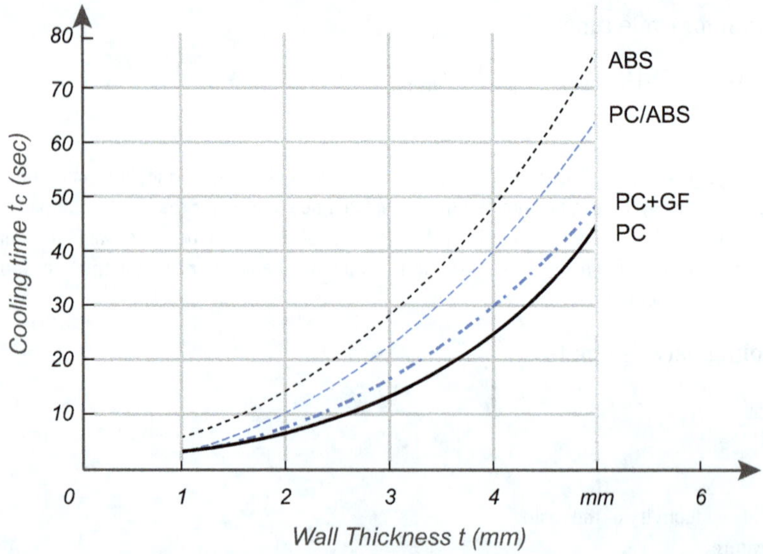

The cooling time variation is quadratic related to part thickness

Figure 6.26 Cooling time ($t_{cooling}$) for ABS and PC variants [3]

$t_{injection}$ + t_{hold} cooling can be obtained from reference charts, such as shown in Figure 6.27 and Figure 6.28. $t_{ejection}$ (mold open time) can be estimated using Table 6.1.

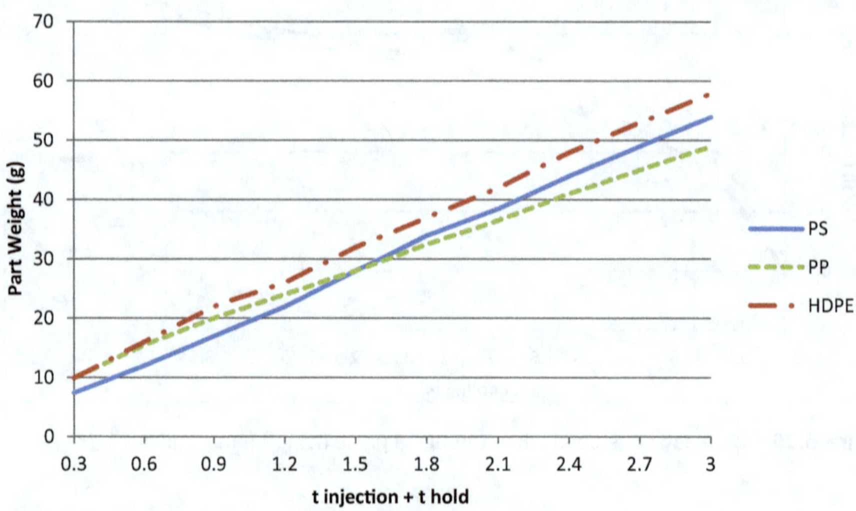

Figure 6.27 $t_{injection}$ + t_{hold} (empirical data from 74 molds)

[3] Source: M. Wang, R. Chang, C. Hsu, *Molding Simulation: Theory and Practice,* Hanser Publishing, 2018 (Figure 8.25).

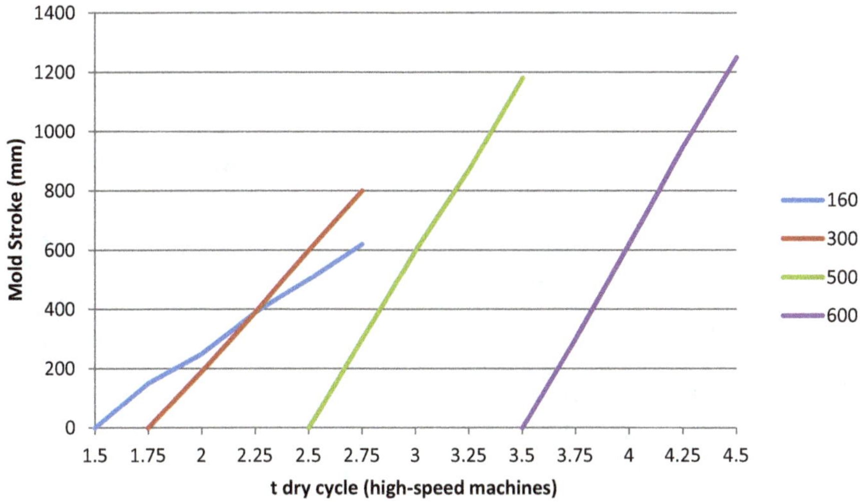

Figure 6.28 $t_{\text{dry cycle}}$ by tonnage size (high-speed machines)

Table 6.1 Suggested Mold Open Times to Add to the Cycle Estimation

Type of ejection	Cycle time added for thin-wall applications (< 1 mm)	Cycle time added for mid-size part moldings (300–1000 tonne m/c)	Cycle time added for large part moldings (> 1000 tonne m/c)
Air ejection	0.3–0.5 s	N/A	N/A
Machine ejection (free fall)	0.75–1.25 stripper rings 1–2 s ejector pins	2–3 s	2–4 s
Robot removal	0.5–1 s high-speed side entry 1–3 s top entry	3–5 s	5–10 s

Computer-aided engineering (CAE) simulation software can be a great help in estimating the cycle times for the part to be molded. The software is designed to be able to detect hot spots that limit the cycle time, such as in Figure 6.29.

Figure 6.29 CAE simulation showing the hot spots that limit the cycle time in a drill housing [4]

Many of the guidelines in this chapter can be used to fine-tune reference cycle components to suit the application being considered. Ultimately, the most reliable method to estimate the cycle time will be to find a similar reference job and make any fine adjustments to the estimate based on knowledge of the previous mold and machine data.

[4] Source: M. Wang, R. Chang, C. Hsu, *Molding Simulation: Theory and Practice,* Hanser Publishing, 2018 (Figure 8.28).

7 Product Cost Estimation

Ultimately, the customer is buying a mold to make a part at an estimated part cost. The mold to be designed must be able to meet the expectations of the customer in terms of output, quality, uptime, and reliability (see the discussion on TEP in Chapter 5). In order to ensure that the mold design is aligned with these objectives, the designer should have a good idea about what goes into the part cost calculation and how the design of the mold may affect this. Additionally, having a relative idea of the cost implications of cavitation and cycle time on the overall part cost can give the designer or project engineer a basis for suggesting design improvements to the mold.

7.1 Machine Hour Cost per Unit Molded

The machine hour cost (C_{mh}) is the *total* of the actual cost of the machine, its accessories, the cost of installation of the machine in its location in the plant, and the connection to the services, plus the cost of money (interest of loans for the purchase of the machine, etc.) spread over the number of hours (or years) the machine is expected to produce:

$$C_{mh} = \frac{\begin{array}{c}\text{Machine}\\\text{purchase price}\end{array} + \begin{array}{c}\text{Accessories}\\\text{cost}\end{array} + \begin{array}{c}\text{Cost of}\\\text{installation}\end{array} + \begin{array}{c}\text{Cost of money}\\\text{(if the machine is leased}\\\text{or a loan taken out)}\end{array}}{\begin{array}{c}\text{\# of years of allowable depreciation}\\\text{by the local tax laws}\end{array} \times \begin{array}{c}\text{Hours of operation}\\\text{per year}\end{array}} \quad (7.1)$$

There is really no definite life expectancy (in years) of a machine, as long as it is well-maintained and as long as the molding technology does not change. In fact, in 2015, the average age of molding machines in North America was 16 years. For practical reasons, and as permitted by tax legislation, the cost of the machine is usually written off in 7 years, even though most machines can (and do) work satisfactorily for much longer before being replaced.

We should also consider the case where a machine could be acquired for a one-time special molding application, and would not be suitable for other molding operations. This machine could possibly be required for a much shorter time and should in fact be considered as an extension of the mold. That means the machine should be depreciated entirely over this single project, and this special product should then be priced accordingly.

For a standard-type injection molding machine, we must first know the actual total price paid and then assume the number of years for writing it off. For example:

Example:

If the total cost of the machine, as defined in Equation 7.1, is $400,000 and we plan to write off the machine in 7 years, what is the machine hour cost? The yearly run in hours is expected to be 6500.

Machine hour cost (C_{mh}) = \$400,000 ÷ (7 × 6500) = \$8.79/h

Obviously, by writing the machine off in fewer years, the cost will be higher. By writing it off over a longer period of time or by using it for more hours per year, the machine hour cost will be lower. The amount is relatively small, and for the simplicity of cost accounting, groups of machines of various sizes and costs, or even the total of all machines in a plant, are lumped together into one machine hour cost for the entire plant.

To the machine hour cost (C_{mh}), the costs of overheads such as cooling water in the plant, compressed air supply, power, plant space, overhead labor, etc., are added to define a plant-wide machine hour cost to be used for costing of any product made there. The machine hour cost used for cost calculations may be \$25 to \$100, or more, even though the actual machine hour cost may only be in the order of \$5 to \$10.

Note that the machine hour cost should really not include the cost of power, compressed air, and cooling water for machine and mold, because these costs are *directly proportional to the number of pieces molded* (the amount of plastic "converted"). However, for simplicity of accounting, it is usually included in the overhead. This (rather artificial) machine hour cost must now be apportioned to the cost of the product.

Typical custom molding rates (which include a profit of 5–15% for the customer molder) can be found in trade magazines such as MAPP's *Plastics Business* [1], *Plastics News* [2], and *Plastics Technology* [3]. They can range from $30 to $160/h, depending on machine size and location. Table 7.1 shows the rates from the 2019 MAPP survey of injection molders.

Table 7.1 Machine Hour Rates for the US in 2019 [4]

Machine size (tons)	Average rate in the US
< 100	$25–30
100–300	$35–40
300–650	$45–50
650–850	$60–70
850–1100	$80–90
1100–2000	$100–110
> 2000	$160+

Example: Machine hour cost calculation

If we have a single cavity mold, running at a 10 s cycle, we produce 3600 s/h ÷ 10 s/piece = 360 pieces/h. Assuming a machine hour cost of $40.00, the machine hour cost per unit molded is therefore $40.00/h ÷ 360 pieces/h = $0.111 per unit. By selecting a two-cavity mold instead, the production doubles to 720 pieces per hour, while the machine hour cost per unit is reduced by half, to $0.0555.

If another machine could operate the same mold at an 8 s cycle, the figures would be different: 3600 s/h ÷ 8 s/shot = 450 shots/h, and with the same machine hour cost of $40.00, the machine hour cost per unit for a one-cavity mold would be $40.00 ÷ 450 = $0.0889 per unit. Similarly, a two-cavity mold cycling at the faster speed would result in a machine hour cost per unit of $0.0444.

From the above example it can easily be seen that the more cavities there are in the mold, the lower the machine hour cost per unit molded. Also, if the machine cycles faster, the lower is the machine hour cost. This change in machine hour cost per unit can be seen in Table 7.2, based on an assumed machine hour cost of $40.00. It can also be seen in Table 7.2 that the incremental gain, i.e., the reduction in machine hour cost per cavity, becomes smaller as the number of cavities

[1] *https://plasticsbusinessmag.com*

[2] *https://www.plasticsnews.com*

[3] *https://www.ptonline.com*

[4] *https://plasticsbusinessmag.com/articles/2019/whats-in-a-machine-rate*

increases and as the cycle time decreases. For example, with a 10 s cycle as above, increasing the number of cavities from one to two, we reduce the machine hour cost from \$0.11 to \$0.055. By selecting four, eight, or 16 cavities, this cost will reduce to \$0.0275, \$0.0138, and \$0.0069, respectively. With more cavities, these incremental amounts are very small, and would in most cases not influence the choice of the number of cavities.

With *large quantities* (100 million or more), the deciding factor will usually be the highest-productivity mold possible.

Table 7.2 Cost of Machine Hours per Unit Molded

Cycle time	Shots/h	# of cavities				
		1	2	4	8	16
12 s	300	\$0.1333	\$0.0667	\$0.0333	\$0.0167	\$0.0083
10 s	360	\$0.1111	\$0.0556	\$0.0278	\$0.0139	\$0.0069
8 s	450	\$0.0889	\$0.0444	\$0.0222	\$0.0111	\$0.0056

7.2 Mold Cost per Unit Molded

The productivity of a mold increases in *proportion to the number of cavities.* But a mold with more cavities will be more expensive and may require a larger machine with more clamp force and more injection and plasticizing capacity. The cost of two identical stacks will be about twice as much as the cost of one stack, but this relation can change. The more identical stacks are made at the same time, the more attention can be given to better "mass-production" mold manufacturing methods by providing special tooling, jigs, and fixtures. Also, programming for computer numerical controlled (CNC) machines is the same whether one or any number of cavities is produced, so that the first cavity will cost more than additional cavities. Obviously, the mold shoe will be larger as the number of cavities increases, but not necessarily proportional to the number of cavities. Only a tentative layout will make it possible to indicate how much larger the mold shoe will have to be.

$$\begin{array}{c}\text{Mold cost}\\ \text{per unit}\end{array} = \frac{\text{Mold cost (\$)}}{\begin{array}{c}\text{\# of pieces expected to be made}\\ \text{from the mold during its lifetime}\end{array}} \tag{7.2}$$

For the purpose of this discussion, we will assume that a two-cavity mold will be 1.8× as expensive as a one-cavity mold for the same product, and we will use the same increase for larger molds. The factor of 1.8 has been arbitrarily chosen from experience and would apply to a fairly complicated stack; this factor *cannot and must not be used* as a universally applicable factor. It is used here only to indicate that doubling the number of cavities does not mean that the mold will be twice as expensive. With simpler stacks, this factor could be smaller, such as 1.6 or 1.5.

Example: Mold cost per unit calculation

A molder wishes to produce a part with the lowest mold cost per part. A single-cavity mold would cost $20,000. Each subsequent doubling in cavitation (to 2, 4, and 8) would be 1.8× the previous cavitation. For quantities of 10,000–10,000,000, Table 7.3 shows the effect of mold cost on the unit cost of the product.

Table 7.3 Relationship between Mold Cost per Unit and Total Production Requirements

# of cavities	Mold cost	Estimated total production			
		10,000	100,000	1,000,000	10,000,000
1	$20,000	$2.0000	$0.2000	$0.0200	$0.0020
2	$36,000	$3.6000	$0.3600	$0.0360	$0.0036
4	$64,800	$6.4800	$0.6480	$0.0648	$0.0065
8	$116,640	$11.1600	$1.1160	$0.1116	$0.0112

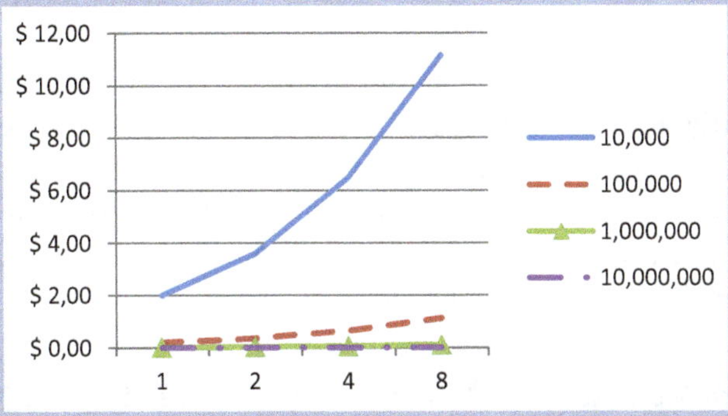

Figure 7.1 Graph of Table 7.3 – Mold cost per unit at increasing cavitation and quantities

From Figure 7.1 it can be easily seen that the larger the number of pieces to be produced, the smaller the mold cost per unit. Table 7.3 shows that the mold cost per unit increases dramatically when the requirements are relatively small and that in such cases a multi-cavity mold does not make much sense, unless the total requirement must be delivered in a very short time. On the other hand, for large quantities, the actual mold cost per unit is very small and a mold with more cavities should be selected. These simple calculations, which take only minutes to perform, must always be made before deciding on the number of cavities planned for a new mold.

In general, it can be stated that **for very large quantities or an indefinite lifespan,** the best quality mold (most solid construction, best cooling, best injection system, best ejection method, etc.) and the largest number of cavities possible is the best choice for a new mold and will yield the *lowest product cost,* even though the initial investment is high.

As we mentioned earlier, in most products, the cost of plastic usually constitutes the largest portion of the piece cost. If, for example, the molds in the above example produce pieces with a mass of 100 g and the cost of the plastic is $1.00/kg, the cost of material would be $0.10 per unit. In most cases, it would not make much difference whether the mold cost per unit of $0.002 or $0.0064 was added to the material cost, but we gain much in productivity of the mold with higher cavitation and at the same time dramatically reduce the total cost by using fewer machine hours.

7.3 Labor Costs

Depending on the level of automation in the work cell, direct labor can be a small fraction or a large portion of the part cost. Table 7.4 shows the average range of cost of direct labor per hour by industry type. This gives the customer a relative idea of the labor burden that would need to be included. Indirect labor in the plant would be added to this to support the direct labor. This would include management, quality, maintenance, shipping, etc.

Table 7.4 Average Direct Labor Rates for Injection Molding in the US in 2019 [5]

Industry	Range of labor cost/h
Vehicles and appliances	$14–17
Food and beverage	$10–12
Medical	$18–21
Consumer goods	$13–16
Industrial	$14–15
Construction and agriculture	$16–17

[5] *https://plasticsbusinessmag.com/articles/2019/whats-in-a-machine*

7.4 Estimating Product Cost

We must always be aware of our goal, which is to arrive at the best-quality product, at the lowest cost to be able to make a profit. It is important to keep in mind that the part cost is typically most influenced by the weight of plastic used in the part. In fact, the plastic itself usually constitutes between 50 and 80% of the part cost. This is another reason why there is such a high focus on reducing the part weight. There is a double impact that occurs when part weight is reduced. Not only is there less material in the part but because this results in a thinner wall section the injection molding systems will be able to run faster, thus reducing the overall cost of the machine and mold per part.

In general, the cost of a molded product consists of these major components:

1. The equipment (machine, automation, auxiliaries) hour cost per unit
2. The cost of the mold, per unit
3. The cost of plastic
4. Labor, direct and indirect
5. Energy
6. Any additional costs, after molding, attributable to the molded piece, such as handling, and post-molding operations, such as printing.

An example of the cost of a typical commodity molded part is shown in Figure 7.2. It can be seen that the plastic cost is 51% of the part cost, followed by indirect costs and then the depreciation cost of the machine and mold. Note that mold depreciation represents 7% of part cost yet it has the largest impact on the overall productivity (or TEP) of the molding workcell.

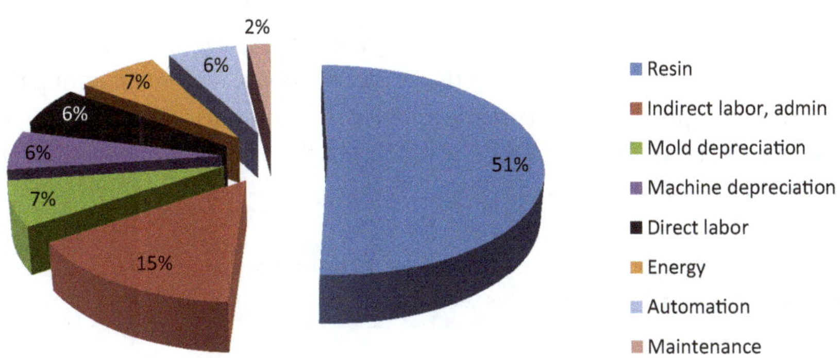

Figure 7.2 Breakdown of typical injection-molded part cost

The conversion cost of the engine cover for an ATV or snowmobile (Figure 7.3) is more impacted by labor than any other cost factor. Note that resin cost is not included in this cost estimate, as it is the conversion cost only, which excludes resin. It is also worth noting that the depreciated cost of the mold is only 5% of the conversion cost.

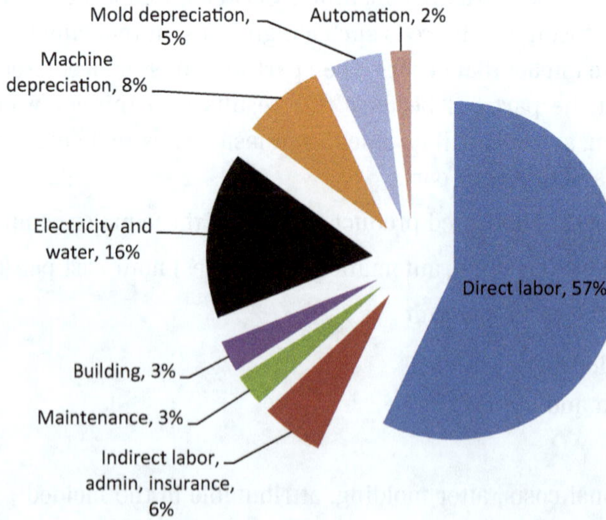

Figure 7.3 Conversion cost of ATV and snowmobile engine covers (resin cost not included)

Automation of the process can have a large impact on the conversion cost. As can be seen in Figure 7.4, the cost of labor is significantly less when automation is added. If the depreciation cost of the automation is justifiable, then automation is an excellent solution to reduce part cost. Usually, automation can be easily justified if the process is highly labor-intensive, as for this cutlery example. Also, automation can be justified pretty quickly if the program extends over many years and the depreciation can be spread over more than 3 years. Again, it is worth noting that in Figure 7.4 the depreciated cost of the mold (which is normally between 3 and 5 years) is only a few percent of the part cost, but the impact of the mold is one of the largest in the entire work cell.

This reinforces the rule of thumb that buying the best possible mold is almost always the best choice for the project.

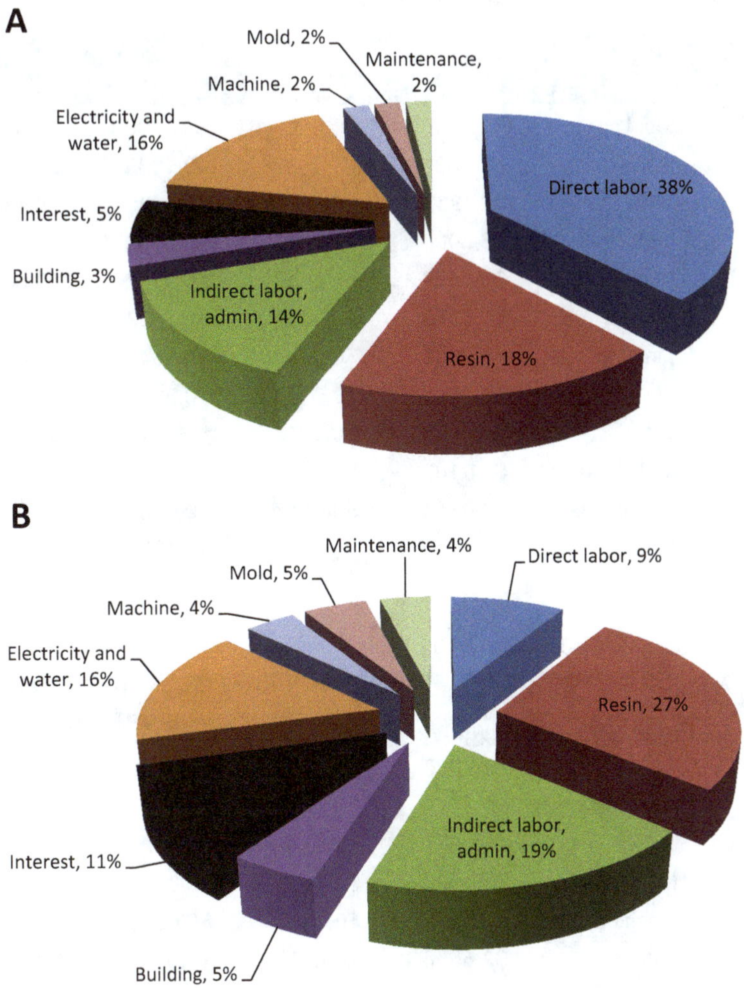

Figure 7.4 Cutlery part cost (A) without automation, (B) with automation

Because both the plastic cost and the cost of direct and indirect labor are constants and have no direct effect on the selection of the cavity numbers, they are omitted in the following tabulations. Remember that the mold and machine hour costs usually represent only a fraction of the total product cost.

Table 7.2 and Table 7.3 demonstrate the effect of the number of cavities on the machine hour time per unit and the mold cost per unit. We can combine these two tables, for each specific estimate; this comparison (Table 7.5–Table 7.7) will greatly help in making a decision on the number of cavities for the planned mold.

Table 7.5 Unit Costs of Product with a 12 s Cycle

# of cavities	Cost per unit	Estimated total production			
		10,000	100,000	1,000,000	10,000,000
1	Mold cost	$2.0000	$0.2000	$0.0200	$0.0020
	Machine cost	$0.1333	$0.1333	$0.1333	$0.1333
	Total	**$2.1333**	**$0.3333**	**$0.1533**	**$0.1353**
2	Mold cost	$3.6000	$0.3600	$0.0360	$0.0036
	Machine cost	$0.0667	$0.0667	$0.0667	$0.0667
	Total	**$3.6667**	**$0.4267**	**$0.1027**	**$0.0703**
4	Mold cost	$6.4800	$0.6480	$0.0648	$0.0065
	Machine cost	$0.0333	$0.0333	$0.0333	$0.0333
	Total	**$6.5133**	**$0.6813**	**$0.0981**	**$0.0398**
8	Mold cost	$11.1600	$1.1160	$0.1116	$0.0112
	Machine cost	$0.0167	$0.0167	$0.0167	$0.0167
	Total	**$11.1767**	**$1.1327**	**$0.1283**	**$0.0279**

Table 7.6 Unit Costs of Product with a 10 s Cycle

# of cavities	Cost per unit	Estimated total production			
		10,000	100,000	1,000,000	10,000,000
1	Mold cost	$2.0000	$0.2000	$0.0200	$0.0020
	Machine cost	$0.1111	$0.1111	$0.1111	$0.1111
	Total	**$2.1111**	**$0.3111**	**$0.1311**	**$0.1131**
2	Mold cost	$3.6000	$0.3600	$0.0360	$0.0036
	Machine cost	$0.0556	$0.0556	$0.0556	$0.0556
	Total	**$3.6556**	**$0.4156**	**$0.0916**	**$0.0592**
4	Mold cost	$6.4800	$0.6480	$0.0648	$0.0065
	Machine cost	$0.0278	$0.0278	$0.0278	$0.0278
	Total	**$6.5078**	**$0.6758**	**$0.0926**	**$0.0343**
8	Mold cost	$11.1600	$1.1160	$0.1116	$0.0112
	Machine cost	$0.0139	$0.0139	$0.0139	$0.0139
	Total	**$11.1739**	**$1.1299**	**$0.1255**	**$0.0251**

Table 7.7 Unit Costs of Product with an 8 s Cycle

# of cavities	Cost per unit	Estimated total production			
		10,000	100,000	1,000,000	10,000,000
1	Mold cost	$2.0000	$0.2000	$0.0200	$0.0020
	Machine cost	$0.0889	$0.0889	$0.0889	$0.0889
	Total	**$2.0889**	**$0.2889**	**$0.1089**	**$0.0909**

# of cavities	Cost per unit	Estimated total production			
		10,000	100,000	1,000,000	10,000,000
2	Mold cost	$3.6000	$0.3600	$0.0360	$0.0036
	Machine cost	$0.0444	$0.0444	$0.0444	$0.0444
	Total	**$3.6444**	**$0.4044**	**$0.0804**	**$0.0480**
4	Mold cost	$6.4800	$0.6480	$0.0648	$0.0065
	Machine cost	$0.0222	$0.0222	$0.0222	$0.0222
	Total	**$6.5022**	**$0.6702**	**$0.0870**	**$0.0287**
8	Mold cost	$11.1600	$1.1160	$0.1116	$0.0112
	Machine cost	$0.0111	$0.0111	$0.0111	$0.0111
	Total	**$11.1711**	**$1.1271**	**$0.1227**	**$0.0223**

To illustrate the points raised, we have assumed production requirements over a wide range and three possible cycle times for a planned product. The same figures for mold costs per unit and machine hour cost per unit as in Table 7.2 and Table 7.3 were used.

Unit costs of product attributable to mold cost and machine hour cost (referring back to mold costs shown in Table 7.2 and based on a machine hour cost of $40.00) are shown in Tables 3.3–3.5.

Instead of the unit values given in Table 7.5–Table 7.7, we could also consider the sum-total of the estimated mold cost plus the cost of producing the total number of products required, by taking the mold cost and then adding the machine hour cost per unit multiplied with the total number of pieces required (in this example, we will only consider the 10 s cycle values (Table 7.8)).

Table 7.8 Estimated Total Cost of Production by Cavitation for a 10 s Cycle

# of cavities	Cost of the total production	Estimated total production			
		10,000	100,000	1,000,000	10,000,000
1	Mold cost	$20,000	$20,000	$20,000	$20,000
	Machine cost	$1111	$11,110	$111,100	$1,111,000
	Total	**$21,111**	**$31,110**	**$131,100**	**$1,131,000**
2	Mold cost	$36,000	$36,000	$36,000	$36,000
	Machine cost	$556	$5560	$55,600	$556,000
	Total	**$36,556**	**$41,560**	**$91,600**	**$592,000**
4	Mold cost	$64,800	$64,800	$64,800	$64,800
	Machine cost	$278	$2780	$27,800	$278,000
	Total	**$65,078**	**$67,580**	**$92,600**	**$342,800**

Table 7.8 Estimated Total Cost of Production by Cavitation for a 10 s Cycle *(continued)*

# of cavities	Cost of the total production	Estimated total production			
8	Mold cost	$116,640	$116,640	$116,640	$116,640
	Machine cost	$139	$1390	$13,900	$139,000
	Total	**$116,779**	**$118,030**	**$130,540**	**$255,640**

To calculate the total cost of the product, we must now add the cost of plastic, the cost of handling directly chargeable to the product, and the cost of any post-molding operations.

Example:

A mold is required for a product with a mass of 40 g, made from PP, at $1.00 per kg. The estimated cycle time is 10 s and the cost of plastic is 40 g × $1.00 ÷ 1000 g/kg = $0.04 per unit. Create a table to examine the impact of cavitation on part cost.

For simplicity, assume that all overheads are included in the machine hour cost of $40.00/h and are the same for all machines. Also assume that there is no direct labor involved with this product.

Table 7.9 Cost of a Product Made with Different Numbers of Cavities with a 10 s Cycle

# of cavities	Cost per unit	For an estimated total production of					
		100,000	% of total	1,000,000	% of total	10,000,000	% of total
2	Mold	$0.3600	78.9	$0.0360	27.3	$0.0036	3.6
	Machine	$0.0560	12.3	$0.0560	42.4	$0.0560	56.2
	Plastic	$0.0400	8.8	$0.0400	30.3	$0.0400	40.2
	Total	**$0.4560**		**$0.1320**		**$0.0996**	
4	Mold	$0.6480	90.5	$0.0648	48.8	$0.0065	8.7
	Machine	$0.0278	3.9	$0.0278	21.0	$0.0278	37.4
	Plastic	$0.0400	5.6	$0.0400	30.2	$0.0400	53.9
	Total	**$0.7158**		**$0.1326**		**$0.0743**	
8	Mold	$1.1160	95.4	$0.1116	67.4	$0.0112	17.1
	Machine	$0.0139	1.2	$0.0139	8.4	$0.0139	21.4
	Plastic	$0.0400	3.4	$0.0400	24.2	$0.0400	61.5
	Total	**$1.1699**		**$0.1655**		**$0.0651**	

# of cavities	Cost per unit	For an estimated total production of					
		100,000	% of total	1,000,000	% of total	10,000,000	% of total
2 × 8 (stack)	Mold	$2.0088	97.7	$0.2009	81.0	$0.0201	29.9
	Machine	$0.0070	0.3	$0.0070	2.8	$0.0070	10.4
	Plastic	$0.0400	2.0	$0.0400	16.2	$0.0400	59.7
	Total	**$2.0558**		**$0.2479**		**$0.0671**	

To arrive at the actual cost of the product, we add the mold cost/unit and machine hour cost/unit from Table 7.9.

Conclusion:

- With *low* quantities, the *mold cost* represents a substantial portion of the product cost; therefore *do not over-build the required capacity*. It cannot be justified economically to make a multi-cavity mold when the total quantities are small, except where such a small quantity is required in a very short time with not enough molding capacity to supply the required quantities on time. Also, two smaller molds are usually faster to build than one larger one.
- With *large* quantities (often disposable items), the *cost of plastic* represents the major portion of the product cost. The incremental increase of the mold cost will be paid back easily when the quantities are high. Therefore, *build to the anticipated capacity*.

However, there are some additional considerations:

- The *machine time* cost can be substantial with a low number of cavities, especially with products with low mass.
- The cost of the product from a *2 × 8-cavity stack mold* ($0.0671) is about the same as the cost of the same product coming from a single-level 8-cavity mold; however, the productivity of the mold-machine system is doubled. This means fewer machines, less plant space, etc.

The tables above apply to relatively complicated molds. However, if the mold is simple, and especially if the product is small, it is usually not much more work to make two, four, or even more cavities and cores without substantially increasing the mold cost, especially with two-plate cold runner molds. In such cases, a multi-cavity solution is preferable, because the mold will use much less machine time while giving much higher production.

For reasons of symmetry in the mold construction, and to avoid unbalanced clamping, single-cavity molds are usually avoided, unless they can be center-gated to ensure proper balancing of the clamping forces. In such cases, a two-cavity mold would often be the preferred choice.

Let us now apply what we have learned so far. The following examples go through the calculations step-by-step:

Example:

In this example, we base our calculations on the following (reasonable) assumptions:

- 72-cavity mold with cooling robot system
- Cycle time: 14 s (used to be in the order of 40 s for molds with standard cooling methods)
- Mass: 26 g (used to be in the order of 40 g)
- Cost of PET: $1/kg
- Production: 7200 h/year
- Write-off time (very aggressive): 5 years for mold, machine, and drying system for PET
- Productivity: 3600 s/h ÷ 14 s/sh × 72 cavities = 18,510 pieces/h × 7200 h = 133,300,000 preforms/year, or 666,500,000 over 5 years
- Cost of mold, including upkeep for 5 years: $1,500,000
- Mold cost, per preform: 1,500,000 ÷ 666,500,000 = $0.00225
- Cost of machine, cooling robot and drying system, including upkeep for 5 years: $2,000,000
- Machine hour cost: $2,000,000 ÷ (5 × 7200) = $55.56
- To this, we add an assumed other $80.00 for overhead (power, water, services, etc.) to get a net machine hour rate of $135.56. If we add to this also a profit of $50.00/h, we get $185.56.

Selling price (machine hour, overhead, and profit) per preform is therefore:

$185.56/h ÷ 18,510 pieces/h =$0.0100/preform

To this we must add the cost of the plastic:

Cost of plastic per preform: 26 g/piece × $1.40/kg PET ÷ 1000 g/kg = $0.0364/preform

	Cost	Proportion
PET	$0.0364	75%
Mold cost	$0.0023	5%
Machine including profit	$0.0100	20%
Total selling price	**$0.0487**	**100%**

From the above it can be seen that:

a) The mold cost is only 5% of the total.

b) One way to reduce the cost of the preforms, besides reducing the cost of the plastic and/or reducing the mass of the perform, is to improve the performance of molds and machines, which is an ongoing process.

By operating 7200 h/year, over 5 years, there will be a profit of 5 years × 7200 h/year ÷ $50/h = $1,800,000 on an investment of $3,500,000.

Example:

The following example compares two molds for the same product; the lower-priced mold has a poor cooling arrangement (it does not use BeCu inserts), resulting in a slow cycle time. In addition, the ejection mechanism fails frequently and causes downtime and scrap.

	Mold 1	Mold 2
Cost of molds	\$10,000 (poor cooling)	\$20,000
Cycle time	10 s	6 s
Up-time	80%	90%
Scrap	5%	2%
Machine time cost	\$40	\$40
Material cost	\$1/kg	\$1/kg
Product weight	100 g	100 g
Productivity (pieces/h)	1152	2160
Machine hour cost per unit (A)	\$0.0347	\$0.0185
Plastic cost per unit (B)	\$0.1000	\$0.1000
Scrap cost per unit (C)	\$0.0050	\$0.0020
Mold cost per unit:		
At total production 100,000 (D)	\$0.1000	\$0.2000
At total production 10,000,000 (E)	\$0.0010	\$0.0020
Unit cost: A + B + C at (D)	\$0.2397	\$0.3205
Unit cost: A + B + C at (E)	\$0.1407	\$0.1225
Total cost, at (D)	\$23,970	\$32,050
Total cost at (E)	\$1,407,000	\$1,225,000
Savings at 10,000,000 production		\$182,000

8 Mold Layout, Drawings, and Inspection

The primary purpose of the CAD drawings and models created by the mold designer is to convey with absolute certainty the details and requirements (such as tolerances and surface finishes) needed to produce the mold. There are normal conventions in the industry that are used to convey these intentions. Following standard conventions makes it easier for all project engineers, machinists, mold makers and operators to understand how the mold functions, how to manufacture it, and how to maintain it. It is therefore essential that every mold designer has a good fundamental knowledge of CAD design and tolerancing. The purpose of this chapter will be to remind the designer of the needs that are specifically related to mold design.

8.1 Steps to Designing the Mold

When given a new mold design, the following steps must be followed:

1. Critique and then complete the plastic part drawing
2. Determine the cavitation and layout of the cavities
3. Concept the stack
4. Establish the part orientation
5. Determine how the part will be ejected (air, stripper, side action, one-stage, or two-stage)
6. Establish the parting line
7. Determine the gate position
8. Are any inserts or core pins required?
9. Determine how the mold will be vented
10. Establish the ejection details

11. Lay out the cooling in the inserts and plates
12. Determine the alignment method between the mold halves
13. Review the above to ensure it is the best method, and examine other possibilities
14. Lay out the machine platen and tie bars, and determine mold shoe size
15. Establish the mold shut height (using the section view assembly)
16. Concept core and cavity plan views
17 Lay out stacks to establish balanced hot runner layout
18. Lay out leader pins, stops, spring returns, linkages, taper locks, shields, and mold supports
19. Show all screws and dowels
20. Make sure assembly and disassembly is possible
21. Review the mold design using the checklist supplied in the job file
22. Call a concept and design review.

Note that this procedure can be short-cut when the design is similar to existing molds. However, you must be sure that all of the above has been considered in the reference design you are using. ALWAYS follow through the mold design checklist, even on repeat or similar jobs.

8.2 Information and Documentation

Before starting the detailed design of a mold, the designer must make sure that all the information is on hand. It is frustrating and wastes valuable time to find during your work that information is missing, or to be asked to make significant changes after starting that affect the concept of the mold.

Are the Tolerances Shown?

Are the dimensional tolerances specified on the drawing the same as when the mold cost was first estimated and the mold price quoted? This can have serious implications, especially if no tolerances were shown when the job was quoted; for example, if a molder requests an approximate mold cost so that he can estimate the final cost of the product for his customer. Unfortunately, sometimes there is not even a drawing, just a sample or model of the product used for the estimate.

While it is desirable that the mold designer is involved in the product design, to ensure that the product can be easily molded and will be satisfactory for the purpose intended, mold designers should not agree to make a product drawing, and if they do, they must insist that it be signed by the customer as acceptable. This will eliminate any possible unpleasantness later on, if the product does not look or function as expected.

Are the Tolerances Reasonable?

Are the requested product tolerances feasible, in view of the size of the product and the plastic specified? This is sometimes overlooked when quoting. As we have seen in Chapter 3, it is nearly always possible to make the steel mold parts accurately, to very close tolerances, this does not mean that the molded part will satisfy often unreasonable and unnecessary requests for close tolerances. If very close product tolerances are wanted, an experimental setup may be required to determine steel sizes, a process that can be very costly and time-consuming. This must be made clear before work is started. Note that in the case of very stringent tolerances, production (the actual molding) can become very expensive, requiring close inspection of the molded products and possibly causing many rejects.

What Is the Expected Cycle Time?

The designer should never guarantee cycle times, and must make sure that the customer understands this. If the customer insists on any guarantee, it could require experimental work (test molds, remaking of mold parts, etc.), which could become very expensive. Any such anticipated costs should be brought to the attention of the customer, and added to the mold price. However, the designer should have some idea of the expected cycle, from past experience with similar products, or should try to get this information from someone with molding experience with such products.

What Is the Expected Production?

The designer must be aware of the total production expected from the mold, and its expected life. There is a significant difference between needing a mold for 1000, 100,000, 1,000,000, or 10,000,000 or more parts. This consideration will affect all aspects of a mold, from selection of mold materials to many mold features selected by the designer.

It cannot be repeated often enough that the mold is only one in a chain of requirements (albeit the most important one) for producing a molded product. The molder, or the final user, should not really be interested in the mold cost, but only in the cost of the molded product. It is the duty of the designer to advise the customer accordingly and build the most economical mold for the intended job.

The following is also a frequent scenario: A new widget is to be marketed. After a few hundred test samples, the customer estimates that during the next year he could sell 10,000 pieces. He does not yet know if the widget will be accepted at large. What size mold will be required? How will the mold cost, divided by this quantity, affect the cost of the widget? Obviously, because of the small quantity, the mold cost will be significant in this calculation. Also, because of the relatively small quantity, there may be only one cavity or at most two or four cavities required. This means low productivity, resulting in a higher molding cost. A simple cold runner system could be suitable and quite inexpensive. But what if the widget turns out to be a success and the required quantities increase to an estimated 1,000,000 over the next 3 years? The first mold probably will not be able to produce these quantities in time. This will then require a new, substantially different mold, with more cavities, a hot runner system, and so on - in short, a more complicated mold, which will cost much more but, despite the higher mold cost, will result in a much lower cost of the molded piece. Which is the better mold? They are both good, and each one is suitable for the specified requirement.

8.2.1 Machine Specifications

Before starting, the designer must know the machine or machines on which the mold is to operate. Chapter 4 covers how to specify the machine for the mold. The following must be checked and determined before the mold design can be finalized and the mold sent into manufacturing.

8.2.1.1 Mechanical Features

1. *Tie bar clearances and platen size.* Will the planned mold fit on the platens? In some cases it is alright to have the mold larger than these dimensions; it may even overhang the platens, as long as the cavities are located within the area between the tie bars. In some cases, it may be necessary to pull one or both top tie bars to be able to install the mold, although today this is rare. If this is required, the designer must find out if the planned machines have provisions for easy tie bar pulling. Sometimes the mold can be rotated 90° to fit between the tie bars and then rotated back for installation.
2. *Locating ring size.* The locating ring centers the injection half of the mold on the stationary (or "hot") platen. Ensure that the size of the locating ring is known before the mold is sent to manufacturing. In some cases there may be locating rings on both halves of the mold, if for instance the mold needs to be installed in two parts for crane capacity limits.
3. *Machine nozzle radius or through hole and melt channel size.* The sprue bushing radius must fit the injection nozzle radius and the size of the hole in the machine nozzle must match that of the sprue bushing in the mold. Failure to have these match can result in burning and degradation problems (if one hole or the

other is too large), hot plastic spraying (if the radii don't match) and filling problems (if the hole is too small). If the machine nozzle is a decompression style (like used in Europe) then the hole in the sprue bushing must match the outer diameter of the machine nozzle. In the case of this style of nozzle, the injection unit must be precisely aligned with the mold to ensure there is no binding when the injection unit carriage moves forward and back.

4. *Mold mounting holes and slot pattern* (Euro, SPI, or other standard?). How will the mold be mounted on the platens? The best method is where the mold halves are directly screwed onto the platens, using standard mounting holes on the platens or clearance holes on the platens with threaded holes in the mold. With this method the full holding force of the screw is utilized. But this is often not possible, especially if the mold must fit several different machines. In these cases, mold clamps are frequently used, with the clamp screws making use of standard mounting holes or slots in the platens. The disadvantage of this method is that only a portion of the holding force of the screw is utilized.
5. *Quick mold change features.* There are a number of commercial and proprietary systems, and the designer must get the specifications to fit the system before starting to design the mold.
6. *Machine ejector.* The ejector force is usually about 10% of the clamp force, which is sufficient for most molds, but there are cases where this is not enough. The mold may have to be equipped with additional ejection means, often built-in hydraulic or air actuators. The machine ejectors are always on the moving platen, but their size and pattern will vary according to the builder's standards (Euro, SPI, other standard?). If the mold is going to make use of the machine ejectors, it is important to know their size and location when designing the ejection mechanism.
7. *Shut height.* This is the total thickness of the mold. This distance must not be greater than the maximum allowable distance between the platen surfaces of the machine when in fully closed position. If the mold is smaller than the minimum machine shut height then a bolster plate may be needed to make up the additional mold thickness to mount the mold in the machine. The machine specifications indicate maximum and minimum shut height. If the laid-out shut height is too great, there are several ways to reduce it:
    a) Investigate whether all the shown mold plates are really necessary. In some molds, for example, the mounting plate under an ejector box can be omitted, by fastening the mold to the machine using the mold parallels
    b) Reduce the thickness of one or more of the mold plates
    c) If neither is possible without compromising the quality (strength) of the mold, a different machine must be selected. This should be discussed with the molder before proceeding.

Conversely, if the shut height is too small, plate thicknesses can be increased, which is not always a good solution because it makes the mold unnecessarily heavy and adds cost to the mold. Some machines are equipped with bolster plates, or bolster blocks, which are mounted on the moving platen in order to decrease the minimum shut height.

8. *Available clamp stroke.* In most machines, the mold clamp stroke is adjustable. For many molds, the suggested minimum stroke should be about 2.5 × the height of the product, to ensure that the molded pieces have enough space to fall free between the mold halves during ejection; however, the stroke should not be less than about 150 mm (6"), so that the mold surfaces can be accessed for servicing while the mold is open. There are exceptions to these two suggested values, for special applications, particularly when using automatic (robotic) product removal methods, which are outside the scope of this book.
9. *Ejector stroke.* This stroke is also adjustable, within the limits of the machine specifications. The designer must make sure that the available ejection stroke is large enough to push the products completely off the cores, in cases where little draft is specified, for example, when molding deep-draw containers. With good draft, it is usually not necessary to do more than push the products some short distance before they fall free, or before air-assist features will blow them away. There are again some exceptions, particularly with robotic product removal methods.
10. *Clamping force.* The designer must make sure that the total projected areas of all cavities, plus the projected areas of any runner system in the same parting plane, multiplied by the estimated injection pressure, will not be greater than the available machine clamping force. As we have seen earlier, the estimated injection pressure depends on the ease of plastic flow (viscosity, temperature) and on the wall thickness of the product. In borderline cases, it is sometimes possible to change conditions, for example, in a very large product, by increasing the number of gates and placing them far apart. It may then be possible to use lower injection pressures, thereby requiring less clamp force.
11. *Auxiliary controls.* Some molds may require specially designed air or hydraulic circuits for ejection, accumulator tanks, and valving for the actuators. Is the machine equipped for such circuits, to be timed within the molding cycles? In some cases, hydraulically actuated side cores may be required. Has the machine a provision for timed core pulls?

8.2.2 Productivity Features

(1) Shot Size (mass per shot)

The total calculated or estimated shot size, that is, the total mass (weight) of the products coming from all cavities, plus the mass of the runner system (in the case of cold runners), should be within 30–90% of the shot capacity of the machine. The shot capacity of a machine is given in g/shot of PS, with a specific gravity of about 1.05. The specific gravity of materials such as PE and PP is less (about 0.90–0.95); that is, the same mass will have a greater volume. Since shot size is rated in grams (or ounces) but is actually a volume (cross-section of extruder barrel times the stroke of the extruder), the shot size of these materials will be less than for PS, by about 10%. These are only approximate figures; exact values should be checked with the material supplier.

What are the practical implications? If, for example, an 8-cavity mold is required to run in a specific machine, but its shot capacity is not large enough, it would not make sense to build it for this machine. This is especially important with cold runner molds, where the mass of the runner can add considerably to the mass of the sum of all molded parts, per shot. A machine could be well-suited for a hot runner mold but be unsuited for a cold runner mold for the same number of cavities. (This is a major advantage of the hot runner system.)

(2) Plasticizing Capacity (kilograms per hour)

Plasticizing capacity is the amount (mass) of plastic a machine can plasticize per hour, that is, melt the cold plastic pellets into a melt of a specific temperature (and viscosity). Plasticizing capacity is usually given as mass for PS, in kilograms (pounds) per hour. Here, the same applies as with shot capacity. The actual mass of other materials, such as PE, PP, or any other, will be different, mostly smaller, and sometimes greater. This should be carefully considered before starting.

But first, the designer must estimate the molding cycle, to find out how much plastic per hour will be required. Dividing 3600 s (1 hour) by the number of the seconds of the estimated cycle will give the number of shots per hour (N). Multiplying the total shot weight S (g/shot) calculated in (1) above, with the number of shots N per hour, we find the total mass W_t in grams per hour required ($W_t = S \times N$). For best quality of the melt (and the molded piece), it is also suggested to use only between 30 and 90% of the rated plasticizing capacity. If W_t is more than the rated capacity, the machine can still be used but the cycle time will have to be lengthened; in other words, fewer shots per hour can be produced than the mold could yield with a suitable, larger size machine.

(3) Injection Speed (grams injected into the mold per second)

This is an important consideration when molding thin-walled products. Because of the narrow gap through which the plastic must flow within the cavity space, the injected plastic will cool rapidly when in contact with the cooled cavity and core walls. As the plastic cools, the gap narrows even more, making it more difficult to fill the mold. To overcome this condition, the melt and/or the mold temperatures could be increased so that the plastic will not freeze before filling the mold.

However, this increase in temperature will also cause an increase in the cooling cycle (and a lengthening of the molding cycle), resulting in a smaller output from the mold. This points to two areas for possible remedy: (a) increasing the injection speed and (b) increasing the injection pressure. But these two are interrelated. The higher the pressure, the faster the melt will be pushed through its paths, from the machine nozzle to the farthest corners of the cavity space. The problem is now that the injection speed depends on the speed with which the hydraulic injection cylinder is filled with pressure oil. Therefore, the speed of the injection cylinder depends on the hydraulic pump output (oil volume per second) entering the cylinder, but it also depends on the size of the associated hardware (hoses, valves, and so on) from the pump to the cylinder.

Most machines for conventional (not thin-wall) products are served sufficiently well by the output of the pump (and the motor driving it). However, the injection speeds required for thin-wall production require the cylinder to be filled more rapidly than what the pump alone can provide. To remedy this, the machine could be equipped with a much larger pump and motor, but in many cases this would be uneconomical or impractical. The preferred solution is to provide the machine injection system with an accumulator, which stores high-pressure oil during the time pressure oil is not used. Additional valving and other hardware is required, which is often sold as an "option" with the machine, called an accumulator package. The accumulator releases the stored high-pressure oil together with the pump output into the cylinder when required for injection. The designer will need to recognize when an accumulator package is necessary for the product for which the mold is to be designed, and must discuss this with the molder to make sure the right machine is available to run the mold.

8.2.3 Additional Requirements

Air Pressure and Volume

Some molds require air pressure for their operation. In general, the designer should be aware that compressed air, especially in large volumes, can be very expensive, especially if it is left to blow for any length of time.

- *Blow-downs* (air jets or air curtains) are often used to help the products clear the molding area rapidly. There are several commercial air jets on the market with low consumption of pressure air. Their initial cost is paid back rapidly by savings from wasted air volume.
- *Air-operated actuators*. The air volume used is usually small, compared with a blow-down. There could be problems with controlling the speed and uniform motion of air actuators, but they are simple and inexpensive.
- *Air required for air ejection*, which is usually activated on demand, for a very short time. Most of the time, the actuation time is controlled from the machine control panel. The designer must make sure that the intended machine is equipped with sufficient controls and hardware (timers, valves, and large enough supply lines). It may be even necessary to add pump capacity, for the added volume of air that will be required for the planned mold. If much air is needed for short blasts, one or several suitable accumulators could be installed near or even on the mold.

Where pressure air comes into contact with the molded products, for example, in blow-downs or in air ejection, the air must be filtered from any oil residues, water (always present in air lines), and so on, before reaching the outlets in or at the mold, to prevent contamination of the products if they are used for food or pharmaceutical purposes. (Unfortunately, most air actuators require lubrication, unless their seals are selected for dry air.) A low-pressure, high-volume blower with its air intake from the shop environment, or better yet, from within an enclosure built around the molding machine when special "cleanroom" requirements are specified, is a preferred solution to ensure that there is no oil or water contamination in the air as it comes into contact with the plastic products. In many cases, such a blower can be directly mounted on the top of the mold. Another advantage is that the power consumption of this type blower is low, on the order of 0.2 kW (0.25 horsepower) or less, and does not require timing or valving.

Auxiliary Hydraulic Supply

For some operations, compressed air may be not suitable: (a) Air cylinders are often jerky in their operation, especially with long strokes: (b) In cases where several air cylinders actuate one large mold member, the forces can be uneven and the member can jam. (c) In most molding shops the compressed air pressure is fairly low, usually about 600 kPa (80 psi), and rarely 900 kPa (120 psi), so large air actuators are needed to produce large forces. It could be difficult to accommodate sufficiently large cylinders within the available mold space, or even outside the mold. In all these cases, the much more powerful hydraulic cylinders would be an alternative. The hydraulic pressure could be taken from the machine system with a pressure-reducing valve, and by providing the necessary safety measures to protect against the very high pressures in that system. A preferred method, however, is to use an auxiliary power supply, usually at a system pressure of about 3500 kPa (500 psi). This is much safer and requires much less expensive hardware (valves, hoses, etc.) than that for higher pressure. The motion of hydraulic operators is smooth and the speed can be well-controlled.

Two points of caution, though. Hydraulic oil (with some special, expensive, exceptions) is highly flammable and there is always the danger of leaks, which could be especially dangerous if the leaks were to occur near heated areas of the mold, as, for example, near a hot runner system. Also, products used in the food or pharmaceutical industry could be contaminated by the oil; this is usually specified as not allowed.

Cooling Water Supply

This is a very important area of concern. There is not much sense in designing the mold with very sophisticated cooling circuitry if the cooling water supply is insufficient in temperature, volume, and pressure. An individual chiller unit may be the answer if the plant supply is too small or has not enough pressure. It is also important that the coolant is clean, that is, with a minimum of minerals or dirt, and is not corrosive. Dirty coolant could gradually plug the water circuits or coat the channel walls with a poorly conducting layer of dirt and lime, thus reducing the cooling efficiency, and could require frequent cleaning of the coolant channels if the mold is expected to maintain high productivity. Corrosive action of the coolant could attack and eat away the mold steels; rust creates insulating layers similar to lime and dirt deposits. It is always good policy for the designer to check with the molder to ensure that there are no such problems with the water supply, and to specify that only clean, non-corrosive coolant is used with the mold.

Electric Power and Controls

The electric power supply in North America and in most developed countries is usually sufficiently stable and uninterrupted, except during natural catastrophes, and of not much concern to the designer. This is not the case in developing countries, where power interruptions occur frequently; the effects of such interruptions on the operation of a mold may cause concern. Typically, in the case of a power failure, a machine using a cold runner mold will just stop, but can resume work as soon as the plastic is again up to molding temperature. However, in a hot runner mold the melt will freeze in the manifold and nozzles, and it may take much more time to restart (in small molds 15–30 min). The expected savings through using a hot-runner mold may be illusory. The controls (breakers, heat controllers) available to operate a mold on a specific machine must be discussed with the molder when designing a mold that will require additional heat controls; typically, such controls are required for hot runner molds. For safety reasons, heaters in molds are rated at 230 V_{AC} or less, and the power consumption may be from as low as 40 W per heater, such as in some nozzle heaters, and up to several thousand watts in hot runner manifold heaters. Since heaters are often bundled in parallel and operated by designated controls, it is important to ensure that adequately sized circuit breakers and so on are available; some can be controlled with time-percentage controllers or variable (voltage) transformers, whereas some will need thermocouples and heat controllers.

8.3 Determining the Mold Cavitation

Many molds, particularly molds for larger products, are built for only one cavity space (a single-cavity mold). However, when volume demands are very large and or economics dictate it, molds can easily be built with two or more cavities (Figure 8.1). The reason for this is purely economical. It takes only little more time to inject several cavities than to inject one. For example, a four-cavity mold requires only (approximately) one-quarter of the machine time of a single-cavity mold. Conversely, the production increases in proportion to the number of cavities. A mold with more cavities is more expensive to build than a single-cavity mold, but (as in our example) not necessarily 4 × as much as a single-cavity mold. But it may also require a larger machine with larger platen area and more clamping capacity, and because it will use (in this example) 4 × the amount of plastic, it may need a larger injection unit, so the machine hour cost will be higher than for a machine large enough for the smaller mold.

Today, most multi-cavity molds are built with a preferred number of cavities to balance filling and respect symmetry in the mold design. Table 8.1 shows the ideal, preferred, and less desirable cavitations. These numbers are selected because the cavities can be easily arranged in a rectangular pattern, which is easier for designing and dimensioning, for manufacturing, and for symmetry around the center of the machine, which is highly desirable to ensure equal clamping force for each cavity. The cavitations listed above are also desirable for creating a balanced filling of each cavity from the runner system. A smaller number of cavities can also be laid out in a circular pattern, and even with odd numbers of cavities, such as 3, 5, 7, and 9. It is also possible to make cavity layouts for any number of cavities, provided such rules as symmetry of the projected areas around the machine centerline (as explained later) and filling balance are observed.

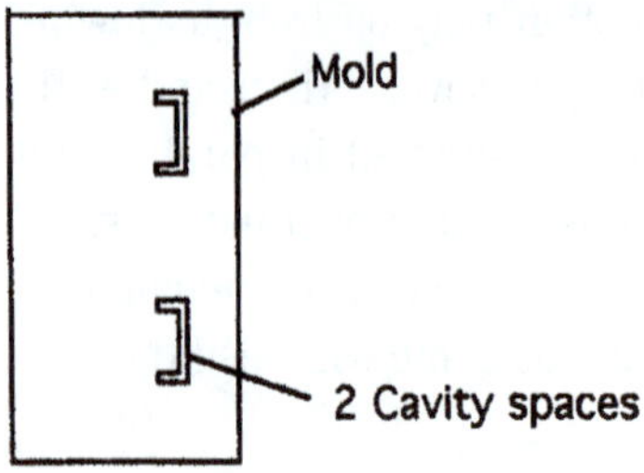

Figure 8.1
Illustration of basic mold with two cavity spaces

8.3.1 Minimum Number of Cavities

To determine the number of cavities required in the mold, the following data (or values based on assumptions) are needed:

- The annual production quantities required of the mold, or the total output of the mold if required only for a shorter period.
- A conservatively estimated cycle time. Factors affecting cycle time have been discussed in Chapter 6.

$$\text{Cavitation required} = \frac{\text{PDA} \times t_{\text{cycle}}}{3600 \times \text{AOH} \times \text{UT} \times \text{Yield}} \tag{8.1}$$

where:
PDA = annual production required (# of units)
t_{cycle} = cycle time (s)
AOH = annual operating hours (h) (typically 7400–7800 used, some dedicated systems 8400)
UT = uptime percentage (# between 0 and 1) (dedicated molding 0.9, custom molding 0.75)
Yield = percentage of good parts as a function of total parts (# between 0 and 1) (typically between 0.95–0.98)
Note that the number that results is just a theoretical value. This needs to be now adjusted to a practical cavitation for a mold design. Normal practical cavitations for mold design are shown in Table 8.1.

Table 8.1 Ideal and Good Cavitation to Select when Designing a Mold

Desirability	Cavitation
Ideal: Easy to balance	1, 2, 4, 8, 16, 32, 64, 128, 256
Good: Can be balanced	6, 12, 24, 48, 72, 96, 128, 144
Not preferred: Unbalanced	3, 5, 10, 14, 18, 20, 36, 56

- The reason these numbers are practical is that they result in an easily balanced layout for the melt distribution systems in the mold, and the layout allows the cavities to be spaced symmetrically to reduce unbalanced forces.

Going to an odd or unbalanced cavitation can result in many molding problems and issues during production, so this should be avoided if at all possible.

- To determine the annual operating hours (AOH) there are two approaches that are generally used. Note that there are 8760 h in a full year (365 days × 24 h). Obviously, there will be non-productive ("lost") hours of inactivity, required for maintenance, or caused by breakdowns. There are two commonly used approaches to determine the actually used hours of a machine.

A more *conservative approach* (often used by a custom molder) would be to use the following logic:

50 weeks of 5 days at 24 h equals 6000 h. By allowing 10% for lost time, there will be *5400 h/year* available per machine. This gives a "buffer" of 8760 - 5400 = 3360 h, in case that working more hours should be required. This buffer can come in handy when deciding on the number of cavities - for example when, by working a few extra days or even weeks, a smaller mold (fewer cavities) could be satisfactory for the expected production requirements.

A more *aggressive approach* (often used by "dedicated molders") plans on using more hours:

50 weeks of 7 days of 24 h = 8400 h/year, and by allowing about 10% for shutdowns, will assume *7560 h/year* available per machine.

Example:

A mold is needed to create 10,000,000 parts per year at a target cycle of 12 s. The machine will be operated for 5400 hours per year. The expected yield is 95% and the uptime is expected to be 90%. What is the minimum cavitation needed?

- PDA = 10,000,000
- t_{cycle} = 12 s
- UT = 90%
- Yield = 95%
- AOH = 5400
- Cavitation required = 10,000,000 × 12 ÷ (3600 × 5400 × 0.9 × 0.95) = 7.219

Eight cavities minimum should be selected

Example:

2 million pieces per year are needed, and the cycle time is estimated to be 30 s with annual operating hours of 7500, an uptime of 95% and a yield of 98%. What is the best cavitation to select for the mold?

- PDA = 2,000,000
- t_{cycle} = 30 s

- UT = 95%
- Yield = 98%
- AOH = 7500
- Cavitation required = 2,000,000 × 30 ÷ (3600 × 7500 × 0.95 × 0.98) = 2.38

Three cavities minimum but four cavities would be the most practical cavitation to use.

These calculations give us the minimum number of cavities required for a job, but they do not tell us whether it is the most economical number of cavities. For this, we must bring in another important consideration: the *machine hour cost* and part cost, which will be covered in Chapter 7 on cycle estimation.

8.3.2 Preferred (Practical) Number of Cavities

Once the minimum number of cavities has been established, we must consider the best cavitation for laying out the mold and for the mold manufacturing process, and which will provide the best flow of plastic to the cavities, whether in a cold runner or hot runner mold. For example, if the above calculations suggest that seven cavities are required, we will select the next larger preferred number, in this case, eight. Note that we will have to *recalculate* to see if the first selected machine will be able to handle the new requirements (clamp force, plasticizing, and injection capacity) for the new total of cavities (see Chapter 4 on specifying the machine for the mold).

Listed in Table 8.1 are the commonly preferred numbers of cavities, although under certain conditions, other numbers are also possible. Essentially, we consider either *rectangular* or *circular* layouts. Each has its own advantages and disadvantages.

For *circular* layouts, symmetrical layouts usually use 3, 4, 5, 6, 8, or 9 cavities, but virtually any number of cavities is possible. However, it could be more difficult to arrange well-balanced cooling and injection. The main advantage of a circular layout is that the runners can be kept very short, especially in two-plate molds.

There are also family molds (see Figure 8.2) in which a number of different products are molded in one shot, as is often the case for toy components, but also, occasionally, for technical products or commodity parts. In these cases, it is recommended to arrange the cavity layout so that the various projected areas are arranged as symmetrically as possible around the center of the mold. These jobs are usually not high-volume items, and a slower cycle can be easily tolerated.

Figure 8.2 Stack family mold layout for two sizes of rectangular container (Courtesy of Husky Injection Molding Systems Ltd.)

8.4 Assembly and Detail Drawings

The purpose of the assembly drawing (including the bill of materials discussed later) is to convey the intentions of the designer to the people involved in purchasing hardware and materials, assembling the mold, and, finally, operating the mold. The assembly drawing of the mold must contain all pertinent information, given in plan and section views and in notes, which are used to explain where the drawings alone could be ambivalent or misinterpreted. Once the assembly drawing is finished, there must be no doubt left about how the mold is to be built and operated. Today, most mold makers depend on machinists specialized in their trade, such as use of a lathe, milling machine, EDM, or other machine tools, to make the individual parts of the mold. The moldmaker will then do the fitting and assembly. These machinists need detail drawings, complete with tolerances and, if deemed necessary, other instructions such as hardness, plating, and finishing. These detail drawings are prepared from the assembly drawings.

8.4.1 Drawings and Views

"How many drawings are really needed?" is the question often asked by a mold designer. The answer is simple: enough to make sure that there is no possible misreading of a drawing. Too few views (or sections) means that in the best case, the machinist will interrupt their work to come and ask for explanations, which costs in lost time. In the worst case, the machinist will not ask, but will proceed in the wrong direction. This could become very expensive if an incorrectly made piece is not discovered until it reaches assembly, and then has to be remade, or it could cause major interruptions until a solution is found to use and repair the wrong part. Sometimes other mold parts have to be altered to make it possible to use an incorrectly made but expensive part. On the other hand, too many (often unnecessary) views make more work for the detailer, and can be confusing for the user of the drawings.

8.4.2 Arrangement of Views

Most molds are laid out by starting from a (significant) cross-section and then drawing to the right of it (as the mold would be when mounted in the molding machine) a view into the cavity half of the mold, that is, into the injection side (see Figure 8.3). The assembly drawing should show above this view words such as "Plan view into cavities".

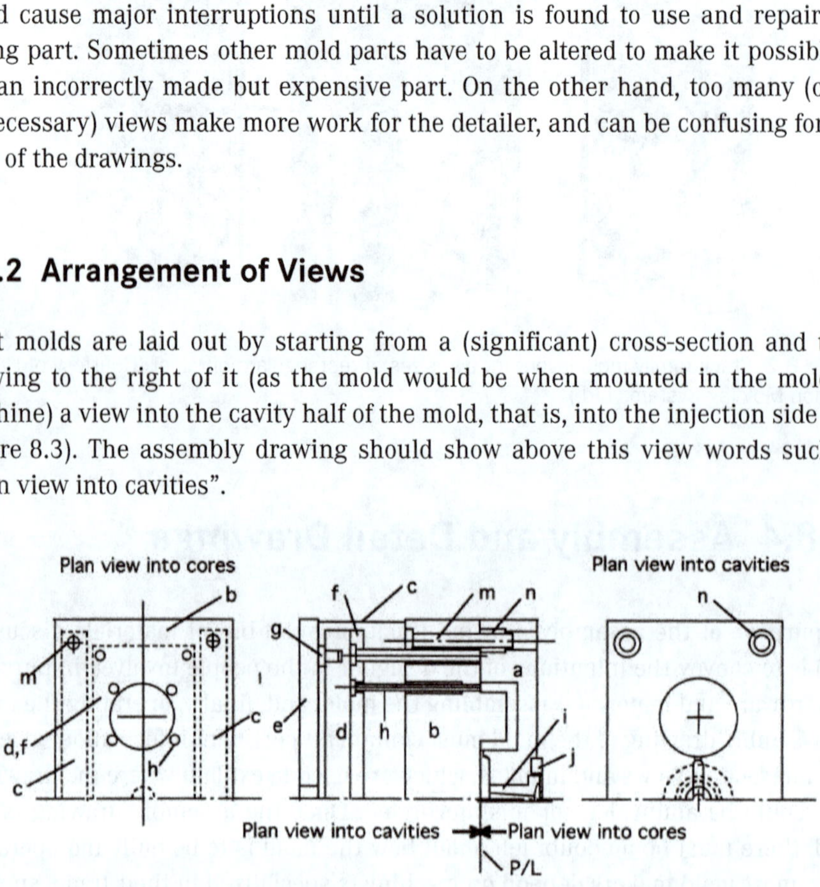

Figure 8.3 Arrangement of mold drawing layout: (a) cavity (plate), (b) core (plate), (c) parallels, (d) ejector plate, (e) mounting plate, (f) ejector retainer plate, (g) stop button, (h) ejector pin, (i) sprue bushing, (j) locating ring, (m) leader pin, (n) leader pin bushing

On the left-hand side of the section view, the core half is shown, as if looking into the direction of the core and the moving platen. The assembly drawing should show above this view words such as "Plan view into cores." The plan view drawings are made so that we see the parting line (plane) as visible, and all plates and mold features behind it as invisible lines. Additional full or partial cross-sections and/or plan views should be added (usually on separate sheets) only when they can add information to the views already shown. The drawing must always be drawn to scale, so that the various parts can be seen in proper proportion. "To scale" in this context means to draw to a selected, set ratio, preferably "to size" (1 : 1), or if this is not practical, in case of large products, smaller, often 1 : 2, or even 1 : 5, or larger, often 2 : 1, 5 : 1, or 10 : 1.

8.4.3 Notes on Drawings

Whenever it is impossible or cumbersome to specify some important information by using standard drawing techniques, a note should be added to express in concise but clear words what is intended. The note should be short, but not so short that it could be open to misinterpretation. Also, the drawing should not be cluttered with too many notes, and must show clearly what the notes apply to.

> A simple example of a note is to mark "Top of Mold" and "Operator Side" for reference orientation.

8.4.4 Additional Information on the Drawings

For more complicated molds, it is good practice to show also, on another sheet if necessary, separate schematic views of (1) all coolant circuits, (2) any air and hydraulic circuits, (3) any special electric circuits, and (4) a sequence of operation of the various mold functions, for example, at what point in the cycle ejection starts, and when air should be activated. This can have legal implications: complete and correct information will protect the designer from any possible future litigation, in case of an accident caused by the mold not being installed and operated as recommended by the designer. Note that the assembly drawings are part of the job and must be shipped to the customer together with the mold. Other drawings may be also sent to the customer if specified in the order.

8.5 Mold Layout and Assembly Drawings

Now the designer has all the basic information about the mold to be built, and can start to consider the basic layout of the mold.

8.5.1 Machine Platen Layout

The platen layout - including tie bar locations - of the machine (or machines) the mold will be used on should be shown first. This will determine the outer limits of the mold and where to place certain mold features. It will, for example, specify where coolant connections must not be located, or any planned auxiliary actuators outside the mold, latches, and so on. The mounting and ejector holes that will probably be used for the mold must also be shown.

8.5.2 Symmetry of Layout, Balancing of Clamp

For multi-cavity molds, it is important that the stacks are positioned such that the projected area of each cavity is as symmetrical as possible about the center of the machine, to ensure that all tie bars are loaded equally as the mold is clamped, thereby providing each cavity with the same preload to prevent flashing. This can present a problem with "family molds", where several different stacks or cavities with different projected areas are used in one mold. A small amount of asymmetry is often acceptable. With edge-gated, single-cavity molds, to balance the load, a pressure pad must be used opposite the stack location to simulate the force of a second cavity. In this case, the cavity itself will see only one-half of the clamping force of the machine. This is important for the selection of the best clamp size for the job. There is no such problem with center-gated, single-cavity molds.

8.5.3 The Views

Start with the significant mold cross-section(s), but always work with all views at the same time; that is, both the plan views of cavity and core will "grow" side by side with the cross-section. This prevents surprises arising when one view is far advanced and then it becomes apparent that it does not go together because another view shows some interferences. Show the selected hot runner hardware, if this is planned to be a hot runner mold. If it is a mold for which the hot runner section is purchased completely assembled by the supplier, show the interface points and dimensions only.

8.5.4 Completing the Assembly Drawing

Everything can now be shown in all views. It is not good practice to show the complete stack in every location, even though it is easy to do with a CAD system. It would make it difficult to read the drawings, especially if there are many other features in the stack. To facilitate the reading of the drawing, the stack should be shown in only one location of each plan view, and just its outlines in all other locations, for example, with heavy, dotted lines. However, important information such as the centers of coolant connections, screws, alignment features, and so on should be identified in all locations with small crosses and/or circles, which can then also be identified with a code, such as S1, S2 for screws and D1, D2 for dowels. Such codes will make it easier to read the drawing; they will be also important when completing the cooling line layout in the plates and the location of plate supports and large screws holding together the various mold plates, where applicable. Now also show the alignment features, the ejection system, the method of mold mounting and any connection (fixed or loose) with machine ejectors, and everything else needed by the detailers to produce the shop (detail) drawings.

At this time, also show where the outside of the mold must be marked (preferably die-stamped) to identify coolant and air connections. There would be a 1 IN, 1 OUT, 2 IN, 2 OUT, and so on, and AIR 1, AIR 2, and so on. The IN and OUT can be important for cooling, because in many cases it does make a difference where the cold coolant should go first (IN). For example, in the core of a container mold, it should first hit the area opposite the gate.

8.5.5 Bill of Materials (BOM) and "Ballooning"

This is also the time to specify the BOM so that all materials can now be ordered and be available when required for the machining operations and the final assembly. The BOM should specify not only the final sizes of steels and so on, but also the hardness of the finished mold part. This is important not only for the buyer, but also for the detailer of the shop drawings.

"Ballooning" is the identification of each mold part on the assembly drawing. Several methods are used, but the preferred one is to show balloons (circles or ellipses about 12–15 mm in size) outside around the drawings. Each balloon contains a number identifying each mold component, but only once, from stack parts to plates to screws, and so on. This number corresponds to a line in the BOM. Each balloon has a leader (line) connecting it with the part identified. Preferably, the balloons should be shown around the main cross-section of the mold or near partial sections; only if these locations would not be clear enough and could cause errors should they be shown in other sections or in the appropriate plan view.

8.5.6 Finishing Touches

Finishing information of the molding surfaces should also be shown - preferably with standard symbols - on the assembly drawing, for future reference, and to be used by the detailer when making the shop drawings. Cross-hatching should be used sparingly, only where it really helps to make the assembly drawing clearer. This also applies to detail drawings. This is also the time to show any notes on the drawing (see also Section 8.4.3).

Usually one "main" title block is shown, preferably on the drawing with the main cross-section; additional, smaller title blocks are on all other drawings. The title blocks identify the mold design office or the mold maker, the project number and drawing numbers, the designer (by name and initials), the checker, and the detailer, if applicable. It will also show any other information pertinent to the product, and will specify for which machines the mold was designed, the types of plastic, and any other information that deserves to be recorded for future use. Tolerances are not shown on the assembly drawings. They are strictly limited to the detail drawings. However, it is a good practice to show fits and clearances where they apply, but only if they are different from standard fits and clearances.

8.6 Inspection Features on Drawings

This book is not intended to be a manual for how to dimension a drawing for manufacturing purposes. However, there are some dimensioning practices used for making molds that are important for the designer to understand. These specific topics related to mold making are covered here as reminders to the designer.

8.6.1 Rollers or Roller Balls

Calculations for dimensions at external and internal taper over "micro" rollers or balls (which are very accurately ground rollers or balls) are shown in Figure 8.4. This is the classical method of accurately specifying the location of tapered surfaces. Today, larger mold shops have measuring machines, based on the same principle.

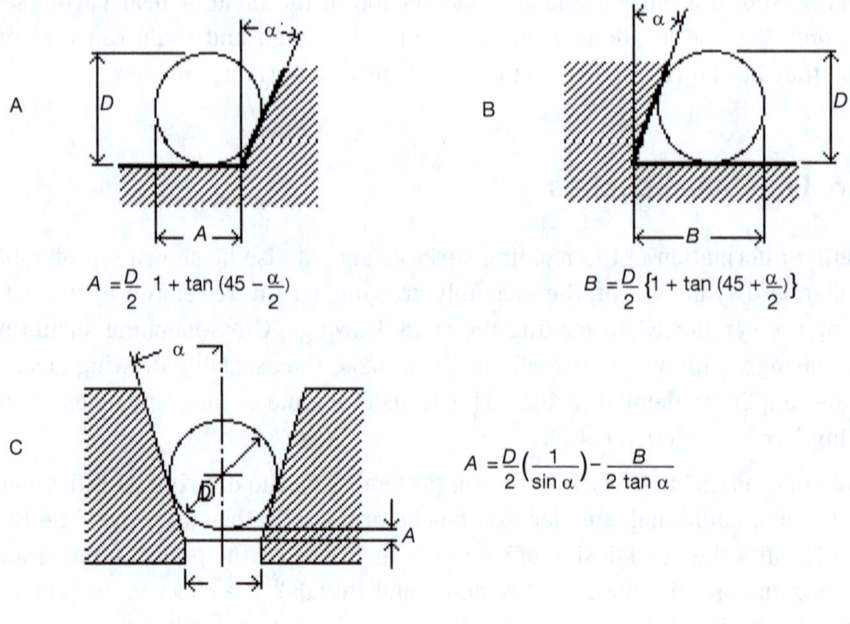

Figure 8.4 Calculating dimensions over rollers or balls: (A) External taper; (B) Internal taper; (C) Small internal taper

8.6.2 How to Inspect Tapers

8.6.2.1 External Tapers

Three alternative methods for inspecting tapers are shown in Figure 8.5. If accuracy is not important, specify taper with the *important diameter d* (in Figure 8.5, the small end was selected "important"), the angle α, and the height of the tapered section h (Figure 8.5A).

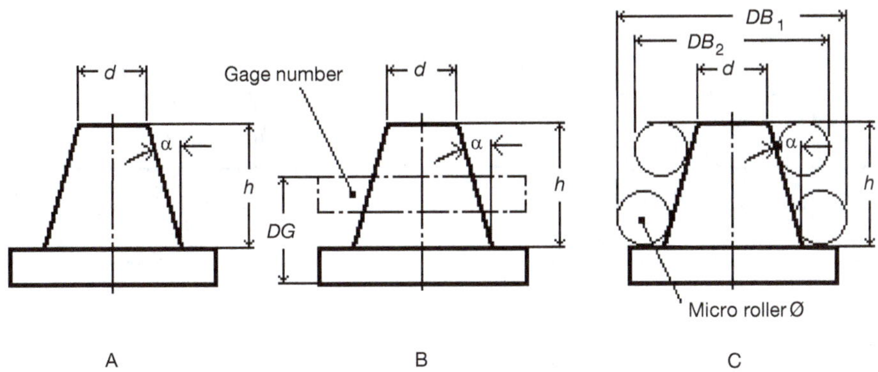

Figure 8.5 Dimensioning external tapers: (A) Simple dimensions when accuracy is unimportant; (B) Use of ring gage for high accuracy; (C) Use of micro rollers also for high accuracy

For high accuracy, use a ring gage. Specify (in addition to d, α, and h) the gage number and the dimension DG (with close tolerance) from the reference plane (Figure 8.5B). For high accuracy if no ring gage is available, specify (in addition to d, α, and h) the size of micro rollers and dimensions DB_1 and DB_2 (with close tolerances).

The plane on which the micro rollers rest must be well-defined; it may be a shoulder of the part, it may be a surface plate, or it may be built up with precision gage blocks (Jo-blocks) from a surface plate. The roller size must be selected so that the roller clearly touches both the plane and the taper.

8.6.2.2 Internal Tapers

Four alternative methods are shown in Figure 8.6 for dimensioning internal tapers. If the dimensions do not have to be very accurate, specify α, d, and (if required) h (Figure 8.6A). For high accuracy, use a plug gage (Figure 8.6B). Specify (in addition to d, α, and h) the gage number and the dimension DG (with close tolerance) from the reference plane.

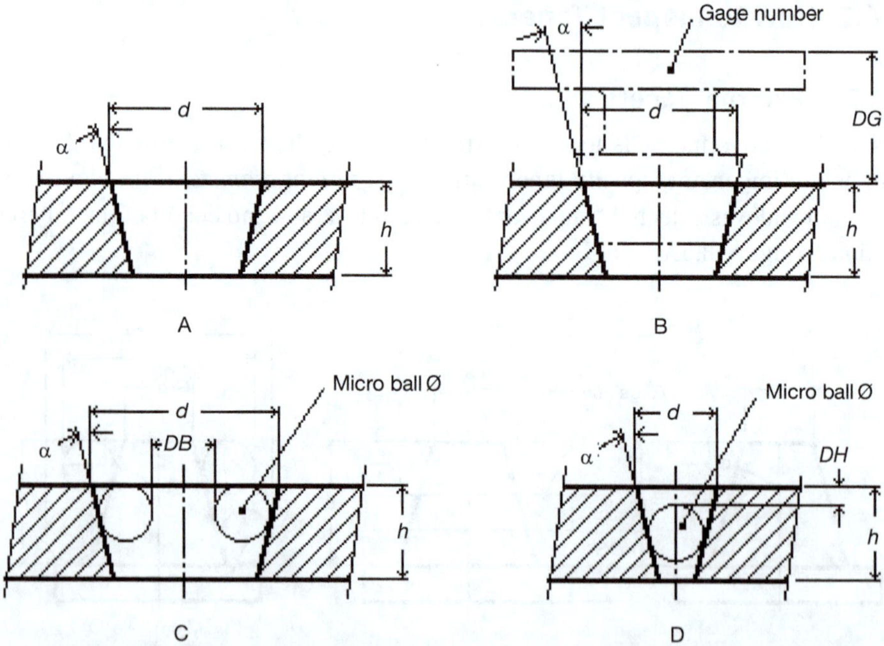

Figure 8.6 Dimensioning of internal tapers: (A) Simple but inaccurate; (B) Use plug gage for high accuracy. If no plug gage is available, then: (C) Use micro balls for high accuracy with large diameters; (D) Use micro balls for high accuracy with small diameters

If no plug gage is available, use micro balls (Figure 8.6C and D). For large diameters, specify (in addition to *d*, *α*, and *h*) the size of micro balls and dimension *DB* (with close tolerance). If the taper is longer, two balls must be specified, similar to alternative C for external taper. For small diameters, specify (in addition to *d*, *α*, and *h*) the size of the micro ball and dimension *DH* (with close tolerance).

8.6.2.3 Angled Surfaces

Four examples are shown in Figure 8.7 for accurate inspection of angled surfaces. For method A, angle *α* and either *h* or *w* are shown. Indicate the diameter of the micro roller and calculate dimension *DR*.

In method B, angle *α* and distance *b* from a ground edge are shown. Indicate the diameter of the micro roller and calculate dimension *DR*.

For a positive dovetail angle *α*, width *w*, and height *h* are shown (Figure 8.7C). Indicate the diameter of the micro rollers and calculate dimension *DR*.

In Figure 8.7D, negative dovetail angle *α*, width *w*, and height *h* are shown. Indicate the diameter of the micro rollers, and calculate dimension *DR*.

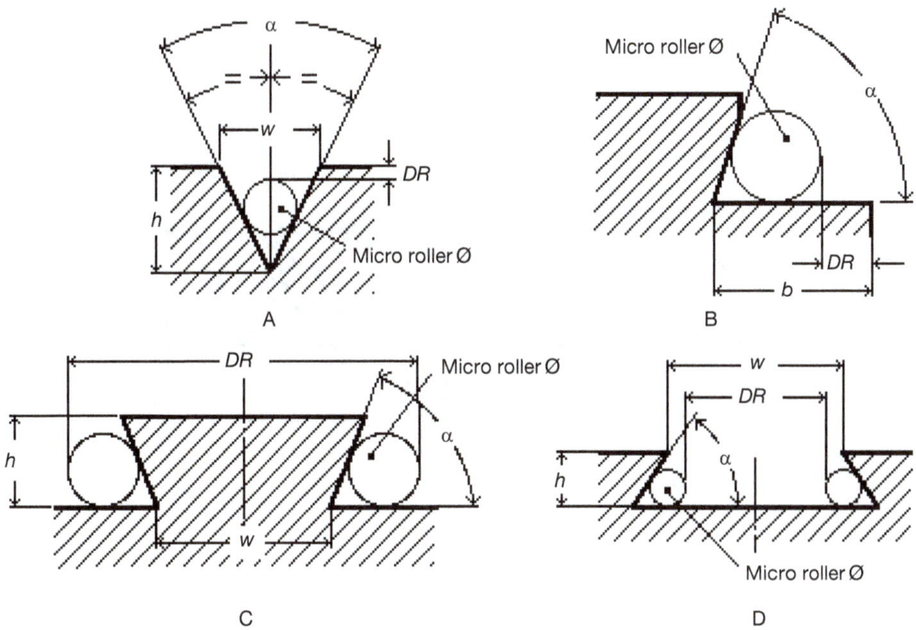

Figure 8.7 Dimensioning of angled surfaces using micro rollers to determine DR: (A) V-Shaped channel; (B) Angled surface; (C) Positive dovetail; (D) Negative dovetail

Note: Either the narrower (as shown) or the wider width of the dovetail can be specified, but not both, unless one is shown as a reference dimension.

8.6.2.4 Construction (Checking) Balls

Where accurate dimensioning (and machining) of angled surfaces or bores is required, an auxiliary hole ("tooling hole") and ball is introduced into the design. The hole is accurately dimensioned (*D* and *C*) from a reference edge (*E*), with close tolerance. A specially constructed ball with stem *A* (specify diameter *B*, and checking ball number) is then inserted into the hole, and the angled surface (α_1) is then dimensioned (*F*) with close tolerance and measured from the ball, as shown in Figure 8.8. An angled bore (α_2) would similarly show its centerline accurately dimensioned (*G*) from the ball. Specifications for some common standard checking balls can be found in suppliers' catalogs, or calculated.

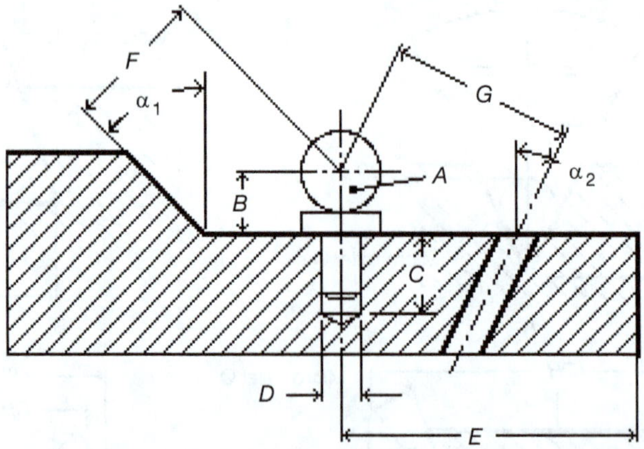

Figure 8.8 Dimensioning using a checking ball

8.7 Chamfers and Radii

8.7.1 General Information

Chamfers are generally specified on all outside edges for safer handling of the mold part; also, chamfers reduce the risk of stress-cracking at the outer surfaces. *Radii* are generally specified on all inside corners to reduce the risk of starting a crack as a result of stress concentration.

The size of desired chamfers and radii can be stated with a general note on the drawing. The relative size of matching chamfer and radius is shown in Figure 8.9 and Equation 8.2. This ensures that the parts will seat properly on the flat surfaces, not on the radius.

Minimum radius for matching chamfer:

$$C_{min} = R + (2 \times \text{tolerance}) \tag{8.2}$$

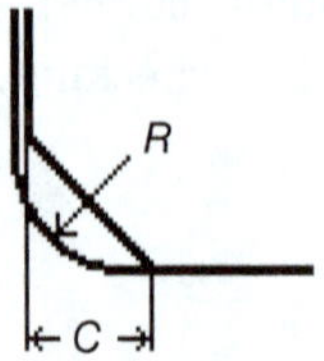

Figure 8.9
Suggested relative size of matching chamfers and radii

All mold plates must be chamfered at least 1.5 mm × 45° on the outside.

All edges of counter-bores and all mold part edges should be chamfered, unless a chamfer would not be permissible because of the product design or the path of plastic flow, to facilitate assembly and to reduce the risk of personal injury (cut fingers) when handling the parts. Any edge which will be dragged over an O-ring or other soft seal on assembly must be indicated with a notation to "round-off and polish".

The *absence* of chamfers must be indicated with the notation "sharp" and a leader pointing to the affected edge. An inside corner must never be sharp. If a sharp corner is necessary for the product design, create it by using separate mold parts (inserts).

Some product designs show a sharp corner (e.g., at the base of a product) where a small radius is perfectly acceptable for the function of the product. In such a case, the designer should request a change in design to permit a small radius; the benefit to the customer would be a longer mold life (less risk of cracking) or savings by avoiding the use of inserts (Figure 8.10).

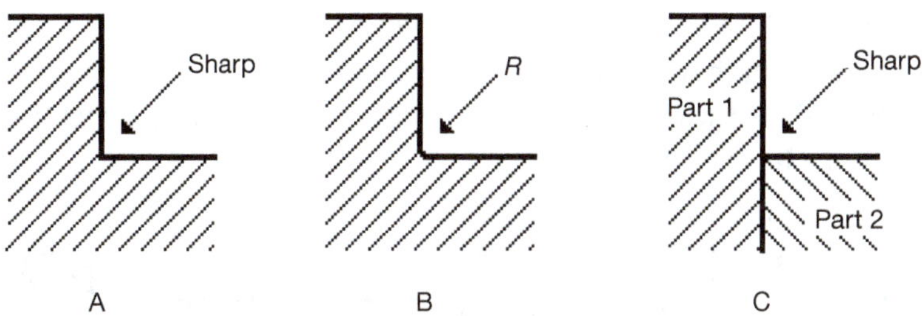

Figure 8.10 Inside corners of mold parts should not be sharp (A): either there should be a radius (B), or the corner must be created by using an insert (C)

8.7.2 Specifying Chamfers

The general method for specifying a chamfer is to show its angle and length (Figure 8.11A). A shortcut method that can be used with 45° angles is illustrated in Figure 8.11B and C. In this case, simply indicate length and angle, with a leader pointing to the chamfer.

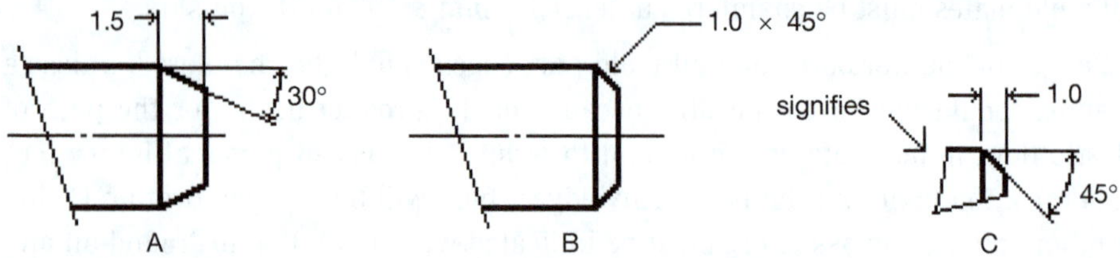

Figure 8.11 Specifying chamfer: (A) show angle and length with dimension lines; (B, C) with 45° chamfers, simply indicate length and angle with a leader

8.7.3 Radii in Corners

Every inside corner, particularly in hardened steel, creates a severe hazard because of the stresses raised when forces are applied. Practically all mold material failures start from cracks originating at sharp corners.

Grinding creates very fine scratches. In hardened steel, in areas where severe loads are expected, such corners must, therefore, not only be specified "ground" but also "polish", to remove any scratches created in grinding.

If it is not possible to have a radius in a corner, as suggested in Figure 8.12, the part must be designed with a "hidden" or recessed radius. Typical high-risk areas where radii must be provided are:

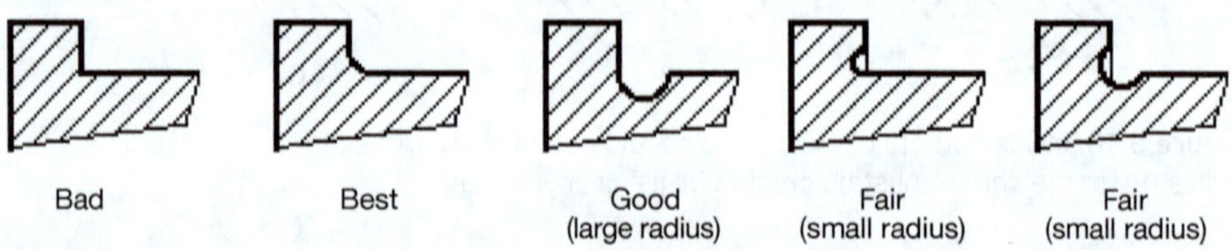

Figure 8.12 Examples of radii in corners

- Any change in a diameter of a mold part
- Inside snap ring grooves
- Inside O-ring grooves
- Roots of cut threads
- Inside counter-bores at the seat of a screw head
- Heels.

9 Mold Shoe Design

A mold shoe is the total of all mold plates making up the mold, including screws and alignment features, but not including the stack, which is the arrangement of all mold parts that touch the injected plastic - typically, the cavities, cores, any inserts in either of them, ejectors, strippers, side cores, and so on. In simple molds (not necessarily low-production molds) the cavities and cores can be machined right into the mold plates. A decision on which way to proceed with the mold shoe should be made only after the product drawing is carefully studied, and one should never lose sight of the expected productivity of the mold. There are several choices for the designer:

1. **No mold shoe used.** The mold may consist of only one plate for the cavity and another plate for the core, with both cavity and core machined right into these plates. Ejection is facilitated by air valves built directly into the core plate. The entire mold, then, consists essentially of only two parts, plus alignment features and air valves. Cooling channels are machined into the plates with the molding surfaces.

2. **Standard mold shoes.** Mold shoes can be bought from mold maker supply houses such as DME, PCS, Hasco, etc. (see Figure 9.1). They have a large selection of standard sizes, with or without leader pin alignment, and with or without ejector plates. All plates are machined and ground square, and are ready for adding the required mold features. Due to the large and expensive machine tools required to manufacture these plates, many mold shops prefer to buy these ready-made mold parts, and instead specialize in the making of the stacks and doing the final mold assembly. These plates are usually available in several qualities of steel:

   a) The first type is an inexpensive, “mild steel”, which is soft, with low strength, and little wear resistance. It is suitable only where the expected forces and wear in the mold are small enough that the steel will not be damaged, for example, by the clamping pressure on a P/L that is too small, or by the hobbing effect, which is the pushing of a supported small insert into a mild steel backing. Also, since mild steels have a low tensile (and compressive) strength, they may permanently deform if loaded beyond their yield point. To protect against corrosion, these plates can be painted with a rust-inhibiting paint.

b) Another common steel supplied for plates is P20. It is treated to a Rockwell hardness of approximately Rc28–35. P20 is produced especially clean, that is without dirt enclosures, which could be detrimental if they appear on a molding surface. These plates are more expensive but cost much less than so-called mold steels. They are very suitable for cutting the cavity or core right into the plates. This is of special advantage for large products where the cost of mold steel would be very high. Mold steel is always supplied very soft (about the same as mild steel), for easy machining. It must be hardened and ground after machining, which represents an additional, considerable expense. In high-quality molds, the stack parts are usually made from steels such as P20PQ (plastic-quality) for large products, and from mold steels for smaller products. To protect these mold plates from corrosion, they are often treated with ENP (electroless nickel plating). Note that this dip process does not full cover the cooling lines drilled into the plates.

c) In high-quality molds, also, both the mold shoe and the stack parts are made from stainless steel (SS). The larger mold parts are then machined from pre-hardened steel, and smaller parts from stainless mold steels. This is helpful in humid climates to prevent rusting of the mold shoe, or where the plastic is corrosive and could attack the stack parts. The higher material cost can often be justified with savings in mold maintenance.

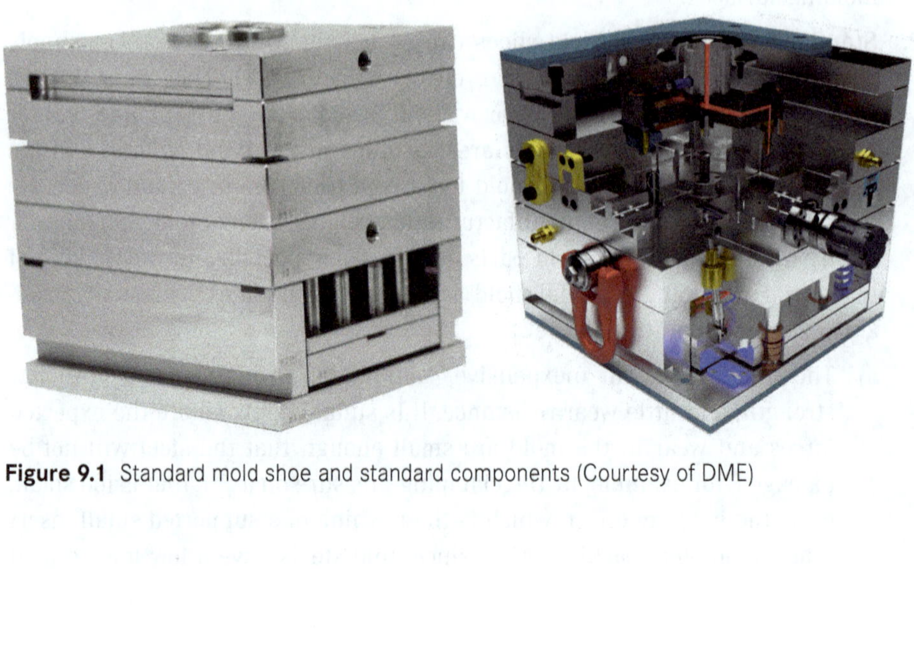

Figure 9.1 Standard mold shoe and standard components (Courtesy of DME)

3. **Home-made mold shoes.** The mold shoes can be made in-house from raw steel plates, which the mold maker can buy from the steel mills or dealers (Edro, Schmoltz and Bickenbach, Krupp, etc.). The mold maker may keep certain plate sizes and thicknesses in stock, and cut and machine them to size as needed. This requires much plant space, heavy lifting and storage equipment, and accurate milling and grinding machines. It is an economic decision that may be different from shop to shop - whether to make the plates or buy them as standard plates or as complete mold shoes. In any case, the same choices of steel apply. Note that often, such in-house made mold shoes or plates are built to the dimensions listed as standard parts by the hardware suppliers.
4. **Special mold shoes.** This applies mostly to special molds for which no suitable standard sizes are commercially available, and to very high-production molds, where the mold shoe is built around the stacks, and optimum layouts are used for all mold features, rather than the stacks being fitted into the space available in standard molds. Some mold makers specializing in certain areas (preform molds, unscrewing molds, etc.) create their own mold shoe standards. In high-production molds, the mold shoe too is usually made from pre-hardened steel. Also, often pre-hardened stainless steel is used for such molds.
5. **Universal mold shoes.** For low-production and relatively small products, a universal mold shoe offers another solution for making a relatively low-cost mold. Universal mold shoes are essentially standard-size chases that are constructed so that different stacks can be easily mounted into them. The mold maker concentrates on making the stacks in sizes and to rules specified by the maker of these universal mold shoes. Mold features such as runners, cooling, and ejection are usually not as efficient as in a mold specifically designed for a product, and the mold will not cycle as fast; but, for the small quantities required this is no problem, and the mold is much less expensive than a complete mold. This makes a lot of sense, especially if a large number of such "inserts" are used or foreseen.

9.1 Mold Plates

A mold plate is any plate required to build a complete mold. The assembly of mold plates (without the cavities and cores) is called the "mold shoe". Figure 9.2 shows the main plates that make up a mold shoe, although many more can be used in complex molds. As a general rule, mold plates have no molding surfaces cut into them. However, in very large molds, or in molds for low production, cavities and occasionally cores are cut directly into the mold plates. In these cases, the quality of steel may be important. It must have good polishability and be free of impurities, which might leave marks on the product.

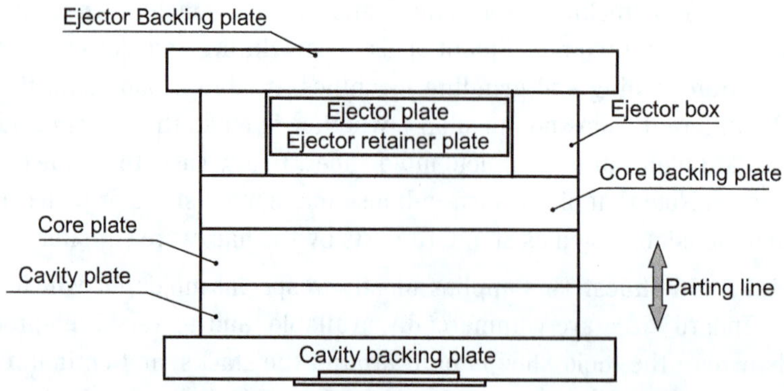

Figure 9.2 Mold plates as part of a mold shoe

In order for mold plates to be properly handled during assembly and maintenance, each plate must have threaded lifting holes on each side and both faces. Each plate must also have pry slots on the edges of the plate to allow for the plates to be separated using pry bars (Figure 9.3).

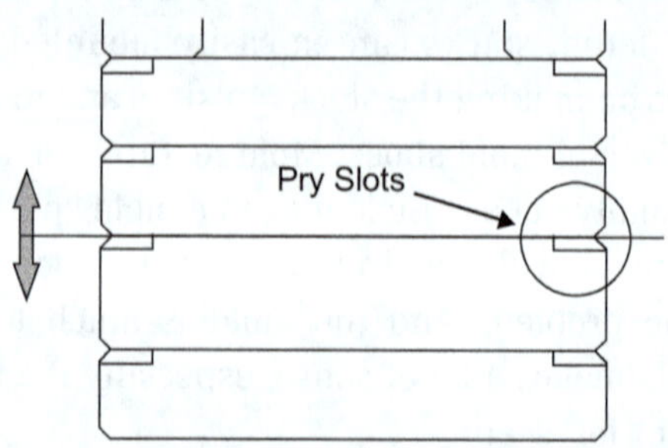

Figure 9.3 Pry slot locations (corners or along edges)

Pry slots should be large enough for standard pry bars to be inserted.

> Typical dimensions of pry slots are 3-4 mm deep ($\frac{3}{16}$" deep) and 25 mm in length (1").

All plates should have chamfers on all edges to eliminate any sharp edges.

> All mold plates should have been stamped with a job number or mold number to ensure that it is identifiable at all times.

9.2 Mold Hardware

Hardware items include leader pins, bushings, ejector pins and sleeves, screws, and many other mold parts that are required for the mold. They are all listed in catalogs issued by the various mold maker supply houses. They are mass-produced, using high-quality materials, and machined to very close tolerances. It is always more economical to buy these parts rather than to attempt to make them in-house. Also, a good mold designer will never modify these products, except for cutting ejector pins and sleeves to length. Also, screws used in molds must never be modified, not even their length. There is always a way to make the design use a standard size, often by just changing the depth of a counter-bore for the screw head. Any modification of a screw will reduce its strength, and it will be difficult to replace in the field.

Every mold requires alignment. The design of alignment systems for the mold is covered in more detail in Chapter 13. The normal method of aligning the two mold halves is with purchased leader pins and bushings (Figure 9.4). In order to prevent any two items from being assembled incorrectly, the mold design will use a simple modification to prevent incorrect assembly. Typically, one of the alignment devices (such as a dowel or leader pin/bushing) is either moved slightly in position relative to the pattern of the others, or one of the leader pin bushings is made slightly smaller. In this way the mold cannot be assembled incorrectly. All of these alignment elements must be held in position tightly and free from rotation to provide the intended purpose of alignment.

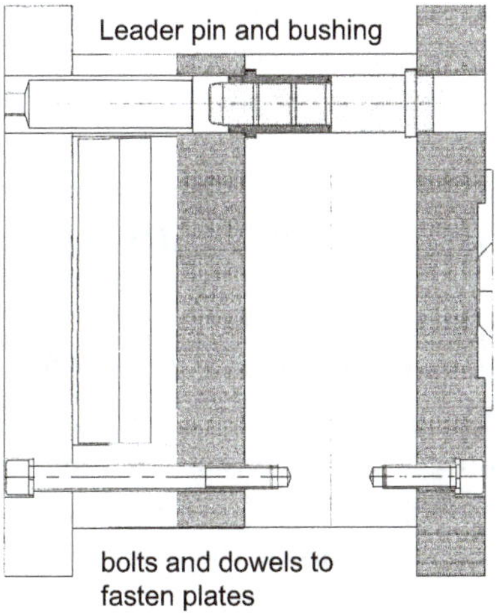

Figure 9.4 Leader pin alignment of the mold halves

Note that when designing any of these elements there must ALWAYS be a way to easily remove them with standard tools available on the shop floor. This can be with extraction holes, pry slots, and punch-out holes or other means. The last thing any designer wants is for an assembler to visit them at their desk and ask them to come out to the shop floor to explain to them how the mold can be disassembled.

9.3 Location of Fittings, Hoses, and Cables

Fittings, hoses, and cables are necessary to feed the mold with required services such as water, air, and electricity. Management of these services should be considered during mold design to make management during use as easy as possible. Electrical connectors and electrical items should come into the mold at the top if possible. They should not be located at the bottom, as water could more easily enter the connectors.

Water lines should be located on the bottom (preferred) and sides of the mold. If they are on the sides of the mold, they should not obstruct part removal or servicing of the mold. The fittings on the mold should be protected from damage by either recessing them or having the mold mounting plates (core backing plate and cavity backing plate) larger in width than the rest of the mold (such as shown in Figure 9.4). This will ensure that when the mold is being handled or mounted into the machine, the fittings do not get damaged.

If possible, the water lines should be internally manifolded in the plates and there should be only one in and one out per plate. This not only eases installation and removal but prevents potential errors when plumbing the mold.

9.4 Forces Affecting Mold Shoes

The plates in the mold shoe are stressed by the clamp force and the injection pressures of the plastic during the injection and hold phases of the cycle. The mold shoe is subjected to mainly compressive and bending stresses. The mold plates must resist these forces to support the cores and cavities. If the designer is unsure if the mold plates will resist these forces, they must calculate the projected deflection and stresses. In order to make these calculations, some information must be assumed about the forces on the mold, such as clamp tonnage and injection pressure. The injection pressures can be estimated using a variety of computer-aided engineering (CAE) software available on the market. The pressure inside the cavity depends on many factors, such as flow resistance within the runner system, plastic viscosity, melt temperature, cooling temperature, injection speed, gate size, etc. Doing a simple CAE simulation will give the designer a good idea of the levels of injection pressure that the mold shoe must resist. Note that the most crucial data needed to model this accurately is accurate material data for the plastic to be injected.

The following sections provide the designer with some basic calculations on deflection and stresses that can be used to assist them in selecting the appropriate plate thickness and designs. However, in general a plate should be made as thick as possible within the constraints of the shut height of the machine.

9.4.1 Deflection of Mold Plates

It is important that the designer becomes familiar with the basics of strengths of materials to ensure that the mold shoe will withstand the forces of the injection molding process. Some important considerations and guidelines for sizing the mold plates are as follows:

1. All plates will deflect under load but the deflection must be small enough (< 0.05 mm or 0.002") that it will not affect the alignment of the mold or cause core shift.
2. If a plate is calculated to deflect more than an acceptable amount, it is preferable to make the plate thicker rather than adding another plate.
3. Only the modulus of elasticity (*E*) affects deflection. The hardness of the steel has no influence on deflection. If a plate is deflected under load below its yield strength it will return to its original shape after the load has been removed. Because the yield strength of harder steel is higher than that of a softer steel, the same load will create a permanent deflection in the softer steel, while the harder steel, under the same load, will return to its original shape after the load has been removed.
   All steels have approximately the same modulus of elasticity *E*, within ±5%, with the exception of tungsten carbides, which may have twice the stiffness of tool steels. Exact figures are given with every material specification sheet. *E* for steel is approximately 200–210 MPa (30,000,000 psi).
4. To prevent failure resulting from cyclical loads, the maximum allowable stress in a plate must *not exceed the fatigue strength* of the plate material at its operating temperature.
5. It is very difficult to calculate the exact deflection of a plate supported in several locations. But for practical purposes, this is usually not required. To ensure that the mold is stiff enough, two assumptions can be made using the worst-case scenario:
   a) Assume single-point loading (at the center of the cavity) rather than spreading the load over an area

b) Assume that the plate is a simple beam, freely supported at two ends only.

With these two assumptions, the estimated deflection is much more severe than arrived at with other, more accurate methods, and provides additional safety against excessive deflection.

6. It is stronger to cut out a plate in the shape required (e.g., for the ejector box or hot runner manifold), rather than using rectangular blocks and adding pillars to create the desired box shape. While it may require removal of considerable amounts of material, it is in the long run less expensive, and the mold will be easier to assemble than if it had many loose parts (pillars, etc.) held together by screws. This does not mean that support pillars must never be used. The decision whether to "hog out" a plate or to provide a simple frame (with parallels and/or large cut-outs) and support pillars rests ultimately with the designer, with input from production engineering. Another advantage of a solid but "carved out" plate, rather than an assembly of many pieces, is that the cooling and air channels are easier to locate, in the absence of the many screw holes.

9.4.2 Calculation of Plate Deflection and Stress

Deflection, Y, of any mold plate should never exceed 0.05 mm (0.002"), or even less for high-performance molds.

Deflection can be calculated using Equation 9.1 and Figure 9.5:

Formula for plate deflection:

$$Y = \frac{P \times L^3}{48 \times E \times I} \tag{9.1}$$

where:
Y = deflection (mm)
P = total load (N)
L = unsupported length (mm)
E = modulus of elasticity (200,000 MPa)
I = moment of inertia ($b \times h^3/12$)
h = beam thickness (mm)
b = beam depth (mm)
Notes:
- Make sure that units are consistent (metric or imperial, etc.)
- The formula assumes a solid plate thickness.

The maximum stress S_{max} experienced by the plate under load P can be calculated using Equation 9.2.

Equation for maximum stress:

$$S_{max} = -\frac{P \times L}{4 \times Z} \tag{9.2}$$

Here, Z is the section modulus, which must be based on the same "unit width" of 1, as explained for the moment of inertia I. Z can be calculated using Equation 9.3:

$$Z = b \times h^2 / 6 \tag{9.3}$$

S_{max} must be equal to or less than the allowable fatigue stress for the steel.

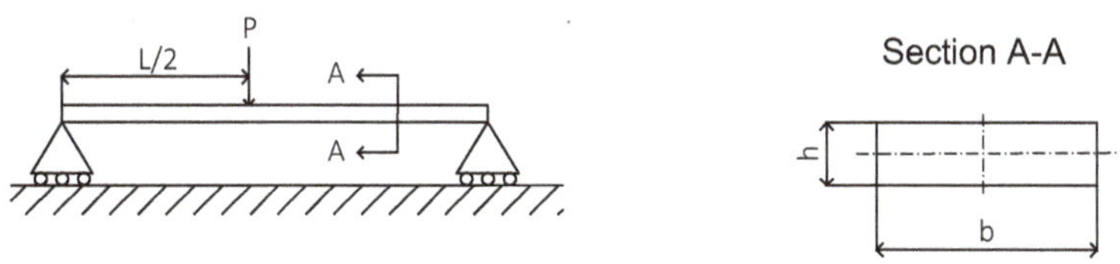

Figure 9.5 Schematic of a simple beam deflection calculation

With the above two very conservative formulae, the designer can easily check whether the plates are thick enough for a specific mold. Minimum deflection results in longer mold life and greater accuracy of products, and is particularly important for thin-walled products.

Example:

The projected cavity area for a rectangular part to be molded is 2" by 4" and the injection pressure is assumed to be 10,000 psi. The concepted plate is 8" wide, 2" thick and the span between supports is 4". The allowable stress on the plate is $S_{allowable}$ = 30,000 psi. What is the plate deflection and is the stress within the allowable limits?

Calculate the force P on the plate:

P = projected area × pressure = 2 in × 4 in × 10,000 psi = 80,000 lb.

From Equation 9.3 the section modulus is:

$Z = (8 \times 2^2) \div 6 = 5.333\ in^3$

From Equation 9.2 the maximum stress in the plate is:

$S_{max} = (80{,}000\ lb \times 4\ in) \div (4 \times 5.333\ in^3) = 15{,}000\ psi$

Since 15,000 psi < 30,000 psi, the S_{max} is acceptable.

From Equation 9.1 the maximum deflection is:

$I = (8 \times 2^3\ in^3) \div 12 = 5.333\ in^4$

$Y = 80{,}000\ lb \times (4\ in)^3 \div (48 \times 30{,}000{,}000 \times 5.333\ in^4) = 0.00066\ in$

The stress in the above example is acceptable, and the deflection within the limit of what is allowable.

9.5 Selection of Materials for Mold Plates

See also Chapter 15.

In all plates used for molds, the main requirements are:

- Good machinability (less machining time)
- Adequate hardness for the requirements of the mold
- Homogeneity of steel (e.g., no hard spots detrimental to machining operations)
- Good corrosion resistance
- Ease of welding for repairs
- Easy availability
- Cleanliness (free from enclosures or dirt) if used for cores or cavities
- Low cost.

An improvement over the mild steels is the use of a "machinery" steel such as AISI 4140, supplied prehardened to about 260–320 BHN (equivalent to about 130,000–170,000 psi tensile strength). This steel meets many of the above requirements. However, it often contains impurities, and rusts easily.

A better steel is a machinery steel that was developed especially for molds, P20 or P20 PQ ("PQ" stands for Plastic Quality). It is essentially an equivalent of AISI 4140 but much cleaner, and can be used also for cavities and cores. P20 PQ meets most of the requirements, but it too rusts easily. To prevent corrosion it must be plated, preferably with the electroless nickel process (ENP) to ensure that the inside channels are also protected. Cost for plating adds to the cost of the steel. Some molders just paint the shoe from the outside with oil paint, but this does not prevent corrosion in the cooling channels.

> Thickness of ENP is normally 0.005 mm (0.0002"). Note that the thickness of coatings must be accounted for in all critical plate dimensions.

Stainless steels such as AISI 420 F Mod., supplied pre-hardened to about 270–340 BHN, do not require protection against corrosion and meet all the other requirements as well. Although the cost of stainless steel is usually considerably higher than that of P20 PQ, by selecting certain standard plate sizes and by committing to buy large quantities, it is possible to reduce the steel price greatly. Also, there are savings because no additional plating is required. To take full advantage of low steel prices, the designer must specify sizes that can be cut from standard mill sizes.

9.6 Guiding of Moving Plates

All plates that move during operation of a mold (opening, closing, sliding) must be properly guided using robust pins and bushings designed for this task. Independent plate guiding must ensure accurate positioning of the non-supporting and guiding parts of the mold such as ejector pins, blades, stripper rings, etc. Whenever one or a number of such components are located on a plate, they must have the ability to float or move, so that each such component can assume its own proper location with respect to the matching mold part in which it slides. If there is misalignment with no float, there is danger of severe side forces, resulting in excessive wear on these mold components, and possible jamming of the motion and breakage. If a sliding mold component is of substantial mass, it must be guided.

Guiding of moving plates is usually achieved by leader pins, support pillars or gibs.

Guiding of plates by their actuators (air pistons, cams, etc.) is not acceptable. Such plates must be supported by independent guides, so that there is no side load on the actuators (as a result of misalignment or the effect of gravity) which could cause premature wear of seals or of the cylinder walls.

> In general, a plate should be supported by three or (better) four pins and bushings, although only two may be acceptable for smaller plates or components.

9.6.1 Gibs

If gibs are used to guide slides, their active length (E) should be at least as large as their distance from each other (W), particularly if there is only one actuator (piston, cam, etc.) to move the sliding plate (Figure 9.6). If $W > E$, two identical actuators should be used, one of each near each one of the gibs, to avoid jamming and premature wear.

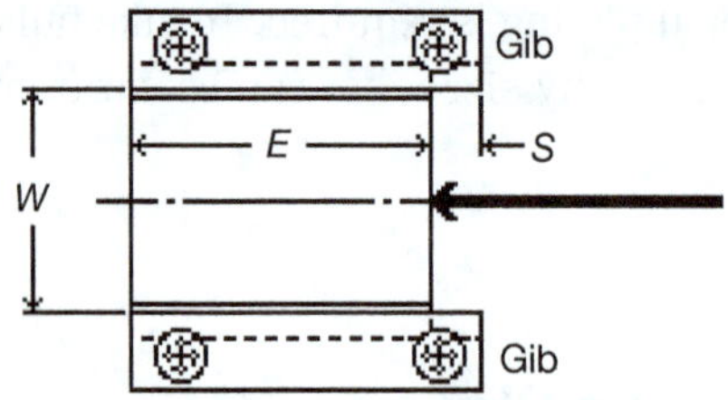

Figure 9.6 Actuator guided by gibs on each side

9.6.1.1 Pillars and Bushings

Figure 9.7 shows guiding for an ejector plate using pillars and bushings. Usually, there should be four such guides. Note that the ejector pins in Figure 9.7 are floating (i.e., free to move sideways) both in the ejector retainer plate and in the core backing plate. The float is usually about 0.5 mm (0.020") radially. Under the heads, there should be a small axial clearance of about 0.025–0.050 mm (0.001–0.002").

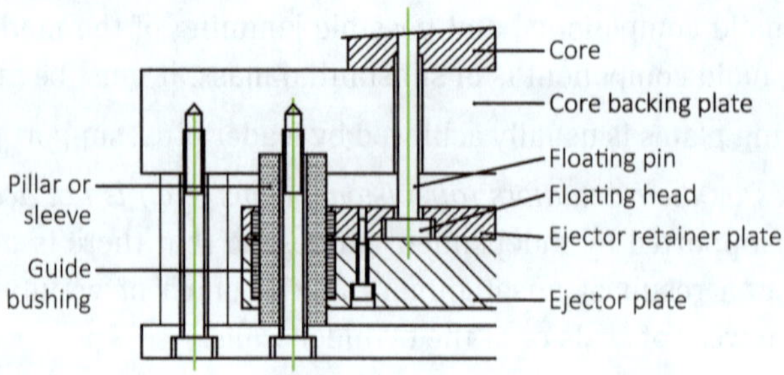

Figure 9.7 Guiding of the ejector plate by pillars and guide bushings

It is important to align the retainer plate with the core backing plate to facilitate mold assembly, and with the ejector plate, to ensure that the ejector pins, etc., maintain their concentricity within the holes of the ejector retainer plate and maintain their float. Alignment with the core backing plate can be done, as shown in Figure 9.7 with short bushings with a radial clearance of about 0.13 mm (0.005") in the guide pillar, so as not to fight the basic alignment provided for the ejector plate. Alignment with the ejector plate also is achieved.

9.6.1.2 Dowel Pins and Bushings

Figure 9.8 illustrates a good and simple design for guiding plates using dowels and bushings. Note that any bushing used for guiding a moving plate must be retained on both ends by recesses, adjoining plate(s), D-clamps, etc. For ease of assembly and disassembly, a removal hole should be drilled behind the dowel so that it can be punched out if needed. A press-fit blind dowel hole is not advised in any situation. In order to ensure that there is enough float in the ejector retainer plate, the following design can be used:

- Two dowels, tight in the core backing plate and loose in the retainer plate. During assembly, these dowels enter the core backing plate while the ejector pins are inserted. The ejector plate is then screwed tight to the retainer plate.
- The bushings in the ejector plate can reach into the retainer plate, and the outside of the bushing is loose with a clearance of 0.13 mm (0.005") in that plate.
- Two dowels, loose in the retainer plate and tight in the ejector plate, ensure that the retainer plate will be in the correct location.

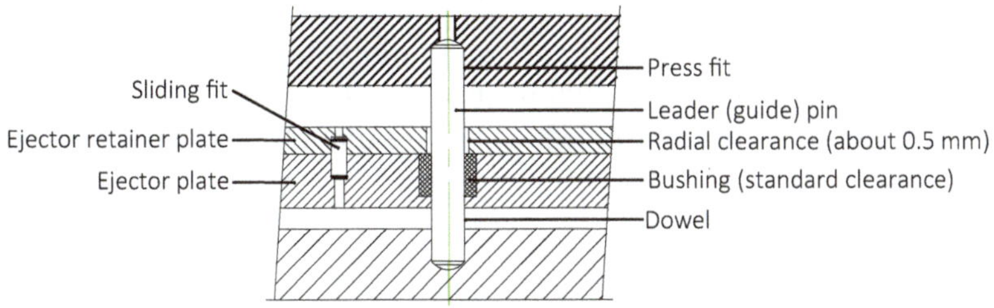

Figure 9.8 Dowel alignment with retainer and ejector plate

9.6.1.3 Leader Pins with Piston and Stroke Limiters

Figure 9.9 shows a typical moving plate with actuator (air cylinder), stroke limiter, and guide pin and bushing. There should be preferably four actuators at the corners (as far apart as possible), to minimize the possible uneven movement and jamming of the plate.

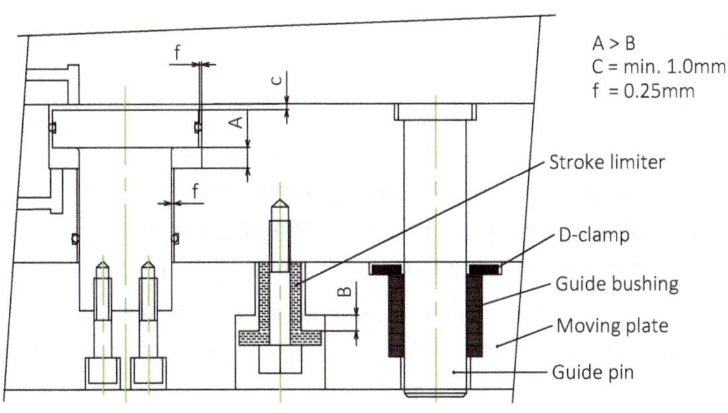

Figure 9.9 Design of a moving plate guidance system with piston and stroke limiter

Note that there must always be a minimum clearance (C) of at least 1.0 mm (0.04") between the top of the piston and the end of the cylinder to allow air to enter and generate force. Alternatively, grooves can be used to allow air to enter the piston area to generate force. Also, the piston should not be used to limit the stroke of the plate. Separate stroke limiters (preferably four) must make sure that the motion stops on these limiters and not on the pistons (A must always be larger than B in Figure 9.9).

The clearance f between air piston and piston rod from surrounding plate must be 0.25 mm (0.010"), regardless of the recommendations by the seal manufacturer, to prevent rubbing and galling between piston, rod, and plate. Note that the diameters of the seal grooves in the plate (for the rod seal and in the piston) and the corner radii must be in accordance with seal specifications to avoid premature failure of the seals.

Guide pins and stroke limiters should be placed as near the actuators as possible. There should be preferably four guides to ensure that there is no side load on the piston rods and seals.

■ 9.7 Mounting Holes and Clamp Slots

The mold is mounted to the platens via the mold plates (typically core and cacvity backing plates). There are two methods to fix the mold to the platens: mounting holes or clamp slots. Chapter 18 shows a variety of mold design examples with both clamps and holes for mold mounting. Depending on the configuration of the machine and the position of the threaded holes in the platens, one or the other could be used. There are two major standards used in the industry for mold mounting and ejector patterns on the machine. They are the SPI standard (imperial) and the Euromap standard (metric). Mounting a mold with screws is always preferred to using clamps.

The mounting screws prevent the mold from falling down or slipping on the platen faces, and resist the opening forces on the mold. Mold opening forces (also called mold break forces) are usually about 10% of the clamp tonnage available on the machine. The screws must be able to resist this in a cyclic loading application.

The screw or clamp must sit firmly on the mold shoe and there must be sufficient clearance for the operator to use an Allen key and torque wrench to tighten-in the screws.

The mold order confirmation must contain all necessary information on the machine mounting and ejector hole locations. It is critical that these are checked and verified before the mold is designed, to avoid significant redesign and rework if the drawings are released to the shop.

9.8 Locating Rings

A method to align the mold in the machine must always be provided. In the majority of cases this is done using a standard-size locating ring. The locating ring, mounted at the center of the mold on the stationary side, aligns the mold in the machine with the extruder. Standard locating rings should be purchased from mold component manufacturers. The locating ring must match the hole in the machine platen, or the mold will not properly align with the machine extruder or the mounting holes in the platens. On large molds that need to be installed in the press in two separate halves, a locating ring on both the cavity side and the core side may be used. Examples of this are shown in Chapter 18.

Occasionally, for specialty applications the mold is mounted on dowels in the platens. It is important as part of the order confirmation to ensure that the mounting method is clarified.

9.9 Mold Handling - Lift Holes, Lift Bars, and Latches

All molds must have lift bars and latches designed to safely allow mold assemblers and service personnel to handle the mold. In addition, each plate or insert in the mold must have lifting holes into which eye bolts can be threaded into. This means all four sides of the plates and the faces of each plate. The location of the threaded holes must be in the center of gravity to make sure they can be lifted level. If this is not possible then multiple holes and a sling between the eye bolts should be used.

9.9.1 Lift Bars

If it is not possible to provide an additional lifting hole to hoist the whole assembly, because of interference with another hole, etc., a lifting bar must be introduced with holes spaced to pick up at least two lifting holes of the parts making up the assembly. The lift bar then has a threaded hole for an eye bolt above the new center of gravity (CG).

The lifting holes should be as close together as possible, to reduce the span on the lift bar, which will bend under the load. If the holes are closer together, the bar can be lighter than if the holes are farther apart.

Figure 9.10 shows a simple calculation for the proper lifting hole location. In this example, this location would be right in the area where two plates meet; therefore, a lift bar must be provided. Occasionally, a lift bar will span the whole length of the mold (particularly in stack molds), and the holes farthest from the center are used to hold the complete mold together. The lift bar must be strong enough to support the whole mold.

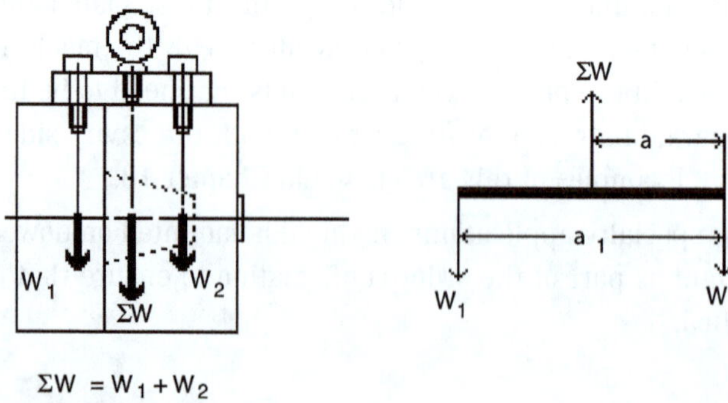

Figure 9.10 Proper lifting: (Left) Mold and centers of gravity (CGs) for masses W_1 and W_2 of the mold halves; (Right) Distance (a_1) of the CGs from each other and the distance (a) of the CG of the whole mold from the CG for W_2

The formula below permits the calculation of a (i. e., where to locate the hole for the eye bolt for the whole mold):

$$a = \frac{a_1 \times W_1}{\sum W} \tag{9.4}$$

The lift bar or eye bolt hole must always be located so that it is possible to handle a complete mold portion (injection side, ejector half, and center section of a stack mold), without the possibility of parts (e.g., stripper plates) falling off during handling. It is sometimes necessary to have more than one lift bar for one mold: one for the whole mold, and one or more for the sub-assemblies. A lift bar must never be screwed to a moving part, such as a stripper plate.

The lift bar (usually cold-rolled steel) must be strong enough that its yield strength is not exceeded; a safety factor of 5 is adequate. The mold number and weight capacity must be shown on the lift bar. This is important, because it will be used only when servicing the mold, and the molder may store it remotely from the mold. All lift bars should be painted "safety yellow". *The lift bar(s) must be shown on the assembly drawing.*

To determine the CG of a complete mold, the designer must use reasonable approximations. Drillings and other hollows within the mold, plates, and other large parts can be assumed solid, to simplify the process of determining the CG. The following two examples illustrate such calculations.

In the "mold" shown schematically in Figure 9.11, to get the total mass M, simply calculate the volume V and multiply it by the specific gravity of the mold material, usually steel. The position of the CG is at $L/2$. For the calculation of the lift bar, the total mass M is used.

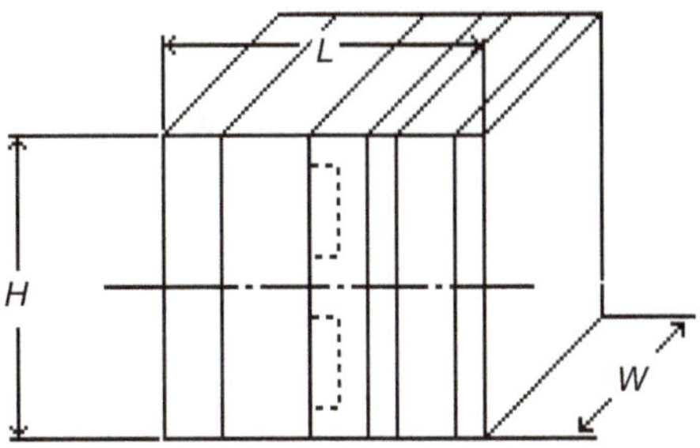

Figure 9.11 The mass M of the mold is the product of the total volume and the specific gravity of the mold material

In the example in Figure 9.12, a "modular mold" is divided into three sections. The volumes are calculated over the distances L_1, L_2 (the volume of the total of all modular parts), and L_3. The volume (mass) is then calculated for each section, and the location of the CG is determined as described above. If a part "hangs" at an angle of not more than 5° from the vertical, this is sufficiently accurate for handling. (This statement should prevent the designer from wasting time on lengthy calculations which will result in little appreciable benefit for handling.)

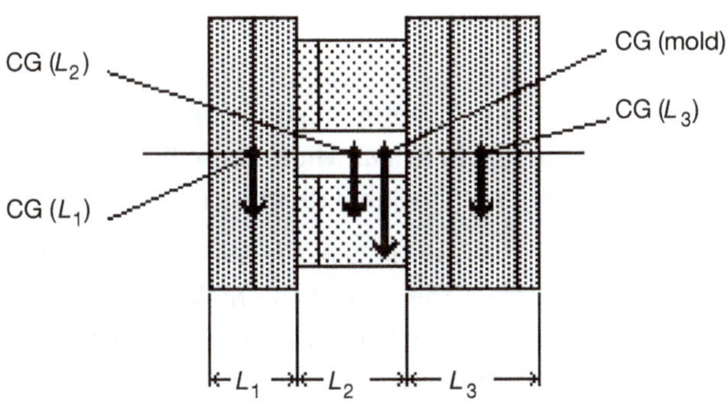

Figure 9.12 The volumes (masses) of each mold section are used to determine CG for the entire mold

If possible, the lift bar should also be kept mounted to the mold. Once the mold is installed, the lift bar can be moved to a parked location. If this is not possible, the latch should be clearly marked with the mold number on it so that it will be easily identifiable.

9.9.2 Latches

Any mold depending solely on tapers for alignment, without the use of close-fitting leader pins, could open up at the bottom during lifting. This could cause damage to the tapers and to the surfaces of cavity and cores, particularly in molds for thin-walled products. This condition, shown exaggerated in Figure 9.13, can be avoided by providing latches.

Even though most molds have leader pins that prevent the mold from opening, it is good practice to provide latches for shipping and handling. This applies also to molds where the lifting bar cannot be mounted directly on the plates but must be spaced with pillars, in which case the screws holding the bars could bend (Figure 9.13B).

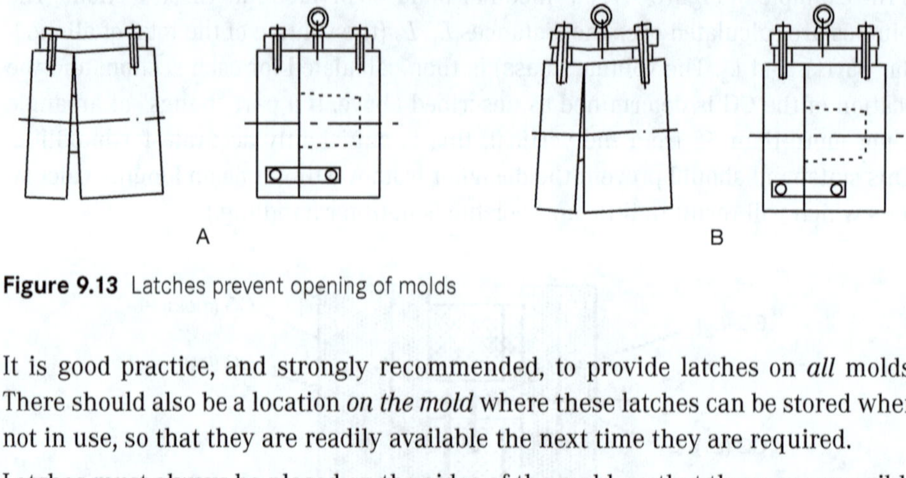

Figure 9.13 Latches prevent opening of molds

It is good practice, and strongly recommended, to provide latches on *all* molds. There should also be a location *on the mold* where these latches can be stored when not in use, so that they are readily available the next time they are required.

Latches must always be placed on the sides of the mold, so that they are accessible from the operator's side and from the rear of the mold when the safety gates are open. Never put latches on the top or bottom faces of the mold, where they may expose the operator to unsafe situations.

9.9.2.1 Latches for Mold Servicing

In some molds, it is necessary to gain access to areas between plates while the mold is mounted in the machine. Typically, it may be required that access be provided to:

1. *Hot runner components,* by separating the cavity plate from the manifold plate to gain access to nozzle tips, band heaters, thermocouples, etc., for inspection, cleaning, or replacement.
2. *Tapers at the base of cores* where the strippers seat for cleaning of vents, etc., by separating the stripper plate from the core plate.
3. *Tapers at the base of cavities* in moving-cavity molds; also for cleaning purposes, by separating the (moving) cavity plate from the stationary cavity backing plate.

As a safety precaution, if two (or groups of) plates as listed in the above examples must be separated while the mold is opened to remove the screws holding these plates together, latches must be attached to ensure that portions of the mold that become loose after removal of the screws cannot inadvertently separate from the plates to which they were held by these screws. These latches must be easily accessible while the mold is in the machine, and not hidden (e.g., behind the tie bars).

The "latching process" is as follows:

1. The latches are attached on that side of the mold where the screws needing to be removed are located (i.e., the stationary side).
2. The mold is opened, and the screws that hold the plates (or groups of plates) are removed. The latches hold the assembly together, and there is no risk of inadvertent separation. In the case of moving plates, such as strippers and moving-cavity plates, the screws to be removed may be stroke limiters, spring return mechanisms, and screws that hold the plates to their actuators.
3. The mold is closed again. Now, the latches are unscrewed from the stationary portion of the mold and mounted to connect the portion of the mold that will be moving away as the mold is opened.
4. As the mold is fully opened, the separation of plates to give access to the desired working area is complete. The latches prevent inadvertent separation from the moving side.

To reassemble the mold, the latching process is reversed:

1. Close mold
2. Switch position of latches and fasten them
3. Open mold
4. Insert and fasten all screws
5. Remove latches.

If possible, keep the latches fastened to the mold in suitable "parking places" so they will not get lost. If they remain on the mold, each latch must be held by two screws to ensure that they will not come loose during the cycling of the mold.

All latches must have the mold part number and mold number stamped on them so that they can be easily identified if kept in a separate location while not in use. They are also shown on the assembly drawing.

The sketches (Figure 9.14) show graphically some of the sequences for removing a plate to get access to the hot runner nozzles. To ensure square pulling off, without cocking, the latches must always be used in pairs, diametrically opposite from the center of the mold. Two latches are usually sufficient; for heavy molds, four latches may be required. The size of latches must be adequate for the size of plates moved. For molds that fit tightly between the tie bars, the location of latches should be planned so that they need not be removed, or do not require pulling of the tie bars when lowering the mold into the machine.

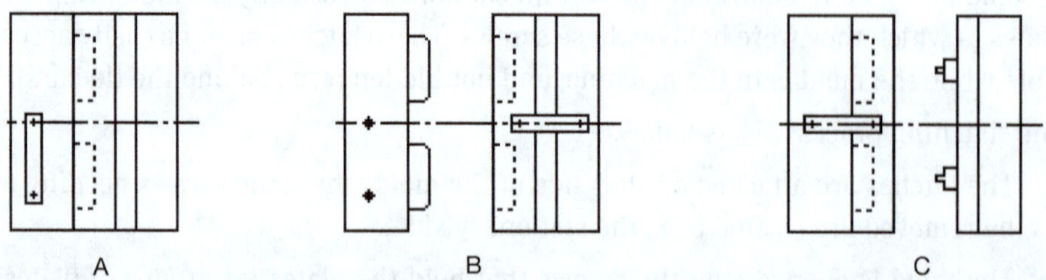

Figure 9.14 Sequence using latches to attain access to hot runner nozzles: (A) Mold closed and latches parked; (B) Mold open with latch holding cavity plate, providing access to screws; (C) After the mold is closed again, the latch is mounted to hold the cavity plate to the moving plate. When the mold is open again, access to the hot runner system is provided

Note: Latches must never be parked by joining two plates that separate during operation of the mold (e.g., between cavity plate and manifold plate in a moving-cavity system, or between stripper plate and core retaining plate). This may seem obvious, but it has been done!

Latches may have two holes if the distance between the two locations can be made the same. If the distance must be different, the latches must have three holes to suit the different distances. If the latches are also required to steady another plate, such as a stripper plate, an additional hole will be required.

Latches should always be kept with the mold, and there should be a latch storage location on the mold so that the latches will not get lost. Typically, this will take the form of additional holes to which the latch can be fastened until needed.

9.10 Mold Identification Nameplates

All molds should have a nameplate permanently attached in a visible location on the mold. This is usually on the operator side of the mold at eye height.

The mold identification tag should contain all key information on the mold. At a minimum, it should contain the following:

- Job number and/or mold number
- Date of manufacture
- Description of the mold (cavitation and part)
- Maximum allowable clamp tonnage
- Mold weight (to ensure crane of lift truck can install the mold). Optional - Weight of A half and B half
- Maximum melt temperature (particularly for hot runners)
- Mold (cooling) temperature
- Voltage and connected load.

Additional items added to mold nameplates are the following:

- Mold cooling line routing diagram
- Mold electrical connections.

10 Cavity and Core Design Overview

The cavity and core form the shape of the molded product and are at the center of mold design. Each unique plastic part requires the customized design of the core and cavity. Many of the most important aspects of mold design are in the core and cavity design. Since these components are so integral to the entire mold, they have a major influence on the design of the mold shoe (Chapter 9), hot runner (Chapter 14), cooling circuits (Chapter 12), ejection mechanisms (Chapter 11), and alignment (Chapter 13) of the mold. Chapters covering these topics cover detailed aspects of core and cavity design.

> All cores, cavities, inserts, and other custom items in the mold should be clearly marked (engraved) with a job number and/or a mold number in a location that can be easily seen. Sometimes the cavity number is also engraved for reference and assembly. However, if the inserts are interchangeable this may not be required.

10.1 Mold Cavity Space

The mold cavity space is a shape inside the mold, excavated by machining the mold materials in such a manner that when the molding material (in our case, the plastic) is forced into this space it will take on the shape of the cavity space and, therefore, the desired product (Figure 10.1). The principle of a mold is almost as old as human civilization. Molds have been used to make tools, weapons, bells, statues, and household articles, by pouring liquid metals (iron, bronze) into sand forms. Such molds, which are still used today in foundries, can be used only once because the mold is destroyed to release the product after it has solidified. In this book, we are discussing permanent molds that can be used over and over. Permanent molds are made from strong, durable materials, such as steel, or from harder aluminum or metal alloys and even from certain plastics where a long mold life is not required because the planned production is small. In injection molding, the (hot) plastic is injected into the cavity space with high pressure (5000–30,000 psi), so the mold must be strong enough to resist the injection pressure without deforming.

By convention, the hollow (concave) portion of the cavity space is called the cavity. The matching, often raised (or convex) portion of the cavity space is called the core. Most plastic products are convex or concave, forming a type of “cup” shape. The outside of the product is formed by the cavity, the inside by what is typically called the core. The alternative to the cup shape is the flat shape. In this case, there is no specific convex portion, and sometimes, the core looks like a mirror image of the cavity. Typical examples for this are plastic knives, game chips, or round disks such as records. While these items are simple in appearance, they often present serious molding problems for ejection of the product. Usually, the cavities are placed in the mold half that is mounted on the injection side, while the cores are placed in the moving half (typically the ejection side) of the mold. The reason for this is that all injection molding machines provide an ejection mechanism on the moving platen and the products tend to shrink onto and cling to the core, from where they are then ejected. Most injection molding machines do not provide ejection mechanisms on the injection or stationary side of the machine.

■ 10.2 Plastic versus Steel Part Drawing

The cores and cavities are made slightly larger than the finished plastic part to account for shrinkage in the process. Plastic shrinkage is covered in Section 2.3.1, and general values are given in Appendix 11. When beginning design on the cores and cavities, the plastic part drawing must be converted to a steel part drawing by adding the shrinkage to the final plastic part drawing. Although it is obvious, it is important to note that steel drawing dimensions will be larger than the plastic drawing. If the mold maker is not guaranteeing the shrinkage of the plastic part, then the steel part drawing (and thus the shrinkage rates) must be signed off by the customer before the design can be finalized and manufacturing started.

Note that in most cases a uniform shrinkage will be used on the part. If there are critical dimensions on the part for fitting to another part, the shrinkages may be adjusted locally in these areas. It is best practice to have the final steel part drawing with associated steel tolerances, or a plastic drawing with shrinkage signed off by the customer, before proceeding with detailed design.

If there are critical dimensions on the part and the shrinkage is in question, then building a prototype mold may be a good idea. If this is not possible then another option is to leave the critical dimensions with some steel left on (called “steel-safe” dimensions) until after the first test of the mold. Once the plastic parts have been measured then the steel safe dimensions can be recut and finalized.

10.3 The Preliminary Stack Layout

By now, the designer will probably have decided what type of mold should be built, guided by the possibilities discussed in previous chapters of this book. This does not necessarily mean that the designer is bound by this early decision. It may become necessary to reconsider their choices as the design is finalized. The designer must always keep an open mind and be ready to revise an earlier idea for a better one. The more time spent on thinking and rethinking the problems at this stage in the design process, the more successful will be the final result, which will save time and money in the long run.

The designer should now take the plastic part model and start to build the core and cavity concept around it. Once this is completed, the designer can begin to evaluate where the mold will be split or separated to eject the part. This may mean creating section views that show all the areas that must be considered when designing the stack. If more than one significant feature cannot be shown in the main cross-section, additional or partial sections may have to be shown. The cross-sections should now be examined and the questions below reviewed and resolved to the designer's satisfaction.

10.3.1 Where Should the Parting Line Be Located?

Figure 10.1 shows the cavity space inside a mold. To be able to produce a mold and to remove the molded parts, there must be at least two separate mold halves. These two sides are called the cavity side and core side. The separation between these plates is called the parting line (P/L). Actually, this is a parting area or plane, but, by convention, in this industry context it is referred to as a line. In a side view or cross-section through the mold, this area is actually seen as a line (Figure 10.1).

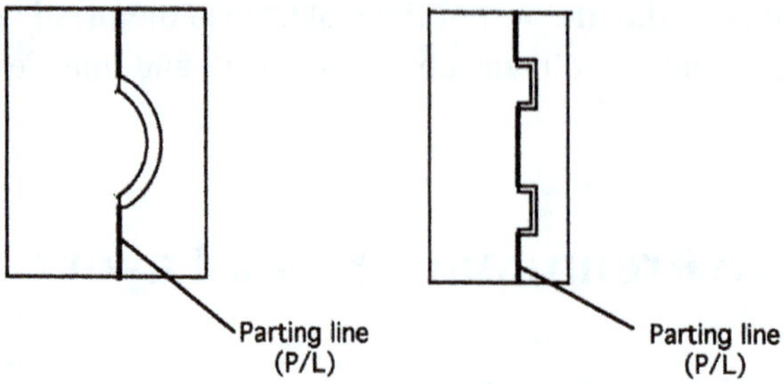

Figure 10.1 Illustration of a basic mold, showing the parting line

10.3.1.1 Primary Parting Line

Before proceeding to concept the stack, the location of the dividing plane (parting line, P/L) between cavity and core must be decided. The parting line can have any shape, but for ease of mold manufacturing, it is preferable to have it in one plane. The parting line is almost always at the widest circumference of the product, to make ejection of the product from the mold possible. With some shapes it may be necessary to offset the P/L, or to have it at an angle, but in any event it is best to have the parting line oriented so that it can be easily machined, and often ground, to ensure that it shuts off tightly when the mold is clamped during injection. If the parting line is poorly finished, the plastic will escape during injection, which shows up on the product as an unsightly sharp projection, or "flash", which must then be removed otherwise, the product could be unusable.

In cup-shaped products, the selection of the primary parting line is usually at the widest portion of the product. As stated earlier, a straight P/L is easiest to produce, preferably, but not necessarily, at right angles to the direction of the mold opening.

Figure 10.2A shows a simple open and closed mold with a single primary parting line. This means that the mold opens without any secondary parting lines. The machine clamping force holds the mold closed at the P/L. Figure 10.2B and C show a part with a potential undercut. However, by moving the P/L to above (B) or at the bottom edge (C) of this feature, the mold becomes a simple open and closed design with a single primary parting line. In this case, the feature is now fully created in the cavity (B) or the core (C). In Figure 10.2D the undercut is more complicated, resulting in the need for a slider. This complicates the design of the mold, increases the stack size and increases the overall cost of the mold.

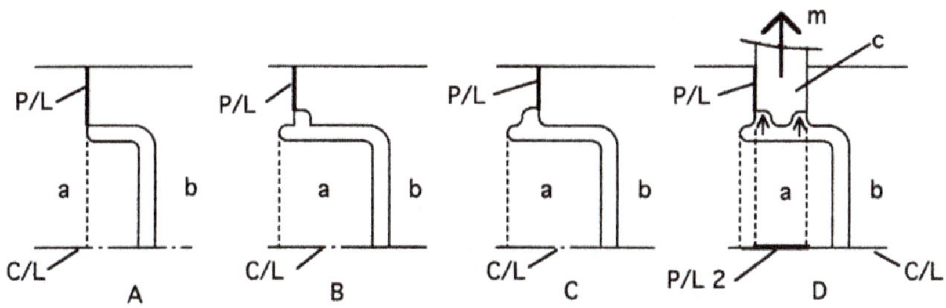

Figure 10.2 Schematic illustrations of location of parting lines (P/L) (only one half of mold shown): (a) core, (b) cavity. (A) Simplest case: P/L at right angles to axis of mold. (B and C) Product with rim but still simple. P/L can be either as in (B) or in (C). (D) Simple product but with rim and projection. Cavity is split, creating an additional P/L

An offset or stepped P/L is sometimes required, due to the shape of the rim. It is also, occasionally, used for molding a large projection, for example, a simple handle of a mug on the outside of the product. Such an offset P/L is preferable to a secondary parting line, which would be much more expensive to build (see Figure 10.3).

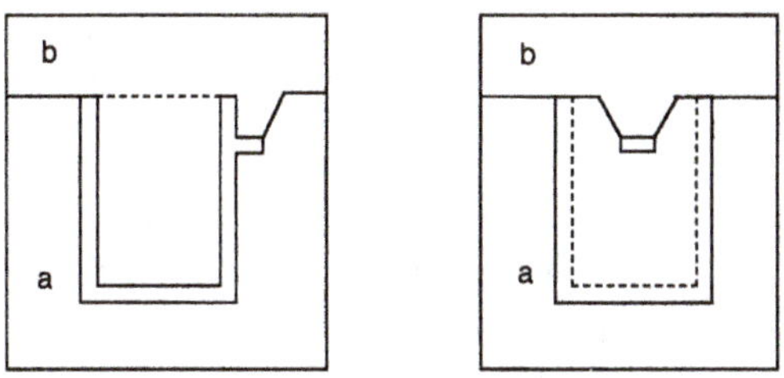

Figure 10.3 Schematic of mug with handle, showing offset parting line: (a) cavity, (b) core

The designer should always consider if a better mold layout could be achieved by turning the product slightly, to achieve with a straight (up-and-down) mold what would otherwise require side cores. For example, in Figure 10.4 if the parting line was flat then a side core (d) would be needed to free the part. However, by tilting the parting line slightly the part can be ejected using a single primary parting line with no side cores.

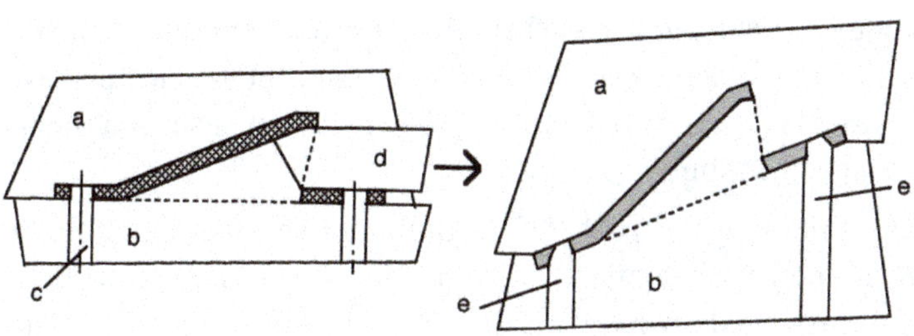

Figure 10.4 Example of a louver mold: (a) cavity, (b) core, (c) round core pin, (d) side core, (e) core pin with shaped tip

Chapter 18 contains examples of molds built with a single primary parting line for reference.

10.3.1.2 Secondary Parting Lines: Split Molds and Side Cores

If the geometry of the plastic part means that a single primary parting line will not reliably eject the part from the mold automatically, then additional, secondary parting lines and splits of the core and cavity may need to be considered. Secondary parting lines are the separating lines between two or more cavity (or core) sections in cases where the cavity must separate (split or retract) to make it possible to eject the molded product as the mold opens for ejection. A good example of a secondary parting line is a split cavity for a handled part. Chapter 18 contains reference designs of molds with secondary parting lines.

Usually, but not always, the split lines are parallel to the axis of the mold, and side cores at right angles to it. Both split cavities and side cores need backing-up and preloading against the forces created by the injection pressure, and some method of operating mechanism, which typically requires significant space in the mold. Operating mechanisms for these side cores can be angle pins (horn pins), or rollers in tracks, both of which translate the opening motion of the mold into a sideways motion. They could also be timed, hydraulic actuators independent of the clamp motion.

In the case of Figure 10.2D, we must add a secondary P/L to eject the part. The secondary parting line in the case of this round part could be a neck ring, consisting of two halves that are retracted as the mold opens to free the part from the cavity. In Figure 10.2D we must consider the injection pressure (P_{inj}) (as shown with small arrows inside the cavity space), which will force the two cavity halves apart. This force also exists in the cases A, B, and C, but the pressure is resisted by the strength of the solid cavity walls, which will expand slightly during injection and then return to their original shape once the injection cycle is completed. Since these side forces can be considerable (see Section 10.6.3), the cavity and/or supporting mold plates must be sufficiently solid to contain these forces and provide the necessary preload to prevent opening of the mold during injection. In the case of Figure 10.2D, which requires split portions of the cavities, these side splits or neck rings must be sufficiently backed-up to prevent movement of these inserts during injection and hold phases. If they move, then the molded part will take the shape of the inserts at the moved location, the mold may flash, and the molded part will be out of specification.

In some cases, rotating the part 90° in the mold concept could also result in a better mold, as shown schematically in Figure 10.5. The right schematic shows the normally expected mold layout, with the center line of the product parallel to the mold axis. Because of the outside shape (e. g., deep engravings or projections) the cavity will have to be split. The projected area (at right angles to the axis) of the product is very small compared to the projected area of the sides of the product. Therefore, the side cores will see considerably larger forces (F_s) at right angles to the mold axis. These forces must be adequately backed-up, robustly supported, and correct preloads created to prevent the splits from being edged open by injection pressure during injection.

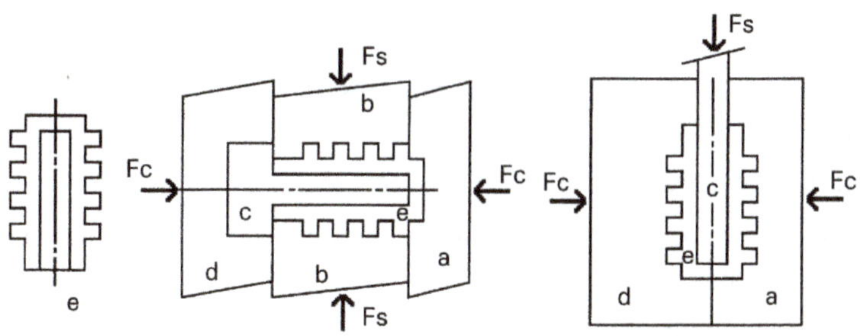

Figure 10.5 Illustration of a product and two possibilities of mold layout: (a) cavity, (b) splits, (c) core, (d) core backing plate, (e) product, (F_c) clamp force, (F_s) force on splits

These backups, especially for large splits, can result in a very bulky mold. But by turning the product by 90°, as shown in the left schematic of Figure 10.5, the primary P/L replaces the split line, and the cavity and core halves are clamped by the machine clamping force F_C. The core must now be withdrawn sideways. It will have a much smaller area exposed to the injection pressure and will need much less backing-up force, but the stroke of such side cores will probably be much greater than the stroke of the split cavities. This could be an undesirable feature, but is often preferred to the alternative of split cavities. Only by laying out to scale these alternatives will the designer be able to determine which is better for the contemplated mold and how to proceed. Note that in the position shown in Figure 10.5, the open end of the product is on top and the product is ejected downward from the (side) core and can fall unhindered. If the core has sufficient draft, air pressure alone could be sufficient to eject the product from the core, which would make for a much simpler mold.

10.3.2 Will the Product Pull out of the Cavity and Stay on the Core?

This point should be investigated first, because it will determine the complexity of the cavity design. This is ultimately determined by the part drawing, but sometimes a small change in the part drawing can help immensely with simplifying the design of the mold, which in turn can speed the cycle time, reduce maintenance, and cut costs.

1. In the case of a simple, cup-shaped product with larger than 1° draft, there is usually no issue with the part staying on the core. Remember from Chapters 2 and 3 that the part shrinks onto the core and grips it firmly.
2. If there is no draft or the part is flat with no features causing it to grip to the core, then additional considerations will need to be taken to ensure the part stays on the core, such as pull rings, a rougher surface finish, or other undercuts.
3. Holes, cutouts, or projections in the side wall of the product may require special attention:
   - Will it be necessary to provide side cores?
   - If side cores are required, should there be one for each hole, or one for a group of holes or projections?
   - Should the cavity have a complete side wall moving? In the case of beverage crates, all four walls may have to move, which will create four vertical split (parting) lines (see reference example in Chapter 18). All this will considerably increase the complexity of the mold, and increase the space required for each cavity and for the stack in general.
4. Are there other projections in the side wall of the product?
   - If they are deep, they will probably be considered like holes. If they are shallow (for example, engraved printing or ornamentation), it will depend on the draft angle of the side wall and the plastic injected. In some cases, typically with draft angles over 5°, shallow engraving could pull out of the cavity, provided there are features (such as undercuts) on the core to ensure that there is enough force on the molded piece to pull it out of the cavity. Cases like this should be discussed with the plastic part designer of the product for which the mold will be built. There may be a good chance that the product design could be slightly changed so that side cores are not necessary at all, thereby saving considerable expense.
   - Other possibilities can be considered, especially for large openings in the side walls, as, for example, the large cutouts in the sides of a typical, large laundry basket, where the cavity and the core can meet at an angle and produce additional, small parting lines, but not require side cores or split cavities.
5. Will the cavity need to be split into multiple sections?
   - The most common case is where the cavities split into two halves, creating two vertical split lines.
   - There may not always be enough space for the long side motions required for two splits, and the cavity will be split into four sections. This is common with pail molds, where four moving side cores are wedged within the

cavity block walls to contain the outward forces of the side cores. Note that in all cases where side cores are used, they must be preloaded and backed-up against the forces generated by the injection pressure.

10.3.3 Will the Product Eject Easily from the Core?

At the very moment that the mold starts to open, there is a tug-of-war between the core and cavity as each part of the mold wants to hold onto the plastic part. Based on the geometry of the part, sometimes this can be a close contest. Therefore, it is the job of the mold designer to consider where it might be necessary to ensure the part stays on the correct side for reliable ejection.

Ejection of the parts from the mold is covered in detail in Chapter 11. However, as the stack is being concepted, the method by which the part will be ejected from the core must be considered. The important question to resolve is whether there are any raised portions inside the product that, when molded, create severe undercuts in the core and prevent the product from being ejected easily?

1. *Does the product have a snap?* (Figure 10.6). A frequently used undercut is a snap feature, which is an undercut on the inside the product, shaped to snap over a similar extension in a matching product, for example, the lid over a can. Provided the shape of the snap rim (its cross-section - tapered and/or rounded sufficiently - and the total circumferential length) is suitably designed for ejection, there is no problem, and a stripper ring will easily eject the product by forcing the rim to expand while ejecting or stripping the part from the core. Stripping with ejector pins, located at some strategic points, may also be used for ejecting, in non-round products. To make the snap ring easier to stretch and to come off the core without breaking, it can sometimes be broken down into several sections, so that there will be, for example, four sections, each covering about 60–70° of the circumference, instead of covering the whole 360°. Note that this type of solution is more difficult and expensive to machine.

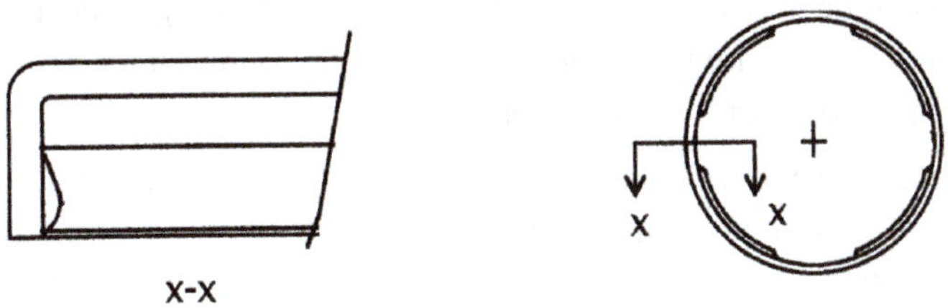

Figure 10.6 (Left) Section through a cap with snap; (right) example of four-section snap

2. *Are there any internal threads?* If the threads are designed suitable for stripping, they can be stripped from the core like the snap rim described above (for more details see Chapter 11 on ejection). It is better if there is not more than one complete thread (360°) on the part. Multiple threads may cause damage to the molded thread projections as they are dragged over the depressions for the successive threads in the core during ejection. In many cases, one thread may be strong enough for its intended purpose. It must also be understood that there is a relationship between the amount the plastic that is stretched radially and circumferentially. For a certain cross-section of the snap rim or thread, and if the product is small, there may not be enough length in the circumference to stretch, and the product will tear.
3. *Does the part need to be unscrewed from the core to eject?* In some cases, the product must be unscrewed from the core, which means a much more complicated and expensive mold. Since this design is quite common, there are standard mold bases available for unscrewing molds. This leaves only the inserts to be manufactured by the mold maker. Note that in order for the part to be unscrewed off the core, there will have to be features added to the part to prevent the part from turning during core rotation, These features are typically indents on the rim of the part crearted by lugs on the stripper ring or stationary part of the core.
4. *Are there sufficient undercuts on the core?* Undercuts are used to hold the product on the core, to ensure proper ejection, especially in cases where the product is designed so that it would stay in the cavity while the mold opens, often held by the vacuum between product and cavity wall or the gate vestige. With many products, there are enough vertical surfaces in the core, such as slots for ribs, or specially shaped slots and holes (as often required in technical products), to hold the product on the core side. The same is true for cup-shaped products with little side draft, where the product will tend to shrink tightly onto the core. If this is not enough to hold the product on the core, carefully designed and placed undercuts should be specified at the time of designing. The proper location for these undercuts (which are usually not specified by the product designer) is (a) near ejectors, or (b) preferably, especially with hot runner molds, near the tip of the core, where the undercuts are more effective because the bottom of the product is stiffer.
5. *Does the part require two-stage ejection?* Two-stage ejection (Figure 10.7) may be a solution for some, somewhat larger undercuts inside the core. This is a more expensive design of the core and the ejection mechanism, but it is frequently used in products that require a snap design inside the product. It is often used for overcaps for spray bottles that are produced in really large quantities. The cooling is less efficient, but it is a well-accepted and reliable design.

6. *Is there deep projection inside the product?* This feature often requires very complicated core design, possibly with moving, retractable core sections, or collapsible cores. Both systems are expensive, difficult to build, and hard to maintain in operation. They are also usually difficult to cool adequately, and thus run much slower than a comparable mold without these features.

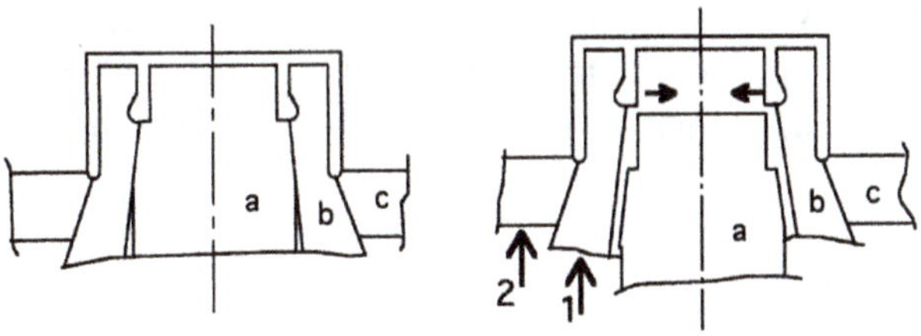

Figure 10.7 Schematic of two-stage ejection: (a) core, (b) sleeve, (c) stripper ring. During ejection, first (1) and (2) move together so that the core can slide out, then the stripper moves up to push the product off the sleeve while the projection on the inner sleeve moves inwards, as shown by the small arrows

10.3.4 Is the Cavity Balanced?

Figure 10.8 shows the elements present in every cavity shape. Within (A), the cavity pressures are balanced. In (B) and (C), the cavity is imbalanced. It does not matter if the side walls are at right angles to the internal pressure. There is always a component of the pressure that will press in the direction at right angles to the mold axis.

Figure 10.8 Schematic of cross-section of: (A) cup-shaped product; (B, C) open-sided product

As can be seen in Figure 10.9 on the left, the pressures inside the cavity push to the left and the right by the same amount, and there will be no force to move the core relative to the cavity. The cavity is therefore balanced.

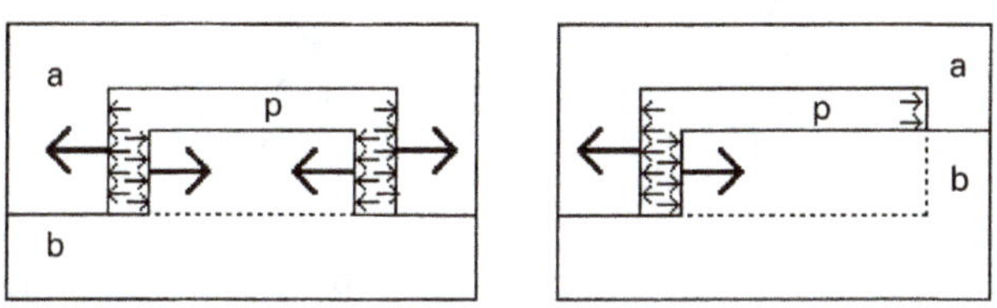

Figure 10.9 Schematic illustration of (left) a balanced mold, and (right) an imbalanced mold: (a) cavity, (b) core, (*p*) internal injection pressure

In the drawing on the right, the pressure p within the cavity tries to separate the cavity and the core by pushing the cavity to the left and the core to the right, as indicated with heavy arrows. This must be taken into account when designing a mold with imbalance in the cavities. The force trying to separate cavity and core can be balanced by placing a second, similar stack near the first one so that the forces are pushing in opposite directions. Failing this, there could be one pair of wedges (similar to a wedge lock) located so that the imbalance is taken up there. If this is not done, the forces of the imbalance will have to be taken up by the leader pins and bushings, which may not be strong enough in some cases, and wear rapidly.

10.4 Determining the Method of Cavity Construction

The designer must begin to think about how the core and cavity might be constructed. However, as mentioned at the beginning of this chapter, the designer must consider many factors such as alignment, cooling, and ejection, which will be covered in subsequent chapters. At this time, they may begin to consider some of these items in the form of a stack concept. Different cooling, alignment, and ejection methods may be considered, such as shown with the simple cup mold in Figure 10.10.

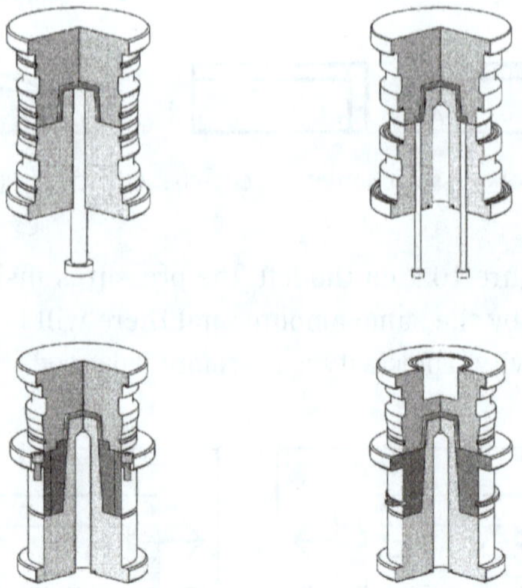

Figure 10.10 Stack concepts incorporating alignment, cooling, and ejection for initial review (Courtesy of Manual do Projectista)

10.4.1 Cavity and/or Core Are Cut Right into the Mold Plate

Some molds have only one or a few cavities cut directly into the mold plate, but the cores are usually separate from and mounted in (or on) the core plate. For practical reasons, one-piece cavity and core plates are often selected for single-cavity molds, mainly for very large products, but there have been molds like this built for smaller products, with more cavities. The problem is to provide the necessary accuracy of machining, especially in the absence of sufficiently large, accurate machine tools. A mold designed like this would consist of fewer parts, and if there are no foreseeable problems with ejection, cooling, and mold life, it is a good method, especially for very large molds. The mold steel selected should be at least a pre-hardened mold steel (such as P20) or pre-hardened stainless steel. Such cavities and cores may still require inserts (usually pins) whenever small holes etc. in the product would require delicate projections in the molding surface. Machining such projections from the solid steel, while possible, would be very costly to repair if they should be damaged.

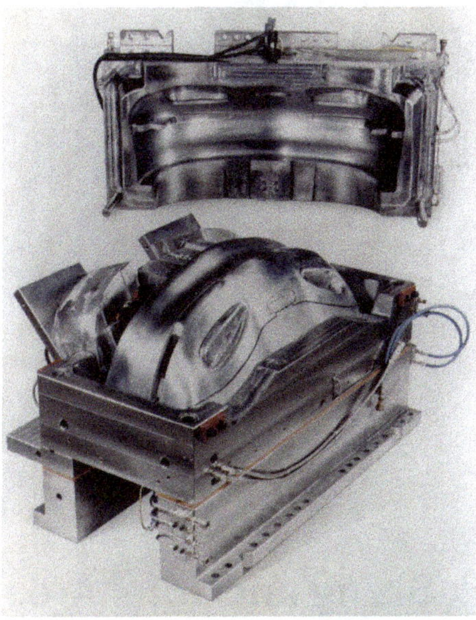

Figure 10.11 Single-cavity car front molding. Core and cavity cut into plate (Courtesy of Accurate Machine and Tool)

10.4.2 Individual Cavities and Cores

This is the most common design for molds. These cavities and cores are then either mounted on top of the cavity plate or inserted into it. Cavities and cores can also consist of an assembly of separate pieces, arranged to form the cavity assembly (see Figure 10.12). If the outside of such an assembly is a (not necessarily round, but sufficiently strong) ring (or "chase") into which the inserts are placed, this assembled cavity can be treated as a solid cavity and mounted on top of the cavity plate or be inserted into it. In some cases, the inserts are directly placed inside the cavity plate, without the need for a surrounding ring. Note that the forces from the injection pressure on the sides of the cavities are considerable, especially if the projected area at right angles to the mold axis is large, that is, where the product is deep. These forces will tend to loosen the inserts and can create gaps between them or between inserts and cavity plate, where plastic can flash into. Properly designed, such inserts must be pressed into their chase (or into the mold plate) to create a preload larger than the expected side forces.

Figure 10.12 Composite core and cavity (Courtesy of Husky Injection Molding Systems Ltd.)

■ 10.5 Stack Sizing and Total Area of the Stack

The total area of each stack is the sum of the space of the cavity (including the ring discussed above, which may also include cooling channels) plus the area (space) of any added features that may be required, such as side core components outside the cavity or core, plus space for their motion, actuation, and the backup. It can be seen that this total space can be much larger than the cavity by itself, and will determine the size of the mold and affect the cavity layout. It is easily understood that a mold without side cores requires much less space, and a much smaller layout, for the same number of cavities.

The size of the module or stack must be sufficient to withstand millions of cycles and repeated pounding from the clamp. The module must also be large enough to prevent undue hoop stress and cavity expansion. Equation 10.1 shows the relationship between the module size and the diameter of the part for round container molds. In general, the following formula can be used to estimate the size of the module from the size of the lip of the part:

$$\begin{aligned}&\text{For cup shaped parts: Module size}\left(\text{mm}\right)=2\times\text{Largest part diameter}\left(\text{mm}\right)\\&\text{For flat parts: Module size}\left(\text{mm}\right)=1.7\times\text{Largest diameter}\left(\text{mm}\right)\end{aligned} \quad (10.1)$$

Note that the size of the module should be rounded **UP** to the next standard module size, to conform to standard pitch and hot runner designs used in house.

10.6 Forces on the Cores and Cavities

As the plastic fills the cavity space under high pressure (P_{inj}), the pressure in the direction of the motion of the clamp will tend to open the cavity at the parting line (see Figure 10.13). The arrow describing projected area in Figure 10.13 really describes an area not a line, as delineated in this section view of the mold. The actual area can be seen (and measured) in a plan view of the mold cavity. From this it becomes clear that the clamping force (F_c) must be at least as great as the separating force, (F_s), to prevent the mold from opening and flashing. The required clamping force calculation is covered in Chapter 4. As shown in Figure 10.13, the injection pressure acts in all directions on the core and cavity, creating high stresses.

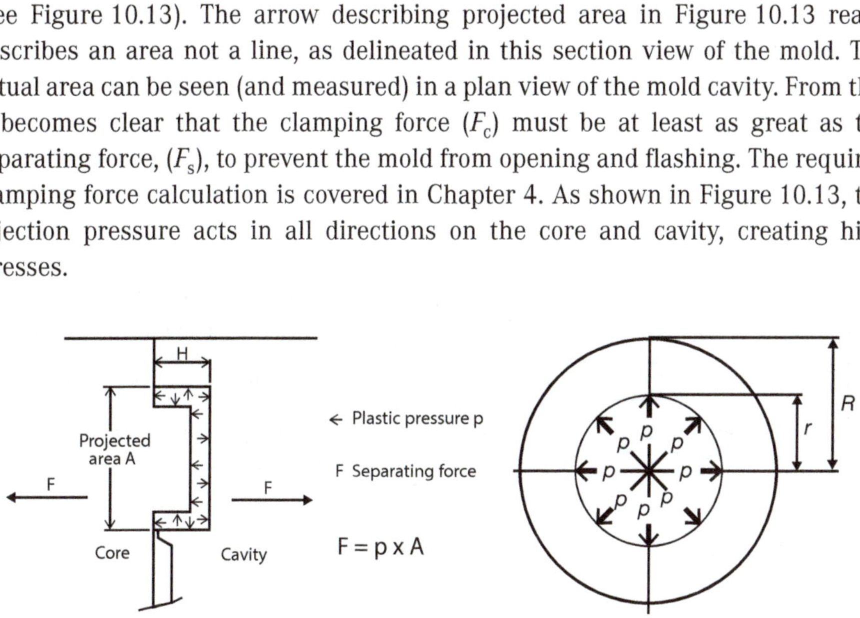

Figure 10.13 Schematic showing a cavity filled with plastic under pressure acting in all directions

> It is always advisable to add the maximum allowable clamp tonnage to the mold identification tag (which is usually mounted on the side of the mold). This is to ensure that mold technician reduces the clamp tonnage if the mold is mounted into a machine with more clamp tonnage than the maximum allowable.

To ensure that the mold and the individual stacks are strong enough to resist the required clamping force, as well as the injection pressure, the design of the mold must be checked against the following conditions:

10.6.1 Checking for Sufficient Steel Shut-off Area in Stack

The stacks are subjected to clamp tonnage on each cycle. The clamp tonnage typically acts only on the areas between the cores and cavities that touch and are loaded. The designer must ensure that there is sufficient shut-off area in the stacks to prevent permanent deformation of the mold.

The shut-off area of steel is the area where cavity and core halves touch as the mold is clamped shut. This is commonly called the parting line (P/L) area. The actual shut-off steel area A_{ss} is shown in Figure 10.14. $A_{ss} = A_{ss1} + A_{ss2}$ + any other areas that are compressed by clamp tonnage, such as tonnage blocks or plate shut-offs. This area under load will be compressed and the amount of compression should be checked to ensure that the individual stacks will not compress plastically and that the stack compression is taken into account when designing the stacks (see Section 10.6.2).

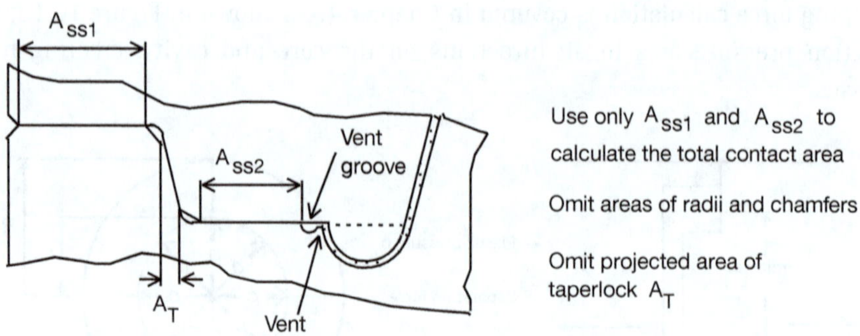

Figure 10.14 Areas that need to resist the applied clamp tonnage in the mold

Once all the shut-off areas have been calculated, then a check can be made to ensure that the shut-off area is sufficient. The stress on the shut-off areas must be less than the fatigue strength of the steel being used. If this is not the case then additional shut-off area must be added in the mold. If this is not practical, load-carrying pads of suitable shapes must be provided outside the molding area to share the load exerted by the clamping force with the parting plane.

Limit of stress on the shut-off area:

$$\sigma_f \geq A_{ss} / F_c \tag{10.2}$$

where
F_c = force of the clamp
A_{ss} = area of steel shut-off
σ_f = fatigue strength of the steel at the shut-off

10.6.2 Checking for Stack Compression due to Clamp Tonnage

Steel is *compressible*. A measure of the compressibility is the modulus of elasticity E, expressed in kPa or psi. All steels have approximately the same value of E, regardless of hardness or composition. The only exception are tungsten-carbides, which may have an E which is 2–3 × that for other steels.

This compressibility affects the mold part dimensions after the full clamp tonnage is applied to the mold. It must be considered when, for example, dimensioning the bottom of plastic parts, especially in molds with deep cavities.

If stack compression is significant, then the cores and cavities must be designed so that, under load, the core and cavity will create a part with the desired wall thickness. If the stack is over-compressed, then the wall thickness in the part can be thinner due to the core moving further into the cavity of the mold. In cases where the wall thickness is critical, this can be significant and must be taken into account at the design stage. If it is not, then a costly recut could be necessary to reach the desired wall thickness in the part.

For a simplified method of calculation of stack compression for a round part, refer to Figure 10.15 and Equation 10.3. When the mold is clamped up, assume a tube defined by diameters D, d, and height h is compressed by the clamp force. The bottom wall thickness size B (after clamp-up) will be smaller (by the amount Δl_c) than the original, unloaded size B_0.

To compensate for cooling lines within the cavity, the calculated amount Δl_c should be increased by about 15% based solely on experience.

Equations for stack compression:

$$\Delta l_c = \frac{F_c \times h}{E \times A} \times 1.15 \quad \text{(see note above)}$$

$$B = B_0 - \Delta l_c \tag{10.3}$$

$$A = \frac{\pi}{4(D^2 - d^2)}$$

where:

Δl_c = stack compression

F_c = stack force (N or lb) (may be expressed in tonnes or tons of force)

h = stack height (mm or in)

D = outer diameter of the stack module (mm or in)

d = part diameter at the parting line (also the largest part diameter on the plastic part) (mm or in)

E = 200,000 MPa or 3 × 10^6 psi

B_0 = clearance between cavity and core on the bottom before clamp-up

B = clearance between cavity and core on the bottom after clamp-up

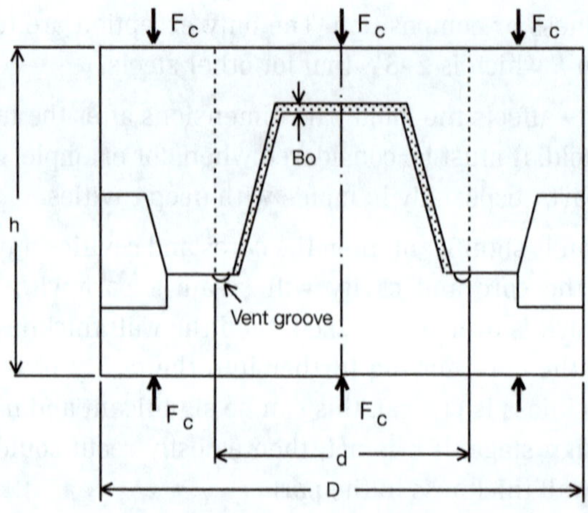

Figure 10.15 Stack compression from clamp tonnage or insufficient shut-off area

The stacking height in the product, and other distances which require accurate dimensions from the open end of the container, must also reflect the size reduction caused by compression, Δl_c of the stack. If the outside of the cavity block is a square or rectangular unit, (e.g., in modular molds), the whole projected area of the cavity block minus the projected area of the product (including any vents) will be used in this calculation. This adds to the area under compression, thus reducing Δl_c. This amount of compression Δl_c must be considered in the dimensioning of the mold parts, especially in products with very thin bottoms.

10.6.3 Checking for Acceptable Hoop Stress in the Cavity

When injection pressure enters the cavity space, it acts at right angles on the cavity walls (see Figure 10.13). This injection pressure, P_{inj}, creates a hoop stress in the cavity, which must be calculated to ensure that the cavity and module size are sufficient to endure the high cyclic loading from injection pressures. The formulae below for thick-walled vessels can be used to calculate this stress in the cavities that are subject to hoop stress.

Formulae for hoop stress calculations:

$$\sigma_{\text{inside}} = P_{\text{inj}} \times \frac{R^2 + r^2}{R^2 - r^2} \leq \sigma_{\text{f}}$$

$$\sigma_{\text{outside}} = 2P_{\text{inj}} \times \frac{r^2}{R^2 - r^2} \leq \sigma_{\text{f}} \tag{10.4}$$

$$\Delta l_{\text{inside}} = P_{\text{inj}} \times \frac{r}{E} \times \left(\frac{R^2 + r^2}{R^2 - r^2} - v \right)$$

$$\Delta l_{\text{outside}} = P_{\text{inj}} \times \frac{R}{E} \times \frac{2r^2}{R^2 - r^2}$$

where
σ_{inside} = inside hoop stress
$\sigma_{outside}$ = outside hoop stress
σ_f = fatigue stress limit of the material
P_{inj} = injection pressure (MPa or psi)
R = outside radius (mm)
r = inside radius (mm)
Δl_{inside} = inside diameter expansion
$\Delta l_{outside}$ = outside diameter expansion
v = Poisson ratio (0.3 for steel)
E = modulus of elasticity

The expansion (which can be calculated using Equation 10.3) can be considerable, and the stress, σ, must be calculated to ensure that the walls are thick enough for the job. Note that stress must be smaller than the fatigue stress σ_f for the cavity material, for long life. Fatigue must be considered, since the stretching (loading) of the cavities occurs cyclically (i.e., at every molding cycle). The fatigue strength of a material is usually around 10% of the ultimate tensile strength.

After the stack has been laid out, the designer must calculate the stresses and ensure that the steel is not only strong enough to reduce the stretching to a minimum but also to prevent the cavity from bursting. Information on the ultimate tensile strength and fatigue strength of steels can be found in the material manufacturer's specifications. In Table 10.1 a few common materials are listed. A more comprehensive list can be found in Appendix 1.

Table 10.1 Ultimate Tensile Strength of Some Common Steels

Material	Tensile strength (kPa)	Tensile strength (psi)
P6	1,160,000	165,000
AISI 4140	1,550,000	220,000
H13	1,760,000	250,000

Example of hoop stress calculation and expansion:

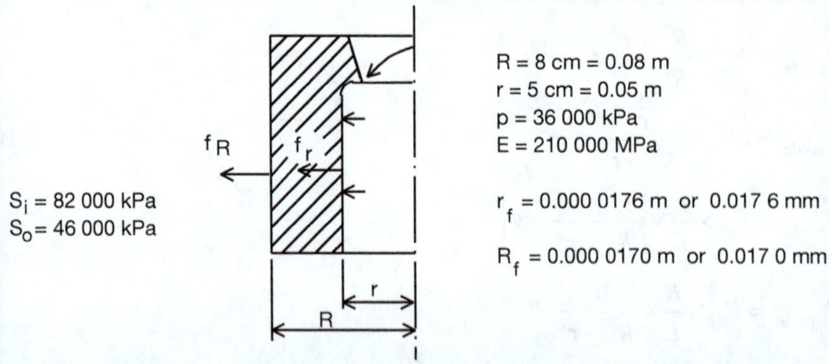

In the example shown above, the stresses are well below the 10% of the ultimate stress value (S_u) of any mold steel selected. Also notice that the inside of the cavity will stretch by the amount r_f = 0.0176 mm (0.0007"), which could be enough for the plastic to flash into the gap indicated with an arrow, or it may make the taper seat lose its fit with the matching mold part. A heavier wall may be indicated even though the stresses are below the fatigue limit.

Under injection pressure and velocity, the centers of the inserts will move outward from the original centers of the bores in the plate. The sketch in Figure 10.16 illustrates this movement. It is important that this movement is considered, especially with high injection pressures and thin walls.

If the core is held solidly but the cavities move away from each other, the wall of the product will be thinner toward the center of the plate and thicker in the direction of the arrow. The use of individual floating core mounting discussed in Chapter 11 overcomes this problem.

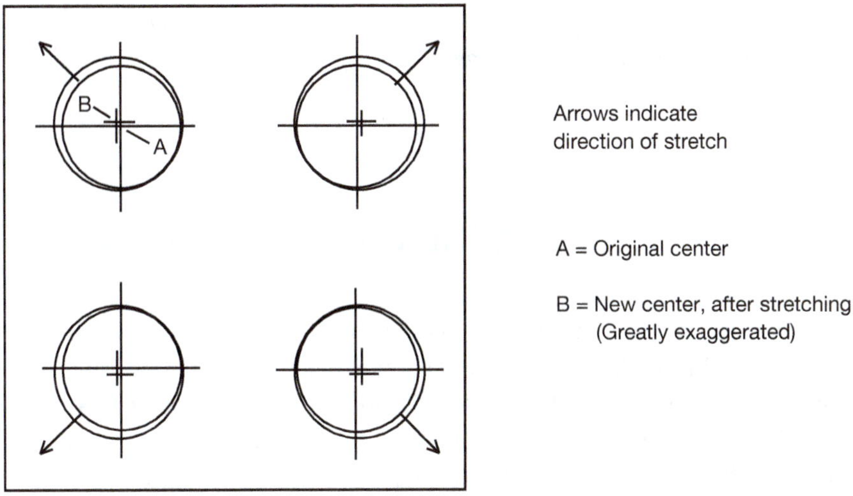

Figure 10.16 Direction of insert movement and expansion under injection pressure and velocity

■ 10.7 Core and Cavity Material Selection

Now that the designer has understood the major forces acting on the cores and cavities, they can begin to consider the materials for construction of the cores and cavities. Using the knowledge gained in Chapter 5 on the factors affecting the design of the mold, the designer can now choose the correct material for the individual inserts. Chapter 15 covers selection of mold materials in more detail, and Appendix 1 contains further detailed information on mold steels.

Table 10.2 Typical Materials Used for Inserts (but not limited to these materials)

Insert	Material	DIN	Hardness
Core, cavity, gate pad, neck ring	H13 420 SS	1.2344 1.2083	49-51 50-52
Core cap	BeCu25	–	36-41
Stripper ring	S7	–	54-56
Wear ring, lock ring	A2	1.2363	56-60

It is critical to ensure that interlocking surfaces that are subject to wear such as tapers have different hardnesses, or there will be fretting wear between the mating surfaces. This will quickly cause major alignment issues and costly repairs.

Mating surfaces that are under load, like tapers, should have a minimum of 4Rc difference in hardness, and preferably 6Rc.

10.8 Determining the Stack Construction

Cores may require quite a number of inserts and even moving parts to allow the part to be ejected from the mold. The injection pressure is usually of little concern (except in some special cases) because this pressure tends to compress the core from all directions, rather than expand it as it does with the cavity. The injection pressure on the core is resisted by the high compressive strength of the core material.

However, there is one serious potential strength problem with the core - relative movement of the core or "core shift". This is a primary concern with long slender cores, when the flow and the pressure of the inrushing plastic can deflect a core, resulting in uneven wall thicknesses around it. This is mostly of concern with thin-walled products, which require higher injection pressures, and where uneven wall thicknesses can create differential pressures on opposing sides of a core, thereby creating forces that deflect (bend) the core during injection. Such deflected cores return to their original shape as soon as the product is ejected, but by that time it already has produced uneven wall thicknesses in the plastic part. Problems like this can be solved by supporting the tip of the core in a matching hole in the cavity when the mold is closed or by using larger more robust module sizes with large tapers and preloads.

For larger parts the cores are usually mounted on top of the core plate, either solidly (the most common method) or floating, which is better, but more expensive. For smaller parts the cores are commonly supported in some way by the core plate using a spiggot or taper inserted into a core retainer plate (see Chapter 13 on alignment for more details).

Examples of various types of stack construction can be found in Chapter 18, and can be used as a reference for deciding on the best stack construction for the part to be molded.

10.9 Venting

Venting is absolutely necessary on all molds. Vents are required to allow air inside the cavity to escape so that the injected plastic can fill the space in its entirety. If air is not properly removed then the air compresses and builds up pressure which in turn opposes the plastic entering the cavity. This results in defects such as shorts, sinks, weld lines and burning due to the dieseling effect of compressed gas. Figure 10.17 shows the basic terminology of vents. The vent itself is the gap or space where the air can escape but that plastic cannot enter into. This is why vents are so closely toleranced. The land is the distance over which the vent exists. The vent grooves are the grooves cut into the core or cavity side that direct the evacuated air to the atmosphere.

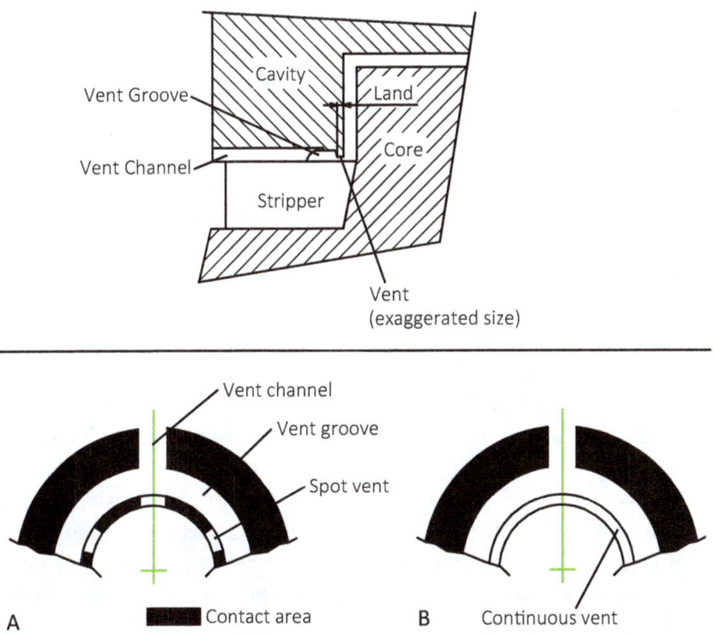

Figure 10.17 Venting terminology and geometry (note that the vent groove is shown on the cavity side as this is preferred, but it could also be on the stripper ring)

Vents are classified into two groups: moving or self-cleaning vents and stationary vents. These groups can be sub-divided into spot vents and continuous vents. When designing vents, the following rules and guidelines should be adhered to:

General venting guidelines

1. Provide as large a vent as is practical for the resin being used. The aim is to provide the least resistance to air escape while avoiding flashing.
2. Place vents at the end of plastic flow paths or where plastic will meet creating a weld line.
3. Vents must be designed so that they can be easily cleaned. Stationary vents are harder to clean and should be designed keeping easy access in mind.
4. Vents must be manufactured to the same tightly toleranced dimensions to ensure equal resistance for air to escape from each cavity in a multi-cavity mold.
5. It is preferred to provide a vent groove connected to all the venting to provide equal air escape (see Figure 10.17A and B). All vents must have a clear and unobstructed path for the air to flow to the atmosphere.
6. Vent grooves should be trapezoidal or preferrably semi-circular for ease of cleaning.
7. Follow the path of the air to the atmosphere, and ensure that air can escape freely (i. e., plates do not shut-off air).
8. Vent at the closed end of all pockets, that is, any area not in the direct plastic stream, into which plastic must flow to properly fill the cavity space. If the pockets (for ribs, bosses, etc.) are created by inserts, it is easy to provide vents where the inserts seat in the main body of the cavity (or core). If the pocket is machined from the solid, vent pins must be provided at the deepest location. Note that ejector pins or sleeves, which should always be placed at the lowest point (of a rib, etc.), provide a natural vent, and usually do not require additional vent pins.
9. If the plastic flow from the gate(s) can circle around an area in the cavity, because it flows easier through wider wall spaces or ribs, such areas must be vented. Because plastic flow always chooses the path of least resistance, it will engulf such areas and trap the air inside it.
10. Wherever possible, use vents that will be operated every time the mold opens and closes, such as ejector pins (or sleeves), two-stage ejection, stroking poppets, or moving cavities. If this is not possible, provide stationary vent pins but design the mold so that these pins can be easily removed for cleaning.
11. It is good practice to pressurize stationary vent pins or slots so that compressed air blows through the vent every time the mold opens. The disadvantage is that it requires additional channeling for the compressed air. It may sometimes be acceptable to leave the air pressure ON permanently, thus not requiring valve operation for this circuit.
12. Ensure that there is sufficient support area remaining on the cavity and core so that the vents will support the clamp tonnage and not compress or collapse over time.

10.9.1 Parting Line (P/L) Venting

The majority of the venting from an injection mold usually comes from the parting lines, as this is normally the end of the flow of plastic and the natural location for the air in the cavity to escape.

The location, depth and land of the vent is dependent primarily on the resin being molded. The following is a guideline for vent depth for some commodity resins. The resin supplier can normally provide additional information on recommended vent sizes. If in doubt about the vent size, consider starting with a smaller vent size for the first test and then increase the size as needed. Although this requires reworking the tool, *it is normally less work to increase a vent size then to decrease it.*

Table 10.3 General Guideline for Parting Line Vent Sizing (for additional details consult the resin supplier)

Resin	Vent depth	Vent land
General	0.01-0.02 mm (0.0004-0.0008")	1-2 mm (0.04-0.08")
PP	0.01 mm (0.0004")	1.5-2.0 mm (0.06-0.08")
PE	0.018 mm (0.0007")	1.5-2.0 mm (0.06-0.08")
PS	0.018 mm (0.0007")	1.5-2.0 mm (0.06-0.08")

10.9.2 Vent Grooves and Channels

The size of vent channels and grooves depends on the volume of plastic entering the cavity and the speed of injection. It is desirable to have large vent grooves and channels but they must not be too wide (*W*), too deep (*DG*), or too close to the cavity (*L*), to avoid weakening the steel. Figure 10.18 shows the typical geometry of a trapezoidal vent groove and channel.

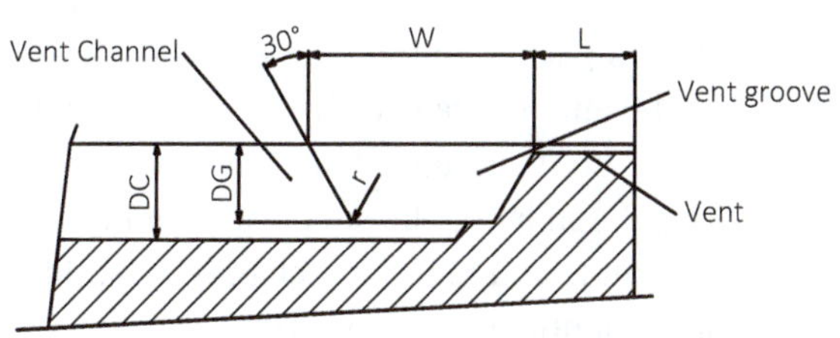

Figure 10.18 Vent groove and channel geometry and nomenclature

The vent channel cross-section should be at least as much as the total of cross-section of the vent leading into that portion of the groove from which the channel leads to the outside. For large cavities (large volume of air to be vented), the number of channels may have to be increased to provide a satisfactory total cross-section. This applies also if the vent channel is created by a chamfer or by a drilled hole. There is nothing wrong with selecting even larger channels if they can be accommodated, because larger tools can be used and the channels are easier to machine. Vent groove and channel dimensions are given in Table 10.4.

Table 10.4 Vent Groove and Vent Channel Geometry[a]

Vent groove or channel	*W* (mm)	*R* (mm)	*r* (mm)	*DG* (mm)	Area (mm^2)
Round	2.00	1.50	–	0.40	0.51
	3.60	2.50	–	0.80	2.07
	5.00	3.70	–	1.00	3.40
	6.00	5.00	–	1.00	4.14
	8.00	6.10	–	1.50	8.22
	10.00	9.00	–	1.50	10.56
	10.00	7.30	–	2.00	13.38
Trapezoidal	2.00	–	0.2	0.60	1.05
	3.60	–	0.3	1.00	3.13
	6.00	–	0.3	1.00	4.55
	8.00	–	0.3	1.50	10.73
	10.00	–	0.3	2.00	17.60

[a] For inches from mm, divide by 25.4, or for square inches from mm^2, divide by 645.

10.9.3 Vent Pins

The standard design for vent pins has a clearance from 0.000–0.005 mm (0–0.0002") in diameter in the portion of the bore below the vent groove to ensure that the vent is concentric (evenly spaced) with the wall of the bore. Figure 10.19 shows the dimensions for the vent and land. Vent pin vents are normally 0.01–0.015 mm deep with a 2 mm land. The suggested vent groove dimensions are $2.5R \times 0.8$ mm deep all around, and the vent channel should be cut down the length of the pin at the same depth as the vent groove.

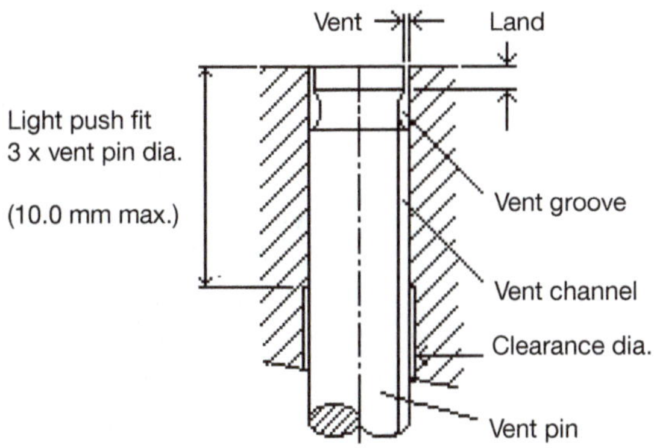

Figure 10.19 Typical vent pin geometry and nomenclature

10.9.4 Venting of Ribs

Venting is absolutely necessary to ensure the filling of any *deep* rib that is more than 1.5 × its width. A sufficient number and size of ejector pins or vent pins at the bottom of the rib must be provided to prevent trapping of air, or an insert must be used.

With regard to dimensions, if feasible the vent should reach over the whole length of the insert to permit maximum air flow. The vent groove and vent channel dimensions should be as large as possible and consistent with the size of the insert. They are usually machined in the insert but could also be in the surrounding part (e.g., the cavity or core). The vent channel can consist of one or more drilled holes, or can be created by a large chamfer in the insert as long as the cross-section of the channel is large enough for the expected flow of air. In all cases, the vent channel must be connected to the outside air or to a pressure air line. For actual vent dimensions, see above discussion for parting line vents.

10.9.5 Venting the Bottom of a Cavity

The purpose of venting in the bottom of a cavity is to vent some of the air ahead of the inrushing plastic stream, similar to the venting of runners. Also, when opening the mold, the vent breaks the vacuum in the cavity, which may otherwise keep (suck) the product into the cavity, and can affect ejection. Such vents may be connected to a compressed air line to assist release of the product from the cavity.

In Figure 10.20A, the vent is in line with the plastic flow and must be within normal vent sizing guidelines. In Figure 10.20B, it is at right angles and, therefore, could be larger than recommended from Table 10.3. The drilled vent channel hole must be large enough, so as not to restrict rapid venting.

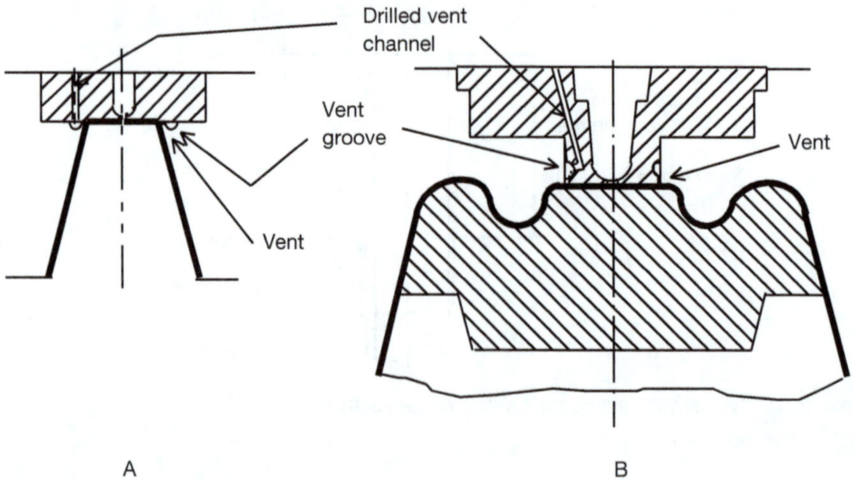

Figure 10.20 Cavity bottom venting

10.9.6 Core Cap and Core Sidewall Venting

Standard stationary core cap vents are perpendicular to the flow of plastic and can be larger than those recommended in Table 10.3 due to the fact that plastic pressure compresses the vent and flows perpendicular to the vent. This makes the likelihood of plastic flowing into the vent lower. As a result, these vents can be slightly larger than the recommended parting line vents (Table 10.5). In the case of air-ejected molds, the core cap and core sidewall vents can be pressurized and used effectively for ejection (see Chapter 11 on air ejection).

Table 10.5 Perpendicular Vents on Core Caps and Other Compressed Vents

Plastic	Perpendicular vent size
PE, PS	0.022 mm × 1.0 mm
PP	0.015 mm × 1.0 mm

11 Ejection

Ejection is a primary requirement of every mold and often one of the most difficult parts of designing a mold. The complexity of any ejection system from the mold is a question of the desired productivity and the required quality. To reduce cycle time, the products should be ejected as early as possible from the mold. Certain ejection methods permit earlier ejection; others depend on the plastic to be stiffer. For example, stripping permits hotter (softer) products to be ejected without damage to them, whereas unscrewing ejection or pin ejection requires the pieces to be more rigid.

At this point in the design process, the designer will determine which method will be most suitable for the shape of the product, the type of mold, and the expected productivity.

Usually, the products are ejected by one of the following methods:

- Ejector pins, blade ejectors, and sleeves (Section 11.6)
- Stripper plate, ring, or bars (Section 11.8)
- Air ejection (Section 11.9)
- Special ejection mechanisms (Sections 11.10–11.12)
- Combinations of the above.

In this chapter we will review these ejection methods plus some specialty applications.

> When designing the ejection system in a mold, the deisgner must always consider the potential conflicts with other critical aspects of the mold design, and ensure that an acceptable solution has been concepted for not only the ejection system but also the cooling, alignment, mold deflection, and serviceability.

Space requirements for ejection mechanisms, including the location of the ejector plate return features, can influence the spacing of the stacks in a multi-cavity mold. The designer must consider that the core must be supported to prevent deflection of the core backing plate during injection. It is often quite challenging to locate the ejector pins in the most effective locations, while allowing sufficient space for the supporting of the core plate and for aligning and guiding the ejector plate. Note that all ejector plates must be guided independently. These plates must not be guided by ejector pins or return pins, because the weight of the plate will tend to bend these pins, resulting in premature wear on the pins, core inserts, and the mold.

Wherever possible, the designer should use one of the simple ejection methods (i.e., ejectors, strippers, or air) in their simplest applications as discussed earlier. However, there are occasions where these methods are not suitable or reliable, and special ejection methods may be necessary.

■ 11.1 Manual or Semi-Automatic Ejection

Some molds are designed for manual or semi-automatic ejection. This requires an operator to stand beside the machine and after each cycle, remove the products from the molding area, reclose the gate, and initiate the next molding cycle. Some lower-quality molds that are not properly finished or have unreliable injection and ejection systems require that the operator reach into the molding area to pry loose a stuck product, and from time to time to lubricate the molding surfaces with mold release agents. All this is not only labor-intensive, adding greatly to the cost of production, but was also very unsafe and the cause of many serious injuries. During the time between cycles, the operator inspects the part or conducts some secondary operation such as degating, de-flashing, or assembly.

Since much of this operation also depends on the acquired skill of the operator - some workers are faster, some slower - and on the time of the day or night, or even on the day of the week, the overall molding cycle time can vary considerably, resulting in quality differences in the product because of different residence times of the melt in the machine; many rejects may result. There is also the problem of absenteeism of the personnel, which often plays havoc with production planning. Much effort can therefore be spent on eliminating operators from the actual molding process.

Due to the above, it is more cost-effective and quality-effective to build a mold that reliably ejects the parts automatically and continues to cycle without operator interaction each cycle. This chapter will focus on automatic ejection only.

11.2 Automatic Ejection

There are many advantages to designing a mold with automatic ejection, and this is always the preferred method of molding. Some of the most relevant reasons are:

- Consistent cycle time
- Consistency of products from cycle to cycle
- Absence of operators interacting with the machine for improved safety of plant operation
- Possibility of cycling faster than with an operator. There is a limit to how fast an operator can handle a semi-automatic operation, without excessive fatigue. With very short cycles, it may be impossible
- Reduced reliance on humans (who are less reliable than a machine in terms of attendance and repeated monotonous functions)
- Reduction in stoppages due to faults and reduction in mold damage.

There are two types of automatic ejection:

- Free fall. In free fall, the products fall freely into a box or onto a conveyor belt (see Figure 11.1)
- Controlled removal. For controlled removal, the products are removed from the molding area with the assistance of a robot or take-off mechanism (see Figure 11.2).

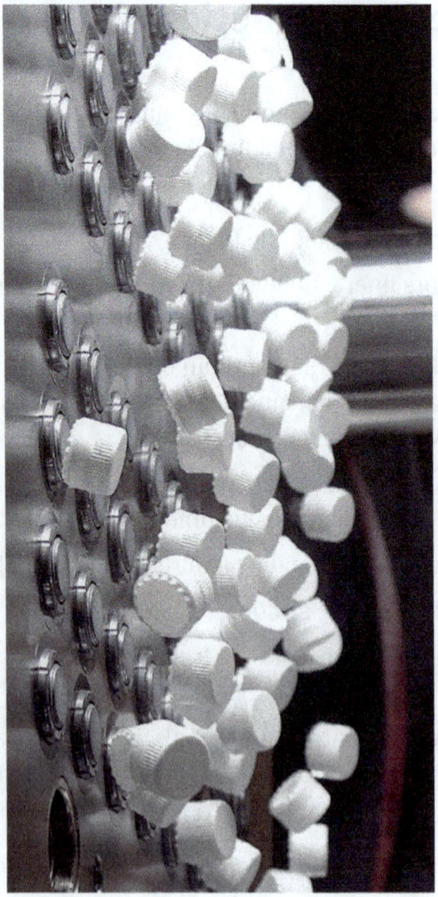

Figure 11.1 "On the fly" ejection of 128 closures (Courtesy of Husky Injection Molding Systems Ltd.)

Figure 11.2 Air ejection into a side-entry robot (Courtesy of StackTeck Systems)

Fully automatic (FA) molding depends essentially on two factors:

- *Reliable injection.* The molding machine must be repeatable from cycle to cycle in every aspect, but especially in dosing the same amount of plastic into the cavities.
- *Reliable ejection.* Every mold can be designed so that there is no chance of the product hanging up and not ejecting. The key to good ejection is that the product always stays on the side from which it is intended to be ejected from.

The designer must select the appropriate method of ejection and make sure that there is enough ejection stroke to clear the products from the cores. This is frequently overlooked, and can also be caused by improper setup of the mold.

11.3 Basic Requirements for Ejection

The one cardinal rule in ejection is: *As the mold opens, the products must stay on the side from which they will be ejected.* Unless the shape of the product (ribs, little draft, etc.) is such that it will stay naturally on the ejection side, features such as undercuts (also called pull rings), different surface finishes, controlled vacuum, or other expensive features may have to be used to ensure that the product stays where required for ejection.

Injection molding machines normally have an ejection actuator on the clamp side of the machine. Most molds, therefore, have the ejection mechanism on the clamp (moving platen) side. There are some exceptions, such as when ejection must be from the same side as injection, for cosmetic reasons (e.g., inside gating of caps, underside of plates, trays, etc.), and in stack molds, where one half of the mold has the cores on the injection side of the machine. In these cases, an additional ejection mechanism as well as its actuation must be designed into the injection side of the mold.

11.3.1 Machine Opening Stroke for Ejection

There must always be enough clearance between the cores and the cavities to permit the products to fall freely, without hanging up. With robotic take-offs, the clearance must be also sufficient to permit the products on the take-off plate to clear the molding area, without interference with cavity and core.

In general, the mold stroke should be at minimum the following:

$$\text{Mold stroke } (S_{\mathrm{m}}) = 2.5H \tag{11.1}$$

where H is the total height of the part or the greatest diagonal distance across the part (Figure 11.3).

When the part free-falls out of the mold it will tumble due to the center of gravity. If the part is robotically removed then the requirement for the mold stroke will be just enough for the robotic take-out to enter, grab the part, and exit from the molding area (see Figure 11.2). However, for free-fall a longer stroke may be required.

> *Example:*
> While the part height of a large lid may be small, the required H to be used in the mold opening stroke calculation will be the diameter of the part to prevent jamming, because the lid may tumble (see Figure 11.3).

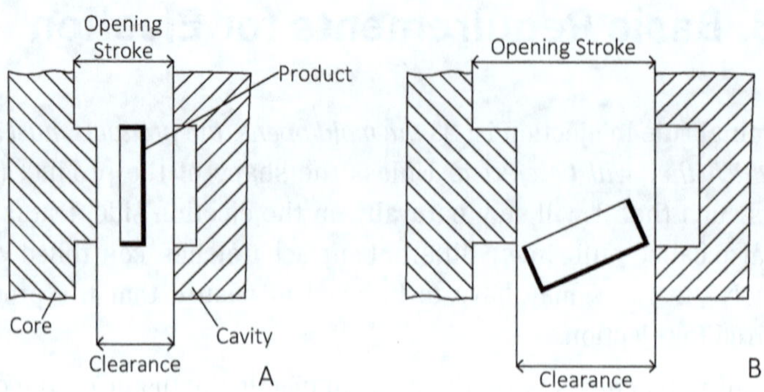

Figure 11.3 Determination of the dimension H to be used in the mold stroke calculation

There are exceptions to this rule. Containers with considerable draft will fall freely off the core much sooner and will, therefore, require less clearance (Figure 11.4A versus B). Also lids may require less stroke to prevent them from tumbling (see Section 11.8.5). Because less clearance is required for the product to fall safely, less stroke is required, and the mold will run faster. The use of a smaller mold stroke than suggested from Equation 11.1 may be possible but must be carefully considered.

The method of alignment can also affect the required stroke. The example in Figure 11.4D shows a method of cavity and core alignment which is sometimes selected, compared with Figure 11.4A and B. Because the face of the cavity is recessed by the height of the alignment taper, the stroke must be **increased** by this amount to prevent the ejected product from catching in the tapered recess in the cavity.

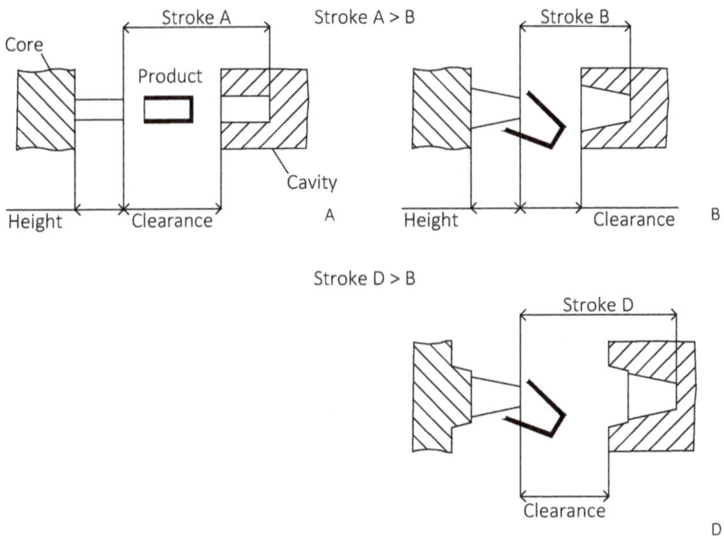

Figure 11.4 Considerations for machine opening stroke length

Excessive stroke usually presents no problems other than extending the cycle time and, therefore, reducing the productivity.

For servicing of the mold, usually during start-up, it is advantageous to keep the stroke large enough, at least 150 mm (6"), to allow the mold technician to have access for viewing (with a mirror) and cleaning such areas as the gates. The larger the molding area, the more clearance should be available. Occasionally, the machine can be set up and operated so that the minimum stroke required for ejection is used to operate the mold at maximum speed, but additional stroke can be selected during set-up and servicing.

In single-level molds, the available opening stroke is usually ample, but with two (or more)-level (stack) molds, the total stroke, which is a respective multiple of a single-level stroke, must be within the maximum available stroke of the molding machine.

An exception to the above stroke requirements occurs in molds where the cavities split in a vertical plane (i.e., where the splits move horizontally). In Figure 11.5, which shows the mold from the top view, the clearance between core tip and surface of cavity is less than the height of the product, but because the cavity is (partially) split, there is enough clearance between the slides to permit the product to clear the mold in vertical free fall. The clearance between the slides (splits) must be sufficiently large for the products to pass without getting scratched on the sharp edges of the slides. The designer must be careful to ensure that part molded at the top does not hit the slides of the cavities below it.

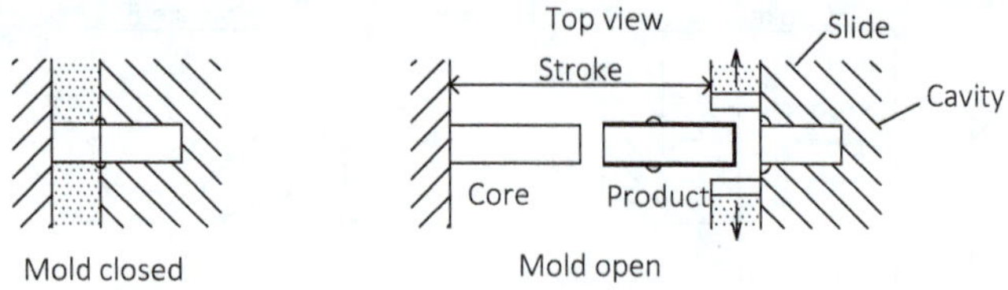

Figure 11.5 Mold cavities split in a vertical plane may allow for smaller mold opening strokes

11.3.2 Venting for Ease of Ejection

Venting should be added to the mold where a vacuum could be created during ejection between core and plastic. This applies to both cup-shaped and flat products.

Figure 11.6A shows that when a straight-walled product is pushed off the core, the vacuum created will deform or even crack the product during ejection. This is more severe with complete absence of side draft, but even with draft angles of several degrees, combined with fast ejection, air to break the vacuum may not enter fast enough at the opened end, and cause the part to implode. If the molding material is flexible (e.g., PE or PP), it usually returns to its molded shape. If the material is brittle (e.g., PS or PC), the bottom of the product may break, stress or permanently deform.

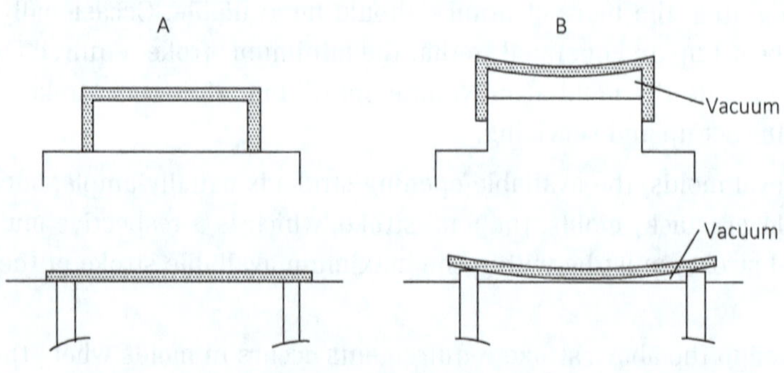

Figure 11.6 Deformation caused by ejection without venting: (A) Vacuum caused by ejection of a straight-walled part. Adding venting to the core will resolve the issue; (B) Ejection on the edges of part may cause deformation due to sticking and vacuum. Adding a vent or another ejector pin in the middle of the part would resolve the issue

Similar conditions may occur with flat products when ejected only near the edge (Figure 11.6B). While the part near the ejectors will be demolded, the portion of the product in the center may just deform and continue to stick to the core. Ejectors near the center would act as a vent to break the vacuum.

11.3.3 Surface Finish Forces on Ejection

Dr. Davide Masato

The finish of the molding surfaces can have a significant impact on the forces required to eject the part from the mold. The frictional force developed at the polymer-mold interface during ejection is mainly determined by the mold surface finish and the polymer type [1]. Injection molding processing conditions have a less significant effect. The understanding and optimization of the effect of surface finish can help reduce the ejection force and minimize deformation or breakage of the molded part [2]. Part and mold designers should be aware of the effect of their choices on the ejection frictions to avoid issues and lower part quality upon processing [3]. It is also essential to notice that some mold geometries, such as tall standing cores, can be more critical than others [4]. Upon molding, the polymer shrinks onto such geometries, leading to an increase in the frictional force that has to be overcome by the ejector pins or other ejection elements [5].

11.3.3.1 Mold Surface Roughness

Mold designers should specify surface roughness for cavity and core surfaces, taking into consideration the effect of the ejection force. Surface roughness is commonly specified using parameters *Ra* and *Sa* for profile and surface measurements, respectively. These parameters measure the arithmetical mean deviation of the assessed surface sample, thus representing the average mold topography. More advanced functional roughness parameters can be used for applications requiring high tolerances, high accuracy, or molding of small geometries [6]. Surface roughness comparators are not recommended for core and cavity inserts, and in general, for surfaces that require acceptable tolerances.

For most applications, the lower the surface roughness is, the lower the ejection force, indicating that a better surface finish will lead to reduced ejection friction. However, it should be noted that highly polished surfaces, such as A1 and A2 surfaces, might lead to higher friction due to polymer stiction to the mold

[1] D. O. Kazmer, *Injection Mold Design Engineering* (2nd edn), Hanser Publishing, 2016.

[2] K. D. Delaney, G. Bissacco, D. Kennedy, A structured review and classification of demolding issues and proven solutions, *International Polymer Processing,* 2012, 27 (1): 77-90.

[3] B. J. Araújo, A. S. Pouzada, Design of ejection systems for injection moulds, *O Molde,* 2002, 54: 36-41.

[4] D. Masato, M. Sorgato, P. Parenti, M. Annoni, G. Lucchetta, Impact of deep cores surface topography generated by micro milling on the demolding force in micro injection molding, *Journal of Materials Processing Technology,* 2017, 246: 211-223.

[5] D. Masato, M. Sorgato, G. Lucchetta, A new approach to the evaluation of ejection friction in micro injection molding, *Journal of Manufacturing Processes,* 2021, 62: 28-36.

[6] M. Sorgato, D. Masato, G. Lucchetta, Effects of machined cavity texture on ejection force in micro injection molding, *Precision Engineering,* 2017, 50: 440-448.

surface [7]. The roughness threshold value can be identified for *Ra* values of about 0.2 µm; differences should be expected for different polymers. Surface finish is discussed in more detail in Chapter 3 of this book.

11.3.3.2 Molding Surface Finish Guidelines for Ejection

The following guidelines should be considered during the plastic part design phase and reviewed during the mold design phase of the project, to ensure the product surface finish will not hinder ejection of the part from the mold:

- When using ejection pins, draft angles should always be specified to facilitate ejection and prevent hang-ups. If draft angles represent a cosmetic or functional problem, other ejection systems should be used, such as lifters, stripper plates, etc. Typical draft angles are 1-3°. When molding high-shrinkage resins, higher draft angles should be used to minimize the effect of surface friction.
- Mold designers should specify surface roughness using *Sa* (preferred) or *Ra*. Draft angles should be increased by 1° for every additional 20 µm of surface roughness.
- For a specific mold surface roughness, higher draft angles should be used for glass-filled or abrasive polymers. Resin suppliers typically provide the information in the property datasheet.
- Ceramic coatings, such as DLC or nitrides, can be used to reduce ejection friction at the polymer-mold interface. Also, they can be used to increase the wear resistance of the mold surface. For some applications, coatings can be useful to minimize the draft angles. Coating suppliers can suggest using a specific treatment based on the mold geometry and the molding resin. Coatings might require additional maintenance over time, as they can wear out.
- Mold makers should consider the effect of the machining process on mold surface roughness and polishability. For example, EDM can result in deep discharge craters that might be more complicated to polish than milling, reaming, burnishing, or grinding marks.

Injection molding simulation currently does not accurately predict the ejection force. However, it can be used in combination with structural analysis to analyze the ejection system design. Moreover, the analysis of shrinkage results can be used to identify areas that can be critical for ejection.

[7] T. Sasaki, N. Koga, K. Shirai, Y. Kobayashi, A. Toyoshima, An experimental study on ejection forces of injection molding, *Precision Engineering,* 2000, 24(3): 270-273.

11.4 Force Required for Ejection

The force required to eject the part from the core can be significant. Most injection molding machines have a rated ejection force that is 10% of the clamp force of the machine. Although it is not common, in some cases additional ejection force may have to be added to the mold to augment this ejection force or to provide ejection that is not inline with the machine ejector stroke direction. A good example of this is the high ejection force requirements on preform molds due to the combination of high cavitation, very shallow draft angles and high-polish surface finishes.

The force required to remove the parts from the core is dependent on a number of factors including the type of resin, molding conditions, surface finish, undercuts, cycle time, shrinkage, draft on the part, and plant conditions. The largest of these forces is the shrinkage holding part on the core and the sidewall draft. Although it might be difficult to calculate the exact force needed, the following equation can help to determine the expected force for ejection:

$$F_t = U \times F_n \times \cos D \tag{11.2}$$

where
F_t = tangential force (ejection force)
U = coefficient of friction for the resin
F_n = normal force (shrinkage)
D = sidewall draft angle

11.5 Where to Eject the Product

Selecting how and where to eject the product is a key part of mold design. A good understanding of the principles of ejection will help in arriving at the most suitable ejection method, sometimes with the help of the product designer.

First some basics:

- In plastics, as in all materials, the shear strength of the material is about half the tensile strength.
- Compressive strength is about the same as (or even higher than) the tensile strength.
- Tensile (and shear) strength varies with temperature. The strength is lower at higher temperatures. In other words, cold plastic is stronger than hot plastic. This seems obvious, but is often forgotten when planning for rapid ejection.

As a result of the above basics, it is always better to compress plastic than to shear it during ejection.

If an ejector mechanism (stripper, pin, and sleeve) pushes on a still-warm (weak) plastic part, it could damage it or not eject at all, pierce the plastic, or shear it off. If the ejection mechanism pushes rather than pulls it, we take advantage of the higher compressive strength (instead of the lower shear strength) and thereby can gain valuable cycle time by being able to eject sooner.

Figure 11.7 shows two typical cup-shaped parts. One of the parts has almost straight sidewalls (little or no draft), the other with good draft (5° or more). The ejection force (F_e) to eject the piece from the core depends on the friction and adhesion between the plastic and the side wall of the core, on the finish of the core or any undercuts where the plastic could "hang up", and to some extent on the force created by the shrinking of the plastic onto the core. It will depend heavily on the angle of the sidewalls, as less force is required with good drafts. In addition, more force is required if there is vacuum trapped between the core and the molded piece.

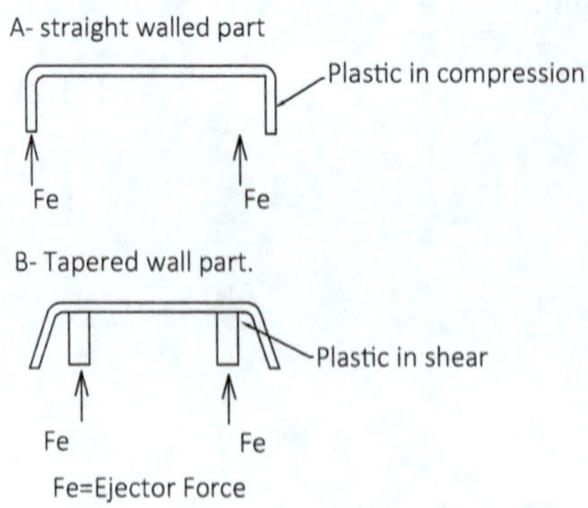

Figure 11.7
Ejection of typical cup (or box) shapes

The choices are to eject either from the rim or from the bottom of the parts, as indicated with arrows in Figure 11.7. Ejecting from the rim (with stripper or with pins) has the advantage that the force is directed against the bottom of the side-wall. In this case, F_e is *pushing*, and the sidewall is loaded in compression. This allows the product to be ejected *sooner*, but we must make sure that there is provision to break the vacuum, especially if there is little or no draft angle.

Ejecting from the bottom means that the product is *pulled* off the core; the force of the ejector pin acts on the bottom wall and loads it in *shear*, since the pins try to pierce the plastic. We must therefore make sure that the plastic is strong (and cold) enough not to be pierced, which means a longer cycle. There is usually no problem with vacuum, because ejector pins or sleeves provide *ideal, self-cleaning vents*. The largest possible ejector pin diameters should always be selected to increase the length of the shear edge created at the circumference of the pin. The cycle time difference between using ejector pins on the bottom and ejecting from the rim could be in the order of one or more seconds.

Another disadvantage other than longer cycles of Figure 11.7B, is the required long length of ejector pins passing through the whole depth of the core. There is also the danger of the product hanging-up on the extended pins at the end of the ejection stroke.

In the case of Figure 11.7, both stripper ejection and ejector pins could work fine. However, the straight-walled part would be best ejected from the rim using a stripper ring, if air could be added to the core to break the vacuum seal between the core and the plastic part. On the other hand, the highly tapered sidewall shown in Figure 11.7B would be best ejected using ejector pins, as the sidewall is highly tapered and the part could be deformed on ejection because the ejector forces do not act in pure compression. In some cases if the rim is not captured by the stripper ring the part could be deformed to be permanently larger on ejection.

11.5.1 Ejection of Deep Cup-Shaped Parts

For deep cup-shaped products it is best to eject at the points where the product is stiffest. The left-hand illustration in Figure 11.8 is preferred because the ejector force is applied to the coolest section of the part and the ejector force compresses the plastic. Remember that plastic is stronger in compression than tension. The image on the right would be less desirable, as it would be putting the plastic in tension and pulling the part out of the mold rather than pushing it. The mold may still run if the ejectors are in the incorrect position but the cycle time may be compromised.

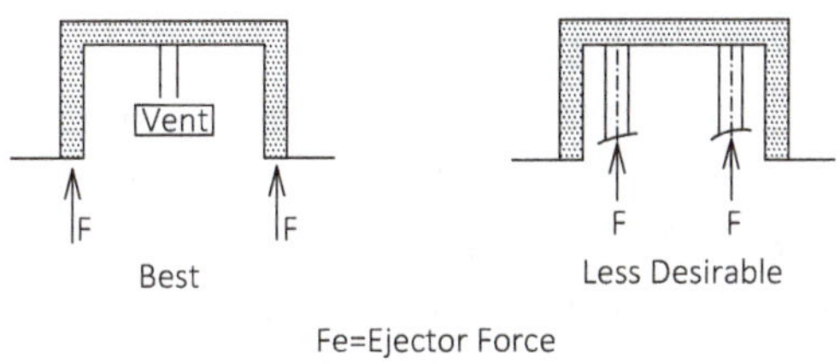

Figure 11.8 Ejection of a straight-walled part. The image on the left shows the preferred option

11.5.2 Ejection of Parts with Deep Ribs

A challenge with solid bosses and ribs is that the plastic forms a large volume at the base of the boss (shown in Figure 11.9 by the inscribed circle at the intersection of the rib and wall section). If this area is too large and the ejector pin is located under the boss, it could deform or puncture the part. Additionally, ribs and bosses can stick in the cavity or core and cause ejection problems if the draft is too small, if there are tool marks left from manufacturing, or if the surface finish is incorrect. The section in Chapter 3 on boss and rib design should be consulted to ensure that the design meets the recommendations for good part design for injection molding.

A radius *r* at the point where the boss joins the wall, while advantageous for strength of the boss, increases the already larger-than-wanted volume at this point. If the radius specified on the product drawing is too large, permission should be requested to reduce it.

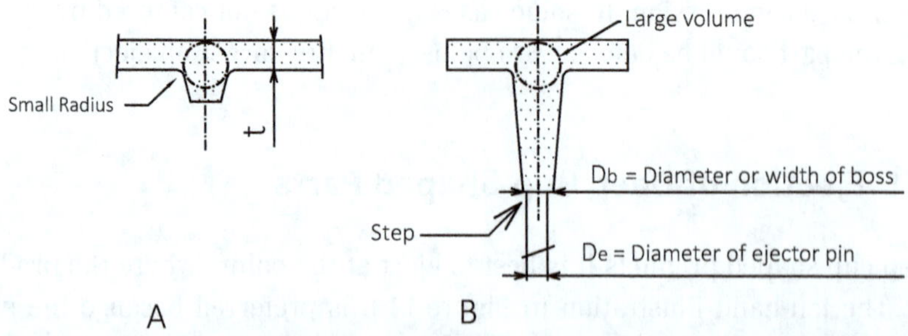

Figure 11.9 Need for ejection of ribs and bosses

Figure 11.10 shows that the ejection of deep ribs is similar to ejecting cup-shaped products. The proper way to eject the ribs is from the deepest point of the rib, for two very important reasons:

- The plastic will fill any rib better if the bottom is well-vented, as it would through the clearance of any ejector pins.
- During ejection, the rib will be pushed, which is desirable because the plastic is under compression. The smaller the draft angle, the more important it is to push, rather than pull. There is less force required to push on a rib because the rib will usually shrink away from the mold wall.
- The smaller the draft angle, especially with thin ribs, the more difficult it is to provide the necessary fine finish for easy ejection. Even though the plastic may shrink away from the wall, it can still hang up there.
- When molding an extended hub such as in Figure 11.10B, the plastic will shrink onto the core pin and will require more force. It should therefore be ejected (pushed) from the bottom, with *sleeve* ejectors (note that a sleeve will result in higher costs). The alternative - pulling the rib (or the hub) from the top - is not recommended, because there is the danger of piercing the top surface and possibly leaving the plastic in deep, inaccessible spots; it could also restrict the space available for good cooling in the core just where it is most required.

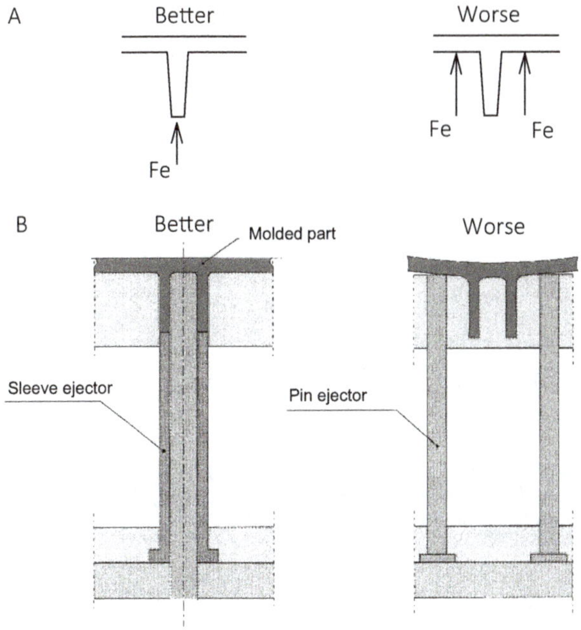

Figure 11.10 Preferred ejector pin locations for ribs and hubs: (A) Rib ejection; (B) Hub ejection (B - Courtesy of Manual do Projectista)

There is usually no need for ejectors for shallow ribs where the height of the rib is less than 1.5 × its thickness, but the need for venting must be considered to prevent implosion of the part or dieseling [8] at the bottom of the rib.

11.5.3 Ejection of Bosses

Ejection of bosses depends on the shape and solidity of the boss. For solid bosses (round or any other shape), there is usually no need for an ejector under the boss if the boss is relatively shallow (i. e., the depth is equal to or less than the diameter or width) and there is good draft angle (at least 5° per side). If the ratio of depth to diameter is larger, or whenever the draft angle is small, an ejector will be required, not only to eject the product but also to ensure good filling of the product, by providing venting at a low spot.

There are two kinds of tubular bosses: one with a through hole passing through it and another with a blind hole (Figure 11.11). A through hole or a deep blind hole

[8] Dieseling is caused by air being compressed by the plastic during injection in a non-vented region of the part. The result is burning of the plastic in contact with the compressed (heated) air.

can help in maintaining the wall thickness uniformity, eliminating the likelihood of sink marks and an extended cycle time. Ejector sleeves, such as those shown in Figure 11.11A and B, can be used for ejecting smaller bosses.

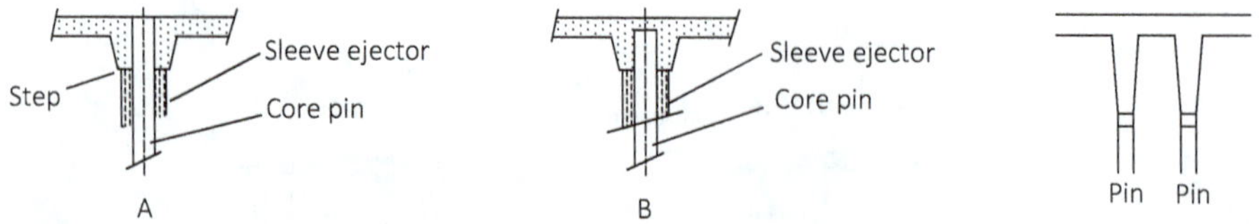

Figure 11.11 Examples of tubular boss designs: (A) Through hole; (B) Blind hole; (C) Using ejector pins rather than sleeves

In general, any boss that has a depth greater than the wall thickness and a draft less than 5° should be positively ejected.

The use of sleeve ejectors for bosses is expensive because it increases the mold shut height by adding at least one plate to the mold (see Figure 11.22). One plate is needed to retain the pin and one for the ejector sleeve. If possible, standard commercially available sleeve ejectors should be specified. The manufacture of small sizes requires specialized equipment. An alternative would be to place a few evenly spaced ejector pins at the bottom of a large boss (Figure 11.11C).

11.5.4 Ejection of Shallow Parts

For ejection of shallow products, with or without a rim, the designer should distribute the ejectors evenly across the part, and strategically in areas needing more force to ensure that the product is lifted without tilting or warping. If there are deep ribs or bosses, ejectors must be placed at their lowest point. Ejector pins and sleeves are "natural" vents (i.e., because of their operating clearances (fit), air can pass through). It is important that the land of any ejectors and sleeves be short enough not to create too much resistance for the air to pass through. However, the land must be of sufficient length to provide adequate guiding and surface area to prevent premature wear. For land length guidelines see Section 11.6.1.

11.5.5 Ejection of Rim Shapes

In general, for round parts a stripper ring is the preferred method of ejection. A stripper ring is a round or rectangular ring that surrounds the core and is attached to the ejection system. The ring has a portion of molding surface rim on it that is used to push the part off of the core. Details about stripper ring ejection can be found in Section 11.8. Stripper ring ejection is effective, relatively low cost, and allows a fast cycle because it pushes on the sidewalls that can be still quite warm but are already strong enough to take the ejection force. With today's technology, it is possible to produce odd-shaped stripper outlines (almost any shape that can be generated on CNC milling machines and grinders and/or with wire EDM). However, due to manufacturing costs, it is much better (easier, less expensive) to use round strippers, stripper bars, or conventional, inexpensive ejector pins.

In Figure 11.12, items 1-6 show simple rim configurations as are often found on utility shapes, ejected with strippers. The corner of the stripper must be about 0.1 mm away from the core (1). If a radius is specified there (2), it should be no larger than 50% of the wall thickness to make sure that the stripper will continue to push under the wall. If the end of the part is round (3) then the stripper ring should have a radius on it. A round design is preferred as it captures the part as the stripper ring moves forward.

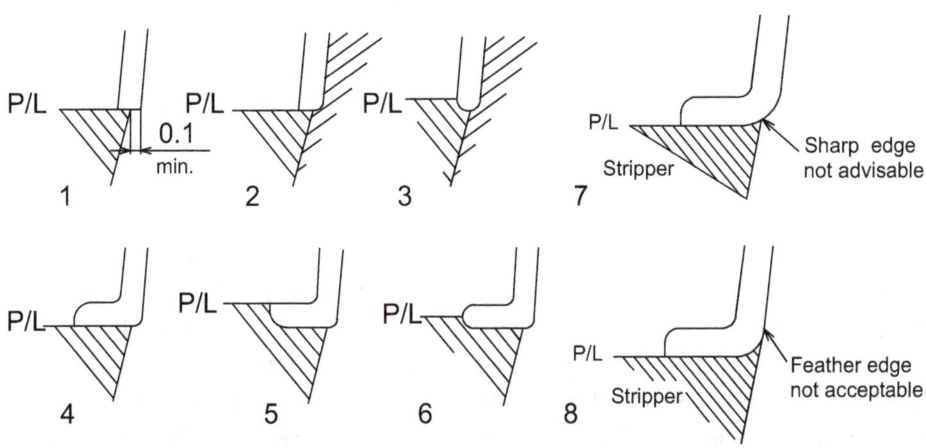

Figure 11.12 (1)-(6) Simple rims and stripper ring locations, (7) and (8) not advisable

If there is a large ledge on the rim, as in Figure 11.12 panels (4)-(6), the stripper ring should contain the ledge and the front of the stripper ring should still push on the vertical wall of the part. This will prevent the ledge from deforming or breaking.

Note that in all cases in Figure 11.12, there will be a witness line from the stripper ring all around the rim. If over time the stripper ring wears, then the interface between the stripper ring and the core may start to flash, causing the rim to be sharp to the touch.

A stripper can have a sharp edge if it incorporates an angle or radius but the stripper must never end in a feather-edge as shown in Figure 11.12 panels (7) and (8). A stripper designed as in these figures will damage easily and the mold will constantly need maintenance. A part design change is highly recommended.

If the radius is larger than the wall thickness (see Figure 11.13), there is the danger that the stripper (or ejector pins) will cut off (or pierce) the rim rather than eject, unless the plastic is well cooled and stiff. A smaller radius should be selected, such as shown in, so that at least part of the stripper (or the pins) will be under the sidewalls.

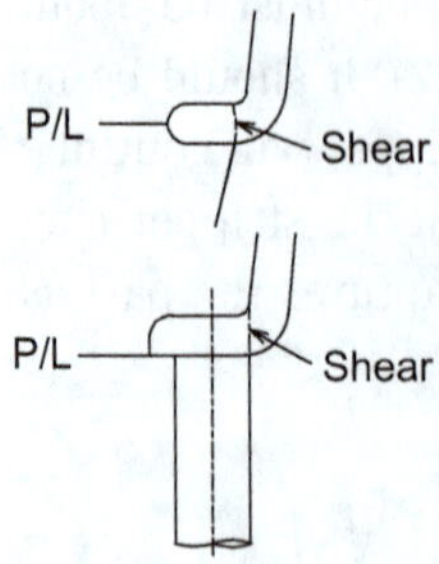

Figure 11.13
A large radius on the top lip should be avoided with stripper rings or ejector pins

Figure 11.14, panels (1)–(6), show similar rims as are shown in Figure 11.12, but this time ejector pins are employed. Note that the pin diameter is much larger than the wall thickness, and will push with a segment of the face of the pin that is exposed to the plastic. This has the advantage of not only using a large, rather than a delicate, small diameter pin, but it also covers a relatively large area of plastic which will therefore see a much lower specific ejection force with less risk pushing into, or shearing the plastic when ejecting. Such large ejector pins will leave an arc-shaped witness line. If the ejector pin contains any part of the molding surface then it will need to be located in position rotationally and it must be designed so that it cannot be installed 180° opposite the intended orientation. Normally this is done by adding a flat face to the ejector pin head and a matching surface in the ejector retainer plate to orient the pin (see Section 11.6.4 for more details on ejector pin alignment).

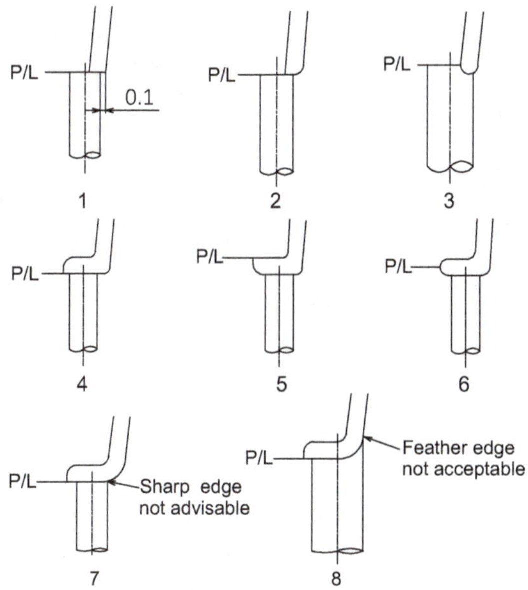

Figure 11.14 (1)–(6) Simple rim and ejector pin locations, (7) and (8) not advisable

The pin should never end in a feather-edge as shown in Figure 11.14 panels (7) and (8), which is difficult to maintain in production. In both cases, the pins must be secured from turning. It should be made clear to the customer that it would be much better to have the design changed so that it is not necessary to resort to cases as shown in Figure 11.14 panels (4), (5), or (6).

11.5.6 Ejection of Cold Runners

For automatic molding, the runner must be positively ejected without a risk of hanging up in the mold and requiring an operator to remove the runner. To ensure that this occurs, runner pullers have been developed to eject the cold runners after the molding cycle has been completed. A runner puller will not work while the plastic is too hot and, thus, too soft. It is, therefore, important to consider the cooling layout in the area of the runner pullers to ensure that the runner will cool down quickly and not cause any undue increase in the molding cycle. It is also important to keep the cross-sections of the runner to a minimum while still allowing proper, fast filling of the cavities. Similarly, it is important to reduce the mass around the pullers to a minimum without causing restrictions in the plastic flow.

The puller arrangement shown in Figure 11.15 reduces the mass of the plastic, brings additional cooling into the center of the junction of sprue and runner, and eliminates the large mass of plastic at the head of the ejector pin.

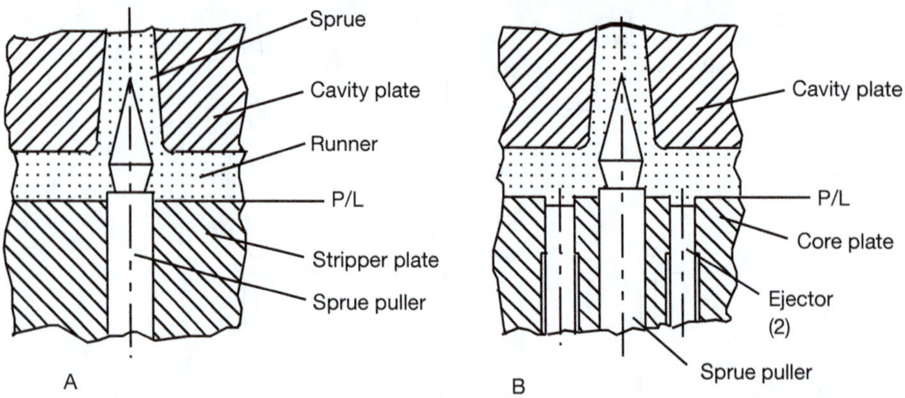

Figure 11.15 Example of runner puller in an intersection to reduce mass and add cooling

The two examples above are intended to show how a typical problem of reducing the plastic mass can be solved. With combination of hot and cold runners in a mold, a heavy sprue is avoided, but the design principle is the same for the runner pullers and/or suckers required in such molds.

There are several methods to eject the cold runner:

1. *Using ejector pins for ejecting runners.* Regardless of the cross-section of the runner, the ejector pin must *never* extend into the runner channel but must be short by the distance *P* (Figure 11.16). The mold parts are toleranced so that *P* will be recessed between 0.0 and 0.2 mm (0.010"). This prevents the runner from molding onto, and sticking on, the ejector pin.

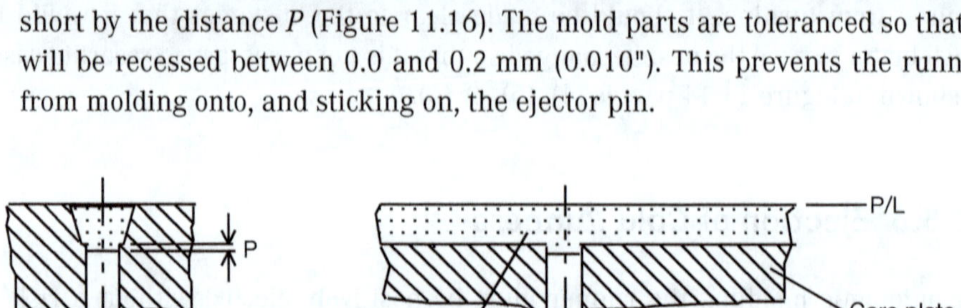

Figure 11.16 Ejector pin used to eject the runner (ejector pin must be recessed)

2. *Pulling the runner with sucker pins.* In many cold runner molds, the runner is retained on the core side by suckers until ejection is commenced. A typical shape of sucker is shown in Figure 11.17. In terms of the dimensions of the sucker, $W > 1.2D$, to let the ejector pass. *A*, the re-entrant angle of the sucker, should be between 5 and 10°.

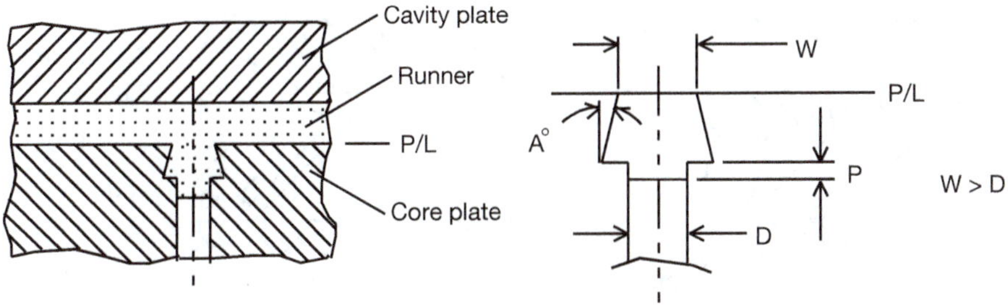

Figure 11.17 Typical sucker pin detail

3. *Pulling the runner with sucker pins* (in three-plate molds, and two-plate molds with stripper plates). As the mold opens, the runner stays on the stripper plate. The stripper plate then strips the runner off the sucker pins. It is important that the runner does not remain with the stripper plate after ejection. Therefore, the sucker pin must enter the runner by between 0.0 and 0.2 mm (0.0–0.010").

Hardened bushings (not shown in Figure 11.18) are recommended around the sucker pins. Occasionally, even for hardened stripper plates, commercial, very hard "drill bushings" are used to minimize wear. The bushings must be held by shoulders, screws, or a retainer plate to stay in place. It is not good enough to press them into the plates. Note that the *pulling force* of the sucker depends on the circumference of the head, and on the angle *A*. This is especially important with three-plate molds, where the resistance against pulling a drop (the runner extension toward the gate) may be considerable, especially where a very small draft angle is selected to keep the mass of the plastic in the drop small.

Sucker pins are usually made from standard ejector pins. The diameter *D*1 should be only slightly less than *D*, and the angle *A* should not be more than 5° (Figure 11.18). It is always possible to increase the pulling force required by increasing the angle, without greatly weakening the sucker pin. It is important to provide a good lead-in for the pins in the stripper plate (or bushing), not only for ease of assembly but also in cases where the stroke of the stripper is so large that the pin disengages from the stripper plate at every molding cycle. Such stroking out of the plate during operation should be avoided, but this may not always be possible.

As with all ejector pins, all sucker pins must be allowed to float freely sideways in the ejector retainer plate to ensure proper alignment without forcing the pins, which would result in rapid wear of pins and/or bushings, or breakage of the pins.

Since it is important to make the sucker pins as large as possible, there is the danger that they may restrict the flow of plastic in the runner. If the pins restrict the flow, then the runner can be widened locally to provide an unrestricted path around the sucker pins.

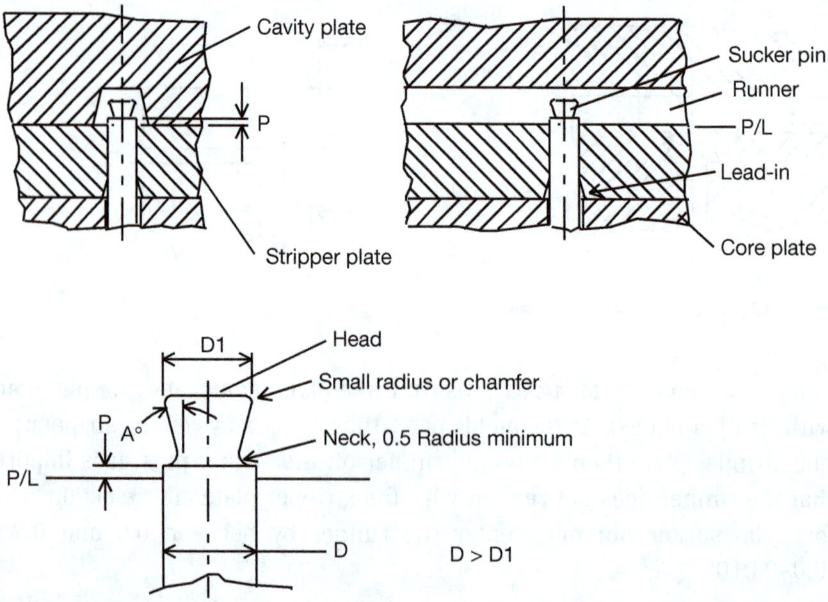

Figure 11.18 Sucker pin design for pulling the runner and ejecting with stripper plate

In general, the stiffer the plastic is (at the moment of ejection), the fewer ejectors are needed. Adding an ejector to each leg of the cold runner is a good place to start. Sometimes the pins can be located at the intersections of runners to avoid adding one to each branch. Since the runners are usually thicker than the product itself, it is important to provide good cooling to remove the heat from the runner. It is good practice to have cooling lines just below the runners; however, this makes it impossible to place ejectors (or sucker pins) right under the runners. Cooling lines are often required on either side of the runners (Figure 11.19, left). Where cavities are fed from a network of balanced runners, it may be possible to place the ejectors (or sucker pins) under the connecting runners (bridge runners), and the cooling directly under the runners (Figure 11.19, right).

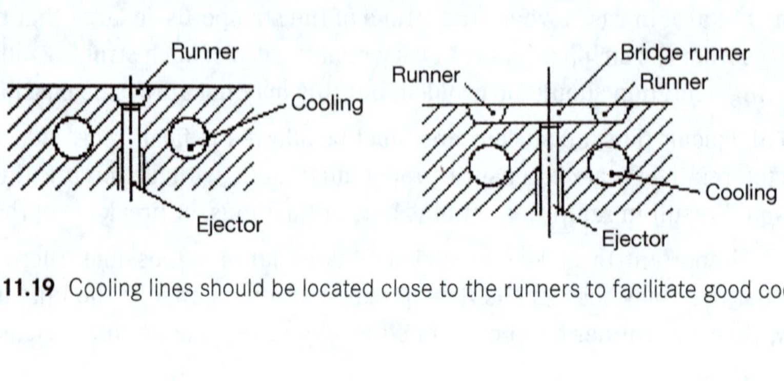

Figure 11.19 Cooling lines should be located close to the runners to facilitate good cooling

11.6 Ejector Pins, Blade Ejectors, and Sleeves

Pins, blades, and sleeves can be used for ejecting any shape of product. The advantage of ejector pins and sleeves is that they are relatively inexpensive to purchase from a standard mold parts company and easy to use. These companies also provide recommendations on the proper design and installation of these devices. Ejector pins, sleeves, and blades provide good natural, self-cleaning venting in areas where the air would otherwise be trapped and would require vent pins.

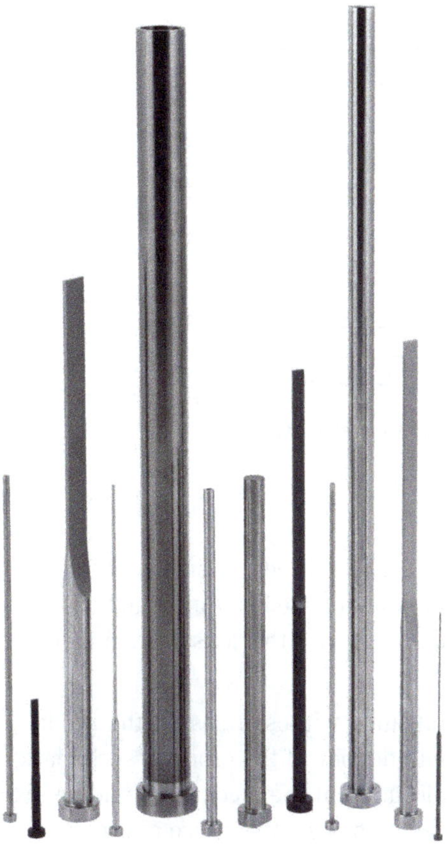

Figure 11.20
Typical injector pins, blades, and sleeves (Courtesy of DME)

The main disadvantage of ejector pins and/or sleeves is that the area where the product is pushed is relatively small, and the product must be well-cooled and stiff enough so that the forces exerted by the pins or sleeves will not deform or pierce the plastic during ejection. Other disadvantages are that these ejectors require an ejector box that can weaken the rigidity of the mold, and the pins often take up precious cooling line space on the core side.

Figure 11.21 shows a section view of a typical ejector pin type mold. In this type of ejection the (mostly) round pins push directly on the plastic part to be ejected. An ejector box, containing the pin, is located behind the core plate. This plate typically interfaces with the machine ejection mechanism. The pins are located strategically to allow the part to free itself from the mold. If the mold must be inserted between the machine platens by sliding it into the machine along the platens, knockout pads (see Figure 11.21) are often used to keep the machine ejectors level with the moving platen surface for ease of mold installation.

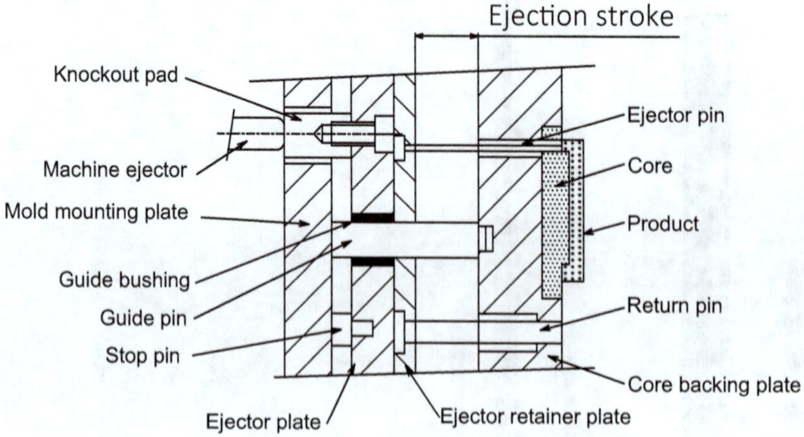

Figure 11.21 Section through ejector pin mold

If an ejector sleeve is needed then Figure 11.22 can be used to visualize the ejector box construction of the mold. Note that an additional plate is required to house the ejector pin at the center of the sleeve ejector. Otherwise the construction is much the same as a pin ejector design.

Ejector pins and sleeves leave round marks called witness lines on the product. Since ejectors are generally inside or on the underside of the product, such marks are considered acceptable. In all cases the location of the ejectors should be discussed with the customer to avoid any issues with the appearance of the molded part.

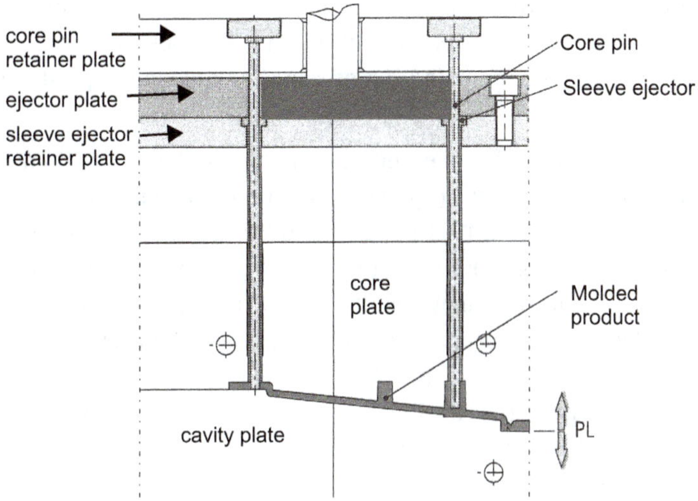

Figure 11.22 Section through sleeve ejector mold (Courtesy of Manual do Projectista)

11.6.1 Ejector Pin Clearance (Fit) and Length of Land

The diameter of the ejector should be a standard ejector pin size and should always be smaller than the bottom diameter (or width) of the boss or rib to ensure that the pin will not scratch the wall when advancing during ejection. The difference per side (step) should be at least 0.13 mm (0.005"). This step helps machining and permits the use of oversize ejectors if required for repairs later on.

In Figure 11.23 the clearances at the molding surface and the length of the land for ejector pins and sleeves should be as specified by the pin manufacturer. In general, clearances correspond to the vent sizes shown in the chapter on venting. There are several reasons for limiting the clearances in the diameter and the length of the land of ejector pins:

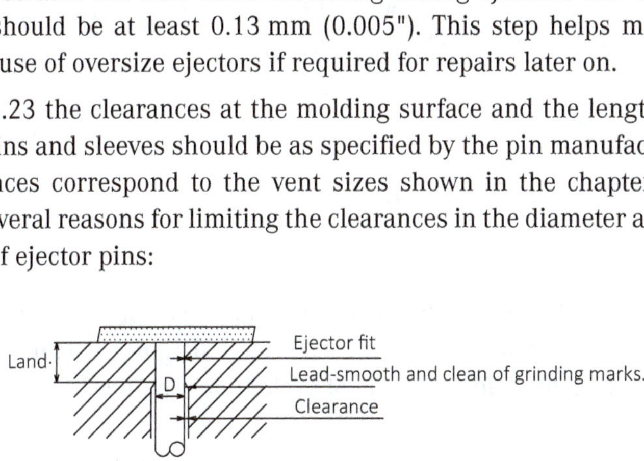

Figure 11.23 Ejector pin detail at molding surface. The land should be <2*D* and must have a polished lead-in to avoid damaging the pin

Clearance of pin in bore: Too tight a fit may bind the ejector, resulting in breakage, or may contribute to rapid wear of pin and/or bore. Too loose a fit may cause flashing, especially when injecting with high pressure and when the plastic is very hot and has a low viscosity.

Length of land: As suggested above, the land should not be longer than 2*D* (the diameter of the pin for small sizes (up to 6 mm), and not longer than 1.5*D* for larger sizes. If the land is too long, it is more difficult and expensive to produce, as it requires more jig grinding or reaming time. A longer land also creates more resistance to the escaping air and may result in poor venting. If the land is too short, there may not be enough surface area to provide adequate guiding for the pin, which can result in premature wear. A land that is too short does not allow enough resistance to keep plastic from flashing, especially when injecting high-flow materials with high pressure. If the land is too short the pin could also buckle when the ejector plate is energized forward.

11.6.2 Blade Ejectors

If very small ejectors are required in locations such as at the bottom of ribs, etc., it is better to provide a flat ejector pin instead of a very small, round ejector pin.

If the width at the bottom of the rib is 1.6 mm, it is better to have a flat ejector, 1.5 × 5.0 mm, instead of a round pin 1.5 mm diameter. The area facing the plastic in the first case is only 1.77 mm²; in the other, 7.5 mm².

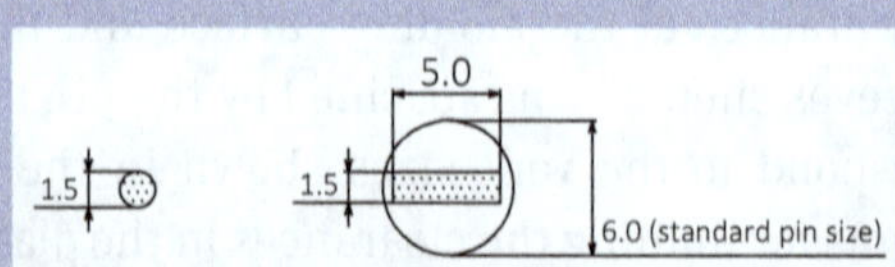

The much larger surface is less likely to penetrate the plastic during ejection, particularly if the plastic is still warm and has little resistance. This may translate to the possibility of saving a few seconds of cycle time. The pin is also stronger and will last longer.

Mold hardware suppliers have listings of standard blade ejectors. Designers should always try to find a standard size before specifying a special part to be made in-house. A special ejector blade would not only be very expensive initially, it would also have a high maintenance cost if it needed replacement. However, keep in mind that sometimes a special ejector may be required - the function of the mold should not be compromised for the sake of a standard part cost.

When designing the slot in the core for the special shape ejector blade, the designer has the option to wire-EDM the slot into the solid core or design a composite core with a two-piece structure to allow for grinding and conventional close-tolerance machining. The clearances and tolerancing for the ejector blade and the length of land should be selected to be similar to the dimensions for a round pin of a size corresponding to the smaller dimension of the slot. This will not only ensure adequate guiding, and enough sealing and venting length, but also facilitate ease of manufacture and avoid excessive length of the special shape. The rear of the core under the slot is bored out for clearance of the shank of the pin from which the special ejector was made (Figure 11.24).

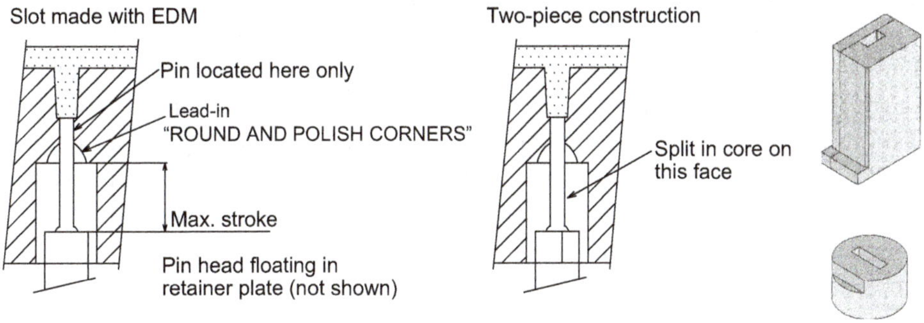

Figure 11.24 Common core construction options for blade ejectors

With a two-piece construction (Figure 11.24, right), a relatively shallow U-shaped groove is machined into one portion of the core. The matching portion is flat. All four sides can be ground if necessary. The counter-bore for the shank of the pin is similar to the solid core.

Do not try to match two U-shaped grooves opposite each other, as shown in Figure 11.25. To ensure perfect matching, the tolerances would have to be unnecessarily tight and costly. Any slight mismatch between the two grooves will result in binding, galling and premature failure of the blade.

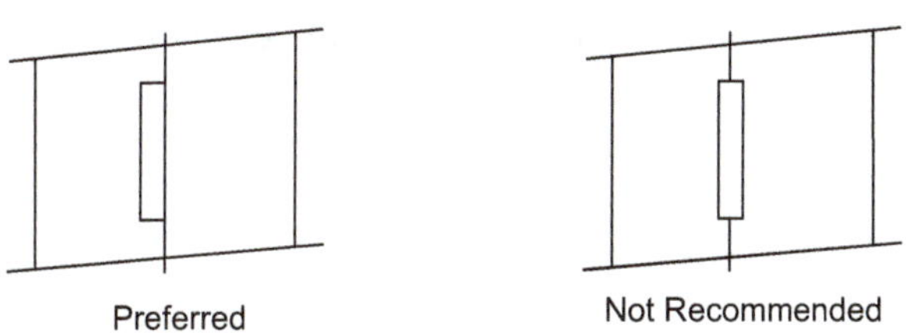

Figure 11.25 Preferred design (left) of a two-piece core construction for a blade ejector. Poor design (right)

11.6.3 Number, Size, and Location of Ejector Pins

There should be as few ejector pins as possible to ease the challenges of other aspects of the mold design and to reduce costs, but good, reliable ejection must be the prime objective. The size, quantity, and location of ejectgor pins is critical to mold function.

11.6.3.1 Ejector Pin Sizing

The designer should always use standard sizes available from part manufacturers. Special sizes are very expensive to make or to procure from suppliers.

As a general rule, make the ejector pins as large as possible. (Note that bore diameters above 6 mm (ca. $\frac{1}{4}$") are cheaper to produce than smaller bores.) Sizes 3 mm ($\frac{1}{8}$") diameter and smaller should be avoided, particularly if the pins are longer than 50 × the diameter. Commercially available ejector pin sizes smaller than 3 mm (2.5, 2.0, and 1.5 mm) are stepped down from a 3 mm ejector pin. The thicker portion extends to a distance 75 mm from the head of the pin. The reason to avoid such long, slender pins is the danger of buckling the pins under the forces from the resistance to ejection. Frequent replacement of these pins is required, as they wear more rapidly.

In order to determine the size and number of ejector pins required, the following equations can be used.

The *compressive stress on an ejector pin* (σ_{pin}) can be used to define the number of pins or their diameter:

$$\sigma_{pin} = \frac{F_{ejection}}{n_{pins} \times A_{pin}} \tag{11.3}$$

where

$F_{ejection}$ = ejection force

n_{pins} = number of ejector pins

A_{pin} = surface area of the front of the ejector pin

Note that the resulting compressive stress should be checked agains the pin steel fatigue limits.

The *shear stress* (τ_{part}) in the part can be used to define the pin diameter or the ejection system parameters:

$$\tau_{part} = \frac{F_{ejection}}{\pi \times n_{pins} \times D_{pin} \times H_{part}} < \frac{\sigma_{plastic}}{2} \tag{11.4}$$

where
H_{part} = plastic wall thickness opposite the pin
$\sigma_{plastic}$ = plastic yield strength
D_{pin} = pin diameter

To ensure that the *pin will not buckle* during ejection, the following equation (for round pins only) can be used:

$$\frac{F_{ejection}}{n_{pins}} < \frac{\pi^2 \times E \times \left(\frac{\pi \times R_{pin}{}^4}{4} \right)}{\left(0.7 \times L_{pin}\right)^2} \tag{11.5}$$

where
E = modulus of elasticity of the pin
R_{pin} = pin radius
L_{pin} = the unsupported length of the ejector pin

11.6.3.2 Ejector Pin Locations

There are some locations where ejector pins are essential and preferred. Pins should be located:

- Where they will provide necessary venting of areas which do not vent to the parting line
- At the lowest points of the product, such as rims, deep ribs, or bosses
- At or near the corners of the product
- Symmetrically and evenly spaced about the product, if possible
- At intersections of rib and rib, or wall and rib. By locating pins at intersections, the ejector size can be increased. By adding a small radius or chamfer in the corners, the pin size could be made still larger, which may be necessary to get to the next standard size pin.

Figure 11.26 provides some recommendations on good and poor placement of ejector pins.

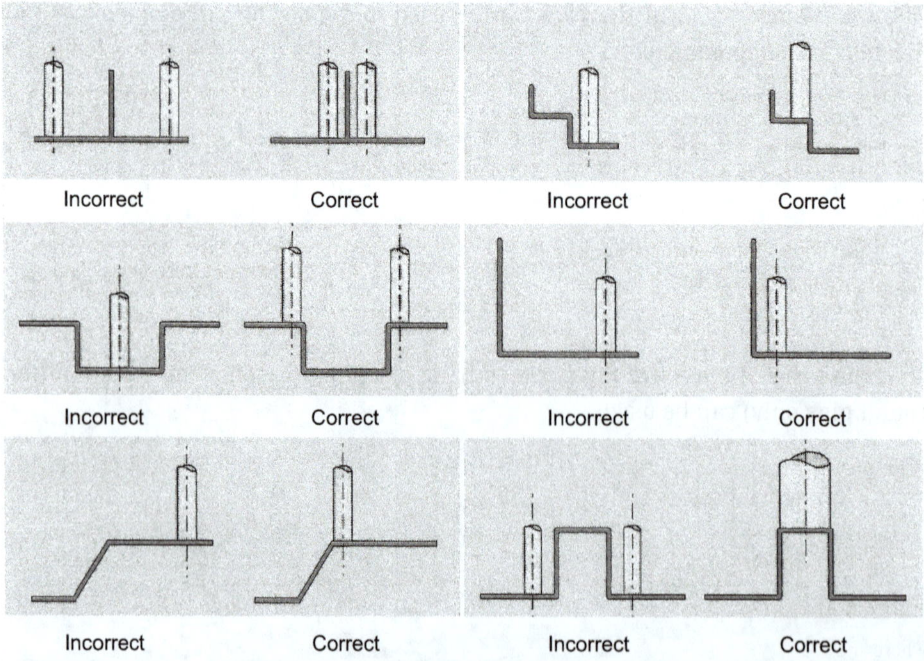

Figure 11.26 Recommended locations for ejection using pins (Courtesy of Manual do Projectista)

11.6.4 Finish of Ejector Pins and Bores

To ensure long wear life, pins are made from H13 or similar steel or powder, hardened and nitrided, with a surface hardness in the order of 70Rc. They are finished to a very high polish and then covered with a solid lubricant.

The bores into which the pins will be inserted must be smooth and free of grinding marks that could act as a file on the pins. This is especially true on the lead-in (see Figure 11.23). A suggested finish is 0.4 µm or better (16 micro-inches). The bores are usually produced by jig grinding or reaming, to create a very straight hole with no flaring or bell-mouthing at the molding surface.

The ejector pin boles should be funnel-shaped from the rear to facilitate entry of the pins (sleeves) during assembly and to provide a large, low-tolerance clearance hole. Pin manufacturers will provide specific details on the design of the installation cores for ejector pins and blade ejectors.

11.6.5 Preventing Ejector Pins from Turning

Ideally an ejector pin should be located where its orientation is not important - for instance, on a flat portion of the part or at the bottom of a rib or boss. This eliminates the need to orient the pin. Ejector pins must be prevented from turning when:

- The front of the pin in the cavity is part of a contoured shape of the product
- It carries engraving, which must stay in a determined relation to the product
- A large pin is used partly for ejection and partly as a return pin.

A preferred method is to machine the head of the pin with one or two flats, as shown in Figure 11.27. If the two-flat method, is used special attention must be given to ensure that the pin is not installed incorrectly, as it would be possible to install it 180° from the correct orientation.

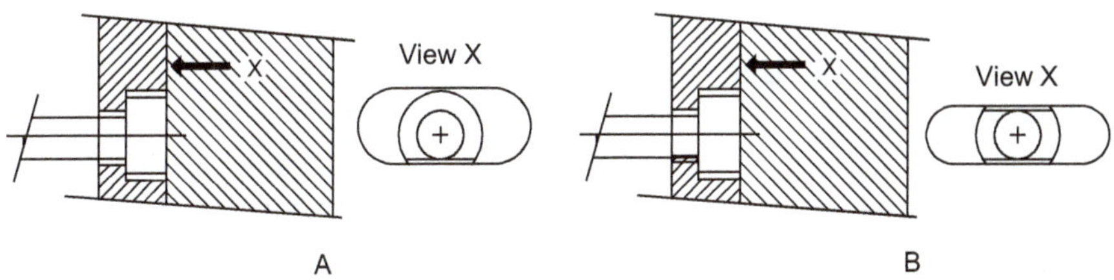

Figure 11.27 Common methods to maintain orientation of ejector pins

As an alternative, an inserted key may be used (see Figure 11.28). The disadvantage of using a key is that an additional mold part is required. The advantage is that one larger key can be designed to orient multiple pins in a row.

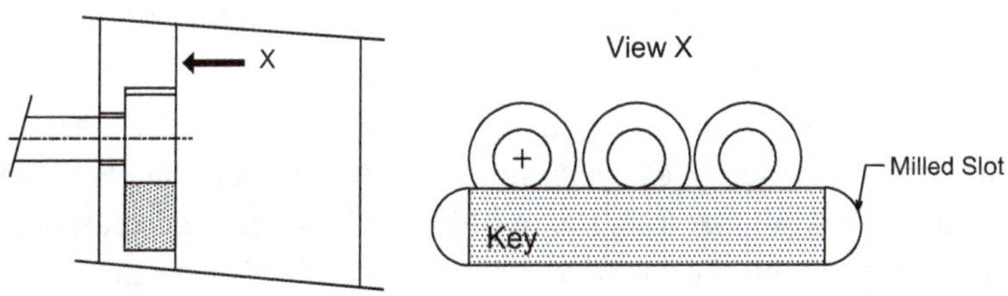

Figure 11.28 Use of a key and a flat on the ejector pin head to prevent rotation

11.7 Ejector and Ejector Retainer Plates

Ejector plates and ejector retainer plates are used to house the ejector pins and blade ejectors. The machine ejectors push against the ejector plate to activate all the ejector pins in parallel. Ejector plates are subject to high forces from the machine and high resistance from the plastic part which is held onto the core or cavity. Ejector and retainer plates need to resist these forces along with the injection forces created from molten plastic entering the cavity.

The ejector plate should be designed to be of a generous thickness, especially if the span between machine ejectors is large, to keep deflection to a minimum. If in doubt about its thickness, calculate the deflection (see Section 11.7.4). Remember, deflection depends on the geometry of the plate and the modulus of elasticity, not on the tensile strength of the steel.

11.7.1 Ejection Forces

The ejection forces acting on the ejector plate can be significant, and they are difficult to estimate. They depend mainly on the following elements:

- *Finish of the core:* In general, the rougher the core, the more force is needed to push the product off the core. However some materials, like polyolefins, release easier with a slight texture (like a sand blast finish) on the cores, as air can travel along the texture to help with release.
- *Draft angle:* The smaller the draft angle, the more force is required to remove the product from the core.
- *Undercuts in the product or added to retain the part on the ejection side of the mold during opening:* Some heavy undercuts may even require additional, high-power hydraulic actuators built into the mold.

All the above forces which resist ejection of the part must be considered. As it is difficult to estimate these forces, we assume that the ejector plate is subjected to the maximum force available from the machine. This is also a good assumption in case the part is frozen on the core and will not eject. In this case the full machine ejector force will be applied to the ejector plate.

> Injection molding machine specifications list the maximum available ejector force. It is usually in the order of 10% of the full tonnage of the machine.

Once the maximum machine ejector force is known, the required strength of the plate can be calculated. This ejection force data is available from the molding machine specifications.

11.7.2 Injection Forces

The ejector pins are loaded at their face with the force generated by the injection pressure during the injection cycle. The greater the area subjected to the injection pressure, the greater the force transmitted through the ejector pins to the ejector plate.

By placing a sufficient number of *stop buttons* under or near heavily loaded pins, the effect of plate deflection due to these forces can be eliminated. Occasionally, the ejector plate of a mold rests directly on the machine platen (no mold mounting plate). In this case, the designer must make sure that the stop buttons will be located so that they are not located over any ejector holes, tapped holes, or slots in the machine platen and, thereby, become at least partly useless.

11.7.3 Number and Location of Machine Ejectors to Be Used

The designer must take relative locations of the machine ejectors into account and must keep deflection as small as possible when concepting the ejector plate. Where possible, the designer should use machine ejector locations symmetrically spaced under the load of ejector pins.

It is preferable to use *four or more machine ejectors if possible.* In small molds, one center ejector is often acceptable.

The use of more than four ejectors is more difficult in mold setup, since all ejectors must "hit" at the same time. However, occasionally it may be necessary if the span between four ejector locations is too large and the resulting plate deflection would be unacceptable.

11.7.4 Calculation of Ejector Plate Deflection

By reducing the ejector plate to a simple beam supported on both sides, the resulting deflection calculated will be slightly higher than with a full plate, but it reduces the formula to a simple calculation shown in Equation 11.6. The resulting higher deflection calculated will push the designer to make the plate thicker than needed, but this is a good design practice anyway.

The plate deflection, Y, must be kept to a minimum. $Y < 0.1$ mm (0.004") is normally acceptable.

Equation for simple beam deflection:

$$Y = \frac{P \times L^3}{48 \times E \times I} \quad (11.6)$$

where:

Y = maximum deflection at the center of the plate (mm)
P = total load (N)
L = unsupported length (mm)
E = modulus of elasticity (200,000 MPa)
I = moment of inertia = $b \times h^3/12$
h = beam thickness (mm)
b = beam depth (mm)

Example: Ejector plate deflection calculation

From Figure 11.29, assume a plate 600 mm (L) × 200 mm (b) × 30 mm (h) with a force of 600 kN acting on it. What is the deflection of the plate with the applied force?

$I = b \times h^3 \div 12 = 200\ \text{mm} \times (30\ \text{mm})^3 \div 12 = 450{,}000\ \text{mm}^4$

Conversions:

$1\ \text{N/mm}^2 = 1\ \text{MPa}$

$Y = P \times L^3 \div (48 \times E \times I) = 800\ \text{N} \times 600\ \text{mm}^3 \div (48 \times 200{,}000\ \text{N/mm}^2 \times 450{,}000\ \text{mm}^4) = 0.04\ \text{mm}$

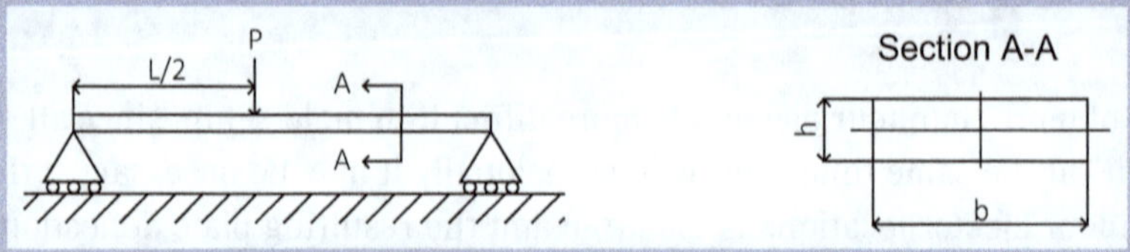

Figure 11.29 Simple beam deflection calculation

11.7.5 Ejector Pin Retainer Plate

The purpose of this plate is to hold the ejector pins, blades or sleeves in position on the ejector plate without restricting their lateral movement.

The heads of the ejector and return pins, or sleeves, should float with lateral play in the retainer plate so that the pins or sleeves find their proper alignment with the bores in the core (Figure 11.30). The forces on the retainer plate are small, usually just the friction of the pins in their bores.

The thickness of the retainer plate at the head (*t*) need never be more than 3 mm ($\frac{1}{8}$"). Axial clearance should be 0.1 mm (0.004"). Radial clearances HC and SC should be at least 0.5 mm (0.020").

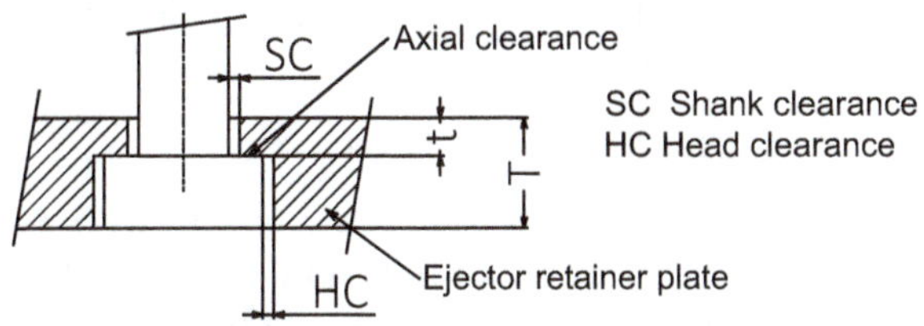

Figure 11.30 Typical ejector pin installation in ejector pin retainer plate

The thickness of the ejector retainer plate does not add to the strength of the ejector plate against deflection. In cases where the thickness or shut height of the mold is critical, or additional thickness in the ejector plate is needed, it may be necessary to place the bores for the ejector and return pin heads into the ejector plate, and reduce the retainer plate to 4 mm ($\frac{3}{16}$") but at the same time increase the stiffness of the ejector plate considerably because of the thickness gained. As the stiffness increases with the third power of the thickness, the ejector plate at the right in Figure 11.31 is much stiffer than the one on the left.

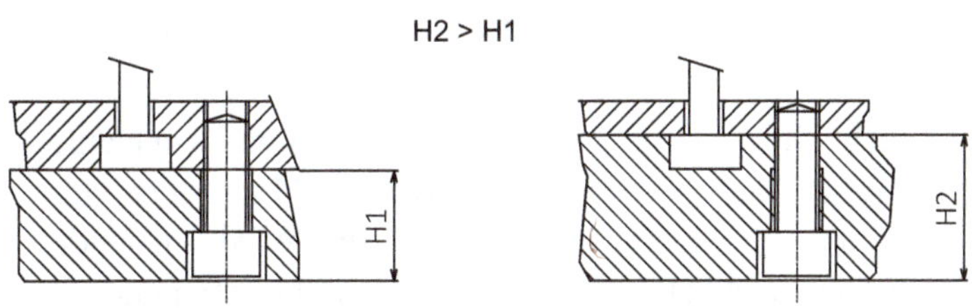

Figure 11.31 Surface-mounted ejector pin (left) and recessed ejector pins (right)

Example:

If the ejector plate thickness H = 25 mm, and the head is 6 mm, the stiffness is proportional to 25^3 or 15,625. If the plate thickness is increased to H = 31 mm, the stiffness is proportional to 31^3 or 29,791. This illustrates that the plate that is 31 mm thick is almost twice as stiff against deflection compared to the 25 mm one.

11.7.6 Ejector Return Pins

Ejector return pins ensure that all ejector pins are returned to their back position by the time the mold closes, to prevent the pins, sleeves or blades from ever touching the cavity surface and causing damage. Return pins must always be provided, even if the ejector plate is positively returned by the molding machine or by any other method, in case the positive return fails.

General guidelines for return pins are as follows:

- The return pin should be sized to return the ejector pins back flush to the molding surface. The return pins can be toleranced to be loaded in the back position by 0.13 mm (0.005") to ensure that the pins are fully back.
- There should always be at least four return pins, evenly spaced, to ensure that the ejector plate moves smoothly and does not jam or twist. If the mold is long and narrow, then six return pins (four corners and two in the center) is recommended.
- Do not position the return pins where they could hit a vent channel.
- Minimum diameter of the return pin should be 12 mm ($\frac{1}{2}$"). Preferred sizes are 16 or 19 mm ($\frac{5}{8}$" or $\frac{3}{4}$"). The larger the pins, the less damage will be done on the opposing plate surface where they hit.
- Always use standard (commercial) pin sizes.
- For proper clearance for a return pin in a mold plate, the passage through the plate can be much larger than that for an ejector pin, to prevent wear and pick-up.

Usually, the return pin meets the cavity at the parting line - see Figure 11.32 (left). If, however, the return pin in its forward position would interfere with a take-off or with the free-fall of the products, the pin must be shortened and another pin placed opposite so that they meet below the parting line, in the core plate, when the mold is closed as shown in Figure 11.32 (right).

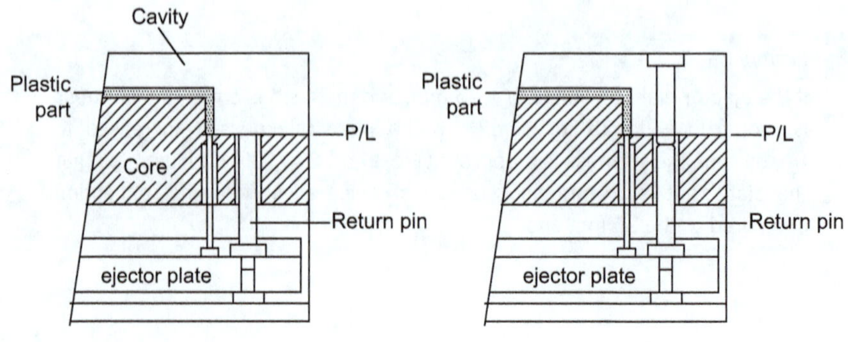

Figure 11.32 Return pin configurations

A combination ejector and return pin may also be used. Occasionally, it is practical to use large ejectors at the rim of a product with an odd-shaped circumference that makes it impractical to provide a stripper plate. This has the advantage that a large segment of the pin acts as an ejector, and the remaining, much larger surface, acts as a return pin (Figure 11.33). In this case, the length of the pin must be so that it stops exactly at the P/L. Note that in these cases the combination ejector/return pin must be keyed to prevent rotation.

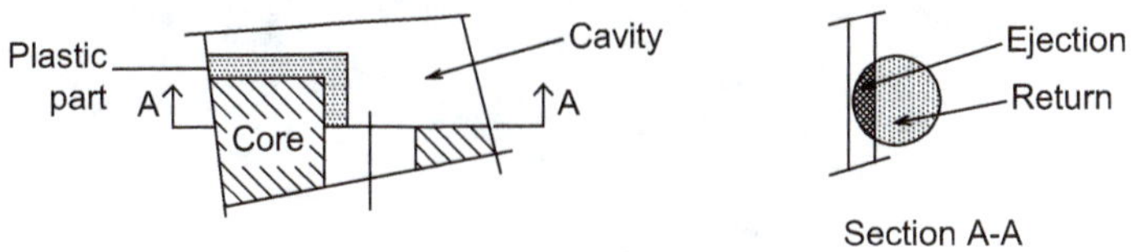

Figure 11.33 Example of a combination ejector and return pin

11.7.7 Ejector Box

The ejector box is that portion of the mold assembly which surrounds, supports, and guides the ejector plate. Traditionally, the ejector box consists of a mold mounting plate, two parallel bars (or four bars making a frame), and additional internal support pillars as needed for support (Figure 11.34 and Figure 11.35).

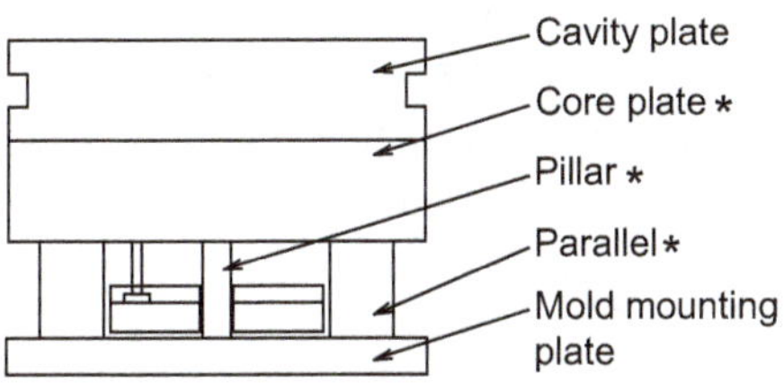

Figure 11.34 Traditional ejector box end view. * Parts of the ejector box - the pillar, parallel and core plate - can be replaced by a single plate with a cutout for the ejector plate

Figure 11.35 Automotive mold with hydraulically driven ejector box (Courtesy of Accurate Machine and Tool)

There are still many molds built this way using complete standard commercial mold sets or using certain standard sizes of rectangular plates, parallels, and round pillars, all held together by a large number of screws.

If the ejector plate must have ejector pins in odd locations, a specially shaped ejector plate, such as shown in Figure 11.36, may be more practical to use with an ejector box machined from a solid plate. Sometimes this can be combined into the core backing plate (the plate that mounts to the moving platen). Combining all the parts of an ejector box into one plate results in a more solid and thinner mold with more space available for cooling and air channels.

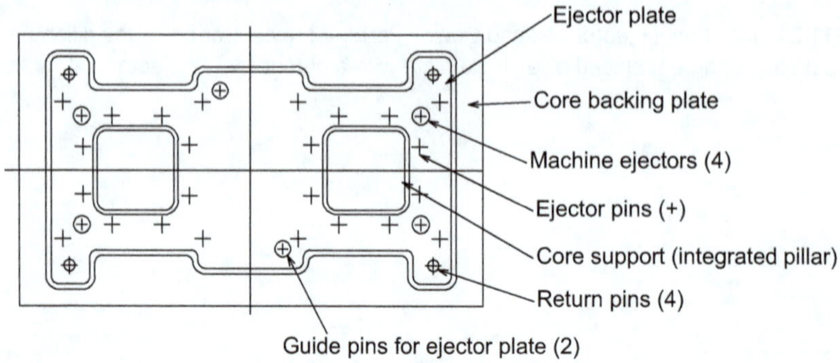

Figure 11.36 Plan view of typical ejector plate layout in a single solid ejector box/core backing plate

For ease of manufacturing the ejector box cutout, the inside radii for ejector plate and for pockets should be as large as possible. Smaller radii (smaller cutters) are of course possible, but they are much slower, and therefore more expensive to use.

The outside corners of the ejector plate should be chamfered generously so as not to interfere with the radii in the core backing plate. The minimum clearance around the ejector plate should be about 3 mm (⅛").

11.7.7.1 Guiding Ejector Plates

Ejector plates must be guided in at least four evenly spaced locations using separate guide pins. Standard leader pin bushings should be used to guide the plate. The clearance *C* in the retainer plate (see detail in Figure 11.37) should be 0.1 mm on diameter (or 0.005").

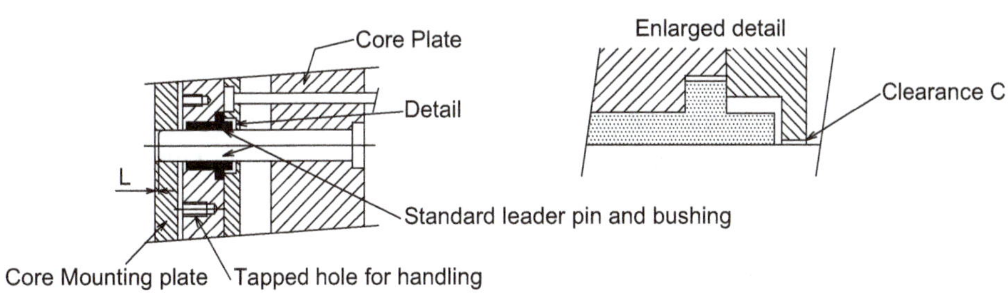

Figure 11.37 Ejector plate alignment and guiding

At assembly, the leader pins are first inserted (tight fit) into the core plate. Then the ejector pin retainer plate is assembled into the leader pins. The ejector pins are then inserted through this plate and into the core plate and cores. Next, the ejector plate is lined up with the guide pins, pushed into position, and all the screws are installed to hold the retainer plate to the ejector plate.

> It is good practice to align the ejector retainer and ejector plate with two dowels at opposite corners.

When the ejector box is made from one piece, there is usually no access from the sides to help in the alignment of the ejector plate during assembly. For this reason, the guide pins must be long enough (see dimension *L* in Figure 11.37) to project beyond the bottom face of the ejector box so that the alignment bushings can enter the guide pins for aligning before the assembler loses hold of the plate. Also, because the plate is recessed inside the ejector box, there must be tapped holes on the underside of the ejector plate to make it easier to pull it out of the ejector box.

In molds without a mold mounting plate (see Figure 11.34 for location of mold mounting plate), the designer must add at least two shoulder bolts from the ejector plate to the core backing plate (see Figure 11.38). This prevents the ejector plate from falling off the ejector guide pins. An adjustable stop button is used to back-up the ejector plate on the platen. Make sure that the stop button locations are not over the threaded holes in the plate. The stripper bolts must never be exposed to any load except when preventing the ejector plate from falling out during handling.

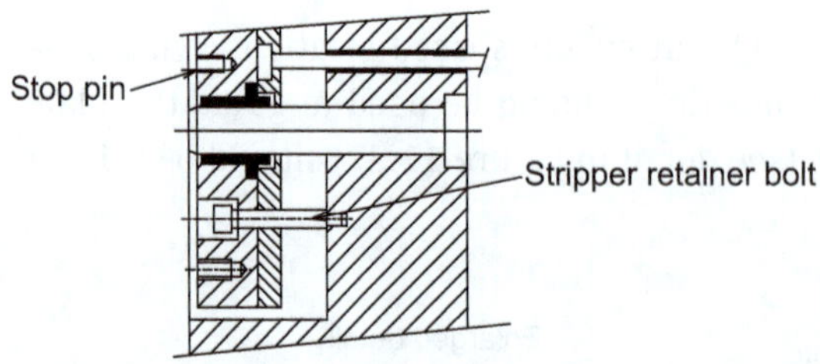

Figure 11.38
Ejector box configuration when no mold mounting plate used

For assembly, a cutout must be provided in the ejector box for access to the hoist hole on the side of the ejector plate, to clear the eye bolt, and to hold the ejector plate for assembly (see Figure 11.39).

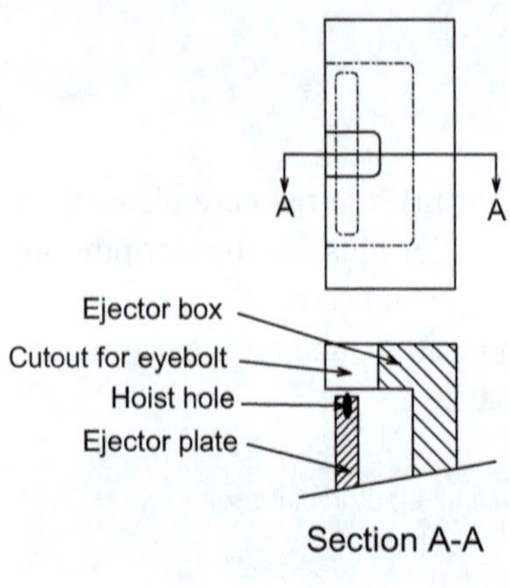

Figure 11.39
Cutout in ejector box for ejector plate handling

11.7.8 Returning the Ejector Plate

Theoretically, with the use of return pins, there is no need to provide additional features to return the ejector plate. As the mold closes, the return pins would drive the ejector plate back into its "rest" position, until the *stop buttons* touch the mounting plate or the machine platen. However, this is only acceptable with slow closing speeds of the clamp, for example if the ejector stroke is small, and where the mold is moving very slowly.

Normally, the ejector pins move so far out that the return pins would strike the cavity side of the mold while the clamp is still moving rapidly. The resulting impact would be damaging to the return pins and the surface where they hit, particularly if the masses of the ejector system are large. For this reason, the ejector plate must be returned by an independent method before the return pins strike the opposing mold half.

The following methods are used to return the ejector plate:

11.7.8.1 Tie Ejector Plate to Machine's Ejector Plate

Called "push-pull", this method is used with the machine ejectors. The ejector rods must be fastened to the ejector plate so that they can push forward and pull back. This normally requires threading the ejector rods into a corresponding threaded stud or ejector plate on the mold side. Care must be taken in setup that the ejector plate is retracted and seats on the stop buttons before the mold is fully closed. The disadvantage of this method is the additional connection of mold and machine, which slows down the mold installation.

11.7.8.2 Linkages Attached to the Mold

Both ejection and return motions of the ejector plate are linked to the mold stroke. To prevent damage to the linkage, this method requires a very exact and repetitive opening stroke. Care must be taken to ensure that the clamp does not open beyond what the links are designed for; machine clamp stroke limiters may have to be provided.

11.7.8.3 Return Springs

Return springs (Figure 11.40) can be placed inside or outside the ejector box (see Figure 11.41). The advantage of springs is that they are inexpensive and simple to use, providing the guidelines in Table 11.1 are followed. In general, springs are quite appropriate for short strokes. Details of spring selection and spring installations can be found in websites and catalogs for standard mold parts. Designers should consult the spring supplier for details, but in general the following tips should be followed:

- In a cyclic loading applications, a spring should be pre-loaded to about 10% of its length
- A spring should be compressed to within its recommended operating range
- The unguided (unsupported) length of any spring should never be more than its diameter. If the distance is larger, the spring must be supported (internally) with a rod.

Table 11.1 Die Spring Operating Ranges (Reference *www.DME.net/*Copyright Raymond®)

Die Spring Color	Duty Rating	Maximum deflection of free length	Efficient operating range of free length
Blue	Light medium	50%	25–35%
Red	Medium	37%	20–25%
Gold	Heavy	30%	15–20%
Green	Extra-heavy	25%	15%

Figure 11.40 Examples of die springs (Courtesy of *www.DME.net/*Copyright Raymond®)

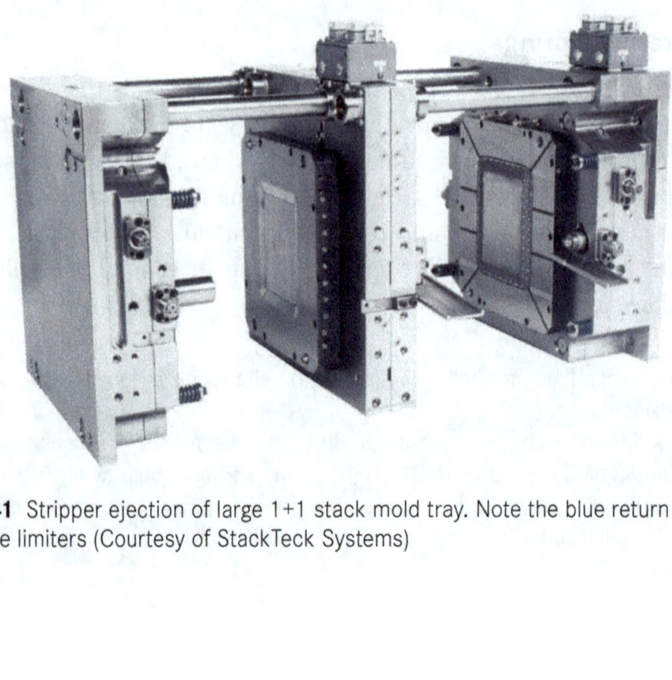

Figure 11.41 Stripper ejection of large 1+1 stack mold tray. Note the blue return springs and stripper plate limiters (Courtesy of StackTeck Systems)

11.7.8.3.1 Internal Return Springs

When using internal springs, at least two stripper retainer bolts must be used to prevent the springs from driving the ejector plate too far back during disassembly of the mold. There must be clearance under the heads of the stripper bolts during regular cycling of the mold to prevent fatigue failure of the bolt. Placing springs between core plate and ejector plate is simple for short strokes (Figure 11.42) but presents challenges with longer ejection strokes (see example below).

> *Example of die spring length*
>
> Given the rules in Section 11.7.8.3 and Table 11.1, the effective stroke of an extra-heavy-duty spring must not be more than about 15% of its free length. This means that for an ejection stroke (E_s) of 30 mm, the return die spring would be 200 mm (8") long.
>
> Free length = Stroke ÷ Efficient range = 30 mm ÷ 0.15 = 200 mm
>
> Such a length is usually difficult to fit in the available space inside the ejector box.

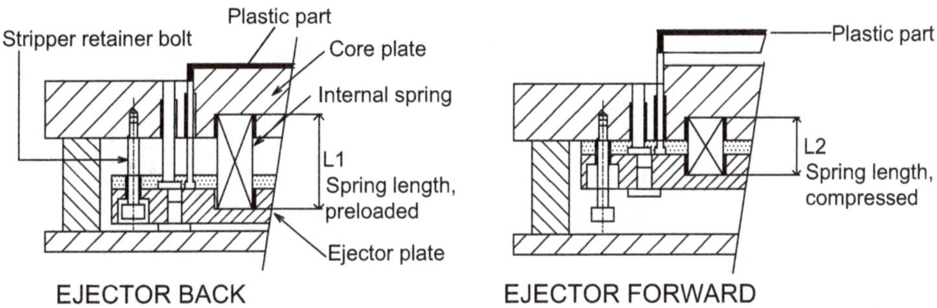

Figure 11.42 Typical return spring internal configuration in an ejector box

11.7.8.3.2 External Return Springs

When using springs with a long ejection stroke, four springs should be placed outside the ejector box, where there is enough space for them and where broken springs can be easily detected. The springs must be covered with a sleeve (for safety in case of breakage) and a label attached to the mold indicating that the springs are used for ejector plate return. The same sizing criteria are used for external and internal return springs.

For small molds a single central ejector return spring may be located at the back of the mold where there is a hole in the moving platen to accept the protruding spring assembly.

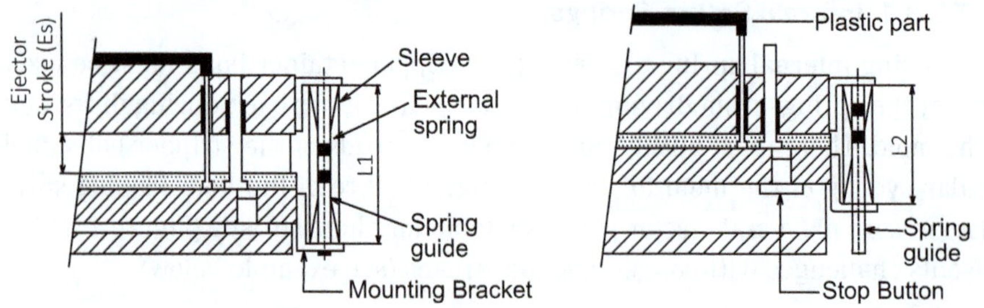

Figure 11.43 Schematic of external return spring mounting (minimum four per ejector plate)

11.7.8.4 Air Cylinders (Air Springs) for Ejector Return

The main advantage of a force generated by a air cylinder is that the force is constant over the whole length of the stroke (a spring is not). The disadvantages are cost, lack of control throughout the stroke, and limited force generated by air. An air cylinder can be bought from a parts supplier and mounted to the ejector plate, or a set of air chambers (usually four on a plate) can be built into the mold rather easily (see Figure 11.44).

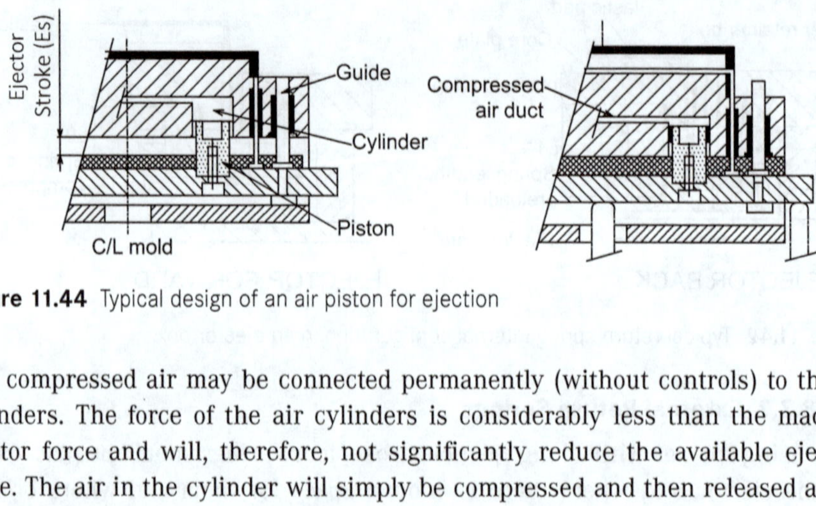

Figure 11.44 Typical design of an air piston for ejection

The compressed air may be connected permanently (without controls) to the air cylinders. The force of the air cylinders is considerably less than the machine ejector force and will, therefore, not significantly reduce the available ejection force. The air in the cylinder will simply be compressed and then released as the machine ejector strokes, ensuring proper return of the ejector plate.

Air cylinders can also be used in place of machine ejectors when it is not possible to use the provided machine ejector positions. Air pistons and machine ejectors should not be combined into one ejector stroke, as the machine ejectors will dominate the stroke and the pistons may become damaged. If multiple ejectors are needed for proper ejection, the combination of machine and an auxiliary ejector mechanism, such as air, is common.

11.7.8.5 Early Ejector Return Systems

Early ejector return systems return the ejector assembly back before the mold completely closes. They are used when ejector pins or other moving components (slides or lifters) interfere with normal closure.

Return systems work by moving the ejector into its back position before the mold fully closes. Most systems use a linkage/cam system to return the ejector plate early. Early return systems can be internal to the ejection or externally mounted (see Figure 11.45). They function as follows. In Figure 11.45B, as the mold opens the machine ejector has pushed the ejector plate forward and the cam fingers in the early ejector return device have slipped inwardly from the counterbore in the bushing. This allows the ejector plate to move forward. In Figure 11.45C, the early ejector return cam fingers have just touched the actuator pin. As the mold continues to close (Figure 11.45D) the ejector plate is pushed all the way back. Once this is done the cam fingers move outward allowing the actuator pin to slip by. The ejector plate is fully retracted before the mold is closed. The clamp continues to the closed position and the mold is then fully closed with the ejector in the return position.

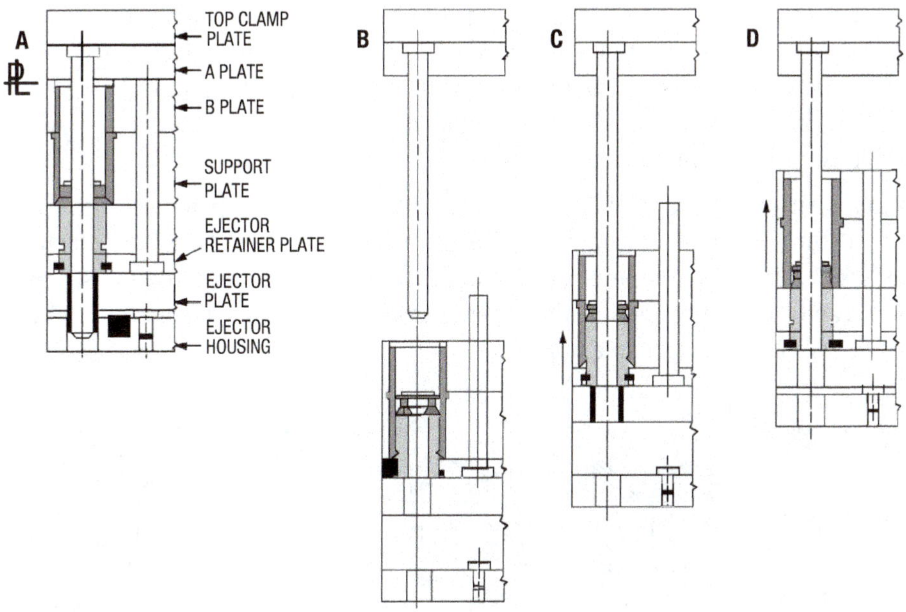

Figure 11.45 Schematic showing the funcation of an early ejector return system: (A) Mold closed in molding position; (B) Mold fully open; (C) Mold closing; (D) Mold continues closing (Courtesy of DME)

A toggle lock is an early return device used to positively return the ejector assembly before the mold completely closes. It is employed when ejector pins or slides interfere with normal mold closure. The external mounting allows more space for cavities, slides and other mold components. The toggle lock system is available in three styles to suit a broad variety of applications. Each style corresponds to a specific mold base requirement.

11.8 Stripper Ejection

Stripper plates, rings, or bars (collectively called strippers as shown in Figure 11.46) are preferable to ejector pins when it is possible to use them. The surface of the product where the stripper pushes during ejection is relatively large, and the ejection force is evenly distributed. Also, the vestige left from the stripper mechanism is usually less noticeable than a pin, sleeve or blade.

Strippers should be used where the product has a shape at the parting line which can easily be generated using conventional machining methods. The stripper must seat on a taper, which must be *preloaded* to prevent flashing into the space between stripper and core. Today, odd-shaped strippers with tapered shut-offs can be efficiently produced using EDM machining, CNC milling and grinding. However, if at all possible, a stripper should have a simple geometric shape (circular or linear), which is easier to produce.

A mold with a stripper plate that is actuated from an ejector plate may be necessary if the ejector plate cannot be directly reached (Figure 11.46) (for example, because of unsuitable locations of the machine ejectors). If, however, the stripper plate can be directly accessed, the expense of an ejector plate and ejector box may not be required (see example in Figure 11.47).

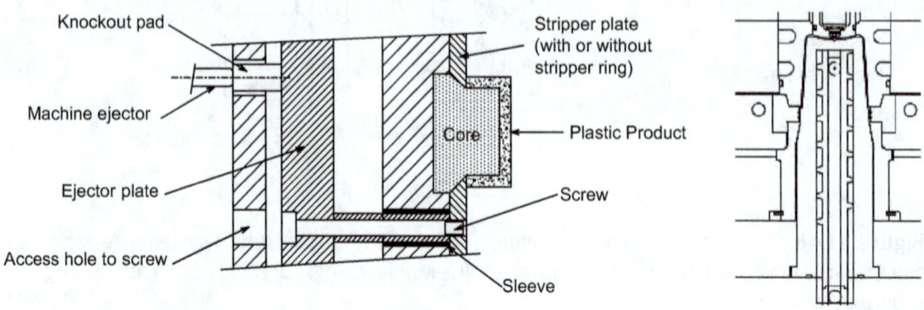

Figure 11.46 Section through stripper ring mold

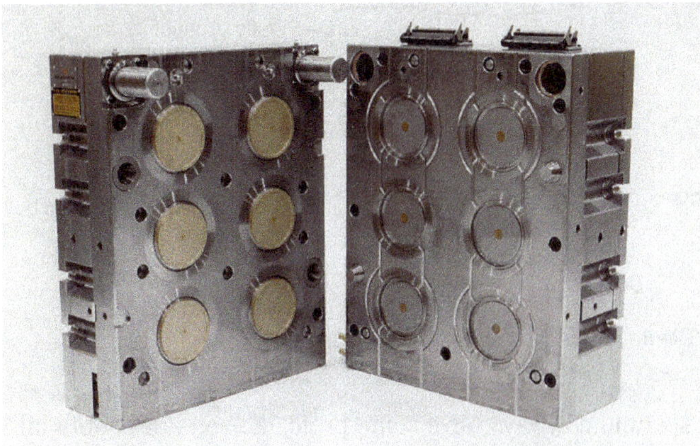

Figure 11.47 Stripper plate ejection without ejector box (machine ejectors used directly on stripper plate) (Courtesy of Husky Injection Molding Systems Ltd.)

11.8.1 General Rules for Stripper Ejection

There are several rules of thumb regarding stripper design in the mold. Many considerations accompany each guideline listed below:

- The stripper must always clear the side of the core at all points of its stroke
- The stroke must extend past the top of the core to clear the parts for free-fall ejection
- Strippers must seat on a taper with the core. Cylindrical shut-off between a stripper and core (no taper) is not recommended, as it will prematurely wear and require replacement
- Tapers of 15° are standard but can be as low as 7° to ensure that there is not a locking taper (tapers less than 7° are locking tapers).

As the stripper moves along the core during the ejection sequence there must be a minimum clearance between stripper and core of 0.25 mm (0.010") (see Figure 11.48). The exception is where the wall thickness is less than this. This clearance prevents scratching and damage to the core and to the sharp edge of the stripper, in case of slight misalignment or excessive play in the guides of the stripper. This is particularly important with no or little side draft of the core.

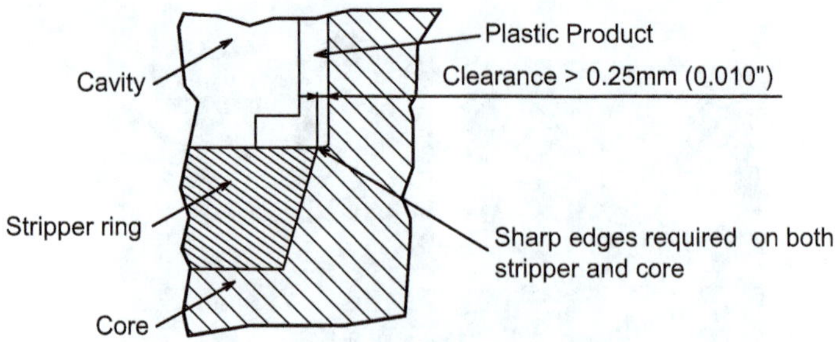

Figure 11.48 Clearance between stripper ring and core

Any stripper seat must always have some preload, even if it is only minimal. Without preload, there will be flashing between the stripper and the core. For further discussion of preload and how to calculate it, see the section on preloads in Chapter 13 on alignment.

A tapered shut-off between the core and stripper is standard practice. Although 15° is the normally preferred angle, smaller tapers may have to be specified if there is not enough thickness in the stripper to accommodate a standard taper without the steel becoming too thin at the end with the wider taper dimension (see Figure 11.49, secondary ejection). The taper ensures that the critical edges of stripper and core will not collide and damage as the stripper moves toward its seat on the core. Also, if the sharp edge of the stripper is damaged, it is relatively easy to recreate a sharp edge by grinding over the surface of the taper or the parting line. However, this may require recutting of the core to avoid a step in the product or grinding the back of the core to drop it down to the same height as the stripper ring (Figure 11.50).

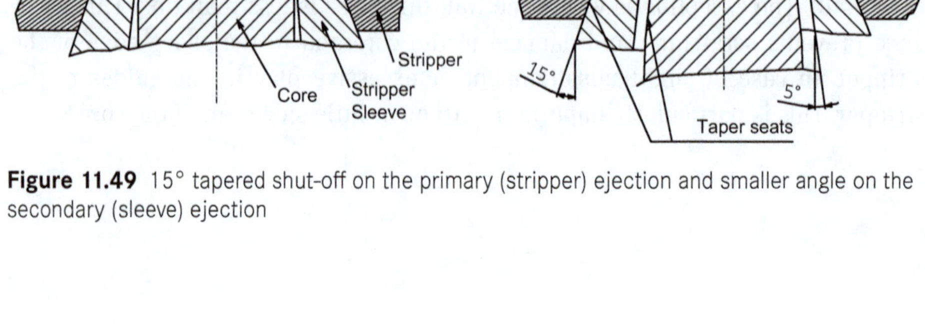

Figure 11.49 15° tapered shut-off on the primary (stripper) ejection and smaller angle on the secondary (sleeve) ejection

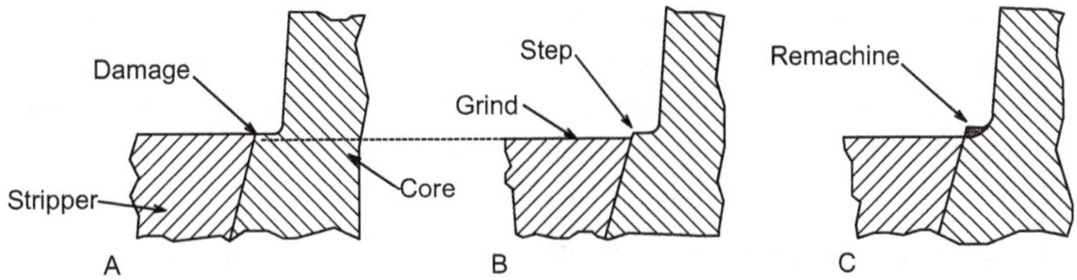

Figure 11.50 Refinishing a damaged stripper ring by grinding the face of the stripper ring and back of core

If the taper is worn loose, it is also relatively easy to reseat it by grinding the underside of the stripper. The effect is similar to grinding the top surface in the previous example. It will create a step that may have to be removed by re-machining the core.

11.8.2 Guiding the Stripper Plate

The stripper must be guided in the axial (moving) direction of the mold throughout its stroke. This ensures that the sharp edge of the stripper does not scratch the side of the core, and that the taper seats properly on the core taper each cycle.

One method to provide this guidance is to use the leader pins of the mold to guide the stripper plate. This applies whether the stripper plate is one piece or carries stripper rings or stripper bars (Figure 11.51). The disadvantage of this method is that the alignment of the leader pins may fight the alignment of the stripper taper. To avoid this, the guide bushings on the stripper plate are made looser than standard, to provide an approximate location for the plate as it returns. The extra clearance in the leader pin bushings must not exceed the total of the positional tolerances of leader pins and cores. Floating core mounting will overcome this problem of aligning, but it could result in uneven wall thickness of the product. This method should only be used if side wall thickness tolerances are sufficiently large.

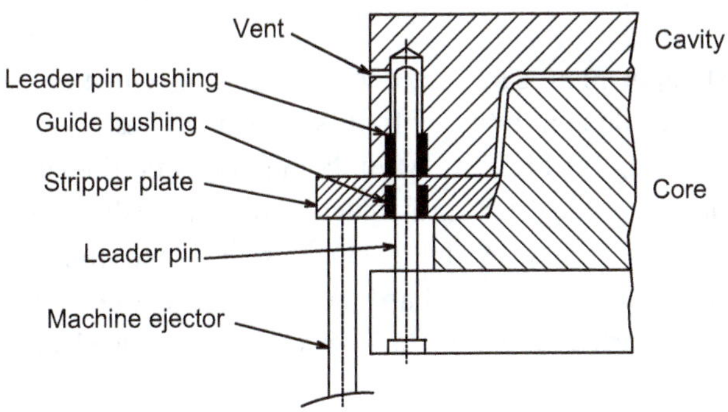

Figure 11.51 Stripper plate alignment with mold leader pins

Because of this added looseness in the bushings, in horizontal machines, the stripper plate "hangs" on the guide pins, and the top of the alignment taper of the stripper will always be hit first as the mold closes and will, therefore, wear sooner. To reduce this problem, the use of direct-acting stripper plates should be avoided and floating stripper rings be used instead.

A better guiding design is to use separate guide pins and bushings for the stripper plate. The mold leader pins, LP1, align the mold (Figure 11.52) but also guide the third plate (cavity plate) and the runner ejector plate. The guide pins, LP2, and guide bushings align the stripper plate and protect the core(s). Floating stripper rings are required with this design; otherwise the taper fit will fight the alignment in the leader pin bushings.

Note that in both examples of stripper guides shown above, there must be stroke limiters to prevent the stripper plate from falling off the leader pins when the plate is pushed out too far (Figure 11.52). This may happen accidentally when cleaning the mold behind the stripper plate or during handling of the open mold. The normal ejector stroke must be less than the stroke limiter stroke so that the limiter bolt will never be stressed by the ejector force.

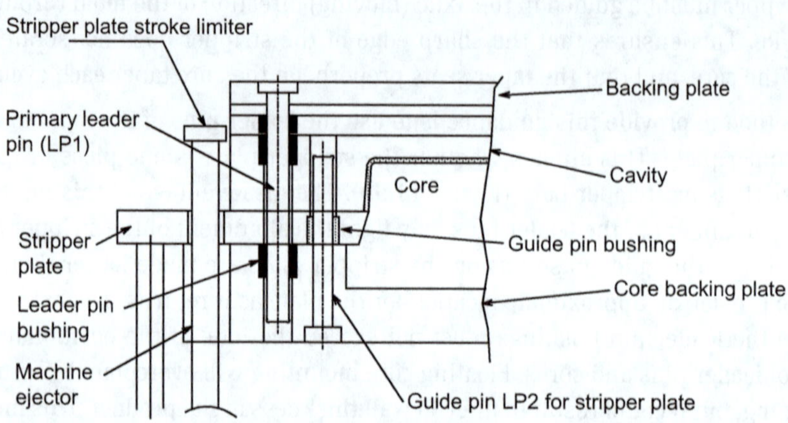

Figure 11.52 Stripper plate alignment with separate leader pins from the mold leader pins and stripper plate stroke limiters (cold runner and direct ejection from machine ejectors)

Figure 11.53 shows how the stripper plate can be guided when an ejector box is being considered in the design of a cold runner mold. In this case, mold leader pins LP1 have the same functions as in Figure 11.52: they guide the mold and the runner plates. The stripper is guided by guide pins in the ejector box and bushings in the core backing plate. The stroke is limited by the ejector box so stroke limiters are not required. The guide must be made from hardened steel, and the guide bushings, which are difficult to lubricate, should be permanently greased ball bushings, or plastic bushings. The sliding fit must be large enough so that the taper fit in the stripper will not fight the alignment in the guide bushing.

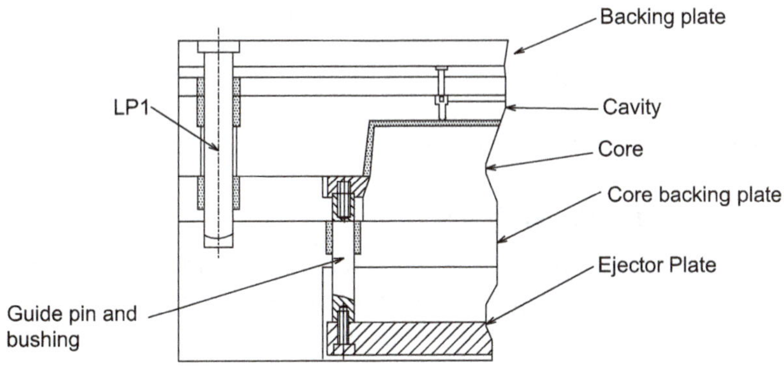

Figure 11.53 Guiding the stripper plate from an ejector box (cold runner mold)

When using a stripper seated on a taper, there is no need for stop buttons. If too high, they would prevent the stripper from seating properly. If they are too short, they would not touch the backing plate anyway.

11.8.3 Stripper Rings

11.8.3.1 Fixed Stripper Rings or Stripper Inserts

Fixed stripper rings are hardened inserts (usually S7 or A2 material) in a soft(er) plate. Their advantage is that they will wear less than a softer plate and are easier to replace in case of damage. By using individual strippers that align to the core there will be no fighting of the tapers on each stack with each other. The disadvantage, as with the solid stripper plate, is that the alignment of the tapers can fight the alignment of the leader or guide pins if they are not allowed to float or align to their matching core. Fixed stripper rings are used primarily in single-cavity molds (e.g., pails, etc.) or in combination with floating cores because of this problem. In Figure 11.54 it can be seen that the stripper ring rests on the stripper plate. The configuration on the right is more expensive but preferred as the taper is longer.

The screws do not see the stripping force, as they are used only to hold the ring on its seat. The stripper inserts are mounted from behind, so that there are no holes on the front face of the stripper where plastic from flashing could enter and cause issues.

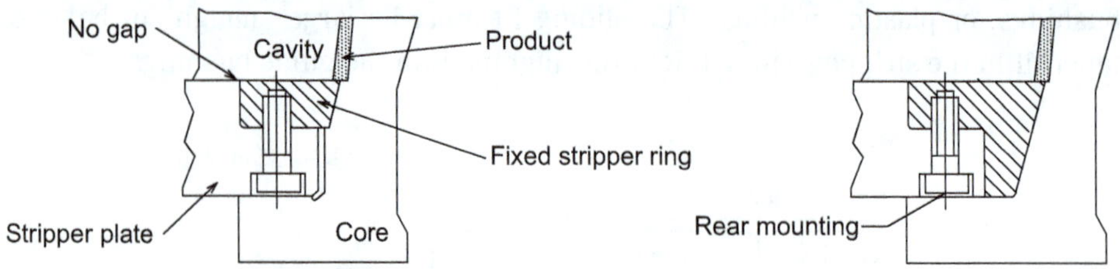

Figure 11.54 Fixed stripper plate insert designs

11.8.3.2 Floating Stripper Rings

Floating stripper rings align with the taper on the core, as they seat without trying to move the much larger stripper plate and fighting the stripper rings on other cavities in the process (Figure 11.55). The masses of the rings to be realigned are much smaller than the mass of the stripper plate, and the forces to realign are small compared to the force required to align the whole stripper plate.

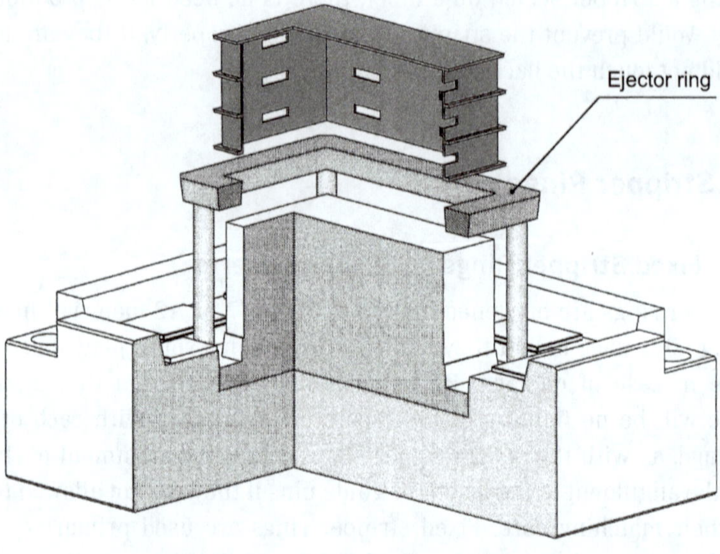

Figure 11.55 Stripper ring ejection of a rectangular container (Courtesy of Manual do Projectista)

The design shown in Figure 11.56 uses a hardened sleeve to allow for floating while allowing the bolt to be properly torqued. This method works well but may be a problem if space is tight. A disc spring can be added between the shoulder of the sleeve and the stripper plate, to ensure that the stripper ring stays loaded in the back position and does not move and align on every cycle. This will reduce the long-term wear on the tapers and sharp edges.

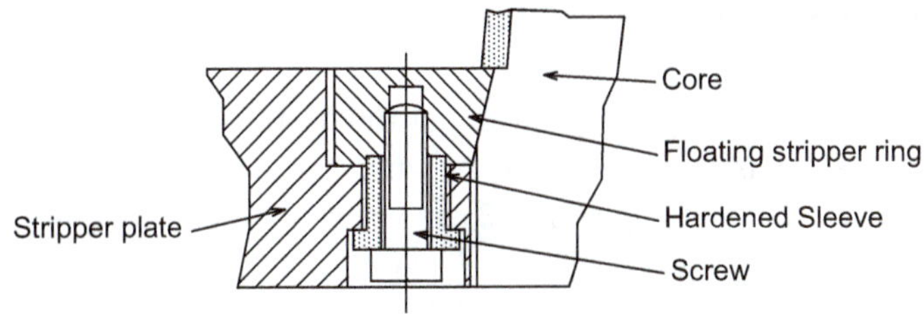

Figure 11.56 Floating stripper design using a sleeve and bolt

A frequently used method for smaller strippers (up to 150 mm or 6" in diameter) is to use a snap ring to hold the stripper ring from lifting. The standard snap ring in Figure 11.57 is inserted after the stripper ring is in position. The clearance bore in the stripper plate must be large enough to accommodate the expanded snap ring for assembly. If you are considering this method of stripper ring retention, follow manufacturers' design specifications for dimensions, clearances, and tolerances. In order to prevent the stripper ring from moving and aligning on every cycle, a dished snap ring can be used to create pre-load between the stripper ring and the stripper plate.

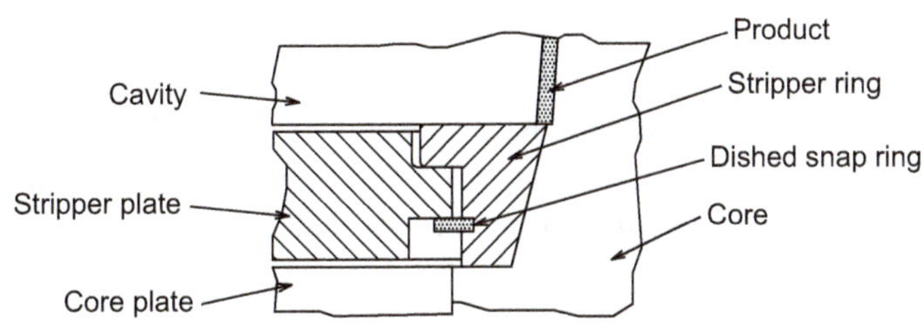

Figure 11.57 Snap ring retention of a floating stripper ring

Note, in Figure 11.57, that the stripper ring seats on the core and on both the flat and tapered areas of the core after overcoming the preload of the taper. The stripper plate is clear from both cavity and core and does not help in supporting the clamp force. If the seating areas of the ring(s) are too small to support this force safely, additional support areas must be added from the stripper plate or other areas such as tonnage support blocks.

The stripper ring float is exaggerated in Figure 11.57. *A typical clearance of 0.03 mm (0.001")* is normally used. If the clearance was larger, gravity would allow the stripper to fall and excessive wear would take place on the upper edge of the stripper and core.

11.8.4 Stripper Bars

Stripper bars can be used when the layout of the parts in the mold align with the stripping action along one axis. Sometimes when the shape of the rim is complex, bars such as those shown in Figure 11.58 can be used to create a number of smaller sections.

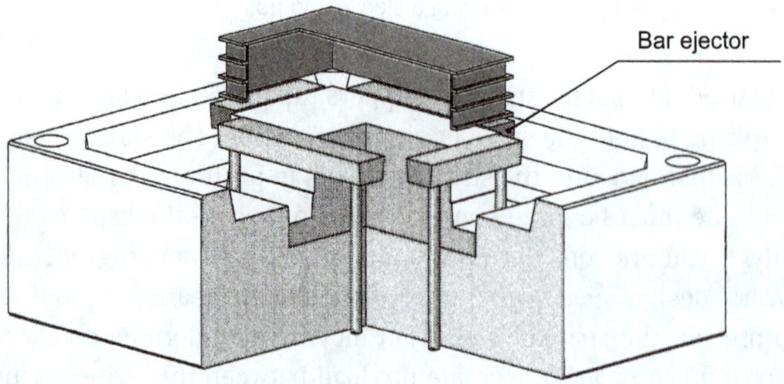

Figure 11.58 Stripper bar ejection of a rectangular container (Courtesy of Manual do Projectista)

All rules for stripper rings apply, including guiding, limiting of stroke, taper seat, etc. However, while the circular ring naturally creates its own backup stress by the core, a bar needs an additional back-up force *F* in Figure 11.59 to make it seat on the core. This force must be supplied by an outside taper that holds and seals the stripper bar on the core.

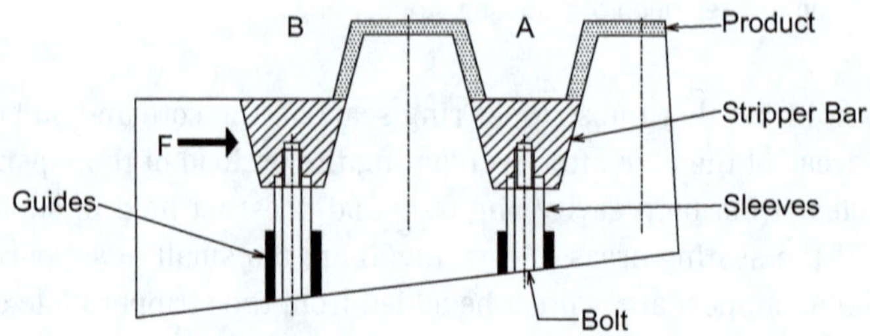

Figure 11.59 Stripper bar configuration when parts are aligned to a common ejection axis

In Figure 11.59, at position A, the stripper bar lifts two products, and the two opposing tapers ensure proper seating on the core. The forces between the two cavities are balanced. At position B, the "outside" stripper bar lifts only one product, and the outside taper provides the necessary balancing force *F*. In this illustration, the taper is cut into the core block, but it could also be a separate wedge, held in the core plate.

Stripper bar ejection is sometimes used where a number of elongated products must be stripped, rather than ejected by pins (e.g., razor handles, toothbrushes, cutlery). The products can be arranged side by side, with the stripper bars between the cores, or with only two outside bars lifting a series of products from their narrow ends.

11.8.5 Stripper Ring Ejection of Lids

The method of ejection for lid molds is almost exclusively stripper ring ejection due to the undercut or snap ring on the skirt of the lid. The actuation method can be hydraulic, from the machine ejector, or air from pistons mounted in the core plate. The preferred method is machine-actuated hydraulic ejection, as it is much more controllable in terms of speed and force.

Design tips for stripper plus air ejection of lids is as follows:

- Use cavity air to avoid a vacuum in the cavity on mold opening
- Core air not required except where a deep-draw lid is being molded
- Blow-down is required to make the lids blow out of the molding area quickly
- The machine stroke for a lid should be 1.5 × the diameter of the lid, to allow the lid to fall between the mold halves but not tumble
- The stroke of the stripper ring or stripper plate must clear the top of the core
- The force required for pneumatic air piston ejection can be determined by completing a force analysis on the lid as a function of the depth of the undercut, the resin used, the height of the lid, and the circumference of the lid. Conversely, this simple rule of thumb has been effective for the vast majority of applications:

$$\text{Piston area/face (in}^2) = \frac{\text{Part diameter (mm)} \times \text{\# of cavities/face}}{120\,[130 \text{ maximum}]} \tag{11.7}$$

Note: Rule is based on an undercut depth of < 0.7 mm and 100 psi air available at the machine.

If using air pistons for stroking the stripper plate, it should stop on stroke limiters, not on the piston itself. This limits the amount of wear on the piston. The stroke limiters should have a 90A durometer polyurethane bumper on them.

11.8.6 Stripping from the Injection (Cavity) Side

There are cases where it is necessary to gate from the inside (or underside) of the product for cosmetic or functional reasons (e.g., some caps and closures, dinnerware, and trays). The ejection practices covered so far in this chapter are similar to those from the core side. However, added plate thickness to the injection side is necessary to accommodate the ejection mechanism. This increases the length of the nozzle drops of hot runner or three-plate molds, or the length of sprues in two-plate molds. With the additional steel and complications in the design, the mold cost will be higher than conventional ejection form the core side.

■ 11.9 Air Ejection

Air ejection can be used for practically any cup-shaped product that does not have large undercuts on the core side (see Figure 11.61). When compressed air enters the space between the core and the product, it does not immediately exhaust into the open, as it would with a flat shape, but has enough time to create pressure under the product to push it free from the core. Air ejection alone can also be used for some flat products. Air ejection is often used in very high cavitation production molds by combining air ejection with integrated take-out robots.

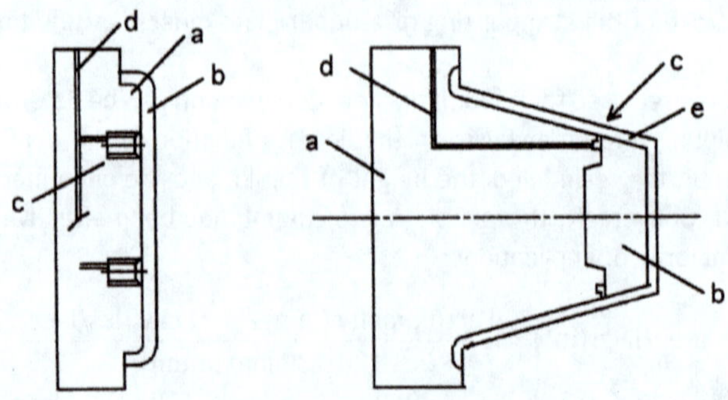

Figure 11.60 Air ejection stack layout. Left: Mostly flat products: (a) core, (b) molded product, (c) air poppets, (d) pressure air supply. Right: Deep cup-shaped products: (a) core, (b) core cap, (c) circular air gap, (d) pressure air supply, (e) molded product

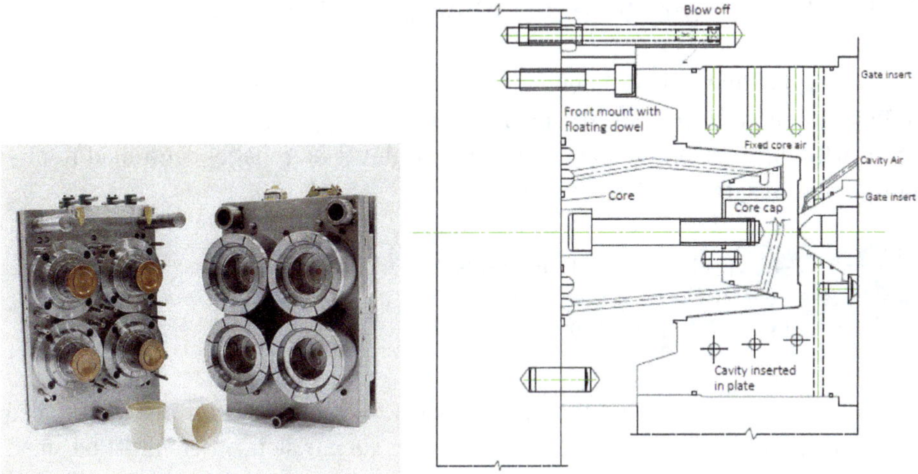

Figure 11.61 Air-ejected 2 lb margarine tub (cavity lock stack design). Note blow-off pins on core side and blow-down bars at the top of the mold (Courtesy of Husky Injection Molding Systems Ltd.)

The advantages of air ejection without mechanical assist are:

- Simpler core half construction
- Fewer masses to be moved (power savings)
- Less shut height
- Less mold mass
- No impact noise
- No wear of ejector mechanism and taper seats
- Less down time and cost required for maintenance, particularly on parting line.

Disadvantages of air ejection are:

- Significant volumes of compressed air are required, which can be expensive
- Existing compressed air supply may not have enough capacity, and accumulator tanks may be needed at the machine
- Balancing of air outlets, particularly in multi-cavity molds, requires intricate channeling in cores and/or core plate. Since this is in addition to the cooling channels, it makes the mold layout more complicated
- Controls such as additional (air) circuits, hoses, and valves are required
- A container with little to no draft (less than 2°) and heavy undercuts holding it on the core may build up a heavy pressure inside the product. The product may burst if it is not strong enough, or it may strike the cavity side and become stuck in the cavity.

11.9.1 Requirements and Timing for Air Ejection

In order to facilitate air ejection, the air supply (volume and pressure) must be sufficiently large to provide the air needed. The supplied air must be clean (filtered) without dust, oil, or water. Air pressure must be reliably controlled to provide continuous, trouble-free ejection.

> A 20–40 L (5–10 gal) tank located at or near the molding machine should be used to provide a sufficient volume of air during ejection at a constant pressure of 5–6 bar (80–90 psi).

The amount of air permitted to escape in each cavity must be properly balanced with small, uniform jet openings at each cavity. The air slots or jets must be small enough to avoid resin flashing into them, but at the same time large enough to ensure sufficient air flow through the slots or jets.

The timing sequence of the components in the air ejection is very important. In a typical setup, before the mold is opened cavity air is turned on for 0.2–0.5 s to ensure the part stays on the core. Then core air (side and optional top circuits) is activated for 0.2 s to break the part free from the core surface. When the mold is opening, the blow-off jet is actuated for 0.3 s to propel the part from the core. Finally, a blow-down jet (from air curtains mounted on the top of the mold) of 0.5 s may be used to accelerate the parts out of the mold. These short blasts are sufficient for good ejection.

> There are limits to air ejection. In general, parts with drafts as low as 1.75° can be ejected and undercuts of 0.25 mm can be air-ejected without any issues.

The designer should be aware of the following when designing an air ejection mold:

- Cup-shaped parts with little draft and heavy undercuts (> 0.25 mm) can get stuck on the core. This will create a buildup of pressure and may burst the part or project it towards the cavity side of the mold. In this situation a stripper ring is needed or both core side air and core top air is required.
- Cup-shaped products with an aspect ratio larger than 1 : 1 should use a moving core ring design (see Figure 11.68) to provide enough air to propel the part forward. Parts with an aspect ratio less than 1 : 1 should use a fixed for cap design (see Figure 11.66).
- If the ledge of the core cap is wide or has shut-offs on it (like in a flower pot mold), consider using an interrupted (spot) vent.

- Based on the geometry of the part, a low-pressure area may form, which will prevent the part from ejecting. For example, on a rectangular part the air may escape from the sides while the plastic still hugs the corners. In this case, move the air as close to the corners as possible.
- Parts with openings in the bottom can only be ejected if the compressed air entering is greater than the air leaving the openings. For example, flower pots require an independent end air poppet to eject.
- There must be sufficient air to eject the part. The core cap should have four air lines leading to the vent. The blow-off should also take place from four positions on the core. Note that the air must be an independent circuit, and ideally is supplied by an accumulator with at least 80 psi.
- Make sure that the air supplied to each position in a multi-cavity tool is balanced or is fed from a large enough feeder line.
- Venting is important for ejection. The designer should avoid a vacuum between the part and the core.

11.9.2 Poppet Air Ejection

A stroking poppet (single- or double-acting) is normally used in applications where there is a very small draft or where there are holes in the bottom of the part (for example, for flower pots - see Figure 11.64). A stroking air poppet can deliver a large amount of air in a short space of time. During injection, the poppet is retracted and seals tight on a taper seat in the tip of the core, which prevents plastic from entering into the air system. After the mold starts opening, the poppet moves forward by the amount *SE* (Figure 11.62 and Figure 11.63). This forward motion of the poppet comes from the air pressure on the underside of the poppet. The stroke *SE* will be small, in the order of 0.25 mm (0.010"), but this allows ample air to escape from the air circuit and propel the part off the core. The poppet is returned to its seat by injection pressure on the top of the poppet. The poppet must be located opposite the gate so that the in-rushing plastic will close the poppet and prevent plastic from entering the air channels.

If the poppet is not opposite the gate, springs can be used to close the poppet to make sure that plastic cannot enter under the poppet. These springs must be properly designed with enough force to hold the poppet closed and with enough length to make sure that the spring will not fatigue.

If the stroke needs to be longer, the air poppet can also be actuated by the ejection mechanism of the mold (combined mechanical and air ejection), but it must be positively returned to avoid hitting the cavity side on mold closing.

In the case of a single-acting air poppet (see Figure 11.62), the stroke *SE* must be less than the 50% of the thickness of the product at this spot. There must be enough space for any cold slug in the gate to eject freely into the space between gate and the top of the poppet.

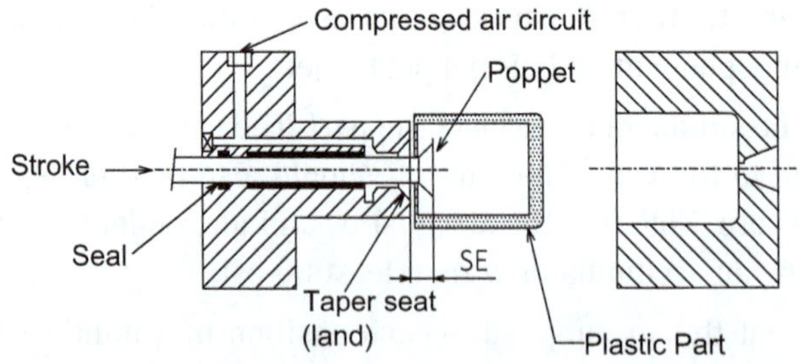

Figure 11.62 Single-acting poppet air ejection

The size of seat (land) of the poppet is similar to any other seat that comes under injection or clamping pressure. The load should be limited to 5% of the yield of the softer of the two materials contacting at the seat. For calculations, use the projected area, not the actual width.

Commercially available air poppets (Figure 11.65), which consist of complete assemblies of poppet and spring in one unit, can be pressed into the core at the desired location (see *www.dme.net*, *www.diemouldequipment.com.au*, or *www.pcs-company.com* for reference designs).

When using air poppets to aid ejection, use the largest possible size, as smaller poppets tend to stick more easily. Commercially purchased air poppets generally cannot be surface-finished as it affects the performance of the poppet and the installation procedures must be strictly adhered to. For rectangular molds, place the air poppets as close to the corners of the core as possible.

For positive return of poppets or if a larger poppet stroke *SE* is required, the poppet must be returned to its seat before the cavity approaches the poppet to ensure that no damage is caused to the poppet or to the surface of the cavity. The forward and return stroke of the poppet can be done either mechanically, by connecting the poppet to the mold ejector system, or by independent double-acting air (or hydraulic) actuators acting directly on the poppet.

A poppet with double-acting actuation is moved by the double-acting piston, which is held by a snap ring. The air sequence is as follows (see arrows in Figure 11.63).

Air 1 moves the piston forward. Air 2 is actuated (mostly in parallel with Air 1) to provide blow-off air. Air 3 is actuated to re-seat the poppet in the back position.

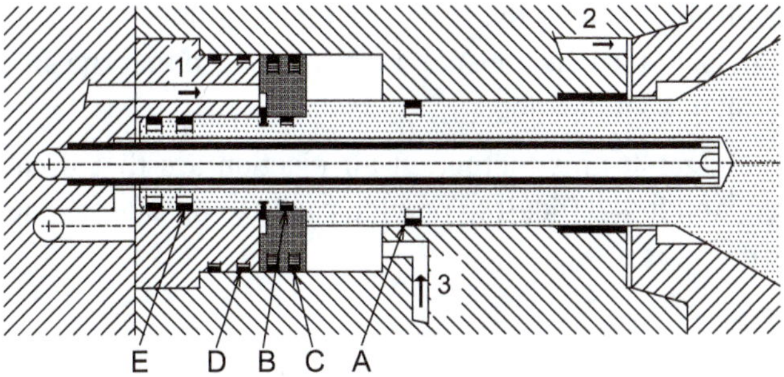

Figure 11.63 Double-acting poppet actuation. Letters (A)–(E) indicate seals: (A) Double delta II ring, (B) O-Ring, (C) Glide ring, (D) O-Ring, and (E) Double delta II ring

Considerations for a dual-stroking poppet design (Figure 11.63 and Figure 11.64):

- Poppet stroke should be maintained at a maximum of 0.25 mm (0.010") for poppets with a 45° taper angle. This is to avoid unnecessary restriction to flow of plastic, as well as to allow enough space for the cold slug to be expelled from the gate. The poppet is seated on a taper to prevent plastic from entering into the air system.
- In order to maintain a consistent stroke, the snap ring installation should be dimensioned from the molding surface to the bottom ledge of the snap ring groove. This will ensure that the poppet does not favor resistance to flow between cavities, by maintaining a consistent stroke.
- The poppet must always be located opposite the gate to allow the force of the plastic to push the poppet closed.
- If a bubbler is added to the poppet for cooling, it should be located in position so that the cooling is the same for each cavity.
- Below the bushing locating the poppet in the core there should be a clearance bore.
- Ensure that the bore for the poppet is stepped to prevent the O-rings from being pinched.
- Ensure that there is a gap between the poppet and the core plate to allow water to enter or escape.
- Use T-seals rather than O-rings for sealing of water or air on the shaft.
- Bore installation for the poppet should have alignment in two locations. One should be close to the top of the poppet where the 45° taper starts, as well as immediately above the snap ring installation.

- For the snap ring not to bend due to insufficient seating when the poppet is stroked fully forward, a gap of 0.01 mm (0.0004") at maximum should be allowed per side.
- When considering the bore diameter surrounding the snap ring, remember to leave enough clearance for the snap ring wrench, so that the snap ring can be removed without complications.
- The disadvantages to this method of air ejection are that it complicates the mold design, and it is difficult to provide cooling opposite the gate, where it is most needed.

Figure 11.64 Air-ejected flower pot stack with custom stroking air poppet. Note the air poppet sitting forward on core in the assembled core on the left (Courtesy of Husky Injection Molding Systems Ltd.)

If an insert cannot be used to provide air to a needed area for ejection, then commerically available air poppets of various sizes can be inserted in strategic locations in the inserts to provide air for ejection (see Figure 11.65).

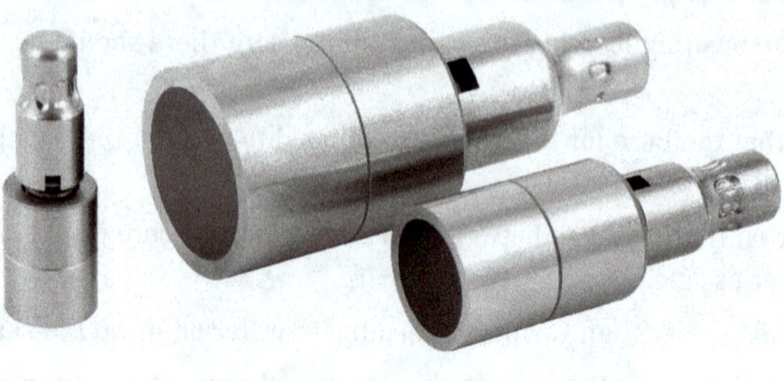

Figure 11.65 Commercially available air poppets (Courtesy of DME)

11.9.3 Core Air

A blow slot or core air is a static gap the size of a vent that allows compressed air to escape and assist in ejecting the part. This air channel can be always open or closed during injection, and open on ejection (see Section 11.9.3.2). The gap or vent is small enough so that plastic cannot enter the gap, but large enough to allow air to escape.

11.9.3.1 Fixed Core Air

In fixed core air the slots stay open during injection even though they are in the cavity space and subject to injection pressure. Advantages are the simplicity of the design and the rigidity of the core. The main disadvantage is that the blow slot has a relatively small opening and is located near the gate, where the plastic is hottest and least viscous. The narrow gap and the small circumference reduce the total cross-sectional area of the blow slot and the amount of air that can pass through it. This method can be used to assist ejection by breaking the vacuum inside the product. It is typically used in combination with stripping or blow-off air. There are three basic types of open blow slots shown Figure 11.66, In all cases the core cap is typically manufactured from a high thermal conductivity material such as BeCu (see Figure 11.61 for reference design). In Figure 11.66A the air gap is on the top of the core. This design provides more velocity to push the part off but is subject to easier flashing, since it is closer to the gate area. In Figure 11.66B the air gap is on the side of the core and the plastic more easily passes this gap without flashing. However, in this design the core cap area is only cooled by a bubbler and the higher conductivity material. Figure 11.66C shows a preferred design which allows for more cooling in the core tip area.

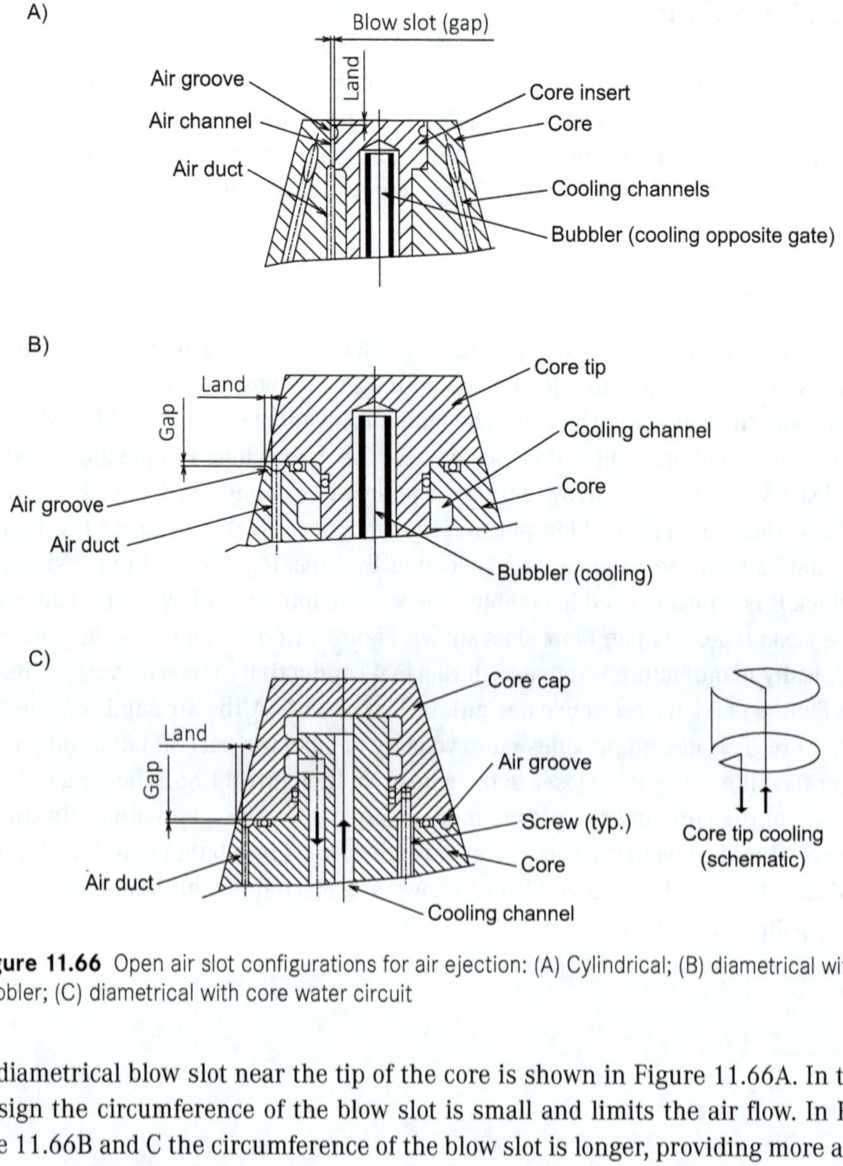

Figure 11.66 Open air slot configurations for air ejection: (A) Cylindrical; (B) diametrical with bubbler; (C) diametrical with core water circuit

A diametrical blow slot near the tip of the core is shown in Figure 11.66A. In this design the circumference of the blow slot is small and limits the air flow. In Figure 11.66B and C the circumference of the blow slot is longer, providing more area for air to escape; in these cases, the plastic has already cooled down somewhat. With less danger of flashing into the slot, it permits a wider gap. This helps to increase the amount of air that can pass through the gap. If more cooling is needed in the core cap, then configuration C in Figure 11.66 could be used. The slot should be in the upper third of the core height. If the slot is too near the open end of the product, the air will escape too soon and fail to eject the product. Suggested dimensions for blow slots are listed in Table 11.2:

Table 11.2 Suggested Dimensions for Open Blow Slots (Core Air Gaps) for PP, PE, and PS Resins

Type of core air	Suggested gap
Cylindrical	0.01 mm (0.0004")
Diametrical, near core tip	0.015–0.020 mm (0.0006–0.0008")
Diametrical, near core base (not recommended)	0.025 mm (0.001")
Land	1.0 mm (0.040")

The size of the gap (open slot) also depends on the type of plastic and the molding conditions (temperatures and pressures). With high operating temperatures and pressures, and with a plastic with very low viscosity, the gap must be smaller than with a plastic with high viscosity and/or with low operating temperatures and pressures. Because this is not always easy to control, the use of a closed or stroking core air blow slot is often preferred (Figure 11.67), even though the system is somewhat more complicated (and costly) and subject to wear.

Figure 11.67 Stack design with fixed core insert air (see BeCu part) and stripper ring ejection (Courtesy of Husky Injection Molding Systems Ltd.)

11.9.3.2 Stroking Core Air

Closed core slots stay closed during injection and open after the mold starts opening. There is no risk of plastic entering during injection, and they can open wider than the open slots to let more air pass through. The disadvantage of closed slots is that they require moving parts, which are subject to wear and cost more to produce. Figure 11.68 shows a schematic of a core with a closed blow slot. It is opened after the mold opens by the blow air, which acts on a large surface in the large groove inside the cap. The air pressure creates sufficient force to push the cap to the end of its small stroke, which is usually 0.125 mm (0.005"). As soon as the core ring lifts, the blow slot is open, and the air enters freely into the space between the core and the product.

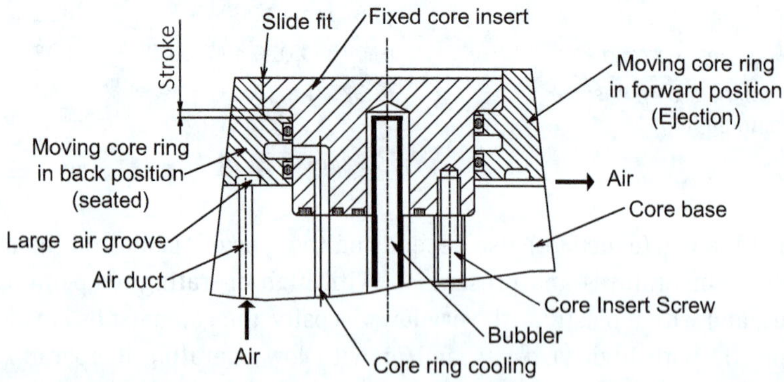

Figure 11.68 Core design with stroking core air

In Figure 11.68 there is also a gap is also between the moving core ring and the fixed core insert. This gap must be small enough that plastic will not flash into it. This gap is commonly used as an additional (open) blow slot, with an air groove connected to the large groove (not shown).

The moving ring is returned to the original position by the pressure of the plastic during injection. The land at the slide fit must be wide enough to prevent plastic from entering the slot before the cap is seated.

11.9.3.3 Combination of Air Poppet and Stroking Core Air

It is also possible to have a combination of stroking core air and poppet ejection (Figure 11.69). There is very little stroke of the poppet (0.25 mm), and the peripheral blow slot is always open at 0.025 mm (0.001"). Air groove sizes are similar to vent grooves.

Because the core cap (from the O-ring outward) overhangs the core base, it is suggested to make the blow slot interrupted (Figure 11.69), so that at least a portion of the core cap can seat on the core base, beyond the air supply groove. This prevents flexing of the overcap and failure at the slot.

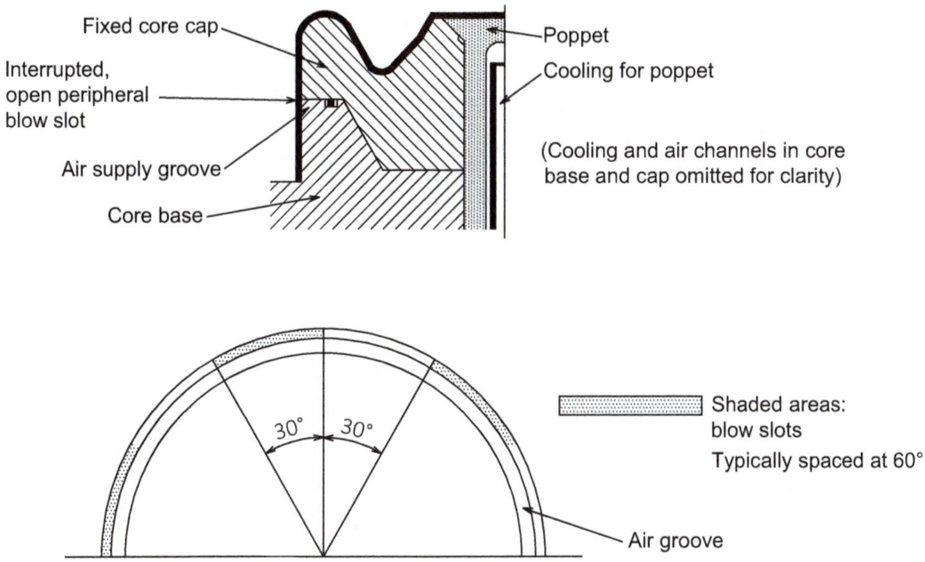

Figure 11.69 Configuration for an open slot and poppet air ejection

11.9.4 Blow-off Jets

Blow-off jets are used to move the part completely off the core after mechanical ejection or core cap air has been used to break the part free from the core (see blow-off pins in Figure 11.61). The force of the air blast coming from a blow jet (orifice) decreases quickly with the distance from the orifice. The orifice must be far enough from the molding surface so that plastic cannot inadvertently enter and plug the orifice or, even worse, the air channel. The orifice should be uncovered only after the cavity moves away, and/or after the stripper moves forward a certain distance.

Blow-off also permits the ejection stroke of the stripper to be limited to less than would be normally required to keep the shut height of the mold small, in case the stripper is actuated from the ejector box. The stripper travels only a short distance, and the final ejection is achieved by blowing air against the underside of the product (Figure 11.70).

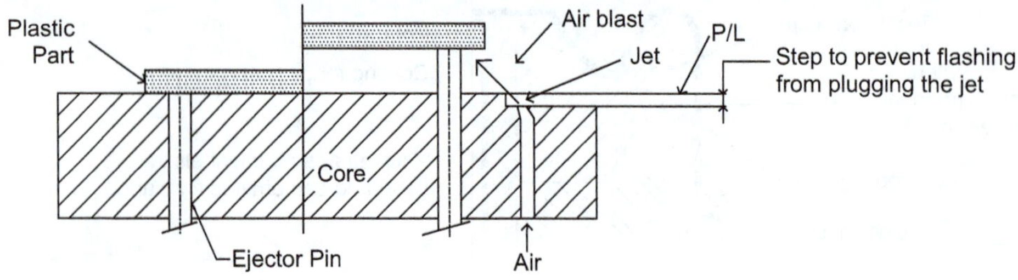

Figure 11.70 Blow-off air schematic

In Figure 11.71 (left) the jet inside the core is fixed while the product moves forward. As the distance increases, the effect of the blow-off decreases. In Figure 11.71 (right) the blow-off is in the taper of the stripper ring and the distance to the parting line does not change as the part moves off the core. This is only practical if the stripper is carried by a stripper plate, where the necessary channeling and sealing for the air can be located.

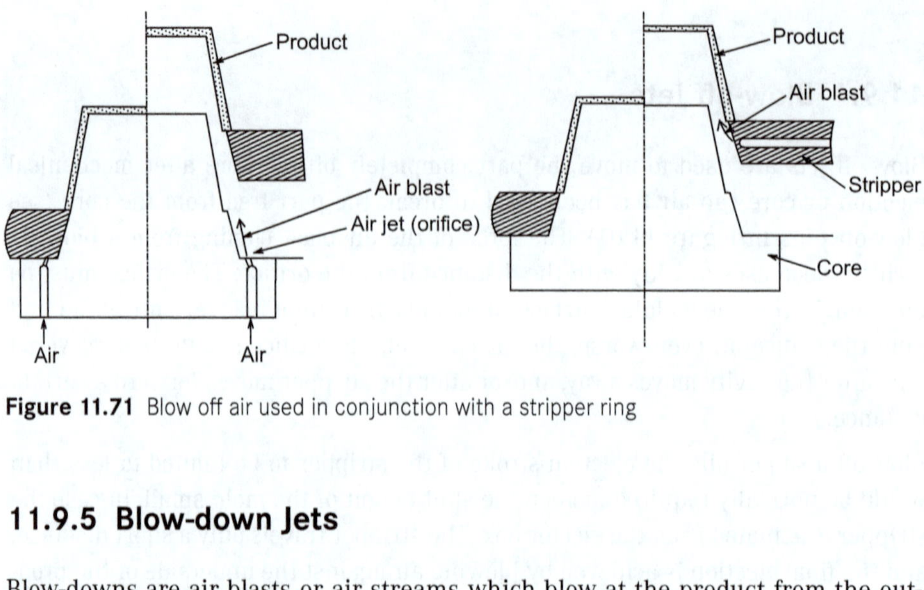

Figure 11.71 Blow off air used in conjunction with a stripper ring

11.9.5 Blow-down Jets

Blow-downs are air blasts or air streams which blow at the product from the outside, after the ejector (stripper or pins) have come forward, to remove the products if they stick to the ejectors. There are two methods of blow-down commonly used: individual open jets or blow-down curtains. Individual adjustably mounted jets are directed against the tip of the product to get the best leverage for knocking the products off the stripper or core. Previous designs used open jets, usually by squeezing the ends of flexible copper tubes. This method is simple but has some distinct disadvantages:

- Safety regulations do not permit pressurized air to escape freely if above 30 psi
- The noise of escaping compressed air is excessive and often above legal limits
- Air through the jets is difficult to control
- High operating cost.

A better system is Venturi-type air accelerator such as the Exair™ nozzle or air knife, which is shown in Figure 11.72. These nozzles utilize the entrainment effect (wall attachment of a high-velocity fluid).

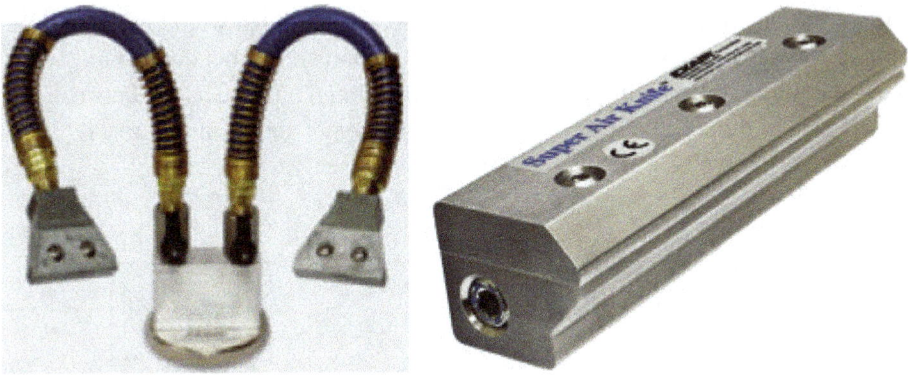

Figure 11.72 Exair™ blow-off nozzle or air knife (Courtesy of Exair)

Exair has shown that these nozzles use as little as 20% of the air used by an equivalent open nozzle. Also, the noise level is below 90 dBA, and the air pressure at the exit is below the permissible 30 psi. For more data on the Exair nozzles, see the manufacturer's data sheets.

The general concern with all blow-down methods is that the products are thrown down with a force greater than that created by gravity alone, and are more likely to be damaged (scratched) than by free-fall alone. As with air ejection, the use of air directed at certain food and medical products may not be acceptable because of the risk of contamination from the dirty shop air. This must be established before designing an ejection system for a mold. Filters, etc., may then be required, or another reliable ejection system must be selected which does not use air.

11.9.6 Air Ejection from the Cavity Side

This type of ejection is simpler than stripper or ejector pin ejection from the injection side, and does not require much (if any) added shut height to accommodate the air channels. All basic rules of ejecting with air from the core side apply. The additional air channels in the cavity side, in addition to the runners and cooling channels, result in a more complicated cavity and gate pad layout. For example, air pressure may be directed on the product in the cavity side with a moving cavity. Figure 11.73 shows schematically how air pressure can be used to assist in freeing a severe outside undercut (rib) at the bottom of the product.

As the mold opens, the cavity follows the core and moves forward sufficiently to permit the outside rib to squeeze between the cavity walls, while holding the product on the core by the air pressure on the injection side. The product will then be ejected (by air poppet) from the core, and the outside rib will return to its molded shape.

The advantage of this design is a much simpler mold, which would otherwise require a split cavity to free the rib. The forward motion should not depend on the air pressure behind the cavities, but actuators (air pistons) must be added between the cavity plate and the cavity backing plate, which are not shown in Figure 11.73.

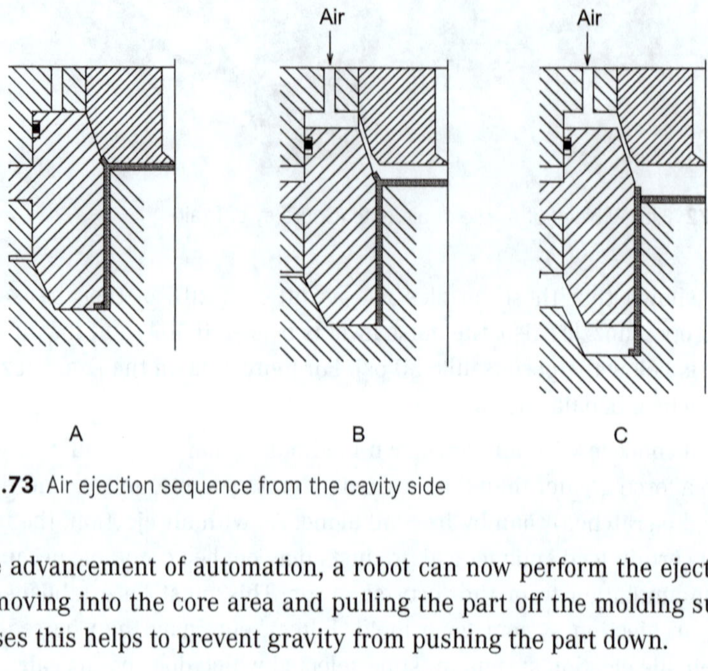

Figure 11.73 Air ejection sequence from the cavity side

With the advancement of automation, a robot can now perform the ejection function by moving into the core area and pulling the part off the molding surface. In some cases this helps to prevent gravity from pushing the part down.

11.10 Ejection of Significant Undercuts

So far in this chapter, the parts to be ejected have not had any major complexity in terms of ejection. There may be some small undercuts on a rim or a thread, but they can be directly stripped off the core. Additionally ribs, bosses and other features which do not eject easily can be pushed off the core with pins, blades or other direct-acting ejection mechanisms.

The methods in the subsequent sections of this chapter deal with undercuts that cannot be stripped off the core without additional movements in the ejection sequence. The parts have features that require additional movement in the ejection process to allow the part be be freed from the core or cavity.

11.10.1 Cam Ejection (Internal and External) Using Neck Rings

The movement of the inserts can also follow a predetermined path from a cam system. This can allow the ejection to follow a path that allows the part to be partially ejected by the insert and then freed from it or vice versa. Figure 11.74 shows an example of a simple ejection system that strokes forward for a distance L_s, and then the movable insert moves outwards a distance M over the remaining stroke L_a.

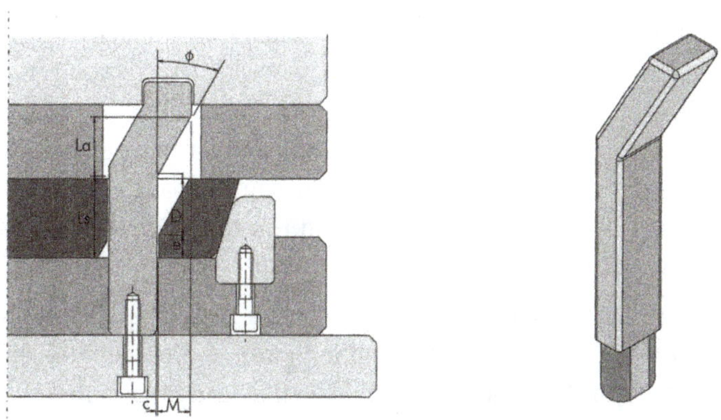

Figure 11.74 Cam ejection allowing for a stroke and then movement of the insert (Courtesy of Manual do Projectista)

A good example of this is a neck ring (or neck splits) consisting of two or more sections of a ring that are split so that when opened, they can free any undercuts or threads (see Chapter 18 for examples). Neck ring ejection is used when an undercut or thread cannot be directly stripped off. In Figure 11.75 the undercuts made by the thread are used to partially strip the part from the core, and then it spreads apart to free the plastic part from the undercuts of the thread. A profiled cam is used on the ejector bars holding the neck rings to guide the movement of the neck rings.

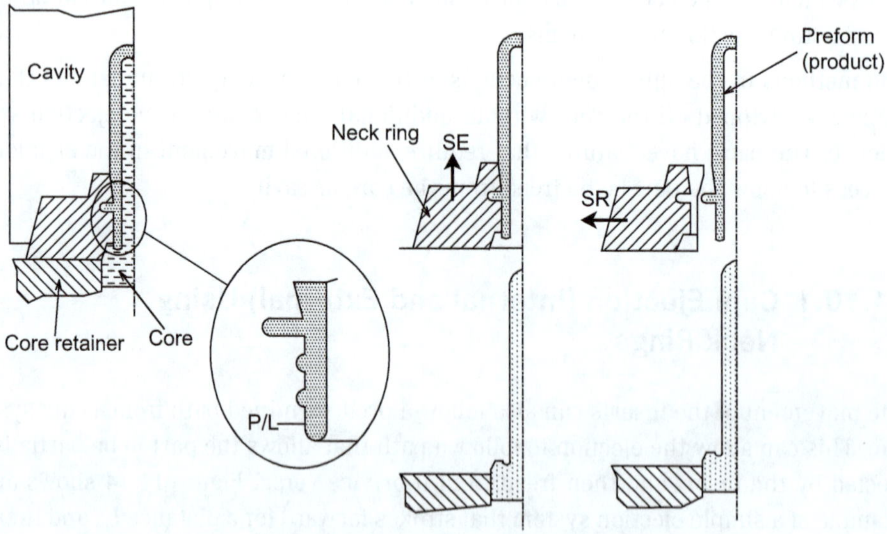

Figure 11.75 Schematic of neck ring ejection (external threads)

Immediately after the mold has opened, the neck rings (which are connected to the ejector mechanism) advance (SE). Cams in the mold first separate the neck ring slightly while the core is still well within the product, to ensure that the product will not stick to the neck ring. As the neck rings advance, they strip the product from the core. Nearing the end of stroke SE, the cams separate the neck rings fully in a radial motion (SR) to free the threads and the ring. The product can then fall freely or be removed by a robot.

To create this path for the neck rings, a cam system is developed to move the slides shown in Figure 11.76 along a predetermined path, as the machine ejectors push the stripper plate with the slides on it (and the neck rings mounted on the slides).

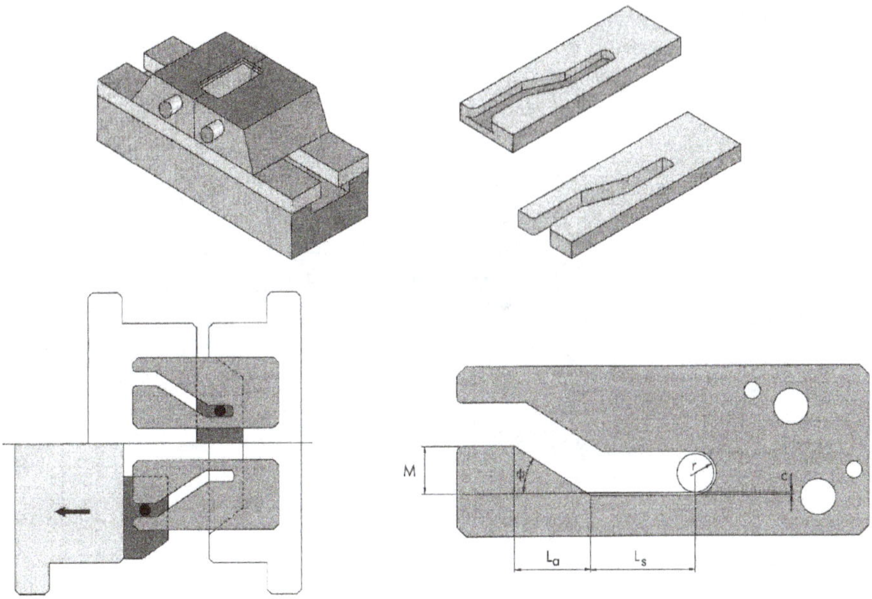

Figure 11.76 Sample cam profile for ejection of a part with undercuts (Courtesy of Manual do Projectista)

11.10.2 Core Pull Ejection

This method is used where the product must be split in a direction where the projected area is large when compared to the other direction, and where ejector pin marks are not permissible or, because of the nature of the plastic, the product would not eject reliably with pins. In the case shown in Figure 11.77 the cores rotate for ejection rather than pulling out, but the principle is the same as the core pull method.

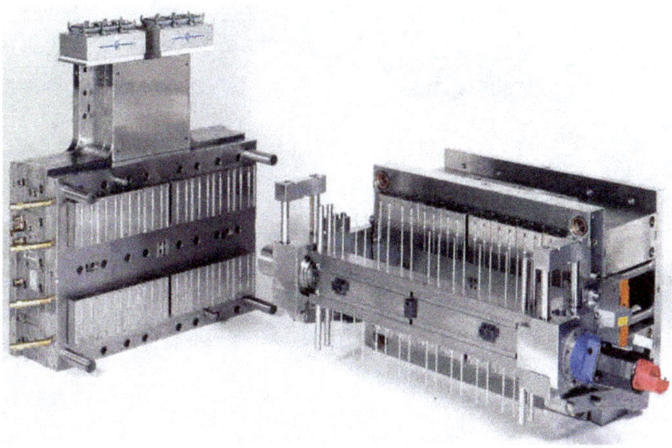

Figure 11.77 Split-cavity/rotating core ejection mold (Courtesy of Caco Pacific)

In the example shown in Figure 11.78, the deciding reason to select this design was that the plastic was soft and rubbery, and ordinary ejector pins into the side of the product would not eject reliably.

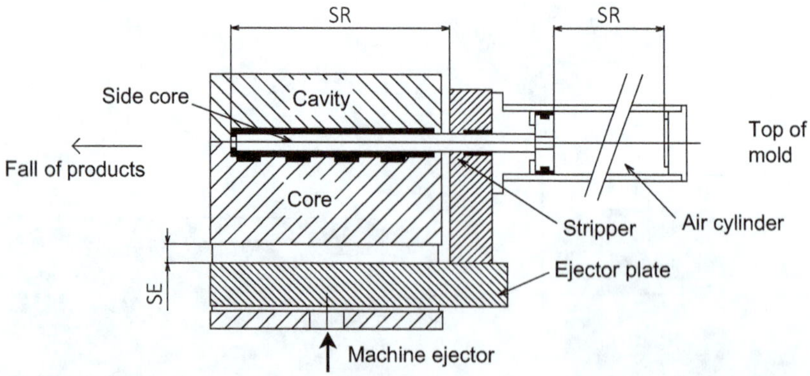

Figure 11.78 Core pull ejection

After the mold opens, the machine ejector drives the ejector plate forward (stroke SE) to lift the products out of the core. Then, the air cylinder strokes to the top, and as the products contact the stripper, they are pulled off the side cores. The mold is very simple, and there are no ejector marks on the product.

Figure 11.79 Typical core pull cylinder (Courtesy of DME)

A similar design can be used for very long, small-diameter spools, which would require high, split cavities. The splits would have to be very solid to resist deflection created by injection pressure.

With the core pull design, the spools would lie in the parting line. When compared with the large side core forces in the split-cavity design, with this design the side core forces are relatively small - proportional to the size of the flange of the spool, or even only to the size of the hole in the spool. Also, there are no ejector marks in the working area of the spool.

11.10.3 Collapsible Cores

Collapsible cores are used where the core cannot be pulled out of the product because there are projections into the core that are too large or too stiff to be ejected by stripping. Typical applications are large-size rigid screw caps, such as for threaded coffee jars, which do not require complete threads going 360° around the rim. The design can also be used for containers or technical products that have deep inside projections. An alternative would be to use a rotating core unscrewing mold design (see Section 11.10.6).

Figure 11.80 shows only the basic principle of collapsible cores, without going into more details such as cooling, the actuation and guiding of the cores, and the stripping of the product. The design of such molds can be one of the more difficult tasks, even for an experienced mold designer. Figure 11.80 and Figure 11.81 show a mold for a threaded cap with deep undercuts. The thread must be modified by interrupting it in two spots because the cap would otherwise be damaged as the center core (CC) withdraws. The interrupted thread can end with a radius, as shown, or it can run out into the center core. This is easier to produce but will create sharp corners where it meets the center core. The thread usually covers only about 65–80% of the circumference, which is acceptable.

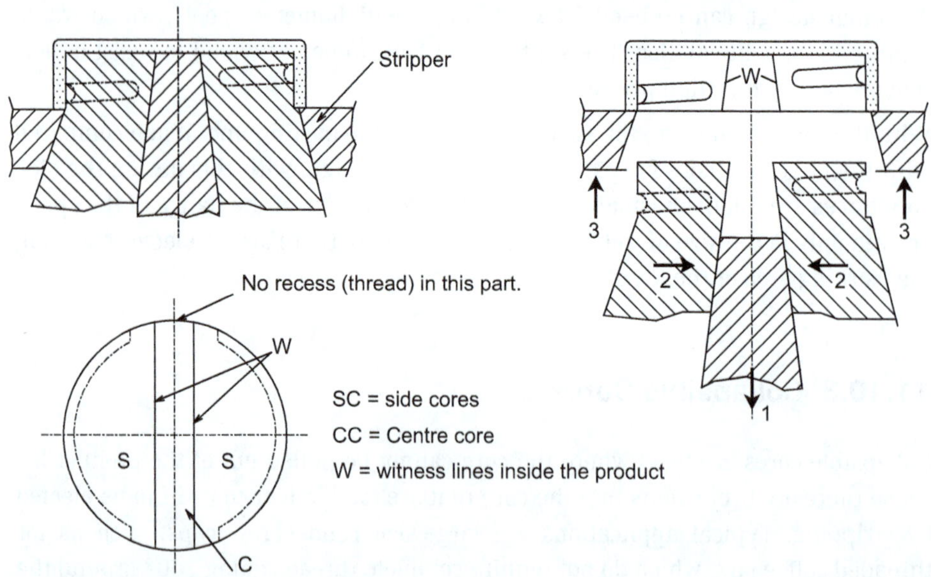

Figure 11.80 Collapsible core ejection for a threaded cap. Step 1 retracts the center core, step 2 contracts the collapsible cores, and step 3 strips the part from the core

Figure 11.81 Example collapsable core driven on dovetail (Courtesy of DME)

In operation, as soon as the mold starts opening, the center core is withdrawn relative to the side cores, which permits them to move toward the center of the core and thereby release the undercut thread. The third motion is the stripping of the product.

In the illustrated example (Figure 11.80), the core consists of three moving parts; other cases may require five, or even more, parts to withdraw an undercut from the complete circumference. This is, therefore, much more complicated, since all these components must slide against each other with minimal wear and lock up completely without gaps where plastic could enter during injection. They must be well-guided and solidly supported to withstand the injection pressure. All corners must be sharp to minimize the appearance of the witness lines where the core components meet. For best molding performance, there should be cooling channels inside each core section.

Some collapsible core molds may be simpler than unscrewing molds. The disadvantages are:

- Complexity of the core assembly, which consists of at least three members
- Difficulties of actuation and locking of the components
- Wear of the sliding surfaces, with the inherent danger of flashing, resulting in high maintenance costs
- Difficulty of providing cooling to these moving core sections
- High mold cost, and slow cycles.

A product intended for molding with collapsible cores may still require some design modifications to suit the planned sectionalizing of the core.

11.10.4 Lifter Ejection

Lifters (sometimes called jiggle pins) are used often in molds where there are one or several severe undercuts on the inside which cannot be stripped with conventional methods. Lifters can be purchased as standard parts from mold parts suppliers. The mold maker only has to design the end of the lifter with the molding surface on it and attach it to the standard lifter. The key for the lifter is the pivot in the ejector plate that allows the lifter to move forward while the angle between the plate and the lifter changes. These pivots come in pin/slider or ball style.

During ejection, the lifter moves forward and sideways, freeing the undercut, while ejector pins move straight forward (Figure 11.82 and Figure 11.84). Both the lifter and the pin are driven by an ejector plate. The stroke of the lifter needs to be just enough to clear the undercut, and the ejectors may continue to move forward to eject the product without it hanging-up on the moving core (this requires two-stage ejection; see Section 11.12). On mold closing the lifter must be positively returned using return pins (see Section 11.7.6).

There must be a minimum 3° difference between the angle of the lifter and the shut-off so that the insert will re-seat in the closed position without causing sliding wear between the insert and the core, and to prevent damage to the sharp edges (see Figure 11.82, right).

To prevent binding and wear on the lifter, the angle of the lift is typically restricted to less than 15° but preferrably less than 12° (see Figure 11.82, right). There are dual lifter systems that allow for angles of up to 30° (by increasing the angle the stroke can be reduced).

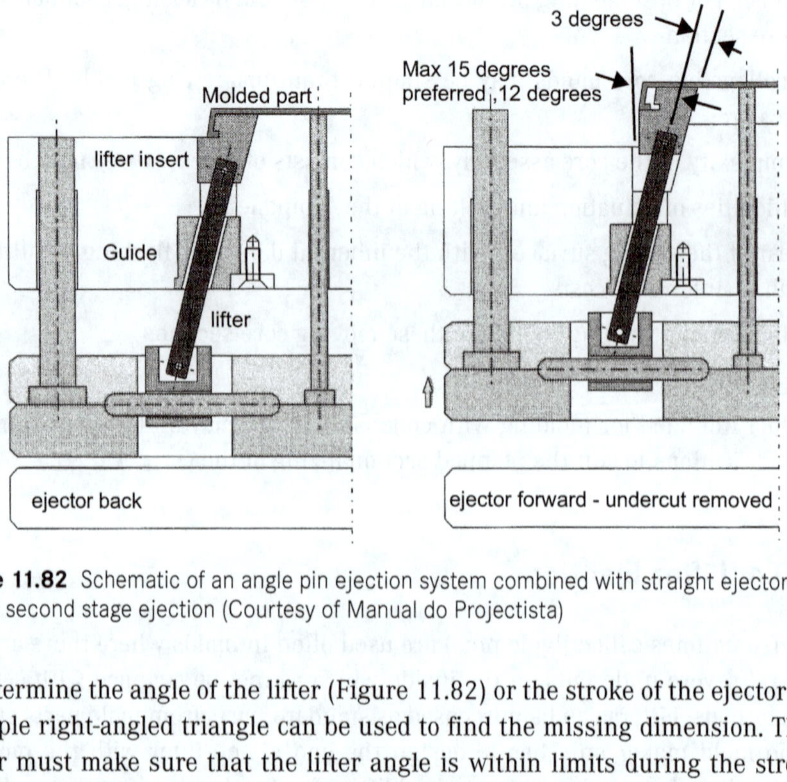

Figure 11.82 Schematic of an angle pin ejection system combined with straight ejector pins for the second stage ejection (Courtesy of Manual do Projectista)

To determine the angle of the lifter (Figure 11.82) or the stroke of the ejector plate, a simple right-angled triangle can be used to find the missing dimension. The designer must make sure that the lifter angle is within limits during the stroke to prevent binding and wear. This can be determined using CAD or the right-angled triangle in Figure 11.83.

Lifters can have cooling integrated into them, or be uncooled, as shown in Figure 11.84.

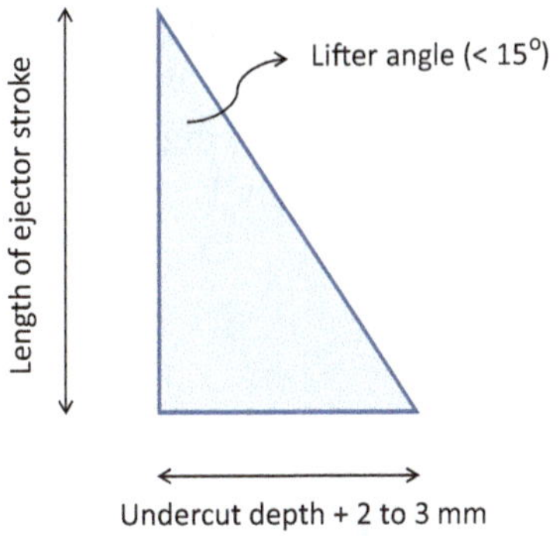

Figure 11.83
Determination of lifter dimensions

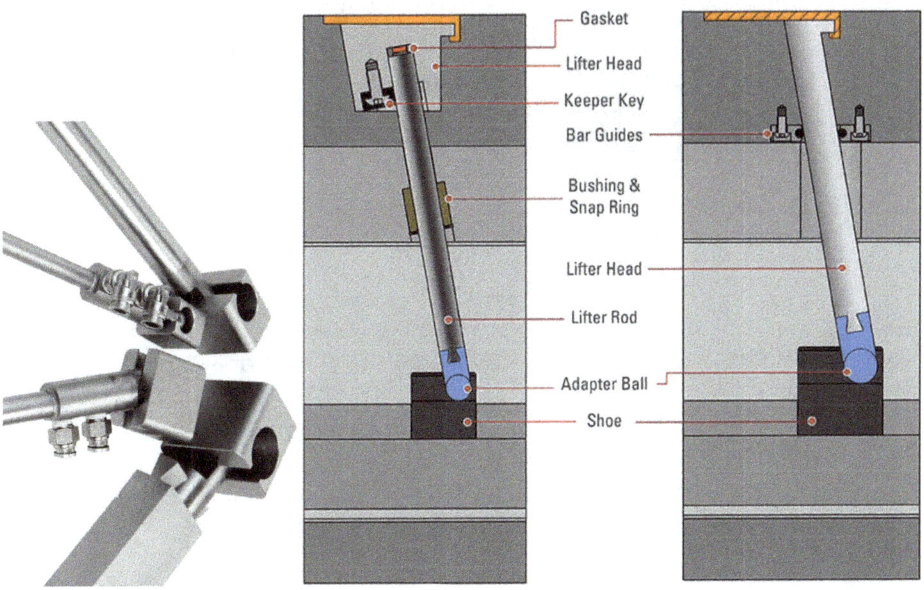

Figure 11.84 Example lifters (cooled and uncooled) (Courtesy of DME)

Examples of molds with lifters can be found in Chapter 18. There are several references showing how lifters have been incorporated into the ejection system.

11.10.5 Sliders

Sliders allow sideways movement of an insert without any forward movement. They are used when the undercut is much larger and cannot be removed with a lifter. The moving insert is held by guides and is allowed to slide as the angled pin (sometimes called a horn pin) drives the insert open during mold opening. The insert with the undercut slides along guided ways that are lubricated and rigidly held in place to resist the ejection forces. Figure 11.85 provides an exploded view of an assembly for an slider system. The moving element (1), which will slide to free the undercut on ejection, is guided by the angled pin (9). The moving element (1) is guided on all sides by lubricated guides (5, 6, 7). The angled pin is contained within a guide (8) or the core plate itself.

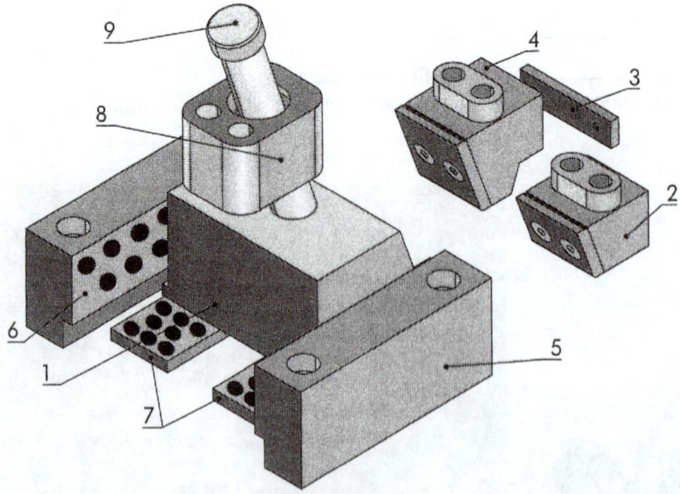

Figure 11.85 Design of retainer for angled pin (or horn pin) in slider ejection (Courtesy of Manual do Projectista)

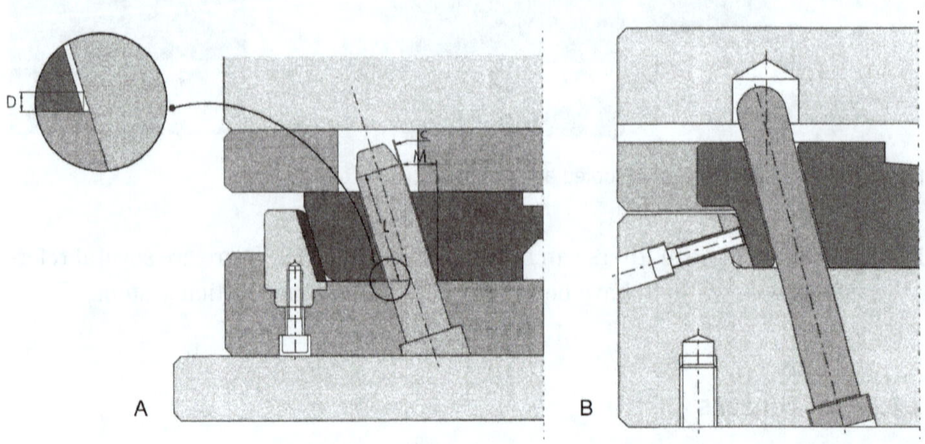

Figure 11.86 Determining the length of the guide for an angled pin ejection and the amount of gap between the pin and guide: (A) Back-up using a wedge; (B) Back-up by plate (Courtesy of Manual do Projectista)

The size necessary for the guides can be calculated using Equation 11.8 and Figure 11.86. Their length (*L*) depends on the stroke required for the movement (*M*), the angle of the pin (j), and the clearance to be applied to the moving element (*c*).

Note that the moving insert must be backed up by a separate wedge bolted to the plate (A) or a recess in the plate (B).

Formula for the length of pin required for side movement:

$$L = \frac{M}{\sin\varphi} + \frac{2c}{\sin 2\varphi} \tag{11.8}$$

where:
L = length of engagement for pin
M = lateral movement of moving insert
c = clearance between pin and guide (typically 0.13–0.25 mm or 0.005–0.010")
φ = angle of pin

To determine the length of the guide for the activation of the moving element, choose the standardized leader pin length with the length just above the calculation in Equation 11.8.

The displacement of the movable element caused by the chosen guide (*M*), see Figure 11.86, can be calculated using Equation 11.9.

$$M = \left(L \times \sin\varphi\right) - \frac{c}{\cos\varphi} \tag{11.9}$$

The movement delay (*D*) of the moving element of the mold, in relation to the opening, can be calculated by:

$$D = \frac{c}{\cos\varphi} \tag{11.10}$$

Note that this delay in opening the movable element, in addition to facilitating the extraction, causes the displacement of the movable element of the lifting bar, thus contributing to less wear.

General guidelines for sliders:

- As with lifters, all shut-off surfaces for sliders must have a minimum of 3° draft in direction of slide travel. This is to prevent damage to sharp edges and to prevent sliding wear between inserts. There must also be a 3° difference between the angled pin and the back-up wedges or support system.
- If the insert to be moved is greater than 150–175 mm (6–7"), then two angled pins (or horn pins) should be used to guide the insert. There must be a 3–5 mm gap between the slider and the molded part when the slider is in the fully open position to prevent any hang-up of the plastic part during ejection.

- If the slide is large then it must have its own independent cooling, or the slide will overheat and impact the molding cycle.
- Any deep ribs in the slider must be vented using the guidelines for venting ribs (see Chapter 10).
- Ensure that the slide face is held back with springs or in a gib to prevent forward motion at all times.
- To prevent galling there should be sufficient difference in hardness between the gib and the slider (4–6 points).

11.10.5.1 Split-Cavity Movement on Angled Pins

If there are undercuts in the cavity, these need to be freed up as the mold opens, before the molded part can be ejected from the core. Sometimes angled pins are used to allow the cavity to open, as shown in Figure 11.87. Examples of this type of mold can also be found in in Chapter 18. As with the angled pins on the core, the moving insert needs to be held rigidly during the clamped and injection phase of the cycle and then allowed to move on lubricated ways during the ejection sequence. The angled pin provides the sliding movement.

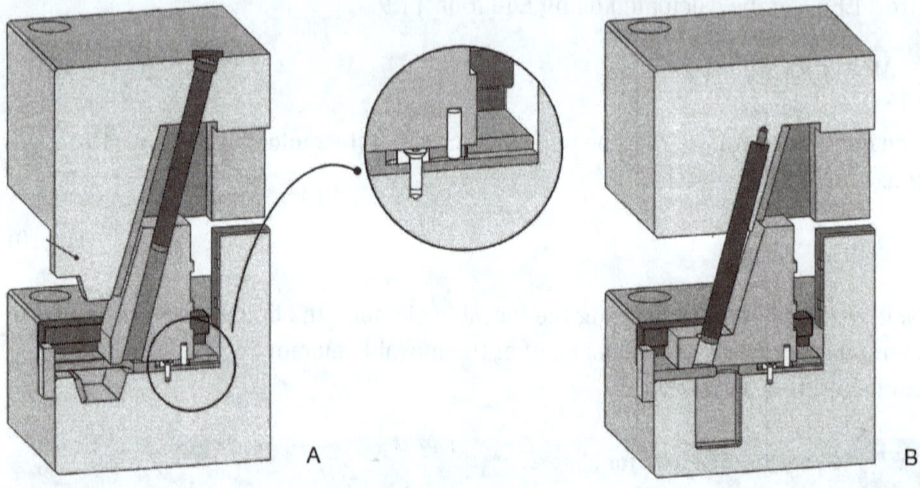

Figure 11.87 Split cavity as part of a two-stage ejection system: (A) Guiding pin located on cavity side; (B) Guiding pin located on core side (Courtesy of Manual do Projectista)

The angled guide pin can be located on the cavity side (A) or the core side (B), as long as it can be firmly held and smoothly guided both forward and back.

The angle of the guide pin must not be too severe as it will generate forces that attempt to pull the pin from the mold. Therefore, the angle of attack for the guide pin should be in the range 10–20°. If there is a taper backing-up the movement, it should be 2–3° more than this angle.

11.10.6 Unscrewing Ejection

Unscrewing refers to removing inside threaded products from the core by rotating the core. In the case of an unscrewing mold ejection, the product is kept from rotating while the core rotates to unscrew itself from the product. Figure 11.88 shows a system using a rack and pinion for the rotation of the core.

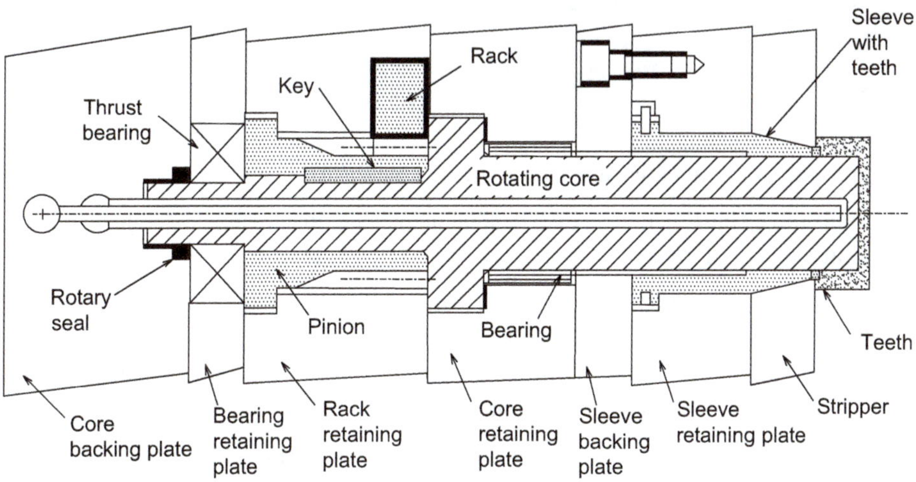

Figure 11.88 Unscrewing ejection using a rack and pinion

As the core rotates, the sleeve-retaining plate and the stripper plate move forward by a light spring pressure under the sleeve backing plate. This motion can also be produced by mechanically synchronizing the forward travel with the travel of the unscrewing motion. In some systems, the core is held in a lead nut below the drive pinion, so that the core is positively retracted, rather than depending on a spring or other synchronized forward motion of the stripper assembly.

The product is kept from rotating by teeth (or other projections) in the end of the sleeve. At the end of the unscrewing, the stripper plate moves forward to eject the product from the teeth, which occasionally may still be keeping the product from falling free. The rack is usually driven by a hydraulic cylinder outside the mold (see Figure 11.89).

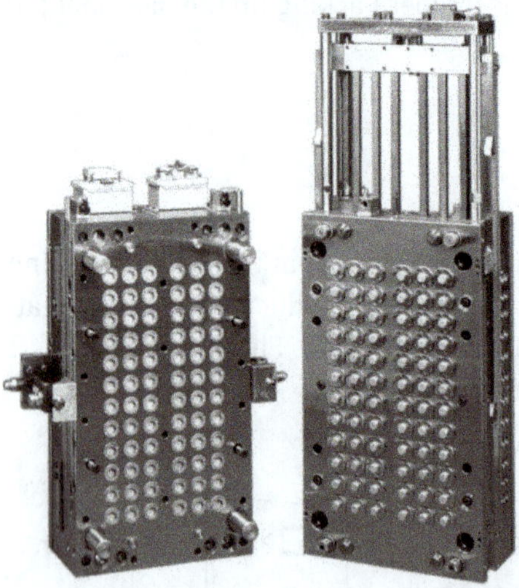

Figure 11.89
72-cavity unscrewing mold using a rack and pinion for unscrewing (Courtesy of Stackteck Systems)

The common challenges with unscrewing systems are:

1. Combining the mechanical drive system with a mold that is subject to temperature variations.
2. The bearings need lubrication, and the sealing gaps must be tight enough so that the plastic will not flash during injection, but wide enough to prevent galling of mold materials that cannot be properly lubricated because of the potential contamination of the products with lubricants.
3. The coolant must be contained with rotary seals, which are, from past experience, a common source of leaks, even after relatively short periods of use.
4. It is difficult to cool the core, and resulting cycle times are slower than a conventionally cooled core.

Despite these problems, there are many unscrewing molds in operation, and some mold makers specialize in this field, using their own standards of core assemblies, drives, etc., so that these molds can be produced on a production basis very competitively.

11.10.7 Cam and Harmonic Linkage Ejection

If very short strokes are required, the movement of the stripper plate can be driven by a linkage system connected to the stationary and moving sides of the mold. An example is shown in Figure 11.90. In this case a cam on the moving side is tied to a stripper plate. As the mold opens, the roller on the stationary or fixed side moves the cam up. This in turn drives the stripper plate forward to eject the shallow parts.

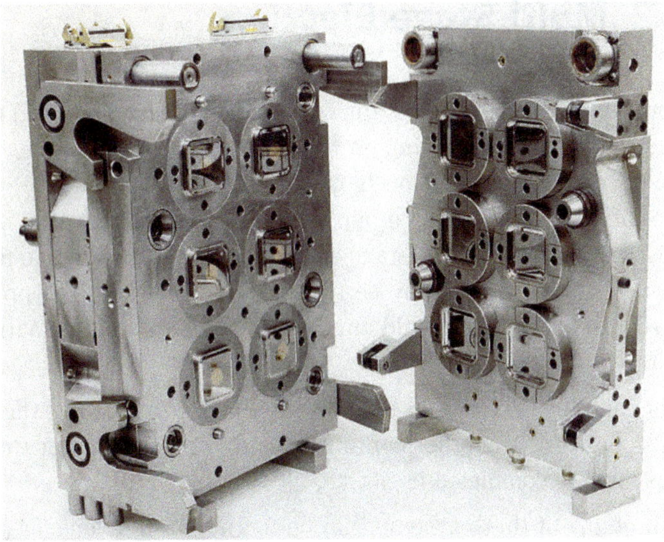

Figure 11.90 Cam ejection (Courtesy of Husky Injection Molding Systems Ltd.)

11.11 Multiple Ejection Strokes

Today, all molding machines are equipped with a feature (standard or optional) that allows the machine ejector to stroke more than once after the mold has opened. Some molders use it to "shake" the product loose from the stripper if the product hangs up occasionally. Machines usually permit up to ten strokes.

In the case of stripper molds, often the need for multiple strokes could be surface tension between the finish on the stripper and the material being molded. This can often be eliminated by reviewing the finish on the stripper and making adjustments to it. The other potential reason for the parts not falling off the stripper readily is a slight step between the molded part and the stripper, meaning that the part is subtly molded into the stripper - this step could be as small as 0.025 mm (0.001"). Both of these potential issues should be checked before accepting multiple ejection strokes.

While multiple strokes can sometimes be successful in running a poorly designed mold fully automatically (without an operator), it is bad practice because the cycle time is lengthened by the additional ejector travel times of stroking forward and returning to shake the product loose, as well as by the additional wear of machine and mold.

■ 11.12 Multi-Stage Ejection

Two-stage (or multi-stage) ejection is the combination of two (or more) mechanical ejection systems in one mold, acting on the same product. Combining blow-off or blow-down with mechanical ejection is not two-stage (or multi-stage) ejection. In this context, the use of robots or other take-offs with molds using single-stage ejection is also not considered two-stage ejection. Occasionally, due to the shape of the product, two-stage ejection may be necessary, even where the product is removed with robots, because the mold must also be able to run automatically when operating without them.

The timing is usually such that (a) both ejection systems start together, and then one of them stops while the other continues, or (b) one starts after the other has started or has already completed its stroke.

The actuation of any of these systems can be by the machine ejector (usually hydraulic), cams, air cylinders, or hydraulic cylinders.

When using all-mechanical ejection, the force available is the machine ejector force. When using air actuators, the required forces must be calculated to provide sufficient force for 100% reliability of ejection.

Hydraulic actuators are used when it is not practical or possible to place strong enough air actuators within the available space in or on the mold. They are also required where it is important to provide smooth operation of the actuator. There is no jerky motion (often associated with air cylinders); also, hydraulics can have accurate speed control. The disadvantages with any hydraulics are higher mold cost and the risk of contamination of the products with oil.

Control of two-stage actuation depends on the methods used. If both stages move mechanically (e.g., with methods using lost motion or dwelling on cams), no additional controls are required to operate the two-stage ejection. This is usually preferred but not always possible. If it is not practical or possible to use a mechanical method for both motions, air or hydraulic actuators will be necessary, with their additional piping and controls.

11.12.1 Why and When to Use Multi-Stage Ejection

Two-stage ejection is always required when the product will not eject freely by being pushed off the core using standard (single-stage) ejection, regardless of whether the ejection is mechanical or uses air. This may be the case for products with undercuts, as in the following example (Figure 11.91). In this example an undercut is trapped in the core. To eject the part, the core is split into a core and a core sleeve and multi-stage ejection is used. The sleeve and stripper move first together (arrows 1 and 2), lifting the product and bringing the inner rim above the top of the core. At this point, the motion of the sleeve (1) stops, but the stripper (2) continues to push the product. This causes the inner cylinder (and rim) to be somewhat squeezed (indicated with arrows), which allows the rim to slide out of the groove where it was formed. This permits automatic ejection of the product without damaging it.

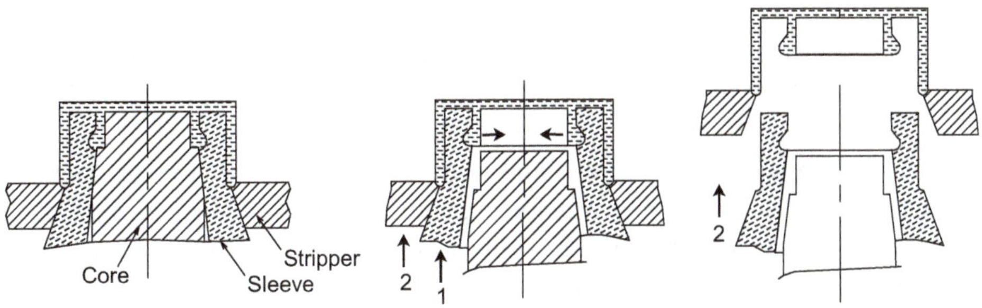

Figure 11.91 Example of product requiring two stages of ejection. Stage 1 uses a stripper + core sleeve to lift the part off of the core. Stage 2 continues with the stripper only to free the part from the core sleeve

Two-stage ejection may be used whenever there is a possibility that stickiness of the plastic (or flash) would cause the product to adhere to the stripper or ejectors. Two-stage ejection ensures that a product which otherwise could be ejected with single-stage ejection will be positively pushed off by the stripper or ejectors.

11.12.2 Typical Two-Stage Ejection Actuations

Figure 11.92 shows a two-stage ejector pin system driven by the machine ejectors. When the machine ejector moves forward, it moves plate B and plate A together. Plate A stops when it arrives at the core backing plate C, but ejector plate B keeps moving forward while compressing the springs. This method is simple, but the springs must be strong enough to move plate A and to overcome the ejection forces. At least four springs must be used and evenly distributed. If the springs are too strong, the machine ejector may have insufficient force to move plate B after A has stopped.

> **Note:** Shoulder bolts in this section are shown for simplicity of illustration only. They must always be the screw-sleeve arrangements. Standard shoulder (or stripper) bolts must never be used in these applications, because the shock forces they are subject to will eventually cause them to fail.

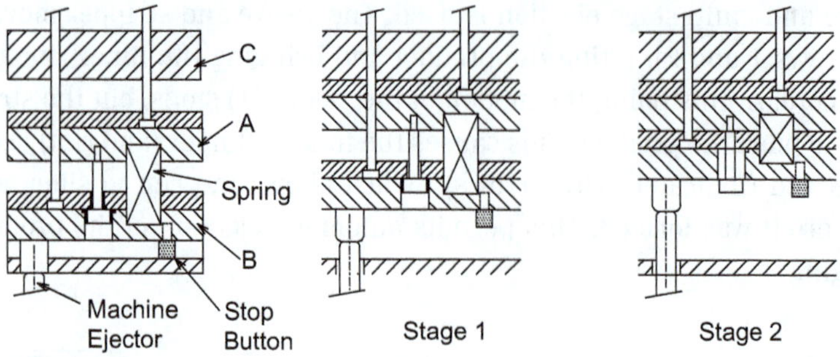

Figure 11.92 Two-stage ejection using springs: (A) Ejector plate; (B) Second ejection plate; (C) Core plate. * Not shown: screws holding retainer plates to ejector plates or return pins

Even though this method is simple and frequently used, it is not recommended because of the inherent danger when using a number of preloaded springs. Also, because stretching/compressing the springs by more than a small proportion of their length will reduce their lifespan, the travel is limited to short strokes only.

When ejection instead uses lost motion and pulling, as shown in Figure 11.93, there is no need for springs. The timing is different from the spring actuation method. The machine ejector drives first plate A, while plate B rests. When the sleeve/bolt assembly (shown as a shoulder bolt) heads engage, plate B is pulled along with plate A until A (and B) stop.

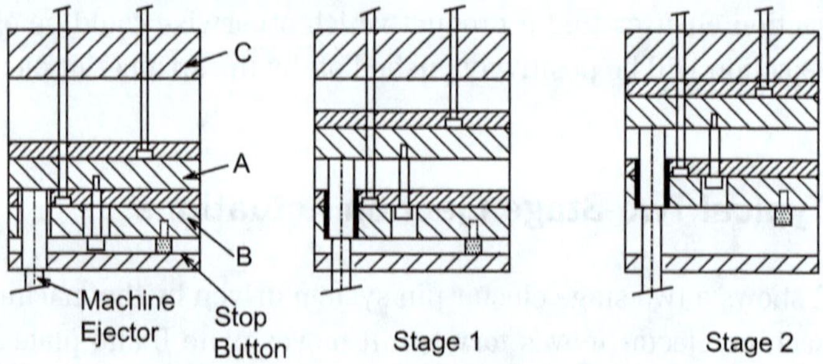

Figure 11.93 Ejection in two or more stages using lost motion and pulling

In all cases, the ejector and/or stripper plates must be positively returned by springs or by air actuators, which is not shown for simplicity of illustration. The return pins provide safety in case of failure of such a return system.

A different method of two-stage ejection involves the use of interlocking cams. There are several mold accessories commercially available that interlock the two ejector plates so that they stroke in sequence. (First, both plates together, then one alone.) The basic operation is similar in all these devices, and it is similar to the arrangement illustrated in Figure 11.94. Because of the high loads on the cams, good lubrication and/or specially selected materials must be provided to prevent early wearing out of the critical cam faces.

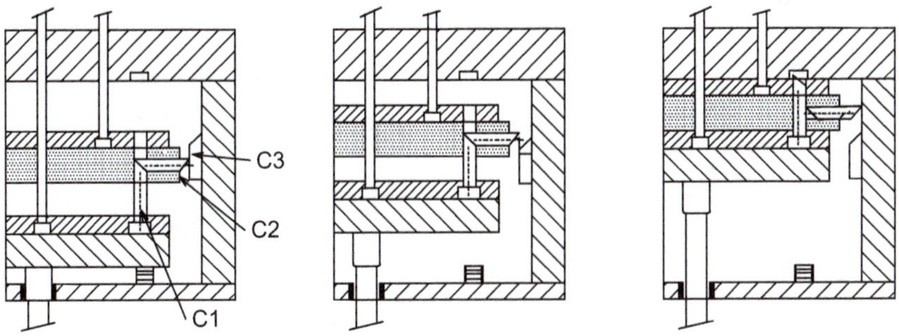

Figure 11.94 Two-stage ejection using cam mechanism

When the machine ejector moves forward, it drives the lower ejector plate, but it also drives the upper ejector plate through the cams C1, C2, and C3. As long as C2 rests on C3, it stops C1 and the upper plate is pushed forward. As soon as C2 comes to the slope of C3, it can move sideways and let C1 pass, so that the lower ejector plate can now approach the upper ejector plate. There must be at least four such cam systems per ejector plate to ensure even advance.

11.12.2.1 Latch Locks (Internal and External)

Latch lock systems can be purchased from mold component suppliers. They can be mounted externally on the outside of the mold for ease of service (Figure 11.95), or they can be contained within the ejection system internally in the mold (Figure 11.96). Latch locks control the sequence of one ejection stroke opening after the first ejector plate has traveled a predetermined distance. The latch lock systems allow multi-stage ejection to be sequenced.

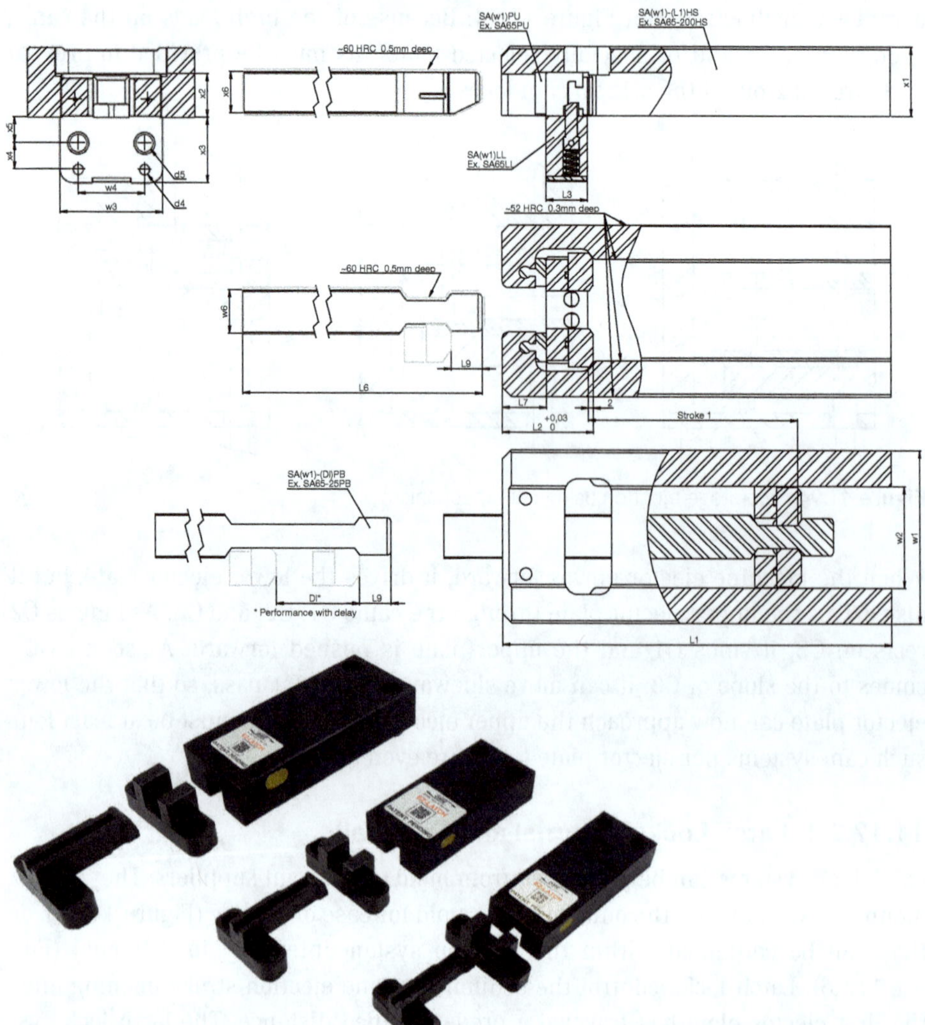

Figure 11.95 External latch lock: Positive and precise positioning of floating plates (Courtesy of DME)

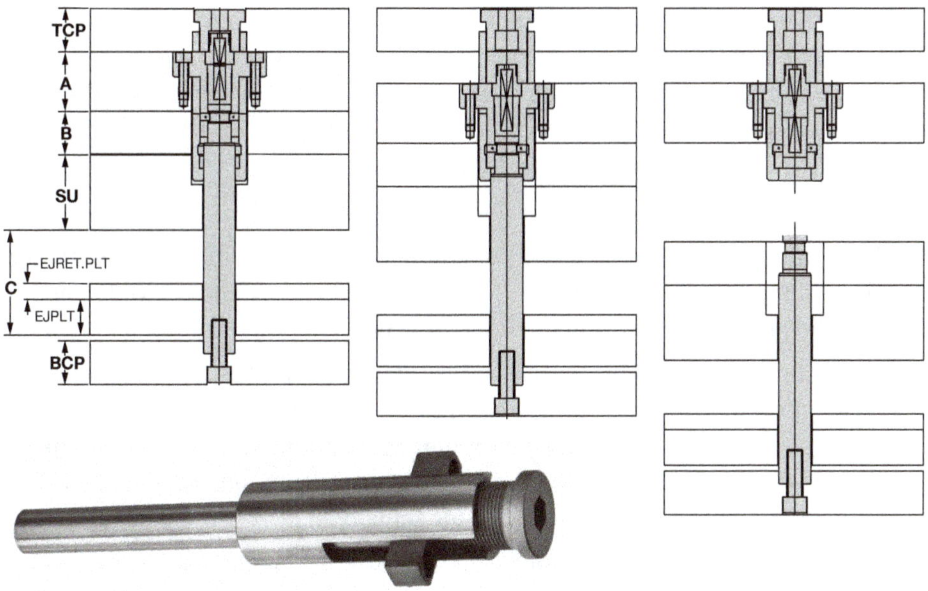

Figure 11.96 Internal latch lock: To float plate A away from top clamp plate while locking plates A and B, the center rod is tied into the bottom clamp plate (Courtesy of DME)

11.12.2.2 Two-Stage Ejection with Levers

Two-stage ejection using lost motion may also operate through the use of levers. Figure 11.97 shows a typical arrangement. There are an infinite number of possibilities using links and levers, but the principle is usually the same.

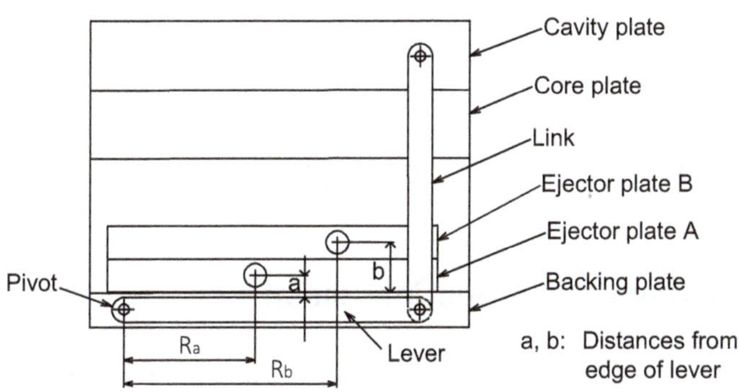

Figure 11.97 Two-stage ejection using levers and rollers

As the mold starts to open at the P/L, the lever, pulled by the link, will contact the roller on ejector plate A and start lifting plate A and B together. As the mold continues to open, the roller in plate B is contacted, and because of the larger distance from the pivot, plate B will move away from the (still advancing) plate A.

Note that by spacing the plates A and B, and by supporting plate B on its own stop in the rear position, the motions could be as follows: Plate A moves first, then plates A and B together, then plate B away from plate A. The shape of lever and link, and the position of the rollers *a*, *b*, R_a, and R_b, depend on the mold requirements and must be developed by the designer.

Rules to observe when working with links and levers:

1. There must be at least one, but preferably two, pairs of links and levers per side, acting on opposite sides of the plate (usually front and back), to ensure that the plates lift parallel to the backing plate.
2. The plates must be returned positively. Since the strokes are usually fairly short, spring return with springs within the plates is practical as long as the springs are properly designed and strong enough for the job. With linkages, there is usually not enough space to place return springs outside the mold. The alternative would be to use (internal) air actuators.
3. The rollers (usually cam followers) must be strong enough for the anticipated forces. If there are two rollers on each plate, each roller could see one-half of the mold opening force. While this is not likely to be the case, the rollers are often under-engineered (too weak).
4. Lubrication of rollers and articulations is very important:
   - Rollers: Use standard cam followers with permanently lubricated needle bearings
   - Pivots: Use permanently lubricated needle bearings. If using bronze or steel bushings, they require lubrication, either through drilled bores in the links and levers, or with separate tubing.

   Lubrication of the faces where the rollers touch the levers is difficult to achieve. The rollers are hardened steel. The levers should also be hardened, at least at the contact faces, to reduce wear, which in the long run could cause imbalance (locking) of the actuation of the ejector plates.

11.12.3 Moving-Cavity Ejection Assist

A moving-cavity ejection system is essentially a limited motion of the cavity with respect to the gate pad, designed to free features of the product trapped in the cavity. This motion frees the product and prevents it from tearing or breaking. It can also used to free a trapped undercut similar to Figure 11.91. A moving cavity design allows air to enter behind the cavity, to break the vacuum preventing the product from staying with the core.

Figure 11.98 shows the essential elements of a moving cavity. As the mold opens (right half of picture), the cavity and cavity plate follows the clamp motion for the distance *S*, which must be sufficient to release that portion of the product that was locked between cavity and gate pad. The motion *S* is generated by internal air pistons (see Figure 11.99) or external pistons acting between the cavity retainer plate and the cavity plate. If there is little or no draft, or if an undercut in the cavity requires the wall of the recessed portion to squeeze to come out of the cavity, the stroke *S* should be slightly greater than *F*. With good draft, the distance *S* need not be as large as the depth of the recess *F*; a small motion of about 3 mm (⅛") will be sufficient.

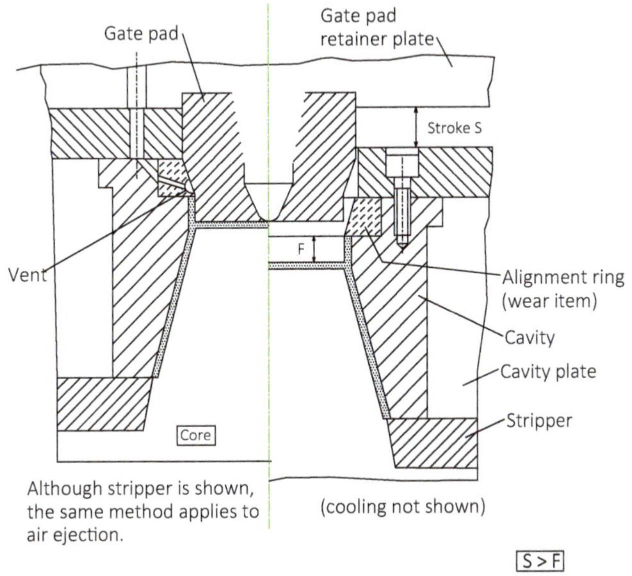

Figure 11.98 Moving-cavity ejection assist

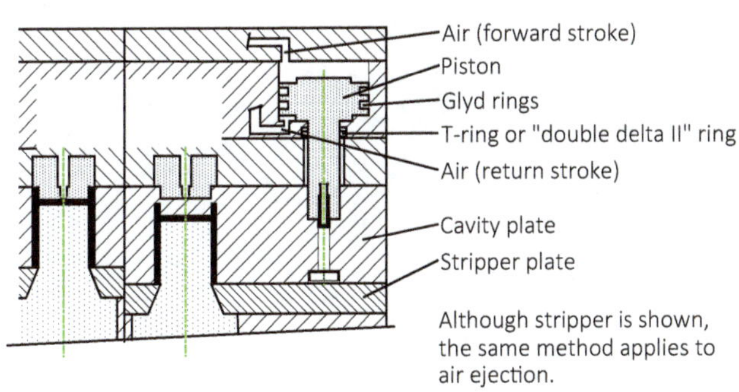

Figure 11.99 Internal air piston layout

On the left in Figure 11.99, the mold is closed. On the right, it is opening and has just reached the end of stroke of the actuator. The cylinders are bores in the manifold or core backing plate (usually at the four corners of the plates), and the piston rods are mounted in the cavity plate. In this design, the piston rods also act as stroke limiters to prevent the cavity plate from moving too far.

Chapter 18 contains an example of a moving cavity air eject mold for reference.

12 Mold Cooling

Cooling is one of the main functions of an injection mold, which acts as a heat exchanger transferring heat from the polymer to the coolant. During and after injection, the hot plastic is cooled so that it is rigid enough to be ejected from the mold. As plastic enters the cold cavity, it beings to cool very quickly once in contact with the molding surface. As it cools, it becomes more viscous and forms a solidified layer (see Figure 12.1). New molten material enters from the remaining channel in the center of the cavity in a "fountain flow" pattern. After the cavity is filled, the velocity drops to zero and conduction to the cooler cavity walls solidifies the part enough to be ejected.

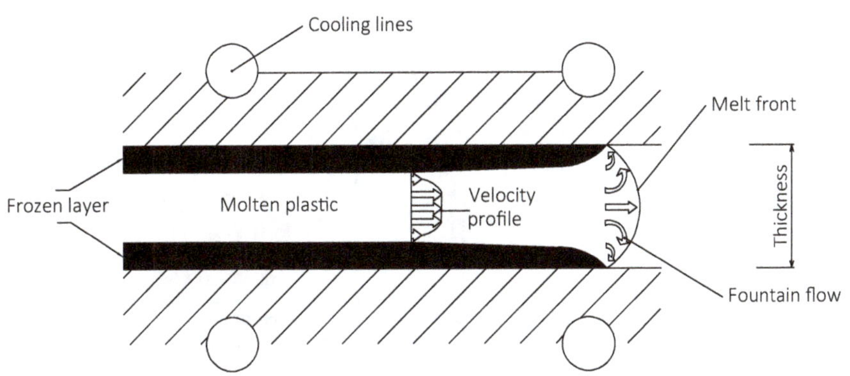

Figure 12.1 Plastic cooling in the cavity space during injection

Cooling is added to every mold in the form of cooling channels that are engineered to allow the plastic to cool evenly and quickly. Most molds are cooled with water, but occasionally glycol or oil is used where the plastic being processed requires different mold temperature conditions than water cooling would create. The designer should be aware that a substantial portion, sometimes *as much as 80% of the total molding cycle, including injection, hold and cooling elements, could be required for cooling*. Therefore, for high-production molds, it is imperative that cooling time is reduced to a minimum.

An injection mold functions as a heat transfer vessel. It takes in the hot molten plastic and conducts, convects, and radiates the heat to cool the plastic until solid (Figure 12.2). In order to design an efficient cooling circuit in a mold, a mold designer must have a solid grasp of *thermodynamics and heat transfer*. The designer must understand that there are significant differences in heat conductivity in the materials commonly used for molds. The designer must also understand that the amount of heat removed per unit of time depends on the distance the heat must travel, the temperature of the cooling medium, and the speed of the water flowing in the cooling channels. Dirt and corrosion in cooling lines act like heat barriers and negatively affect the heat transfer from the mold to the cooling medium. In some cases, it may be almost impossible to provide cooling lines in small mold parts. Special mold materials, such as beryllium-copper (BeCu) and other alloys, provide about 4 × better heat conductivity than steel, and are often used for delicate mold parts where adding cooling channels is not practical. The heat will then move faster than in steel to reach a neighboring well-cooled mold part or plate. Certain larger parts in high-speed molds are also made from BeCu even though such parts can be well-cooled by cross-drilling or are surrounded by coolant channels. This method is often used in mold parts opposite the gate, where the hot plastic hits first as it comes out of the gate, and in parts surrounding the hot gate.

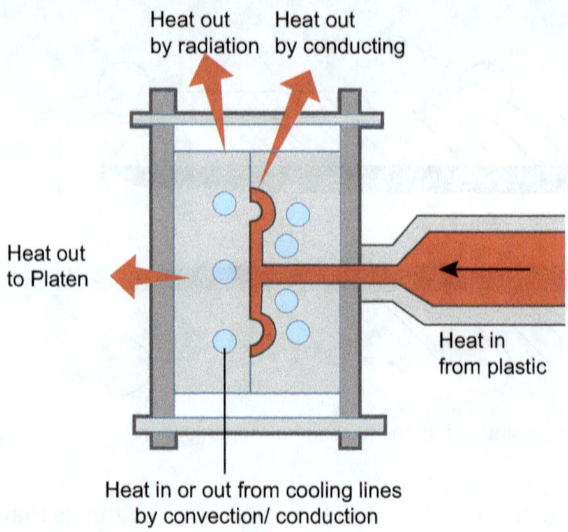

Figure 12.2 Heat transfer in an injection mold [1]

[1] Source: M. Wang, R. Chang, C. Hsu, *Molding Simulation: Theory and Practice,* Hanser Publishing, 2018 (Figure 8.4).

The higher the productivity that is expected from a mold, the better the cooling must be. Since the molder is primarily interested only in getting the best product at the lowest cost, for anything except for really low production requirements there should be no limits to ingenuity when designing the cooling channel layout or selecting the mold materials for good conductivity and mold life.

Plastics have a relatively low thermal conductivity, and so act as an insulating layer, preventing the center of the injected material from cooling. As soon as the plastic touches the walls of the cooled cavity space, it freezes and creates a boundary layer, which makes it more difficult for the following layers to give up heat to the mold wall or insert (Figure 12.1). This is significant for products with heavy walls, and will increase the cooling time regardless of how well the wall (or mold part) is cooled. Plastics, too, have different heat conductivities and this can affect the rate of transfer of heat to the mold. There is a significant difference between the ways that semi-crystalline and amorphous plastics cool. Semi-crystalline plastics (e.g., PE or PP) contain more heat and give it up more slowly to the coolant than amorphous plastics (e.g., PS or PC). More energy (heat) is needed to melt semi-crystalline plastics, and more energy (in cooling) is needed to cool them down again. In practical terms, it takes longer to mold a product from PE than from PS.

The cycle time of the injection molding process is directly proportional to the square of the wall thickness.

Also, as soon as the plastic begins to cool, it will start to shrink; this happens in most cases in the direction away from the cavity. Because the shrinking plastic will start to hug the core, there will be a better contact with the core and a space will be created between the plastic and the cavity wall. This space contains a vacuum or (if not properly vented) air. Both air and vacuum are ideal heat insulators, and reduce the heat flow from the plastic to the cavity wall. In most such molds the cavity does not need as much cooling as the core. In fact, *about 70% of the cooling from a mold comes from the core and 30% from the cavity side*. Unfortunately for the designer, there is usually much more space in the cavity where cooling is not needed as much, while the core, which does more of the cooling, is much more difficult to cool. This is especially true when there are also ejectors, moving parts, and/or air channels in it. Many existing molds have lots of unnecessary cooling in the cavity, and it should be noted that cooling lines reduce the structural performance of mold plates.

Sometimes molders might extend the cycle time to compensate for incorrect shrinkage used in the mold design. Ultimately, the additional design and engineering time taken to optimize the cooling system to achieve the intended shrinkage and avoid the additional cycle time is easily paid back over the lifetime of the mold. With the creation of CAE simulation tools such as those shown in this chapter, it has never been easier to design cooling circuits that produce the right size molded part art the optimal cycle time.

■ 12.1 Injection and Mold Temperatures for Common Plastics

Approximate injection and mold temperatures for some materials are listed in Table 12.1. The table shows that plastics are generally processed in the 150-300 °C (300–600 °F) range, whereas molds are cooled to 5–140 °C (40–280 °F). While these temperature differences are a challenge to the designer to maintain, there are effective and efficient ways to create the right conditions for processing these plastics. A more extensive list of the processing conditions for plastics can be found in Appendix 11.

Table 12.1 Typical Injection and Mold Temperatures of Plastics

Material	Injection temp. (°C)	Injection temp. (°F)	Mold temp. (°C)	Mold temp. (°F)
Polyethylene (PE)	150-260	300-500	5-70	40-160
Polystyrene (PS)	175-260	350-500	5-60	40-140
Polyamide (Nylon)	225-290	440-550	40-120	100-250
Acrylonitrile-styrene	195-215	380-420	50-80	120-170
Polycarbonate (PC)	270-295	520-560	85-140	180-280
Polyacetal (Delrin)	175-225	350-440	70-130	150-260
Polypropylene (PP)	190-290	370-550	5-80	40-170
ABS	195-260	380-500	50-120	120-250

12.2 Coolant Requirements for a Mold

First and foremost, a designer must understand the desired cycle time for the mold. This will ultimately decide how complex the cooling circuit needs to be and how much coolant will need to be supplied. After that, the designer must specify the coolant flow rate and at what temperature the coolant will be. This will determine if a cooling tower, a chiller, or a thermolator will be needed. The cooling requirements for the mold will also dictate to some degree the size of the cooling lines needed to feed the mold. Finally, the designer must understand the details of ejection to make sure that the cooling system does not interfere with the ejection of the parts from the mold or the ejection system inside the mold.

A mold is a heat exchanger. Energy (in the form of heat and electric energy) put into the plastic on its way from the machine hopper to the machine nozzle is used to bring the plastic from room temperature (T_r) to the temperature required for injection (T_i). This energy (in the form of heat) must be removed from the plastic while it is cooling in the mold so that the finished product becomes stiff enough and can be ejected.

If we assume that there are no heat losses due to radiation and conduction away from the machine, the actual amount of energy used for plasticizing can be measured and calculated. Important considerations for such calculations are:

1. The rated wattage (kW) of the heaters installed at the extruder (plus the shooting pot in a two-stage injection system).
2. The amount of energy (in kWh) actually used in the extruder motor and the heaters to plasticize the resin during the time the heaters and the drive motor are "on" in each cycle, or more accurately, during one hour of operation, once the machine is "on cycle" (i.e., running and producing without further adjustments).

Such a calculation is quite easily taken from an electrically driven extruder by reading the current to the electric motor. This is more difficult for a hydraulic-motor screw drive because of the losses between the hydraulic pump motor in the pump and in the hydraulic screw drive motor. It would be wrong to use the total connected load of heaters plus motor in a calculation without allowing for the time they are "off". Most of the energy is supplied to the plastic by the screw drive motor, and only a small amount by the heaters surrounding the extruder barrel. This ratio depends to a great extent on the design (configuration of flights) of the extruder screw. The better the screw design, the less heat will be required from the barrel heaters.

Note that the heater capacity in the hot runner system (manifold and nozzle heaters) must not be added in these calculations. In well-designed hot runner systems, the heaters are used only to maintain the melt temperature on its way from the machine nozzle to the gates. (The reason for the often large ratings of manifold and nozzle heaters is to be able to start up the mold faster, from cold. Once the mold is running, the heaters use less than 10% of their rated capacity, and in fast-running molds, they can even be turned "off" altogether.)

Note that the above measurements and calculations can only establish the power consumption and heat load for a cooling system *after* the mold has been built.

12.3 What Affects Mold Cooling Performance?

Many factors affect the cooling performance of a mold. A list of the factors is shown below with the most important listed at the top:

- Thermal conductivity of mold parts
- Temperature of the cooling medium
- Uniformity of cooling. Cooling channel layout and size of cooling channels in the components in contact with the molded product
- Flow rate of coolant through the mold including layout, size (diameter) and number of supply and return lines (hoses)
- Temperature increase (ΔT) of the coolant, from "in" to "out"
- Chemical composition of coolant
- Plastic material characteristics
- Runner system (size, layout) and type of runner system (hot or cold)
- Machine parameters like clamp speed, injection heaters, nozzle design, screw design, etc.
- Plastic material pre-conditioning (e.g., drying).

In this chapter, guidelines for cooling will be presented and reviewed. An overview of heat transfer and thermodynamics will be provided to the extent needed to make required calculations for mold design. If more background is needed, please refer to textbooks on thermodynamics and heat transfer.

12.3.1 Temperature of the Cooling Medium

The cooling medium, usually water, should be conditioned so as to minimize contamination of the cooling system with lime deposits or other substances present in the water that can corrode the cooling channels. These contaminations also create deposits on the walls of the cooling channels, which reduces the heat transfer and can also block the passages, thereby reducing the flow.

Where water temperatures are low or the supply lines are exposed to freezing temperatures, antifreeze is added. However, in most cases, the cooling water is kept at or above 5 °C (40 °F) and antifreeze is not required. The addition of antifreeze changes somewhat the capability of the water to remove heat, but this can usually be ignored.

In most molding installations, the coolant to the molds is supplied either from a water reservoir and a cooling tower, or from a water chiller system. Ideally, a whole plant would run on one cooling water temperature, supplied from one source. However, in many cases this is not practical because a plant may be molding products from plastics requiring different cooling conditions. The designer should know the cooling system used in the location where the mold is going to be installed.

Note that with some materials, such as ABS or PC, the cooling water is actually heated to the required temperature, usually by using portable heating units. In some cases, especially with higher temperatures, oil or a non-inflammable oil substitute is used instead of water.

> It is important to remember that the temperature of the cooling medium is NOT the same as the mold temperature. The mold itself will be warmer that the coolant, as it is in contact with the hot plastic and subject to the laws of thermal conductivity, convention, and radiation.

12.3.2 Achieving Steady-State Temperature in the Mold

It is important that the mold is designed to quickly reach a steady-state temperature cycle when operating in a production mode. If this is not achieved, then the molded products will not meet the required tolerances set out in the plastic part specification until this state has been reached. These parts are either scrapped or subject to secondary inspection.

An injection mold operates in a transient mode during start-up, and the temperature constantly cycles up and down during every phase of the process. Heat is extracted from the mold throughout the total cycle time, not only during the cooling time (Figure 12.3). At start-up the mold is at the temperature of the coolant.

During the injection cycle, the mold temperature will rise due to the hot plastic being injected while coolant continues to remove heat from the mold. At the end of injection and hold, no additional heat will be added to the mold, but the coolant continues to remove heat from the mold until the next injection takes place. It may be that the amount of heat removed is less than the heat added for a number of cycles. The mold will eventually reach equilibrium, where the amount of heat entering the mold is the same as the heat removed from the mold (Figure 12.4). This will be at a temperature slightly above the coolant temperature. This is the point at which the mold is consistent and in a steady state. Note that if the mold stops for some reason, then when the mold re-starts the same equilibrium process will take place. Consequently, the first shots from the mold after start-up are typically discarded until steady state is reached again.

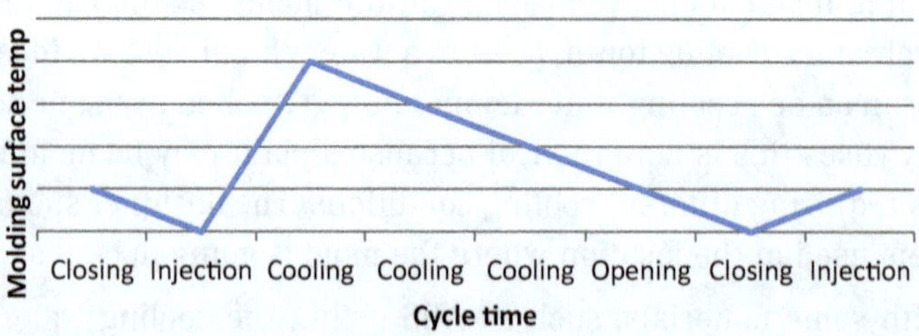

Figure 12.3 Schematic showing the cooling cycle in a mold

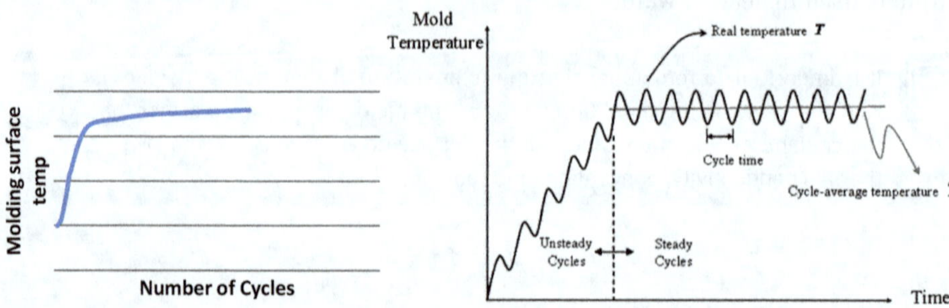

Figure 12.4 Mold surface temperature build up from start-up until steady-state

Note that heat is not only removed by the coolant but also by the products as they are ejected from the mold, because they are ejected at a temperature that is above room temperature but below their heat deflection temperature (HDT). Also, a certain small amount of heat is radiated from the surface of the mold into the plant and by conduction into the machine platens.

If the cycle time is too short to remove the heat or the amount of cooling is insufficient, the temperature of the mold will gradually rise and will eventually be too hot. The mold will not reach a steady state, and the products will become unusable or of poor quality. The parts will be too soft to be ejected and/or the parts will warp or be out of dimensional tolerance when cool. Cooling time must be added, or cooling capacity (more flow or lower temperature) added.

If a steady-state is not reached, then alterations must be made to the process to find the point at which a heat balance occurs. There are various options available to the processing engineer to make this happen. Some of them are related to machine controls and some are related to coolant control in the mold. If the mold is designed properly, the process engineer should be able to find the steady-state processing conditions relatively quickly. The mold designer must design the cooling circuits so that the mold can achieve the steady-state heat balance without too much trouble. Overall, the process engineer who is optimizing the mold in the machine has very little flexibility to optimize the cooling compared with the mold designer. The mold should have a wide "operating window" to make a good part. If a mold can only make a good part within a very narrow window of processing conditions, then the designer has not done a good job in their design.

> It is not the purpose of this book to cover the techniques used by the process engineer to make an acceptable molded part. However, the mold designer should work closely with the process engineering team to make sure that their knowledge and requirements have been considered in the design. It is ALWAYS a good idea to get input from the people that will ultimately need to run the mold and maintain it. They will have a critical eye for things that matter in the design for them that differ from the mold designer.

12.4 Principles of Heat Transfer and Thermodynamics

A critical function of any injection mold is to efficiently and homogenously transfer the heat in the molten plastic to the cooling medium. It is important for all mold designers to understand the fundamentals of heat transfer and thermodynamics. It is not the purpose of this book to teach these engineering principles. Instead, we focus on ensuring that the mold designer understands how to apply them to make good design decisions.

During the design of the mold, the mold designer is primarily concerned about two types of heat transfer - convection and conduction. Although radiation also plays a role in heater transfer (to the atmosphere and between the hot manifold and plates), it is generally not considered when designing an injection mold. Conduction happens between the plastic and the mold steel of the inserts (Figure 12.5). It also occurs between plates and the machine platens. Convection occurs between the steel and the cooling medium (usually water).

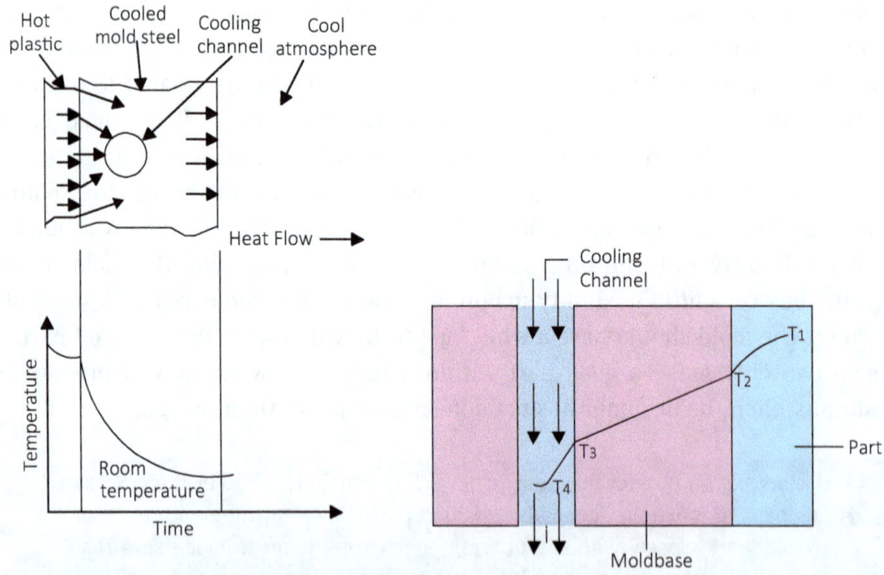

Figure 12.5 Temperature gradient from the plastic to the mold coolant

12.4.1 Heat Transfer Basics

Temperature and heat are sometimes confused with one another, but they are not the same thing. Temperature is a measure of how hot something is, measured in °C or °F, while heat is a measure of the thermal energy contained within an object measured in joules, calories or British Thermal Units (BTU) (see conversion in Table 12.2). One calorie (1 cal) is the amount of heat required to raise the temperature of 1 g of water by 1 °C at atmospheric pressure.

As heat is added to a substance in the solid state, the molecules of the substance gain kinetic energy and the temperature of the substance rises.

Table 12.2 Units of Heat

Unit	Conversion
1 cal	4184 J
1 BTU	1055 J

To calculate the energy (in the form of heat) required to raise the temperature of a known mass of a substance, Equation 12.1 can be used:

Energy required to heat a substance: (12.1)

$$Q = m \times C_p \times \Delta T$$

where:
Q = energy transferred in joules over the desired timeframe (example per hour or per cycle)
m = mass of the substances in kg, in the timeframe of interest (for instance per hour)
C_p = specific heat capacity in J/kg·°C
ΔT = temperature change in °C across the area of interest

Example:

A 24-cavity mold is producing a 16 g part made from PP. The injection temperature of the PP is 260 °C and the ejection temperature is 90 °C. How much heat did the mold remove from the plastic before ejection? The C_p of PP is 1920 J/kg·°C.

$$Q = m \times C_p \times \Delta T$$

$$Q = 16 \times 24 \times 1920 \times (260 - 90) = 110 \text{ kJ}$$

The rate of heat (q) flowing from one location to another is governed by Fourier's law of heat conduction. This is expressed in the following simplified formula:

Fourier's law of heat conduction: (12.2)

$$q = -k \times A \times \Delta T / L$$

where:
q = rate of heat flow (J/s)
k = thermal conductivity of the material (W/m·°C)
A = cross-sectional area of the body (in the direction of heat flow)
ΔT = temperature change in °C across the area of interest
L = length over which the temperature change happens
The negative sign is used to indicate the direction of heat flow.

Heat always flows from hot to cold. In a mold, the heat from the plastic flows into the mold, then into the coolant and finally into the air and into the machine platens.

Example:

The temperature of the molding surface of a PP part is 200 °C. The temperature of the water coolant, which is 20 mm away from the plastic, is 5 °C. The cross-sectional area of flow is 500 mm². What is the rate of heat transfer from the hot plastic to the cold water? The thermal conductivity of the steel is 14.4 W/m·°C.

$500 \text{ mm}^2 = 0.0005 \text{ m}^2$

$20 \text{ mm} = 0.02 \text{ m}$

$q = -k \times A \times \Delta T / L$

$q = -14.4 \times 500 \times 195 \div 20$

$= -14.4 \times 0.0005 \times 195 \div (0.02\ \times) = 70\ W$

12.4.2 Melting and Cooling Behavior of Plastics

To raise the temperature of any material, a certain amount of heat (energy) must be introduced into it. The amount of energy required is different for different materials. To get the plastic to its molding temperature, a certain amount of heat energy must be introduced into the plastic. This is done by the injection unit (consisting of the screw, barrel, and heaters) of the injection molding machine. This heat content flows with the plastic into the mold cavity and must then be removed so that the plastic returns to its rigid state for ejection.

The amount of heat required to melt the plastic is directly proportional to the mass of the plastic to be melted. For the same ΔT, the amount of heat to be removed from the molded plastic by cooling is identical to the amount that was introduced into the plastic. However, most of the time, the product is ejected fairly hot and, therefore, still contains some of the heat added during plasticizing. This remaining heat will eventually be convected and radiated from the products into the surrounding plant air.

Amorphous plastics, such as PS, vinyl, and ABS, have a randomly arranged molecular structure, and soften at a glass transition temperature rather than melting suddenly. They start melting proportionally to the heat added, and cool down proportionally to the heat given up to the coolant (Figure 12.6). Semi-crystalline polymers, such as PE, PP, nylon and PET, have a regular order to their molecular arrangement, and a sharp melting point. They require a certain amount of heat at a certain temperature level just to melt the crystals without increasing the temperature of the plastic. Only after the crystals have melted will the plastic continue to increase its temperature in proportion to the heat input.

The cooling process is similar. As the plastic passes through the crystallization range, heat is given up while the temperature remains the same. Only after crystallization will the plastic cool down further (Figure 12.6). For practical purposes, this means that a crystalline material needs more heat (energy) to raise a certain mass to the desired melt temperature than an amorphous material and, therefore, also more cooling (energy) to remove its heat content.

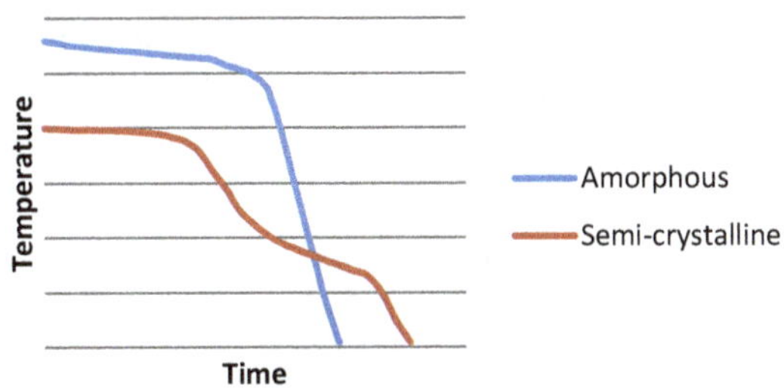

Figure 12.6 Cooling behavior of amorphous versus semi-crystalline plastics

12.4.3 Conduction and Thermal Conductivity

Heat conduction is the transfer of heat between two objects in contact with each other. Heat energy moves from the hotter object to the cooler object. Eventually, the two materials will even-out in temperature as the heat keeps on transferring until the temperatures are the same. This is equilibrium.

Thermal conductivity (k) is a measure of the rate at which a material conducts the heat from hot to cold. This rate is different for various materials and can be looked up in reference books. *Polymers have lower thermal conductivity, while metals are characterized by much higher thermal conductivity.*

When the plastic comes into contact with the cold steel from the mold, it freezes almost instantaneously and forms an insulating layer between the still-molten polymer and the mold steel. This influences the rate of cooling to a much greater degree than the temperature or materials of the mold (see Figure 12.5). This is of particular significance when designing molds for products with heavy walls, or when there are isolated areas in the product which are thick and cannot be cooled properly.

Steady-state conductive heat transfer in one dimension can be calculated using the following equation:

1D Steady-state conductive heat transfer:

$$k \times \frac{d^2T}{dx^2} = 0 \quad (12.3)$$

where
k = thermal conductivity of the resin (W/m·°C)
T = temperature (°C)
x = dimension through the wall

Using this simple formula can provide the designer with a good estimate of the effectiveness of the planned cooling. However, using 3D finite-element modeling software would be able to provide an excellent view of the overall transient cooling of the mold and the impact on shrinkage and warpage. It is highly recommended to use this type of software in applications where there is not a good reference design to compare to. Figure 12.7 shows an example of the cooling of a part in the mold. This information can be used to optimize the cooling circuit and also to anticipate warpage in the molded part after ejection.

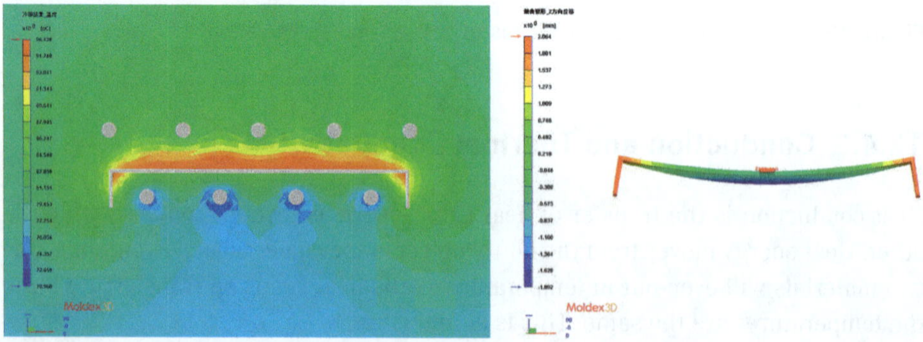

Figure 12.7 Mold temperature distribution and resulting warpage analysis using CAE simulation [2] (Courtesy of Moldex3D)

12.4.3.1 Thermal Conductivity Values

Each material has a coefficient of thermal conductivity. Shown in Table 12.3 are approximate values of some materials for comparison purposes only. More information on plastic and mold material properties can be found in Appendices 8–11. Data sheets from the material supplier should be used to obtain more accurate coefficients of thermal conductivity for detailed analysis.

Table 12.3 Approximate Coefficients of Thermal Conductivity

Material	*k* (W/m·°C)
Pure copper	350
Aluminum	270
Beryllium copper (depends on alloy)	100–250
Brass and bronze	50–100
Nickel	50
Steel (mild)	40–50
Steel (alloyed), hardened	10–30
Stainless steel (40% nickel)	10

[2] Source: M. Wang, R. Chang, C. Hsu, *Molding Simulation: Theory and Practice*, Hanser Publishing, 2018 (Figure 8.12).

Material	k (W/m·°C)
Titanium alloys	2-4
Glass	1.0–1.5
Plastics (not burnt or degraded)	1.0–1.5
Insulating boards	< 1

12.4.4 Convective Heat Transfer

Various factors affect the flow of the coolant through the mold. The amount of heat removed from the mold is proportional to the ΔT (temperature difference) between the temperature of the plastic and the coolant. But unless the heated coolant is removed and replaced by fresh coolant, there is no cooling effect. Generally, the following factors affect the flow of the coolant:

- Reynolds number (flow rate of coolant)
- Pressure differential (Δp) of coolant between supply and return lines
  - Cross-section of passages (area available for flow)
  - Length of cooling lines and their layout
  - Viscosity of the coolant
- Condition of the coolant channels.

12.4.4.1 Reynolds Number (Re)

The Reynolds number (Re) is a dimensionless number that is used to characterize a liquid flow by defining the ratio between inertia and viscous forces. Re measures the effect (on the cooling) of the velocity of the fluid in a channel, of the diameter, and of the specific density and the specific viscosity of the coolant. It has been found that the heat transfer from the walls of the channels to the coolant is greatly improved when the Reynolds number is above 3200. Below 3200, fluids have laminar flow, and above 4000 they have turbulent flow. Figure 12.8 shows that in the laminar flow only the coolant next to the wall removes heat, while the remainder of the flow just passes through the channel. With turbulent flow, the entire flow of coolant participates in the cooling created by eddy currents.

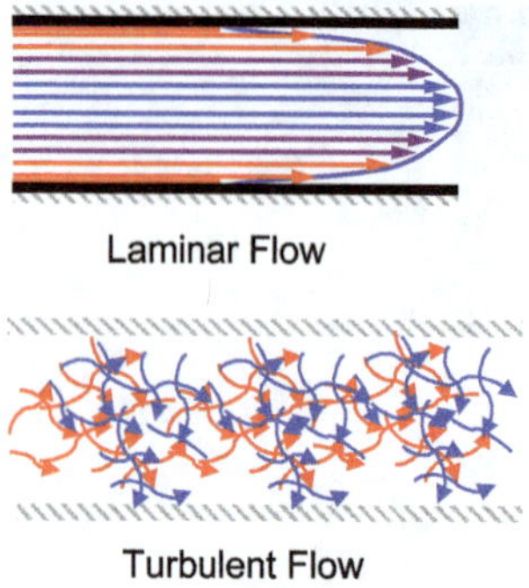

Figure 12.8
Schematic of laminar versus turbulent flow [3]

The Reynolds number can be calculated using the following equation:

Reynolds number:

$$Re = \frac{4 \times Q \times \rho}{\pi \times D \times \eta} \tag{12.4}$$

where
Re = Reynolds number
ρ = coolant density (water = 1 kg/m^3)
η = coolant viscosity (water at 10 °C = 1.308 mPa·s)
D = channel diameter (mm)
Q = volumetric flow rate of the coolant (m/s)

A mold designer should always design the cooling circuits to achieve turbulent flow near the molding surfaces for significantly better heat transfer. A Reynolds number of more than 4000 designates turbulent flow.

The Reynolds number is indirectly affected by the surface condition in the flow channel. The rougher the surface, the higher the Re number, which is desirable for good cooling. However, this roughness is also conducive to corrosion and retention of sediment. Generally, a bore should be left as drilled and preferably protected against corrosion by plating, such as by the electroless nickel process (ENP), or by selection of stainless steel for the mold parts.

Cooling channels in cores and cavities, as well as inserts, must have turbulent flow. In large-diameter passages, the flow may become laminar, but these larger passages are used mainly to distribute the coolant to the actual cooling channels, which are then much smaller and provide the desired turbulent flow.

[3] Source: S. Kulkarni, *Robust Process Development and Scientific Molding*, 2nd ed., Hanser Publishing, 2017 (Figure 11.1).

12.4.4.2 Pressure Drop of Coolant Supply and Return Lines

The difference between the inlet pressure and the exit pressure at the mold is expressed as Δp (delta p). The pressure in the return lines is usually only slightly above atmospheric pressure, just enough to ensure the flow back to the source (reservoir). Usually, the coolant supply at the inlet of the mold is in the order of 400–500 kPa (55–70 psi). Higher pressures would provide higher flow, but for practical reasons such as pump cost seal design and hose construction the industry has adopted this range.

12.4.4.3 Cross-Section of Channels

A larger cross-section lets more fluid flow than a smaller one. The flow rate is roughly proportional to the square of the diameters, that is, a 10 mm diameter channel passes roughly 4 × as much fluid as a 5 mm diameter channel ($10^2 = 4 \times 5^2$).

It is very important to consider this when laying out cooling channels. Designers and molders are mostly concerned about the quantity (flow) of coolant in the vicinity of the cavity space to remove the heat in that area. There must be enough coolant where it is needed, and it must be evenly distributed. Starving the supply to some areas results in hot spots or uneven cooling in the mold.

12.4.4.4 Length of Cooling Lines

The effect of length on the flow is linear, that is, the resistance against flow is proportional to the length of the channel. However, except in rare cases, the influence of length can be ignored because, except in very large molds, the total length of channels in a mold is not very great. Changes in direction of flow, such as those created where bores of coolant lines intersect, add to the flow resistance and reduce the coolant flow. The "equivalent length" means that, for example, a 90° bend in the flow (most common) is the equivalent of a certain length of straight flow path; this equivalent length is dependent on the diameter of the channel and the volume of flow. It is good to remember that fewer changes of direction of flow are better for the flow, especially with small cross-sections of the flow path.

12.4.4.5 Coolant Viscosity

The effect of the viscosity of the coolant, although theoretically affecting the flow, especially with a high percentage of antifreeze or other additives, can for practical purposes usually be ignored.

12.4.4.6 Condition of Channels

Dirt and contamination in the cooling channels can seriously affect the cooling efficiency. It can be easily understood that such deposits have a doubly negative effect on the performance of the cooling system. First, the deposit reduces the cross-section of the passage and thereby reduces the flow of coolant. Secondly, the deposit usually has a much lower heat conductivity than the mold material and reduces the amount of heat passing from the plastic to the coolant. This is of particular importance where small (round or flat) cooling channels are used. They should then be designed so that the channels can be cleaned from time to time.

■ 12.5 Calculation of Cooling Requirements for a Mold

Using the above basics in thermodynamics and heat transfer, the mold designer can now determine how much heat is needed to condition the plastic for injection and then how much cooling is required before the part can be ejected.

12.5.1 Heat Input Required to Condition the Plastic for Injection

The method explained in the example below is very accurate but depends on several factors which have to be assumed, mainly from experience, with similar products in similar molds and plastics. Equation 12.1 can be used to calculate the heat inputted into the mold. Note that the hot runner system is not added to this calculation, as it is never cooled or melted. The hot runner system maintains the temperature achieved in the extruder.

Example:

A product made from PS (C_p = 0.34 cal/g·°C, melt temp. 280 °C) weighs 35 g. It is to be molded in a 16-cavity hot runner mold at a 6 s cycle. The plant is at 22 °C ambient temperature. What is the heat input into the mold per hour?

$$\text{Heat input } (q) = m \times C_p \times \Delta T \qquad (12.1)$$

First calculate shot size: 35 g/cavity × 16 cavities = 560 g

Shots/hour: 3600 s/h ÷ 6 s/shot = 600 shots/h

Mass molded/hour: 560 g × 600 shots/h = 336,000 g/h (or 336 kg/h)

= 336,000 × 0.34 × (280 - 22)

= 29,473,920 cal/h × 4.184 J/cal (Table 12.2)

= 123,401,408 J/h

= 123 MJ/h

One kWh (kilowatt-hour) = 3.6 MJ, therefore the total energy required is 34.17 kWh.

If the designer wanted to know the heat inputted per cycle, the above example would be calculated on a per-shot basis and the result would be 0.73 kW per cycle.

12.5.2 Cooling Required to Remove the Inputted Heat

A significant portion of the heat added to the plastic calculated in Section 12.5.1 must be removed from the plastic during the molding process, so that the sufficiently cooled molded products may be removed from the mold. Equation 12.1 can be used for the calculation with one important difference. It is not necessary to cool the plastic down to ambient temperature but only to a suitable ejection temperature T_e. The T_e can be much higher the ambient temperature of the plant as long as the plastic is stiff enough to resist the forces of ejection and will not lose its shape after being ejected while still warm. Therefore, not all of the heat added to the plastic will be removed within the mold by the cooling water. Most of the remaining heat will be dissipated into the surrounding (plant or storage) air space. Some of that heat will be radiated by the mold and the machine into the plant air space.

A good target for an ejection temperature (T_e) is the heat deflection temperature (HDT) of the plastic. HDT values can be found in the material supplier's literature; general numbers are also given in Appendix 9.

Note that the hot product will continue to shrink after ejection until it is fully cooled. This will affect the dimensions of the product. The desired ejection temperature (T_e) may be critical in cases where accurate product dimensions are important.

To continue the example above:

The 336 kg/h of hot plastic, now in the form of products, has cooled from a T_i of 280 °C to a safe ejection temperature T_e of 50 °C:

$\Delta T = T_i - T_e$, or 280 °C - 50 °C = 230 °C.

As before, calculate the heat removed in the mold:

$$q = m \times C_p \times \Delta T \qquad (12.1)$$

$= 336{,}000 \times 0.34 \times (280 - 50) \times 4.184 \div (3.6 \times 10^6)$

$= 30.55$ kWh or 110 MJ/h (or 0.5 kW/cycle)

In North America, in molding and refrigeration technology, the amount of heat is often indicated in "tons" of refrigeration. This is defined as the amount of heat required (in BTU) to melt one ton (2000 lb) of ice at 0 °C into water at 0 °C. The conversions are:

1 ton refrigeration = 12,000 BTU/h = 3.516 kW (1 kW = 3413 BTU/h)

In the example above, we would need 30.55 kW ÷ 3.516 kW/ton refrigeration = 8.7 tons refrigeration.

An alternative to the above calculation is to use the method from Chapter 5 (Chiller sizing guidelines).

12.5.3 Temperature of Cooling Water

The heat to be removed from the plastic within the mold must be carried away by the coolant flowing through the mold and back to the cooling system, such as a chiller and/or cooling tower. The temperature difference, ΔT, of the coolant between the IN and OUT ports of the mold can be measured after the mold is built, but is typically assumed as an input for calculation purposes, and is based on empirical knowledge.

In many cases, a ΔT from inlet to outlet of 5 °C (9 °F) is a good choice, but occasionally a ΔT of 2 °C (4 °F) may be necessary for critical parts. In some molds, a ΔT of up to 10 °C (18 °F) is quite acceptable (if there are no critical dimensions).

The nominal temperature of the coolant on the mold has only a minor impact on the cycle time. The flow of heat from the hot plastic to the coolant depends largely on the temperature difference between the injected plastic and the coolant, on the design of the cooling channels, and the material selection for the mold components. Assuming (as in the example) T_m = 280 °C:

- With coolant temperature T_c = 5 °C, the ΔT (melt to coolant) = 280 °C - 5 °C = 275 °C

- With coolant temperature T_c = 10 °C, the ΔT (melt to coolant) = 280 °C - 10 °C = 270 °C.

This represents a difference of only about 2% in ΔT. However, when at the end of cooling (i.e., when the plastic has its ejection temperature, T_e), the differences are more meaningful:

- With coolant temperature T_c = 5 °C, the ΔT (ejection to coolant) = 50 °C - 5 °C = 45 °C
- With coolant temperature T_c = 10 °C, the ΔT (ejection to coolant) = 50 °C - 10 °C = 40 °C.

This represents a difference of only about 10% in ΔT.

In fact, using higher cooling temperatures will only slightly increase the time required to cool the plastic, and in many cases, the added cost (equipment and power) of using very cold coolant cannot be justified economically.

12.5.4 Quantity of Cooling Water Required

The quantity of cooling water can have a major impact on the cycle time, as it is the method by which the heat is removed. Continuing the calculations using the same formula as before, but inverted:

To continue the example above:

It is known that 110,000,000 J (or 110 MJ) heat must be removed from the mold every hour. If the coolant is assumed to enter the mold at 5 °C, and to rise in the mold to 10 °C, ΔT = 5 °C. How many liters of water are required per hour for this example?

Note C_p of water = 1 cal/g·°C and density of water is 1000 g/L.

$q = m \times C_p \times \Delta T$ (Equation 12.1), which can be rearranged to:

$m = q / (C_p \times \Delta T)$

From above:

q = 110 MJ or 26,275,200 cal/h (4.184 conversion)

= 26,275,200 cal/h ÷ (1 cal/g·°C × (10 - 5 °C))

= 5,255,040 g/h and since 1000 g of water equals 1 L

= 5254 L/h of water or 76 L/min

To show the influence of the ΔT selected, consider for a moment if the ΔT desired was only 1.5 °C instead of 5 °C in the example. Since the mass flow rate of water needed is directly proportional to the ΔT the required water would go from 76 L/min to 76 × 5 ÷ 1.5 = 253 L/min!

If on the other hand the selected ΔT were 10 °C, the required amount of coolant would be smaller by the ratio of 5 : 10, or 38 L/min.

Similar considerations apply to the ejection temperature, T_e. If T_e were higher, say 60 °C instead of 50 °C, less cooling water would be required. Conversely, if T_e were lower, say 40 °C, more cooling water would be required.

Plastics with higher specific heat factors, such as PP and PE (C_p = 0.50–0.55) require about twice the amount of coolant as plastics such as PVC (C_p = 0.25).

If the cycle were not 6 s as estimated in the example above but 9 s, the number of shots/hour would be only 400 instead of 600, and only two-thirds of the heat input would need to be removed. Conversely, if the mold proved to run faster, say at a 4 s instead of a 6 s cycle, there would be 900 shots/hour, and the plasticized material would have a mass of 504 kg, or 50% more than estimated, and a correspondingly higher water flow would be required to remove the heat.

12.5.5 Efficiency of Cooling on Required Water Flow

The theoretical amount of heat to be removed from the mold calculated in Section 12.5.4 is the true amount of heat which must be removed from the coolant, regardless of the quantity of coolant flowing through the cooling system of the mold. But it indicates the correct quantity of coolant only if the transfer of the heat from plastic to coolant takes place in the stipulated cycle time. Even in a mold with a good cooling layout, it is unrealistic to expect that the theoretical amount of coolant, as it flows through the channels, will pick up all the heat generated in the anticipated cycle time. Therefore, the amount of coolant required depends very much on the cooling efficiency of a mold (E_{cm}) to remove heat from the mold.

E_{cm} depends on various characteristics of the mold construction, the chemistry of the cooling medium selected, the conductivity of the coolant, and on the speed and flow behavior (turbulent or laminar) of the coolant within the mold. Note that the flow (speed) of the coolant depends primarily on the pressure drop between the pressure of the incoming coolant and the pressure of the return line back to the source of the cold coolant.

It is not possible to assign definitive values to the cooling efficiency of a mold, E_{cm}. We only know that a definite (calculable) amount of heat enters the mold with the hot plastic, and that another definite (calculable), usually smaller, amount of heat must be removed by the coolant. In practice, the flow of the coolant through the mold will be larger than the calculated, theoretical amount. An actual E_{cm} could be determined by testing various completed molds, by calculating the theoretical amount of coolant for a certain material and cycle time, and then measuring the required amount of flow through the mold once it is on cycle at the stipulated cycle time.

Note that the effect of increasing the volume of the coolant will reduce the temperature increase of the coolant between IN and OUT, but as stated earlier, the amount of heat transferred to the coolant remains the same as calculated.

12.6 General Cooling Design Rules

The mathematics of cooling is quite complicated in the injection molding process, since everything happens in a very short time with dynamically changing conditions. Computer simulation programs can determine the appropriate means of cooling a particular mold, after input of the geometry of the product and the mold, and based on temperatures of melt and coolant, flow patterns and sizes of the cooling channels, and other variables, such as heat characteristics of the coolant and the mold materials. Given these accurate inputs, the simulation program can determine the best cooling layout for a mold after the mold is designed. If the design is poorly conceived, then the design will have to be changed and the simulation rerun until a good design is achieved. Time should be allocated in the design process for this iterative process, in order to avoid costly mistakes. There are several tips, based on experience, to help the designer create a design to be modeled. This will perhaps reduce the number of iterations needed to reach the final design.

- **Tip 1**: All cooling channels must have flow. Convective heat transfer is much more efficient than conductive transfer of heat. Stagnant coolant in ends of channels, or in any pocket, does nothing for cooling, as the stagnant coolant heats up and the heat has nowhere to travel to.
- **Tip 2**: Cooling channel layouts should be as uniform as possible. *Uniformity is the golden rule for cooling design.*
  - All cavities (and cores) in a multi-cavity tool must be cooled with the same coolant flow at a temperature that is only little different from cavity to cavity (or core to core)
  - The cooling between the core and cavity side should be as uniform as possible
  - When cooling is required in the core or cavity, the distances between cooling channels and the mold walls should be constant.

As long as the temperature difference, ΔT, between the first and the last cavity in one group of cavities (or cores) is 1-5 °C (2-9 °F), the system will work effectively when equilibrium is achieved. For fast cycling multi-cavity molds below 10 s, the permissible ΔT should be 1-3 °C (2-5 °F).

- **Tip 3:** The amount of heat removed depends on the quantity (volume) of coolant flowing through the channels in cavity and core. The faster the coolant flows, the more efficiently it cools because a greater volume will flow through the channels, and there will be a lower temperature rise of the coolant from the first to the last cavity or core.
- **Tip 4:** The coolant must flow in a turbulent flow pattern near the molding surfaces ($Re > 4000$; see Section 12.4.4.1), rather than in laminar flow ($Re < 2300$). Turbulence within the flow causes the coolant to swirl around as it flows, thereby continuously bringing fresh, cool liquid in contact with the hot metal walls of the cooling channels, and removing more heat. By contrast, laminar flow moves along the channel walls relatively undisturbed, so that the outer layer of the coolant in touch with the metal will heat up, but the center of the coolant flow will remain cold, thus doing little cooling. Thus, for cooling in the cavities, cores, inserts, side cores, and other inserts with molding surfaces on them, small-diameter channels and fast-flowing coolant are very important. Most cooling lines for cavities and cores are supplied from channels in the underlying or surrounding plates. Since they are not close to the molding surfaces, they can be much larger to reduce pressure drop in the cooling circuit. This is usually satisfactory because these plates do not need as much cooling as the stack parts, which come in contact with the hot plastic.
- **Tip 5:** It does not matter whether the coolant follows a serial flow, that is, from cavity to cavity or core to core in sequence (Figure 12.9), or whether the flow is split so that the coolant flows in a parallel pattern (Figure 12.9), as long as each branch has the same flow and enough coolant flow is maintained so that turbulent flow conditions are achieved near the molding surface. In many multi-cavity molds, the cooling channels are arranged so that they are partly in parallel and partly in series (Figure 12.10). Often, in the same mold, cavities are in one arrangement of series, parallel, or both, and cores, inserts, or side cores are in another arrangement, whichever is more suitable for the layout. There is no rule for which way to go, as long as the flow rules are followed and as long as the resulting ΔT is within the acceptable limits.
- **Tip 6:** Never starve a branch of the required flow. The channel sizes must be calculated so that there is always more than enough flow capacity in a preceding section to feed equally all the channels in the following split, parallel sections.

For example, if there are four parallel channels of 40 mm^2 cross-sectional area each, the (preceding) feeder must have at least $4 \times 40\ mm^2 = 160\ mm^2$ cross-sectional area. In some molds, there are four or more points where the cross-sections step down in the cooling system. It does not matter if the preceding section is greater than the calculated minimum value, but it must not be smaller, if the coolant is to flow equally through all subsequent channels. Coolant, like plastics, always takes the path of least resistance.

For example, if the preceding cross-section is 3*x*, and each of four succeeding parallel cross-sections are *x*, there will not be enough coolant, and one of the four channels will see little or no flow through it. Unfortunately, this is often missed in designs, with the result that the mold does not function properly.

- **Tip 7:** The difficult-to-cool areas in the mold must be considered first. These are typically all delicate mold features, such as thin and slender core pins, blades, and sleeves. "Slender" signifies, in this context, that the ratio of length over the narrow bottom dimension or diameter of a pin or insert is more than 2:1.
- **Tip 8:** Study the product to locate heavy wall sections of the plastic. The heavy areas of the product are always the limit for cycle time and cooling, even where it is easy to provide good cooling, because of potential shrink and sink marks. Heavy sections are particularly bad if they are toward the end of the flow of plastic where there is less pressure left to ensure good filling. The mold designer should discuss this problem with the product designer. There may be the possibility of a minor alteration of the product design to avoid heavy sections so that not only is plastic saved but also cooling time is reduced. For example, the heavy, solid handle of a coffee mug could be redesigned by coring it from both sides. This could add to the mold cost, but would greatly reduce the cycle time (see Chapter 3 on plastic part design for mold designers)

12.7 Cooling Channels in Plates

The layout of cooling channels is governed by the need to provide equal and ample flow to each cavity, the accuracy of the manufacturing equipment, and the desired cooling in the core and cavity. Given that these plates are often part of the alignment and ejection mechanism of the mold, they must be kept at a steady and even temperature. Although most of the heat in the mold is removed by the cooling of the cavities and cores, a certain amount of heat travels to the plates. Mold plates need to be cooled while also feeding the cores and cavities with sufficient cooling. Differences in the temperature between plates can seriously affect the alignment of leader pins and bushings or taper locks, and result in rapid wear of these components or the mold itself. If the cavities are fed directly (by hoses) and the leader pins and bushings are mounted on uncooled plates, the designer must make sure that there will be no undesirable temperature differences between the two mold plates.

Considerations for mold plate cooling are:

1. Symmetry of the circuits.
2. Number of circuits required to ensure even cooling.
3. Size and available space on the machine. The size of the cooling layout is often limited because of the size of the machine platen and the location of the tie bars. The points where the cooling lines enter and leave the mold must be accessible when the mold is installed in the machine.
4. Number and size of screws used in the plates. The designer must ensure that the proper number of screws is used to mount the cores and cavities, and to hold plates together to prevent any bending or deflection by heat. The large number of screws needed can sometimes make laying out the cooling circuit a challenge. *However, the number of screws required should never be compromised to ease the cooling circuit layout.*
5. Position and size of leader pins and bushings. The leader pins and bushings can also sometimes impede the layout of the cooling channels. Again, these are of primary importance in the mold design and should not be compromised by cooling channel layouts in the plates.
6. Cost. As pointed out in the introduction of this chapter, all parameters must be considered to decide what quality of cooling will be provided and at what cost.
7. Use of standard available drill sizes from the tool shop and standard fitting sizes for hose connections.

12.7.1 Series and Parallel Plate Cooling

Figure 12.9 shows a typical arrangement of a plate cooling circuit in series and in parallel. The flow depends only on the pressure differential ΔP (Delta P) between the IN and OUT ports, and on the resistance within the channels. The longer the circuit, the more resistance there is. The greater the channel diameters in the circuit, the less resistance there is.

The coolant, like any fluid, always flows along the path of least resistance.

As the coolant flows through the channels, it picks up heat from the surrounding metal and thereby increases in temperature. Therefore, the cooling efficiency is greater near the inlet port (greater ΔT from hot plastic to cold metal) than near the outlet, where the coolant is warmer. This is the reason why it is recommended to keep the maximum temperature difference between IN and OUT to a minimum (see Section 12.5.3).

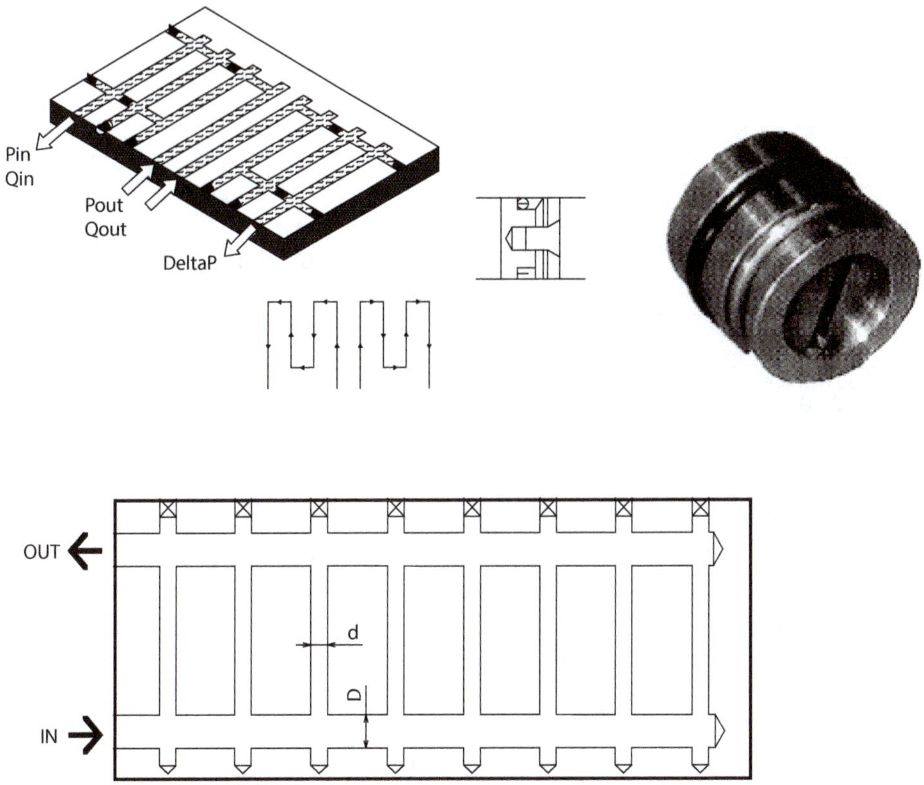

Figure 12.9 Typical layout of a series cooling circuit (top) and parallel circuit (bottom) in a plate

For a parallel arrangement of cooling channels, it is important that the cross-section of the IN and OUT channels (D) is sufficiently larger than the sum of the cross-sections (d) of the parallel branches. If d were the same as D, the coolant would take the shortest route (first branch) from IN to OUT and provide no cooling for the rest of the plate. The channel sizes can be calculated using Equation 12.5 (refer to Figure 12.10).

Calculation for line sizing in a plate:

$$D_2 = \sqrt{\frac{D_1^2}{N_2}} \quad \text{and} \quad D_3 = \sqrt{\frac{D_1^2}{N_2 \times N_3}} \qquad (12.5)$$

where

D = line diameter

N = number of parallel lines in the branch

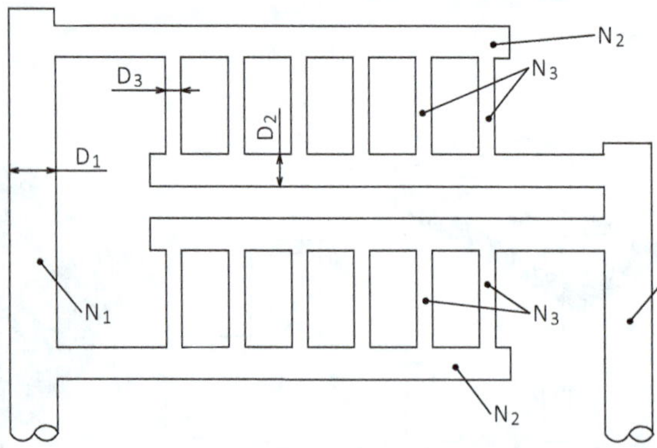

Figure 12.10 Example of a combined series and parallel cooling circuit

Example:

Figure 12.10 shows a combined series and parallel cooling layout. If D_1 = 20 mm Ø, what size should D_2 and D_3 be?

In this example, the number of the branches N_1, N_2, N_3 is 1, 2, and 6, respectively. To arrive at the approximate diameter of the two smaller lines, D_2 and D_3, use Equation 12.5.

D_2 would be 14.14 mm Ø (select 15.0 mm Ø standard drill)

D_3 would be 5.77 mm Ø (select 6.0 mm Ø standard drill).

Parallel and series cooling are used in combination in many molds. For example, the plates may be all series-cooled, and the cavities and cores may be series- and/or parallel- cooled.

The cooling in the plates is only part of the circuit. The flow of the water will enter and exit each core or cavity from the plate cooling circuit and travel through the insert and then exit back into the plate cooling circuit. The pressure drop and flow must be considered for the entire system.

Figure 12.11A shows a series cooling arrangement in a plate including a cavity. Provided there is enough flow to keep the ΔT between IN and OUT to about 5 °C (10 °F), this circuit is acceptable and simple to manufacture.

Figure 12.11B shows a similar mold with a series+parallel cooling layout. The feeder and return channels must be larger than the branch channels to provide equal flow through each of the branches. This method requires more drilling, but the cooling efficiency is better than with series cooling, and so this approach is often used for multi-cavity molds.

Figure 12.11C shows a fully parallel circuit. Each insert receives the same amount of water at the same temperature. The water enters the insert from one feeder line and exits into a different return line. This is how parallel flow to each cavity is achieved. The water from one insert never enters another insert. As can be seen, this circuit is much more complex and should be used only when the ΔT between IN and OUT is a significant concern.

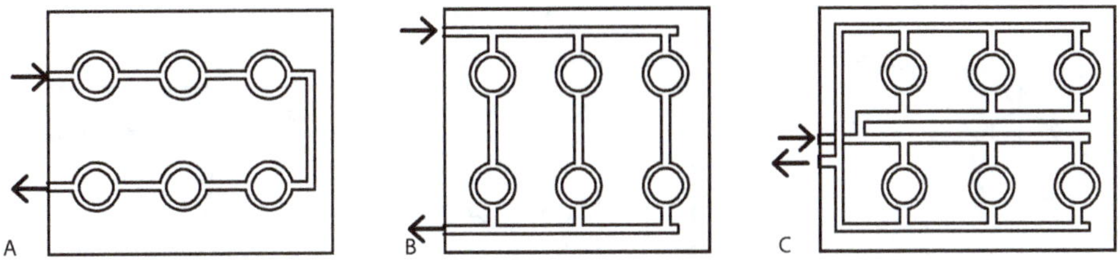

Figure 12.11 Cooling layout for inserts: (A) Series; (B) Series+parallel; (C) Parallel

It should be understood that the cooling arrangement of the cavity and the core side do not need to be the same. It is often quite easy to provide parallel cooling for the cores, while it is more difficult to do so for the cavities. A very common approach is to use series or series+parallel cooling for the cavities and parallel or series+parallel cooling for the cores.

Restrictors are sometimes used to achieve equal flow through a number of parallel branch channels. Generally, the use of such restrictors can be avoided by properly sizing the flow channels in the first place. However, this is not always possible, and restrictors may be the only way to correct a bad flow situation. The restrictor is always inserted at the outlet end of the branch of the flow path to ensure a back-pressure to the coolant flow. This, in turn, ensures that the coolant in the channels stays in full contact with the channel walls.

12.7.2 Distance of Waterlines from Holes and Surfaces

The longer the length and the smaller diameter of the drilled hole, the more the drill could wander off its intended path. Figure 12.12 provides the minimum recommended distances between holes and features. At the start of the hole, this distance may be less. This applies also to long holes produced with deep hole drills, even though they stay in line much better than twist drills.

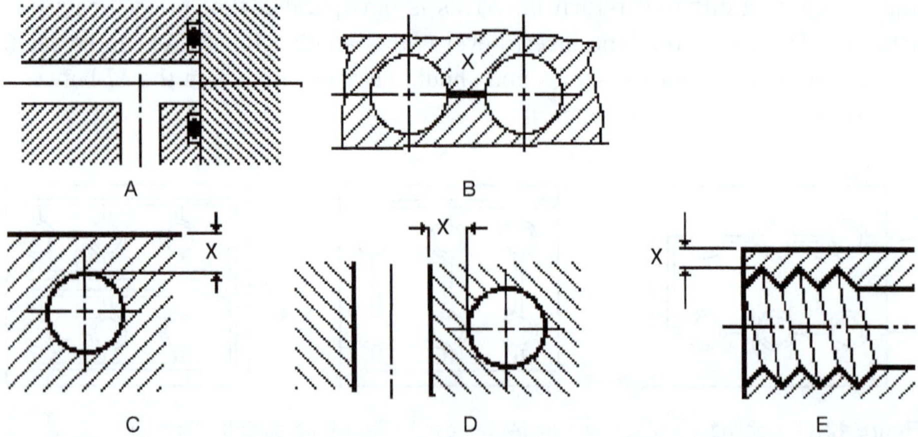

Figure 12.12 Minimum distances between cooling line and features in mold plates: (A) >1 mm from O-ring groove; (B) $x > 5$ mm; (C) $x > 5$ mm; (D) $x > 5$ mm; (E) $x > 4$ mm

Using gun drilling reduces the risk of wandering drill bits considerably. If the distance x from Figure 12.12 is too small, the cyclic stresses induced during molding can be very high. Corrosion inside the bore may cause stress raisers and cracks in the steel. Also, in such thin spots, heat cracks may develop during quenching in the heat treatment of the mold steel.

12.7.3 Plugging of Cooling Channels

The size of the cooling channel (bore) is usually selected to be suitable for tapping the standard pipe thread sizes (see catalogs from parts suppliers for plug thread sizes and drilling recommendations). The most common method of plugging coolant lines is to use standard tapered pipe plugs (NPT or BSP). As a rule, they are installed flush (level) with the surface of the part or with the surface of the intersecting channel (Figure 12.13).

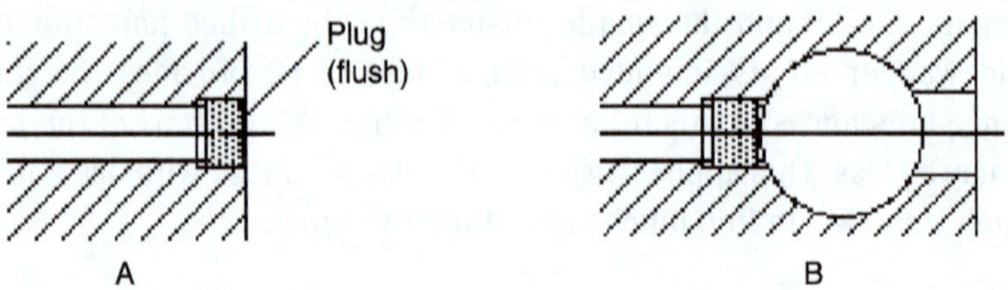

Figure 12.13 Plug installation in cooling channels

Brass (rather than steel) plugs are used because they are softer than the material they thread into. They deform more easily to fit the thread, and will not overstress the mold part by any wedging action when tightening, especially if the wall surrounding the plug is thin. Another reason for using brass plugs is that they do not rust and can be fairly easily removed, even after a long time.

For **Standard National Pipe Thread (NPT)** tap drill sizes and cross-sectional area of bores, follow recommended standards found in machinery handbooks and parts suppliers, etc. The NPT standard thread taper is 1:16, corresponding to a 1°47' angle with the centerline of the bore. This taper is desirable to reduce stresses in hardened steels and to provide cleaner threads.

British Standard Pipe (BSP) threads differ significantly from NPT in that the tapped bore is straight and the pipe thread is tapered. The number of threads per inch is different, and the threads are *not interchangeable with NPT* threads.

German Standard VSM 51100 for pipe threads *(Rohrgewinde)* is interchangeable with BSP, but *not* with NPT.

12.7.4 Plugs and Baffles in Plates

Figure 12.14 shows the end of a plate with plugs at the end of bores and baffles to direct the flow. In the case of Figure 12.14A, the baffles are fastened to the rod with pins, screws, or solder. The rod ensures that the baffles stay in position. This method is relatively inexpensive, and standard commercial parts are available from parts manufacturers.

A disadvantage of baffles and rods is that their fit in the bore must be loose so that they can be assembled. As a result, the bore is gradually covered with sediment or rust, and the baffles are difficult to remove for cleaning after the mold has been in use for some time.

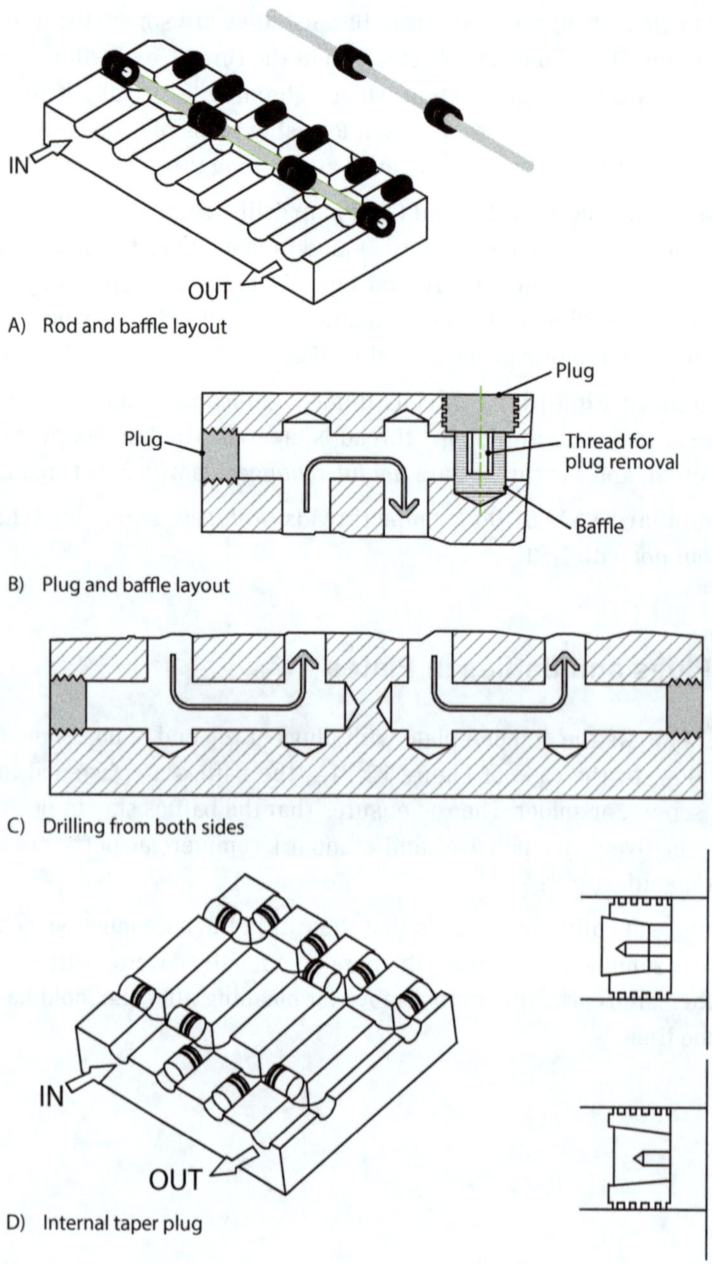

Figure 12.14 Typical arrangements of plugs and baffles in plates to create a cooling circuit (Courtesy of Manual do Projectista)

Figure 12.14B shows a different method for installing baffles that eliminates the problem of removal of the rods shown in A. The baffle is positioned so that its centerline approximately intersects the centerline of the cooling channel. By selecting the diameter and the length of the baffle to be somewhat larger than the diameter of the channel, the flow is completely shut off. A threaded hole in the baffle facilitates removal. A plug in the side of the plate closes the access to the baffle and eliminates a potential leak.

Figure 12.14C illustrates the commonly used practice of drilling from both sides of the plate to avoid inserting baffles. This method should only be used if there are other holes to be drilled from both sides, as this saves the extra machine set-up time, which may cost more than a baffle.

Figure 12.14D shows another variant using plugs that have an expansion feature to lock them in position inside the cooling channel core. While these plugs are convenient and provide a lot of flexibility, they must be fastened tightly or they will move in the cooling channel and potentially block or short-circuit the cooling path.

If a critical element of the mold such as a leader pin, bushing, ejection mechanism, or screw is in the way of a cooling line layout, then Figure 12.15 shows a commonly used drilling arrangement. It is not absolutely necessary that the centerline of the cooling bore and the centerline of the opening (in the illustrated case, the bore for the leader pin bushing) intersect.

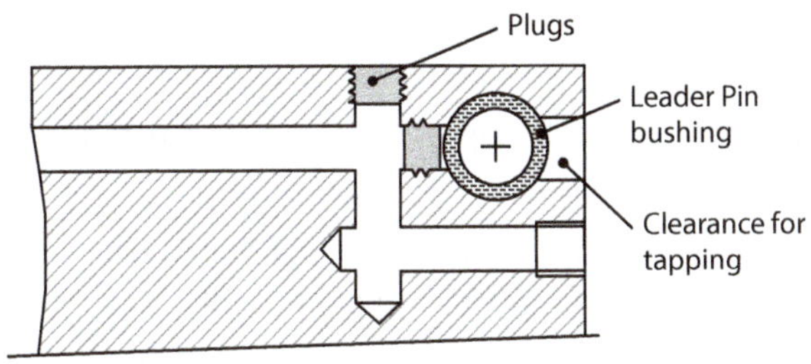

Figure 12.15 Typical cooling line arrangement to avoid a mold plate feature

12.7.5 Cooling of Hot Runner Plates

Hot runners are designed to maintain the manifold, nozzles, and sprue bushing at the same temperature as the processing temperature of a specific resin. The manifold is housed within the manifold plate and the manifold backing plate of the hot runner. Although the contact area between the manifold and these plates is minimized to reduce heat transfer, a lack of cooling or insufficient cooling of the hot runner plates could adversely relax the preload (or sealing force) on the nozzle stack. This could result in plastic leakage within the hot runner. Other problems that could arise include conduction of the excess heat to the cavity plate, causing it to bow or to misalign with the moving side, and premature wear of the valve gate piston O-rings and hot plates. All these issues are potentially hazardous to the operator, and therefore it is important to carefully consider the cooling in the hot runner plates.

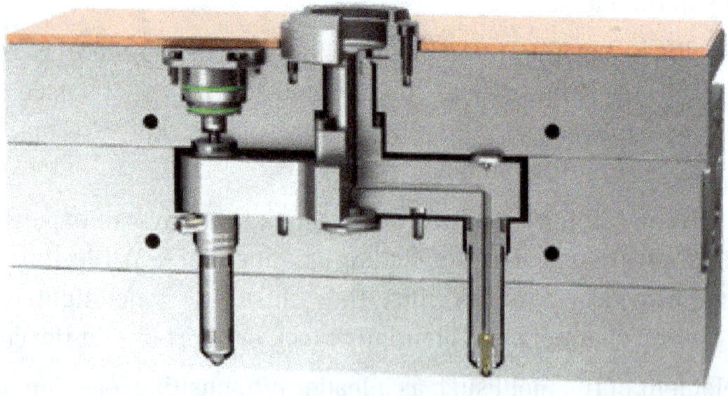

Figure 12.16 Hot runner plate cooling and heat sources (Courtesy of Mold Masters)

The design of the cooling circuit route should follow the same guidelines as normal plate cooling. Note that the cooling around the nozzle housings and valve gates should be symmetrical in the layout, so as not to create an uneven heat transfer condition to some drops.

12.7.5.1 Calculation of the Sizes of Cooling Channels for Hot Runner Plates

The wattage installed in the hot runner can be obtained from the electrical schematic drawing of the hot runner. If this is not available, it can be estimated using the volume of the steel manifold and an approximate rule of 25 W/in^3. This value is the effective power deliverable during the start-up phase of the mold, and is not used in cooling calculations for the steady-state case. Instead, *a duty cycle of 40% of the total wattage should be applied.* The wattage of the band heaters should not be included in the total power calculation, since the gate insert cooling in the cavity plates is assumed to be the primary heat sink for this heat source. Therefore, the heat to be removed from the hot runner plates during steady-state operation is equal to 40% of the total heater wattage deliverable.

Heat to be removed from hot runner:

$$Q_r = 0.4 \times (\text{Wattage of manifold} + \text{sprue}) \quad (12.6)$$

where

Q_r = heat to be removed (in W or J/s)

To remove this quantity of heat via water cooling, a certain water flow rate must be maintained around the hot manifold. This water flow rate is given by Equation 12.7, and is the requirement for the entire hot runner.

Total required hot runner coolant flow rate:

$$\text{Since } Q_r = m_{HRT} \times C_p \times \Delta T, \quad = \rho \times v_{HRT} \times C_p \times \Delta T \tag{12.7}$$

$$\text{and } v_{HRT} = \frac{Q_r}{\rho \times C_p \times \Delta T} = Q_r \times 7.9523 \times 10^8 \ (\mathrm{m^3/s})$$

where

m_{HRT} = total required hot runner water flow rate in kg/s
v_{HRT} = total required hot runner water flow rate in m^3/s
C_p = specific heat of water (4182 J/kg·°C)
ΔT = temperature difference between the manifold and the cold plates (°C)
ρ = density of water (999.2 kg/m^3)

To determine the water flow rates to the individual plates, the following assumptions are made:

Split of flow between plates:

Flow to manifold plate = 0.55 × v_{HRT} (12.8)

Flow to manifold backing plate = 0.45 × v_{HRT}

The header diameter required for the respective flow rate in each plate can then be determined using Equation 12.9:

Calculation of header diameter (mm):

$$\text{Since } \mathrm{Re} = \frac{V \times d_H}{\upsilon}, \text{ we can rearrange to obtain:} \tag{12.9}$$

$$d_H = v_{HRT} \times \text{Plate type} \times \frac{4}{\pi \times \upsilon \times \mathrm{Re}}$$

where

d_H = header diameter (mm)
Re = Reynolds number desired (> 3000 = turbulent)
V = velocity in cooling line (m/s)
u = kinematic viscosity for water (1.311 × 10^{-6} m^2/s)
For plate type, see Equation 12.8

Example:

Consider a hot runner plate with an installed power of 7500 W for the manifold and sprue. What should the flow rate of water be through the plates to keep them at the desired temperature, and what should the header size be in the manifold plate if a Re of 10,000 was desired?

$$Q_r = 7500 \times 0.4 = 3000$$

$$V_{HRT} = Q_r \times 7.9523 \times 10^{-8}\ (m^3/s) = 0.000238\ m^3/s$$

$$d_H = V_{HRT} \times \text{Plate type} \times \frac{4}{\pi \times \upsilon \times Re} = 0.000238 \times 0.55 \times 97.12 = 12.7\ mm$$

12.8 Layout of Cooling Channels in Inserts

In order to produce the fastest possible cycle for any mold, the cooling design of the stack must be very efficient. Cooling only in the plates will not be effective in providing efficient cycle times. Cooling the plates only provides good cooling for the plate, but at the intersection between the stack and the plate there is contact resistance, which impedes thermal conduction. Additionally, the cooling water will be far from the hot plastic.

As mentioned previously, it is possible to divide molds into three basic groups, according to the shape of the molded product: flat, cup-shaped, and a combination of flat and cup-shaped products. Cooling of these types of products will be considered in this section for simplicity, as there are endless possibilities of plastic part designs.

The layout of cooling lines in the inserts (Figure 12.17) may take considerable design time but it is always worth the effort as cooling represents the majority of the molding cycle. It may also require going back to the stack layout and changing the ejection layout to arrive at a good compromise in locating both ejection and cooling features.

Figure 12.17 Cooling channels in a thin wall container stack for fast cycling (Courtesy of Husky Injection Molding Systems Ltd.)

There are two main methods for supplying water to the stack components:

- *Individual hoses for each core and cavity from the machine header.* This is fairly inexpensive, but not very good for molds with more than four cavities because of the large number of hose connections required. Remember that every cooling circuit has an IN and an OUT connection (two hoses), often there are several cooling circuits per cavity, and, similarly, often more than one cooling circuit per core or cavity. All this can add up to a very large number of hose connections (Figure 12.18, left) which makes installation difficult and leakage a strong possibility.
- *Cooling supplied from the core and cavity mounting plates.* This method is more complicated, but reduces the number of hoses and leak points to a minimum (one IN and one OUT per plate) (Figure 12.18, right). The mold plates are cross-drilled with channels of various (larger) sizes to supply the coolant and to return it. These sizes should be calculated and located so that all cavities or cores will be fed with the same amount of coolant.

Regardless of which of the above methods are used, the designer must now consider where the coolant connections are located in those stack members that will be cooled.

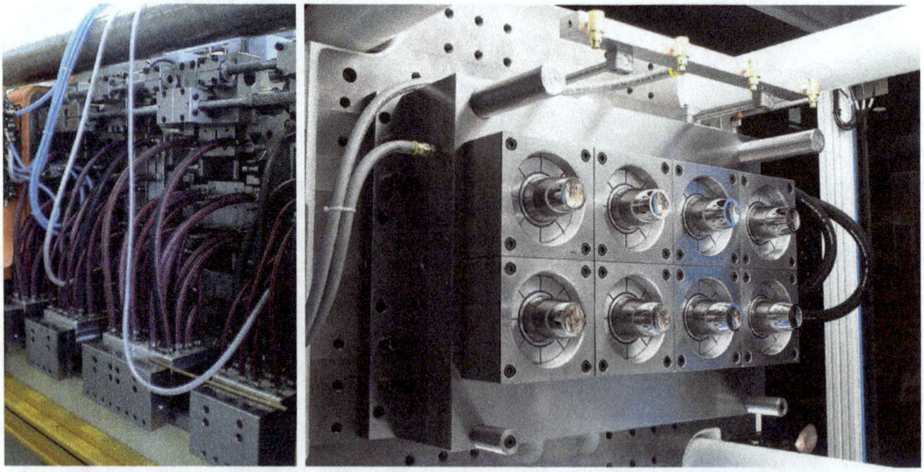

Figure 12.18 (Left) A large number of cooling connections and hoses make installation and maintenance more difficult but may be unavoidable. (Right) One IN and one OUT cooling line per plate (Courtesy of StackTeck Systems, left, and Husky Injection Molding Systems Ltd., right)

Tips for insert cooling:

- Parallel cooling should be used whenever possible for equal cooling of the stacks of multi-cavity molds.
- Always make sure that all inserts can only be installed one way, so that the cooling circuit is not blocked. Add an offset dowel or flat to make sure the insert is installed correctly.

12.8.1 Shrinkage of Plastic and Cooling Layout

As the hot plastic enters the cavity, it is hot (expanded) and under more or less high pressure (compressed). As the injection is cut off and the plastic cools down, the plastic shrinks to return to its smaller, cold volume, and the pressure on the plastic also decreases.

Usually, the effect of shrinkage is the greater of these two influences, with the result that the plastic, as it cools down, loses its close contact with the cavity walls but maintains its contact with the core as it "shrinks on the core". This condition has a great influence on the cooling of the molded product.

When designing cooling for a flat product, it is important that both sides of the product are equally well cooled. This can be seen in Figure 12.19, which shows one side of a flat product with less cooling than the other side. To achieve short cycle times, the product is usually ejected as soon as it is rigid enough for ejection. At that moment, the side in contact with the poorer cooling is still hotter than the other side. As the product cools down outside the mold, the warmer side still continues to contract, and warps the product.

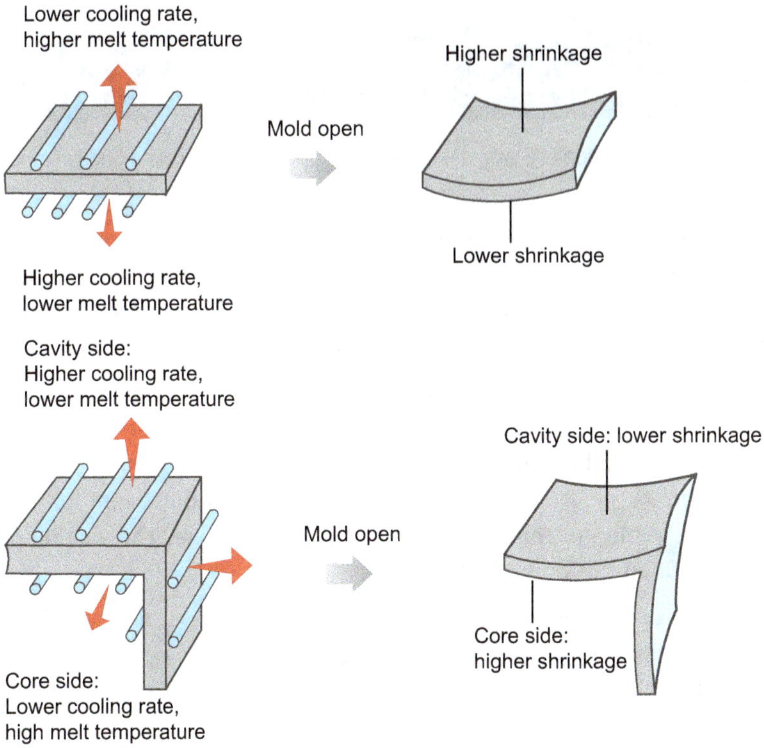

Figure 12.19 Effect of cooling line layout on shrinkage and warpage [4] (Courtesy of Moldex 3D)

There are cases when such differential contraction is designed into the mold, such as in the case of snap-on lids. The surface must be flat, but the rim, being heavier, cools down slower and creates a reverse taper which is desired for the product. The amount of "toeing-in" can then be controlled with the amount of cooling and the cooling time. Also, by cooling the rim more than the panel of the flat product, the rim acts to constrain the shrinkage and results in a flat part. If too much cooling is added to the panel, the lid forms a "potato chip" as shown in Figure 12.20.

[4] Source: M. Wang, R. Chang, C. Hsu, *Molding Simulation: Theory and Practice,* Hanser Publishing, 2018 (Figures 8.9 and 8.10).

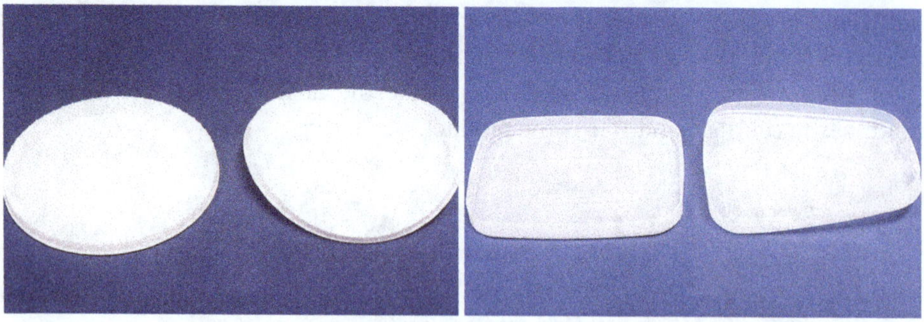

Figure 12.20 Results of poor cooling design (and part design) on a round and rectangular lid

With thin-walled products, the difference in efficiency between cavity and core is not as noticeable because the cooling time is very short. Shrinking onto the core ensures that the product will stay on the core for ejection purposes and remain in its intended shape. Both cavity and core should be equally well-cooled.

For heavy-walled products, the conditions are different. The product, as it cools, loses contact with the cavity wall so that the cavity cooling becomes less efficient. The molding cycle will be much more dependent on the quality of the core cooling. This is unfortunate, because it is usually much easier to find room for cooling in the cavity, which is larger than the product, than in the core. The smaller the core is, the more difficult the cooling becomes.

Making the appropriate decisions up front on part design (see Chapter 3) will alleviate a lot of problems in the mold design stage, especially in cooling design layout. Always remember that good mold design starts with good plastic part design for injection molding.

12.8.2 Preventing Water Leakage in the Mold

When water passes across any joining inserts or plates in the mold, the circuit MUST be sealed against leakage. This is commonly done by inserting O-rings at the sealing points. This applies to all locations in a mold. All O-ring applications for cooling circuits are considered static seals. For O-ring groove dimensions and finishes, use the established standards supplied by the O-ring manufacturers. O-Rings are used for flat seals or circumferential seals, as shown in Figure 12.21.

O-Rings located between flat surfaces are compressed because the plates are joined. Figure 12.21 shows the preferred arrangement. Some material (at least 1 mm or 0.040") is left standing between the O-ring groove and the coolant passage. Figure 12.21A also shows an arrangement where two or more passages are sealed by a single O-ring. This is quite acceptable because any leakage from one passage to the other can usually be ignored.

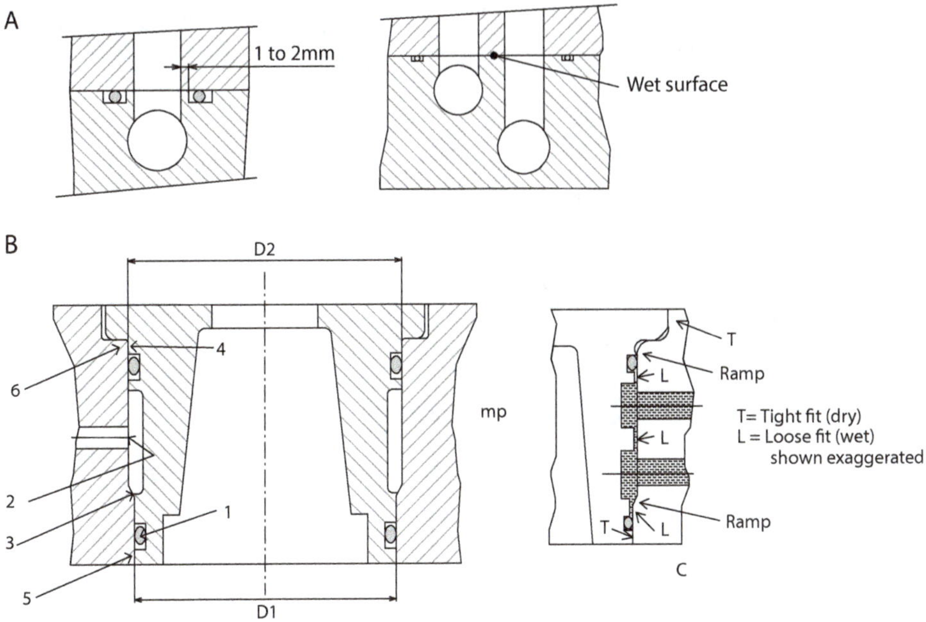

Figure 12.21 Typical configurations for flat and circumferential O-ring seals: (A) Flat seal configurations; (B) Circumferential seal configurations. Note the step at point (3) and chamfer at point (6)

> Screws within a "wet" area must be sealed with O-rings to prevent leakage and corrosion of the threads.

O-Rings are always needed between cylindrical surfaces, and become compressed when the insert is assembled. The main problem is the possible damage to the O-ring during assembly. Any sharp edge, for example, at the entry of a drilled passage into the bore for the cavity insert, which lies in the path of the O-ring during assembly, must be well-rounded or chamfered. Figure 12.21B shows the correct method of using two diameters to fit the insert into the plate. D_2 is large enough for the uncompressed O-ring in groove (1) to pass almost to the end of its insertion travel without contacting the surrounding wall of the bore or the entry of the supply channel (2). A chamfer (3) ensures smooth transition from D_2 to D_1. There must be a radius under the flange and a chamfer in the bore (6). The accurate location areas of the insert within the bore (4) and (5) are thereby kept dry and free of corrosion.

To calculate the size of the ramp (*R*) in Figure 12.22, the minimum step required is the maximum *actual* O-ring cross-sectional size *D* (*not* the nominal size), less the *minimum* depth of the required groove *d* for correct sealing (obtained from the manufacturer), then rounded up to the next one- or two-tenths of a millimeter.

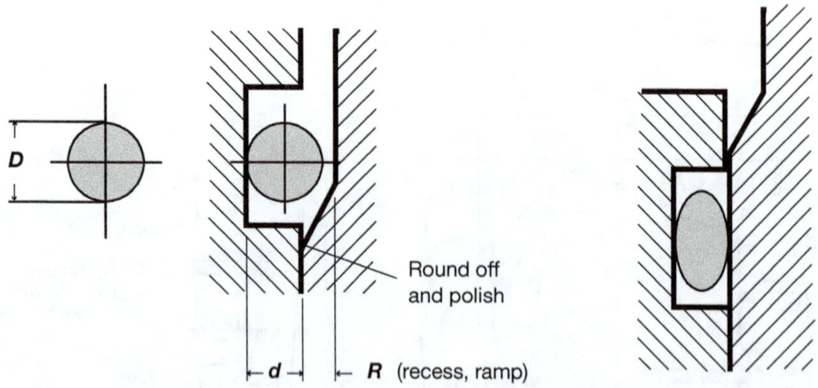

Figure 12.22 Determination of ramp size for cylindrical O-ring installation

Example:

What size should the step be for a nominal ⅛ size O-ring ($D = 3.53 \pm 0.10$ mm and $d_{min} = 2.87$ mm)?

$D_{max} = 3.63\ (3.53 + 0.1)$

$D_{max} - d_{min} = 0.76$ mm

Rounding to the next one-tenth means that the *R* selected = 0.8 mm.

12.8.3 Cooling Channel Distances in Inserts

Distances of cooling channels from the molding surface and from each other must be considered for optimal and even cooling of the molding surface (Figure 12.23). The recommendations shown in the tips below are based on experience of what is needed to achieve even surface temperatures. As the cooling channels are moved closer to the surface, the cooling effect on the surface is greater near the channels and smaller in the space between them. If the channels are moved closer together, the cooling will be more even, but the added cooling channels will increase the cost of the mold and require more (and possibly unnecessary) flow of the coolant.

The stack parts must withstand the forces created by the plastic pressure during injection. These forces are very high and can be in the order of 50–100 MPa (7000–15,000 psi) and even higher in thin-wall molding. Also it is important to note that these forces change from zero to full to zero at every molding cycle, so fatigue is a critical consideration in material selection and positioning of cooling lines.

Under cyclical conditions, the safe, permissible stresses in the steel should be held below 10% of the yield stress of the mold material. When following the recommendations for steel shown in Figure 12.23, it will not be necessary to calculate the stresses. For other materials and other proportions, stress calculations should be done to avoid insert failures.

In molding inserts (round holes):

Recommended distances between cooling lines and molding surfaces (see Figure 12.23) are:

- $B \approx 2.5 - 3.5D$
- $A \approx 0.8 - 1.5B$.

For molding inserts (square and round-milled cooling channels, see Figure 12.23):

- Recommended dimensions are $S \geq L$ and $A \geq 0.66L$ for a square channel
- Recommended dimensions are $S \geq L$ and $A \geq 0.6L$ for a round channel.

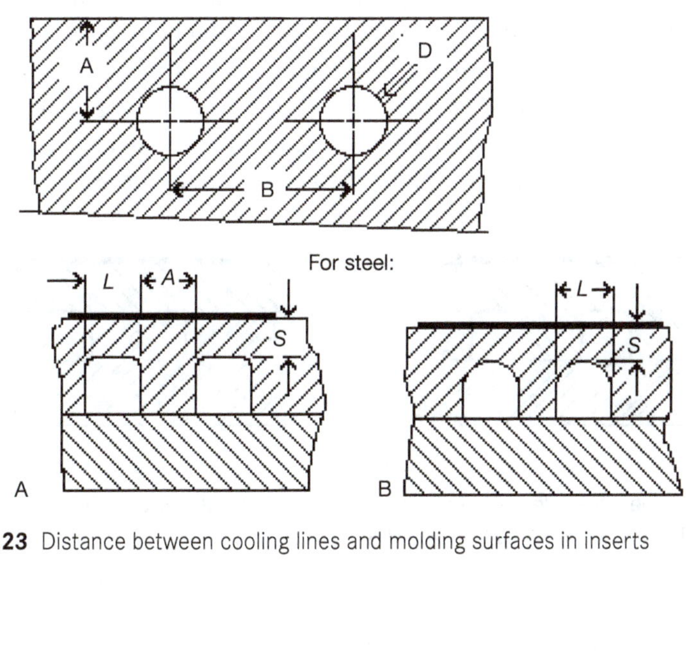

Figure 12.23 Distance between cooling lines and molding surfaces in inserts

12.8.4 Distance to Pipe Threads and Fittings

For straight adaptors and plugs, the minimum distance from obstacles is governed by the size of the wrench required to tighten the adaptor and the swivel hose fitting. For elbow adaptors (fittings with 30 or 45° angles), consider the swing of the elbow when screwing it into the mold plate.

It is important to remember that the outside diameter of the thread is larger than the bore or through hole. The combination of notching caused by the cutting of the threads and the wedging action of the plug when tightening can exert tremendous forces that tend to split the steel if not enough material is left around the bore. The minimum distance between the thread with the largest outer diameter in the insert or plate and a nearby feature should therefore be the same as in the tip above.

The installation dimensions for the fittings or plugs to be used should follow the hardware manufacturer's recommendations.

12.8.5 Cooling Channels Split between Inserts

Figure 12.24 shows the same channel cross-section and the same distance from the molded product in both A and B. However, the active cooling surface in Figure 12.24A is about 3 × the active surface shown in Figure 12.24B. The reason for this is that in A there is about 3 × as much cooling channel surface in the insert with the molding surface cut into it, but in B only one side of the channel is in direct contact with the steel of the mold cavity, and the heat must travel through the interface between the two parts. The conductivity at the contact area of two pieces is at best only 50% of what it would be if there was no separation. In general terms, it means that the *cooling lines should be, wherever possible, in the mold parts that are in contact with the plastic, rather than in adjoining plates.*

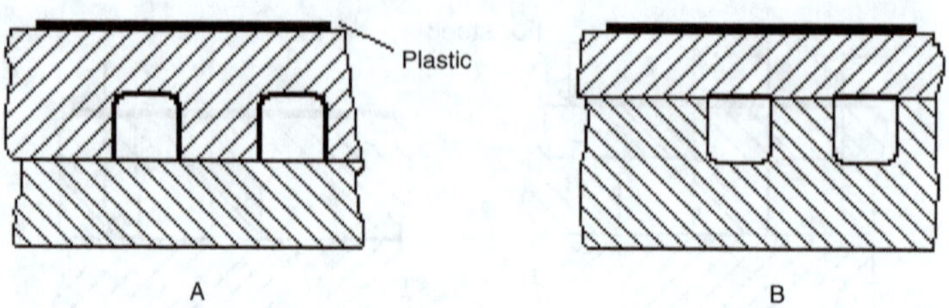

Figure 12.24 When possible the cooling channel should be cut into the part of the insert in contact with the molding surface, i. e., (A) rather than (B)

12.8.6 Design of Insert Cooling for Flat Products

Any cooling arrangement for a relatively flat product must satisfy the requirement that the heat will be equally well-removed from both sides of the molded product, to prevent unwanted post-mold warpage (see Section 12.8.1). Difficulties achieving this are usually due to:

- The gate area being a source of additional heat input (especially in hot runner molds).
- The ejector mechanism, whether stripper plate or ejector pins, requiring space that competes with cooling lines on one side of the mold.
- Inserts and/or core pins often making it difficult to provide a symmetrical, equal, and balanced cooling layout.
- Localized heavy cross-sections in the molded product.

12.8.6.1 Drilled and Milled Cooling Circuits in Flat Parts

The decision of which cooling channel layout to choose rests usually with the designer, and it depends on the available mold-making equipment and the expected productivity of the mold. As a rule, the better the cooling, the more complicated and expensive it is to fabricate the mold.

Figure 12.25 shows a frequently used arrangement of concentric cooling circuits. The circular grooves are easily produced on a lathe or milling machine. The crossover channels are milled. If the rings are turned, it is necessary to insert baffles (shown schematically) to direct the flow. With milled grooves, such baffles can be avoided by milling only approximately 330–340° of the circumference.

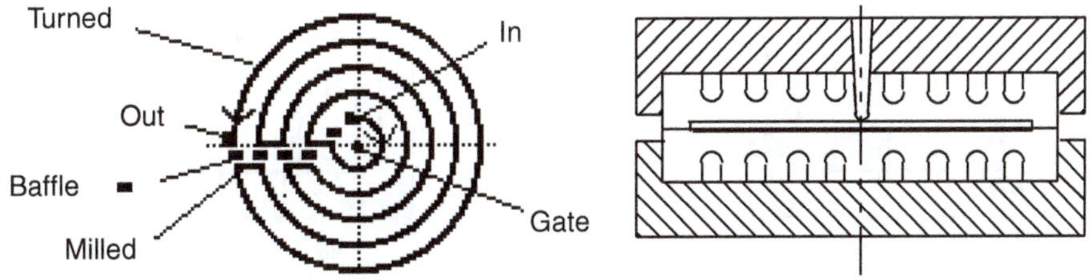

Figure 12.25 Circular cooling pattern for a flat part

Figure 12.26A shows a single spiral. A small disadvantage of both circular and single-spiral grooves is that the coolant enters offset from the gate location. Figure 12.26B shows a double-spiral layout. In this arrangement, the coolant inlets are symmetrically located about the gate and provide better cooling than the single spiral. The number of spirals and their distance from each other depends on the cavity size, on the strength of the material, and on the number, size, and location of screws. The use of milled slots also permits the designer to create detours around screw locations, but in critical molding applications this may affect the evenness of cooling desired.

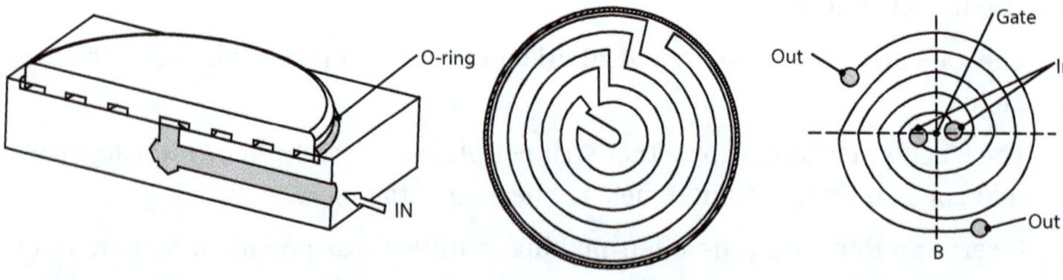

Figure 12.26 Single and double spiral cooling for a flat product (Courtesy of Manual do Projectista)

In all molds, the cooling should enter near the gate location and spread out toward the outside of the product, since the plastic is hottest where it enters and will give up more of its heat content more quickly, because the coolant at this spot is coolest (ΔT is at a maximum). At the same time, as the plastic flow is cooled by the mold, it becomes more difficult to push between the cavity walls toward the edges of the product, and it therefore makes mold filling easier to provide less cooling near the end of the flow path of the plastic. Figure 12.27 shows two commonly used layouts for rectangular parts. In case (A) the cooling enters closer to the gate, but this design is more complicated and costly than the design shown in (B).

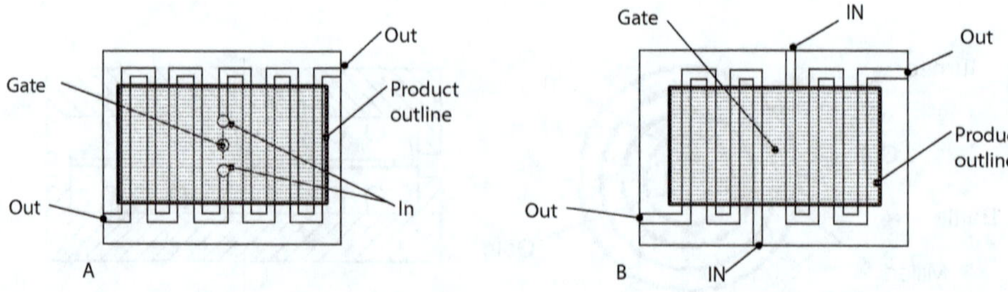

Figure 12.27 Common cooling layout for rectangular-shaped flat parts (baffles and plugs omitted in schematic)

Core cooling for flat products is essentially the same as the cavity cooling. However, an added difficulty arises if ejector pins must pass between the cooling lines. If the cooling channels are milled, the ejector pins must be sealed against the coolant with O-rings.

It is desirable and often mandatory to provide additional spot cooling opposite the gate, especially if the cooling circuits cannot be routed close enough to the gate area. Such spot cooling is usually done with bubblers.

12.8.7 Cooling Design for Cup-Shaped Products

Cup-shaped products typically cannot have the same cooling pattern for the core and cavity side due to geometrical differences. Since the parts typically shrink onto the core, the cooling on the core side is the dominant cooling in the stack, contributing up to 70% of the effective cooling.

12.8.7.1 Cavity Cooling in Cup-Shaped Products

Cavities are typically mounted onto the cavity plate or inserted into the plate. For cooling design, the main difference is in the method of supplying coolant to the cavity blocks or inserts for these types of mountings.

12.8.7.1.1 Cavities Inserted into Plates

Figure 12.28 shows a common method of effectively cooling cavities inserted into a plate. A circular groove around the cavity insert divides the flow of the coolant into two branches. It is important that the sum of the cross-sections of the branches is equal or less than the cross-section of the drilled lines, to prevent one half being less well-cooled than the other. There may be more than one cooling groove around the cavity, and the feeder and return channels may be at different levels, as shown in Figure 12.28. Milled slots and baffles in the insert cooling grooves and in the drilled channels of the plate are usually required to direct the flow as desired into the different levels. As a general rule, if multiple levels are used the flow should start at the bottom of the cavity near the gate, and end at the lip or end of flow.

A further modification to the multi-level grooves is a spiral as shown in Figure 12.28 with single or double spiral grooves instead of ring grooves. This eliminates the need for milling connecting grooves (risers) from ring to ring or the insertion of baffles. Figure 12.29 shows double-spiral cooling. Both the inlet and outlet are at the same level in a double spiral, while in a single spiral the groove would enter near the gate and exit near the lip.

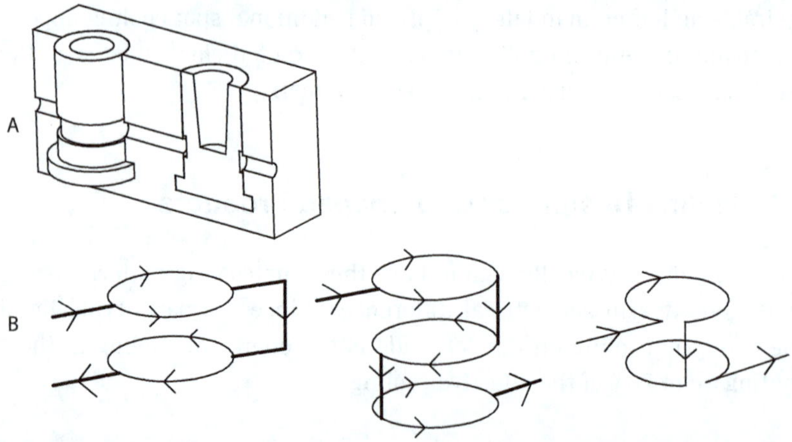

Figure 12.28 Cavity cooling for inserted cavity: (A) Single-level circuit in series from one cavity to the next around a groove; (B) Multi-level circuits (Courtesy of Manual do Projectista)

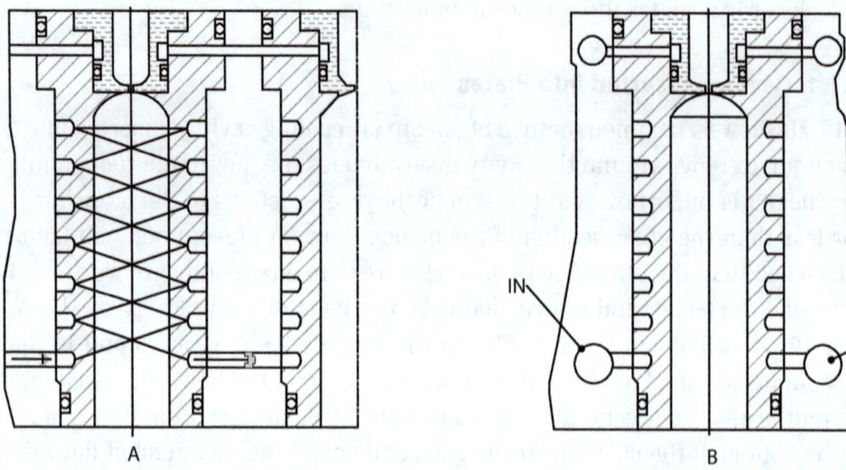

Figure 12.29 Double spiral cooling of a cavity: (A) Cavities in series; (B) Cavities in parallel

Occasionally, cavity inserts are cooled with drilled channels. This could be the case if the cavity is offset within the cavity insert, to provide optimal cooling as may be required for a non-symmetrical part like a disposable cup with a handle (Figure 12.30).

> Make sure to provide dowel pins or keys to locate the cavity inserts so that cooling channels in the plates line up properly with the channels in the inserts.

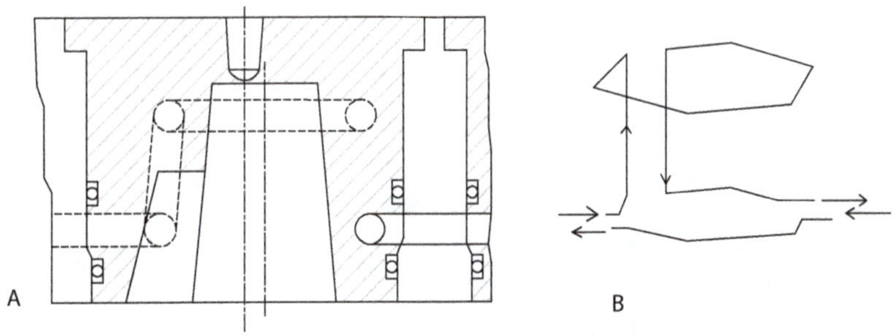

Figure 12.30 Inserted cavity with drilled cooling lines due to offset cavity configuration (cup with a handle)

12.8.7.1.2 Surface-Mounted (Modular) Cavities

Cavities can also be mounted on a backing plate to reduce the cavity plate thickness. Each block (module) contains at least one cavity. Coolant supply can be accomplished with hoses directly to each cavity or via the cavity backing plate.

Coolant Supply with Hose Lines

This method is used with a smaller number of cavity blocks, usually one to four blocks. A possible advantage is the cost saving by avoiding drilling in the backing plate, which can then be thinner, resulting in a saving in weight and shut height of the mold. Also, it permits individual flow control for each cavity if necessary. The disadvantage is the large number of hoses limiting access to the mold, and the time required to make the connections at setup. Also, the number of rows of modules limits the number of cavities that can be connected in this way: where there are more than two rows, the center rows cannot be reached easily (or at all) from the outside with piping.

Coolant Supply through the Backing Plate

Connections are with drilled passages in a backing plate, sealed with O-rings. The backing plate is cross-drilled for all supply and branch connections. The flow to the cavities can be arranged either for series or parallel cooling (see Section 12.7.1). There are two common methods of cooling the modular blocks:

- *Channels are parallel to the parting plane* (Figure 12.31), typically at one or more levels. When several layers of cooling channels are used, the connections (risers) from one layer to the other are with vertical bores, with baffles and plugs where required. These connections can create additional uneven cooling, but in many cases it is only important to keep the block at an average, even temperature.

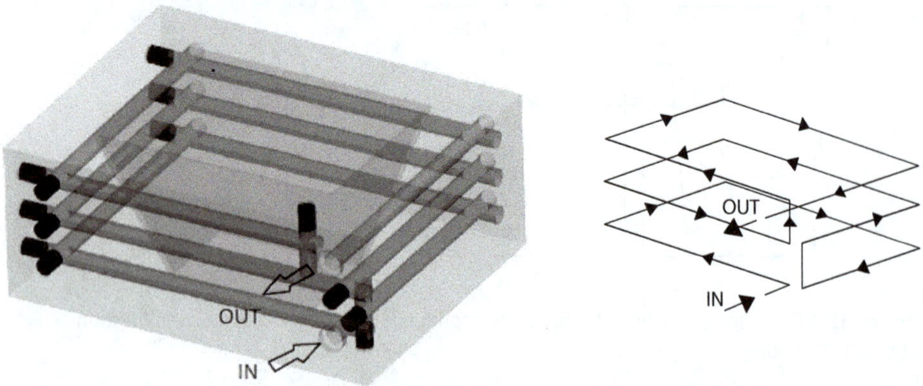

Figure 12.31 Surface-mounted cavity with cooling channels drilled parallel to the parting line at several levels (Courtesy of Manual do Projectista)

- *Channels are approximately vertical to the parting plane.* This method (Figure 12.32) gives better cooling, especially for deeper cavities such as containers, but is not always practical to achieve, and may not even be necessary.

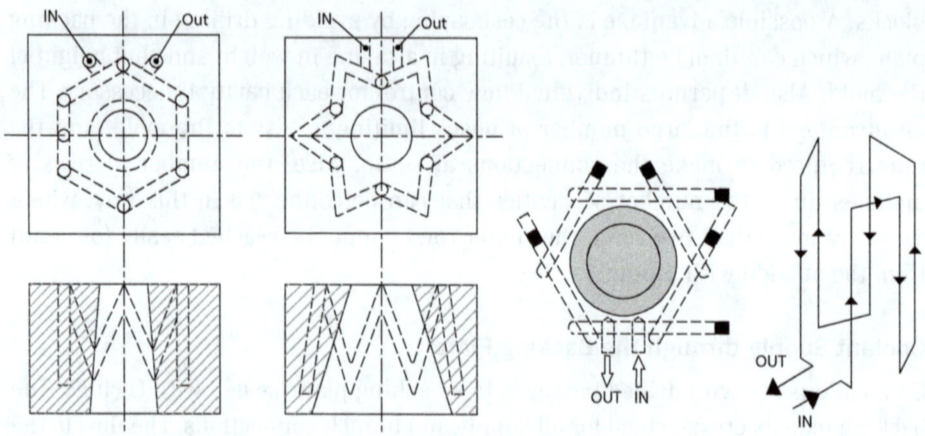

Figure 12.32 Cavity cooling drilling patterns

In Figure 12.32A, the channels are at right angles to the parting plane and are, therefore, further from the plastic near the closed end of the product and closer near the rim of the product. This method is simpler than the one shown in Figure 12.32B, as there are no compound angles for drilling. In Figure 12.32B, the cooling channels run parallel to the cavity wall (tapered container, pail, etc.) and are at an angle to the parting plane. This arrangement provides better, closer spacing of the coolant over the whole length of the cavity. In the case of Figure 12.32C, the cooling lines still run at right angles to the parting line, but are connected at the top and bottom to create a single cooling circuit around the part. The disadvantage of this design is that the water cooling the part at the IN line will be cooler than the water at the OUT line. For some parts, this may cause some issues around dimensions and warpage. In less critical cases, like thicker parts, this may be perfectly fine.

Another option to reduce the drilling and eliminate the need to connect the drill lines is to insert a baffle inside the drilled holes so that the water flows up one side of the hole and down on the other (see Section 18.8 for cavity lock container design). This can reduce the amount of drilling in the cavity, which can sometimes weaken the strength of the cavity.

As a rule, the drilling lines in the cavity should not remove more than 40% of the material through the centerline of the drilling.

Another alternative is to add baffles (examples shown in Figure 12.33) to the drilled cooling lines to create higher flows and to reduce the number of cooling channels to drill. Figure 12.34 shows how baffles can be used to create a cooling circuit with cooling lines generally perpendicular to the parting line. A groove is added to the back of the cavity to allow the cooling water to flow out one side of the baffle and into the next baffle. The baffles must be anchored in the bottom of the cavity groove to prevent them from moving forward and short-circuiting the cooling path. Other examples of baffle cooling in the cavity can be found in Chapter 18 and also in Figure 12.42.

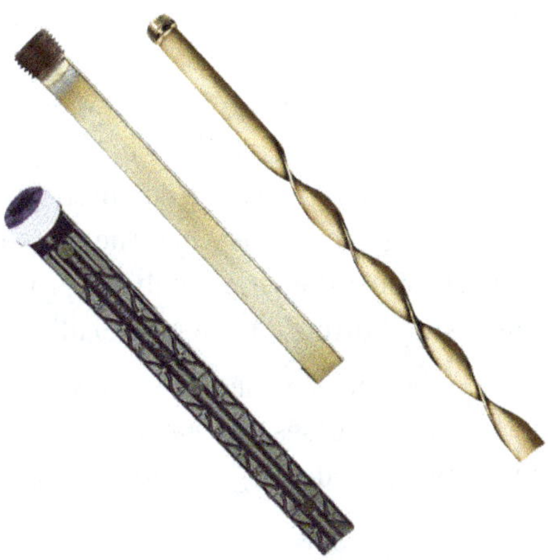

Figure 12.33 Standard baffles for use in core and cavity cooling (Courtesy of DME)

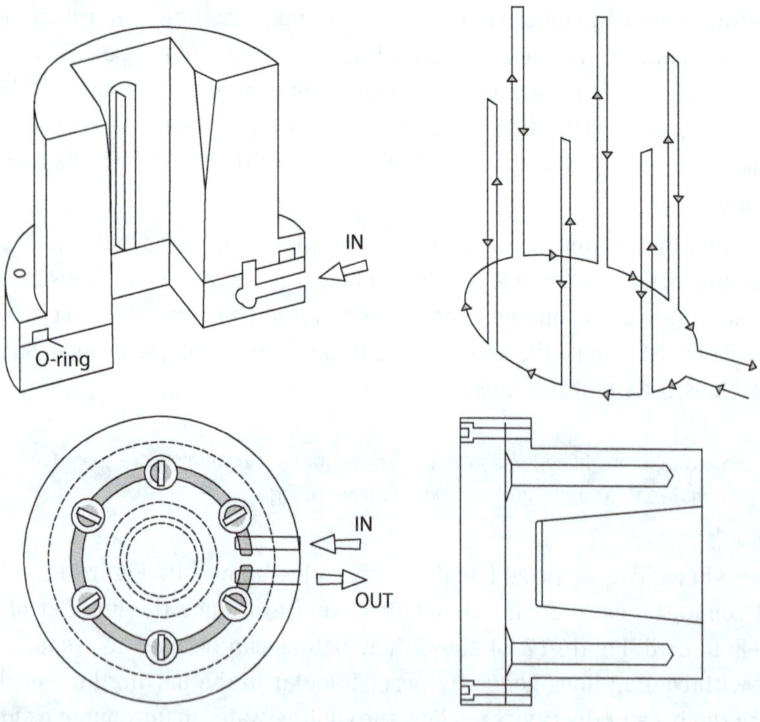

Figure 12.34 Baffle cooling in cavity

12.8.7.1.3 Gate Pad and Gate Insert Cooling

The gate pad or gate insert is an insert in the cavity, surrounding the gate area. Typically, a gate pad or gate insert is used when additional cooling is needed near the gate. The most common reason for this is when a hot runner is used as a melt distribution system. The hot runner keeps the material hot up to the gate and it needs to seal on the cold plates of the mold. To avoid a hot gate area, the cooling near the gate needs to be especially efficient in a hot runner. The most efficient way is to add an insert into the cavity to create a comprehensive cooling circuit for the gate area. A gate insert or gate pad also has the added benefit of being an inexpensive replaceable part. Since the gate s subject to wear and damage, it is always advisable to create a small insert for the gate to allow easy replacement.

For some products, it may not be permissible to have a witness line where the gate pad is located in the cavity. In such cases, the cavity must provide sufficient cooling in the gate area, usually by cross-drilling or with circular grooves.

There are several reasons why a gate pad is desirable:

- *Wear of the gate.* It is usually more economical to replace the smaller gate pad than the whole cavity, although it is often feasible to repair a damaged gate by welding and re-machining.

- *The material used for the gate pad* can be different from that used for the cavity. For example, to improve cooling, the gate pad can be made from BeCu, while the cavity may be made from hardened mold steel.
- *Venting.* Cavity venting can be easily added when a gate insert or gate pad is designed into the cavity space.
- *Separate cooling circuits in the cavity and gate areas.* With multiple circuits, the cooling for the cavities and gate pads can be separately controlled at start-up and during operation. The warm water from near the gate area can be separated from the cooler water needed to cool the cavity.

Figure 12.35 illustrates examples of gate pad and gate insert cooling. Figure 12.35A shows a gate pad with a cooling groove. This works well for round inserts and provides efficient cooling with no cross-drilling. Figure 12.35B and C show a gate pad with cross-drilled cooling channels. Figure 12.35D shows a drilling pattern entirely in the gate pad being fed from the cavity plate. The pattern is similar to cavity cooling, as shown in Figure 12.32.

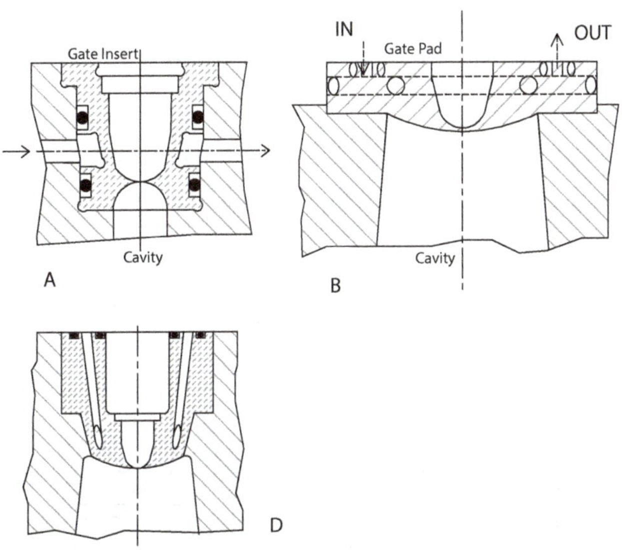

Figure 12.35 Gate pad and gate insert cooling circuit examples

12.8.7.2 Core Cooling of Cup-Shaped Products

Core cooling for cup-shaped products is very important, since in these parts the core cooling contributes up to 70% of the overall cooling in the mold. Unfortunately, core cooling of cup-shaped products is often quite challenging due to ejection mechanisms and the shape of the core.

The area opposite the gate must always be properly cooled while providing adequate cooling for the rest of the core. Wherever possible, the cooling supply for the core cooling should be directed first against the gate area before being channeled to cool the rest of the core. If for some reason this is not possible, a separate bubbler should be directed at the gate area.

When laying out the cooling channels inside the core, allowance must be made for vent pins and/or vent slots, and for the connecting air lines between the drilled coolant lines or open channels if they are needed for ejection or filling. Space must also be allowed for sealing with O-rings to prevent leaks into the vent lines.

Thick products or products with bosses, which preclude fast cooling, will usually not need an elaborate cooling layout. As long as there is enough cooling capacity in the circuit to remove the heat of the plastic during the projected cycle time, the mold will work well without designing an excessively complicated and expensive cooling arrangement.

12.8.7.2.1 Core Cooling Circuits

Figure 12.36 shows three simple methods to provide cooling in a core. A series of baffles such as shown in Figure 12.36A is often used around the periphery, near the molding surface, on the inside of the cavities, or the outside of the cores. There can be practically any number of such baffles, and they can all be in series, parallel, or a combination of both. It is very critical that these baffles be located in the correct position. If the baffle is rotated even slightly, it will affect the cooling in that entire circuit, so they must be rigidly fixed in position. If the baffle is rotated 90°, the water flow will simply bypass the baffle path and travel along the path of least resistance.

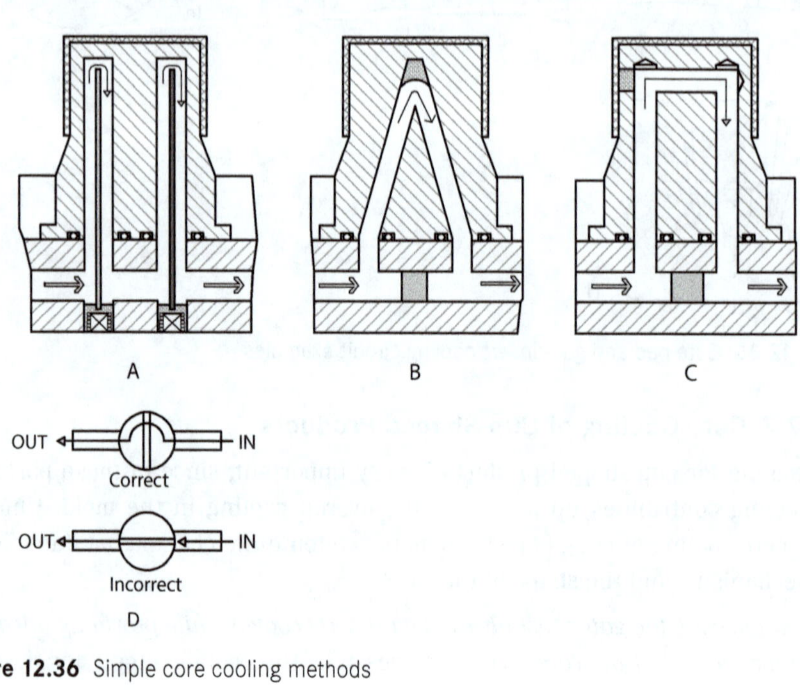

Figure 12.36 Simple core cooling methods

Figure 12.36B illustrates a commonly used method, with two drilled channels converging toward the gate area. The advantage of this method is that the gate area is better cooled than with bubblers, provided the enclosed angle between the two drilled holes is not too acute. If this is the case, the main flow of the coolant will find the path of least resistance, and flow around the lower portion of the intersection of the two bores, leaving the coolant stagnant near the gate (as shown). Also, the total channel surface exposed to the coolant is smaller than in the other two examples shown.

Figure 12.36C shows another good method of providing cooling channels in such a core, but it requires blocking of a channel (usually by welding or soldering a plug) on the molding surface, which may leave a witness mark on the inside surface. This may be acceptable in parts where the inside of the part (the core) is not a visible surface. Automotive panels are a good example of this.

In all three cases, individual O-rings are shown at each passage, which is good practice to reduce the wet area exposed to corrosion. However, in small cores there may not be enough room for two O-rings, and so one O-ring surrounding both passages could be used.

12.8.7.2.2 Small Cores with High Aspect Ratios

Because of the aspect ratio of products such as the test tubes shown in Figure 12.37, the choice of cooling is very restricted on the core side. In most applications, it is possible to cool such cores with bubblers.

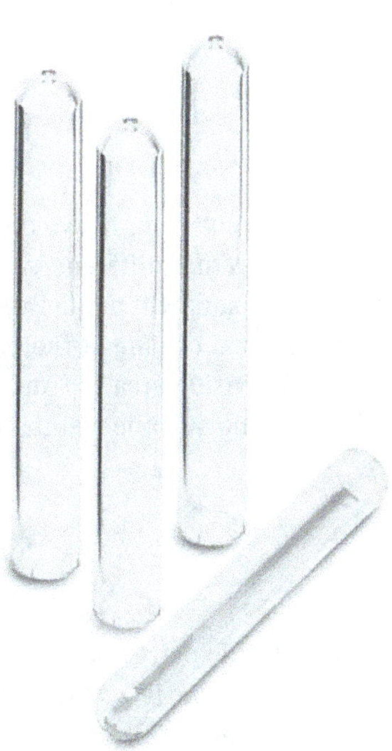

Figure 12.37
Long slender parts such as these test tubes lend themselves to bubbler cooling

Baffles should be avoided in long slender cores, as the water coming up one side of the baffle will be colder than the water flowing down the other side. This will result in a part that warps into a "banana shape" after cooling.

Figure 12.38A and B show a typical bubbler arrangement in a small, elongated core. The bubbler allows water to flow up the tube inside the core and then cascade down the outside of the tube in the hole. This type of cooling provides a very even cooling around the core. The bubbler itself must be held in the center of the drilled hole in the core with standoffs or another supporting means at the tip of the bubbler. Figure 12.38C shows a variation of a bubbler where the coolant is forced to return along a helical path, which is dimensioned so that the area A of the tube is equal or slightly larger than the area F of the helix.

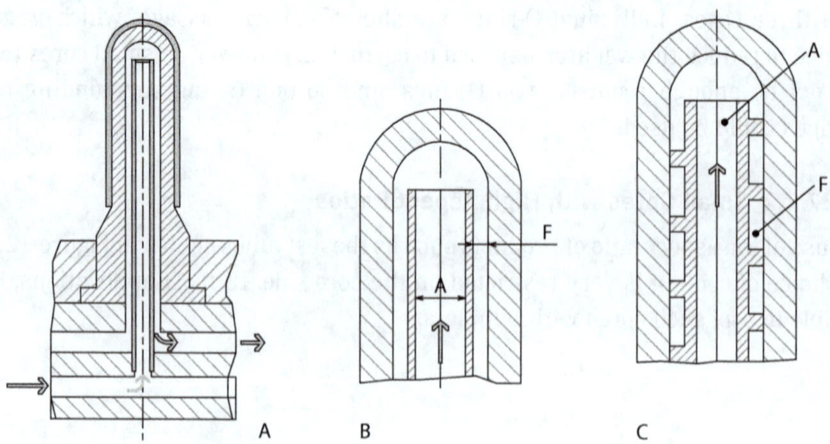

Figure 12.38 Core cooling using a bubbler

It is very important to ensure that the flow down the core is even, to avoid core shift and uneven post-molding shrinkage caused by temperature differences from one side of the core to the other. If this is neglected, the coolant will not flow symmetrically (cascade) around the tube and will create unequal cooling of the product. In order to ensure that the resistance to flow on the inside area (A) and the outside area (F) of the bubbler is the same, the following ratio must be maintained:

Bubbler diameter ratio:

$$\frac{\text{Inner diameter of bubbler}\left(\text{mm}\right)}{\text{Outer diameter of bubbler}\left(\text{mm}\right)} = 0.71 \tag{12.10}$$

Alternatively, if the core is too large for a bubbler, a double-helix spiral can be created inside the core and the cores can be aligned in series. The flow of the water comes in one helix, flows to the top, and returns down the other.

12.8.7.2.3 Single-Piece Cores

One-piece cores are generally used where a witness line on the core is not acceptable. An internal plug is manufactured from preferably stainless steel and provided with tapped holes to facilitate removal for maintenance and repair. Figure 12.39A is a simple design, similar to the bubbler method. The groove in the plug can be a helix, or consist of a number of circular grooves connected with milled slots and provided with baffles to create the desired flow pattern. Figure 12.39B is a reversal of Figure 12.39A that provides improved cooling efficiency but at a higher cost. The benefits of the cooling grooves in the outer core are the same as those described in Figure 12.24. Note that in both cases of Figure 12.39, the cross-section of groove must be equal to or smaller than the inlet drill line (similar to the ratio described in Figure 12.38 and Equation 12.10).

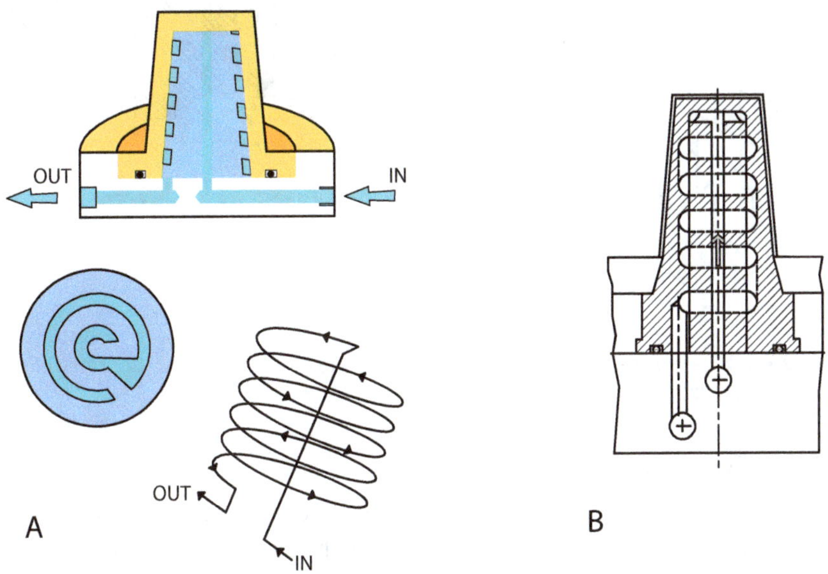

Figure 12.39 Single-piece core design. Design A has the grooves in the plug while design B (preferred) has the grooves in the core to improve heat transfer (Courtesy of Manual do Projectista)

If the sidewall of the core has a significant taper, then a tapered plug can be used to keep the cooling lines in close proximity to the molding surface for the entire sidewall.

Note that the tapered plug must fit snugly, to assist in the support of the core between the grooves. The suggested fit of the plug is as follows:

> The fit of the plug in the core of Figure 12.39 is a light sliding fit. Clearance *F* (per side) = 0.010–0.015 mm (0.0004–0.0006").

Figure 12.40 shows the common approach in large cores of cross-drilling cooling lines. In this case the drilled cooling lines enter and exit through the molding surface. The plugs on the molding surface of the insert must be welded in or press-fitted. The plugs on the molding surface will create a slight witness line at the plug location, but this may be perfectly satisfactory on the inside of the part.

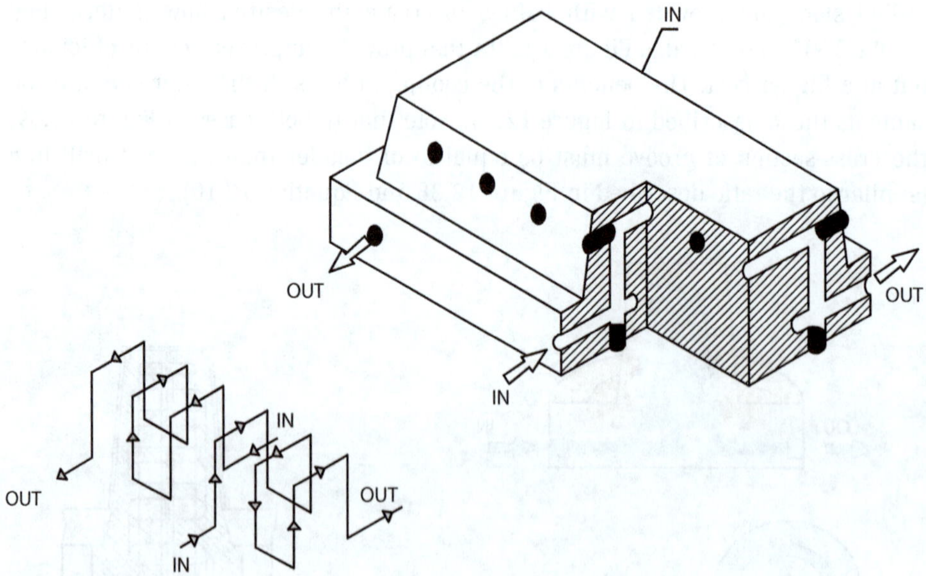

Figure 12.40 Single-piece core with cooling plugs on the molding surface (Courtesy of Manual do Projectista)

12.8.7.2.4 Two-Piece Cores

Two-piece core designs are used where witness lines are acceptable on the inside of the product and where additional cooling at the top of the core is essential to meet the cycle time requirements of the mold. Figure 12.41A shows a core tip insert made from beryllium-copper that is held only by a light press-fit. The pressure of the coolant is not enough to dislodge it, and the injection pressure holds it in place. If the core is large enough for bolts, then they can be inserted to hold the insert in if needed. Figure 12.41B and C show a two-piece core cooling design that is quite common and effective. A threaded tube is used to supply the coolant and to fasten the tip to the core. This tube is subject to high stresses and may require frequent replacement. Care must be taken to ensure that the distribution grooves in the center of the beryllium-copper insert do not weaken it too much.

A still better design (see Figure 12.41D) uses a bolt to hold the core cap in place and two grooves at the bottom of the core to feed water. One groove is used for the inflow and one groove is used for the outflow. This avoids the need to introduce water up the center of the core where the bolt is.

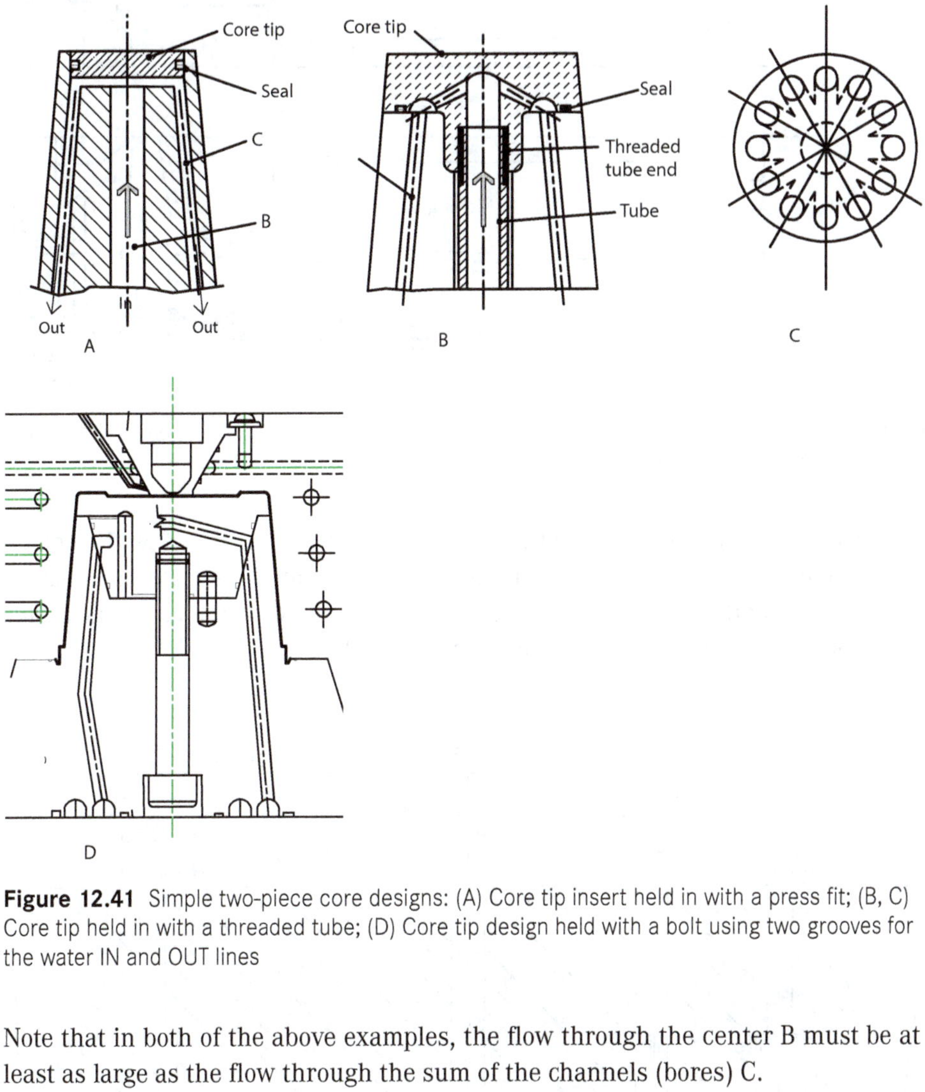

Figure 12.41 Simple two-piece core designs: (A) Core tip insert held in with a press fit; (B, C) Core tip held in with a threaded tube; (D) Core tip design held with a bolt using two grooves for the water IN and OUT lines

Note that in both of the above examples, the flow through the center B must be at least as large as the flow through the sum of the channels (bores) C.

Figure 12.42 shows a layout that uses a core cap or a core insert with two grooves at the bottom of the core. A solid bolt instead of a tube holds the core cap. The flow through the channels is divided into two groups: one set is for the inlet and the other is for the outlet. Notice the two grooves at the bottom of the core for the IN and OUT lines.

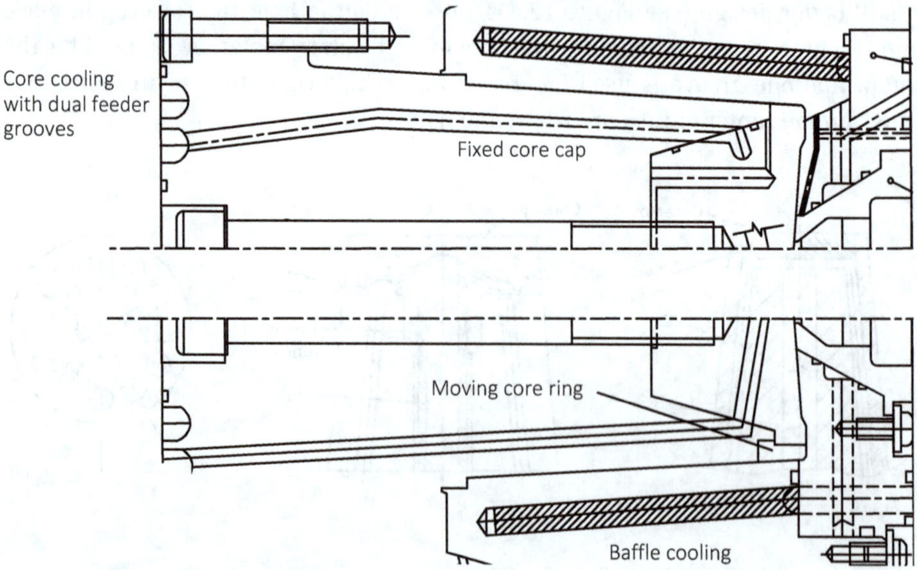

Figure 12.42 Two-piece core design with fixed core cap (top) and moving core ring (bottom). Note that the flow is divided into two cooling circuits on the bottom of the core

Figure 12.43 shows the design of a core with a complete overcap with cooling grooves cut into it. It is usually made from a high conductivity material (such as BeCu25). The pattern of the grooves can be decided based on the geometry of the part. Note that the grooves are cut into the cap itself to improve cooling efficiency.

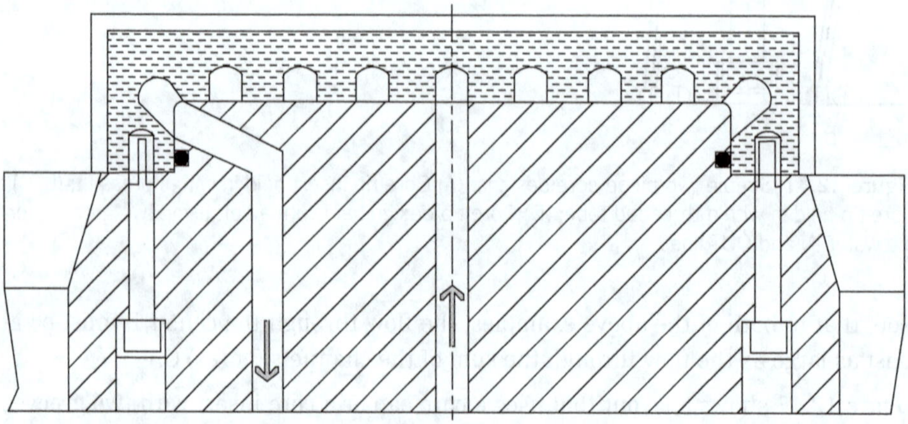

Figure 12.43 Two-piece core design with an overcap

Figure 12.44 shows a two-piece design using an inserted plug. The inserted plug allows the designer to develop a cooling circuit right by the lip of the part where the cooling needs to be well-controlled. The insert may be cross-drilled in a pattern such as shown in Figure 12.44B. Note how some of the zigzag lines converge close to the gate area, to cool this area well. An alternative design could be to add an additional insert opposite the gate with a groove on it. However, the additional cost of this insert must be considered in the design.

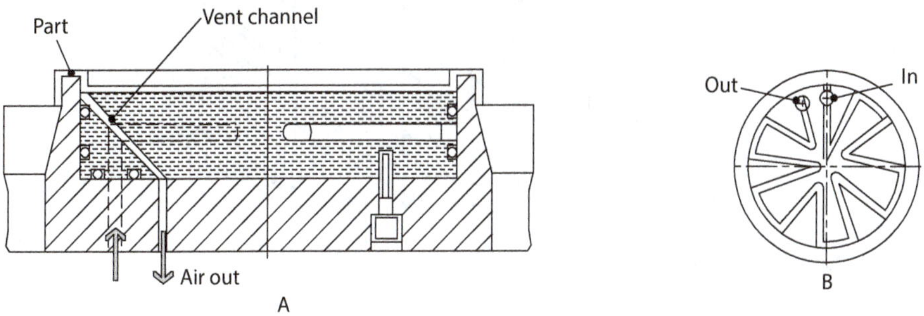

Figure 12.44 Two-piece plug design for a core

> Whenever a material with high thermal conductivity is being used in a stack component, this part should not be on the parting line or in contact with a moving taper such as a stripper ring. Soft high-conductivity materials will wear quickly or hob, and quickly cause molding issues.

If the part is internally gated, then provision must be made for the nozzle as well as cooling of the core. This can be a significant challenge, especially on smaller parts. Section 18.7 shows how a gate insert is inserted into the core cap with a cooling circuit, with the core tip itself having a separate circuit. The core tip is held in place with bolts. Note also that the cavity design uses a cavity insert with cooling pointed at the gate to keep this cycle-limiting section cooled.

Figure 12.45 shows a typical overcap for a larger product. Coolant flows spirally in concentric grooves, with connecting slots and baffles from the center toward the rim of the product. Venting is provided by the vent pin that passes through the core structure to the top of the core. This type of design allows the designer to provide cooling evenly at all points on the core. In this case, ample cooling can be provided opposite the gate. A core insert or a bubbler could also be inserted opposite the gate if cooling in this area was not sufficient.

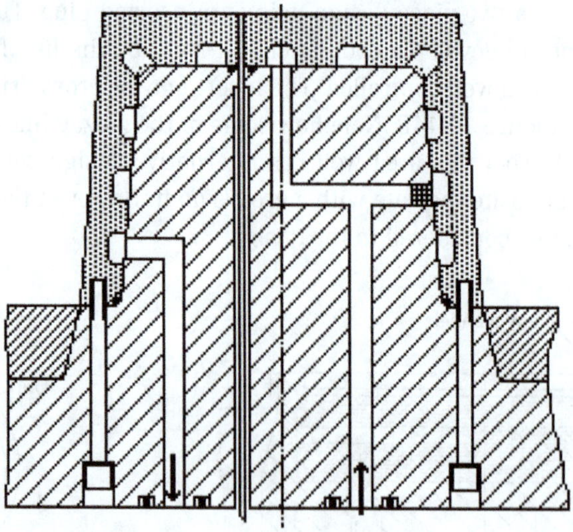

Figure 12.45 Two-piece overcap core design for a large part

12.8.7.3 Cooling of Inserts

In some cases, small features on the part are best created with separate inserts on the core or cavity side. Designing a separate insert may be more cost-effective to machine than making the core or cavity out of a single solid piece of steel. The insert can be easily changed if damaged and modified if necessary after the mold is built. An insert is often difficult to cool, but since it is a small mass, it will rise in temperature quickly if not cooled. Where size permits, cross-drilling or bubblers should be used, similar to the cooling of small cores described in Section 12.8.7.2.2. Beryllium-copper alloys for inserts are sometimes used to help remove the heat from the insert, but at the cost of lower strength and higher wear. If space is very restricted, the designer might consider the use of a heat pipe. For more information on heat pipe theory and application, the designer should consult suppliers of this technology.

Figure 12.46 illustrates various methods of cooling small inserts. Figure 12.46A shows cooling by conduction to a cooled plate. The cooling is somewhat limited because the conduction needs to pass over the interface between the insert and the adjoining part containing the direct cooling. The method in Figure 12.46B is better, as it cools the pin directly and the surface area of cooling is about 3 × that of Figure 12.46A, with the coolant flowing around the base of the insert. Figure 12.46C and D show a bubbler inside the base of the insert, reaching as close as possible toward the hot area. This is the most efficient method, but the size and strength of the insert may not permit it.

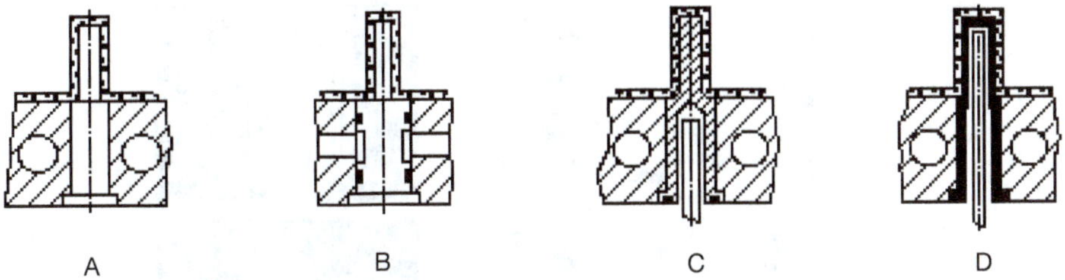

Figure 12.46 Methods of cooling small inserts in the mold: (A) Passive cooling by conduction; (B) Direct cooling at the base of the insert; (C) Bubble cooling of the base of the insert; (D) Full cooling of the insert with a bubble or baffle

Using CAE cooling analysis, the impact of direct and passive cooling in inserts can be shown. For example, in Figure 12.47 the directly cooled insert is 18 °C cooler than the passively cooled part. Note that in place of the bubbler a baffle or a heat pipe could be used. If direct cooling of the insert is being considered, it is important to make sure that there is enough flow available through the small passages in the inserts to cool the rest of the circuit. The cooling through the insert should not cause a pressure drop (restriction) to the flow that causes the heat transfer efficiency to drop.

12.8.8 Use of CAE Molding Simulation for Cooling Design and Warpage Analysis

CAE analysis tools are extremely useful for evaluating the impact of cooling water design. If the design of the cooling circuit or the specific insert cooling is in question, running an analysis on the design and conducting simulations on variants of the design is both cost- and time-effective. Providing that accurate inputs can be provided to the model, the equations for transient heat transfer are well-defined and accurately predict the cooling of a mold.

Figure 12.47 shows an example of a molded product that cannot be easily cooled using traditional methods. Traditional methods such as a baffle or a cooling tube (a and b) still leave the top of the part hot. In this case, a conformal cooling design approach was used to bring the cooling right up to the tip of the part. Alternatively, an insert with high-conductivity material such as BeCu could have been added to take away more of the heat from the plastic.

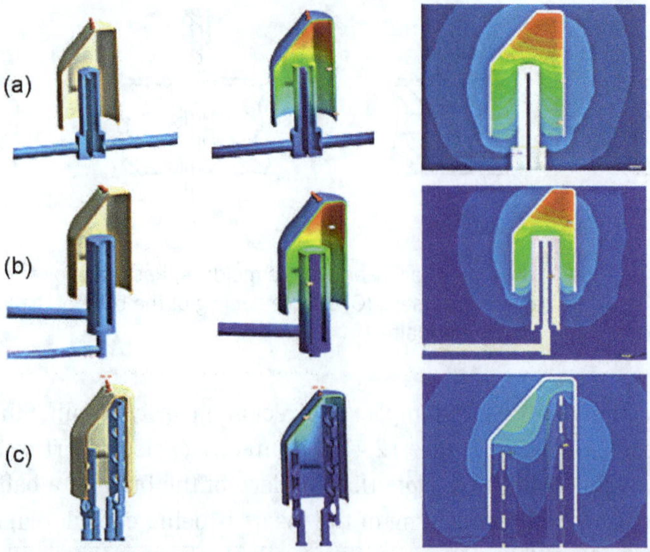

Figure 12.47 Cooling insert analysis using CAE analysis [5] (Courtesy of Moldex3D)

Figure 12.48 shows how CAE analysis tools can be used to optimize the cooling of a part. A conventional cooling circuit and a conformal cooling circuit were designed, and a cooling analysis shows clearly that the conformal cooling provides more effective and consistent cooling.

[5] Source: M. Wang, R. Chang, C. Hsu, *Molding Simulation: Theory and Practice,* Hanser Publishing, 2018 (Figure 3.46).

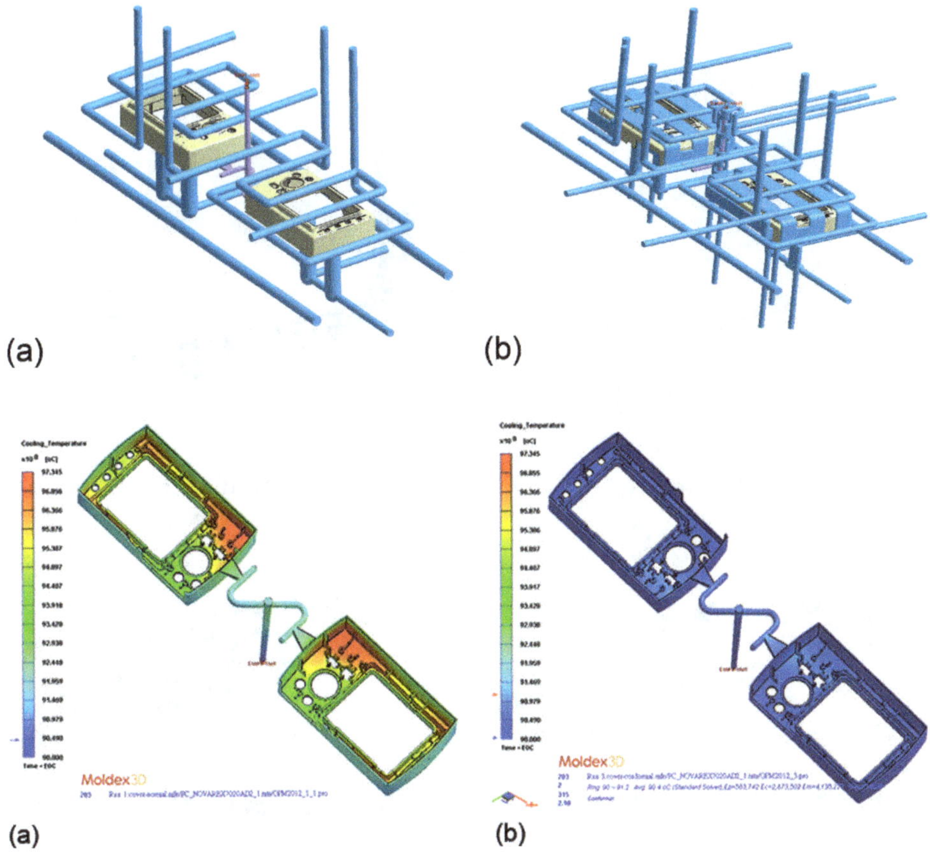

Figure 12.48 Impact of using conventional cooling and conformal cooling on a housing [6]: (a) Conventional cooling; (b) Conformal cooling. Top: Layout of cooling lines; Bottom: Cooling analysis (Courtesy of Moldex 3D)

Based on the results of the cooling analysis, CAE simulation tools can also predict how the part will warp during cooling once outside the mold. As discussed in Chapter 3 on part design, changes in thickness can impact the warpage of the part. This, in addition to how the mold is cooled, can be predicted using the analysis tools available to designers today. Figure 12.49A shows an example of a frame which warps outwards after cooling. Figure 12.49B shows a different frame with and without metal inserts for improved cooling. Note in the figure that the cooling inserts actually produce a larger warpage even though the designer thought they were improving the cooling and therefore the quality of the molded parts. This is where using simulation tools can identify less obvious results and greatly benefit the designer.

[6] Source: M. Wang, R. Chang, C. Hsu, *Molding Simulation: Theory and Practice,* Hanser Publishing, 2018 (Figure 8.38 and 8.39).

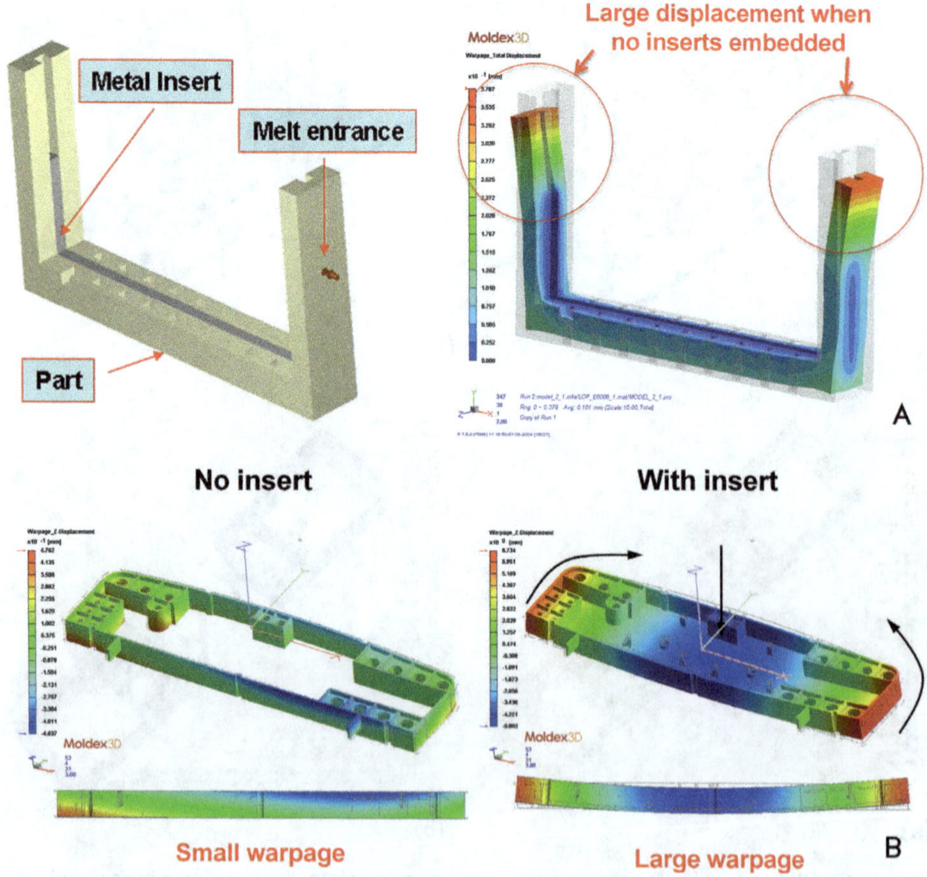

Figure 12.49 Warpage simulation on molded parts with and without cooling inserts [7]: (A) Warpage analysis on a part with no cooling inserts; (B) Warpage analysis on a molded part with and without cooling inserts (Courtesy of Moldex3D)

■ 12.9 Supplying Water to the Mold

Up to now, we have considered only the cooling of the various mold elements. We shall now look at the whole mold. The mold plates and inserts need to be fed with coolant through cooling hoses. Typically, on an injection molding machine there are cooling manifolds mounted on the moving and stationary platens. These can be used to feed the mold with water. If these are not present, then water manifolds can be built into the mold itself to evenly distribute the primary water entering the

[7] Source: M. Wang, R. Chang, C. Hsu, *Molding Simulation: Theory and Practice,* Hanser Publishing, 2018 (Figure 9.28 and 9.29).

mold. Water manifolds have the advantage of eliminating the "spaghetti mess" of hoses sometimes seen in production facilities, making installation faster, improving access to the mold, and eliminating the many sources of leakage from independent cooling lines.

A separate controllable cooling circuit should be used for each area of the mold requiring different cooling criteria (flow or temperature). However, the number of cooling circuits should be kept to a minimum. Typically, the three to four circuits are created:

- Hot runner
- Gate insert or gate pad
- Cavity
- Core.

> If at all possible, water connections at the top of the mold should be avoided, as any leaking water will enter the mold. The preferable location for hose connections on a mold are on the non-operator side and the bottom.

12.10 Thermal Expansion

An often-overlooked fact is that mold cooling is not only for cooling the plastic, but also for cooling the various mold plates that are close to areas heated by the plastic, such as the hot runner systems. As is explained in Chapter 15, all materials expand when heated. In many molds, the mold plates are essential for the alignment system because they carry the leader pins and bushings or other alignment members. If the mold plates are at different temperatures, they will expand differently from their original cold state, and cause misalignment between the alignment elements.

Thermal expansion (ΔL) can be calculated using the following equation:

Thermal expansion: (12.11)

$$\Delta L = K \times L \times \Delta T$$

where:

K = thermal expansion (mm/°C):

Steels = 0.000011-0.000013

Coppers = 0.000016-0.000017

Brass = 0.000018

Aluminum = 0.0000224

L = length of the area of interest (plate or across and insert)

ΔT = temperature difference across the length L

Example:

Assume that the distance of two leader pins in a mold is L = 400 mm and that a temperature difference of ΔT = 10 °C (18 °F) exists between the two plates carrying the pins and bushings. Calculate the thermal expansion of the plate.

$$\Delta L = L \times \Delta T \times 0.000011 = 400 \times 10 \times 0.000011 = 0.044 \text{ mm } (0.00173")$$

Considering that the standard diametrical clearance between leader pins and bushings is only 0.025 mm (0.001"), the example shows the pins will bend at every cycle, wearing the leader pin bushings or binding in the bushings. This points to the importance of ensuring during mold design that both mold halves are kept as close as possible to the same temperature.

13 Mold and Stack Alignment

The main reason for any alignment method between cavity and core is to keep the centerlines of cavities and cores in line. Any deviation from the actual centerlines of cavity and core from the "true" centerline will result in thickness variations of the sidewalls of the product, as shown in Figure 13.1, as well as premature wear and potential damage to the molding surface components.

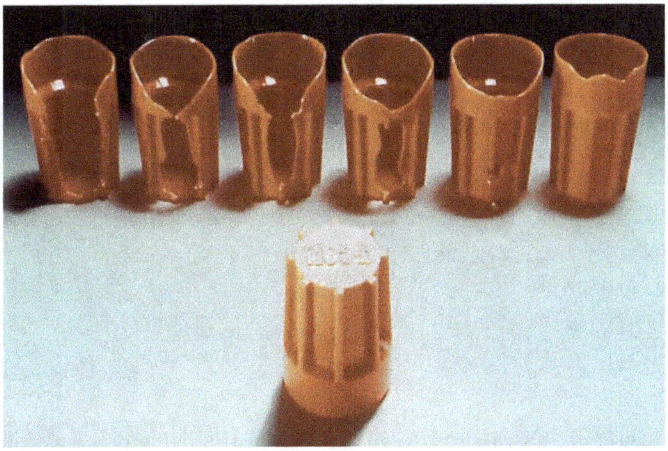

Figure 13.1 Uneven filling and wall variation on these molded parts is created by poor alignment of the stack

There are several choices available to align the mold halves and the individual cores and cavities:

1. *Leader pins and bushings between mold plates* (see Figures 13.2, 13.4, and 13.9). This is the oldest and most common method used, and is suitable for any size of mold, and for any number of cavities. This is the lowest cost method of alignment.
2. *Taper interlocks or straight interlocks between mold plates* (see Figures 13.6 and 13.8). This method usually also requires at least two or more leader pins to

protect the core(s) from damage and to facilitate handling, assembly, and servicing of the mold outside the machine. The designer can choose from standard round taper pins (less expensive), straight interlocks or standard wedge locks (adjustable). The advantage of this type of alignment system is that the parts are standard and can easily be replaced when wear occurs.

3. *Taper lock between each cavity and core* (locking achieved by round tapers or flat wedges). This style of alignment provides the best alignment, but it also increases the mold cost and the cost of repairing any worn tapers.
4. Or any combination of the above.

Deciding the type of alignment mechanism to use depends on a few factors, including the tolerance of the plastic parts, the cavitation, and the wall thickness to be molded. Table 13.1 provides general guidance on which method to select. If any of the criteria falls in a more complex solution then this alignment should be chosen.

Chapter 18 contains many reference mold designs with the above three types of alignment. The reader is encouraged to consult these to determine how and when to apply each type of alignment system.

Table 13.1 General Decision Tree for Type of Ejection to Be Used

Factor	Leader pins only	Plate alignment with straight or taper interlocks	Individual core/cavity alignment
Shape of part	Flat	Dished	Deep cup-shaped
Wall thickness	> 2 mm	1-2 mm	< 0.8 mm
Wall thickness tolerances	Wide	Narrow	Critical
Cavitation: cup-shaped	≤ 4	≥ 2	≥ 8
Cavitation: flat	≤ 8	≥ 2	> 16

The shape of the part has a significant impact on the decision of which alignment method to use. A flat part stack design can generally use a simple alignment of the mold plates, as misalignment will not impact the part significantly. Alternatively, a deep cup-shaped part, as shown in Figure 13.12, can be significantly affected by core shift in molding, and so individual stack alignment is generally used. If the part design is such that round tapers cannot be used to align the stack, then a wedge lock design is preferred such as shown in Figure 13.13 and Figure 13.15. The advantage of a wedge lock type alignment system is that adjustable thickness spacers can be used behind the hardened wedges to correct any core shift encountered when molding.

If the sidewall tolerances are large, which is often the case with heavy-walled products, any possible misalignment between cavity and core is usually insignificant, and alignment with leader pins is perfectly viable.

13.1 Leader Pin and Bushing Alignment

Leader pin and bushing alignment between mold halves is shown in Figure 13.2–Figure 13.4. This alignment method is quite accurate, and is used when there is really no need for significant alignment within the mold. The main reason to have leader pins in these cases is to protect the projecting cores from physical damage, when handling the mold. In theory, only two leader pins are ever required to ensure proper alignment. The fact that many molds use four pins is mainly to protect the cores during mold servicing. Leader pins and bushings are manufactured by mold hardware suppliers in various styles and materials to suit the specific needs of the mold (see Figure 13.5). Designers should attempt to always use standard hardware for these items, rather than creating custom components in order to facilitate easy replacement and maintenance.

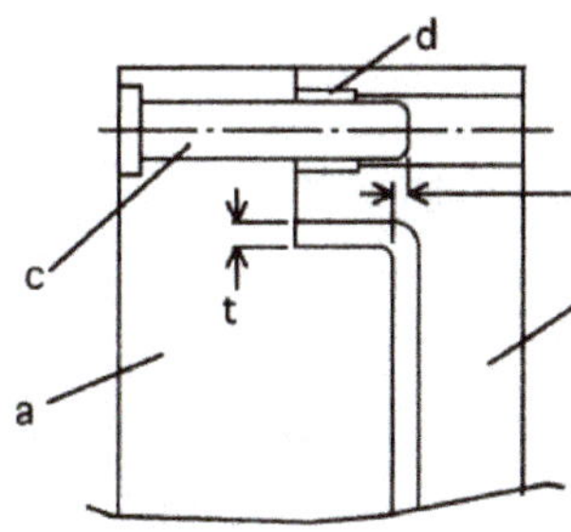

Figure 13.2 Typical mold with leader pin and bushing alignment: (a) core plate, (b) cavity plate, (c) leader pin, (d) leader pin bushing, (s) safety distance of pin above core, (t) wall thickness of plastic product at parting line

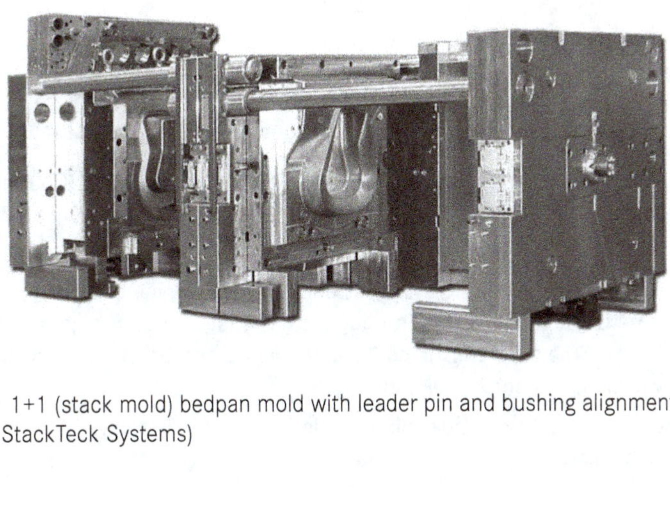

Figure 13.3 1+1 (stack mold) bedpan mold with leader pin and bushing alignment (Courtesy of StackTeck Systems)

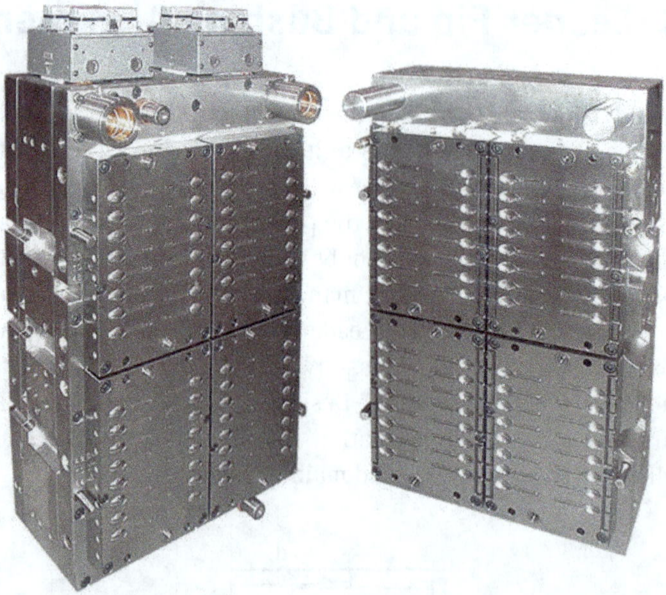

Figure 13.4 Ice cream scoop mold with flat parting line and leader pin alignment (Courtesy of StackTeck Systems)

Figure 13.5 Examples of commercially available leader pins and bushings (Courtesy of DME)

It is always better to use the largest possible leader pins on your mold to ensure robust mold robust. If in doubt, go to a larger size pin and bushing combination. The difference in cost is minimal but the added strength in the mold will provide much better support.

For example, a 24 mm leader pin provides 4× the cross-sectional area of a 12 mm leader pin.

The average clearance between leader pins and leader pin bushings (standard hardware) is about 0.03 mm (0.001").

The protection of the cores by use of leader pins applies also to all other mold alignment methods. Wherever leader pins are used, they should be placed on the same side of the mold as the cores, and be longer than the longest projection of the cores to protect them from damage (see dimension *s*, in Figure 13.2).

Example:

Consider dimension *t* in Figure 13.2, and assume a wall thickness *t* = 1.50 ± 0.05 mm (0.060 ± 0.002"). With standard commercial hardware, the leader pin is usually nominal size minus 0.025 mm (0.001"), and the bushing is nominal size plus 0.025 mm (0.001"). Therefore, with one set of pins and bushings, the maximum clearance, in the highly unlikely worst case, between one set of leader pins and bushings could be 0.05 mm (0.002") on the diameter, so the centers would be misaligned only half that amount. By having at least two (but usually four) sets, the total clearance between the pins in all the bushings would be even less. In the worst case, the possible play and misalignment would be well within the tolerance limits specified in this example, and therefore acceptable.

13.2 Taper or Straight Interlock Alignment between Plates

If the part has some critical dimensions, has a significant cup shape, or if the cavitation is relatively large, then the designer should consider some additional alignment between the mold plates. In this case, adding some standard catalog taper pins or wedges (see Figure 13.6 and Figure 13.10) would be a satisfactory solution to ensure alignment. These alignment features should be located in four positions on the mold, as shown in Figure 13.7, to ensure they will be able to align the mold halves without premature wear. Figure 13.8 is a good example of a mold with both leader pins as well as opposing taper locks on the mold plates. In general, the largest taper pin or wedge lock should be used to provide the most robust alignment.

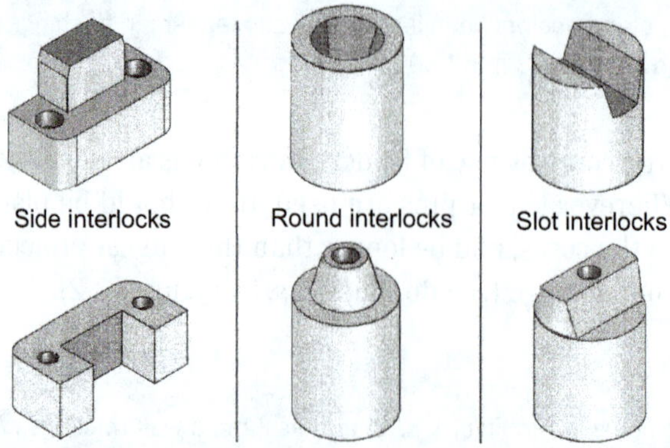

Figure 13.6 Types of taper and straight interlocks used on molds (Courtesy of Manual do Projectista)

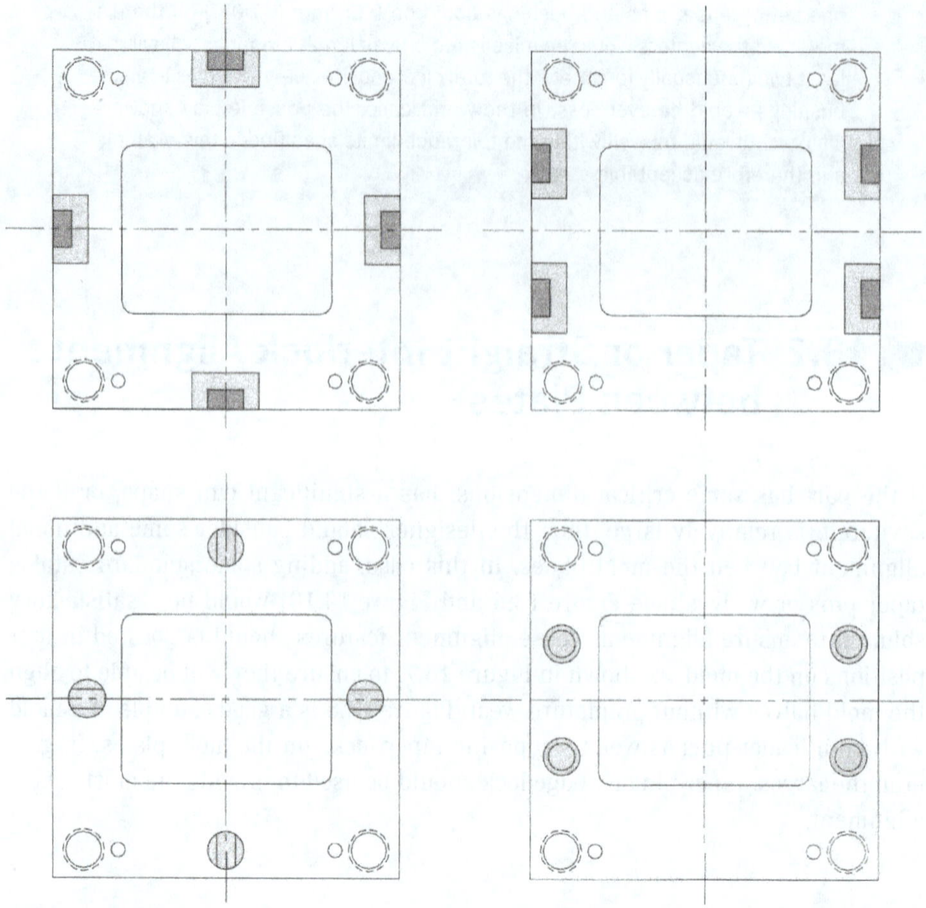

Figure 13.7 Suggested placement of interlocks on mold plate faces (Courtesy of Manual do Projectista)

It is always better to use a lower number of larger-size taper interlocks on the mold to get the desired alignment. For example, it is preferred to use two 25 mm (1") interlocks rather than four 12.7 mm (½") interlocks, as the two larger interlocks provide more than 30% more surface area for alignment than the four smaller ones do.

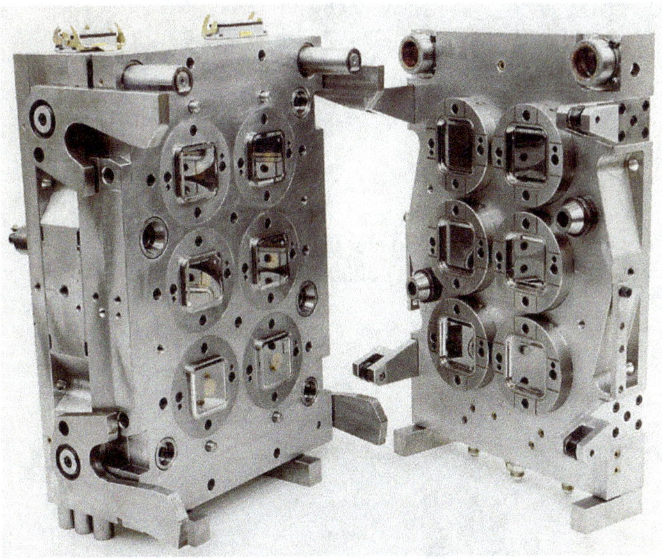

Figure 13.8 This lid mold has leader pins and round taper lock alignment, while the modules have no alignment mechanism. This works well for shallow parts (Courtesy of Husky Injection Molding Systems Ltd.)

The preloads between a taper pin and bushing should be in the range 0.013–0.025 mm (0.0005–0.001").

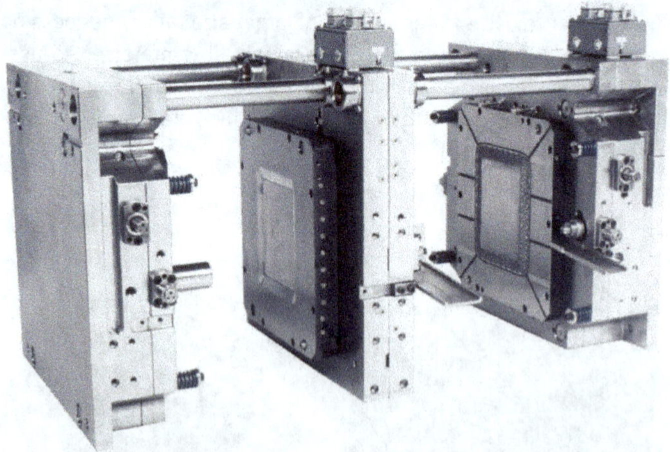

Figure 13.9 This 1+1 cavity mold uses leader pin alignment and round taper locks to keep the mismatch to an acceptable level (Courtesy of StackTeck Systems)

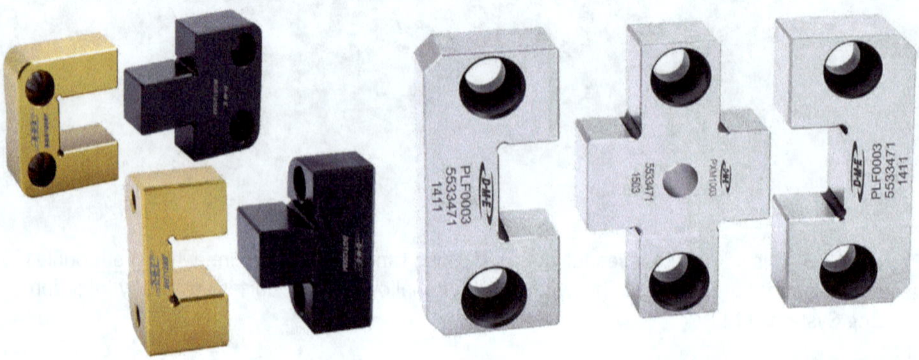

Figure 13.10 Examples of side interlocks for plate and mold alignment (Courtesy of DME)

■ 13.3 Taper Lock between Each Cavity and Core

If the walls are thinner than 1 mm and the tolerances are tighter, alignment with leader pins may not be sufficient to ensure that the variations fall between the allowable limits. In these cases, individual taper locks are required.

Round tapers are relatively easy to manufacture, but require high accuracy to ensure concentricity with the center of the cavity and to ensure that the proper preload is achieved. The basic requirement of any taper fit is the preload between the matching faces. Without any preload between the tapers, the alignment is useless. The biggest problem with round taper fits is that the tapers wear with time, and need to be reset or replaced.

Wedge locks are more expensive than round tapers, and are used mainly for molds where accuracy is very important and where the higher cost can be easily justified over long periods of use. It consists of two or four opposing pairs of matching hardened wedges, at 90°. The advantage is that the wedges are easily accessible and can be adjusted (by shimming and/or grinding) or replaced with little cost. The main disadvantage of wedge locks is that more space is required to accommodate them than for round tapes, thereby making the mold larger and more expensive.

In both round and wedge lock style alignment systems, the length of the taper should be sufficient to provide the required alignment but not so large as to add unnecessary cost and shut height to the mold. A good rule of thumb is to have the taper height roughly 20% of the height of the part, as shown in Figure 13.11. The angle of the taper must not be less than 7° (locking) and should ideally be 15°. Table 13.2 is also a good reference if the 20% rule of thumb cannot be achieved.

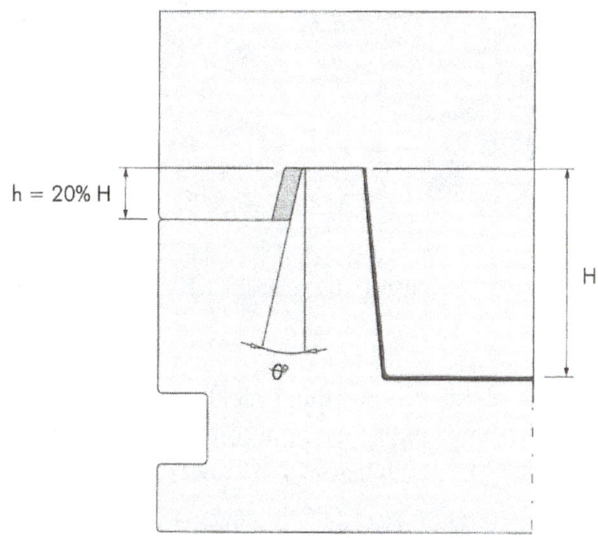

Figure 13.11 Size of taper for good alignment strength (Courtesy of Manual do Projectista)

Table 13.2 Recommended Alignment Taper Lengths

Part shape	Length
Shallow part (length or diameter bigger than height)	10-15 mm
Deep-draw part (length or diameter smaller than height)	20-30 mm

Figure 13.12 shows an example of a stack for a container with circular alignment tapers (A). Note that the cavity (B) is set into the cavity retainer plate (not shown), while the core (C) is mounted on top of the core backing plate (not shown) to ensure proper alignment. Note the absence of a stripper ring, as this product is air-ejected from the core, making for a much simpler mold. For best cooling efficiency, there is a beryllium-copper alloy (BeCu) core cap (D), and a BeCu gate insert (E) in the cavity bottom plate (F). Note the intricate venting channels in both cavity and core to ensure fast filling of the cavities. This type of stack usually has cycle times of 6.0 s or less.

Figure 13.12 Modular mold for container (round taper interlock) (Courtesy of Husky Injection Molding Systems Ltd.)

Figure 13.13 shows an eight-cavity modular mold for rectangular containers. The cavities are set into the hot runner plate; the cores are mounted with float on the core backing plate. Each core is aligned with its cavity with wedges; the containers are ejected by air.

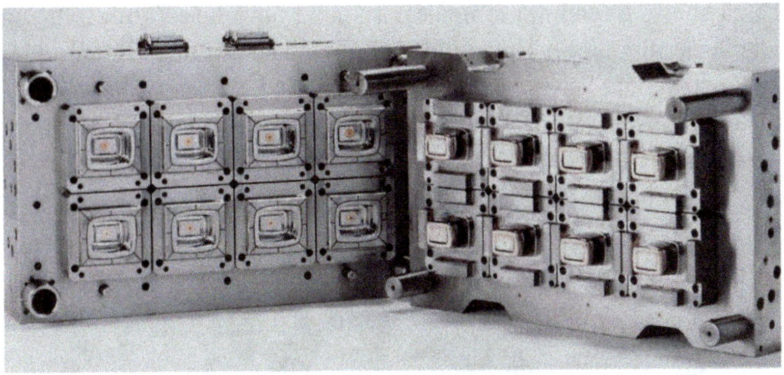

Figure 13.13 Eight-cavity mold for rectangular containers (Courtesy of MULLER Technology Group)

Note that when wedges or any alignment device is used to align cores and cavities they should be backed-up by the core or cavity, or spigotted into the plate, as shown in Figure 13.14.

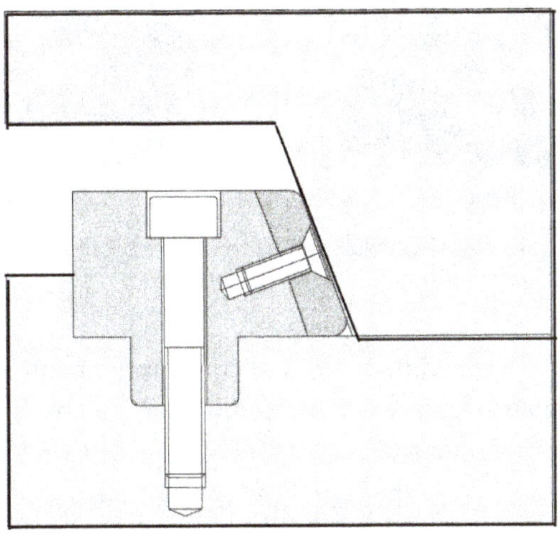

Figure 13.14 Wedge design for core cavity taper locks (shown with back up and spigot) (Courtesy of Manual do Projectista)

Figure 13.15 shows a modular single-cavity mold for a large, thin-wall container, with wedge lock alignment. This mold also operates with air ejection only, and is simple in construction. For the most effective cooling, there is a BeCu core cap (A) and a BeCu gate insert (B). The jaws (C) for the wedge lock are easily adjustable. The two leader pins (D) are basically only for mold handling and for protection of the cores, and fit only loosely in the leader pin bushings (E).

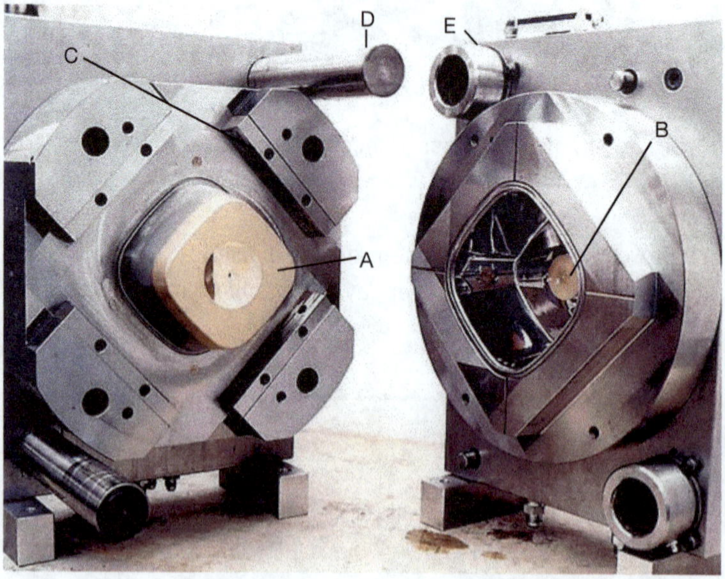

Figure 13.15 Modular single-cavity mold for large thin-walled container (square lock alignment) (Courtesy of Husky Injection Molding Systems Ltd.)

Examples of molds with individual core and cavity lock alignment can be found in Chapter 18. These molds can be used as references for the designer to use for their mold design.

13.3.1 Backing up a Taper

When a locking taper is used between the cavity and core, it is subject to very high stresses due to injection pressure and clamping. In order for the taper to work correctly when aligning the core to the cavity, it must be very rigid. The taper should therefore be backed-up by being buried in a plate or healed into a plate, as shown in Figure 13.16. This additional back-up ensures that the locking taper will not lose its preload during injection and clamp-up.

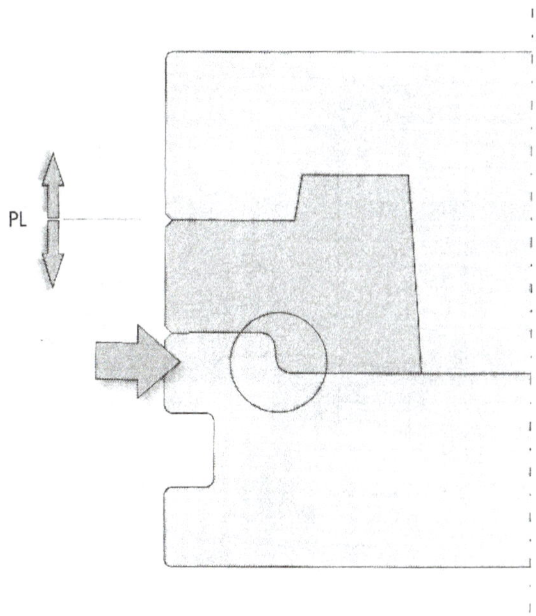

Figure 13.16 Secondary back-up for tapers, providing rigidity (Courtesy of Manual do Projectista)

13.3.2 Core Lock and Cavity Lock Alignment

A core lock alignment system means that the core has the female taper and the cavity has the male taper. A cavity lock alignment system means cavity has the female taper and the core has the male taper (see Figure 13.17). Both of these air-eject stack designs work well, but the core lock system is stronger and is used for the most aggressive molding applications. Sometimes a wear ring is added to the core of a cavity lock system if replaceability of worn components is important. In both cases, the cavity is fixed and the core is allowed to float slightly to align to the cavity.

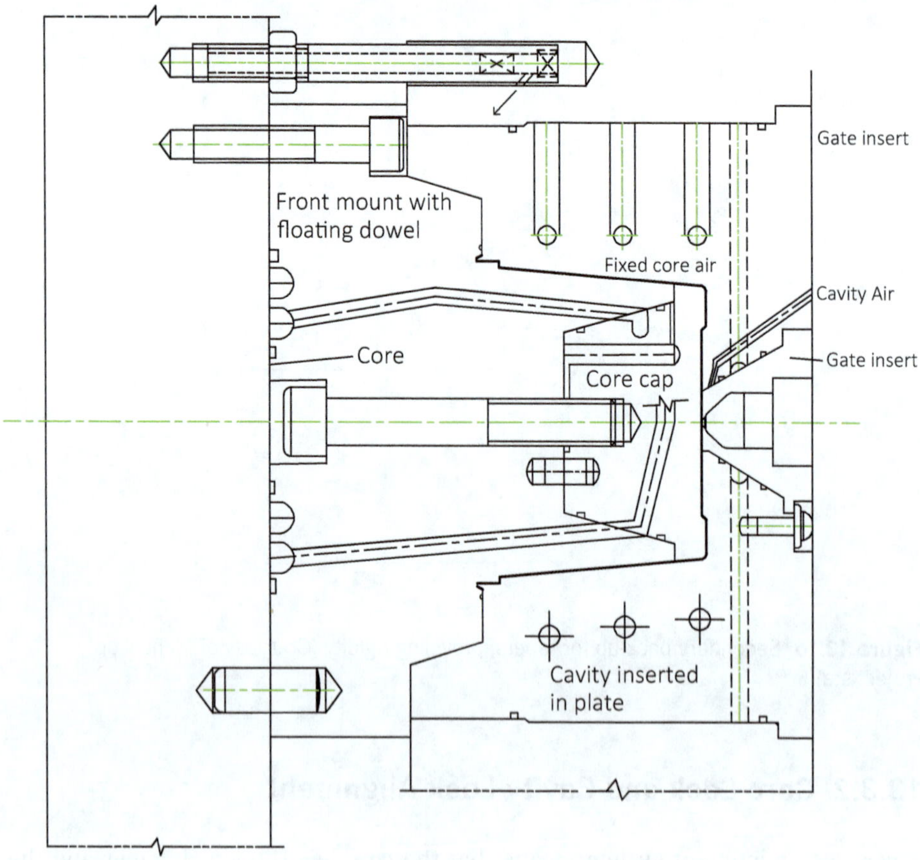

Figure 13.17 Typical cavity lock stack design

In the core lock illustration shown in Figure 13.18, the core is pressed into the lock ring from behind to make the two pieces act as one. In some designs, the core and the locking ring shown in Figure 13.18 are made from one piece of steel. However, this is expensive, difficult to make, and nearly impossible to adjust in case the taper fit is lost. Recommended preloads between the alignment members of these stacks are covered in Section 13.4.

If a stripper ring is needed to eject the part, then a core lock configuration such as shown in Figure 13.19 can be used. In this case, the (floating) core is aligned to the cavity via the lock ring taper (a). The stripper ring is also floating, and seats on the core taper (c). The surfaces (b) and (d) are fully loaded when the mold is clamped up, and both the lock ring and the stripper ring are stretched to ensure good fit.

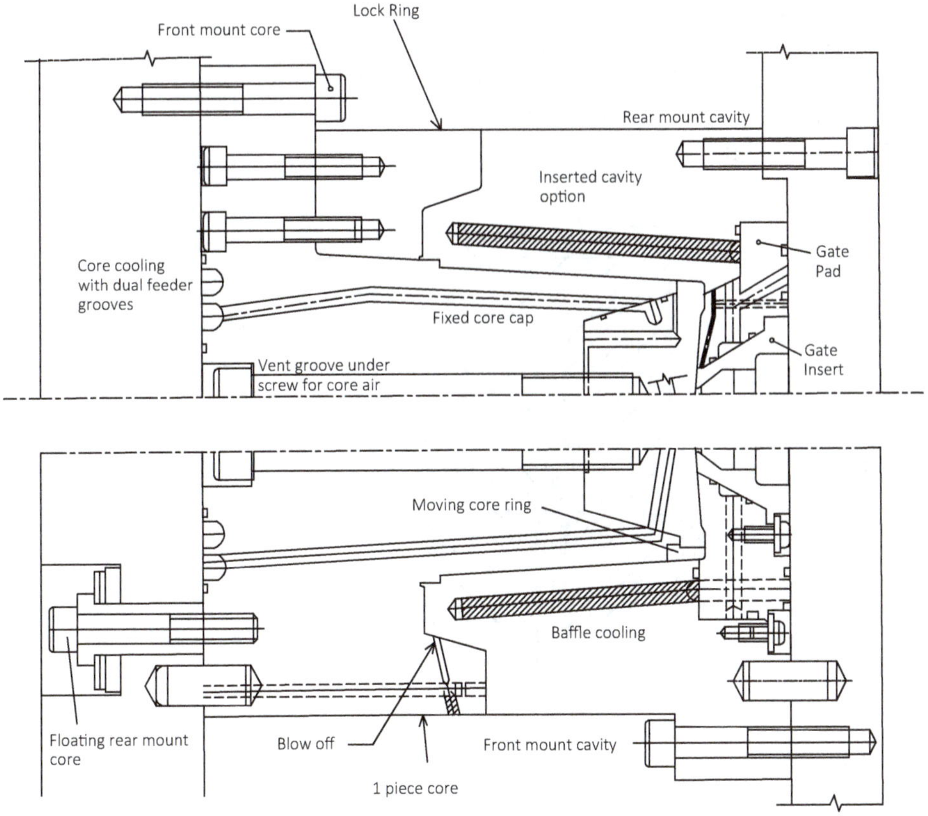

Figure 13.18 Typical core lock stack design with options

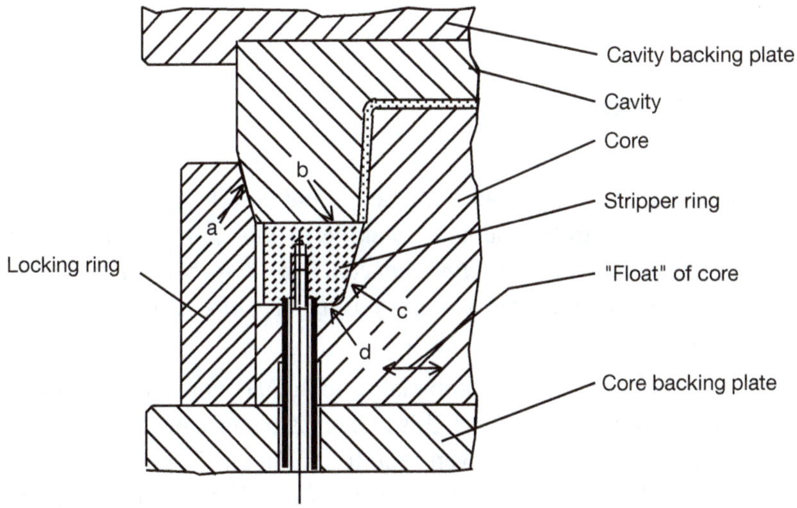

Figure 13.19 Core lock alignment with a stripper ring ejector

In the arrangement in Figure 13.20, the stripper ring has two tapers: (a) seats in the cavity, and (c) on the core. Both core and stripper must be able to float sideways to permit alignment. This design is quite common where the stripper ring must split to release the product (e.g., a thread, as with preforms) but must be held with preload during injection.

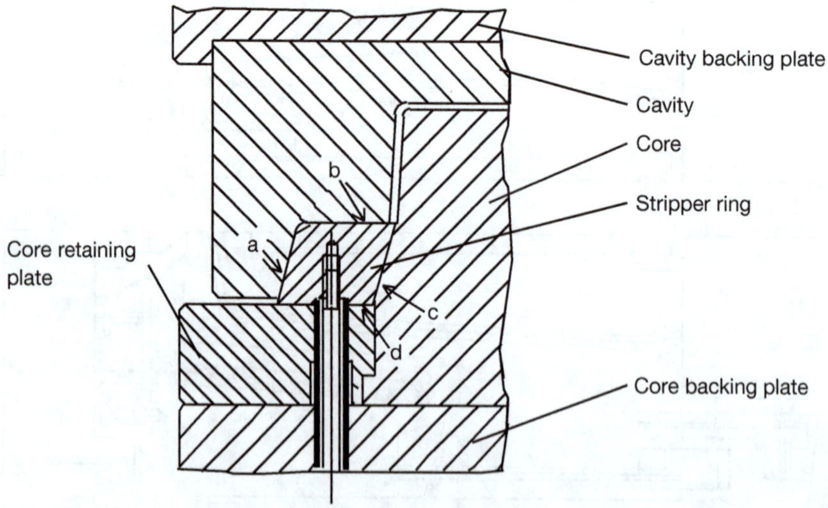

Figure 13.20 Cavity lock alignment with stripper ring

Detailed examples of air eject core and cavity lock designs, as well as a core lock stripper ring tumbler mold, can be found in Chapter 18.

■ 13.4 Preloads

In cases where alignment between each cavity and core is needed, preload is required to make the tapers functional. *Preload, even though it is strictly speaking a force, is usually defined as the additional distance (x) that the male part of a taper needs to move in order to be seated on the opposing face, after initial contact with the female part of the taper (see Figure 13.21).*

In Figure 13.21 (left) the tapers in both male and female members match perfectly. Because of manufacturing tolerances, this is impossible to achieve except, perhaps, by individual fitting of parts, and even then it is difficult. To solve the problem of providing proper alignment, the matching parts are dimensioned such that the male member is slightly larger than the female member, and the female member will be slightly expanded from the moment the mold halves touch, until the mold is fully clamped. The amount that the pieces stay apart before final clamping (x) is called preload in Figure 13.21 (center). This amount x is very small, and depends on the length of the taper and on its angle. It must be greater than zero. In Figure 13.21 (right), the female member is larger than the male member. This taper lock is useless because the tapers do not touch due to the clearance (c), and no force is generated to pull the mold halves into alignment.

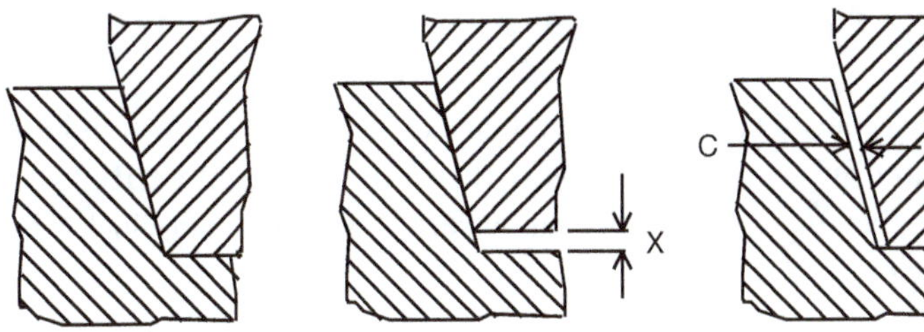

Figure 13.21 Taper (or wedge) lock: (Left) Ideal condition (difficult or impossible to achieve). (Center) Correct application, with a preload x. (Right) Useless taper.

Preload is usually indicated either with dimension lines or with a note and a leader pointing to the taper (see Figure 13.22).

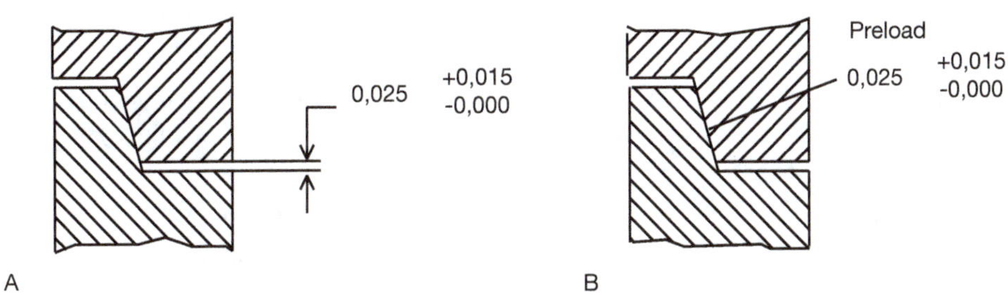

Figure 13.22 Expressing preload on an engineering drawing

> Common taper angles are 7, 10, and 15°, and the length of the taper should be substantial enough to align the core to the cavity.

If the angle is self-locking (< 7°), there will be a resistance to closing, as with other tapers. However, the increase in stress will rise quickly as the preload closes. With a self-locking angle will also be a resistance to *opening*. The smaller the locking angle α, the more force is required to open the taper lock. Too small a taper angle may cause locking and separation difficulty because of friction in the tapers. Too large a taper requires too much force to close. Obviously, to move the tapers for the preload distance x, until they seat properly, means that the matching, female taper will have to be spread. This requires considerable force. When considering the clamp force of the machine, this must be considered and the forces calculated, especially with multi-cavity molds in which every stack is aligned with taper locks. If too much force is required for closing the mold, there may not be enough clamp force left to hold the mold closed during injection.

> Generally, taper lengths vary between 10 and 40 mm depending on the space available and the aggressiveness of the application.

There are three types of core-to-cavity taper lock designs as shown in Figure 13.17, Figure 13.18, and in Sections 18.13 and 18.15. In general, a cavity lock design is the least expensive to produce, followed by core lock, with a core lock with stripper ring as the most expensive.

General preloads (dimension x in Figure 13.21) used in taper or wedge lock applications are as follows:

Table 13.3 Recommended Preloads by Lock Type

Lock type	Preload between:	Preload *x*
Core lock	Lock ring and core	0.103 mm (0.004")
(Figure 13.18)	Lock ring and cavity (15° taper angle)	0.013 mm (0.0005")
(Figure 13.17)	Lock ring and cavity (10° taper angle)	0.019 mm (0.0007")
Cavity lock (Figure 13.17)	Core and cavity (15° taper angle)	0.013 mm (0.0005")
(Figure 13.17)	Core and cavity (10° taper angle)	0.019 mm (0.0007")
Stripper lock (Figure 13.19) 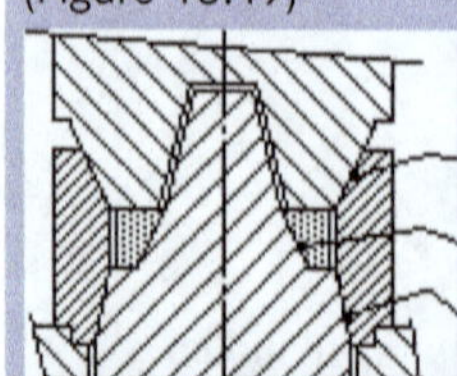	Stripper ring and core	NO PRELOAD 0 to -0.013 mm (0.000" to -0.0005")
General taper lock on mold plates 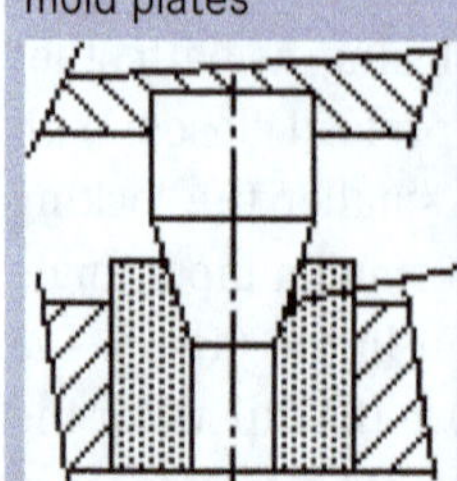	Male and female	0.013-0.025 mm (0.0005-0.010")

Forces, in the form of stored potential energy, are generated by the following in a mold:

1. *Screws.* As a screw is tightened, it is preloaded to a recommended torque, which generates force as it stretches (see Chapter 16 on fasteners). The resulting force (or holding power) is used to keep parts firmly together in the mold, with no other forces being exerted.
2. *Clamp force.* As the mold is clamped up, force is generated by eliminating the preload dimension on the tapers, wedges, stripper rings, and so on.
3. *Thermal expansion* in the hot runner due to the heating of the hot elements in the melt distribution system. A preload is used to maintain a seal between elements of the melt distribution system.

Taper fits are used for other areas as well, such as the accurate alignment of gate inserts in gate pads, or the alignment of gate pads in cavities. In this latter application, a preloaded taper is especially important to prevent flashing after the gate pad and the cavity separate a short distance S at each cycle, as shown in Figure 13.23 for moving cavities.

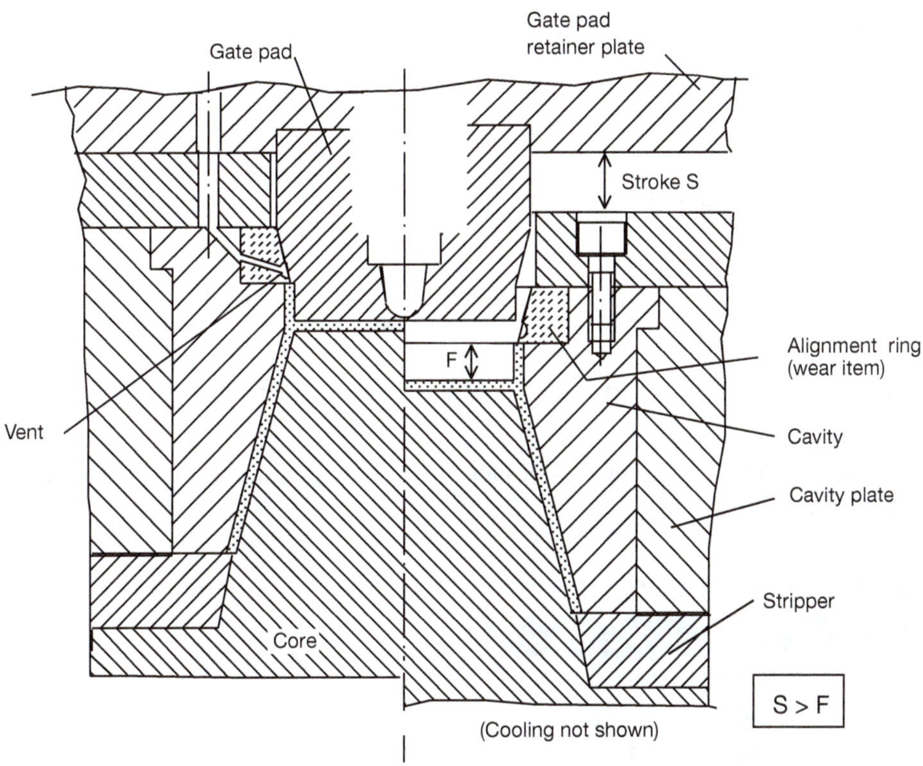

Figure 13.23 Tapers on the gate pad and stripper ring prevent flashing and maintain alignment

If the tapers (or wedges) are shiny all around, they are working well. If they are rusty, or just dirty, they do not work, and the mold probably depends on the machine tie bars and tie bar bushings for alignment, or on the mold leader pins and bushings.

Working tapers are subject to severe wear and they will eventually require reworking, even though they are manufactured from hardened steels. Tapers with heavy preloads will wear more rapidly than those with lower preloads, since the abrasive effect increases with the force with which the surfaces contact each other.

Taper fits can be fairly easily repaired (reset) if they are worn and have lost their preload. Such repair usually requires grinding on one of the faces which are at right angles to the taper axis, and shimming or regrinding a face on the mating part, to re-establish the original preload.

13.4.1 Restoring Preload by Grinding of the Tapers

When grinding any surface, it is not practical to make cuts smaller than 0.0025 mm (0.0001"). When removing this minimum amount of a taper, it will change the dimension (preload) X by the amount ΔX (Figure 13.24).

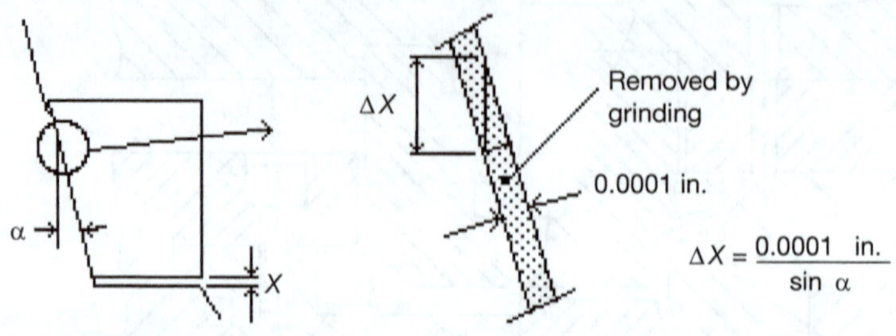

Figure 13.24 Impact of grinding tapers on preload (0.001" removed in this case)

For example, Table 13.4 shows ΔX at four different angles, α, for every 0.0025 mm (0.0001") removed.

Table 13.4 Impact of Grinding 0.0025 mm (0.0001") on Preloads

Angle α	ΔX
3°	0.049 mm (0.0019")
5°	0.029 mm (0.0011")
10°	0.015 mm (0.0006")
15°	0.010 mm (0.0004")

It is, therefore, not practical to ask for tolerances that are smaller than the effect of grinding would permit. For this reason, preloads of rings with small taper angles are more difficult to produce. Similarly, in a multi-cavity mold, manufacturing variations during the final grinding steps will mean that so-called "identical" parts will have different taper seats, and thereby create different preloads.

13.4.2 Preload Stresses

The forces generated by compressing preloads creates hoop and contact stresses in the mold components. The designer must ensure that the mold components can withstand these forces on a cyclic basis for the expected life of the tool. This life should be assumed to be infinite for calculation purposes. Chapter 10 covers the forces on the inserts, and the designer must ensure that their design is within acceptable limits for the materials being used.

13.4.3 Mounting of Stacks to Mold Plates

With the alignment of the stack established, the next step is to design how these stacks will be mounted onto the plates in a fashion that will allow them to hold their position, while remaining flexible enough to adjust to heat expansion and misalignment.

Regardless of the robustness of the alignment and mounting elements used in mold design, *it is always difficult to fight heat expansion. It is therefore critical to ensure that both mold halves are at the same temperature*, particularly the mold plates carrying the alignment elements. A temperature difference of only 20 °C between two plates, on a distance of 400 mm, causes an expansion of 0.091 mm (0.004 in). This can result in a serious misalignment. If the mold depends on leader pins for alignment, they will deflect and/or wear rapidly, as will the bushings. If taper elements are used, they too will wear out rapidly and lose their usefulness.

Even with the mold halves at the same temperature, there may still be some minor misalignment due to manufacturing tolerances. This must be accounted for in the design, or otherwise the cores and cavities will fight each other and wear prematurely. Therefore, *molds are usually designed with the cavities rigidly fixed in position and the cores floating slightly to align to the cavities.* There are several reasons for this. First, the stationary side contains the runner system, and a fixed gate location prevents any leakage from the runner system. Secondly, the cavity side generally has fewer ejection functions, and as a result it can be fixed more easily. Finally, the cavity is subject to expansion stresses during injection, and having the cavity backed-up by rigid mounting and the cavity plate assists in keeping the mold aligned.

Since the cavities are rigidly fixed in location, the cores (with their taper alignment) are designed to float 0.15 mm (0.005") on their mounting surface (plate) to find the matching position taper in the cavity. There are two methods commonly used to achieve this:

- Large dowels can be used to restrict the floating motion. Each core or cavity should have two dowels, one offset to prevent incorrect mounting of the core or cavity. The cores are screw-mounted to the backing plate, with the screws accessible from the parting line (front-mounted cores - see Figure 13.17). The mold is assembled completely, but these screws are at first not tightened fully, so that the mold, the first time it is closed, will push the cores into proper relation to the cavities. After the mold is opened again, the screws can be fully tightened by the setup technician to be ready for production.
- Mount the cores on bevel spring washers and a hardened sleeve to allow them to move into position (see Figure 13.18). Once in position, the spring force in the washers will keep them rigidly fixed.

If inserts such as cavities are backed-up in a plate, then the fit must be tight enough to provide back-up while loose enough to allow assembly, disassembly, and ease of manufacture. The fit for cavities in a plate is normally 0.05 ± 0.03 mm (0.002 ± 0.001").

For surface-mounting cores to plates, two large dowels (bigger than the mounting screw sizes) should be used for location. The dowel should be tight in the plate and slightly looser in the core to allow for float. Normally the hole dimension in the plate would be 0.03 mm larger than the dowel diameter + 0.03 mm tolerance (0.001" + 0.001"). The hole dimension in the core should be 0.15 mm larger than the dowel dimension + 0.03 mm tolerance (0.005" + 0.001"). This allows limited float for alignment of the core and cavity.

The figures in this chapter and examples shown in Chapter 18 provide the designer many references showing how cores and cavities can be mounted into or onto mold plates.

13.5 Over-Constrained Alignment

Typically, cavities and cores can be aligned by leader pins and bushings, or taper (or wedge) locks. Where high accuracy in alignment is required, taper (or wedge) locks are the preferred choice. However, they do not ensure that the mold halves will stay together when handling the mold; there is always the danger that the cores and cavities could be damaged if the mold halves should separate and bang together once the taper engagement is lost. It is therefore necessary to equip the mold with leader pins and bushings, in addition to the taper locks. Since the tapers will determine the final alignment, the leader pins must fit only loosely in their corresponding openings (or leader pin bushings) without actually contributing to the final alignment of cavities and cores. Quite often, even for large molds, only two such pins need to be provided, usually located at the top of the mold on the core side.

Similarly, some multi-cavity molds are built with small leader pins (usually only two) and bushings for each set of cavity and core, and are mounted on the stack plates; they ensure the final alignment of each stack. In addition, two or four large leader pins are used to align the complete mold halves, but these pins also must be "loose" in their bushings, to prevent "fighting" between the two separate sets of alignments. An exception to this rule of loose pins is when a more expensive but superior method is used: the cores are mounted so that they can move slightly (float) on their backing plates; as the mold closes, the final alignment (tapers or pins) will move each core into position relative to its cavity. In this case, the leader pins mounted in the mold shoe (on the core side) will have their regular, standard clearances.

14 Melt Distribution and Gate Design

In order to fill a plastic injection mold with molten plastic, provision must be added for bringing the plastic into these cavity spaces. This must be done with enough pressure so that the cavity spaces are filled completely before the molding injection plastic cools so much that the plastic can no longer flow (called freezing). The flow passages are:

- The sprue, from where the machine nozzle (see Figure 14.1) contacts the mold
- The runners, which distribute the plastic to the individual cavities
- The gates, which are (usually) small openings leading from the runner into the cavity space.

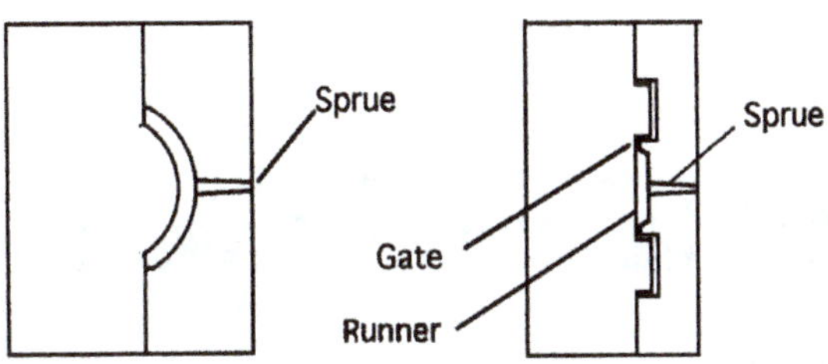

Figure 14.1 Illustration of schematic mold, showing cold sprue (left) and cold runner (right)

The left part of Figure 14.1 shows the simplest case of a single-cavity mold, with the plastic injected directly from the sprue into the cavity space. This is a frequently used method, mostly with large products. It is inexpensive, but requires the clipping or machining of the relatively large (sprue) gate. The right drawing is of a typical two-plate cold runner system, with the plastic flowing through the sprue, the runner and the gate then entering the cavity space through relatively small gates. The runner will easily break off of the part after ejection. Instead of the two cavities as shown here, there can be any number of cavities supplied by the cold runners.

Figure 14.2 shows the three of the basic types of runners used in injection molding. In the following sections we will discuss each of these types of runner systems in more detail. Section 14.2 describes an important variation of a cold runner called an insulated runner.

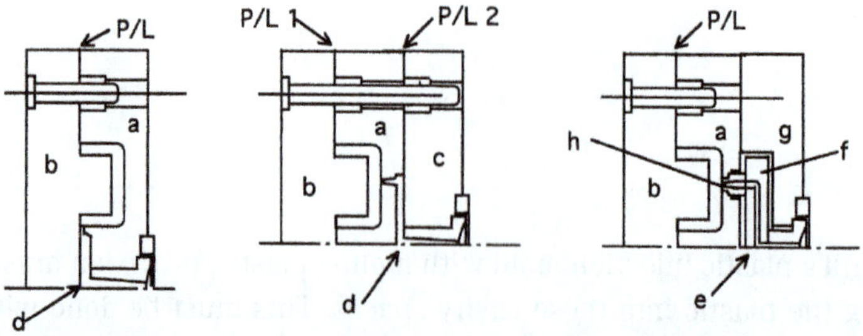

Figure 14.2 Schematic illustrations of: (Left) A two-plate mold; (Center) A three-plate mold; (Right) A hot-runner mold. (a) Cavity plate, (b) core plate, (c) third plate, (d) cold runner, (e) hot runner, (f) hot runner manifold, (g) hot runner backing plate, (h) nozzle. P/L = parting line

As can be seen in Table 14.1, there are many advantages to a hot runner system. However, there are some cases where the cost of the hot runner cannot be justified by the quantities and price of the molded product. Cold runners therefore remain effective and economic solutions for small-run, small-volume, and low-cost applications.

Table 14.1 Advantages of Different Types of Runner Systems

Attribute	Cold runner (two- or three-plate)	Insulated runner	Hot runner
Cycle time	Long	Moderate	Short
As molded gate cosmetics	Poor	Good	Very good
Gate placement options	Limited	Limited	Many
Amount of scrap/regrinding	High	Moderate	Low
Capital cost	Low	Moderate	High
Color changes	Fast	Fast	Moderately fast
Energy use	Low	Low	Moderate
Clamp tonnage required	High	Moderate	Moderate
Extruder size required	Larger	Smaller	Smaller

14.1 Cold Runners

14.1.1 Cold Runner, Single-Cavity Molds

The arrangement, as shown in Figure 14.3, is simple, effective, and good for very large products, but is also sometimes used for smaller ones. The disadvantage is that the gate is large and must be cut or even machined if appearance is important.

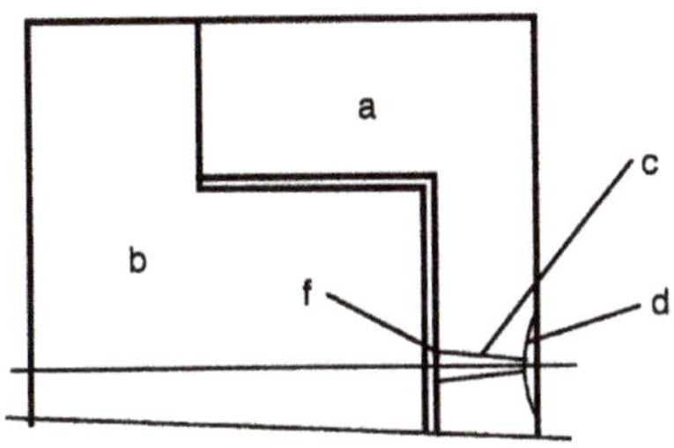

Figure 14.3
Schematic of (large) single-cavity mold: (a) cavity, (b) core, (c) sprue, (d) nozzle seat, (f) gate

14.1.2 Cold Runner, Two-Plate Molds

The mold on the left in Figure 14.2 has only one parting line as is known as a two-plate mold. An example of a two-plate mold is also shown in Figure 14.4 and in Section 18.1. An edge-gated arrangement is shown in both figures. The products and the runners stay together when ejected and must be separated after molding. There are other methods of gating, some of which are self-degating as the mold opens, but products and runners are still mixed together and require separation.

The *advantages* of the single-parting-line cold runner system are (1) simplicity and (2) low cost. Also, (3) color changes are easy, and (4) the system is not sensitive to dirt in the plastic. If a gate is blocked, it is cleaned again after the runner is ejected.

The *disadvantages* are (1) these molds usually have longer molding cycles because of the longer time required to cool the often large runners. (2) The mass of the runners must be added to the mass to be plasticized for the products, therefore, energy is wasted, first in plasticizing, then in cooling. In some cases, the mass of the runners is as great as the mass of the products, or even greater. (3) Although in many cases the runners can be reused, this requires more handling (costs), energy is needed for regrinding, and there is always a danger of contamination of the plastic. Also, losses of plastic (maybe 10% of the scrap) in the course of this process are unavoidable. Even so, two-plate molds are used in many multi-cavity molds. (4) The additional projected area adds to the clamp force needed to keep the mold closed during injection. This may require that the mold is run in a larger, more expensive machine.

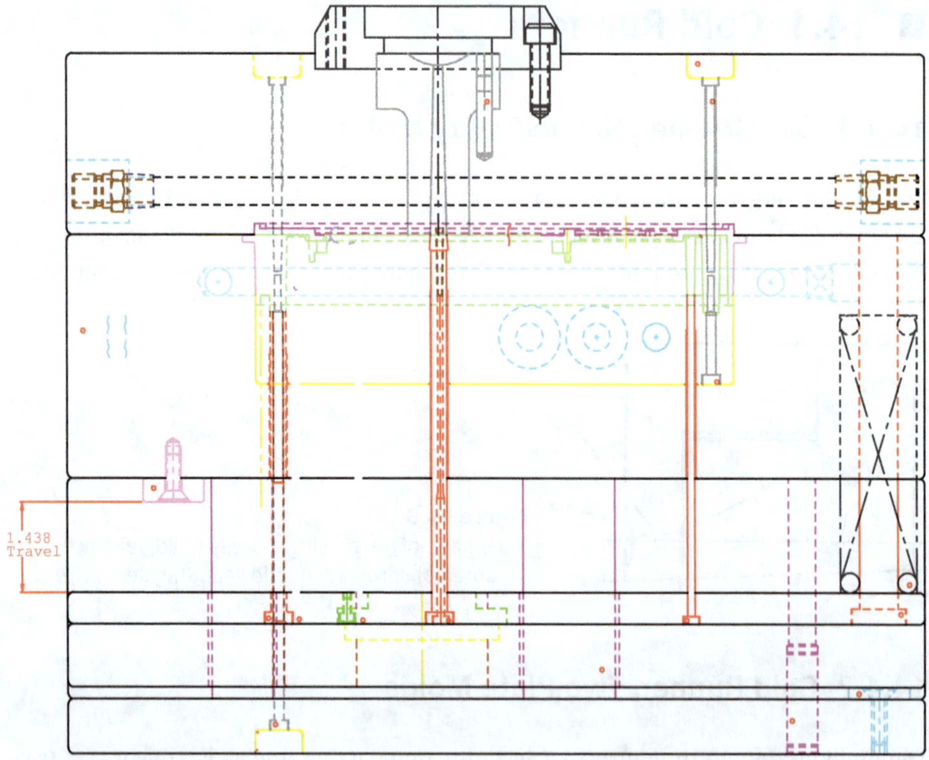

Figure 14.4 Two-plate cold runner mold (Courtesy of Don Smith)

In two-plate molds such as shown in Figure 14.2 (left) the plastic is injected either into the cavity itself (large, single-cavity mold, which will be discussed later) or into a distribution channel – the "cold-runner" system, consisting of a sprue and a runner which delivers the plastic to each cavity through (usually just one) gate. This cold runner is located within the parting plane of the mold and will be ejected together with the molded products, when both products and runner are cool enough (see Figure 14.5).

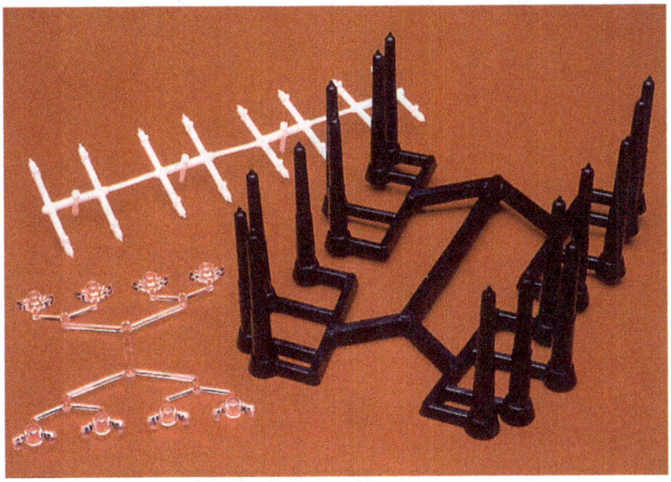

Figure 14.5 Examples of various two-plate cold runner systems

14.1.3 Cold Runner, Three-Plate Molds

Three-plate molds are also cold runner molds, but the system is inherently self-degating. The mold in the center of Figure 14.2 has *two* P/Ls. As the clamp opens, first, the cavity plate travels with the moving mold half; as soon as the cavity plate has reached a limited distance, the moving mold half (the core, with the products still on them) continues to move away from the cavity plate and the product can be ejected, after P/L 1 is opened. At the moment when the product, still on the cores, starts pulling out of the cavity, the plastic in the gates is severed as shown in Figure 14.6. Then, by a mechanism designed into the mold, P/L 2 separates and permits the ejection of the runner system in a separate plate.

Advantages: (1) the products can be center-gated, or gated anywhere on the top surface. (2) Due to the absence of runners in the P/L, the cavities can be closer together and more cavities can be placed in a mold of comparable size; see the difference between the left and center (or right) illustrations in Figure 14.2. (3) The gate vestige is usually very small, with excellent appearance. (4) Color changes are easy. (5) The system is not sensitive to dirt in the plastic. If a gate is blocked, it is clean again after the runner is ejected.

Disadvantages: (1) Three-plate molds are much more complicated and expensive (Figure 14.6 shows an example of the runner from a three-plate mold). (2) It is very difficult to guarantee 100% automatic ejection of the runner system. There are numerous systems, with links, chains, air actuators, and so on to provide the necessary motions. (3) With three-plate molds, too, the runner mass can be greater than the total mass of the products; the same comments regarding productivity apply as for two-plate molds.

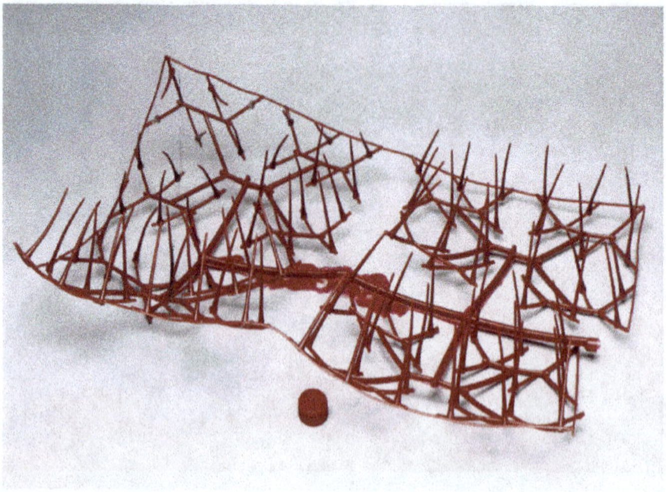

Figure 14.6 Runner for a 72-cavity, three-plate mold for a small cap

A three-plate mold is sometimes used for very small products, such as screw caps, overcaps, and so on, which should be center-gated for best molding flow, such as in containers, and where the appearance of the top surface of the molded piece is important.

For more information on runner layouts for three-plate molds, see Sections 14.3.3 and 14.1.5.

14.1.4 Cold Runner Gate Types and Configurations

14.1.4.1 General Features of a Cold Runner Gate

The following discussion highlights certain features associated with cold runner gates. Even though most high-production molds built today are hot runner molds, the designer may require the use of a cold runner or a combination "hot-to-cold" runner.

Small gates are important for the appearance of the product. However, there is a limit. If the gate is too small, it will slow the filling of the cavity or even make it impossible to fill before it freezes. Short lands are required to reduce the restriction to flow. However, there are limits because of the strength of the cavity material. If the runner ends too close to the cavity, the wall may break. Tapered lands are important to ensure that the plastic will break where it joins the product.

A sharp corner at the entry into the cavity will ensure that the gate breaks cleanly at that spot, leaving no or only a minimum projection (vestige). In most molds, a very small vestige projection is acceptable, and a short, straight section at the end of the gate of between 0.1-0.2 mm (0.004-0.008") is suggested (Figure 14.7) to provide an "almost" sharp corner, which ensures that all gates in a multi-cavity mold are of uniform cross-section.

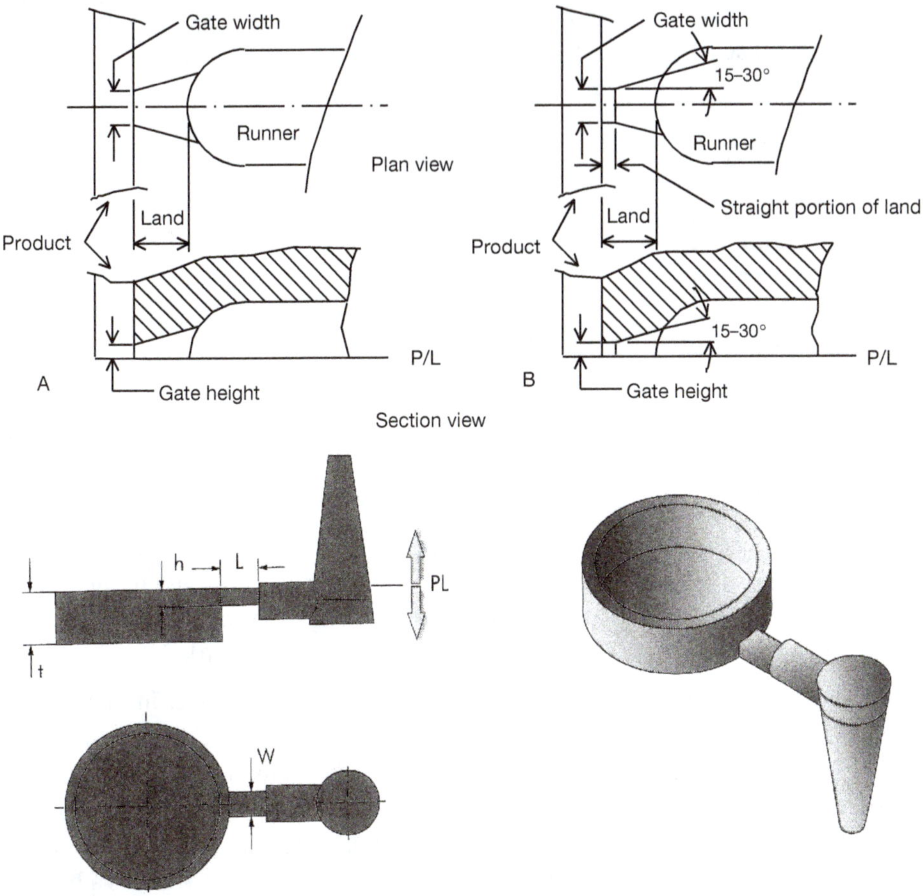

Figure 14.7 Typical cold runner gate configurations (bottom image Courtesy of Manual do Projectista)

14.1.4.2 Edge, Fan, Tab, and Diaphragm Gates

An edge gate is the simplest form of gate and is used wherever the product can or must be gated from the parting line and where self-degating is not required or practical. The sketches in Figure 14.7 illustrate the ideal, sharp, configuration and the practical gate configuration with a short, flat portion (Figure 14.7). Suggested dimensions for general applications of edge gates are outlined in Table 14.2.

Table 14.2 General Guidelines for Edge, Tab, Fan, and Diaphragm Gating

Product size	Mass (g)	Height *h* (mm)	Width *W* (mm)	Cross-section (mm^2)	Land (mm)
Very small	0–5	0.25	0.75	0.19	0.5–0.8
Small	5–40	0.50	1.50	0.75	0.5–0.8
Medium	40–200	0.75	2.25	1.69	0.8–1.5
Large	200+	1.00	3.00	3.00	1–2

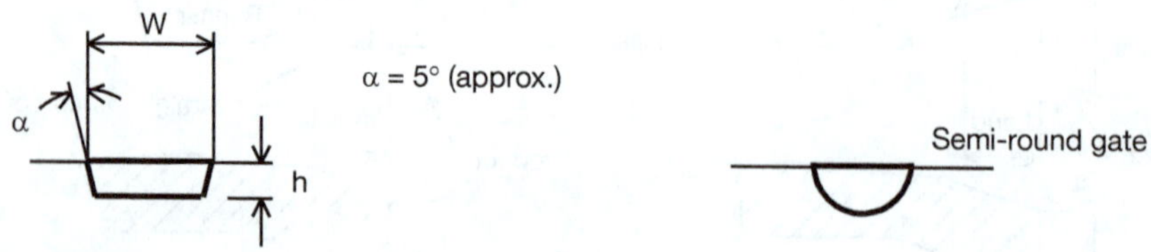

Figure 14.8 Suggested dimensions for Table 14.2 ($W = 3 \times h$)

The dimensions in Table 14.2 are only *general* guidelines. Gates can be calculated using flow analysis software. In case of doubt, it is better to start with small gates and, if required, open them up after sampling the mold.

Fan gates are a variation of the edge gate. Gate width W is much greater than $3 \times h$. The gate height h may be only 0.1 mm (0.004"), and W may be 10 mm or more. Fan gates are used where edge gating is appropriate, except that the typical gate vestige should be avoided. The fan gate vestige looks like a flash mark.

Diaphragm gates (Figure 14.9) are another variation of edge gates. In fact, a diaphragm gate is a circumferential fan gate. In its simplest form the gate is a disk, where the width W equals the length of the inside circumference. The gate height is usually about 0.1–0.15 mm (0.004–0.006"), and the shape is similar to the fan gate, with an angled passage from a disk or circular runner, ending in a short straight section. An ideal application for this type of gate is where there is a hole in the part or around the outside of a round part. A great advantage of this type of gating is that the product will remain round and not become distorted, as it would be with a single gate or with a small number of gates.

Because the plastic enters the cavity all around through a diaphragm gate, there are no weld lines (weak spots), as are unavoidable with one or several distinct edge gates. In Figures 3.24, 3.29, 3.31, and 3.32, there are as many weld lines as there are gates. Since the plastic flow divides at each gate into two streams, this creates another weak spot at each gate which must be considered.

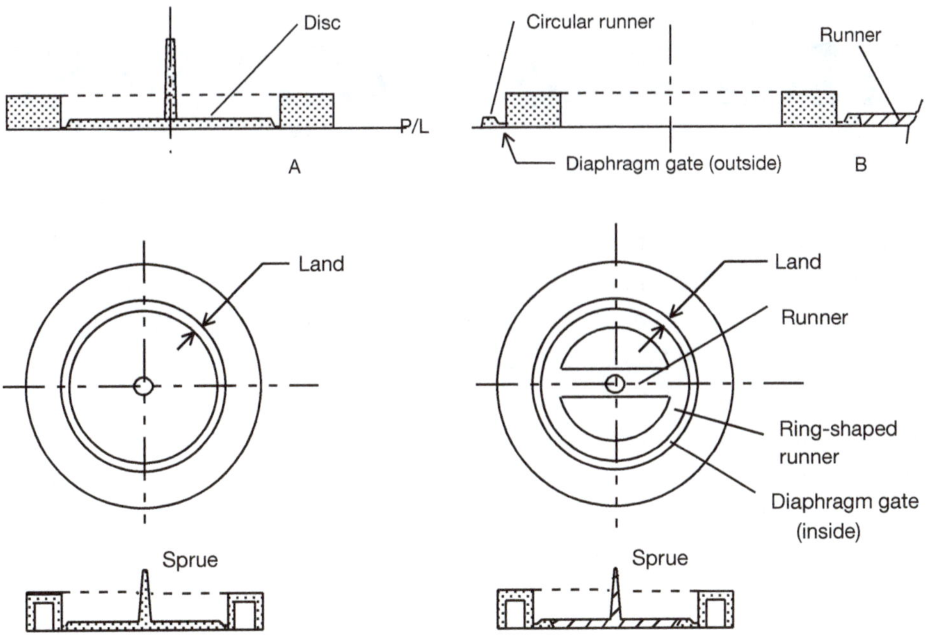

Figure 14.9 Example of diaphragm gates

14.1.4.3 Tunnel Gates

Tunnel gates (Figure 14.10) are commonly used in two-plate molds to provide automatic, in-mold separation of the product from the runner. In most molds, self-degating is desirable.

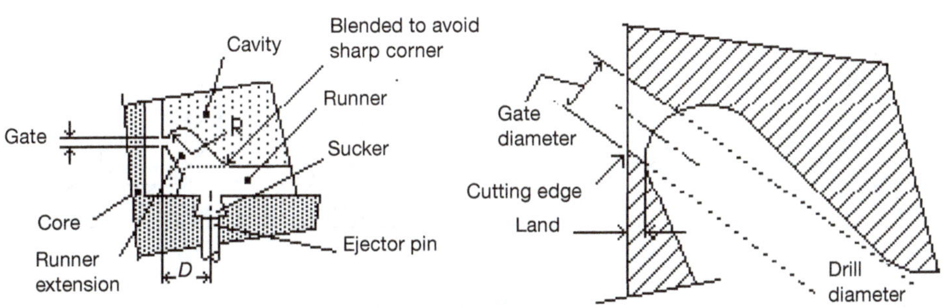

Figure 14.10 Example of a cold runner tunnel gate layout

The cross-sectional area of the tunnel gate is similar to edge gates, except that tunnel gate passages are usually circular (Table 14.3). The dimensions provided are only general guidelines. Gates can be calculated using flow analysis software if available.

Table 14.3 General Guidelines for Tunnel Gate Sizing

Product size	Mass (g)	Gate diameter (mm)	Cross-section (mm^2)	Land (mm)
Very small	0-5	0.5	0.19	0.5-0.6
Small to very small	5-20	0.62	0.31	0.5-0.6
Small	10-20	0.75	0.44	0.6-0.8
Small to medium	20-40	1.00	0.78	0.6-0.8
Medium	40-100	1.25	1.23	0.8-1
Medium to large	100-200	1.50	1.76	0.8-1
Large	200+	2.00	3.14	0.8-1

It is important to keep the tunnel gate diameter as small as practical to facilitate the shearing-off (degating) without creating an ugly vestige or even deformation of the product. The runner extension should be rounded to provide enough bulk in the plastic to prevent the gate from freezing too soon. Tapered extensions, even though easier to produce, are not recommended because they freeze off easily near the gate.

Figure 14.11 shows how the ejection sequence of a tunnel gate works. The product stays with the core as the mold opens and shears the gate at the sharp corner where the gate enters the cavity space Figure 14.11B. After the cavity has opened, the ejector can move forward and eject the part and the separated runner.

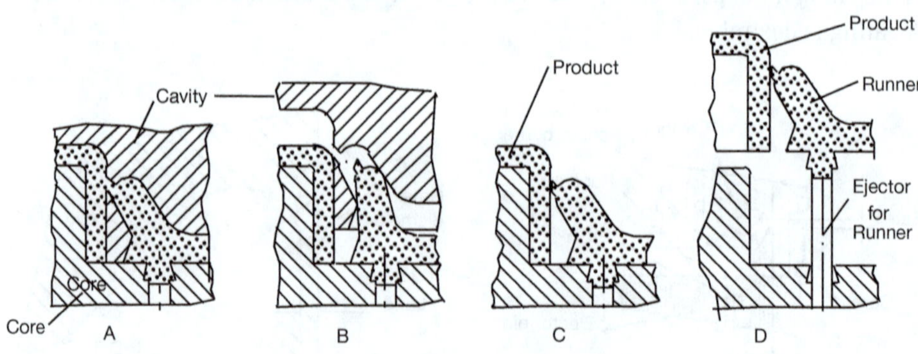

Figure 14.11 Schematic of tunnel gate ejection sequence

Note that the steel between the cavity wall and the runner extension is very thin and, therefore, easily damaged, particularly if the gate is close to the P/L. Damage is often caused by operators who try to remove a frozen, broken slug from a cylindrical gate passage; this is another reason why the land should be tapered. When molding stiff and/or reinforced, abrasive materials, the land should be longer than the values in Table 14.3 to provide more strength (the values in the table are meant primarily for commodity resins such as PE and PP).

Mold steel selection for tunnel gating may differ from that for other gates. Because the tunnel gate area is so delicate, steels selected must be tough rather than hard. H13 is a good selection when hardened up to 46–49Rc. A few more examples of tunnel gates are described and illustrated in the following paragraphs for different types of products.

Depending on the space available, the runner feeding the tunnel gate can be cut into either the core plate or the cavity side plate.

14.1.4.4 Multiple Tunnel Gating

Typical applications of multiple tunnel gating are slender products such as vials or hypodermic barrels, with the end remote from the P/L closed or open. To prevent core deflection, it may be advantageous to provide two opposing tunnel gates on the part for filling, as shown in Figure 14.12.

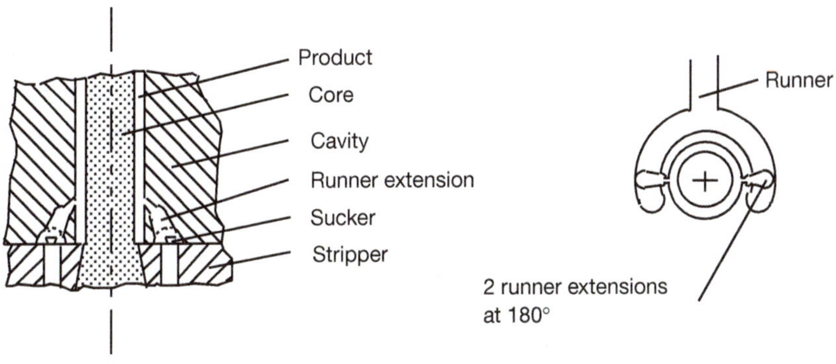

Figure 14.12 Multiple tunnel gates on one part (e.g. cylindrical barrels)

14.1.4.5 Curved or Submarine Tunnel Gating

This method could be selected where gating into the side of the product is not acceptable. The runner must extend to the distance *E* within the ejector hole (Figure 14.13) to ensure that the curved runner is fully ejected by the time the ejector is out of its hole in the core. This method is rarely used, because it is essentially restricted to flexible materials. It is also very expensive because of the need for precise core inserts, which are split to allow the machining of the curved runner extension.

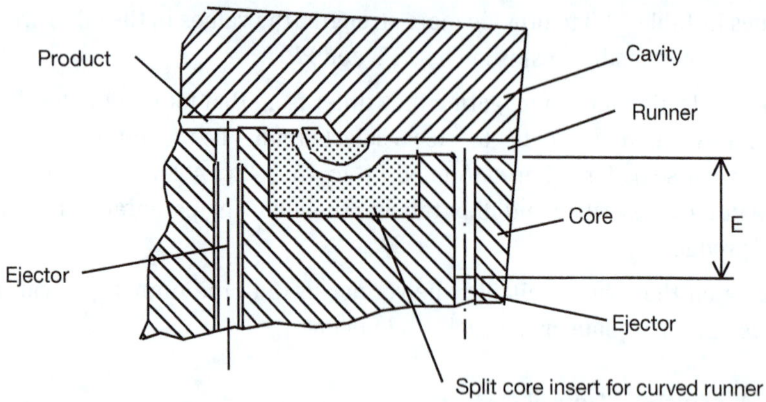

Figure 14.13 Example of submarine gating

The gate is similar in shape and size to that of a three-plate, pin-point gate, tapered, and with a sharp, well-defined break point.

14.1.4.6 Three-Plate Gates

These gates are inherently self-degating, since the gate breaks off as soon the product is pulled out of the cavity. It is important to keep the gate as small as possible to minimize the vestige and any possible projection, as would be the case with an excessively large gate (Figure 14.14).

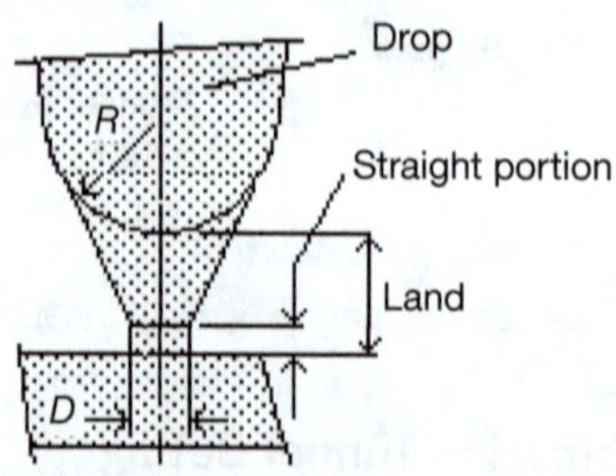

Figure 14.14 Recommended configuration of a three-plate runner gate

The three-plate gate design shown in Figure 14.14 creates a good cutoff point and is easier to make. The included taper angle should be between 60° and 90°. Recommended gate sizes are similar to sizes for tunnel gates listed in Table 14.3, or even smaller.

14.1.4.7 Optimizing the Cold Runner Gate Placement

Finding the optimal location for the gate on the part can be obvious or in many cases unclear. Chapter 3 covers suggestions for gating many common types of parts. If the part is not common, or if there are new features on the part that change the filling pattern, then a simulation of the mold filling is strongly recommended. The filling simulation will allow the designer to run iterations with different gate locations to find the optimal number and position of the gates. Additionally, the simulations will allow the designer to identify potential problems in the existing plastic part design that could be addressed at the earliest possible stage of the mold design process.

Figure 14.15 shows an example of an electrical housing design filled with one and two gates. The two-gated part results in a more even volumetric shrinkage and resulting warpage. It also identifies an opportunity to modify the part before the mold design has begun, to eliminate a potential problem area for the resulting mold. By making this slight modification, the resulting part warpage and shrinkage is even further reduced. Additional iterations could be undertaken to continue to optimize the design and the gating without significant additional cost to designing the mold.

> The gate should be positioned to inject directly (ideally perpendicularly) onto a wall to avoid jetting and other flow defects. By impinging on a wall, the flow is disrupted and it flows more evenly in all directions.

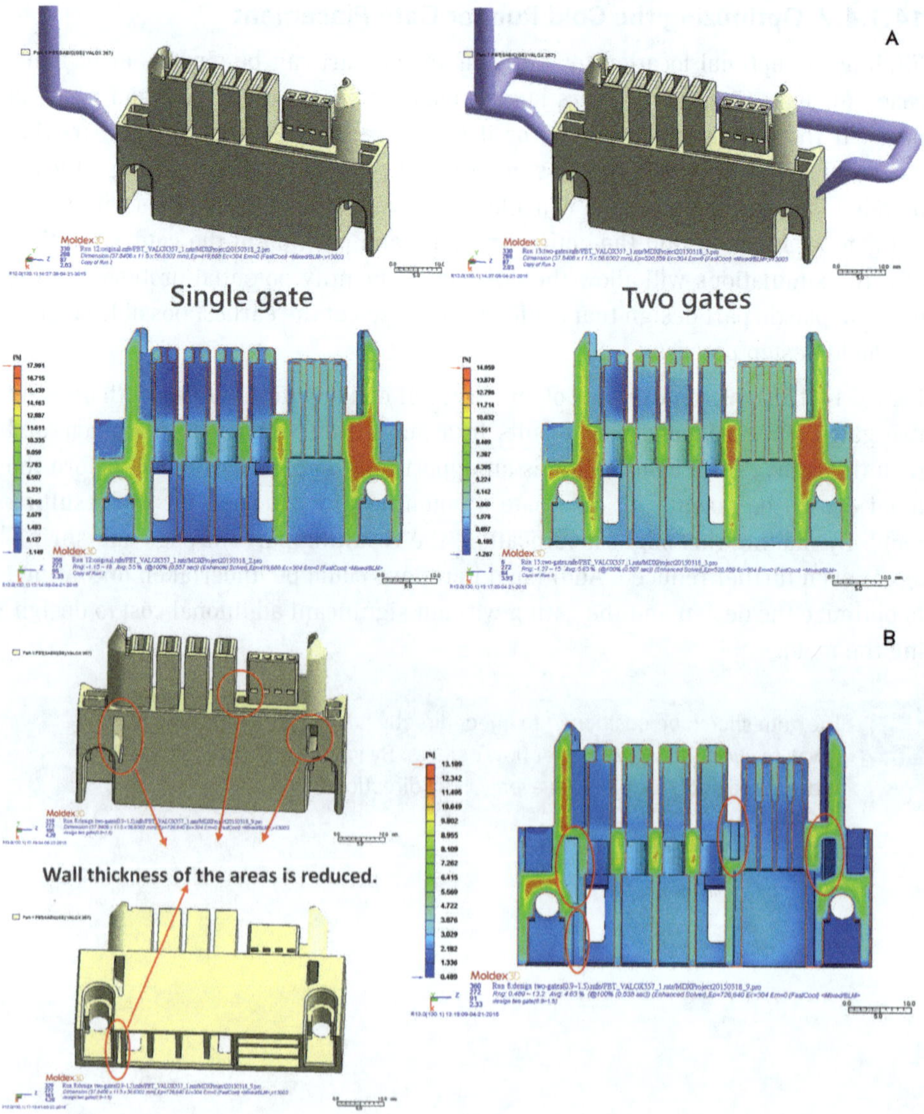

Figure 14.15 Impact of single versus multiple gating locations on an electrical housing [1]: (A) Simulation of one versus two gates; (B) Modification of the plastic part to improve the molding conditions even further (Courtesy of Moldex3D)

[1] Source: M. Wang, R. Chang, C. Hsu, *Molding Simulation: Theory and Practice,* Hanser Publishing, 2018 (Figures 9.35 and 9.37).

14.1.5 Layout and Balancing of Cold Runner Melt Channels

Chapter 8 on cavitation provides information to determine the correct number of cavities and the preferred number of cavities for optimal mold function (see Table 8.1).

> A well-designed runner system (hot or cold) should have the same rheological history in each cavity. This means the same pressure drop (distance and diameter), the same number of turns, and the same shear history.

While it might seem logical to lay out a runner system like the branches off the trunk of a tree, this causes the branches near the sprue to fill before the ones further away. The length of flow to each cavity would not be equal, and the result will be uneven filling of the cavities in the mold. Those cavities closest to the sprue will fill and cool first and those further away will fill and cool later. The result will be a mold with uneven shrinkage, packing, and many other molding defects. It is therefore absolutely critical that the runner system, hot or cold, be balanced. Keeping the tip above in mind will help immensely in determining the best way to gate the parts in a multi-cavity mold.

If a two-plate cold runner is being used, then the cold runner will be on the same parting line as the part. In this case, the runner system must be cut into the space between cavities. If a hot runner or three-plate mold is being used, then the runner layout is independent of the space taken by the cavities. This gives the designer another level of freedom in how to arrange the runners, as can be seen in Section 14.3.3 (where an example of a three-plate runner layout is described).

In order to understand how to lay out the cold (or hot) runner system, it is important to review how shear-sensitive materials flow, and the relationship between viscosity and temperature. Chapter 3 provides an introduction to non-Newtonian fluids (to which category plastics belong). It also discusses the impact of shear and temperature on viscosity. Since we know that higher temperatures reduce viscosity and that shear increases temperature due to friction between layers, we can conclude that the material in the center of the flow will flow easiest. The plastic near the cooled walls will cool and become more viscous, and so will slow down and eventually freeze on the melt channel walls, as shown in Figure 14.16. This is important, as it determines how to best lay out the cold runner feed system.

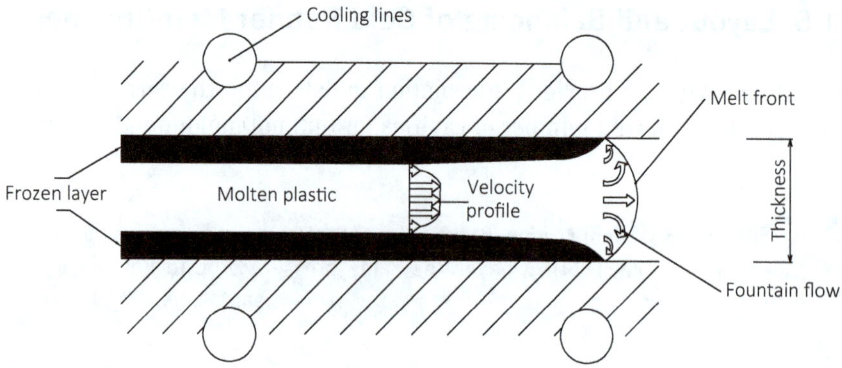

Figure 14.16 Fountain flow of plastic in cold runners and in cavities and cores

It is also important to understand what happens when a runner system branches into two or more sub-runners. At the point of intersection, the molten plastic splits into two flows. However, due to the laminar viscous flow characteristics of plastic, the material in the center of the channel flows to the outer edge of the intersection and the material nearest the melt channel walls stays on the nearer wall (Figure 14.17). This creates a new shear, velocity, and temperature profile in each of the two new secondary branches. At further intersections (tertiary branches in Figure 14.17) these profiles are further modified, causing differences in filling from cavity to cavity. For critical molding applications, this may become important. It is therefore important to keep the number of intersections to a minimum.

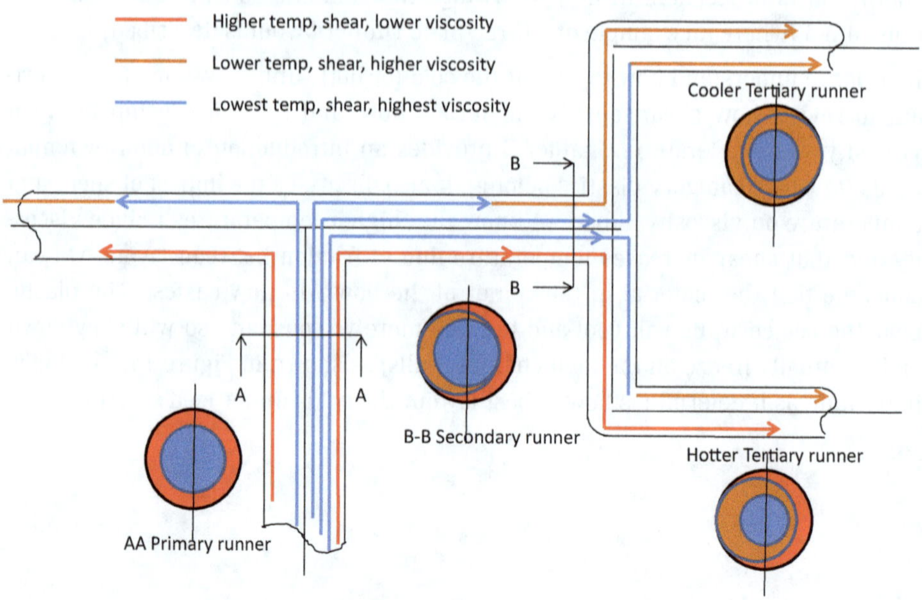

Figure 14.17 Impact of branched flow in a runner system

There are several ways to reduce the impact of this shear-induced balance issue. It is less of an issue in hot runners, as the boundary layer build-up does not occur as it does in cold runners and three-plate runners. Hot runners also have the advantage of being able to change levels, which reduces the imbalance at the intersection points. The runner sizes can be adjusted to help to offset the impact of this effect. A full 3D flow simulation would be needed to determine the required runner sizes. Finally, a simple device called the MeltFlipper® [2] can be used to reverse the shear impact.

Figure 14.18 shows three layouts of cold runners. None of the runners shown in this figure provide an equal shear history to each cavity. The middle picture is the most balanced, as the channel sizes and distances to each gate are equal. However, the non-Newtonian flow behavior of plastics described in Figure 14.17 will create some flow imbalance, and this may become important in thin-wall or critical-dimension applications. The middle layout would be preferred, with the use of a MeltFlipper®-type technology to reduce the high shear history build-up to some cavities.

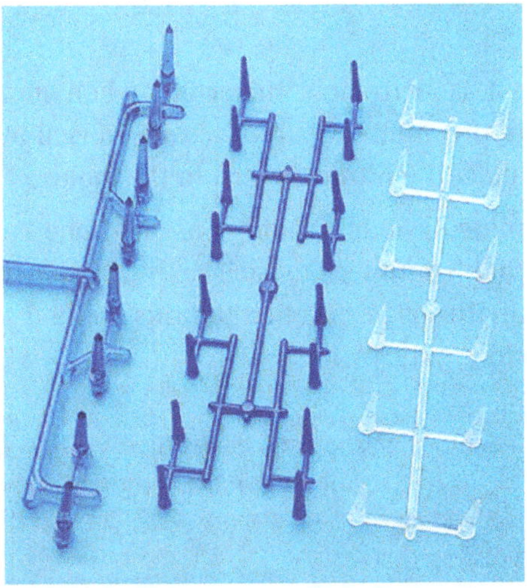

Figure 14.18 Cold runner layouts. The middle layout is the best, although all of these layouts may have some filling issues due to the characteristic non-Newtonian flow of plastics

For more details on the topic of plastic flow, the reader should review the principles of fountain flow, non-Newtonian plastic behavior, and laminar fluids in Chapter 3 and in fluid mechanics textbooks.

[2] MeltFlipper® was developed by Professor John Beaumont, and is commercially available from the INCOE Corporation.

14.1.5.1 Multi-Cavity Runner Layouts (One Gate per Part)

For molds with multiple cavities and one gate per part, the guidelines above remain the governing rule for laying out the runner system. Each cavity should receive molten plastic with exactly the same shear and temperature history.

A circular runner layout, as shown in Figure 14.19, is a good option if this layout works for the mold design. However, this is often not the case because cooling, ejection, and alignment are more challenging with these layouts.

Figure 14.19
Circular runner layout (eight cavities shown)

Balanced runner layout is extremely important when molding thin-wall parts (< 0.8 mm thick) where the differences in viscosity caused by differences in shear and temperature will affect the filling pattern in the mold.

One- to four-cavity runner layouts should consist of one center sprue and then equal-diameter and equal-length legs to each cavity, as shown in Figure 14.20. If a circular layout of the cavities is preferred, then this method can be used to gate all the cavities, as shown in Figure 14.19.

> In order to keep the forces balanced on the platens and thus prevent molding problems such as flashing, vent hobbing, shorting and other issues, *the cavities must be spaced equally and symmetrically* around the center of the stationary platen.

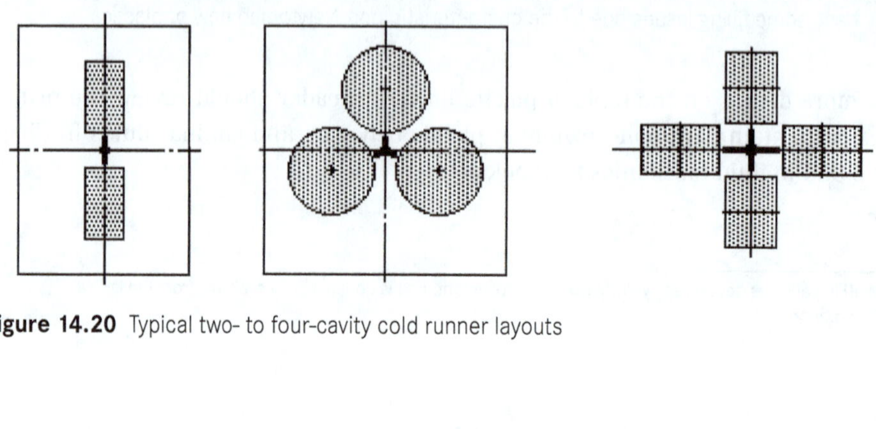

Figure 14.20 Typical two- to four-cavity cold runner layouts

Above three cavities, the runner layout may require branches and sub-branches. At the point where a branch divides into a sub-branch, the runner diameter should be decreased to keep the flow and shear rate approximately the same. Common branching options for four, six, and eight cavities are shown in Figure 14.21. The branches can be created with a T (branch into two), a Y (branch into three) or an X (branch into four). It is better to reduce the number of branching intersections to a minimum, as each branch builds up a condition of uneven shear history due to the splitting and fountain flow effect shown in Figure 14.16. Methods have been developed to lessen these effects. The above-mentioned MeltFlipper® technology is a good example of how the effects of shear build-up can be overcome. In hot runners this effect can be overcome by using multi-level hot runner systems (see Section 14.3.3).

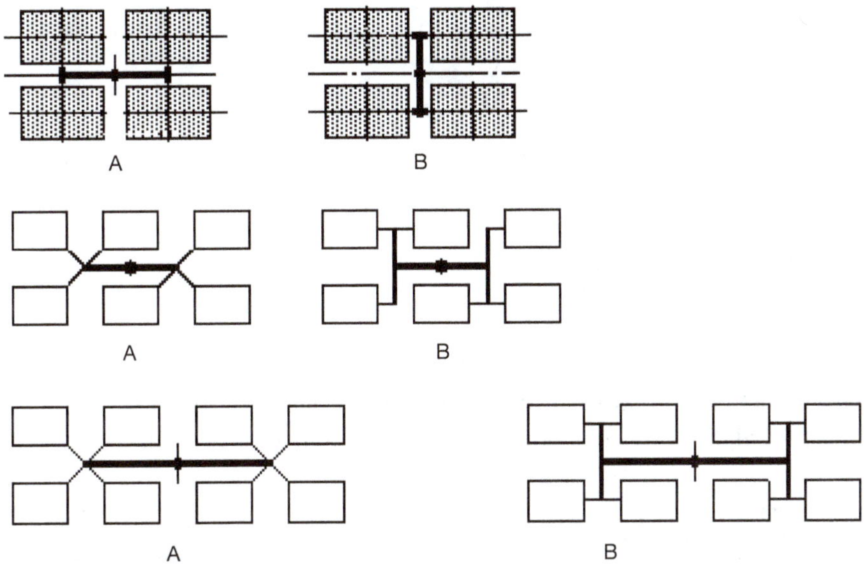

Figure 14.21 Typical cold runner layouts for four to eight cavities (option A or option B)

When laying out the arrangement of cavities in the mold, it is important to consider how the runners will be positioned. Since the cold runner will likely dictate the cycle time for the mold, it should be as small and compact as possible.

As can be seen by the line widths in Figure 14.21, each time the runner system is branched the runner size is decreased to ensure that the shear rate is maintained and the runner system is properly sized. A good way to calculate the correct sizing is to calculate an equivalent hydraulic diameter (d_h) for each successive branching. The total of d_h for the sub-branches should remain the same as for the larger channel feeding them. This can be used for round, rectangular or trapezoidal runners.

$$d_h = 4 \times A / p \tag{14.1}$$

where

d_h = hydraulic diameter (mm, in)

A = area section of the duct or pipe (mm^2, in^2)

p = "wetted" perimeter of the duct or pipe (mm, in)

Example of runner sizing:

An eight-cavity mold is to be fed with a T-shaped cold runner system (refer to Figure 14.21, bottom right). The runner at the gate will be a half-circle with a diameter of 6 mm (d_3). What size should the runners be that feed this branch?

$d_3 = 8 \times 0.5 \times (\pi \times d_3^2 \div 4) = 113\ mm^2$

$113 = 4 \times 0.5 \times \pi \times d_2^2 \div 4$

$d_2 = 8.5$ mm

$113 = 2 \times 0.5 \times \pi \times d_1^2 \div 4$

$d_1 = 12$ mm

The following re-arranged formulas can also be used to calculate the correct diameters if d_1 is known.

$$d_2 = \sqrt{\frac{d_1^2}{N_2}} \quad \text{and} \quad d_3 = \sqrt{\frac{d_1^2}{N_2 \times N_3}} \tag{14.2}$$

For even higher cavitations, the guidelines above can be used, because each successive branching will be a T, Y or X subset such as shown in the previous figures. Examples of higher-cavitation layouts can be seen in Figure 14.22 and Figure 14.5.

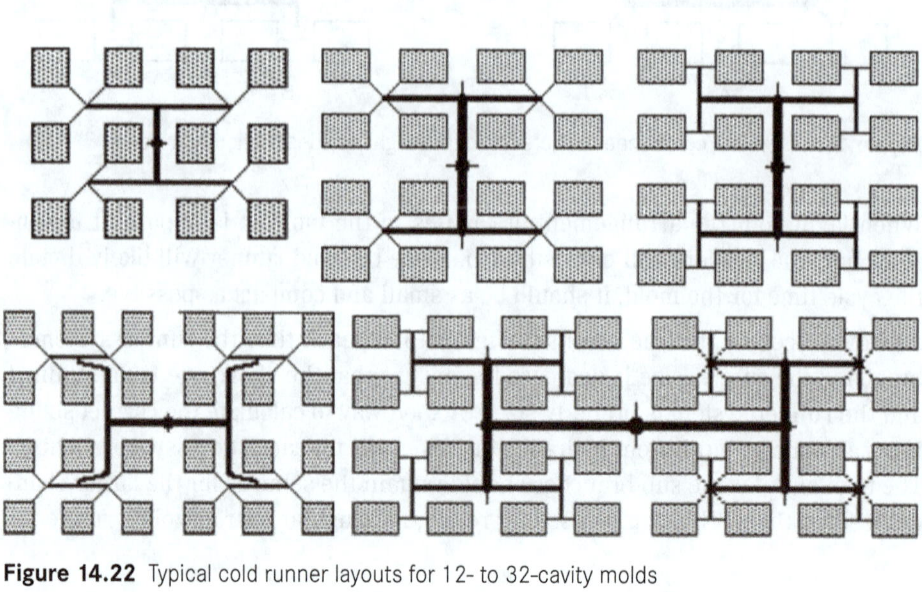

Figure 14.22 Typical cold runner layouts for 12- to 32-cavity molds

14.1.5.2 Single-Cavity Multi-Gate Runner Layouts

If a part is planned to have multiple gates, then the runner layout, gate sizing, and runner sizes should be determined by completing a CAE flow simulation, such as shown in Figure 14.15 and Figure 14.35. Since the most important information required will be where the flow fronts meet in the part and how the air is vented out, a flow analysis is the best method to ensure that no significant recuts are needed to the runner system after the first test.

14.1.6 Cold Runner Ejection

See Section 11.5.6.

14.2 Insulated Runners

Figure 14.23 (left) shows a mold similar to that in Figure 14.3, but the sprue is an insulated runner. The plastic within the runner will stay hot long enough that the material injected during the next cycle will "shoot through" the still-hot plastic, even pushing the (by then) frozen gate out of the way, but only if the cycle is not too long. Cycle times of up to 30 s can be successfully handled, and with some materials even longer, before the plastic freezes. If the plastic freezes, the now-frozen runner is easily ejected after retracting the machine nozzle. The runner ejection is usually a separate ejection movement activated only when the runner is frozen or a color change is needed. As soon as the machine nozzle is again in position, the next cycle can be started. This method is simple, inexpensive, and reliable, and it is not sensitive to dirt. If some dirt blocks the gate, it can be easily removed, as if the gate were frozen. The gate can be very small but must be properly designed for shape and size.

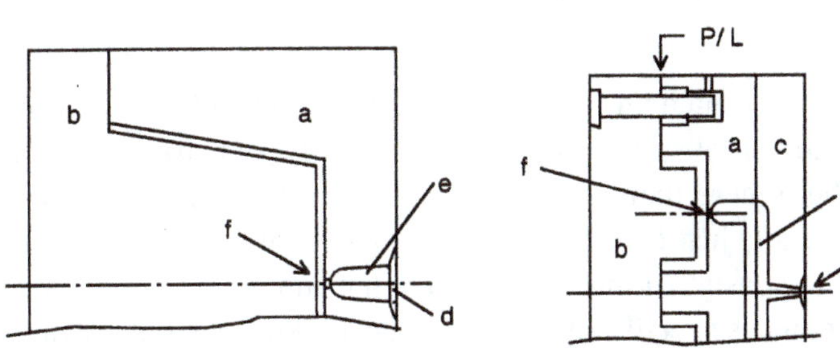

Figure 14.23 Schematics of single and multiple cavity insulated runner molds: (a) cavity, (b) core, (c) backing plate, (d) nozzle seat, (e) insulated runner, (f) gate.

Figure 14.23 (right) shows a mold similar to the schematics in Figure 14.2 (right). The hot runner is replaced with a much simpler insulated runner channel (e). In this system, the plastic in the center of the runner remains hot, even though the plastic close to the cooled walls will freeze; successive injected plastic will be able to flow through the hot core of the runner system without any added heat (see Figure 14.24). It works well at cycle times up to 30 s, and even longer, depending on the plastic used. This system is very inexpensive, simple, and reliable, but the start-up procedure is somewhat awkward and possibly dangerous if performed by operators not skilled in this system; often several starts are needed before the mold will run on-cycle. If the runner freezes, the mold must be split open between the cavity plate (a) and the backing plate (c), the frozen runner must be removed, and the plates locked together again before restarting. Molds with up to 16 cavities have been built and run successfully, but it is better to use no more than six cavities. Color changes are very easy. Without ever stopping the machine, by just changing to a new color in the extruder hopper, after about 15 shots, pieces with the new color are produced.

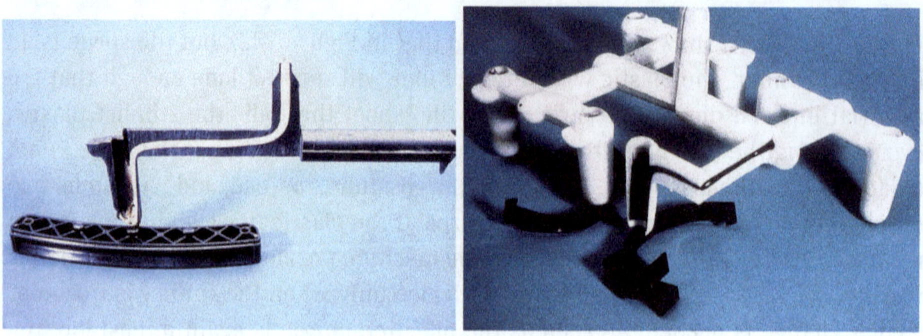

Figure 14.24 Sectioned insulated runners showing color change

■ 14.3 Hot Runners

Hot runners have become an increasingly popular choice for runner systems. Invented in the early 1960s, they are used in almost every market segment today. In most cases where more than 50,000 or 100,000 parts are needed in a year, a hot runner can be easily justified. Due to the popularity of hot runners and the added complexity in the design, a number of global hot runner suppliers have emerged. These suppliers focus specifically on this aspect of the mold and provide a number of different solutions for hot runner systems. *The purpose of this chapter is to highlight what is important to know about hot runners for mold design.* It is not intended to be a design manual for hot runner systems or components, as this is best left to the experts in this technology. If the designer is looking for more details on hot runner design, an excellent reference book is available [3].

Hot runners can be purchased as a complete unit with plates called a "hot half", as an assembled melt delivery system with wires and cables routed, or as components to be assembled into a mold by the mold maker (see Figure 14.25 for examples of the three purchase options for hot runners). The decision on which hot runner type to purchase normally depends on the customer's comfort with machining the plates and the complexity of the hot runner solution. Usually for simple, small-cavitation molds, customers are comfortable with purchasing the components only, machining their own plates, and assembling the hot runner in them. For large-part moldings, such as automotive parts, many customers prefer the pre-wired and assembled hot runners that can be inserted directly into their plates as a unit. Finally, in cases where there are a larger number of cavitations (more than eight) and more complex technologies, the customer frequently chooses the complete "hot half" solution shown in the left in Figure 14.25.

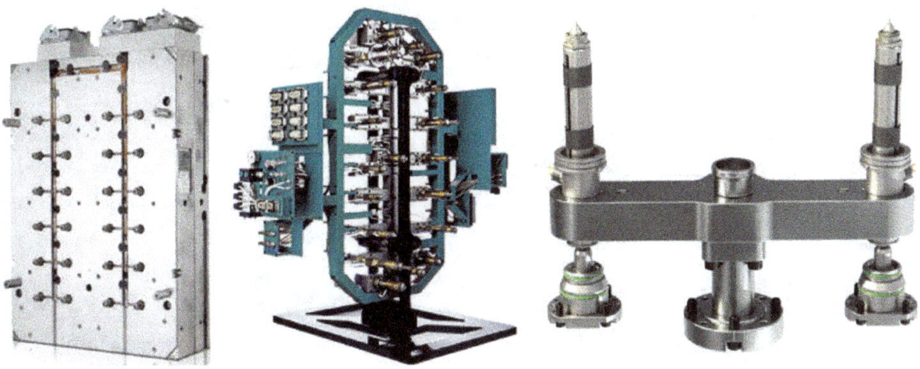

Figure 14.25 Three hot runner purchase options: (Left) Hot half; (Middle) Pre-wired assembly; (Right) Hot runner system components only (Courtesy of Mold-Masters)

Figure 14.26 shows a schematic of a typical hot runner system consisting of a manifold to distribute the melt to the gate and one of each of two types of nozzles: an open-type nozzle (bottom), and a valve gate-type nozzle (top). The various elements are labeled. Normally, these two types of nozzles would not be used in the same manifold.

[3] J. P. Beaumont, *Runner and Gating Design Handbook* (3rd edn), Hanser Publishing, 2020.

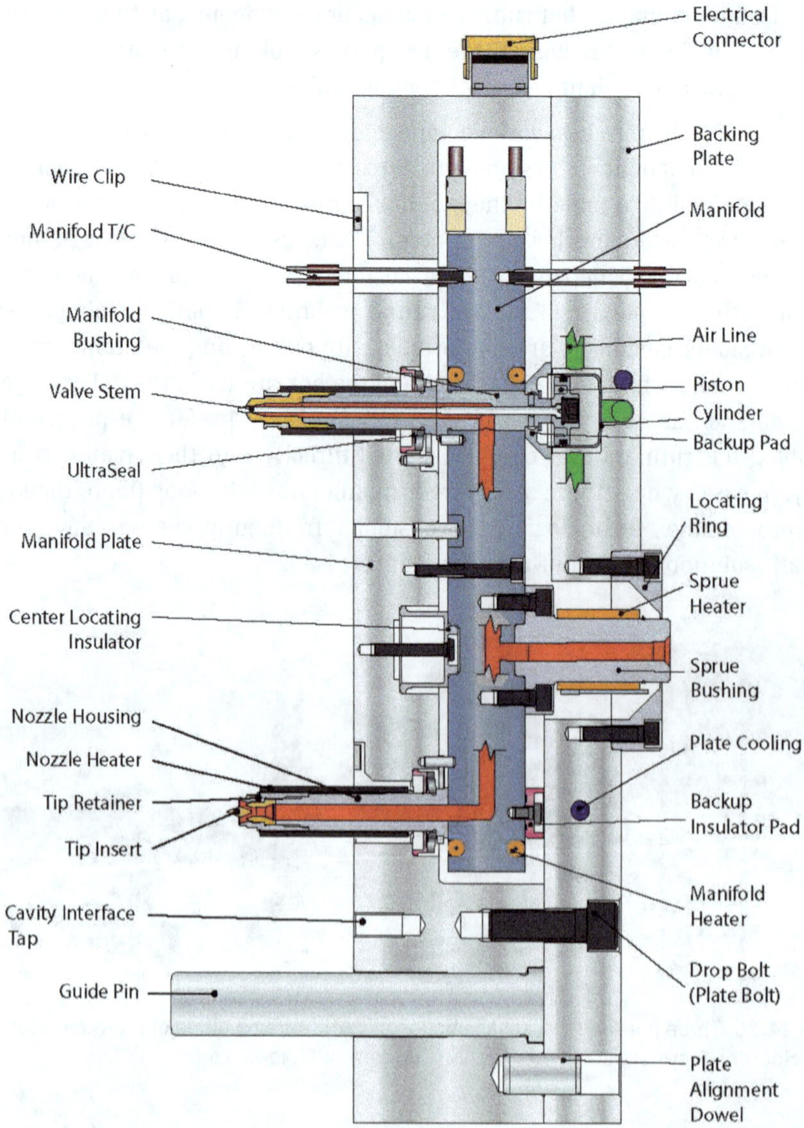

Figure 14.26 Schematic of a typical compression fit hot runner system: (Top) Hot tip; (Bottom) Valve gate (Courtesy of Husky Injection Molding Systems Ltd.)

The *advantages* of hot runner molds are many, and they have changed the economics of injection molding. A few of the many advantages are listed below:

1. Cycle times are much faster than cold runners, since the runner is no longer the limitation to the molding cycle.
2. There is additional freedom in gating location, since the gate can be located directly on the part, rather than into a runner on the edge of the part.

3. Current gating technologies allow for almost invisible gate marks on the part.
4. The cavity spacing can very close, making good use of the available space on the mold platen.
5. There is no need for regrinding, except for scrap during start-up.
6. The extruder size can be smaller, since the runner is not ejected after each shot and the shot size can be smaller. This could result in the ability to increase cavitation in the press and increase output.
7. The clamp can be smaller, since the projected area of the runner does not affect the required clamp size.

Despite all the advantages of hot runners, there are some potential *disadvantages* the designer should be aware of when considering using them. A few of these are listed below:

1. Higher mold cost than a cold runner, but not much different from a three-plate mold.
2. Longer color changes than a cold runner system. The plastic within the hot runner system must be completely clean before a new color can be used.
3. Very sensitive to dirt in the plastic. If there is a gate blocked by dirt, the nozzles must be accessed for cleaning, which may take an hour or two of stoppage. In some cases it may be necessary to remove the mold from the machine. A good mold design makes sure that this cleaning can be easily performed while the mold is in the machine.
4. The sensitivity to dirt also suggests that the molder should use virgin plastics rather than regrinds, which are more likely to be contaminated. This will affect the cost of the product.
5. There may be problems with molding very small products, because of the long residence time of the plastic in the runner system, which can cause the plastic to degrade.
6. The shut height of the mold will likely be larger than with a two- or three-plate cold runner. The designer should make sure they always consider the hot runner thickness in the total shut height of the mold.

Example of hot runner advantage:

Referring back to Figure 14.5 and Figure 14.6, in each of these cases the runner makes up as much plastic as the part to be molded or more. The cycle time is controlled by the cooling time of the (clearly visible) heavy distribution runners. A hot runner system for such a mold would reduce the injected mass by half, and the mold would cycle twice as fast, but would be more expensive to build. This is a typical example where both methods must be considered, in view of the total requirements of the product.

The answer is mostly a matter of economics. We must consider the *price of about $2000 to $3500 per hot runner drop*. On the other hand, the gain in cycle time and productivity by using hot runners can be considerable. A hot runner mold for these parts could cycle about twice as fast with hot runners. Also, the cost of recycling the runners and the percentage of plastic lost during recycling must be considered. A smaller injection unit will be required with hot runners, because all of the injected plastic will be converted into products. There is also a saving in energy, because less plastic material is processed for the same output. The mold sizes are about the same, whether a three-plate or a hot runner mold is used. Therefore, it is just a question of the quantities to be produced. If the quantities are large (typically more than than 30,000 per year), the savings in the cost of the product can easily justify the higher cost of the hot runner mold.

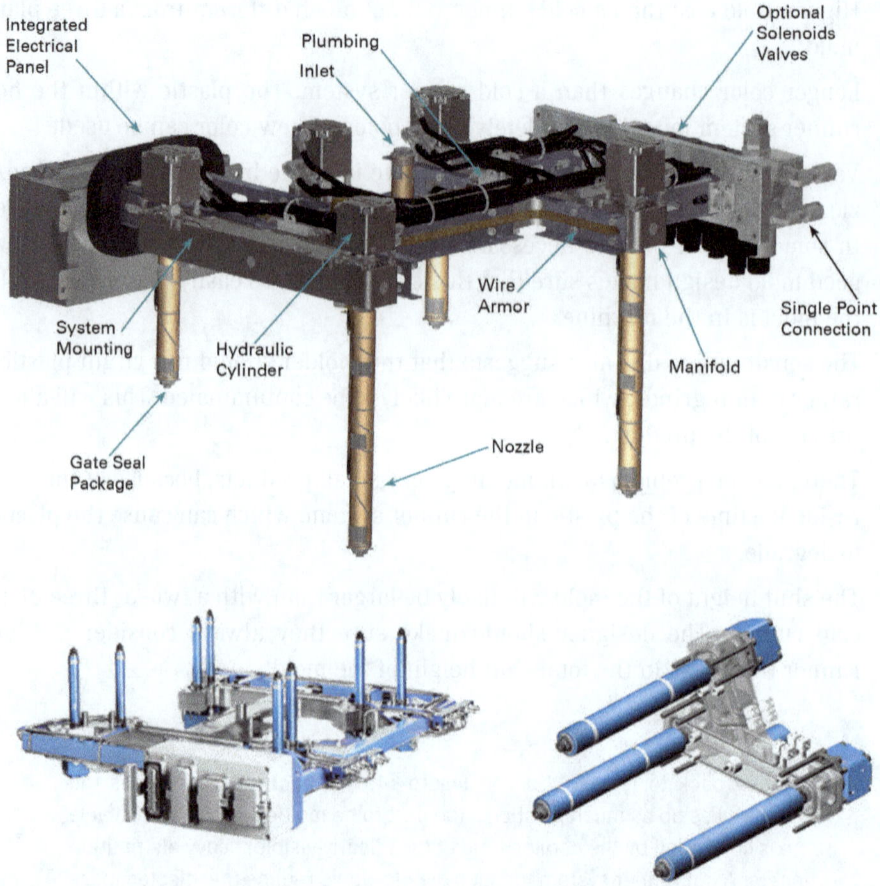

Figure 14.27 Typical pre-assembled, pre-wired hot runner system (Courtesy of Mold-Masters (top) and Mold Hotrunner Solutions (bottom))

14.3.1 Considerations for Specifying a Hot Runner

Given that the hot runner will probably be purchased as components or as a complete unit, there are some areas that need to be considered to ensure that it will perform as expected.

1. **Pressure drop.** There should be a minimum pressure drop between the machine nozzle and the cavity space, after the gate. This affects selected runner lengths, runner cross-sections, and gate size. The longer the runners, the smaller the runner cross-sections, and the smaller the gates, the higher will be the pressure drop; therefore, less pressure is available to fill the cavities. This means that often the length and thickness of runners must be increased, thereby increasing the inventory (see rule 2) in the case of hot runner molds but not the cycle time (as would be the case for cold runner molds because the entire runner needs to be cooled before the mold can open).

> *As a rule of thumb, the injection pressure used in the hot runner system should be no more than 25% of the machine injection pressure capacity.*

2. **Plastic inventory.** In hot runners, there should be as small a volume (inventory) of plastic as possible in the system between machine nozzle and cavity. The larger the cross-section of the runners the less pressure drop, (see rule 1) but the greater will be the inventory, and the longer the plastic will be exposed to the heat in the hot runner manifold. Extended times in a high-temperature environment can degrade the plastic within the runner system. The time at a given temperature before the plastic starts to degrade is different for each plastic, and is shown in graphs that can be obtained from plastics materials suppliers. Some plastics are very heat-sensitive; others are not.

> *For most plastics, it is desirable to have the inventory not larger than 2–3× the total mass of one shot, so that the plastic within the manifold is continually replaced, thereby reducing the time during which it is exposed to heat.*

   This is especially important with slow cycles, where the plastic resides for a long time in the manifold; the same applies when molding very small products.

3. **Heat loss.** There should be a minimum heat loss between machine nozzle and cavity. Heat loss affects the melt temperature and increases the viscosity of the plastic, making it harder to inject.

> *The temperature difference across the manifold should be less than 5% of the rated hot runner temperature.*

4. **Insulation.** The hot runner system should be well heat-insulated from the surrounding plates. Some heat losses are unavoidable because the hot runner manifold must be well-supported (against injection pressure) by its backing plate, and by the features necessary to locate it within the mold. These necessary areas of contact conduct heat away from the hot runner to the surrounding cooled plates. Heat may have to be added through the hot runner manifold heaters to make up any heat losses, thus increasing the electric power used. Also, the heat loss into the surrounding plates can raise their temperature and affect the mold alignment; good cooling is necessary for these plates.
5. **Cooling.** There must be ample cooling in the cavity (or core) near the gate because the nozzle of the hot runner seals-off in this region to prevent leakage of plastic out of the hot runner. This creates a large heat increase in the core or cavity in the area of the nozzle shut-off. This heat must be removed to prevent the hot runner from increasing the necessary mold cooling time.
6. **Balanced runners.** In any runner system, the pressure drop from the machine nozzle to each cavity space (gate) should be the same. Pressure differences from cavity to cavity will affect the amount of plastic entering the gate before it freezes, the density of the plastic in the cavity space, and thereby the strength and quality of the molded piece. It will also result in differences in the surface definition, and appearance of the product. It is not always possible to follow rule 6 completely, but every effort should be made to do so. In some cases, individual adjustments to gate sizes may help to ensure greater product uniformity.
7. **Number of gates per cavity.** Wherever possible, there should be only one gate per cavity. There are exceptions to this rule: (1) where core shift could be a serious problem, two or more gates may be located symmetrically around a delicate core to equalize pressure and flow around the core; (2) where the flow length L from the gate to the farthest corner (or rim) of the molding is very great. This applies especially to large moldings.

14.3.2 Overview of Hot Runner Design

It is not the intention of this book to teach the mold designer how to build their own hot runner. However, the mold designer should understand the fundamentals of hot runner design to ensure that the mold will perform as intended. Also hot runners can be purchased as complete assembled units or as heated components that need to be placed into the mold or hot runner plates. A designer should therefore have a good understanding of how a hot runner is designed and how to select the best solutions for their mold.

14.3.3 Layout and Balancing of Hot Runner Melt Channels

Most of the principles discussed in cold runner melt channel layout (see Section 14.1.5) also apply to hot runners, but with some exceptions, which will be discussed in this section. As the plastic flows through the manifold, friction with the walls of the channels causes a pressure drop. But also, every turn in the flow at intersections causes a pressure drop equivalent to a certain amount of straight length of channel. Depending on the number of cavities and complexity of the manifold layout, there could be a number of different turns, some 90°, some 45° and some at 135°. To ensure that the pressure drop is the same for each cavity on the way from the machine nozzle to each gate, there must be the same number of the different bends, and the same lengths of channels of the same diameter (if there are different diameters). This last requirement makes it often necessary to have the channels *at different levels within the manifold,* with connecting bores or channels between the levels (see Figure 14.30).

Table 8.1 in Chapter 8 shows the preferred cavitations for a balanced layout of a hot runner. Any cavitation that is a multiple of 2 (2, 4, 8, 16, etc.) or 3 (6, 12, 24, etc.) is a good choice for a hot runner cavitation, as these can be easily balanced. Hot runner manufacturers have ways to "artificially" balance cavitations that are not naturally balanced, but this is less desirable.

Due to the laminar flow of the highly viscous plastic in the melt channel, the material near the walls of the channel is subjected to more shearing (and heat) than the material that flows in the middle of the melt channel. At every intersection of flow, the plastic gets split. If the hot runner designer is not careful in their layout, the material with more shear heat history may end up largely in one leg of the intersection (see Figure 14.17, Figure 14.28, and Figure 14.29). It is therefore *always advisable to use multi-level manifolds and the minimum number of intersections (X is better than a Y, which is better than a T),* such as shown in Figure 14.30, to ensure that the shear history of the plastic is evenly distributed in the channels after two, three, or four intersections. Figure 14.31, Figure 14.32, and Figure 14.33 show typical examples of layouts for multi-cavity three-plate and hot runner molds with one gate per part. One of the big advantages of layouts for hot runner molds is that they are not constrained by the layout of the cavities in the way that a two-plate cold runner would be. A hot runner layout can have a shorter path to the gate because it does not share a parting plane, and of course the part can be gated anywhere on the cavity, rather than having to be at the edges of the part (as for two-plate cold runners).

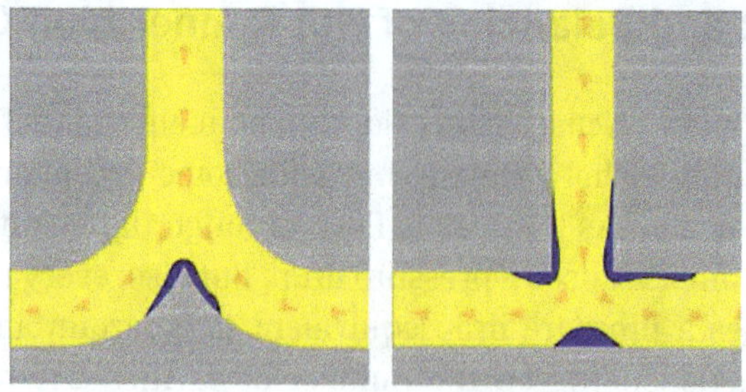

Figure 14.28 Stagnation points at intersections in the melt distribution system

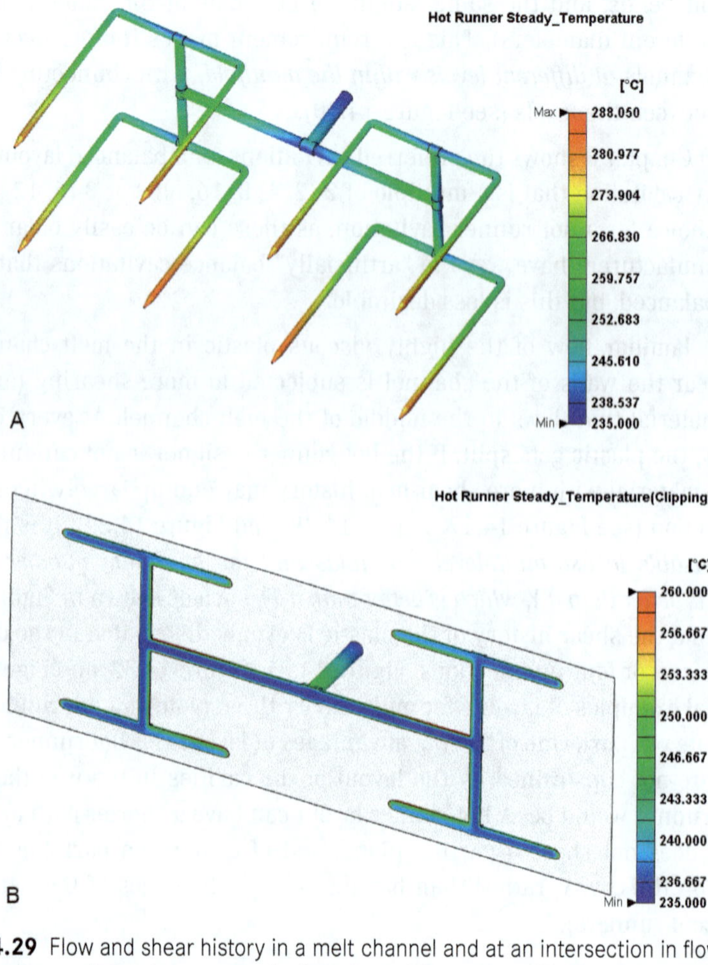

Figure 14.29 Flow and shear history in a melt channel and at an intersection in flow: (A) Temperature variation in a geometrically balanced hot runner; (B) Viscosity variation in a geometrically balanced hot runner (Courtesy of Moldex3D)

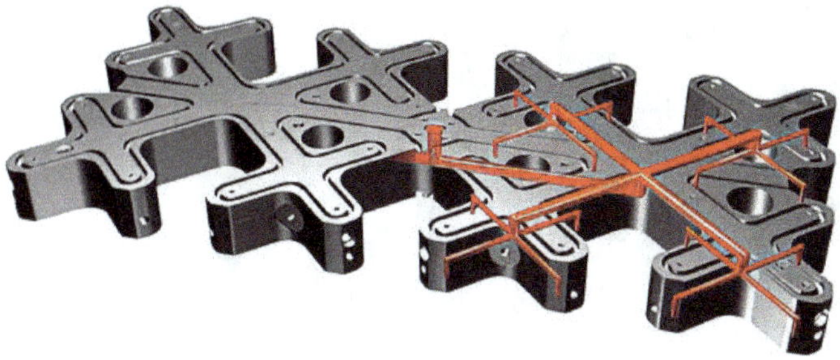

Figure 14.30 Typical layout of 16-cavity multi-level manifold (Courtesy of Husky Injection Molding Systems Ltd.)

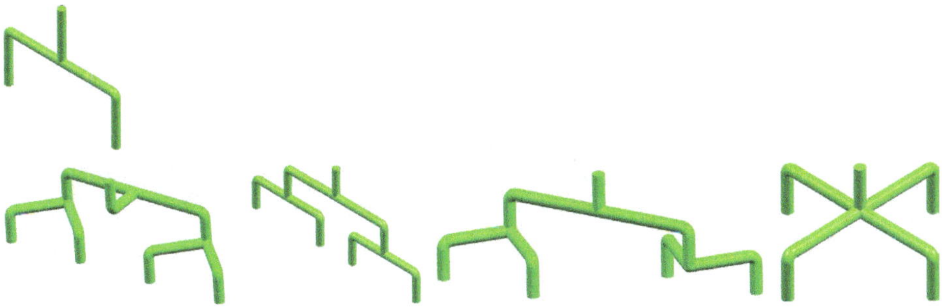

Figure 14.31 Typical one- to four-cavity three-plate or hot runner melt channel layouts (Courtesy of Husky Injection Molding Systems Ltd.)

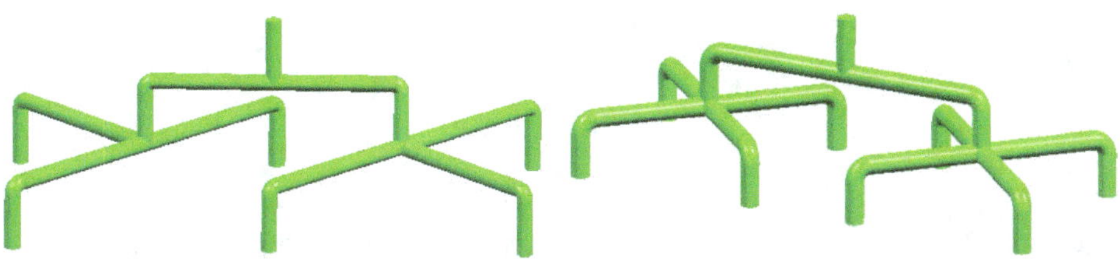

Figure 14.32 Typical six- to eight-cavity three-plate or hot runner melt channel layouts (Courtesy of Husky Injection Molding Systems Ltd.)

A

B

C

D

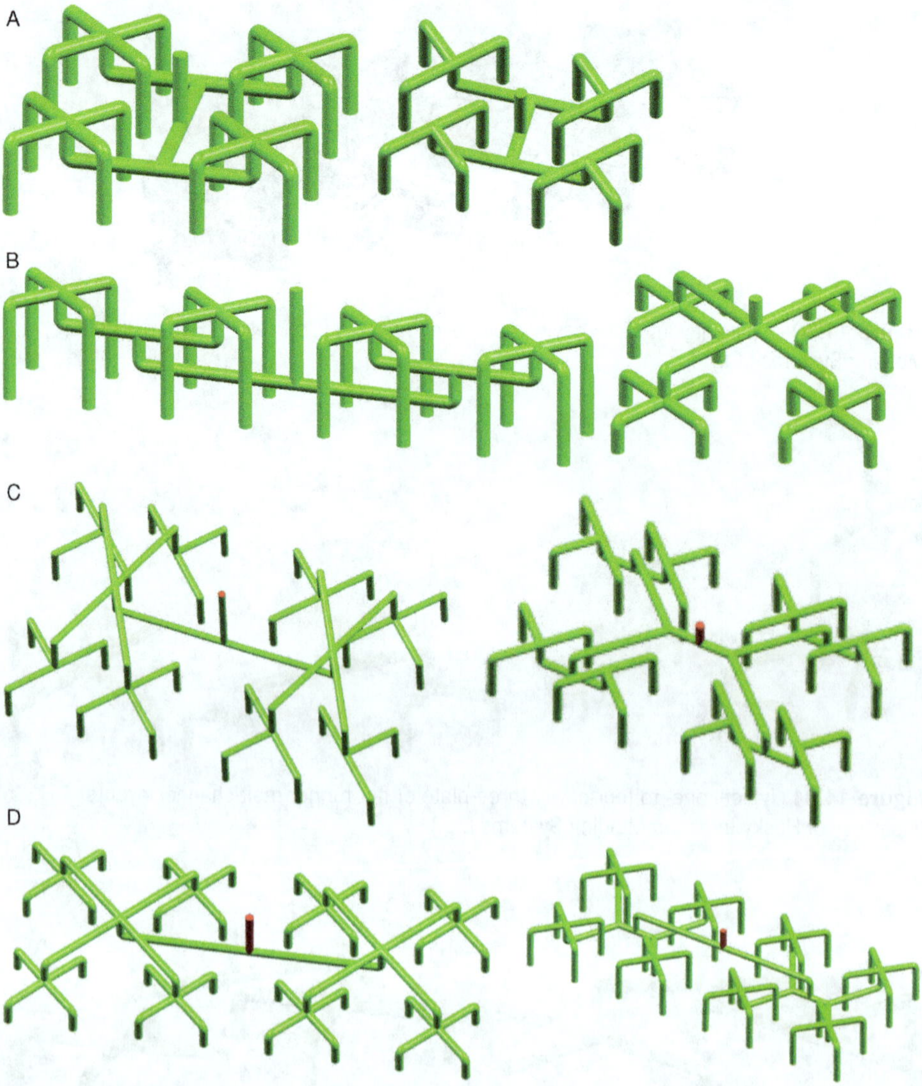

Figure 14.33 Typical hot runner melt channel layouts: (A) 12 cavities; (B) 16 cavities; (C) 24 cavities; (D) 32 cavities (many more options are possible) (Courtesy of Husky Injection Molding Systems)

At the intersection points of melt channels in a hot runner, it is advisable to change levels in the hot runner, as shown in Figure 14.38.

The melt channel layouts are created by drilling a solid block of steel, or milling in the matching surfaces of two manifold halves and then brazing or diffusion-bonding them together (see Figure 14.34). If the manifold is made in two or more pieces, then brazing or diffusion-bonding is normally used to permanently join the pieces.

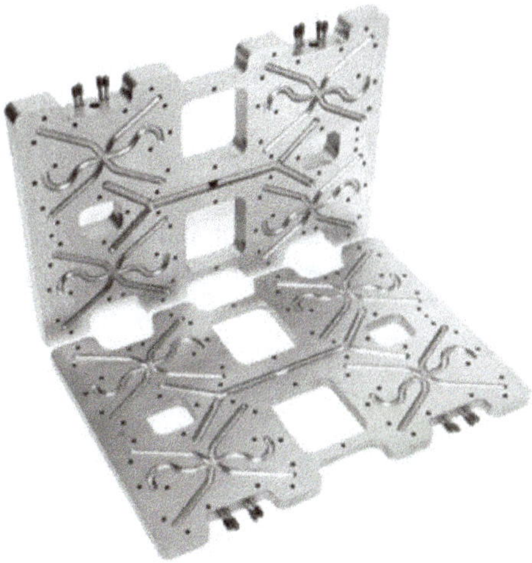

Figure 14.34 Two-piece brazed melt channel layout for a 24-cavity main manifold (Courtesy of Mold-Masters)

When drilling the channels, the holes intersect with others to create the desired pattern. It is important that the points of intersection do not create steps that could create hang-up. One disadvantage of drilled holes is that the opening must be securely plugged up to prevent hot plastic escaping under very high pressure. Some use screw plugs of various designs, so that, if needed, the channel can be cleaned out. Other suppliers weld the entry solidly. The other disadvantage is that the gap in the bore between the plug (or weld) and the next intersection will fill up with plastic, which will degrade. To prevent this, a shaped insert backed by the weld (or screw) fills this gap. The other end of the insert is shaped to the desired bend in the flow. The insert must be well-fitted and oriented (there are different designs to solve these problems). This is also a very important safety consideration. An improperly assembled and secured screw subjected to the internal manifold pressure may be ejected like a projectile, and cause injury. Milling of the channels has the advantage that sharp corners and bends can be avoided, for better plastic flow and so that the melt channel lines do not have to be straight as when using a drill. The milled manifold process is gaining popularity as the demands on the hot runner continually increase.

With either gun-drilled or milled channels, the inside of the channels must be very smooth to reduce the friction of flow, but also to prevent even minuscule "hang-ups" of plastic in machining grooves, where it would gradually degrade and then from time to time re-enter the plastic stream, leaving unsightly marks in the product. To create this smooth finish, the melt channels are normally purged with a thick slurry (trade name Extrude Hone®) and then cleaned.

Laying out the melt channel path can be straightforward when the cavitation is naturally balanced (refer to Chapter 8 on cavitation). When the cavitation is not naturally balanced or when there are multiple gates per part, then CAE simulation software can help to optimize the channel layout and the channel sizing if necessary to "artificially" balance the mold. Figure 14.35 shows an electrical lamp housing with six gates that is (a) artificially balanced with different melt channel sizes and (b) naturally balanced with a more complex manifold layout.

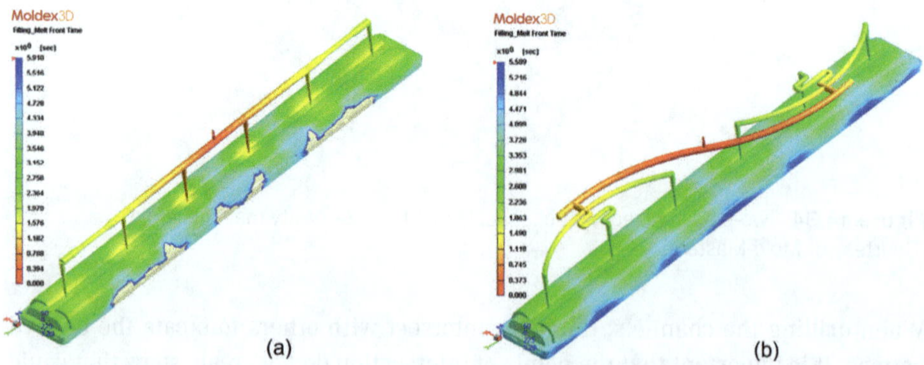

Figure 14.35 Flow simulation of a part with six gates [4] (Courtesy of Moldex3D)

Note that while a brazed or diffusion-bonded manifold allows limitless freedom of melt channel layouts, such as that shown in Figure 14.35, the halves to be brazed or bonded must be closely matched so there is no mismatch and resulting hang-up. Also, since this type of manifold is permanently bonded together, any cleaning of the manifold must be completed with chemical or thermal cleaning solutions.

The resulting design of the hot runner manifold can be tested for its effectiveness in balancing the fill of the mold by intentionally creating "short shots". This will show how the mold fills before the boundaries of the cavity space stop the flow in some cavities. Figure 14.36 shows the fill balance of a 48-cavity part. The testing procedure for this balance test is to create short shots until just one of the cavities is filled. The resulting parts should be collected and weighed. The level of imbalance can then be calculated by using the fill balance formula below:

$$\text{Fill balance} = \frac{\text{Heaviest part} - \text{Lightest part}}{\text{Average full part weight}} \times 100\% \qquad (14.3)$$

A well-balanced mold should have a balance of 90%. Anything above 90% would be considered excellent.

[4] Source: M. Wang, R. Chang, C. Hsu, *Molding Simulation: Theory and Practice,* Hanser Publishing, 2018 (Figure 7.27).

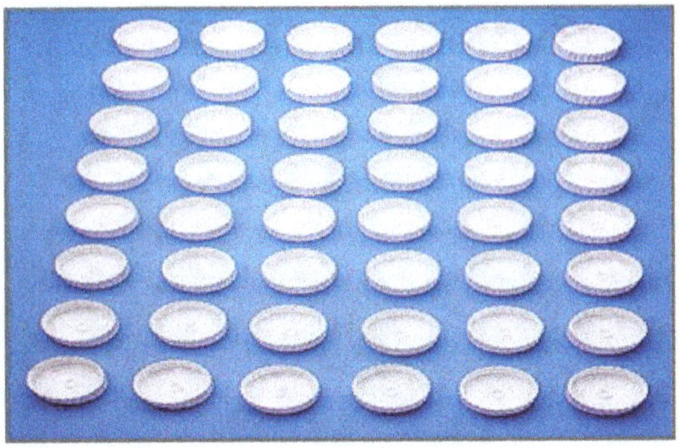

Figure 14.36 High-cavitation fill balance assessment using short shot weights

14.3.3.1 Bridge and Sub-Manifolds

In some cases a single manifold cannot be used to create an even melt distribution path for the plastic to flow into the cavities. In cases such as this, bridge or cross-manifolds are used to join the melt flow to a series of main manifolds, as shown schematically in Figure 14.37 and in the 96-cavity hot runner shown in Figure 14.39.

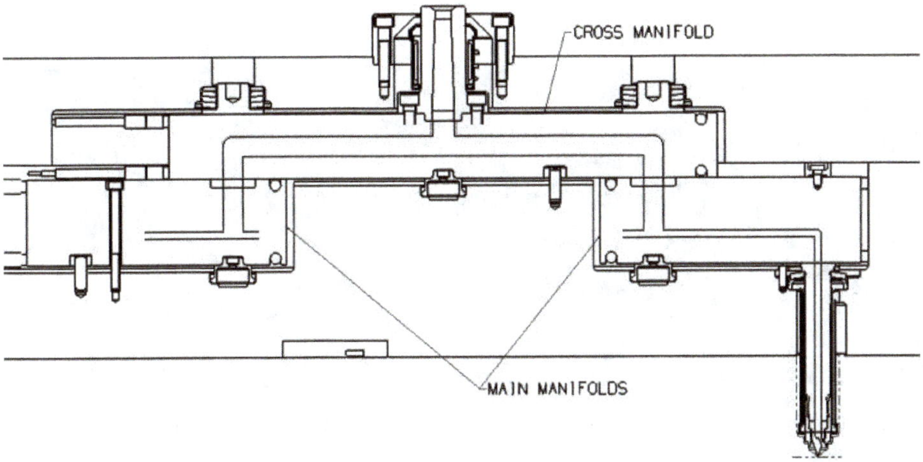

Figure 14.37 Bridge (or cross) manifold layout in a hot runner system (Courtesy of Husky Injection Molding Systems Ltd.)

The use of bridge manifolds can reduce the size of the main manifolds and more effectively deal with thermal expansion, by allowing some thermal expansion to occur between the main and bridge manifolds. The downside to this is that there are additional locations where plastic can leak from the manifold system into the manifold pocket. Typical layouts of bridge and main manifold systems are shown in Figure 14.38.

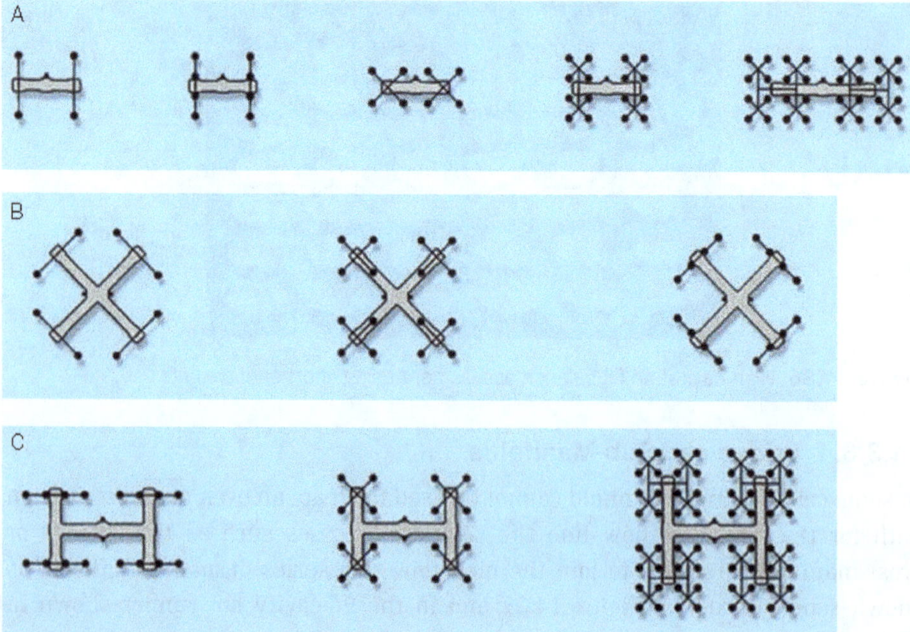

Figure 14.38 Manifold layouts using bridge manifolds: (A) Two-drop bridge examples; (B) Four-drop bridge examples (Courtesy of Mold-Masters)

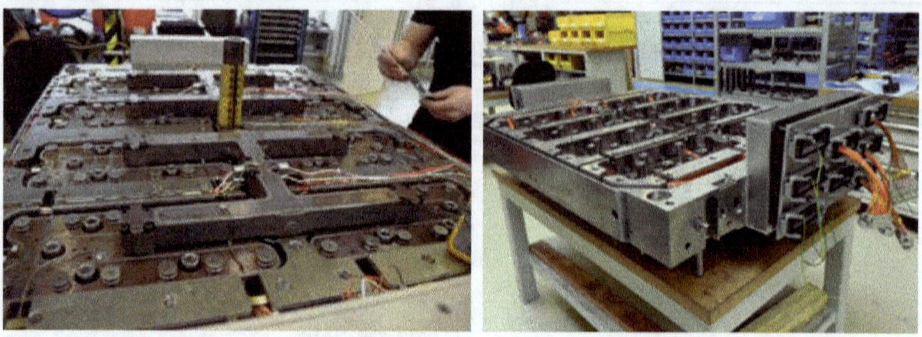

Figure 14.39 96-Cavity hot runner system with bridge manifolds (two X-shaped and one two-drop bridge)

14.3.4 Melt Channel Sizing

The melt channel sizing in the hot runner is very important, as it is a balance between keeping the inventory of melted plastic low to prevent degradation, keeping the shear rate high for good flushing, and keeping the pressure drop within reasonable limits to allow the cavities to fill. In some cases, there must be a trade-off between the inventory and the pressure drop in the hot runner. This is particularly important when it comes to fast color changes. Since smaller channels facilitate faster changes, these are desirable. However, smaller channels also create more pressure drop in the hot runner, therefore leaving less available to fill the cavities.

In general, the pressure drop in the hot runner should not constitute more than 25% of the available injection pressure in the machine. The temperature rise due to shear in the hot runner should be less than 10 °C. The shear rate in the hot runner should be kept as high as practical (within the pressure drop and temperature rise limits above) to facilitate good wiping of the melt channel walls for color change and reduced degradation. This means a shear rate of around 1000 s^{-1} and an inventory of plastic in the runner of less than three shots.

Suppliers of hot runner system have programs to calculate the channel sizes and find the right balance. In the end, the purchaser may need to decide which compromise they are willing to live with. A mold designer can also use one of the many computation fluid dynamics programs available, such as Moldex3D, to conduct their own calculations.

14.3.4.1 Plastic Inventory in the Hot Runner

We define inventory as the total amount of plastic (in grams) within in the manifold and the nozzles.

To prevent degrading, plastic should be exposed for as little time as possible to high (but necessary) molding temperatures. This would suggest making the channels as small as possible. However, this would increase the pressure drop toward the gate and create problems in filling the mold.

Example 1:

A manifold has an inventory of 320 g. The shot weight is 160 g and the cycle time is 6 s. That means that every two shots (cycles), the inventory is completely renewed, and the plastic is exposed to critical temperatures for only 12 s. This is a good condition.

Example 2:

The same manifold is used, but the product is thick, the shot weight is only 80 g, and the cycle time is 30 s. In this case, the inventory is four shots, and the plastic is exposed to these hot conditions for over 2 min. This may be too long, and unacceptable for some heat-sensitive plastics. A different hot runner channel size may be required.

14.3.5 The Hot Runner Manifold

The manifold is the component of the hot runner system that distributes the melted plastic equally to each cavity with minimal impact on its rheological state. It connects to the hot runner nozzles, the sprue bushing, and the cold plates. It is subjected to high stresses generated by heat, expansion, injection, and clamp pressure. The manifold must provide ample backup to the cavities to prevent any deflection in the mold, while keeping the plastic at the ideal molding temperature between the machine nozzle and the gates, despite the fact that it is held in cold plates. It is a challenging engineering application.

14.3.5.1 Locating the Manifold

The manifold is subject to heat expansion as it heats to the melt temperature relative to the cold plates supporting it. The manifold must therefore be held in position and prevented from rotating. This is usually done by the center back-up insulator opposite the sprue bushing. We must also be sure that the manifold will not lose this central position. This can be achieved by placing a dowel or similar element into the manifold right opposite the machine nozzle or somewhere on the manifold. However, any method to keep the hot manifold from rotating out of position must also ensure that minimum contact is made between it and any cold plate, and that it does not hinder expansion of the manifold in all directions. Figure 14.40 (left) shows a center back-up insulator opposite the sprue bushing. Figure 14.40 (right) shows the expansion of this four-cavity manifold outwards from the center locating insulator. Also shown is an anti-rotation dowel to ensure that the manifold stays located in the center of the manifold plate pocket.

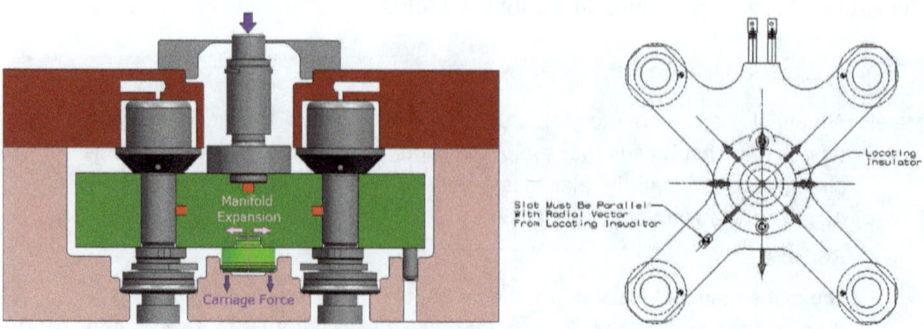

Figure 14.40 Manifold back-up and locating insulator shown with a valve gate pistons for each nozzle (Courtesy of Husky Injection Molding Systems Ltd.)

14.3.5.2 Thermal Expansion of the Manifold, Sealing the Hot Runner Systems, and Nozzle Tip Position

There is no practical method to oppose heat expansion as the forces are tremendous. A hot runner supplier therefore needs to work with it, and there are two ways of doing so. One is to have the manifold expand and slide over the fixed nozzles during heat-up (see Figure 14.26 and Figure 14.49), and the other is to screw the nozzles into the manifold (see Figure 14.27).

Sliding nozzle sealing: In the cold condition there is a calculated cold clearance between the manifold and nozzles. As the manifold heats up and expands, the cold clearance is eliminated and the manifold is free to expand lengthwise and through its thickness. The backup insulators and the nozzle housings are built to accept some of the pressure from heat expansion, and this creates a seal to eliminate any leakage during high-pressure injection. Some hot runner manufacturers add bolts between the nozzle and manifold to keep them in contact at all times and force the cold clearance to be between the backup insulator and the manifold. Other manufacturers use a spring system to keep the nozzle and back-up insulator firmly compressed to the manifold at all times (see Figure 14.40).

Example:

If the manifold is heated to 230 °C, the manifold will expand in all directions from its fixed center. Suppose we have a two-cavity mold, with the cavities spaced 150 mm apart, i.e., 75 mm right and left of the center. With the mold at 21 °C, we have a temperature difference of ΔT = 209 °C. The manifold will grow over the distance of 75 mm by the amount of heat expansion (see Section 12.4):

$$\Delta L = k \times L \times \Delta T \quad \text{(Equation 12.11)}$$

$$= 0.000012\ °C^{-1} \times 75\ \text{mm} \times 209\ °C$$

$$= 0.18\ \text{mm}\ (0.0068")$$

The difference shown in the above example is significant, even in a large mold. To overcome this misalignment, when machining the cold parts, the distance in the mold parts will be different, so that at the transfer points the hot parts will align properly with the cold part when in operation.

We can now see three immediate challenges affecting the mold design:

1. The manifold must be allowed to expand freely between its supports, but must be centered in the mold and kept from rotating
2. When dimensioning the mold components, we must make allowances for the heat expansion

3. To build the hot runner accurately, we must know its operating temperature beforehand, even though it might change in operation. Normally a temperature range of up to ±50 °C (90 °F) from the name-plate temperature is permissible for a hot runner to function correctly.

Screw-in nozzle sealing: The other method of sealing is to screw the nozzles into the manifold. This reduces the number of sliding sealing faces in the hot runner system and reduces the number of heat loss locations, since the nozzle does not require a seat to seal on (see Figure 14.50).

The position of the nozzle tip in the gate area is affected by thermal expansion. In the case of a hot tip application, the position of the nozzle tip can have a significant impact on the gate quality, as can be seen in Figure 14.41. If the thermal expansion is calculated incorrectly, then the nozzle tip could be too far forward, creating stringing or plugged gates. If the tip is too far back, the gate may freeze-off and cause incomplete filling of the parts. If the nozzle tip is in the correct position (normally flush with the molding surface or slightly protruding), then the gate mark will be very small and consistent. The position of the gate is particularly important in a high-cavitation tool. The position of the nozzle tip must be very even from cavity to cavity, to prevent imbalanced filling issues and inconsistent gate quality.

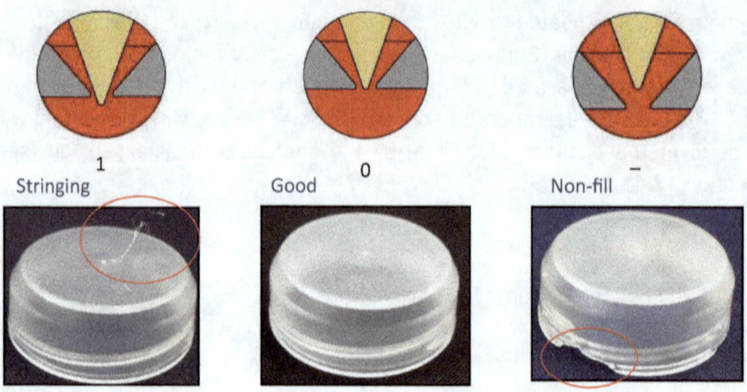

Figure 14.41 Thermal expansion and impact on nozzle tip position (Courtesy of Husky Injection Molding Systems Ltd.)

14.3.5.3 Manifold Supports (Back-up Insulators)

In order to support the manifold between the cavity or cavity backing plate and the mold mounting plate, we must provide supports. Since these supports contact the hot manifold and the cold plates, they must also act as insulators. These supports, or back-up insulators, serve the following purposes:

1. To add to the support of the cavity plate, in case the surrounding plate is insufficient to prevent the cavity plate from deflecting during injection

2. To provide pressure to seal (a) the contact areas between the manifold and (b) the nozzles, to prevent leaks of plastic at the transition points
3. To insulate the hot manifold from the cold plates.

We have seen earlier that the heat flow is directly proportional to the temperature difference ΔT and the cross-sectional area through which it flows, and inversely proportional to the length of the flow.

This would suggest thin, long supports. This is not always practical, especially because the length is often restricted by the mold shut-height limits.

Instead of thin "columns", it is now practice to use hardened, tubular, precision-ground supports. To increase the seating area to prevent the harder tubes from permanently pressing into the softer plates and the manifold, the ends of the tubes are flanged. The tubes have a wall thickness in the order of 1–2 mm, and diameters of typically 15–25 mm. We use similar tubular supports as part of the nozzles, and these will be discussed in more detail later in this chapter. When installed, and in operation, the ends touching the cold plate are cool, and the other ends (where the manifolds can slide) are hot. For calculations, we use the mean value of these temperatures.

In Figure 14.40 and Figure 14.42, there are three areas for the support shown:

1. A center support (Figure 14.40) that counteracts the force of the machine nozzle pressing on the sprue bushing, or directly on the hot runner manifold. It is located between the manifold and the cavity backing plate. This force is usually not more than 10% of the rated machine clamping force, and is used to ensure that the plastic will not leak during injection where the M/C nozzle contacts the sprue. If this force, if not backed-up, could unnecessarily deflect the manifold and create leaks or cracks in the manifold.
2. Manifold back-up support. While injecting, and especially at the end of injection, there will be a plastic force pushing the nozzle back against the manifold. This force must be contained by the back-up insulator. This back-up insulator also acts as an opposing force to the nozzle skirt, to provide a sealing force between the nozzle and the manifold.
3. A nozzle support spaces the manifold from the cavity plate and opposes the back-up insulator to provide sealing force. This ensures the proper distances for the ends of the nozzles and the gates within the cavity well where the nozzles are located.

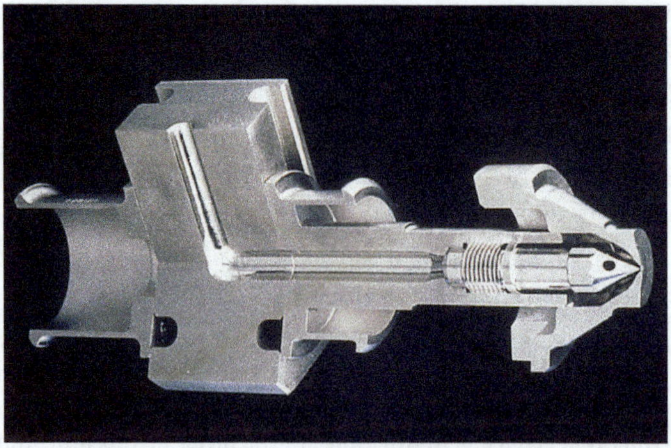

Figure 14.42 Section view of typical hot tip hot runner system showing back-up insulator and nozzle skirt used to support and insulate the hot manifold (Courtesy of Husky Injection Molding Systems Ltd.)

14.3.5.4 Manifold Plate and Manifold Backing Plate

As mentioned in the previous sections, the manifold is located on insulators to keep the cold plate from affecting the temperature of the manifold. The plate supporting the manifold could be a separate plate or frame in the mold (called a manifold plate), or the cutout for the manifold could be incorporated in the cavity or cavity backing plate. Whichever design is chosen, the manifold plate must provide excellent support throughout so that the mold remains rigid enough to prevent deflection from the molding forces and to resist the expansion forces of the manifold. The manifold plate must also be well-cooled, to prevent it from expanding due to the heating of the manifold.

The major design elements of the manifold plate are shown in Figure 14.43. As can be seen, the manifold plate must be very rigid to support the injection and clamp forces. There must be excellent support throughout the plate to ensure that the mold does not deflect under the loading conditions of the process. Since the hot manifold located in the manifold plate will expand, the bolts holding the manifold plate and manifold backing plate together must be sufficient to prevent the plates from separating (see Chapter 16 on fasteners). Channels for the heating and thermocouple wires must be created so that the wires can be routed to the electrical box mounted on the mold. The wire channels must be sufficiently deep to prevent the wires from being pinched. The wire channels must also be routed so that the replacement of heaters or thermocouples does not require complete disassembly of the hot runner. For example, as shown in Figure 14.43 (right), there is a wire channel for routing the thermocouple and heater wires. This allows the nozzle heater or thermocouple to be replaced in the machine if it fails.

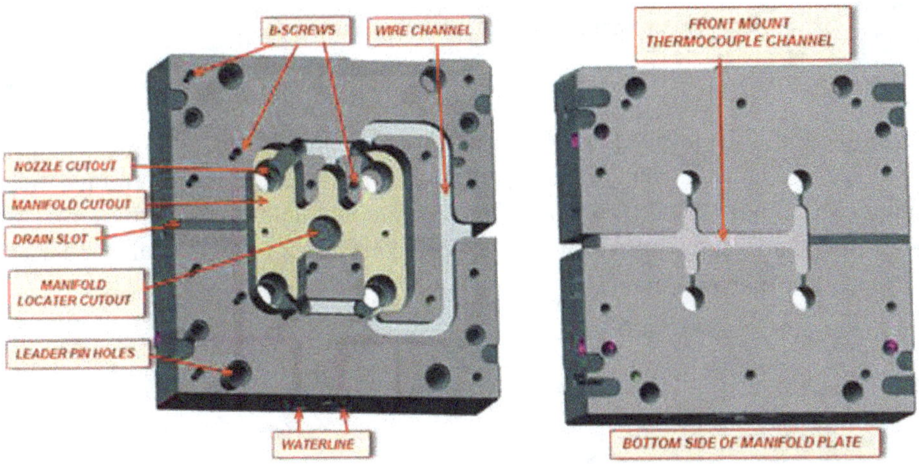

Figure 14.43 Manifold plate cutout for hot runner manifold (Courtesy of Mold-Masters)

14.3.5.4.1 Manifold Plate Cutout or Manifold Pocket

In order to house the manifold and nozzles, the manifold plate must be cut out. This pocket into which the manifold is placed is designed to be slightly larger than the manifold, by about *3–10 mm all around,* to provide good clearance and air insulation, as shown in Figure 14.44. It is best practice for the manifold pocket cutout to follow the contour of the manifold profile to retain as much support as possible. Additional space must be cut out for the heater and thermocouple wires to be routed to the electrical connection boxes on the outside of the mold.

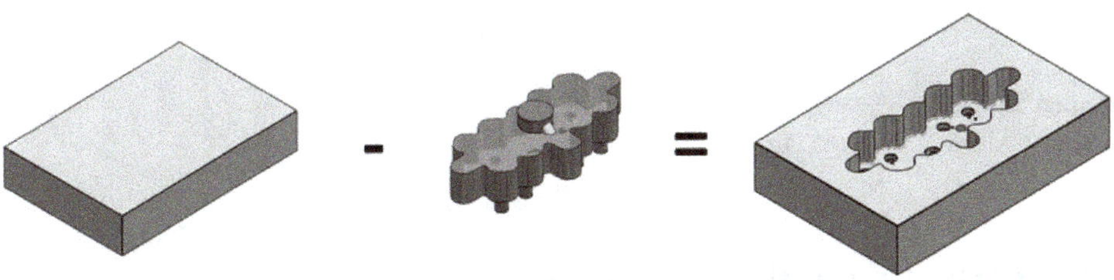

Figure 14.44 Manifold pocket cutout in manifold plate (Courtesy of Husky Injection Molding Systems Ltd.)

The frame provides good support for the cavity plate, except for the area of the well, but we must be sure that the cavity plate is also well-supported. Ideally, the manifold cutout should follow the layout of the manifold so that the maximum amount of plate steel remains. Although the manifold is well-supported by the nozzle skirts and the back-up insulators, solid plate steel is still the best back-up support, so as much steel should be retained as possible.

> There should be no more than 100 mm (4") of unsupported length in the manifold plate.

If, due to the pitch spacing and layout of the manifold, the amount of remaining manifold plate steel is compromised, then consideration must be given to adding pillars into the manifold plate and cutouts in the manifold (see Figure 14.45). As with the tip above, the pillars should be located such that there is no unsupported span greater than 100 mm (4").

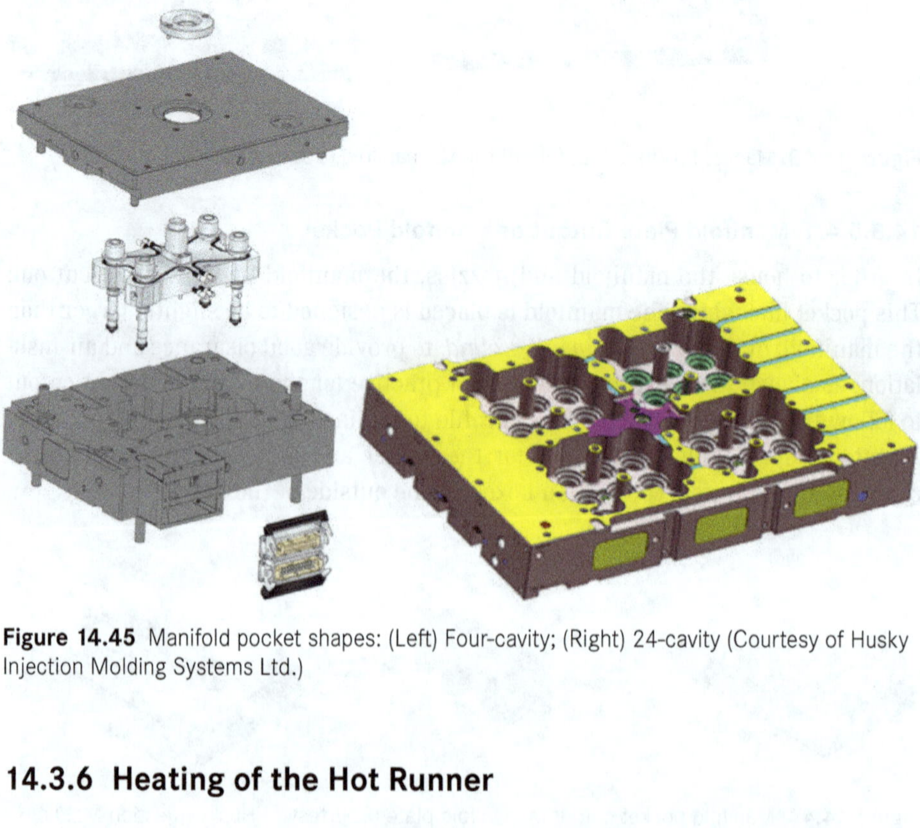

Figure 14.45 Manifold pocket shapes: (Left) Four-cavity; (Right) 24-cavity (Courtesy of Husky Injection Molding Systems Ltd.)

14.3.6 Heating of the Hot Runner

A key role of the hot runner is to maintain the temperature of the molten plastic in its ideal state uniformly at each cavity until it is ready to be injected into the mold. Also, the hot runner must be able to be heated at start-up to re-melt the solidified plastic in the hot runner from the last molding run. The hot runner components must therefore be efficiently heated. As stated earlier, the heaters are required for two reasons:

1. To facilitate fast re-starting of the mold. The plastic within the channels should be at a temperature high enough so it can be injected, preferably, within 10 min or less.

2. To replace all heat losses to the surrounding (cooled) plates, by (a) conduction through the supports and the locating device to keep the manifold in the center, and (b) radiation to the plates.

Today, in the plastics industry, electric resistance heaters are used almost exclusively for heating a hot runner system.

14.3.6.1 Basics of Resistance Heaters

Resistance heaters are based on the Ohm's law. An electric supply with a voltage differential V (in volts, V) is applied to the ends of a wire with a resistance R (in ohms, Ω). This creates a current I (in amperes, A). In doing so, the work E (in watts, W) of the electric current I is transformed into heat (measured in watts, British Thermal Units (BTU) or in calories).

Ohm's law can be expressed in three formulae shown in Equation 14.4:

$$\begin{aligned} V &= I \times R \\ E &= I \times V \\ E &= V^2 / R \end{aligned} \tag{14.4}$$

The resistance R is directly proportional to the length L or the coil, and inversely proportional to the cross-section Q of the wire, and proportional to a factor f specific for each material. This can be written as

$$R = f \times L / Q \tag{14.5}$$

We can conclude form the above that the heat generated by a heater is proportional to the square of the voltage applied. In most countries today, 220 V is used for heaters, but this is just a "rated" or "nominal" figure. In real life, this can be anywhere from 200–240 V, depending on the source of the power supply.

Usually, heaters are rated at 230 or 240 V (even though the line voltage is normally 220 V), and the "wattage" (or heat rating) is based on that figure. This ensures that the heater is safe at that voltage, and that it would not burn out at a slightly higher voltage. However, it would also provide less heat should it be used at a lower voltage, such as 200 or 220 V. For example, a heater rated 1000 W at 230 V would deliver only 870 W at 200 V.

14.3.6.2 Heaters

The following sections will provide the designer with a brief introduction to the heater types. For more detailed information, the designer should contact their heater or hot runner supplier.

There are essentially two types of hot runner heaters: cartridge heaters and tubular heaters. In both types, a "resistance" wire is connected to an electric power source. The electrical energy is transformed into heat, which is then given off to the surroundings. The resistance wire is usually drawn very thin (usually round, occasionally flat), and then wound into very small spiral coils 1–2 mm in diameter. These coils are in some cases again wound into larger coils. One or the other of these spirally wound coils are then used in the heaters. (Thicker wires are usually wound only once.)

14.3.6.2.1 Tubular Heaters

The vast majority of the heating in a hot runner is done with tubular heaters. These tubular heaters are similar to the heating elements on a stove. A tubular heater is a coil or resistance heater that is embedded in silica and then placed in a tube. The ends of the tube are sealed and leads are attached via a cold junction point. This allows the heater to be bent into a pattern needed to provide even heat to the hot runner. The wattage of a heater should be limited to a maximum of 10 W/cm^2 (65 W/in^2) of heater surface area. This figure applies to good heat sink conditions, that is, where the heater is cast into a heated body or pressed into the steel of the hot runner manifold. It should be less, 5 W/cm^2 (33 W/in^2), when the tubular heaters lie in a groove, with good contact on about 50% of its surface, and poor or no contact with the rest of the surface.

Tubular heaters can be easily formed to virtually any shape, by using appropriate bending tools, and by observing certain rules. There are limitations on the minimum bend radius allowed, and there must not be any bends in the area of the heating wire where it is joined to the connecting wires. If a bend there is unavoidable, a special length for the unheated length must be requested with a special order. Some heater suppliers make tubular heaters that are more flexible (see Figure 14.46).

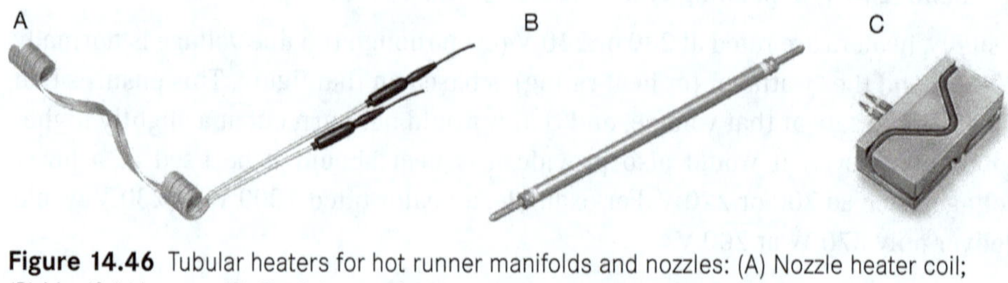

Figure 14.46 Tubular heaters for hot runner manifolds and nozzles: (A) Nozzle heater coil; (B) Manifold heater; (C) Flexible manifold heater (Courtesy of Hotset GmbH)

The fact that the terminals are on the ends of the heater can sometimes make it more difficult to wire the mold. This is less of a problem in manifolds, where the heaters can usually be bent so that the terminals are near each other.

After the shape of the heater is established, a groove is cut into the manifold. It is now important that good contact with the manifold is ensured. There are several methods used by the various manufacturers. Fill the groove with special, heat conducting cement, push the heater in, and smooth the surface flush with the manifold surface. This is often adequate, and is not expensive. If required, the heater can be easily removed. Alternatively, the heater can be brazed into the groove. This ensures perfect contact, but the heater is now part of the manifold, inseparable, and would have to be machined-out for replacement.

14.3.6.2.2 Cartridge Heaters

The use of cartridge heaters should be avoided if possible, mainly because of poor heat distribution within the heater and difficulties in achieving uniform heat transfer to the surrounding steel. There are also other mechanical and insulating problems with these heaters which make them more susceptible to damage. In general, they have a shorter life than tubular heaters but are less expensive and easy to obtain, as most dealers carry a wide variety in size and wattage of these heaters.

A cartridge heater is constructed as follows. A length of coil (with terminals welded to both ends) is wound through and/or over a highly heat-resistant ceramic insulating support, with both terminals ending at one end of the insulator. This is then inserted into a metal sheath (usually made from stainless steel or a copper alloy), and then crimped to seal the assembly. In its application, the cartridge is placed into a suitable bore. The fit must be tight enough to ensure good contact with the heated mass, but it should not too tight, to permit easy removal of a burnt-out cartridge. It is helpful if the bore is a through-hole, to permit pushing the cartridge out when needed.

One advantage of cartridge heaters is that both power terminals are at the same end of the heater, which can simplify wiring. There are, however, several disadvantages to cartridges. If contact with surrounding steel is poor (poor machining, or rust), hot spots can cause premature burnout of heaters. Accumulation of dirt around the terminals, especially in dusty, humid surroundings, causes electric flashover and short circuits.

Cartridge heaters usually have a surface wattage of about 5 W/cm^2 (33 W/in^2), and are used where use of a tubular heater is not viable.

14.3.6.3 Heat Input per Mass of the Hot Runner Manifold

A simple, useful rule for designers of hot runner manifolds is that the heater capacity required for a hot runner system (provided it is well insulated from, and has little contact with, the surrounding cooled mold plates) is about 2–2.5 W/cm^3 (30–40 W/in^3) of total mass of manifold. Using a larger wattage than suggested above may create hot spots and overshoot, and is usually not required. Also, it could increase considerably the cost of the system by requiring larger controllers.

In a well-designed hot runner system, once the plastic in the manifold and in the nozzles is at the molding temperature, the only heat supplied by the hot runner heaters (manifold and nozzles) is to make up for the heat losses to the surrounding mold plates. The plastic is heated in the extruder to the required melt temperature, and its temperature is maintained as it flows through the hot runner system, without requiring any additional heat input from the hot runner heaters.

In order to prevent cold spots in the manifold, it should be evenly and well-covered by the manifold heaters, preferably on each side (injection and clamp side).

14.3.6.4 Thermocouples (T/Cs)

A thermocouple (or T/C) is a sensor that measures temperature (Figure 14.47). It consists of two different types of metals, joined together at one end. When the junction of the two metals is heated or cooled, a voltage is created that can be correlated back to the temperature. Thermocouples are widely used in injection molding wherever temperature needs to be controlled. This is particularly true in hot runners, where thermocouples are used in the sprue, manifold, and nozzles. There are two main types of thermocouples used by hot runner suppliers, J-type and K-type.

The differences between J- and K-type thermocouples are the metals used, the operating range, and the sensitivity. K-type thermocouples are made with chromel-alumei, have a range from -200 °C to 135 °C, and a sensitivity of 41 µV/°C. J-type thermocouples are made from iron-constantan, have a working range of –40 °C to 750 °C, and a sensitivity of about 50 µV/°C [5].

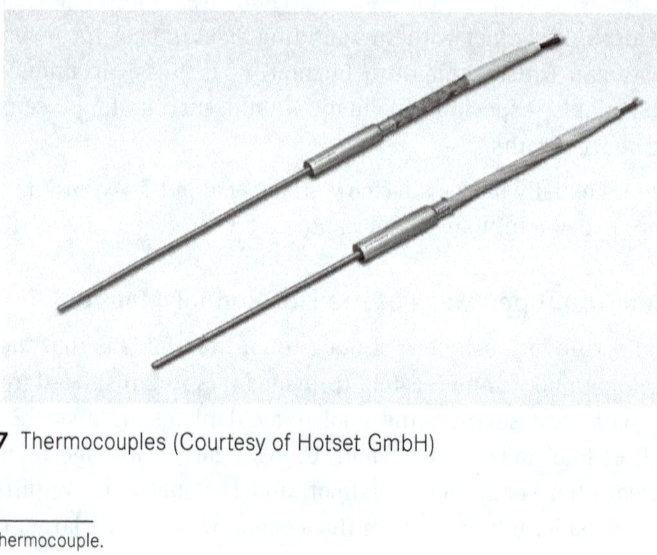

Figure 14.47 Thermocouples (Courtesy of Hotset GmbH)

[5] Wikipedia – Thermocouple.

The thermocouples and heaters are connected to a controller that adjusts the heat inputted to the device based on the feedback from the thermocouple. A thermocouple must therefore be strategically placed in the area to be heated, so that it accurately measures the temperature with a quick response time. The thermocouple should be as close to the melt as possible and in an area that represents the average temperature of that heat source. Incorrect location of the thermocouple, a ground fault, a pinched wire, or the thermocouple loosening in its core are the main reasons for poor readings.

Ideally, each heater zone should have its own thermocouple and a back-up in each location. For a more accurate reading, a probe thermocouple should be used rather than a surface thermocouple. Hot runner nozzles and nozzle tips

The nozzle connects the manifold to the nozzle tip, and the nozzle tip controls the temperatures and flow in the gate area. Usually the nozzle tip is removable from the nozzle using a thread for ease of maintenance in the machine. Almost all hot runner nozzles are heated by a resistance heater and controlled by a thermocouple. In some cases, the heater is removable while in others it is brazed into the nozzle material itself, or into the tip in special circumstances and when temperature uniformity is critical. The hot runner should allow in-machine servicing of the nozzle tip, thermocouple, and heater (if removable), as shown in Figure 14.48.

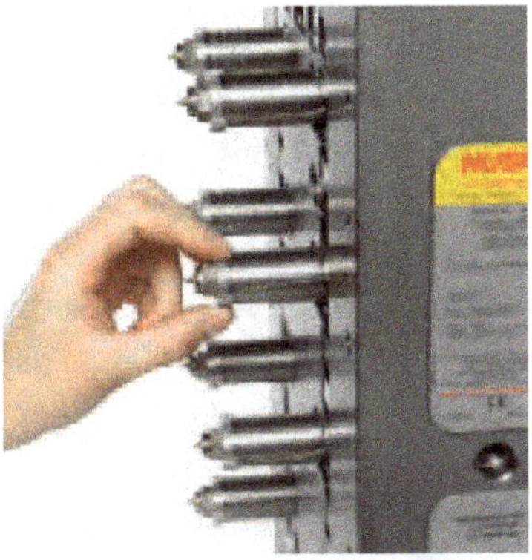

Figure 14.48 Nozzle tips, thermocouples, and heaters (if removable) should be accessible in the molding machine for servicing (Courtesy of Husky Injection Molding Systems Ltd.)

14.3.7 Hot Runner Nozzles

There are two types of nozzles: sealing nozzles and a screw-in nozzles. Sealing of these types of nozzle was discussed in Section 14.3.5.2.

The *compression sealing nozzle* (typical example shown in Figure 14.49) has a skirt on it that supports the manifold, generates forces to seal the interface between the manifold and nozzle, and restricts heat loss. The nozzle remains fixed in the manifold plate while the manifold slides over it during heat-up and expansion. Generally, this type of nozzle is used for short nozzle sizes (< 180 mm), small-diameter nozzles (< 25 mm diameter), and high-cavitation molds.

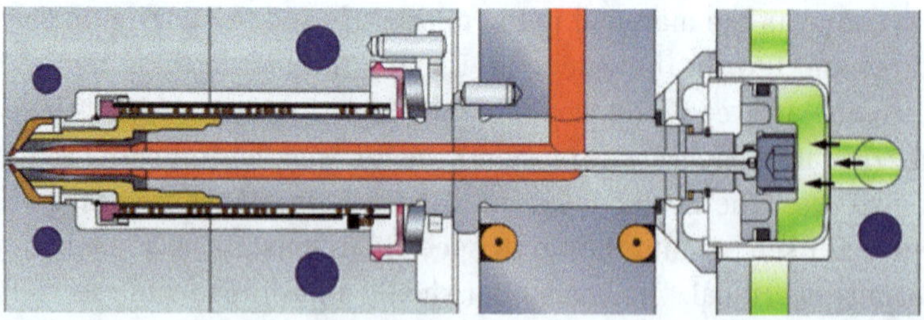

Figure 14.49 Compression-sealing nozzle design (Courtesy of Husky Injection Molding Systems Ltd.)

A *screw-in nozzle* threads into the manifold to create a seal, and moves at one end with the manifold as the manifold expands when heated. This type of nozzle is used frequently for large nozzles and long nozzles, as the thread creates a better sealing force for larger channels and the long nozzles have the flexibility to bend slightly as the manifold heats. A threaded-in nozzle creates a fixed joint between the nozzle and the manifold. This can be a challenge, as the nozzle is fixed in the gate insert and the manifold moves as it heats and expends.

Screw-in nozzles should only be considered in the mold design when the total length of the nozzles is **more than half of the distance** between the nozzle position and the manifold locator. Otherwise the threaded interface with the nozzle will be subjected to excessive stresses that could cause leakage or breakage.

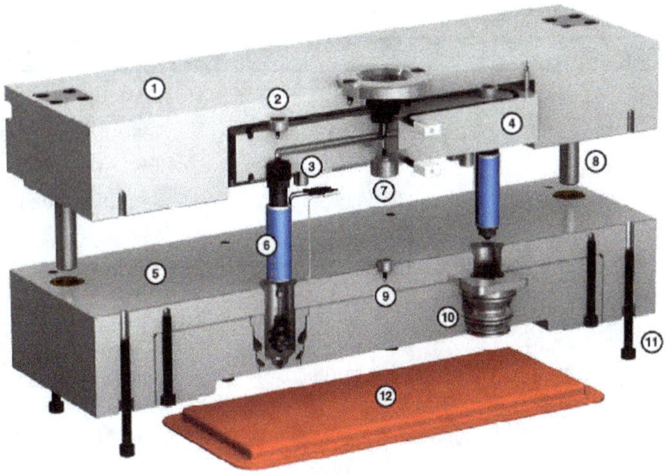

Figure 14.50 Typical screw-in nozzle: (1) Manifold plate; (2) Manifold back-up insulator; (3) Manifold support pillar (screw-in nozzles only); (4) Manifold; (5) Cavity plate; (6) Threaded nozzle; (7) Manifold support locator; (8) Guide pin; (9) Manifold center locator; (10) Gate insert; (11) Interface screws; (12) Molded part (Courtesy of Mold Hot Runner Solutions)

14.3.8 Nozzle Tips and Gating in Hot Runners

The gate is where the hot runner and the mold meet. It is an area where the resin transitions from a hot melt to a cold solid. During injection, it is also an area subject to high pressures, high shear rates, and large temperature increases. While all this is happening, the resulting gate must be for all intents and purposes invisible. There are two types of gates: hot-tip (or open) gates and valve gates. A summary of the advantages and disadvantages is shown in Table 14.4.

Table 14.4 Advantages of Hot-Tip and Valve Gating

Feature	Hot-tip (open) gating	Valve gating
Cost	Low	Higher
Shut height	Less	More
Gate vestige	20-50% of the wall thickness	Zero
Gate size	Small (< 2 mm)	Large
Pressure drop	Higher	Lower
Shear rate in gate	Very high	Lower
Cycle times	Longer (wait for gate freeze)	Shorter
Maintenance	Low	Higher
Processing (for gate vestige control)	Harder (stringing, high gates)	Easier (fewer gate defects)

Table 14.4 Advantages of Hot-Tip and Valve Gating *(continued)*

Feature	Hot-tip (open) gating	Valve gating
Decompression	Needed	Not needed
Gate wear	Some wear due to high shear	Wear due to valve pin over time
Ease of manufacture (mold cost)	Relatively easy	High tolerances needed in gate
Importance of thermal profiles	High importance	Less importance
Use of fillers	Harder	Easier
Use for engineering grade materials	Low usage	High usage
Energy usage	Lower	Higher (air and oil for piston)
Need for controls	Less (temperature only)	More (temperature and air/hydraulic valving)

14.3.8.1 Hot-Tip Gates

In the case of a hot-tip (or non-valved) gate, the gate seal-off is created by thermally cycling each shot. At the start of injection, the gate is barely frozen. The injection pressure forces this cold slug out of the gate and into the part. The high shear rates re-melt this slug of partially cooled plastic as it enters the cavity. During injection and hold, the gate is wide open to allow for filling and packing of the part. During cooling, the gate area starts to cool and solidify. During this time, the hot runner channels are decompressed by screw pull-back and decompression. At the end of the cooling section of the molding cycle, the gate is again frozen-off enough and decompressed so that plastic does not drool out of the gate.

The main challenge in all hot-tip applications is to keep the gate small (so it can freeze-off) while still being able to rapidly fill the mold. These gates work well with commodity resin types such as PP, PE, and PS.

The shape of the gate, such as diameters, any tapers, and the length of the gate (the gate land) are critical, and must be identical for all cavities in a multi-cavity mold. Also, the temperature of the gate areas must be the same for all. The temperature of the nozzle, the mold cooling, and the design of the nozzle tip affect this temperature. There is a lot of very close-tolerance machining work required in both these areas.

The set-up of the mold (proper adjustment of the molding parameters) has a direct influence on the performance. In proper operation we expect that the gate freezes just enough by the time the mold opens for ejection to stop any residual pressure of the plastic inside the hot runner system from pushing plastic into the cavities and cause "drooling". At the beginning of the next injection, the frozen plug in the gate is pushed into the cavity. In some cases (e.g. with clear plastic), the plug can sometimes be seen somewhere in the product. But it can also melt and mix with the next shot.

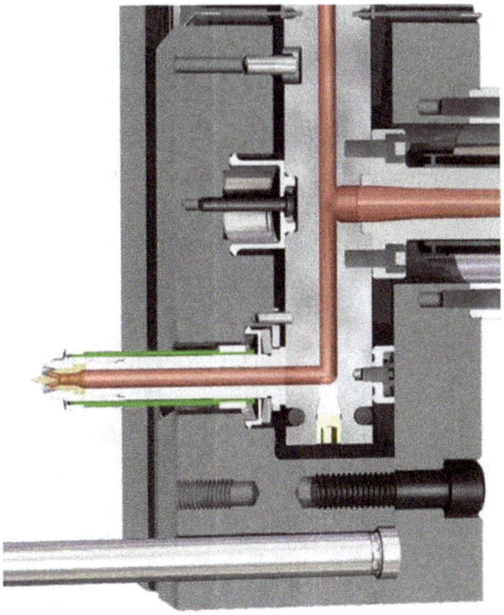

Figure 14.51
Hot runner hot-tip gate system (circular and annular) (Courtesy of Husky Injection Molding Systems Ltd.)

If the gate area is too hot, plastic will drool into the (now open) cavity during ejection and create a partially solidified slug of plastic. This glob can be cool enough when the mold closes again, and could leave a bad mark at the gate area of the product, it could prevent the cavity from filling at the next cycle, or in some cases may even activate the mold protection by causing it to sense a foreign object in the mold.

If the gate area is too cold, it cannot be pushed into the cavity during the next cycle. The cavity does not fill, and if not cleared the mold will continue to operate with fewer cavities, thus seriously affecting the production.

There are three main variations of open hot-tip gates: circular, annular and edge gates.

Circular Hot-Tip Gates

A circular gate is a round hole with a diameter D at its narrowest point, and a length (of the hole) of L, called the land.

The best way to make sure that the gate always breaks at the same spot is to make the gate length L tapered from both the nozzle and cavity side; the plastic will always break where the tapers meet (at D). The gate vestige will be then a small, short cone. There are two basic designs of circular hot-tip nozzle tips. Figure 14.52 shows typical examples of the most common designs.

Figure 14.52A and B illustrate the most common configuration. The nozzle tip does not form part of the molding surface, but is held inside a plastic bubble formed by the gate insert near the gate. The advantage of this design is that the hot nozzle tip does not need to form part of the molding surface, and so does not affect the cycle time. The disadvantage is that this insulating bubble formed by the molded plastic can hinder color changes. The alternative typical design is shown in Figure 14.52C, D, and E. Here, the nozzle tip forms part of the molding surface, eliminating the insulating bubble. This makes color change much faster and prevents any degradation in the partially cooled bubble. However, this design creates a hot molding surface in the gate area, and so is not normally used in fast-cycling applications (< 15 s).

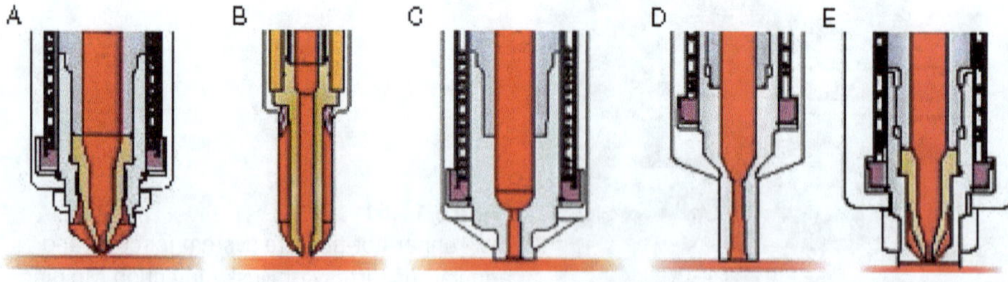

Figure 14.52 Typical circular hot-tip gate designs (Courtesy of Husky Injection Molding Systems Ltd.)

Annular Hot-Tip Gates

These gates (Figure 14.53) were developed to keep the gates from stringing or freezing. The point of the nozzle tip must be accurately positioned in the center of the gate. The plastic flow from two (or more) holes in the side of the nozzle rejoins to create a tubular flow where it enters the cavity space. The advantage of this arrangement is that while the total cross-section of the plastic flow can be relatively large, the actual wall thickness of the "tube" can be very small. This permits the flow close to the outside to be cooled rapidly, while a very thin "skin" hugging the nozzle point remains fairly hot. If properly dimensioned, these nozzles do not string, and do not drool.

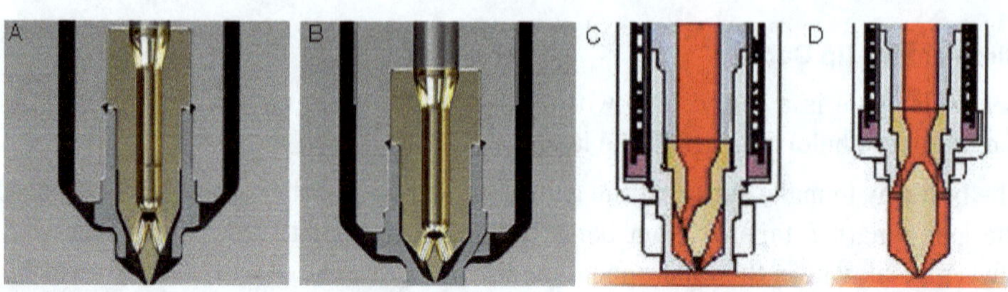

Figure 14.53 Typical annular hot-tip gate designs (Courtesy of Mold-Masters (A, B) and Husky Injection Molding Systems Ltd. (C, D))

Edge (or Side) Hot-Tip Gates

These gates (Figure 14.54) are used mainly for products for which gating into the top-center is not permitted, such as hollow parts without a bottom face.

The main difference in an edge-gated part is that they are naturally self-degating. After injection, when the mold opens to eject the product, the gate is shorn off at the cavity wall, leaving a small oval, dull vestige. The cold slug remaining in the gate will be driven into the cavity space during the next injection and melts with the incoming plastic. The mold designer must make sure that the length of the slug is shorter than the distance to the opposing core wall, so that the slug will not stay in the gate and plug it.

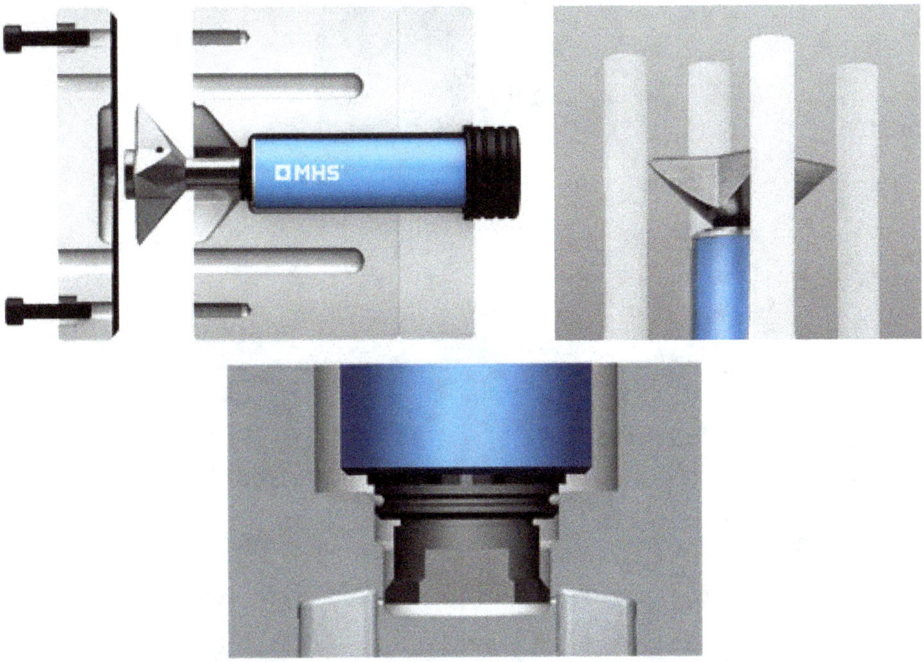

Figure 14.54 Typical edge hot-tip gate design (Courtesy of Mold Hot Runner Solutions)

14.3.8.2 Valve Gates

Valve gates (Figure 14.55) are circular gates with a stroking pin that fit closely inside the gate hole to control the plastic flow between injection cycles. Proper (timely) opening or closing the gate does no longer depends on the accuracy of the temperature and pressure settings of the machine, or on the shape of the gate. For the reasons listed in Table 14.4, a valve gate is often the option of choice for mold designers.

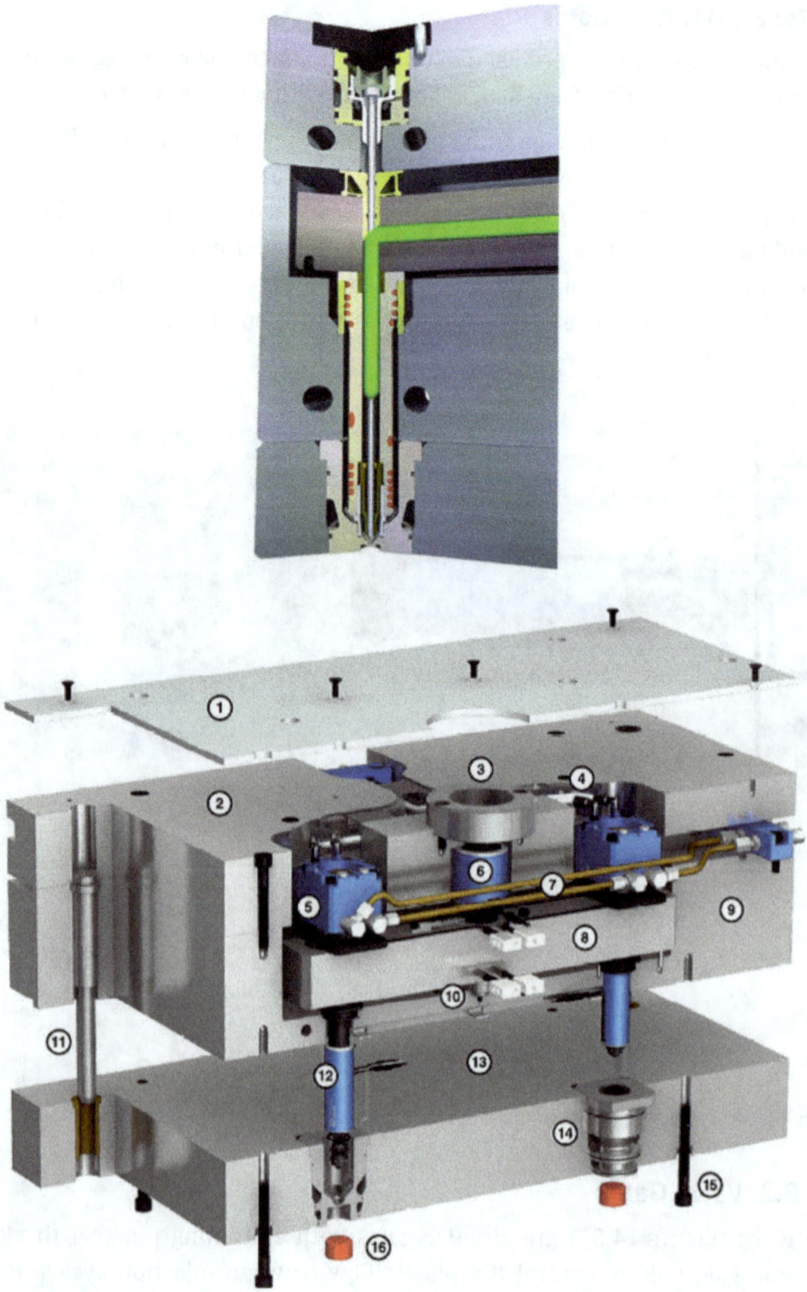

Figure 14.55 Typical hot runner valve gate nozzle stack designs. Top: Piston in backing plate design. Bottom: (1) Insulating plate; (2) Top clamp plate; (3) Locating ring; (4) Air/hyd supply line); (5) Valve gate piston; (6) Sprue bushing; (7) Actuator cooling; (8) Manifold; (9) Manifold plate; (10) Manifold support locator; (11) Guide pin; (12) Sealing nozzle; (13) Cavity plate; (14) Gate insert; (15) Interface screws; (16) Molded part (Courtesy of Mold-Masters (top) and Mold Hotrunner Solutions (bottom))

There are two types of valve gate shut-offs with the gate insert: a tapered shut-off and a cylindrical shut-off (see example configurations in Figure 14.56). A tapered shut-off has the advantage of being self-centering and the shut-off taper cools the valve pin in each cycle. However, in some cases plastic material can be caught between the pin and the colder gate insert, causing high gates and sometimes partially frozen material being injected into the part. A cylindrical shut-off is often chosen when the gate vestige and gate appearance is critical. The pin and gate must be closely toleranced to allow the pin to cool in each cycle, and the pin must be guided into the gate hole so that neither the pin nor the gate insert is damaged. Eventually, gate wear may occur, and the valve pins may need to be replaced and the gates refinished. However, this should not occur before millions of cycles are completed by the mold (between 2 and 5 million cycles).

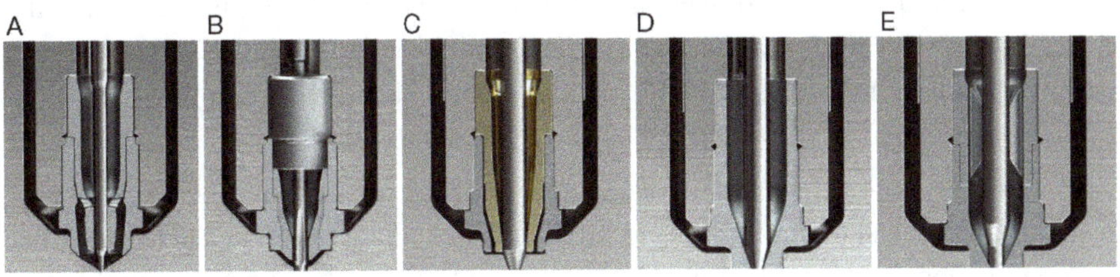

Figure 14.56 Tip styles for valve gating: (A) Accuvalve CX; (B) Accuvalve MX; (C) Bimetallic C valve; (D) Hot valve; (E) Cylindrical hot valve (Courtesy of Mold-Masters)

The gate size for a valve gate can be much larger (2–9 mm diameter) than is possible with open gates. This is of great advantage when filling large volumes of plastic rapidly. Large gates are more suitable for heat-sensitive and shear-sensitive plastics because the plastic is not forced at high speed through a small opening. Valve gates are not sensitive to dirt in the plastic; the machine will rarely stop producing because of plugged-up gates, which is sometimes the case with hot-tip gates.

Although a valve-gated hot runner is more expensive, the advantages of ease of processing, predictable vestige, low gate vestige, and reliability make it a popular choice with mold designers.

Valve Gate Pistons

There are several design options to operate the valve pins. In most cases the valve pins are controlled with air pistons or hydraulic pistons, although electric pistons are becoming more popular as the price and complexity of electrics decreases. The piston can be located on the hot manifold or in the manifold backing plate (see Figure 14.55). Both solutions work well, and are compared in Table 14.5.

Table 14.5 Valve Gate Piston in the Back Plate versus Piston on the Manifold

On manifold	In manifold back plate
Hotter piston requires high temperature O-rings and periodic maintenance	Cooler piston uses common O-rings and requires minimal regular maintenance
Piston moves with manifold expansion to achieve alignment. Valve pins are tight in the piston.	Piston is fixed in location. Valve pins float in piston.
Back plate must be removed to access pistons and pins	Valve pins accessable via back plate covers. Valve pins must be removed prior to hot runner dissassembly

The air or hydraulic oil is controlled by valves on the machine or in a standalone controller. Usually one valve controls some or all of the pistons. It is not typical to have an individual control for each nozzle piston. Instead, all the pistons on a main manifold are typically grouped together and controlled by a single valve. If there are several main manifolds joined by bridges, then each main manifold may have its own valve for control to keep the circuit size manageable.

In some cases, when the spacing of the nozzle drops is very close, a single piston or plate can be used to drive multiple valve pins (see Figure 14.57). These systems, typically called synchro plate designs, increase the shut height of the mold, add cost, and work best when all cavities are running all the time (i.e., no shut-off cavities).

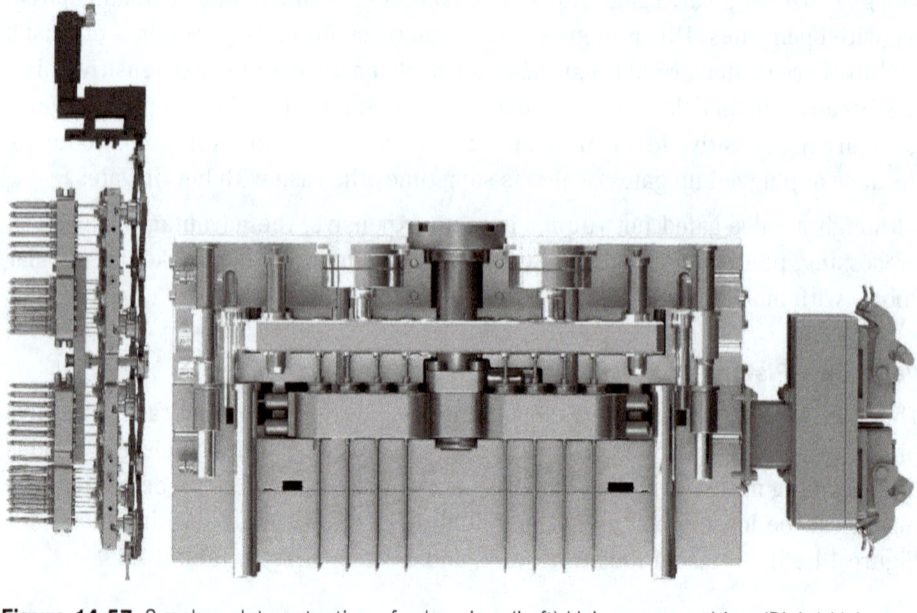

Figure 14.57 Synchro plate actuation of valve pins: (Left) Using a servo drive; (Right) Using a hydraulic piston (Courtesy of Mold-Masters)

14.3.8.3 Slot Gating

All hot runner gating configurations up until this point have been round or annular. A slot gate allows for rectangular or slotted gating, similar to tab or edge gating in cold runners. The benefit of this style of gating is it allows for gating on ribs and bosses and in locations that a round gate would not fit. Since it can be almost any configuration of length × width, the area available for flow can be much larger than a round or valve gate, resulting in lower pressure drops, less shear heat generation and shorter cycle times. Wear and service requirements may be higher for this type of gating.

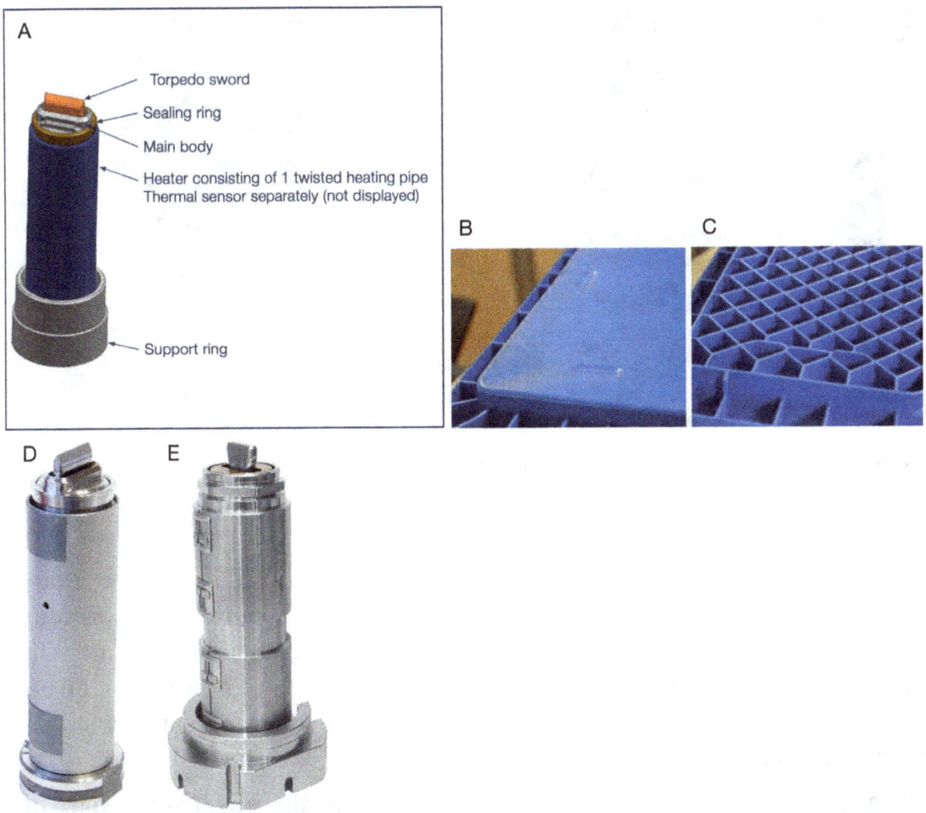

Figure 14.58 Slot or flat die unit gating: (A) Nozzle description; (B) Box bottom; (C) Crate bottom; (D) FDU open nozzle; (E) FDU valve-gated nozzle (Courtesy of FDU Hotrunner)

14.3.9 Hot Runner Temperature Controls

The manifolds, sprue bushing, and heaters on the nozzles need to be controlled so that the temperature remains constant in the nozzle, does not create hot spots, and has enough power at start-up to heat the solidified plastic. This can be accomplished by using mold temperature control zones on the machine, but more often by using a standalone hot runner temperature controller like those shown in Figure 14.59.

Figure 14.59 Hot runner temperature controllers (Courtesy of Mold-Masters (left, center) and Husky Injection Molding Systems Ltd. (right))

The controls for hot runners require a more precise algorithm than those for a machine barrel or other machine components. Standalone hot runner temperature controllers have been specially designed for the heating requirements in the hot runner using PiD or PiD^2 control algorithms. They also have special heat-up and cool-down sequences to ensure that the heaters have a long life, that they have no moisture, and that there are no ground faults.

The manifold heaters can be paired together with other similar zones to reduce the number of control zones needed, as long as the required electrical current is within the range of the zone. The nozzle tips can be individually heated and controlled, or in some cases grouped together and controlled by a single zone. While this saves money on the number of control zones and amount of wiring, it reduces the control of the mold in a critical area. To give the molding technician as much control of the process as possible, individual zones for each nozzle are highly recommended.

14.3.10 Hot Runner Gate Location

There are computer programs designed to provide guidance to the mold designer on where to gate. An example of this is shown in Figure 14.60. The gate size and geometry itself is normally recommended by the hot runner supplier. Many hot runner suppliers will also conduct a flow analysis to identify the optimal gate position as an integral part of the service they offer, but ultimately the gate placement will be the responsibility of the mold maker or customer.

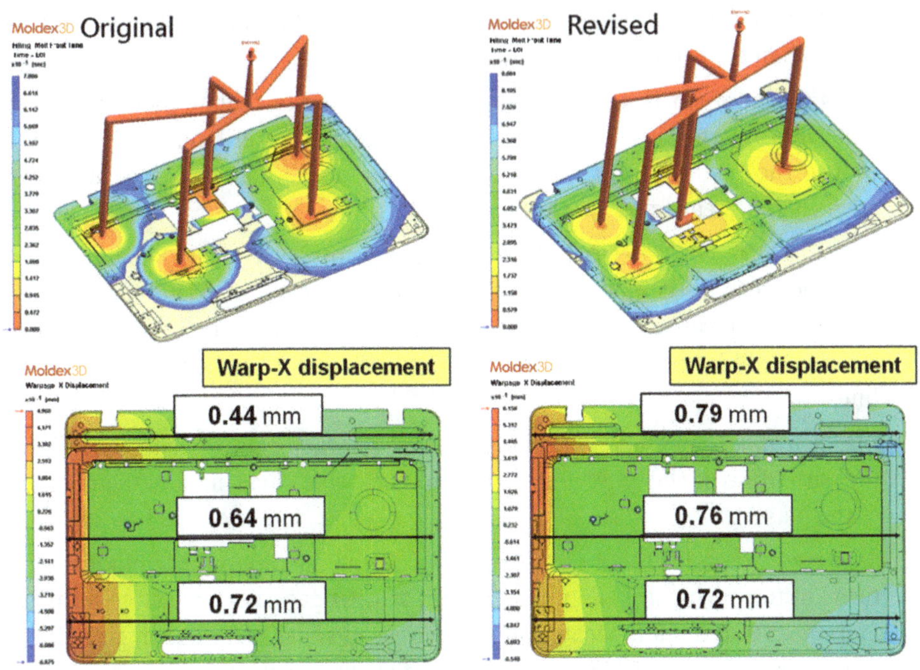

Figure 14.60 Hot runner flow balancing of a computer housing using CAE software [6] (Courtesy of Moldex3D)

14.3.11 Hot Runner Gate Sizing

The determination of a suitable gate size is part of the design of the complete hot runner system. The gate size affects the plastic injected, the stresses created within the product, and the gate appearance or vestige. There are a number of factors that can affect the choice of the best style and size of gate.

[6] Source: M. Wang, R. Chang, C. Hsu, *Molding Simulation: Theory and Practice,* Hanser Publishing, 2018 (Figure 7.31).

1. *Part mass and size:* The longer the flow length and the larger the cavity surface, the larger must be the required gate to reduce the fill pressure and to ensure a sufficient flow of plastic to prevent premature freeze-off of the gate.
2. *Product wall thickness:* A product with a large wall thickness requires a larger gate to supply fresh plastic to the shrinking product during the holding (packing) stage of the injection cycle. An undersized gate would produce a warped or unfilled product. Typically, the gate diameter is smaller than the wall thickness, because once the cavity is full and cooled, the gate must also freeze to allow ejection of the product without creating an ugly vestige. However, for thin-walled products (wall thickness $t < 1$ mm), the gate diameter is often larger than t to reduce the pressure drop at the gate.
3. *Resin:* If the resin is very viscous (i.e., has a low melt flow index and high molecular weight distribution), a larger gate and a shorter land are required to reduce the restriction in the gate area. See Table 14.6 regarding suggested permissible shear rates for various materials.
4. *Location of cooling lines (mold temperature):* Improperly located cooling lines, e.g., too close to the gate, will cause premature freeze-off. If too far from the gate, they will cause drooling. Gate diameter selection can be used to compensate. Select a larger gate for close cooling lines and a smaller gate for distant lines.
5. *Injection time:* Very fast injection requires a larger gate to reduce the local pressure drop at the gate, and to prevent excessive shearing and degradation of the plastic.
6. *Melt temperature:* If the melt is processed at the maximum possible temperature but the cavity still cannot be filled, a *smaller* gate can be used to transmit more heat to the resin because of the increased shear heating, which reduces the viscosity further and makes the resin flow more easily.
7. *Entrance effects:* Sharp corners or restrictions impede the flow of the resin and can cause shear-induced degradation. Use a generous radius on the cavity side of the gate to help provide smooth laminar flow and to prevent jetting or blushing.
8. *Nozzle tip position and tip type:* If the tip is closer to the gate, the gate is less likely to freeze-off prematurely. As a result, a smaller gate can be used, provided the maximum shear stress and pressure drop are not exceeded.

14.3.11.1 Gate Sizing Methods

There are three methods commonly used for estimating hot runner gate size:

1. Use past experience from earlier designs or from information given by experienced designers or molders. This is the most common method, and is often

practical, but it may be misleading when applied to a mold which is (erroneously) believed to be similar to another mold.

2. Use a computer analysis program, as described in Section 14.3.11.2. This (better) method is used today by many mold makers.
3. Use the empirical approximation based on experiments described by R. W. G. Pye [7], and described in Section 14.3.11.3. This method may yield a good enough gate size, and can be used when the necessary software is not available.

14.3.11.2 Gate and Melt Channel Size Calculation

The gate is part of the hot runner, and should be designed together with the rest of the hot runner system. The complete system can then be modeled using CAE simulation analysis. Pressure, shear stress, and temperatures can be checked for the entire system, and the impact on part cooling and warpage can also be simulated (see Chapter 12 on cooling).

The following limitations apply when determining the correct channel sizes in the hot runner manifold:

- Pressure drop (Δp): $\Delta p < 40$ MPa or 6000 psi ($\Delta p < 35$ MPa or 5000 psi for general-purpose (GP) plastics)
- Melt temperature rise (ΔT): $\Delta T < 15$ °C or 60 °F, but less for shear-sensitive plastics
- Shear rate ($\dot{\gamma}$)
  - For manifold: $\dot{\gamma} = 1000\ s^{-1}$ (to allow easy color changes)
  - For gate: See Table 14.6
- Shear stress S_s: See Table 14.6 for *gate and manifold.*

Table 14.6 Suggested Maximum Shear Stress and Shear Rates in a Hot Runner

Material	Max. shear stress S_s (MPa)	Max. shear rate $\dot{\gamma}$ (s^{-1})
PP	250,000	100,000
HDPE	80,000	40,000
LDPE	80,000	40,000
PS	250,000	40,000
HIPS	300,000	40,000
SAN	300,000	40,000
ABS	300,000	50,000
PPS	345,000	50,000

[7] R. W. G Pye, *Injection Mould Design: Manual for the Thermoplastics Industry* (4th edn), Halsted, 1989.

Table 14.6 Suggested Maximum Shear Stress and Shear Rates in a Hot Runner *(continued)*

Material	Max. shear stress S_s (MPa)	Max. shear rate $\dot{\gamma}$ (s^{-1})
Nylon	500,000	60,000
PET	500,000	6000
PUR	250,000	40,000
PBT	400,000	50,000

The critical fill rate of the cavity must not exceed the maximum permissible shear stress (or shear rate) for a material. If this value were reached, flow-induced brittleness (melt fracture) may result, causing molding difficulties and poor product quality. These values can be calculated for the entire hot runner system using flow analysis for hot runners.

14.3.11.3 Empirical Calculation of Gate Size

If a computer program or advice from the hot runner supplier is not available, the designer can use Equation 14.6 to find a reasonable hot runner gate size.

$$D = C \times N \times \sqrt[4]{A} \tag{14.6}$$

where

A = approximate total (not projected) product area in mm²

C and N are experience factors that depend on the average product thickness (Table 14.7) and the type of molded plastic (Table 14.8).

Table 14.7 Variation of Experience Factor *C* with Wall Thickness *t*

Wall thickness *t* (mm)	*C*
0.75	0.178
1.00	0.206
1.25	0.230
1.50	0.242
1.75	0.272
2.00	0.294

Table 14.8 Variation of Experience Factor *N* with Resin Type

Resin	*N*
PE, PS	0.6
PC, PP, Acetal	0.7
Nylon	0.9
PVC	0.9

14.3.11.4 Gate Land Length

The land should be as short as possible to achieve lower pressure for filling and to improve degating by reducing the height of the vestige. The only argument against short lands is the loss of strength of the steel in this area, which may cause the steel to break.

> Land lengths should be in the range 0.00–0.13 mm (0.000–0.005").

14.3.12 Gate Inserts (Gate Pads)

The simplest method is to machine the gate right into the cavity bottom. However, when molding abrasive plastics containing glass fibers or other fillers, the gates will be rapidly eroded. But even unfilled plastics contain additives (colorants, etc.), which combined with high temperatures, pressures, and flow speeds, will over time wear down the gate, and disturb the required balance of gate size and machine settings. More details about the design of the gate insert and gate pad is covered in Chapter 10 on cavity and core design, and in Chapter 12 on cooling.

The usual cavity material is steel, and to cool the gate area, cross-drilling cooling channels close to the gate is a good solution. This is often done and quite successful, but quite a challenge to the mold designer. If a mold is to be used for many millions of cycles, the problem of gate erosion still remains. If the gate erodes, it could be welded and re-machined. However, this is as a rule not recommended. Gate inserts (gate pads) as shown in Figure 14.42 and Figure 14.61, provide the proper answer. There are several alternatives, and the decision of which one to select depends on calculating the product cost over the expected life of the mold, using the various alternatives.

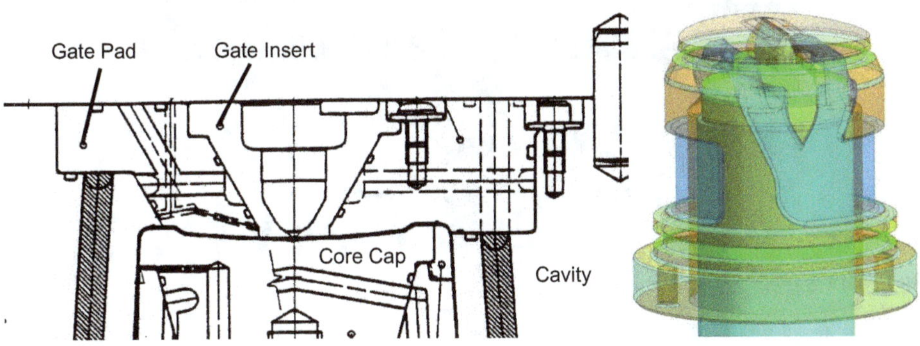

Figure 14.61 Gate inserts (see also Chapter 12 on cooling) (Image on right courtesy of FDU Hotrunner)

Some options for materials for gate inserts are:

- BeCu. This could easily reduce the cycle time by 20% (large gain in productivity, especially with short cycles), but the inserts would have to be replaced quite regularly to maintain the high product quality required. If the inserts wear after time, they are easily replaced.
- Hardened steel similar to the cavity steel. This allows designing the cooling around the gate more efficiently (and at the same time making it less expensive) than by just cross-drilling. The better cooling should also result in some saving in the cycle time.
- Special very hard, abrasion-resistant steels. This still makes it easier to cool the area and to cost-effectively replace the simple inserts, but such steels are usually poorer conductors than mold steel. They may require some additional molding time, but they practically never require replacement.

14.4 Hot and Cold Runner Molds in Combination

Combinations of hot and cold runner molds are usually selected for cases where cold runner molds (edge- or center-gated) would have advantages over hot runner molds. This is done sometimes for very small products to avoid excessive residence time in the hot runner system, or for very large products to reduce the pressure drop from machine nozzle to the gates. If the distances from nozzle to gates are very large, a hot-to-cold runner system would significantly reduce the amount of cold runner ejected on each cycle.

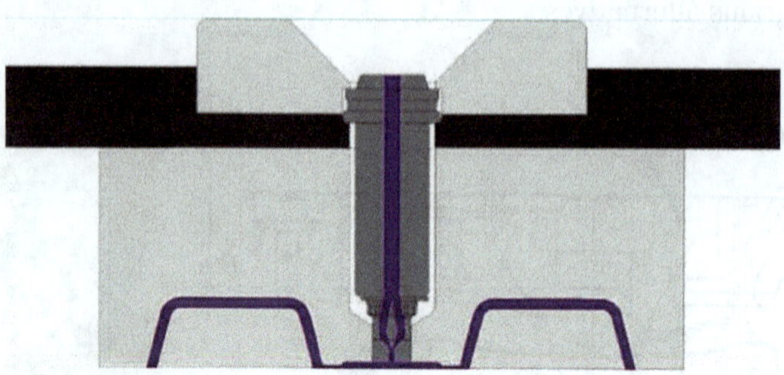

Figure 14.62 Hot-to-cold runner layout

Two typical examples are shown in Figure 14.63 and Figure 14.64. Many such combinations are possible.

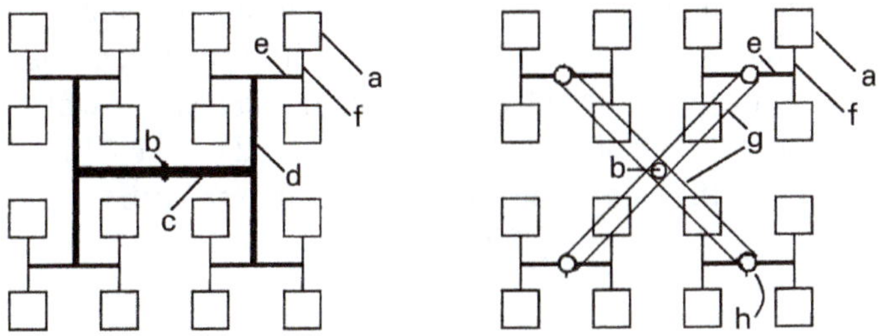

Figure 14.63 Schematic of a 16-cavity mold: (Left) Common cold runner mold; (Right) Combination hot-and-cold runner mold. (a) Cavity, (b) sprue, (c–f) cold runners, (g) hot runner

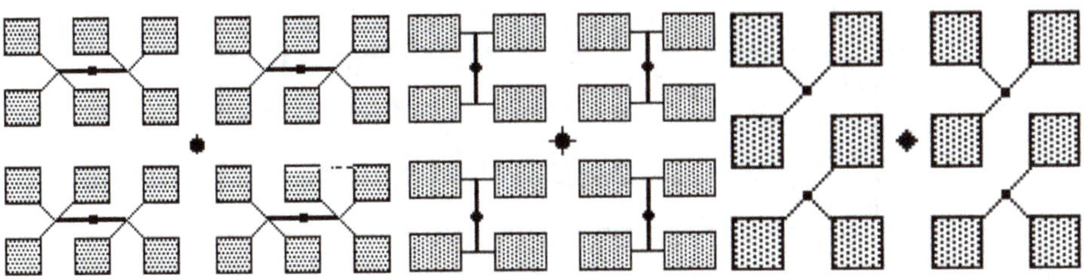

Figure 14.64 Typical hot-to-cold runner layouts (24, 16 and 12 cavities)

For multi-cavity molds (Figure 14.63, left), the runner system could become quite large. To prevent large pressure drops while avoiding unnecessary large masses of plastic, the runner channels taper down from a heavy cross-section where the plastic enters the mold at the sprue, becoming gradually smaller every time the runner splits, until it arrives at the gates (symbolized in Figure 14.63 by the line width of the runners). This will ensure that all cavities are filled properly. However, the heavy runners are difficult to cool, and add much to the plastic that must be recycled; this is wasteful, as explained earlier.

Figure 14.63 (right) shows a simple, schematic example for a similar mold in which the total runner system is divided into a (four-branch) hot runner system, with each branch then supplying a cold runner system, two-plate in this example. The pressure drop in the hot runner channels is small, and the final branches of the cold runner can be kept as small as they would be with the common runner layout shown in Figure 14.63 (left).

For very large products (Figure 14.65) as found, for example, in the automotive industry, the product should be edge-gated. The edge gates are located where best suited for the product, but instead of having a large cold runner to supply these edge gates, a (usually non-standard) hot runner manifold, or other method of ducting the hot plastic, will bring the plastic to the gates (or group of gates), without significant loss of pressure and without the need to reprocess the heavy runners. Another advantage is that the product can be placed approximately symmetrically around the center of the machine, for a balanced clamp force. This is possible with standard three-plate molds, but not with two-plate molds because the machine nozzle is in the center of the platens. (*Exception:* Special molding machines are equipped with an offset extruder, or one that can be located outside the platen and injected in the side of the mold, or even into the P/L.)

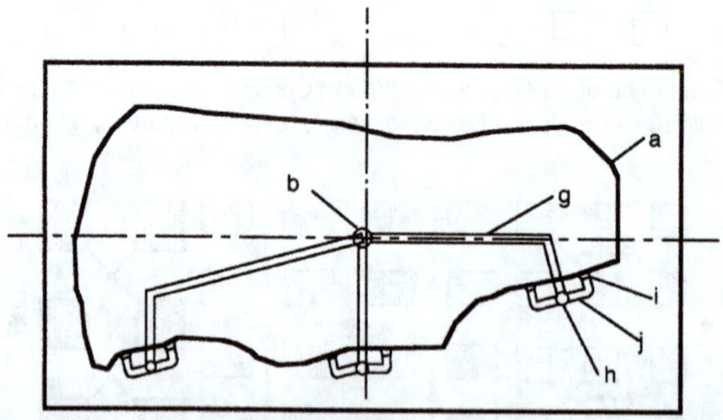

Figure 14.65 Schematic of a large molding, with six cold runner edge gates and a hot runner system with nozzles into each of the cold runners: (a) product, (b) sprue, (g) hot runner branch, (h) hot runner nozzle, (i) gate, (j) cold runner

■ 14.5 Selection of Hot Runner or Cold Runner System

For both products in Figure 14.67(a) and (b), the amount of plastic in the runner system is large in relation to the mass of products molded (25% in one, almost 100% in the other case). What could we gain by using hot runners?

The answer is mostly a matter of economics. We must consider the price of about \$2000 to \$3500 per hot runner nozzle drop (plus the price of plates if a hot half is being considered). On the other hand, the gain in cycle time and productivity by using hot runners can be considerable. A hot runner mold generally cycles about twice as fast with hot runners! Also, the cost of recycling the runners and the percentage of plastic lost during recycling must be considered. A smaller injection unit will be required with hot runners, because all of the injected plastic will be converted into products. There is also a saving in energy, because less plastic material is processed for the same output. The mold sizes are about the same, whether a three-plate or a hot runner mold is used. Therefore, it is just a question of the quantities to be produced. If the quantities are large, the savings in the cost of the product can easily justify the higher cost of the hot runner mold.

Figure 14.66 shows a runner system for a 32-cavity, two-plate mold for caps. The mass of the runner system is about 25% of the mass of all the caps per shot. The runner system is fed in the center by a hot sprue (not shown) to avoid an otherwise large, cold sprue. The cavities are tunnel-gated, and the products and runners are separated after ejection.

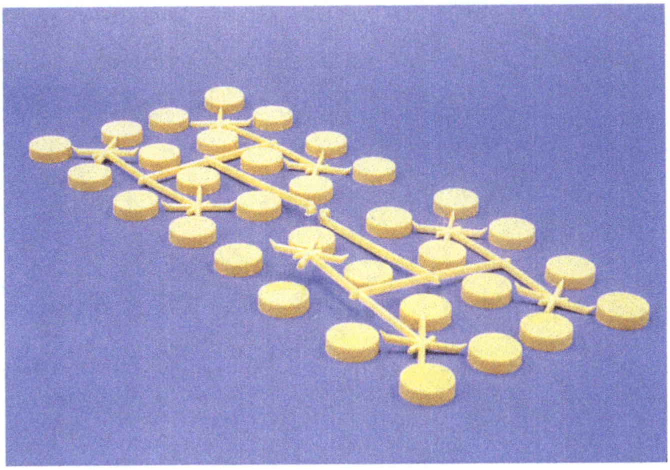

Figure 14.66 Runner system for a 32-cavity, two-plate mold for caps

Figure 14.67(a) shows a runner for a 72-cavity, three-plate mold for a small cap. The runner takes almost as much plastic as the products themselves. The cycle time is controlled by the cooling time of the (clearly visible) heavy distribution runners. A hot runner system for such a mold would reduce the injected mass by half and the mold would cycle twice as fast, but would be more expensive to build in the first place. This is a typical example where both methods must be considered, in view of the total requirements of the product. Figure 14.67(b) shows the runner and products from a very simple four-cavity, two-plate mold for a simple but heavy product. Note the heavy runners and the cold sprue. This too is a candidate for hot runners (valve-gated into the product) provided the quantities justify the greater expense.

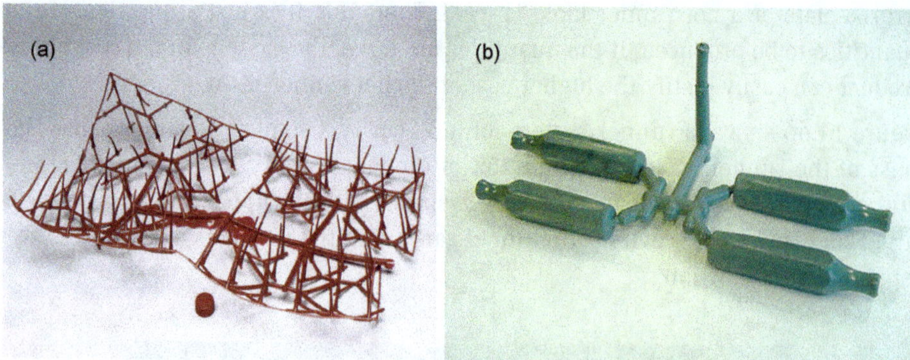

Figure 14.67 (a) Runner for a 72-cavity, three-plate mold for a small cap; (b) Runner for a simple but heavy product

15 Selection of Mold Materials

The designer must pick the appropriate materials for each component of the mold so that it meets its intended purpose. To do so, the designer must always clearly understand the purpose of the mold. Will it have a long life, or a short run and then be discarded? Will the mold run thousands or millions of cycles in a year? What will be the pressures and temperatures during molding? What plastic material is being used? What environment will the mold run in? What will be the price of the mold? What machining capabilities are available in the machine shop? These all impact the best choice of materials for construction of the mold.

Production molds are almost always made from steel, except for certain mold parts requiring materials with high thermal conductivity. Sometimes bronzes or even rigid plastics are used for cams and areas where moving parts cannot be lubricated, or must not be for sanitary reasons. Experimental molds may use softer materials such as aluminum.

The types of steels used depend on the requirements for each application. There are several points to be aware of:

- The material cost represents about 10–15% of the total mold cost.
- Steel costs vary widely, depending on the annual requirement of the mold maker. The more volume is contracted to purchase over a certain period, the lower the base cost of the steel.
- The weight of steel purchased has a significant bearing on the cost per unit of mass: large pieces are cheaper than small ones.
- There are weight, cutting, and other charges, so that even though the base price may appear to be low, by the time the piece is cut and delivered, the price is much higher.

It is always worthwhile to get feedback from your purchasing department and to contact steel suppliers to get a detailed quote before deciding on the materials to be used.

In general, steels, particularly mold and tool steels, are sold by brand names, different for each steel mill, but it is better to specify steels by their generic names or numbers. In general, there is little or no difference between steels of the same specification originating from different suppliers. However, new mold steels are constantly developed for "better" characteristics, and it may become necessary to re-evaluate and update the lists of steels used by the designers from time to time.

Mold designers should ensure that they consult with machine shop team members, and bring them in for design reviews, to ensure that the mold design can be effectively machined in the shop.

■ 15.1 Forces on the Mold Affect Material Choices

Injection molds are subject to many forces. These can be broken down into compression stresses, expansion stresses, and thermal stresses. The mold must be designed and constructed to withstand all these stresses for the duration of its expected life. The designer must know what forces are present within the mold when deciding on the materials to be used to resist these forces. The most important forces acting within the mold that affect material choices are:

1. *Tension:* the forces created by the injection pressure of the plastic inside the runner system and in the cavity space, usually requiring high tensile strength materials. As the hot plastic is injected into the cavity space, it exerts tremendous forces on the mold. These forces are resisted by the clamp pressure and the mold itself. As can be seen in Figure 15.1, the injection pressure acts equally and at right angles to all open spaces in the mold. These forces will tend to stretch and deflect the cavity walls outward. The greater the height *H* of the product, the greater this force will be, and the stronger the walls surrounding the cavity must be.
2. *Compression:* the compressive strength required to counteract the clamp force of the machine, typically, the forces on the P/L, and the forces seen where inserts are supported by plates, and so on. The clamping force is applied as soon as the mold closes. At this moment, the whole clamp force is resisted by the primary and secondary shut-off surfaces, which is the area surrounding the cavity that touches the core side. If this area is too small, inserts will be crushed and damage the sealing-off surfaces of the parting line, eventually ruining the mold. Proper sizing of the shut-offs and correct materials and hardness (steel, etc.), or other measures to counteract the clamping forces are the solution to this problem.
3. *Bending (or deflection):* the forces seen by cores, and by all plates, especially the ejector and stripper plates.

4. *Wear:* the forces created by wedge action, as in stripper rings and so on, or tapers and wedges for alignment, which create wear on the matching surfaces.
5. *Torsion:* the forces seen by screws, coil springs and in mold features, such as unscrewing, or in some robots.
6. *Shear:* forces seen by dowels, or by the backup of wedges.

Note that in many cases, a mold can be subject to combinations of any of the above forces.

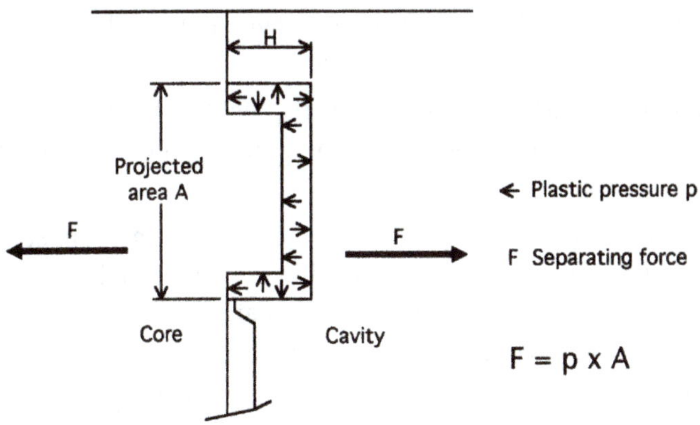

Figure 15.1 Portion of a schematic mold, showing a cavity filled with plastic under pressure acting in all directions

15.1.1 Characteristics of Steels and Other Mold Materials

For every mold part the following factors, in order of importance, must be considered. Unfortunately, some of these characteristics directly oppose each other (e.g., toughness and hardness) and compromises are necessary.

1. **Availability:** Is the material or hardware or standard mold component to be selected readily available in the required sizes within the required lead time? Many items are often shown in catalogues or other listings as standard, but this does not always mean that they are readily available, on the shelf, in the desired size, and in the quantities needed.
2. **Strength of material:** "Strength" is usually used to mean tensile strength; compressive strength is often (but not always) about the same. Shear and torsional strength are about one-half the tensile strength. The designer should always get the exact values from the supplier.
3. **Wear:** Some materials are better for wear than others. Lubrication (or the lack of it) can be a decisive factor. Wear points could be steel on steel, steel on

bronze, steel on hard plastic, and so on. Hard steels are always better, but the designer must never use the same alloy for both members rubbing against each other, as in wedges or taper locks, except if the wear points can be lubricated. Each alloy has a distinct, different grain structure, and surfaces with identical grain structures will lock (seize) when sliding under pressure, and damage (tear) the surfaces. Hardness differences alone are no substitute for different grain structure, except where one of the rubbing surfaces is treated with methods such as nitriding. In nitriding, very hard nitrogen compounds enter between the grains and alter the surface of the steel. Lubrication in molds is never permitted where it could contaminate the molded products, especially for pharmaceutical and food use.

4. **Corrosion:** It is important to decide how the mold will be protected from corrosion if the mold will be operated and stored in a humid environment. This can affect the mold cost.

   A common approach in many shops is to protect the molding surfaces before the mold is put into storage by using silicon spray ("Mold Saver") or to just apply plain, clean machine oil. Many shops paint the outside of the mold shoe with a permanent oil paint to protect the outside of the mold against corrosion.

   The mold shoe (mold plate) itself can be protected against corrosion with electroless nickel plating (ENP) rather than an oil-based paint. ENP has the additional advantage that it also protects some of the inner surfaces of the mold shoe, which would not normally be covered when the mold is just painted on the outside. ENP also enters the cooling channels to some extent and protects them against corrosion caused by the coolant, but the penetration is limited, and does not cover the walls of the channels deep inside the plates. ENP is hard (70Rc) but thin, and not resistant to scratches and wear.

   The best method may be to make the entire mold shoe from stainless steel (SS). The basic cost of SS is higher than the cost of mild steels or pre-hardened machinery steels. However, when SS is bought in large quantities, the cost difference can be much less. When molds are expected to run for a long time, the advantage of SS over other steels can justify the higher cost. We must not forget that chrome plating or ENP also cost money. We must also consider the costs of transportation to and from the nickel- or chrome-plater, the additional time required for these operations, the lack of control over the transport, and the dependence on an outside supplier.

   The stack parts can be protected from corrosion by flash-chrome-plating them or making them from corrosion-resistant steel such as stainless steel. A problem with chrome plating is that any change (requiring remachining) of a chromed surface requires that the chrome must first be removed from the steel part. This requires shipping the part to the plater so the coating can be removed by a process similar to plating. After re-machining, the changed part must again be shipped to be plated. This is an expensive and time-consuming procedure.

The use of fully hardened (or pre-hardened) SS for cavities, cores, and inserts is quite common today, even though the steel cost is higher. When considering the expenses and risks with chrome-plating of mold steels and the time saved, the total cost could be more than using SS.

5. **Fatigue:** The strength figures for steel and other metals are arrived at from stressing a test sample, for one cycle only. The results of such tests are satisfactory for steady loads, such as seen (for example) by preloaded screws, but molds often operate many, sometimes millions of cycles. If there are more cycles, the rated strength gradually declines. This decline is usually shown, as in Figure 15.2, in logarithmic graphs, as a straight line declining from the rated strength (e.g., tensile or yield strength) for one cycle to a point where the value remains the same regardless of the additional number of cycles; for all steels this occurs at about 2 million cycles. The strength of the material after 2 million cycles (the fatigue strength) depends very much on the material and hardness selected, but also on features such as notches, holes drilled into it, and surface finish. The fatigue strength can be as low as 15–20% of the yield strength (yield, in hardened mold steels, is only a little less than the tensile strength; many datasets use yield strength rather than tensile strength). Note that so-called machinery steels, but also the related P20 or P20PQ, do not lose as much strength as hard mold steels.

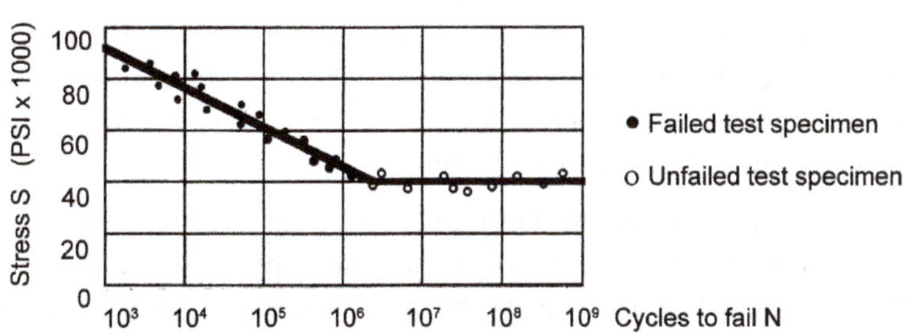

Figure 15.2 Typical fatigue graph for a machinery steel

The fatigue strength is equivalent to the safety factor (frequently, 5) often used by designers when calculating the strength of a part. The problem is that all force calculations depend on an assumption of the injection pressure, as discussed in Section 4.1. But we know that the forces will be greater for thin-wall molding, and since most of them are designed for a very large number of cycles, selecting only the very best materials with appropriate strength and hardness is suggested.

Note that springs inside molds (sometimes specified for ejector plate return) are especially sensitive to cycling. When designing for springs, use the manufacturer's suggested values for maximum compression and load of the selected spring.

> Always watch whether the values given are in metric or imperial systems. The strength values are given either in kPa (kilopascal) or in psi (lb/in^2).

15.2 Steel Properties

Earlier molds used mild steels even for stack parts, but they did not last for long production runs. Mold makers were gradually switching to the types of hardened steels that were used in tool and die making; however, these steels were often too brittle or otherwise unsuitable for mold applications, so the steel industry began to develop steels specifically designed for the plastics industry. The important features for mold makers are essentially as listed below.

1. *Yield strength.* One of the most important properties is the yield strength of a material. If a tensile load is applied on a component, the material is put under stress, and a deflection per unit of the original length (or strain) occurs. As the stress is increased, the strain also increases linearly until permanent deformation occurs, and this is the yield strength. At stresses below the yield strength, the material will return to its original size when the load is removed. At stresses above the yield strength, the component becomes permanently deformed. Mold components should be designed so the stresses in the material are well below the yield point. The ultimate strength is the maximum stress the material can withstand before complete failure occurs.
2. *Impact strength.* Another strength property which is important for mold components is the impact strength or material toughness. It is the amount of energy required to instantaneously break the material. Impact strength is very sensitive to the condition of the part surface. Smooth unnotched surfaces have higher impact strengths than rough, notched surfaces. There are several different standard tests with various types of notches and configurations. One of the most used is the Charpy V-notch test.
3. *Tensile (or compressive) strength.* This is important for long life of the components because they are subjected to high stresses within the mold, particularly those created by high injection pressures and large clamping forces.
4. *Modulus of elasticity.* This quantifies how much a part will deform under an applied force. It is determined from the slope of the stress-strain curve. If we

look at the equation for determining axial deflection, it is apparent that the amount of deflection is inversely proportional to the modulus of elasticity. Therefore, the greater the modulus of elasticity, the less the component will deflect under load. The modulus of elasticity is only valid if the stress on the component is below the yield strength of the material. Above the yield stress, the stress-strain curve is not linear, since permanent deformation occurs.

5. *Hardness*. The hardness of a material measures its ability to resist an indentation. The most common method for measuring hardness is Rockwell hardness. There are many different Rockwell hardness scales. The one most commonly used for mold steels is HRC (or Rc) for steels and HRB (or Rb) for softer metals such as copper. A diamond cone indenter is loaded onto the sample and a "minor" load is applied which penetrates further. The difference in penetration is used to determine the hardness. In general, the higher the hardness of a material, the higher the strength and lower the toughness. Hardness is often used to quantify a material's heat treatment procedure.
6. *Toughness*. This is especially important for long, slender cores and inserts subjected to side forces that deflect them during injection.
7. *Wear resistance*. There are two main types of wear which occur in molds:
   - Abrasive wear can result from abrasive resins acting on nozzle tips, gate holes, and molding surfaces. The abrasive particles may be glass fiber, minerals, etc., which dig into the material and push it away. Abrasive wear can also occur between sliding surfaces when foreign particles or particles of the sliding materials get forced into and along the surface. The properties required to minimize abrasive wear are high hardness and a low coefficient of friction. The high hardness prevents the abrasive particle from penetrating the surface, and the low coefficient of friction helps the particle to slide along the surface rather than dig in.
   - Sliding wear without abrasive particles can occur on components such as mold tapers between valve stems and bushings, etc. It results from an adhesion or microwelding of the sliding materials. As with abrasive wear, a high hardness and a low coefficient of friction is beneficial, as well as surface lubricity and use of dissimilar materials. Surface lubricity lowers friction between the two surfaces. Some metals, such as bronze or 06 steel, contain lubricants in the material. It is also important to use dissimilar materials (and materials with different hardnesses) for components that slide together, as similar materials tend to adhere more to each other. Also, "gummy" materials such as some stainless steels should be avoided in sliding applications. When using similar materials, surface treatments may be used to change the properties of the sliding surfaces, but the treatments should not be the same. A good combination is hard chrome sliding on a nitrided surface.

8. *Hot hardness.* This is of special importance for hot runner components, but also for molds for some plastics that are molded at high temperatures. Note that many hardened tool steels start to anneal at temperatures lower than the melt temperatures sometimes required for injection.
9. *Corrosion resistance.* Some plastics attack (corrode) steels and other mold materials. In a high-humidity environment, molds corrode (rust) because of the high water content in the air. In all such circumstances, the mold parts and plates should be chrome- or nickel-plated, which can be quite expensive, especially when considering the often high handling costs for plating by outside suppliers. These stack parts and plates can also be made from stainless steel, which is more expensive than other steels, but – considering the overall cost connected with chrome-plating – the difference may not be that big, especially if the stainless steels can be bought in large volume.
10. *Thermal conductivity.* The thermal conductivity of a material is its ability to transfer heat. The equation for one-dimensional conductive heat transfer shows that the heat transfer (Q) is directly proportional to the conductivity (k) of the material. For molds and hot runners, copper alloys are used in areas where heat transfer is required. All steels typically have very similar conductivities. Titanium, plastic, or air are often used as insulating materials due to their low conductivities. This can be important with high-speed molds. However, keep in mind that, in many cases, good cooling can also be achieved with steel by using a better layout of the cooling channels, and thus avoiding the use of the softer copper alloys, which require more upkeep than steel.
11. *Thermal expansion.* The coefficient of thermal expansion quantifies how much a material will grow or contract with changes in temperature. As seen from the thermal expansion equation, the coefficient of thermal expansion is directly proportional to the amount of expansion. Therefore, the greater the coefficient, the more it expands. In designing hot runner or other heated components, it is important to consider how much the components will expand and ensure the possible stresses generated are below the yield point of the material, to prevent permanent deformation.
12. *Ease of machining.* This is an important consideration. The addition of certain alloying elements to the steel makes it much easier to cut chips; this can make a big difference in the time and the cost of machining.
13. *Ease of polishing.* Some steels are not well-suited for polishing and will not permit or maintain the high surface polish often required. Do not forget, however, that high polish is often not required.
14. *Ease of nitriding.* Nitriding is a surface treatment applied on top of an already well-hardened and otherwise finished part to provide a very hard surface. It is used mostly to improve the wear characteristic of the steel. To nitride on top of

a soft base does not make any sense: the hard (nitrided) surface will collapse under any heavy load because the supporting steel is soft.

15. *Ease of welding.* In some cases, it may be important to be able to repair a worn mold part by welding. While this, in general, is not good practice and should be done only in exceptional cases, it may be used as a "quick fix" to keep a mold running until it can be properly repaired.
16. *Cost.* We stated earlier that materials constitute a substantial portion of the mold cost, but cost alone must never be the reason to select any steel. There is only one goal for the mold maker and designer: to produce the best mold for the specified purpose, that is, the mold that will produce the lowest cost of the product for the specified production requirements.

Table 15.1 and Appendix 1 give the designer an overview of some common mold materials. The data is approximate, and may vary somewhat from one manufacturer to another. Therefore, the designer must always check with the manufacturer to obtain the correct material property data sheets.

Table 15.1 Commonly Used Mold Materials

Type	AISI designation	DIN material no.	Steel code	Recommended hardness (Rc)	Used for
Pre-hardened	4140	1.7225	42CrMo4	30–35	Plates, leader pins
	P20	1.2330	40CrMnMo7	30–35	Plates, pistons, stroke limiters
Stainless steel Pre-hardened	420SS	1.3216	X42Cr13	30–35	Plates
Carburizing steels	P5	–	–	59–61	–
	P6	1.2735	–	58–60	–
Oil hardening	O1	1.2510	106WCr6	58–62	–
Air hardening	H13	1.2344	X40CrMoV5-1	49–51	Cores and cavities
	A2	1.2363	X100CrMoV5-1	56–60	Wear surfaces, lock rings, stripper rings
	D2	1.2379	X155CrVMo12-1	56–58	–
	S7	–	–	54–56	Wear rings
Stainless steel (SS)	420SS	1.2083	X42Cr13	50–52	Cores/cavities
High-speed	M2	1.3343	S-6-5-2	60–62	Wear surfaces
Beryllium-copper	BeCu25	–	–	36–41	High thermal conductivity inserts

By studying the various steels, it can be seen that any given steel will have only a few of the characteristics required for a certain purpose. Typically, a steel may be very tough but not be very hard, or it may not readily accept nitriding. Therefore, the designer will always have to find the most suitable compromise when selecting a steel for a mold part.

Some very expensive steels are occasionally used in molds such as tungsten carbides. They are very hard and three times as stiff as steel, but also very brittle and make it difficult to produce mold parts. New steels are constantly being developed by steel manufacturers, with better properties than before, for general mold making and for new applications, to keep pace with the development of new plastics and with new methods of use in mold making.

15.2.1 Types of Steels

15.2.1.1 Pre-Hardened Steels

Pre-hardened steels are steels that are hardened by the steel supplier prior to machining. These steels are usually hardened between 26–40Rc and can be machined without too much difficulty. The greatest advantage of these materials is that heat treatment is not required. When components are made from materials that require heat treatment, extra stock must be left on the component in the rough machining stage to allow for distortion and size changes that occur during the heat treatment process. The extra stock must then be ground off after heat treatment. The heat treatment process is also very time-consuming, taking typically 15–30 h. There is, however, a trade-off in using pre-hardened materials. The strength of lower-hardness materials is generally considerably lower than the higher-hardness materials, but nevertheless sufficient for many components such as mold plates, manifolds, etc. Some of the most common pre-hardened steels are P20, 4140 and NAK55. The components in molds requiring high hardnesses and strengths, such as molding surfaces, nozzles, etc., are usually made from heat-treated tool steels.

15.2.1.2 Tool Steels

The steels used for components requiring high hardness and strength are typically tool steels, which require heat treatment as part of the manufacturing process. There are many different grades which are used. There are usually trade-offs when selecting one grade over another grade, such as higher strength and hardness but lower toughness, or better wear resistance but more difficult machinability. The most commonly used material is H13 (DIN 1.2344). It is a hot-worked steel designed for die-casting molds. It is easy to machine in its annealed state, is easy to heat-treat, and has good strength and toughness properties at high temperatures.

Higher hardnesses and better wear resistance can be achieved with other grades but at the expense of toughness.

15.2.1.3 Stainless Steels

Stainless steels are used in applications requiring corrosion protection. The most commonly used stainless steel is 420SS (DIN 1.2083). It has similar hardness and strength properties to H13, but lower toughness, which makes it more susceptible to cracking. 420SS is also used in the pre-hardened condition (approximately 36Rc). For higher strength, 440CSS is sometimes used, but is fairly brittle. The 300 series stainless steels are not commonly used, since although having better corrosion resistance, they have much lower strength.

15.2.1.4 Copper Alloys

Copper alloys are used in applications where high thermal conductivity is required. The ones most commonly used are BeCu3, BeCu25 and Ampco 940. Typically, the higher the conductivity, the lower the strength of the material.

15.3 Quality of Materials

The quality of materials used in molds is just as important as the type of material selected. There are several aspects that should be looked at when determining if a material's quality is acceptable. The proper chemical composition is essential to ensure the required properties are obtained. There are standards for each type of steel that indicate the chemical compositions and tolerances. The correct microstructure is also very important in determining a material's quality. In order to look at the microstructure of a material, a sample piece is polished then etched with a reagent that allows visual differentiation of various parts of the microstructure using a microscope. In this way, is possible to determine grain size and shape, and the distribution of various phases. The micro-cleanliness is also quantified by inspecting the number, size, shape, and type of inclusions. These characteristics of the microstructure are compared to standards for the specific type of material. The material quality can also be determined by mechanical testing. Since the strength properties are often the main criteria for selecting a particular material, this is a good method for quantitatively determining a material's performance. Basic mechanical tests are carried out by the material supplier at the mill to verify the quality of the material, but if stricter quality control is required, independent testing can usually be provided by a third-party source.

15.4 Manufacturing Considerations for Choosing Materials

15.4.1 EDM

During the EDM process, the outermost layer of the steel is melted or vaporized at temperatures up to 50,000 °C. This affects the microstructure, hardness and stress condition at the surface. The overall end result is a very brittle and highly stressed surface. The harmful effects of EDM can be minimized by finishing with a slow removal rate (low current) and high frequency, grinding or polishing the surface to eliminate the brittle layer, and/or stress relieving (15 °C below tempering temperature) after the EDM process.

15.4.2 Welding

Welding also significantly affects the properties of a material. In order to minimize the stresses resulting from welding, it is important to preheat the material before welding. This reduces the rate of cooling of the weld and brittleness. It is also important to use the correct filler metal for the base material, and use dry electrodes, as moisture in the weld causes hydrogen cracking. The most common welding methods used are shielded metal arc (SMAW) and gas tungsten arc (TIG).

If the welded part has been previously hardened and tempered, it should be tempered again or stress-relieved after welding. This will reduce the stresses and reduce the hardness of the weld to a level similar to that of the base material. If the welded part is in the annealed condition, it should be re-annealed and then hardened and tempered. The annealing process anneals the weld to the same condition as the base material, and the properties become similar.

15.5 Heat Treatment

The steel structure of certain steels can be changed by heating and subsequent chilling of the work piece to increase the hardness of the steel from a level that is suitable for machining (usually soft) to the hardness that will provide good working life of the steel under repeated exposure to heat, high pressure, and wear. In general, only steels with a carbon content of at least 0.35% can be hardened. So-called mild steels, with lower carbon content (usually in the range of 0.1–0.3%), cannot be hardened. However, to use these relatively inexpensive steels for mold parts that need good hardness, the surface of such mild steels can (after machining to their shape) be provided with a carbon-rich skin by the process of carburizing – that is, subjecting the work piece at high heat, for about 24–48 h, to a carbon-rich atmosphere (using charcoal or carbon monoxide). This causes the surface to absorb carbon to a depth of usually 0.5–1.5 mm (0.020–0.060"). The work piece can then be hardened as for through-hardening tool steels, by heating to a high temperature, quenching in water or oil, and then tempering (reheating to a lower temperature than before quenching), and finally cooling in air.

The *advantage* of these mild steels is their relatively low cost, ease of machining, and availability. The *disadvantages* are the costs of carburizing and the subsequent heat treatment. During carburizing, the steel often distorts and even grows slightly. The art in using these steels is to foresee such changes and to allow enough material for grinding after heat treatment to arrive at the final mold dimensions. Since the carbon content diminishes with its depth, which is dependent on the time required for carburizing, the danger is that too much grinding allowance can make the hardened skin disappear during grinding, and in such areas the surface hardness is then as soft as the base steel. Tool and mold steels are "through-hardened", and the amount of grinding to size will not affect the surface hardness. For these reasons, over the years, mild mold steels have been used less and less.

Today, with the development of better machining methods and more rugged machine tools, larger mold stack parts are made mostly from pre-hardened mold steels, which are supplied from the steel mills and the supply houses at a hardness of about 30–33Rc. This allows the finish machining of most parts without the need for any heat treatment after machining. Note that very large mold parts may require three steps for finishing: (a) pre-machining to remove the bulk of the outside and any large openings or cutouts, which may cause stresses within the steel that distort the work piece; (b) stress-relieving to reduce any locked-in stresses from the pre-machining; (c) finish-machining to the desired close dimensions.

After hardening, it may still be necessary to grind or otherwise machine the work piece to the final shape before polishing. Heat treatment is usually done by specialists, thus requiring shipping of the parts to and from them, adding time and cost to the heat treatment. By standardizing a small number of different mold steels requiring hardening, costs can be reduced; for larger mold makers, this may make it economical to provide in-house heat treatment.

15.6 Overview of Surface Treatments

Surface treatments can help enhance the performance of the base material used for the mold component. Surface treatments can be used to increase the hardness, build back a sharp corner, add lubricity and reduce wear. Table 15.2 gives a general overview of some surface treatments. The suppliers of the coatings can provide much more extensive details on the process and the benefits. This table is intended to give the designer a basic idea of surface treatments that are available.

Table 15.2 Overview of Surface Treatments

Name and process	Properties	Uses
Electroless nickel plating (ENP): Chemically deposited nickel and phosphorus without electrical current	▪ Very uniform ▪ Holes can be coated ▪ Typical thickness 0.005 mm (0.0002") ▪ 47-50Rc (can be baked to 70Rc) ▪ Good corrosion and wear resistance ▪ Coefficient of friction 0.2	▪ Core and cavity cooling channels, O-ring grooves
Nickel plating: Electrochemically deposited nickel	▪ Builds up on corners ▪ Fixturing required for uniform thickness ▪ Up to 0.13 mm (0.005") thick ▪ 48-94Rc ▪ Coefficient of friction 0.22	▪ Typically used for build-up ▪ Some corrosion resistance
Flash chrome or hard chrome plating: Electrochemically deposited chrome	▪ Builds up on corners ▪ Fixturing required for uniform thickness ▪ Typical thickness 0.002-0.8 mm ▪ 68-70Rc ▪ Coefficient of friction 0.15	▪ Used to build-up surfaces ▪ Good wear resistance ▪ Some corrosion resistance
Nitriding: Surface hardening by introducing nitrogen into steel surface	▪ Parts heated to 500-560 °C in NH_3 atmosphere ▪ Types: gas nitriding, ion nitriding, nitrocarburizing ▪ 70Rc ▪ Typically 0.1 mm deep ▪ Good wear resistance ▪ Parts may grow or warp slightly in process ▪ Polished finishes require re-polishing	▪ TiN aids release of parts ▪ Wear resistance

Name and process	Properties	Uses
Physical vapor deposition (PVD): Material (Ti, Cr, etc.) is evaporated in an atmosphere of N_2 gas, and is then attracted to the part by a voltage difference	▪ 260–510 °C process ▪ Typically 0.002–0.005 mm thick ▪ 80Rc ▪ Very little part distortion due to low process temperature ▪ Line-of-sight process ▪ Coefficient of friction 0.4–0.7	▪ Can coat TiN, CrN, TiAlN, etc. ▪ Good abrasion resistance
Tungsten disulfide (WS_2)	▪ 0.9005 mm thick ▪ Coefficient of friction 0.03 (very low) ▪ No part distortion ▪ Line-of-sight process	▪ Improves lubricity ▪ Reduces wear
Ion implantation: Ionized gas bombards into surface	▪ < 150 °C process temperature ▪ No dimension change ▪ 80Rc ▪ Can use N, B, C, Ti, Cr, etc. ▪ Line-of-sight process	–

15.7 Hardness of Mating Parts

As a general rule of thumb, if the designer is considering using dissimilar material in a contacting condition, the difference in hardness should be 4–6Rc if possible. If the metals are similar, the difference should be 6–10Rc.

16 Fasteners

Proper selection of and use of fasteners is critical to the function and maintainability of every injection mold. Since fasteners basically hold the whole mold together, if the number, location, position or specification of the fasteners is incorrect the whole mold could fail to meet its objectives.

Fasteners are used to hold (clamp) together two or more parts of an assembly. Two types of fasteners are used in mold construction: *screws* thread directly into the part, while *bolts* require a nut to complete the assembly (Figure 16.1 and Figure 16.3).

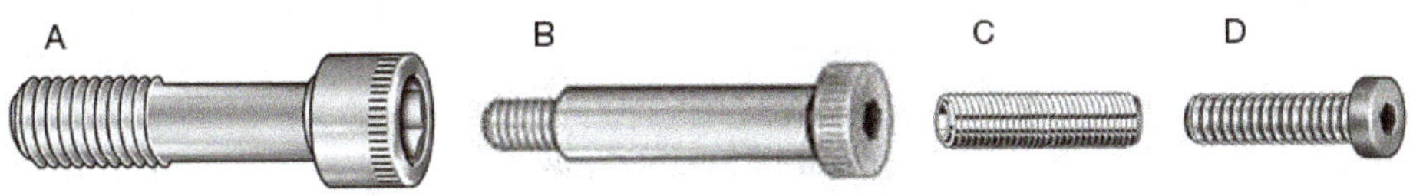

Figure 16.1 Fasteners typically used in molds: (A) Socket-head cap screw (SHCS); (B) Shoulder bolt; (C) Set screw; (D) Low-head cap screw

Socket-head cap screws (SHCS) are preferred to hex-head bolts for increased strength, reduced spacing requirements, and ease of torquing. Low-head cap screws are used only when spacing does not permit the use of a full headed screw. Most bolts some in alloy steel and stainless steel as well as coarse and fine threads. Typical dimensions for a SHCS can be seen in Figure 16.2.

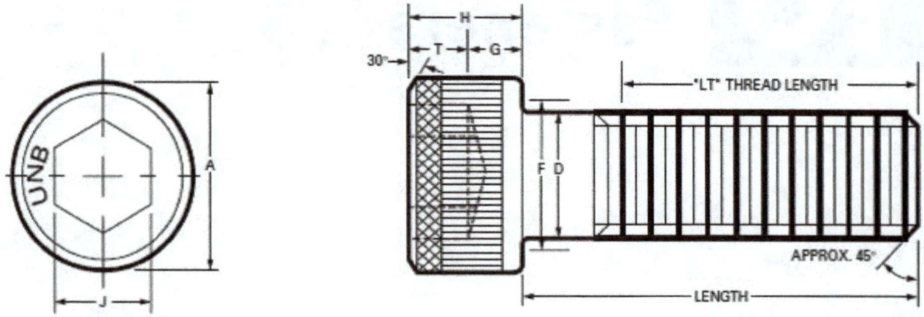

Figure 16.2 Dimensions of an SHCS [1]

In the inch-based measurement system, the thread pitch is given by the number of threads per inch (TPI). For example, a screw thread specified as ½-13 means that the major screw diameter is ½", and there are 13 threads per inch. This results in a pitch of 1" ÷ 13, or 0.077".

In the metric system, the pitch is given in millimeters. For example, a screw specified M12-1.75 means that the major screw diameter is 12 mm, with a pitch of 1.75 mm.

Screws are also defined as either coarse or fine threads. Coarse threads are those with larger pitch (fewer threads per axial distance), and fine threads are those with smaller pitch (more threads per axial distance). Coarse threads have a larger thread form relative to screw diameter, where fine threads have a smaller thread form relative to screw diameter. ½-13 belongs to the UNC (Unified National Coarse) series, and ½-20 belongs to the UNF (Unified National Fine) series. Similarly, M10 (10 mm nominal outer diameter), as per ISO 261, has a coarse-thread version at 1.5 mm pitch and a fine-thread version at 1.25 mm pitch [2].

Figure 16.3 Typical fasteners used in molds (Courtesy of DME)

[1] Source: Unbrako Engineering Guide.

[2] Source: Wikipedia.

16.1 How Screws Work

Every screw consists of a head and a threaded shank. When the screw is turned into the receiving (female) thread in a part (or in the nut), there is at first no resistance except friction, until the joint touches and the head comes to rest on the part. Further turning of the screw begins to stretch the screw, as a result of the wedge action exerted by the threads. Figure 16.4 illustrates the common terminology of a screw. As the screw is torqued (additional force applied), the wedge action pulls on the shank, creating force between the threads and head of the screw (Figure 16.5).

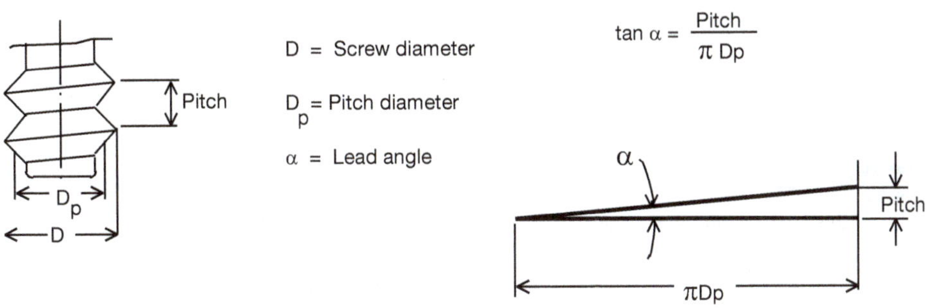

Figure 16.4 Screw thread terminology

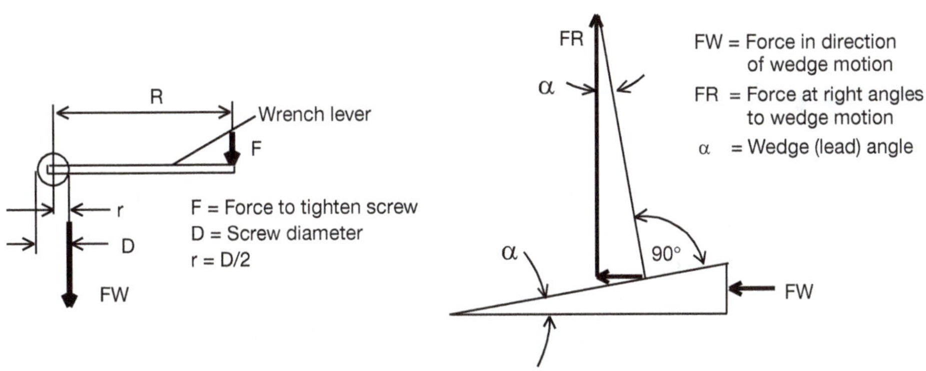

Figure 16.5 How torque creates force in a screw

Using Figure 16.5, the following equations are derived for torque, the force in the direction of wedge motion (F_W), and the force at right-angles to wedge motion (F_R):

$$\text{Torque, } \tau = F \times R \tag{16.1}$$

where:

τ = torque (N·m)

F = force (N)

R = radius or rotation

F_W is the tangential force (at radius r) exerted on the screw by the torque used to tighten the screw. The smaller the screw diameter, the less torque is required to tighten a screw.

$$F_W = \frac{\tau}{R} = \frac{F \times R \times 2}{D} \tag{16.2}$$

The smaller the wedge angle, the greater is the axial force F_R, which acts in the screw. In other words, with a finer thread (on the same diameter), a greater force can be exerted on the joint.

From the relationships shown in Figure 16.5 between the wedge angle and wedge motion (using Pythagoras' theorem), the equation for the axial force F_R (or *holding force*) is:

$$\text{Screw holding force, } F_R = \frac{F_W}{\tan\alpha} = \frac{2 \times F \times R}{D \times \tan\alpha} \tag{16.3}$$

All screws used for fastening have a lead angle or wedge angle smaller than the friction angle between steel and steel, even when lubricated, and the screw is inherently self-locking (i.e., it will under normal circumstances not unscrew after tightening). However, there are conditions that do require securing screws after tightening; they will be discussed later.

16.2 Rolled versus Cut Threads

Rolled threads are made using rollers to force (squeeze) the steel into the required thread form (Figure 16.6). This produces stronger threads, with good flow of grain within the steel. The threads do not have tool marks which could induce notching and premature failure. Before rolling, the *shank diameter* of the length of the screw to be threaded is approximately the same as the *pitch diameter*.

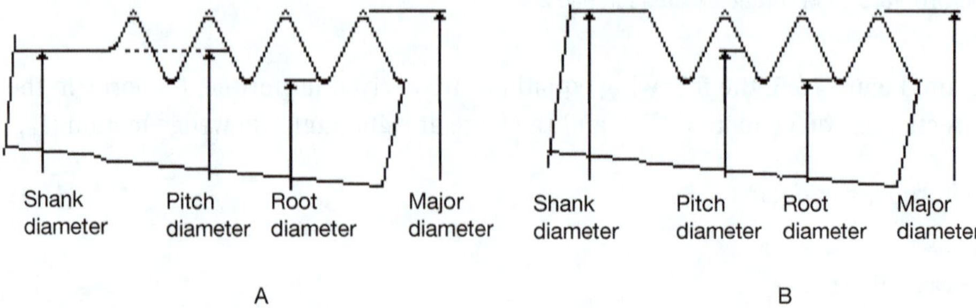

Figure 16.6 (A) Rolled threads versus (B) cut threads

All class 12.9 screws (recommended for molds) up to 20 mm diameter and up to 150 mm long have *rolled* threads. All screws over 20 mm diameter and, also, all those over 150 mm long *may* be rolled, cut, or ground at the manufacturer's discretion.

Cut threads are machined into blanks. There is always the danger of tool marks (scoring) at the root of the thread and at the run-out, which may initiate progressive failure of the screw. The *shank diameter* is approximately the same as the *major* thread diameter (outer diameter) of the screw.

As a rule, the designer should always use standard, easily available screws. The use of specially made screws is not recommended. In molds, generally, only coarse thread screws are used.

16.3 Holding Action and Preload of Screws

Figure 16.7, and Equation 16.4 and Equation 16.5, will help the designer understand the principle of fastening with screws.

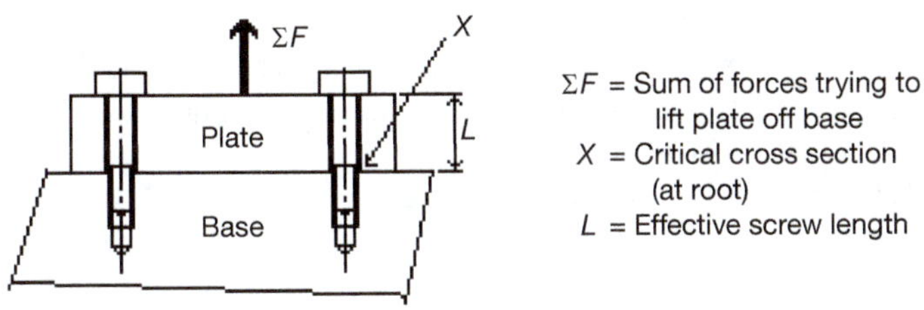

Figure 16.7 Schematic of screws holding a plate to a base

The stretching of the screw, Δl, can be calculated using Hooke's Law:

$$\text{Screw elongation, } \Delta l = \frac{L \times \sum F}{n \times X \times E} \tag{16.4}$$

where:

Δl = elongation (mm)

$\sum F$ = sum of the forces pulling the two pieces apart (N or lb). Forces could be pulling the plates apart or generated internally by pressure

L = the length of the bolt before the thread (the stretched length) (mm or in)

n = the number of bolts holding the pieces together

X = cross-sectional area of the bolt at the root (mm^2 or in^2)

E = modulus of elasticity (N/m^2 or psi)

Elongation (Δl) is solely related to the modulus of elasticity (E) and to the sum of forces $\sum F$. Hardness or tensile strength does not affect elongation. (Soft screws will stretch as much as hard screws.) However, the yield strength of hardened screws is much higher than that of soft screws, so that hard screws can be preloaded much higher than softer ones.

The tensile stress in the screw, σ, is calculated using the formula:

$$\text{Screw tensile stress, } \sigma = \frac{\sum F}{n \times X} \tag{16.5}$$

where:
σ = stress (N/m^2)

The total stress, σ, must be less than the permissible stress, σ_p ($\sigma < \sigma_p$). The permissible stress can be found in the supplier's catalog (see Section 16.6). The permissible stress is usually selected to be about 75% of the yield stress of the screw material used. The recommended torque for each size screw can be found in hardware suppliers' data sheets (see Section 16.6).

Example:

Eight M12×60 SHCS are holding a 30 mm thick plate to another plate. The plate is subject to a load of 5000 N.

a) How much does the applied force stretch each bolt when hand-tightened?
b) How much holding force does each bolt have when torqued to the recommended setting of 110 N·m?
c) What is the stress in each bolt in part (a)?
d) Are the bolts correctly sized for the load?

Answers:

a) Since elongation, $\Delta l = (L \times \sum F)/(n \times X \times E)$

= (5000 N × 30 mm) ÷ (8 × (π × 12^2 ÷ 4 mm^2) × 200 GPa)
= (5000 × 30 mm) ÷ (8 × 113 mm^2 × 200 × 10^3 N/mm^2)
= 0.0008 mm

Each bolt stretches 0.0008 mm.

b) Since torque, $\tau = F \times R$. Rearranged, $F = \tau / R$
F = 110 N·m ÷ 0.006 m
= 18,333 N

Each bolt can hold 18,333 N. Eight bolts can hold a load of 146,000 N.

c) Since stress, $\sigma = \sum F/(n \times X)$

= 5000 N ÷ (8 × 113 mm^2)
= 5.5 × 10^6 N/m^2
= 5.5 MPa

d) According to the Unbrako catalog, the maximum tensile load the screw can take is 110 MPa, therefore the bolts are oversized for the application.

To prevent the plate from lifting, the screws are tightened before applying load $\sum F$. By use of a wrench, the screws will be torqued to a suggested value, which in fact stretches the screws within the effective length *L*. This stretching is equivalent to a preload on the plate.

As long as the preload is greater than the load $\sum F$, the plate will *not* lift off the base. If the force $\sum F$ reaches the preload, the forces in both directions are balanced. When $\sum F$ becomes greater than the preload, the plate will lift off the base.

The critical point (*X*) of a screw is the cross-section of the first thread root, as indicated on Figure 16.7. The whole screw force goes through there, and this is the section that will break when the screw is overloaded.

It is important that the designer has a good estimate for the amount of force ($\sum F$) that must be secured by screws. Often, there are more screws specified than really necessary, which unnecessarily increases the cost of the mold.

If more than one screw is used ($n > 1$), it is important that all screws are torqued equally to share the load. If not, some screws may bear more than their share of the load and be stretched beyond their yield points. Screws stretched beyond their yield point will be permanently stretched and lose their holding force. The other screws will, therefore, have to take more than their share and may also fail.

It is also very important to locate screws as symmetrically as possible around the load, to prevent uneven loading of these screws. Forces acting on a plate are usually distributed evenly on the plate. By adding extra screws at the point of maximum deflection, the plate deflection can be greatly reduced, as shown in Figure 16.8. Note that the plate thickness has a large influence on the selection of number of screws.

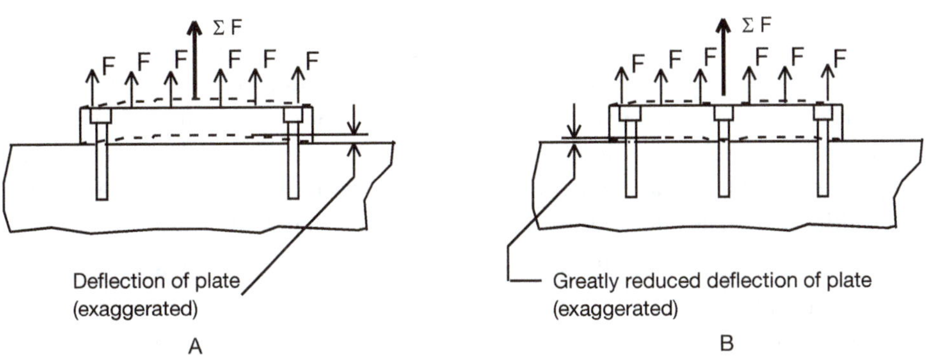

Figure 16.8 Impact of adding more screws on plate deflection

From Section 9.4.2, we notice that the deflection of a plate is inversely proportional to the third power of its thickness t, and directly proportional to the distance M between supports or screws (see Figure 16.9).

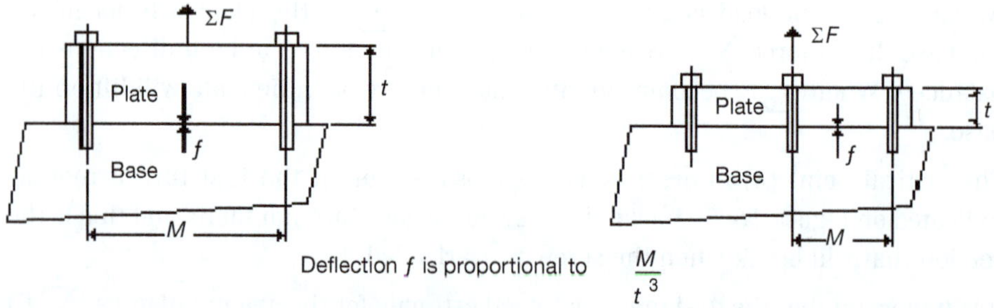

Figure 16.9 Deflection in thick versus thin plates (and the resulting need for more screws)

Example:

$M = 80$ and $t = 40$ for the thick plate in Figure 16.9

$f_{thick} = 80/40^3 = 1/8$

$M = 40$ and $t = 20$ for the thin plate

$f_{thin} = 40/20^3 = 1/2$

Deflection will be four times less on the thick plate, even though there are fewer screws and a greater distance between the screws. This emphasizes the fact that the plate thickness t, when raised to the third power, is more important than the distance between screws. Trying to resolve an issue with a plate that is designed too thin with more screws is difficult. This is why a designer should always use plates as thick as possible.

Note that the cost of a screw joint is not only the cost of the screws, which is usually not very high, but also the cost of drilling and tapping one plate, drilling (and counter-boring) another plate, and, possibly, drilling additional plates held together by the same screws. In addition, the space for unnecessary screws could be used to simplify locating other mold parts or water lines. All these points reinforce the fact that a thicker plate is almost always a better solution than more screws.

A larger screw takes only a little more space than a smaller one. For example, by selecting 12 mm instead of 10 mm screws, the number of screws needed to hold two pieces together could be reduced from 6 to 4, from 12 to 8, or from 18 to 12 screws, which would constitute considerable savings.

16.4 Influence of Temperature on Screws

Temperature has a meaningful impact on screw selection and screw design. The mold designer should be aware of these influences and use good screw design principles to ensure that temperature does not negatively impact the holding force of the screw.

The working range of class 12.9 bolts is -50 to 300 °C. Above this temperature, the yield strength can diminish rapidly; therefore, either the rated load on these screws must be reduced proportionately, or special high-heat screws must be specified. Such screws should be used only in *exceptional cases*, and when they are physically (and in appearance) *distinct* from the regular screws within the mold, because of the possibility that the customer may mix the screws when servicing the mold. *This should also be shown prominently on the assembly drawing.*

The effect of heating up at different speeds will be clear with an example, shown in Figure 16.10. The screw is heated by the surrounding hot body; however, the heat flow is restricted to pass through the small areas at points a and b, while little (radiated) heat passes through the gap around the screw, over the length L. This creates a slow rise in temperature of the screw, while the heated body heats up and expands more rapidly.

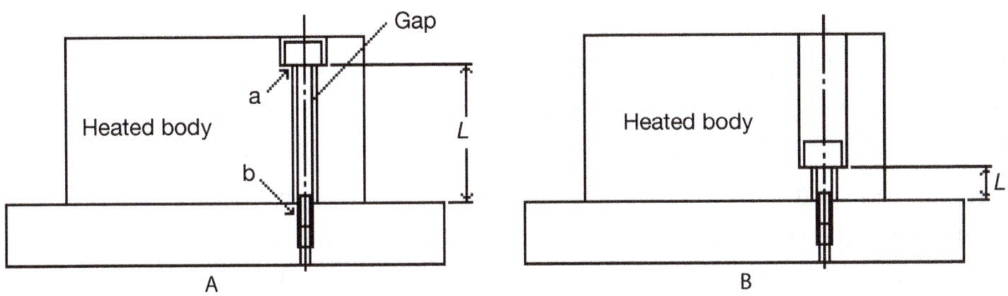

Figure 16.10 Screw configuration in a heated body: (A) Not recommended; (B) Recommended

If the screw was tightened with recommended torque, the additional stretching caused by the expansion of the body can raise the stress in the screw beyond the permissible limit, and the screw will fail. To preload the screw to a less-than-recommended value will not help, since once the screw reaches the temperature of the body; there will not be enough preload to hold the heated body on the base securely.

In this case only, the recommended solution is to shorten the screw, so as to reduce the (non-contacting) length L, which can then more rapidly follow the temperature rise in the body. This is particularly important if the heating-up time is short, as is sometimes the case in hot runner systems.

If the whole system heats up slowly, it is considered to be equivalent to a cold system, except that high-temperature screws may be required if the temperatures are at or above the annealing temperature of standard screws.

16.5 Effect of Cyclical Loads on Screws

When the plates in the mold are subjected to a large clamping force F_c, while the mold is closed, F_c will compress the plate(s) by the amount Δl_c. If this compression (Δl_c) becomes equal to or larger than the stretch caused by preload (Δl), the screws will become temporarily, partially, or entirely unloaded. The screw head may even lose contact with the body held down. If the screws are not loaded, the self-locking mechanism of the screws will be lost and the screws are free to gradually loosen. Also, the screws are now subjected to a cyclical load and can fail much sooner, since they are less tightened.

To prevent this full unloading and subsequent unscrewing, the designer must make sure that the screws are long enough to stretch when torqued (Δl) more than the expected compression (Δl_c); or $\Delta l > \Delta l_c$. This is even more important when the loading is applied suddenly (e.g., by shock). If shocks are unavoidable and of a magnitude that may unload the screws, the screws must be secured from unscrewing by some reliable method. Short screws will loosen more readily than longer ones because they cannot be stretched enough. Heavy vibrations have a similar effect to that of frequent shocks and may cause the screws to loosen, particularly if the screws are short and it is not possible to stretch them sufficiently.

In molds, the plate sizes and cross-sections A_p are usually in proportion to the clamp size and force, and the effect of plate compression can be ignored, but with the full clamp force acting on smaller units such as cavity modules, the compression Δl_p may be significant.

16.6 Screw Standards and Properties

There are many good standards for screws set by ISO and DIN. Mold designers should familiarize themselves with standards such as DIN 912 and ISO 4762-2004.

Specifications for socket-head cap screws can be found in suppliers' catalogs. The following are some companies that can be searched to find the appropriate dimensions and strength information.

For critical applications such as a mold the screws should be the highest quality available. Typically grade 12.9 screws are used for the highest tensile strength.

Screw suppliers used by the mold making industry:

- Unbrako: Preferred by many mold makers. There is an excellent product guide available [3]
- Fastenal
- McMaster-Carr
- Grainger
- Holo-Krome
- Fuller Fasteners
- TorqBolt.

There are standards for many types and qualities of screws, but the designer is essentially concerned with screws used in molds. Today, most molds use only metric screws to ISO standards of quality, or better. The screws are made from medium carbon steel with additives, quenched and tempered to a SI standard hardness of 38–43Rc (high-grade 40–43Rc). Steel (property) class is expressed with a number that must be stamped in the screw head. It is recommended to use socket-head cap screws (SHCS) with a property class of 12.9. The first number (12) equals 1/100 of the tensile strength, expressed in N/mm^2 (e.g., 12 means a minimum of 1200 N/mm^2, equivalent to about 120 kg/mm^2 or 170,000 psi). The second number (9) indicates 10 times the ratio between tensile and yield strength. In the example, class 12.9 means that the yield strength is 90% of the tensile strength, or 170,000 psi × 0.9 = 153,000 psi.

In wet areas, stainless steel screws should be used, which have lower yield strength but prevent corrosion. Areas where the required screw strength is high must be kept dry by use of O-rings. The finish of SHCS is preferably black, and SHCS heads are preferably knurled.

16.7 Length of Thread Engagement

The strength of the screw joint is also determined by the length of the thread engagement with its mating part. The threads themselves, at the bottom of the thread (root diameter), are mostly loaded by shear, with a very small amount of bending. In general, a screw is twice as strong in shear as in tension.

[3] *http://www.unbrako.com/images/downloads/Unbrako_US_Product_Guide.pdf*

The length of the screw engagement in a mold part (with a hardness of at least 30Rc) should be at least the same as the diameter (D) of the screw, but not more than $1.5D$.

The length of the first thread (pitch) of both screw and female thread is usually lost because of chamfers in both parts to permit easier assembly. Excessive engagement is costly because of added, unnecessary machining time and higher screw cost.

Exception: If the female thread is cut in soft materials (aluminum, copper alloys, cast iron, etc.), with a yield strength substantially less than that of the screw, the engagement must be proportionately larger to avoid deforming or stripping of the softer thread when torquing the screw to its limit. An engagement of twice the screw diameter or even more may be required.

It should also be noted that most screw joints will corrode, especially if they are not opened for a long period of time, and become difficult to unscrew. In such cases, excessive length of engagement aggravates the situation and may cause the screw to break when attempting to unscrew it. It may then be difficult and costly to remove the broken piece of screw.

16.8 Additional Recommendations Regarding Screws

16.8.1 Always Use Standard Available Screw Sizes and Lengths

Use standard size screws … period. It is costly and, therefore, bad practice to modify a screw to fit an application.

- Never cut a screw down from a standard length. Usually, by changing the depth of the counter-bore for the head, a suitable standard length can be found.
- Never machine the thread to increase the length of the threaded section. This is bad because cut threads are weaker than rolled threads. The threaded portion of a screw is always at least 2.5 × larger than its diameter and should always be suitable. If necessary, use a larger screw or redesign this area.
- Never reduce the size (diameter and/or height) of the screw head. This can seriously weaken the screw joint. Reduce the screw size and increase the number of screws, or redesign the area requiring the smaller-than-standard head.

16.8.2 Increase the Depth of the Counter-Bore

The depth of the counter-bore (C/B) should be approximately 1 mm more than the height of the screw head so that the head will not protrude above the surface of the mold part. It can be deeper to allow use of a standard length.

Allow for Space for Wrenches

Most screws in mold assemblies are tightened with free access and present no problem. However, if a screw must be tightened after other parts are in place, and/or where an overhang covers some of the area above the screw head, the design must provide sufficient vertical and lateral clearance so that a "hex" (hexagonal) driver can be inserted and rotated by at least 90°. Note that ratchet-driven hex drivers require more space than the slower-to-operate L-shaped (hex) keys.

Use of Shoulder Screws (Stripper Bolts)

The use of shoulder screws is discouraged except in applications where they may be used to limit a motion (e. g., of a plate) *without* contacting the plate at every cycle during normal mold operation. They are meant to limit the stroke only during servicing of the mold (e. g., to prevent a plate from falling during handling of the mold). The reason is that these screws cannot be tightened sufficiently to prevent them from coming loose is because there is no substantial length L to be stretched (see Figure 16.11A). It is good practice to use Loctite® to prevent a shoulder screw from loosening. If cyclical action of a shoulder screw is required, a combination of sleeve and long screw must be used to achieve a long L (Figure 16.11B). In this case, the screw can be fully preloaded, and will withstand cycling loads.

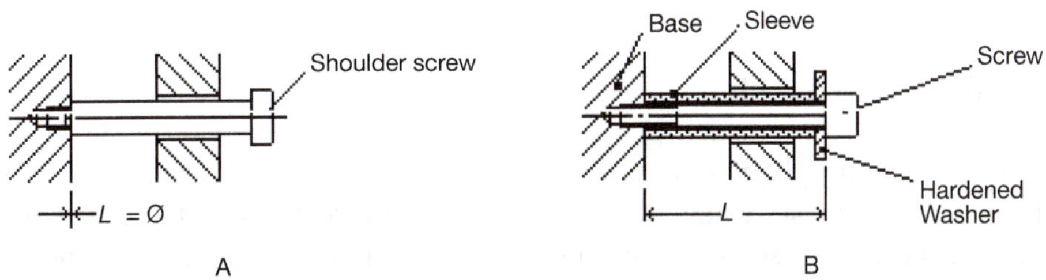

Figure 16.11 (A) Method of using a shoulder screw, and (B) preferred method using a sleeve

16.9 Use of Set Screws

Set screw joints may be used when it is not practical to use a standard SHCS (e.g., when fastening a rod to a plate). Because there is no length L to be stretched (Figure 16.12A), it is good practice to counter-bore one side (or both) so that the threads are free (not engaged) for a distance of at least 3-4 mm ($\frac{1}{2}$-$\frac{3}{16}$"), thus creating a length $L > Ø$ that can be stretched (Figure 16.12B). This greatly improves the strength of the joint. It also prevents the screws from breaking in the event of the rod being hit from the side, as shown with the heavy arrow.

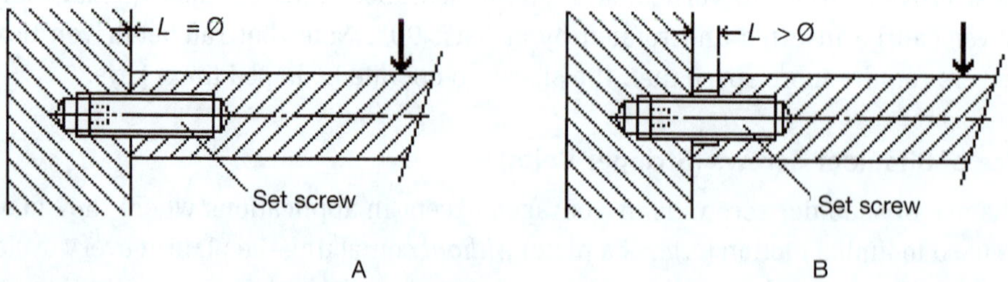

Figure 16.12 Set screw installation

Note that this method for reducing the thread engagement can also be used with a standard SHCS if the length L appears to be insufficient to ensure a good joint.

16.10 Holding Screws in Place (if Needed)

16.10.1 Lock Washers

Lock washers, including "star washers" may be thought of as "spring" washers. The principle of these spring washers is that they are compressed to a flat shape under the head of the screw. Teeth - one pair for an ordinary lock washer, many for star washers - dig (bite) on one side of the washer into the part to be fastened, and on the other side into the screw head (or nut), so that they oppose the unscrewing of the screw (or nut).

To be effective, the hardness of the screw and the fastened part must be less than that of the washer. Since the hardness of lock washers is 44-51Rc, and of star washers 44-47Rc, they will not bite into the underside of a standard (property class 12.9) SHCS, nor into a hard (49-51Rc) mold part, although they would bite into a pre-hardened (≈ 30–35Rc) plate. They are, therefore, not recommended for use with 12.9 SHCS, or on hardened steel parts.

16.10.2 Screw-Securing Compounds

These products are anaerobic liquids which are applied to the (clean, oil-free) threads at the moment of assembly, instead of oil. After application, the liquid hardens and bonds to the threads in the mold part and the screw. This bond is sheared off as the screw is removed, and the liquid must be reapplied for the next assembly.

This ensures a safe joint. It is important that the mold assembly drawing contains a clear note (warning) that Loctite must be applied to the designated screws when the mold is assembled and serviced. Table 16.1 lists some compounds, with their characteristics:

Table 16.1 Typical Compounds for Holding Screws in Place (check with suppliers for up-to-date details)

Compound	Characteristics
Loctite #242 (blue)	Medium strength; minimum thread length L = 1.5D
Loctite #271 (red)	High strength (2.5 × #242); use for studs
Loctite #262 (red)	15% more strength than #271
Loctite #290 (green) or Porosity sealant #290	Same strength as #242 but has "wicking" property (i. e., it will enter and lock an already tightened thread). This is a preferred product

16.10.3 Screws Secured by Plastic Inserts (Nylok™, etc.)

Screws are secured by a small plug (cylindrical or rectangular) inserted into the thread section of the screw (Nylok™ screws, etc.). These screws are standards that can be purchased from hardware suppliers. When the screw is turned into the matching part, the (slightly projecting) plastic is deformed by the female thread; this deformation creates an area of high friction, which prevents the screw from loosening. Such a screw will stay in any position where it is placed, even without being tightened, despite vibrations or shocks.

These screws are often quite useful, but to ensure that they function properly, they *must* be replaced every time they are removed, especially if they have been in place for a long time, because the permanent set in the plastic may reduce the friction force. The assembly drawing must carry a note to this effect.

Typical applications of these screws are in locations where short screws, such as round- or button-head screws etc. are required, but where these cannot be secured sufficiently by torquing (e.g., when fastening sheet metal enclosures subject to heavy vibration or shock). An alternative would be to use Loctite® or an equivalent.

17 Dimensioning and Tolerancing

James D. Sykes, GDTP-S, Profile Services
(www.profileservices.ca)

Along with the assembly and detailed drawings of the mold discussed in previous chapters of this book, the mold designer communicates their intent by dimensioning and tolerancing of the drawings and the 3D models they have created. It is critical that the designer communicates this information in a clear and concise way to ensure that the mold will be manufactured to the specifications intended.

Historically, mold production relied on skilled tradesmen, called mold makers. In that system, tolerancing of mold components was not necessary if you had a small cavitation mold that would be manufactured by a master mold maker, and if you would never need replacement parts to be shipped from the supplier. Today, components are produced by high-precision CNC machines in large quantities, and field replacements cannot accommodate hand-fitting. Reflecting the shift from skilled trades to a large-scale industrial basis, molds are now put together by mold assemblers and the term "mold manufacturing" has largely replaced "mold making". Today, effective tolerancing in the mold design is critical to proper mold functionality and the mold supplier's commercial success.

There are two methods for tolerancing workpieces: *conventional plus-minus tolerancing*, and geometric tolerancing, which is often referred to as *geometric dimensioning and tolerancing (GD&T)*. Designers, machinists, and inspectors in the mold-making industry are familiar with conventional tolerancing methods. To explore the value of GD&T in a precision manufacturing environment, the limitations of conventional tolerancing will be explained and GD&T's benefits will be discussed.

Most mold shops have shifted to CNC machining, with tool paths programmed directly from 3D CAD models. As they improve their machining capabilities, they improve their inspection capabilities as well, adding contact and vision coordinate measuring machines (CMMs), which typically work in a GD&T environment. Mold design and production businesses often try to exist in a hybrid of conventional and geometric tolerancing, thinking that they will have the benefit of GD&T on *important* features, and the convenience of conventional tolerancing on everything else. This inevitably leads to "what to do in this situation" decisions in the shop; and it means that the product definition is unclear and may not result in the expected product.

Increasingly, engineering design is moving away from creating and issuing drawings, instead issuing CAD model files with the geometric tolerances, surface finishes, and other attributes embedded into the model. This is model-based definition (MBD). This allows the computer-aided manufacturing (CAM) software to recognize the tolerances embedded in any given feature and compensate its cutting path. MBD also allows inspection systems to automatically register the tolerances on each feature and program the CMM to verify them. The mold-making industry is quickly moving away from conventional tolerancing and into MBD as a means of improving productivity and efficiency, by reducing costs related to drawing creation, interpretation, and programming.

There are significant elements of *common knowledge* and *common practice* inherent in conventional dimensioning and tolerancing practices. Five attributes define the features on a workpiece: location, orientation, size, form, and surface finish. The first four fall into the domain of dimensioning and tolerancing, and will be addressed here. Surface finish is addressed in Chapter 3 on plastic part design, and in Chapter 11 on ejection. Conventional practices typically address only the location and size of features. Form is rarely indicated beyond a flatness requirement, and perpendicularity or parallelism is often over-controlled as an indirect control for flatness. Industry has historically relied on best efforts in the machine shop. It is difficult to effectively control form (flatness, straightness, circularity, and cylindricity) of a feature with conventional tolerances, because there is no definition for what flatness or cylindricity means, much less any control for irregular surfaces. Shops tend to work to their tightest process capabilities, rather than to the tolerances needed for part functionality, and this drives up costs unnecessarily. Orientation is rarely controlled beyond parallelism or perpendicularity between surfaces. Conventional tolerances for angled features produce their own significant issues. Specified as plus-minus degrees, conventional angular tolerances result in wedge-shaped tolerance zones, allowing the least deviation from nominal for geometry closest to the inflection point of the tolerance zone, and the greatest allowable deviation at the other end of the toleranced feature. Section 17.2 examines the limitations of conventional tolerancing in greater detail.

17.1 Conventional versus Geometric Tolerancing

Dimensions provide the nominal (or target) values for a feature's geometries, size, and location. Tolerances provide the allowable deviation that a feature's actual geometries, size, and location may have from the nominal value, while still providing the expected functionality.

Conventional tolerances can be shown as equal bilateral (e.g. 10.7 ± 0.3), unequal bilateral (e.g. $2.45^{+0.03}_{-0.01}$, $6.250^{+0.035}_{-0.028}$), unilateral (e.g. $4+0.03$, $4^{+0.03}_{0}$, $8^{0}_{-0.01}$), as a limit-based dimension (e.g. $\begin{matrix}10.385\\10.375\end{matrix}$), or as a standard fit (e.g. 40H7/g6). Conventional tolerances are typically attached to the dimensions or provided as a general title block tolerance (Figure 17.1) based on the number of significant digits in the nominal dimension.

UNLESS OTHERWISE INDICATED:		
LINEAR TOLERANCES	X	± 0.5
	0.X	± 0.1
	0.XX	± 0.05
	0.XXX	± 0.005
ANGULAR TOLERANCE	X°	± 2°

Figure 17.1 Conventional general title block tolerances

Geometric tolerances, as defined in ASME Y14.5 [1], are comprised of twelve geometric controls (Table 17.1) provided in a feature control frame (Figure 17.2) and applied to a feature (Figure 17.19).

Table 17.1 Geometric Controls as Defined in ASME Y14.5-2018

Symbol	Name	Controls	Datum referencing
⌓	Profile of a surface	Location, orientation, size, and form of SURFACES	Dependent on application
⌒	Profile of a line	Location, orientation, size, and form of 2D contours	Dependent on application
⌖	Position	Location and orientation for features of size	Required
⌰	Total runout	Location of cylinders about an axis of rotation	Required
↗	Circular runout	Location of any circular element about an axis of revolution	Required

[1] *Dimensioning and Tolerancing* ASME Y14.5-2018.

Table 17.1 Geometric Controls as Defined in ASME Y14.5 *(continued)*

Symbol	Name	Controls	Datum referencing
∠	Angularity	Orientation	Required
⊥	Perpendicularity	Orientation	Required
//	Parallelism	Orientation	Required
▱	Flatness	Form of a nominally flat surface or center plane	Prohibited
—	Straightness	Form of a nominally straight surface line element or centerline	Prohibited
⌭	Cylindricity	Form of a nominally cylindrical surface	Prohibited
○	Circularity	Form of a nominally circular element	Prohibited

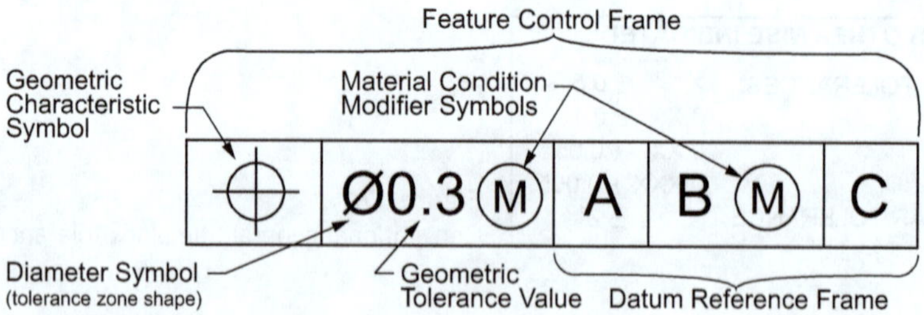

Figure 17.2 Feature control frame

■ 17.2 Limitations of Conventional Dimensions and Tolerances

The underlying issue with conventional dimensioning is that there is no standard definition of what it means, how to interpret it, or how to measure it. This is evident in the way that conventional dimensioning and tolerancing address the following:

- Origins of measurement
- Repeatability of sizes and centers
- Orientation and angles
- Tolerance accumulation.

17.2.1 Origins of Measurement

It seems reasonable to interpret the dimensions of 450 mm and 150 mm in Figure 17.3 to mean that the cutout for the electrical outlet is located 450 mm up the wall, and the outlet cutout is 150 mm tall. Unfortunately, it is not that easy to interpret the intent of conventional dimensions. The question arises, which is the origin of measurement; from the bottom of the wall, or from the floor? Does it make a difference?

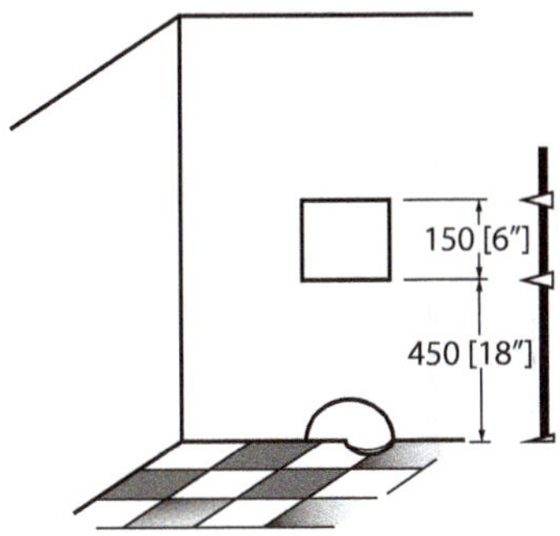

Figure 17.3 Conventional location dimensions

Conventional dimensions are point-to-point, meaning across directly opposed points. Since the square cutout is part of the wall, one interpretation would have the measurement start at the bottom of the wall. However, if you measure from the bottom of the wall to the cutout, you must include the mouse hole at the base of the wall in your measurements (Figure 17.4a). A second interpretation could argue that you would measure from the floor up to the cutout, because you stand on the floor when you reach for the switch (Figure 17.4b). Because the mouse hole extends into the floor, the point-to-point measurement would introduce an error in the measurement (Figure 17.4c). Another interpretation would be to measure from the junction of the wall and the floor, and simulate an equivalent by using a straight edge, essentially establishing a de facto origin of measurement (Figure 17.5a). A final interpretation would argue that all outlets in the room should be at the same height above the floor, no matter where in the room, so you should measure from the flat net equivalent of the entire floor (Figure 17.5b). This is not a valid dimensional interpretation because it is not point-to-point. Ironically, while this net equivalent is not a legitimate interpretation of the point-to-point specification, it does represent the design intent. Unfortunately, there is no effective way to communicate this with conventional dimensioning. So the answer to the question is yes, the origin of measurement *does* make a difference if you want a single interpretation and repeatable results.

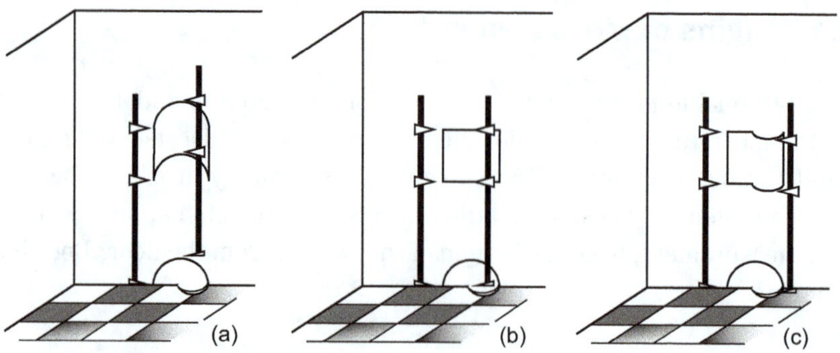

Figure 17.4 Point-to-point measurements

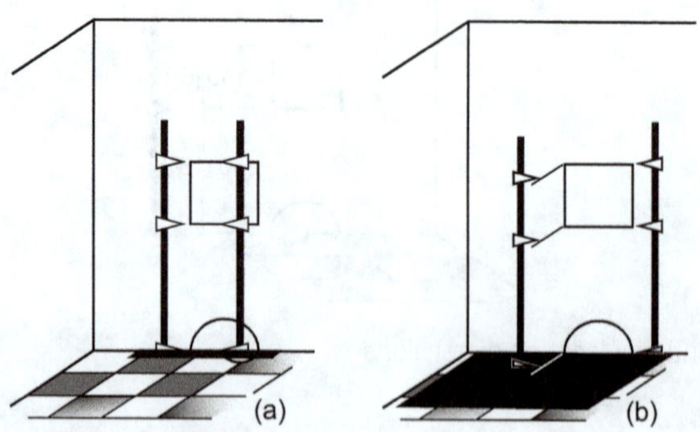

Figure 17.5 De facto origins of measurement

17.2.2 Non-Repeatable Sizes and Centers

In conventional dimensioning and tolerancing, the sizes and centers cannot be found accurately or repeatably. Consider a simple cylindrical solid (Figure 17.6a). Common practice to measure the hole size would be to take two or more measurements of opposed points (Figure 17.6b), then average the measurements to establish a 9.8 mm *actual* size. Note that the largest *measured* size is 10.3 mm diameter, which is larger than the reported size of 9.8 mm. Consider if this were a pin that had to precisely fit into a hole. The local or cross-sectional size is irrelevant to the fit. Instead, you need to know how large the pin is *acting;* that is, what is the equivalent size of a cylinder of perfect form? You can visualize this equivalent (or functional) size as the smallest ring gauge (a perfect cylinder) that can encapsulate the pin (Figure 17.6c). Unfortunately, an averaged measurement will always change depending on where you take your measurements, and will always be smaller than the functional size for a pin, or larger than the functional size for a hole.

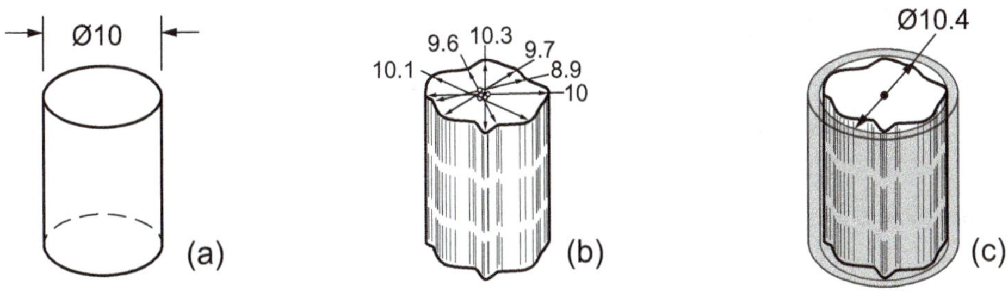

Figure 17.6 Conventional determination of size and centre

Similarly, it is common practice to determine the center of a geometry by halving a measurement between opposed points, or for a series of measurements. Unfortunately, this yields a *cloud of center points* rather than a single true center. Again, each set of opposed points will yield a different, non-repeatable center (Figure 17.6b). Functionally, the center of the equivalent perfect cylinder (Figure 17.6c) is repeatable and of importance when designing mating components. The underlying issue is that conventional dimensioning does not consider the form errors in a feature, and there are no conventional tolerances for form.

17.2.3 Orientation and Angles

Machinists understand that conventional tolerancing does not effectively deal with orientation. Consider a peg that is intended to be perpendicular to the top surface of a plate (Figure 17.7a). Assuming that line-to-line contact would fit, the Ø10 peg should fit into a Ø10 hole that is perpendicular to the surface of a mating plate while the faces of the two plates contact fully (Figure 17.7b). No manufacturing is ever perfect, so even a perfectly-sized peg will be at an angle to the first plate's surface (Figure 17.8a). If we try to engage the Ø10 hole in the mating plate, the parts will fit together, but the plates will not mate flush (Figure 17.8b). With conventional dimensioning and inspection philosophies, the size and location for the peg would be measured, but the orientation of the peg would rarely be inspected unless it is visibly out of perpendicular with respect to the plate surface. The typical resolution for this issue is to rework the hole until it allows the peg to fully engage in the hole (Figure 17.9a), and the plate faces to mate (Figure 17.9b).

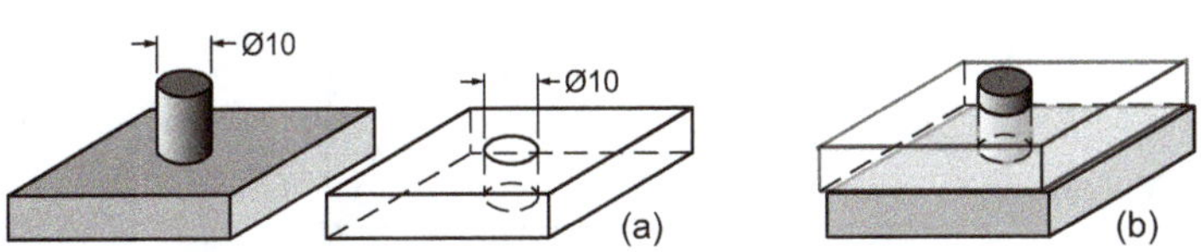

Figure 17.7 Mating with perfect boss orientation

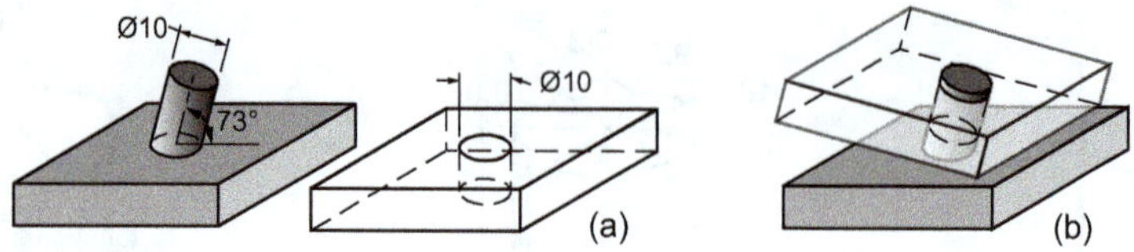

Figure 17.8 Mating with boss orientation error

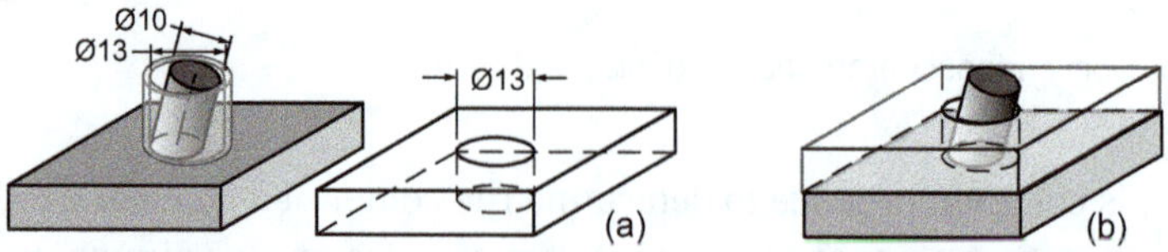

Figure 17.9 Typical mating feature adjustment to compensate for boss orientation error

Angular tolerances (± degrees) present two problems of interpretation; wedge-shaped tolerance zones and how to distribute the tolerance. First, consider an included angle between two sides of a triangle (Figure 17.10). Angular tolerances create wedge-shaped tolerance zones (Figure 17.11a). Notice that the width of the tolerance zone increases as you move further away from the inflection point of the tolerance zone. This means that the longer the surface, the greater the allowable error on it. Now, consider that the toleranced angular dimension applies to all three vertices of the triangle. The result is that the inflection point of the tolerance zone wedge can be at either end of any side of the triangle (Figure 17.11b). This is a valid interpretation of the tolerance callout, because no standard exists for interpretation of conventional tolerances. Furthermore, the inflection point can be anywhere along the side of the triangle, resulting in even more confusing but valid interpretations (Figure 17.11c).

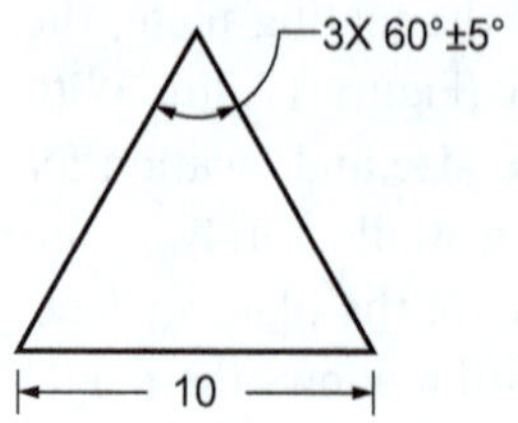

Figure 17.10
Conventional angular tolerance

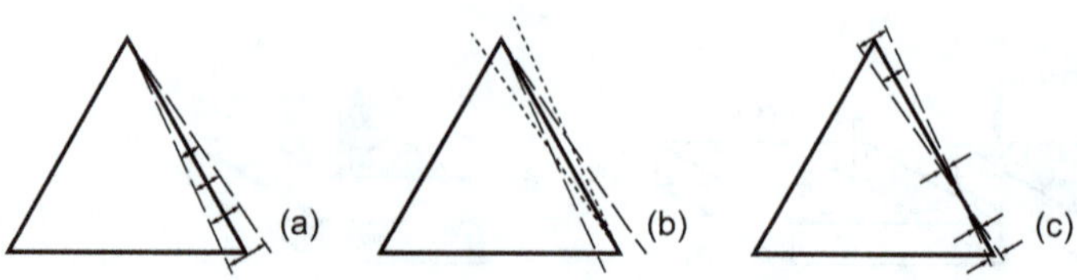

Figure 17.11
Wedge-shaped tolerance zones

Three valid interpretations for this angular callout are shown (Figure 17.12), and other interpretations may be just as valid. The difference between the interpretations shown is how the angular tolerance is distributed. The tolerance can be assigned to one side or the other, or it could be distributed equally or unequally between the two sides. Again, there is no standard by which to interpret this callout.

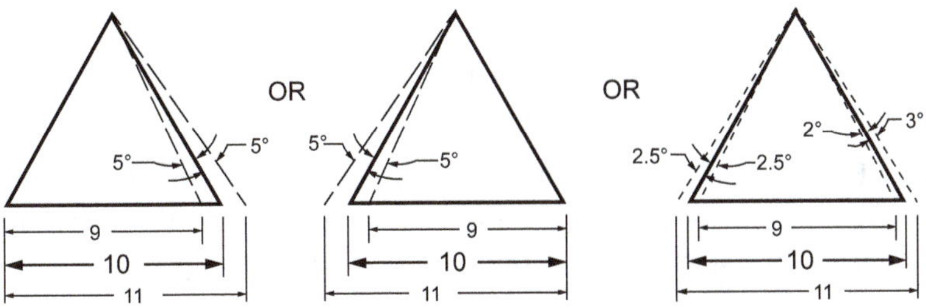

Figure 17.12 Valid interpretations for distribution of an angular tolerance

17.2.4 Tolerance Accumulation

With conventional dimensioning and tolerancing practices, tolerance accumulation is sensitive to how you lay out the dimensions, as well as the tolerances applied to each dimension. Three dimensional layouts are shown in Figure 17.13, Figure 17.15, and Figure 17.17 to demonstrate the issues of tolerance accumulation. For each layout, the tolerance zone on each feature in the dimension path is provided for visualization (Figure 17.14, Figure 17.16, and Figure 17.18, respectively). The tolerance (A, B, and C, respectively) on the length of the center horizontal section will be determined.

For Layout A (Figure 17.13), there are two dimension-vector paths from the left-hand side of the part to the right-hand end of the center horizontal segment; these are designated $A_{1\text{-}2\text{-}3\text{-}4}$ and $A_{5\text{-}6}$ in Figure 17.14. For tolerance analyses, tolerance values add together and do not cancel each other out.

Tolerance $A_{1\text{-}2\text{-}3\text{-}4} = (\pm 1)_{225} + (\pm 1)_{25} + (\pm 1)_{R25} + (\pm 1)_{R25} = \pm 4$.

Tolerance $A_{5\text{-}6} = (\pm 1)_{75} + (\pm 1)_{75} = \pm 2$.

Clearly the results conflict, which means that the specification is invalid.

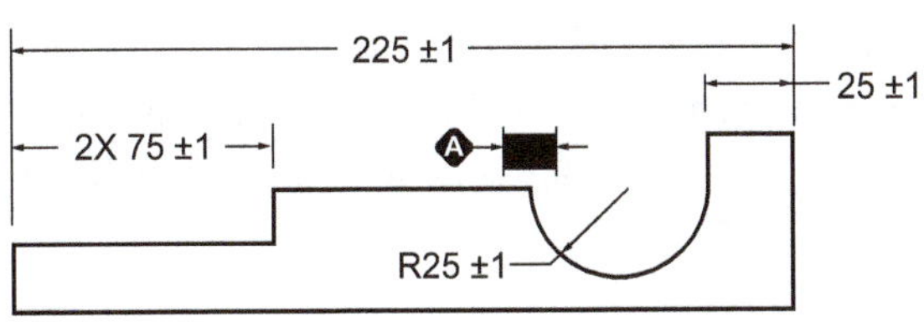

Figure 17.13 Dimensions for Layout A

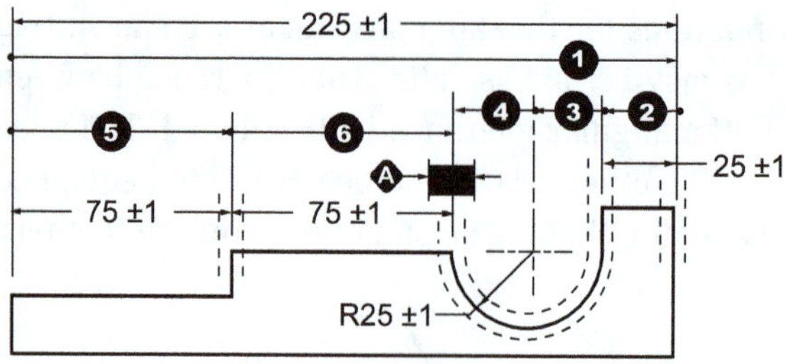

Figure 17.14 Tolerance zones and vectors for Layout A

For Layout B (Figure 17.15), the length of the center segment is provided as a reference dimension; therefore, the tolerance applicable to the location of the right-hand end of the segment is entirely based on chain $B_{1\text{-}2\text{-}3\text{-}4}$.

Tolerance $B_{1\text{-}2\text{-}3\text{-}4} = (\pm 1)_{225} + (\pm 1)_{25} + (\pm 1)_{R25} + (\pm 1)_{R25} = \pm 4$ (Figure 17.16).

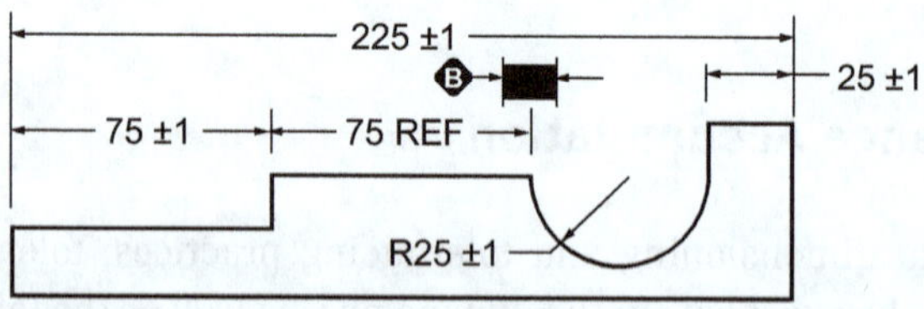

Figure 17.15 Dimensions for Layout B

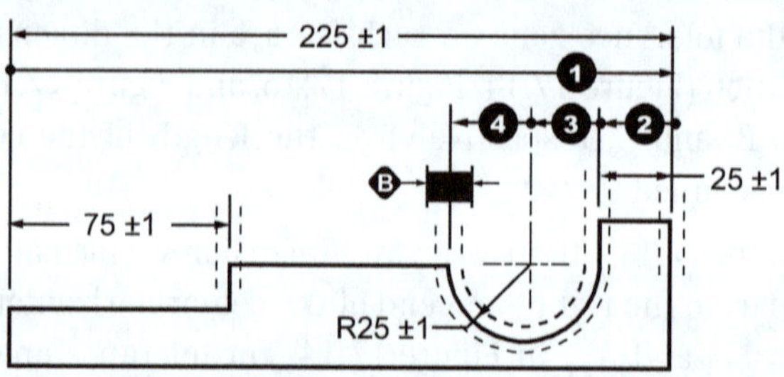

Figure 17.16 Tolerance zones and vectors for Layout B

For Layout C (Figure 17.17), the length of the far-right-hand horizontal section is provided as a reference dimension. The tolerance applicable to the location of the right-hand end of the center segment is:

$C_{1\text{-}2} = (\pm 1)_{75} + (\pm 1)_{75} = \pm 2$ (Figure 17.18).

Though both dimensional layouts B and C are valid, comparing tolerances B and C establishes that the feature's tolerance is dependent on the dimension layout.

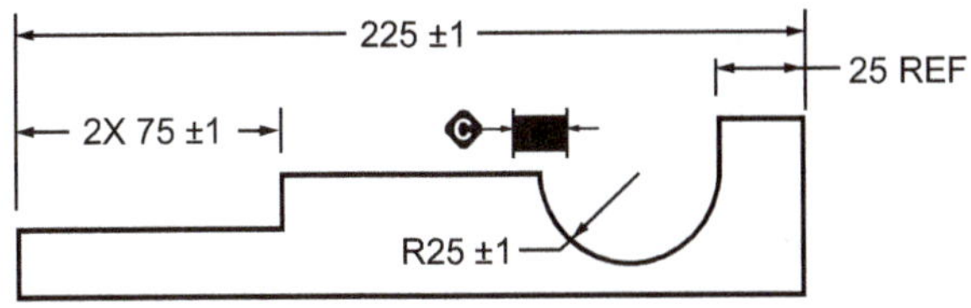

Figure 17.17 Dimensions for Layout C

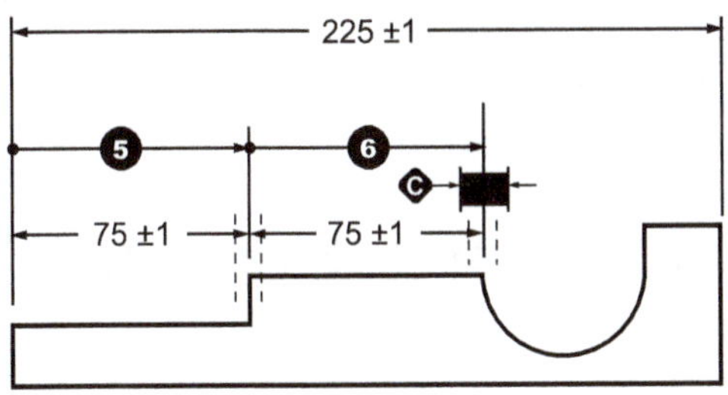

Figure 17.18 Tolerance zones and vectors for Layout C

17.3 Geometric Dimensioning and Tolerancing

Geometric dimensioning and tolerancing (GD&T) addresses the limitations of conventional dimensioning and tolerancing practices described in Section 17.2. There are two international standard series:

- American Society of Mechanical Engineers (ASME) standard Y14.5 [1]
- A collection of standards comprising ISO's Geometrical Product Specifications (ISO-GPS).

While these are substantially similar, because the ISO standards originally derived from the ASME standard, the intent of each standard has grown different over time. In the ASME standard, the definition of each geometric control is rooted in the functional (design) intent of the control, whereas the definitions of the ISO standard are based on the metrology of each geometric control. Though the symbology of both are substantially the same, the two standards are not interchangeable. Therefore, the applicable standard must be indicated on the drawing, and that standard is the legal basis for interpretation of that drawing. For this text, GD&T according to ASME Y14.5-2018 will be discussed.

The primary benefits of geometric tolerancing over conventional tolerancing are:

- Workpiece set-up and origins of measurement are clearly identified using datum feature identifiers

- Geometric controls apply to the features, not to the dimensions
- Addressing all feature attributes: location, orientation, size (geometry) and form
- The feature's sizes and centers are repeatable
- Tolerance accumulation is minimized
- Tolerance zones are uniform rather than wedge-shaped
- Control of patterns of features
- It provides a single interpretation.

17.3.1 Three Core Steps for Applying or Inspecting GD&T

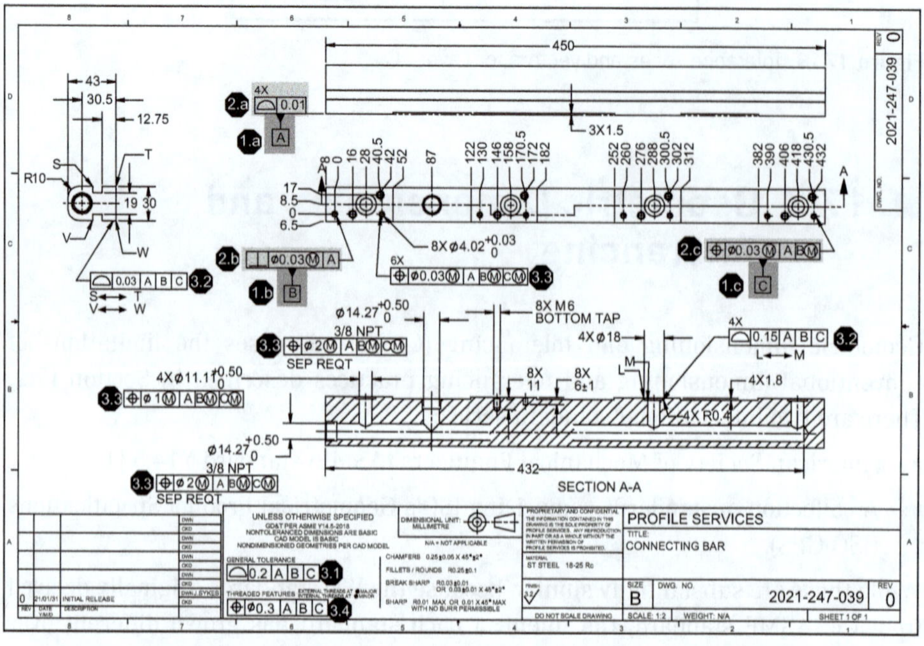

Figure 17.19 Sample drawing toleranced in accordance with ASME Y14.5-2018

Step 1: Identify Datum Features

Datum feature identifiers (Figure 17.19, labels 1.a, 1.b, 1.c) indicate physical elements of a workpiece that establish dimension origins on engineering drawings. While datum features are imperfect physical features, the datums (or origins of measurement) that they will establish are perfect and theoretical. Datums are simulated by engaging near-perfect fixturing components to the datum features; the fixture components are the origins of measurement for the associated geometric controls. Using the datum features directly as the origin of measurement would introduce any deviations in the datum feature as an error in associated measurements. Datum features progressively constrain up to six degrees of freedom, typically reflecting how they engage another workpiece. *Up to three datum features can be referenced in a geometric control.*

Step 2: Control Datum Features

Contemporary mold design is based on mating of perfect CAD-modelled components, but nothing is perfect. The designer uses geometric controls (Figure 17.19, labels 2.a, 2.b, 2.c) to limit how far from perfect the datum feature can deviate while still performing as required.

Step 3: Control the Location of All Other Features with Respect to the Datums

ASME GD&T establishes a hierarchy of feature attributes - location, orientation, size, and form - in order of importance (Figure 17.22). Features fall into two categories, namely surfaces and features of size. Surfaces can only be located with the "profile of a surface" control. Depending on the functional intent for a feature of size, you can locate it with position, runout, or profile of a surface.

In the sample drawing (Figure 17.19), note that non-toleranced dimensions are basic and a general profile of a surface tolerance (labeled 3.1) is provided for all surfaces not directly controlled. Profile of a surface controls (labeled 3.2) are applied to specific areas to override the general profile tolerance. Position controls (labeled 3.3) have been applied to all features of size, and a general position tolerance (labeled 3.4) is provided for threaded features.

17.3.2 Geometric Tolerancing Benefits versus Conventional Tolerancing

ASME geometric tolerancing is a robust system of geometric controls for the four primary attributes of every feature on any workpiece: location, orientation, size, and form. The twelve controls have defined meanings, with rules and syntax for application, which mitigate the many issues of conventional tolerancing practices discussed in Section 17.2.

17.3.2.1 Clear Setup and Origins of Measurement

GD&T uses datum feature identifiers to show which features to use for inspection setup, and the datum reference frame of each feature control frame indicates the sequence in which to engage the datum features to set up a workpiece for inspection. Typically, this reflects how the part mates within an assembly. Datum precedence refers to the sequence in which datums are referenced in a feature control frame. Alphabetical order has no relevance to datum precedence, though common practice for machined workpieces is to designate the primary datum feature as datum feature A, the secondary as datum feature B, and the tertiary as datum feature C.

The feature control frame on the pattern of four counter-bored holes (Figure 17.20) indicates a datum precedence of A | B | C, meaning that you engage datum feature A on its datum feature simulator first, engage datum feature B second, and finally engage datum feature C. For inspection purposes, common open-setup fixturing components such as V-blocks, surface plates, keys, collets, and angle blocks can be used to engage the datum features. For the illustrated datum feature simulator (Figure 17.21a), the keys used to simulate datum features B and C will have to be movable so as to engage their respective datum features in sequence. The resulting three mutually-perpendicular datum planes are the origins of measurement (Figure 17.21b) for the specified datum reference frame. Because datum features are imperfect and datum planes do not physically exist, we measure the location and orientation of controlled features from the datum feature simulators (Figure 17.21b). These simulators are adequately precise as to not introduce errors into the measurements, and to repeatably establish the origins of measurement for inspection.

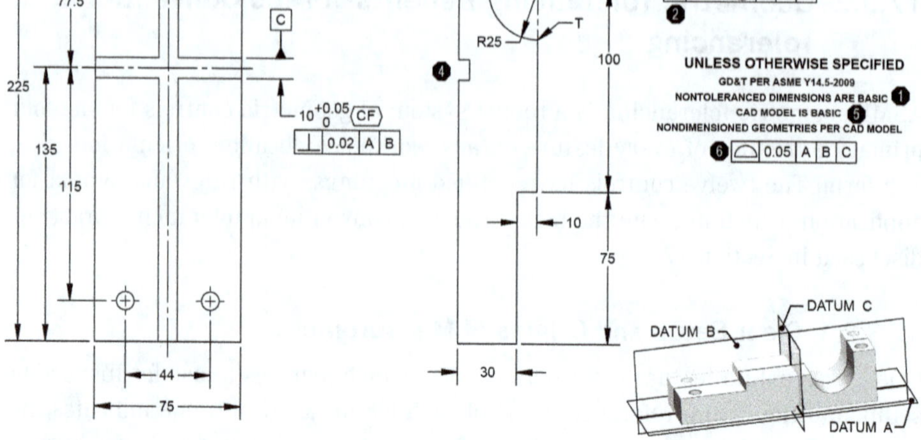

Figure 17.20 Datums as origins of measurement

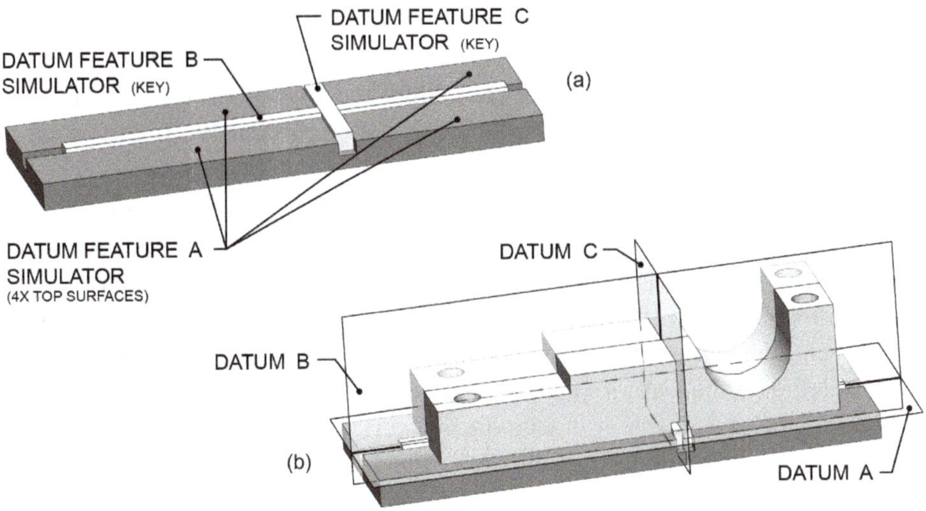

Figure 17.21 Datum feature simulators and resulting datums

17.3.2.2 Geometric Controls Apply to Features, not to Dimensions

Conventional dimensioning and tolerancing ties each feature to a preceding feature in a chain of tolerances and irrelevant relationships. Beyond unwanted tolerance accumulation, the greater concern is that conventional practices do not establish a relationship for each feature directly to its intended functionality. *GD&T addresses four feature attributes: location, orientation, size, and form.* Of these, location and orientation are controlled relative to the datums, establishing a repeatable, independent relationship for each feature to the origins. GD&T drawings indicate the basic, or theoretically perfect, locations for features. Basic dimensions are typically shown in boxes. An alternative practice is to include a general statement on the drawing to indicate that all non-toleranced dimensions are basic (Figure 17.20, labeled 1). In a model-based definition (MBD) environment, where manufacturing and metrology will be working directly from the CAD model, it is common practice to omit basic dimensions for some features (Figure 17.20, labeled 4), instead indicating that the CAD model is basic, and directing the user to the CAD model for further information (Figure 17.20, labeled 5). Once the location of a surface or feature of size is established with basic dimensions, the feature is controlled with a directly-applied geometric tolerance, or with a general geometric control (Figure 17.20, labeled 6).

Features of size have directly opposed points and a limit on the size. A feature which has an enclosed boundary may be considered either a feature of size or a surface. If you are concerned separately about the actual size of the feature, then you will treat it as a feature of size by applying a tolerance to the size dimension and controlling its location with either a position or runout control. Otherwise, your primary concern is that the feature's geometries do not exceed prescribed inner and outer boundaries. Treat such features as surfaces, specifying the geometries with basic dimensions, and controlling the geometries with profile of a surface. In practice, most features should be treated as surfaces. In either case, however, the geometric control must reference the datum(s).

Position (Figure 17.20, labeled 2) can control the center (plane, axis, or point), inner boundary, or outer boundary of a feature of size. Many mold shops are familiar with hard gages, which is a specialized application of position controls. The position tolerance zone is centered on the basic, or ideal, location of the feature. The default condition for profile of a surface establishes an equal-bilateral tolerance zone centered on the feature's nominal geometry. The inclusion of a circle-U (Figure 17.20, labeled 3) modifier in a profile of a surface control allows the tolerance zone to be distributed either unilaterally or unequal-bilaterally from the feature's nominal geometry.

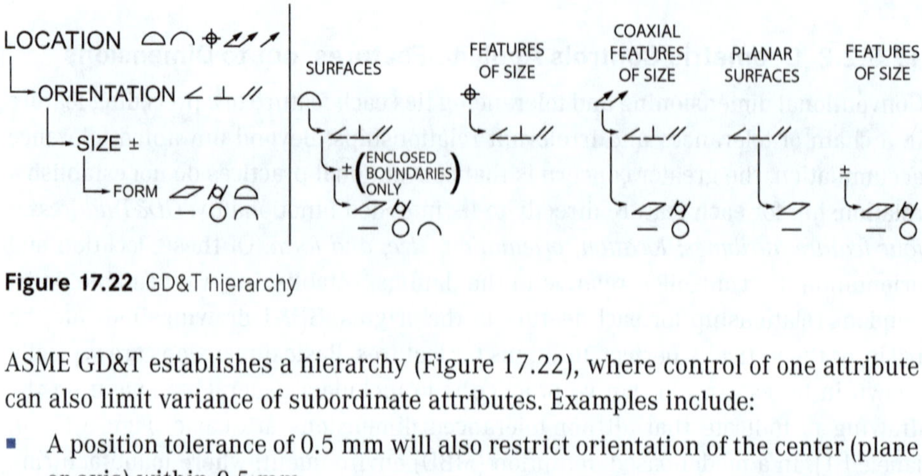

Figure 17.22 GD&T hierarchy

ASME GD&T establishes a hierarchy (Figure 17.22), where control of one attribute can also limit variance of subordinate attributes. Examples include:

- A position tolerance of 0.5 mm will also restrict orientation of the center (plane or axis) within 0.5 mm
- A perpendicularity tolerance of 0.15 mm on the face of a plate will also restrict flatness and straightness of that face within 0.15 mm
- ASME Y14.5 specifies that by default, the limits of size also limit the form. Therefore, a shaft with a size tolerance of ±0.25 mm can have a maximum cylindricity error, or circularity error, of 0.5 mm when the shaft is produced at its smallest size.

Profile of a surface is the most robust geometric control in ASME GD&T. Its primary use is to control the location of surfaces. However, the profile of a surface also controls orientation and form, and can control the size of an enclosed geometry. Applied to a square pocket in a plate, a profile of a surface tolerance of 0.5 mm will control the location of each face of the pocket within 0.5 mm, the orientation of each within 0.5 mm, and the flatness of each surface within 0.5 mm. Indirectly, it also controls the width of the pocket within ±0.5 mm.

17.3.2.3 Repeatable Sizes and Centers

ASME Y14.5 is based on the communication of design intent, and uses geometric controls to limit how far from perfect a feature can vary while still providing the expected functionality. ASME Y14.5 specifies that a feature's size and center are determined by engaging its actual mating envelope (AME), also called its true geometric counterpart. Open-setup tools such as ring or pin gauges, machinist's vices, gage blocks, chucks, and collets are commonly used to simulate AMEs. As described above (Section 17.2.2, Figure 17.6), a pin that is measured as Ø9.8 by conventional methods *acts* as large as Ø10.3 when inspected using the feature's AME (Figure 17.6). Because the AME is a precise cylinder, measuring across any opposed points will yield a repeatable size measurement. Similarly, the repeatable axis of the AME is the axis of the pin.

17.3.2.4 Tolerance Accumulation Is Minimized

Basic dimensions are theoretically perfect, meaning that there is no tolerance applied to them. Therefore, chaining basic dimensions does not accumulate tolerances. Instead, each feature is directly controlled for location with respect to the datums.

According to the **Unless Otherwise Specified** statement (Figure 17.20, Figure 17.23), non-toleranced dimensions are basic, meaning theoretically perfect. The basic location of the vertical segment starting at point S is found by adding vectors one through four: 25 + 100 + 75 - 135 = 65 basic from datum C (Figure 17.23). The location of the vertical segment starting at point S is controlled by the profile of a surface tolerance applied between S and T. The total tolerance zone is 0.5 mm wide. The circle-U symbol indicates an unequally disposed tolerance with 0.4 of the total 0.5 mm tolerance zone to the side of the nominal geometry that adds mass to the part. The tolerance zone on the indicated segment is illustrated.

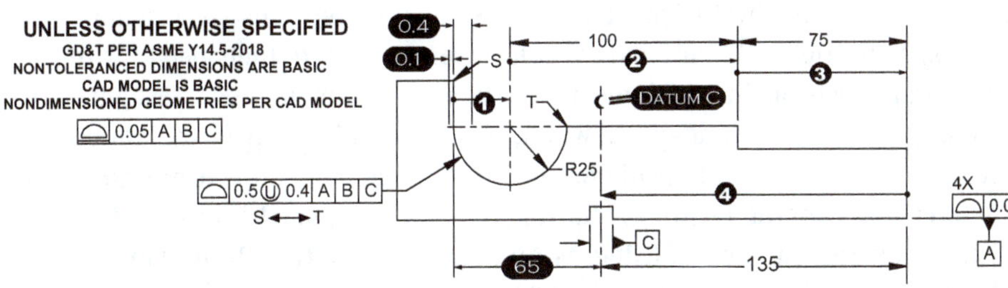

Figure 17.23 GD&T tolerance accumulation

17.3.2.5 Tolerance Zones Are Uniform

For surfaces, design engineers typically envision tolerance zones that are uniformly offset from the nominal geometries. Whether the nominal geometry is flat, curved, complex, or angled, the profile of a surface will establish a uniform tolerance zone offset from the nominal geometry. However, the circle-U modifier (Figure 17.20 label 3, Figure 17.23) allows the specification of a non-uniform tolerance zone when appropriate to the design intent.

17.3.2.6 Control of Patterns of Features

One of the most significant advantages of geometric tolerancing over conventional practices is that it offers flexibility in addressing patterns. Patterns can be comprised of multiples of the same feature, different feature geometries, or combinations of these. A pattern's features can be restricted to locations on one surface, or multiple surfaces. Controlling patterns can allow a comparatively large overall locational tolerance of the entire pattern while controlling the relationship of the features within the pattern more tightly to each other, as is commonly needed in mold design. When appropriate, pattern control can be more restrictive in one direction than another.

17.3.2.7 A Single Interpretation

GD&T is a language, and will effectively communicate the design intent only when both the creator and user of the engineering drawing interpret the GD&T in the same way. ASME Y14.5 provides the definitions, interpretations, and rules of application of the twelve geometric controls. The result, when the standard is followed, is a single interpretation of workpiece functionality.

■ 17.4 Implementation of Geometric Tolerancing

Repeatability of any level of precision, coarse or fine, requires clear communication of specifications. When mold manufacturers recognize their need for repeatability and precision in a volume production environment, they recognize the value of geometric tolerancing. However, a strict requirement for a defined payback period is a common impediment to GD&T implementations. **Understanding geometric tolerancing requires training of designers, machinists, and metrologists by an instructor qualified as an ASME Geometric Dimensioning and Tolerancing Professional (ASME GDTP).** An effective understanding of GD&T requires an iterative and specialist-supported approach of training, application, and validation of the knowledge and skillsets involved. The introduction of GD&T is often synchronized with equipment upgrades in the shop and in inspection. While increased throughput of the equipment and better manufacturing capabilities can be measured, improvements directly attributable to implementing geometric tolerancing are largely intangible for a long period. Over time, however, continued championing of geometric tolerancing skillset development across the enterprise will result in increased manufacturing efficiencies and capabilities, reduced rework and scrap, reduced field service and warranty costs, as well as improved reputation and customer satisfaction.

As companies move from conventional drawing practices to geometric tolerancing, their perspective on geometric tolerancing, and therefore their implementations, typically evolve. GD&T implementations fall into three general categories: GD&T-lite, GD&T integration, and model-based definition. Once geometric tolerancing is effectively understood, further gains are made by introducing advanced elements including manufacturing capability studies and statistical process controls.

17.4.1 Implementation of GD&T-Lite (not Recommended)

Initially, most implementations will take a GD&T-Lite approach, using GD&T only for critical features, meaning those that have very tight tolerances. This is rooted in an unfounded belief that GD&T only conveys precision. In fact, geometric tolerancing communicates feature relationships and how critical each attribute (location, orientation, size, form) of every feature is to the design functionality; it eliminates guesswork. While the relatively important features are more effectively identified and controlled in the GD&T-Lite approach, the comparatively less important features are still subject to interpretations of vague conventional tolerances. The greater issue is that GD&T-Lite often results in multiple tolerances applied to some features, and an unclear definition of intended functionality. Companies using a GD&T-Lite approach will typically convert GD&T controls to conventional tolerancing for manufacturing and metrology purposes. In this way, the value of the GD&T implementation investment is largely lost.

17.4.2 Implementation of GD&T Integration

Whether a decision from the outset, a progressive recognition of the residual issues of conventional tolerancing, or a necessary step to the integration of new technologies, the next phase in a GD&T implementation is a complete transition to a full GD&T integration.

In a fully compliant GD&T drawing, conventional plus-minus tolerances are only applied in two scenarios: to feature of size dimensions, and as a control for *junk* features such as fillets/rounds, chamfers, counter-sinks and non-critical counter-bore and spotface depths. Even then, the location, orientation, and form of the face of a counter-bore or spotface is often important and will be controlled with geometric controls. Companies integrating GD&T into their business will typically provide significant training across the entire enterprise, often including their supply chain. Manufacturing and metrology technologies and processes are typically selected to facilitate the use of GD&T, though these technologies typically require the manual transposition of the geometric tolerances from the drawing into the applicable software. In this way, the value of investing in GD&T implementation is realized, though the process is not the most efficient, and has a risk of miscommununication.

17.4.3 Implementation of Model-Based Definition

Model-based definition (MBD) is a product definition approach where all attributes are embedded in the three-dimensional CAD model. Attributes include datum feature identification and control, sizes and basic location dimensions, geometric controls, and surface finish characteristics. Though some companies using MBD will still produce traditional drawings, the same information is associatively annotated within the model, which can eliminate the need for drawing creation. The greatest value of MBD is that it can directly convey all the attribute data to compatible CAM (computer-aided manufacturing) and metrology software packages, which improves work throughput and mitigates the risk of miscommunication. MBD also introduces the capacity to use automated tolerance analysis software. As mechanical systems become increasingly complex, as the need for controlled precision increases, and as new production technologies introduce new process capabilities, tolerance analyses gain importance, and the capability to analyze "what if" scenarios gains value. The embedded geometric tolerances in MBD allow analysis software to quickly run scenarios without having to restructure manual calculations.

Initially, MBD users may feel overwhelmed at the three-dimensional nature of the information provided. However, users quickly find that feature interrogation reciprocity can be easier to use than reading a conventional drawing. This interrogation reciprocity means that when you click on a geometric control, any features associated with it are highlighted, and when you click on any feature, all geometric controls and datum feature identifiers related to it are highlighted. This single functionality makes it significantly easier for manufacturing and inspection users to accurately interrogate the CAD model and retrieve the specific information they need.

17.5 Considerations for Tolerancing

At first glance, tolerance selection would appear to be an easy task. Essentially, the mold designer just establishes the sizes that ensure the parts will fit together. That is reasonably simple if the assembly is only two pieces that mate in a simple pin-and-hole configuration, or with two nominally flat faces. Molds are not that simple. They have planar and non-planar surfaces that must mate intimately. There are locating features such as leader pins and bushings that must be precisely located with respect to each other while also simultaneously precisely oriented with respect to each other and to the mold plates. All of these interactions must be repeatable, whether for accommodating initial assembly of high-cavitation molds or for field replacement of mold spares.

Geometric tolerance determination focuses on design intent but must include other considerations. For a mold, the decision of nominal CAD model size is the first additional consideration and is guided by the functional fits between components. Material characteristics and manufacturing processes are secondary considerations that will impact the achievable tolerances, while inspection processes may limit the ability to validate tolerances. These are just a few considerations when determining mold tolerances. A final consideration to be discussed relates to both the nominal CAD geometries and the selected tolerances; thermal expansion may become an important consideration when dissimilar materials are used and operating temperatures deviate significantly from ambient conditions. In practice, the final tolerance determination often comes down to the cost and value of the tolerance being applied.

17.5.1 CAD Modelling Practices

One of the first tolerancing considerations is actually a CAD modelling decision; at what size do you model each feature? This decision impacts tolerance application and analysis, as well as manufacturing and inspection interpretation of the drawing. A feature is modelled in CAD at its nominal location and/or nominal size. Typically this will be at the feature's maximum, minimum, or median (middle) size, though in cases of some clearance and interference fits, the nominal size may be outside the tolerance range. Mating of features on different workpieces is frequently the basis for sizing.

Fits-based tolerancing, called "limits & fits", is a common way of determining nominal sizing and applicable tolerances. Fits are broadly categorized as clearance, transition, or interference. ASME Limits & Fits [2] describe ten characterized inter-

[2] *Preferred Metric Limits and Fits* (B4.2), ASME, 1978.

actions from loose running to force fit, and specify them with a nominal size, hole, and shaft fit class. An ASME limits & fits specification of 40H7/g6 yields a hole size of $^{40.025}_{40.000}$ and a shaft size of $^{39.991}_{39.975}$, both of which may be modelled at 40 mm nominal size. Notice that the nominal hole size is at the lower limit of the tolerance band, while the nominal shaft size is not in the range of allowable shaft sizes. Understandably, this may lead to issues with finite-element analyses and tolerance analyses, but perhaps more significantly, model-driven CNC programming may not detect this difference.

Limits & fits is a useful design tool, best addressed by conversion to a median-nominal with equal bilateral tolerances; 40.0125 ± 0.0125 for the hole and 39.983 ± 0.008 for the shaft. Limits & fits, developed before geometric tolerancing, address only size, leaving location and orientation to be controlled with geometric tolerances appropriate to the application.

In general, preferred CAD modelling practice has shifted from maximum or minimum limit to median nominal for all geometries, regardless of the type of tolerance or tolerance value applied.

17.5.2 Materials, Manufacturing, and Inspection

Material properties, including composition, quality or purity, additives, and hardness, affect the chip-forming process, or machinability. Machinability, in turn, will affect feed and cutting speeds, and may limit the tolerance capabilities available to a workpiece. In many cases, material suppliers are able to provide grade-specific information on material characteristics, including machinability; however, specific tolerance capabilities are co-dependent upon the material removal process. Engineering and machinist reference materials may provide general tolerance capabilities for common machining processes for cutting common-grade materials, and machine suppliers will typically provide machine-specific general capabilities. Notwithstanding the value of these resources, they are generalizations of the process capabilities rather than specific to the mold material, cutting tools, and machining center that will be used. Process capability studies, based on analysis of geometric characteristics of a machining center for specific material grades, will provide meaningful data that can be used by engineering in refining mold design.

Throughout the history of manual machining processes, preference was given to *material-safe* conditions, wherein a feature was machined to just within the tolerance limits so that it could be reworked if needed. This bias is evident in limit-based tolerancing in the ASME limits & fits tables, and made it easier for machinists to work with the drawings. The material-safe bias of manual machining processes creates a skewed distribution of manufactured sizes and locations. In most metal removal processes, the migration to CNC in conjunction with various machine tool encoders have made the machines more accurate in terms of feature location and size, and have shifted to a normal process distribution. Normal distributions provide an equal-bilaterally-disposed tolerance centered on a median nominal dimension, another compelling reason to model at median-nominal regardless of the total tolerance value.

17.5.3 Thermal Expansion and Tolerancing

One critical consideration frequently overlooked in mechanical design is thermal expansion, which needs to be addressed in the nominal geometries and in the tolerancing. The ASME Y14.5 geometric tolerancing standard indicates that dimensions and tolerances apply at 20 °C (68 °F) unless otherwise specified, consistent with typical controlled shop environments. However, injection molds generally run at elevated temperatures, which induce thermal expansion in the mold components. Many materials have the same thermal expansion factors, but there can be significant differences between adjoining dissimilar materials such as steel and copper alloys, which are common in mold design. Because of this discrepancy, the fit of the mating components of dissimilar metals may change as the mold temperature shifts from ambient to operating temperature. Mold components should be designed to permit expansion starting at their constraint or mounting location, then compensating for thermal expansion of both the nominal geometries and the tolerances of size and location to ensure that anticipated fit is achieved at operating temperature. Thermal expansion factors are available in Appendix 9, and the impact of thermal expansion on the design of the mold is discussed in the chapters on melt distribution (Chapter 14) and alignment (Chapter 13).

■ 17.6 Application of GD&T to Mold Components

The benefits of ASME Y14.5 geometric tolerancing have been discussed, and the core steps of GD&T have been illustrated. What remains is a demonstration of the application of GD&T to mold components. The following examples, an injection mold core and lock ring, and their associated core plate, illustrate the three core steps (Section 17.3.1) in the application of GD&T. A fourth step, the addition of auxiliary datums, is particularly applicable to patterns of features in mold plates of multi-cavitation molds.

Note: Drawings are intentionally incomplete. Dimensions and controls for most features have been omitted to clarify key concepts. Geometries, dimensions, geometric controls and tolerance values are provided for illustration purposes only, and may not reflect actual designs.

17.6.1 GD&T Applied to an Injection Mold Core

Following Section 17.3.1, the **first step** in application of geometric tolerances is to *identify the datum features*. For the core (Figure 17.24), its engagements with the mating components (core plate, lock ring) are considered. The flat surface is seated on the face of the core plate and is located by the simultaneous engagement of the core's shoulder diameter and conical taper to the lock ring. There is clearance between the spigot diameter and the bore in the core plate, indicating that the interface of the core with the lock ring will be the dominant consideration in the core's location. Based on these interfaces, the flat surface is identified as datum feature A, establishing a plane as the primary datum. The shoulder diameter is identified as datum feature B, and the conical taper as datum feature C. Because datum features B and C are engaged simultaneously, they are used together to establish datum axis B-C, perpendicular to the primary datum plane. Datum precedence |A|B-C| constrains five of the six degrees of freedom. Because there are no features that require control of their angular location about the datum axis B-C, a tertiary datum is not required.

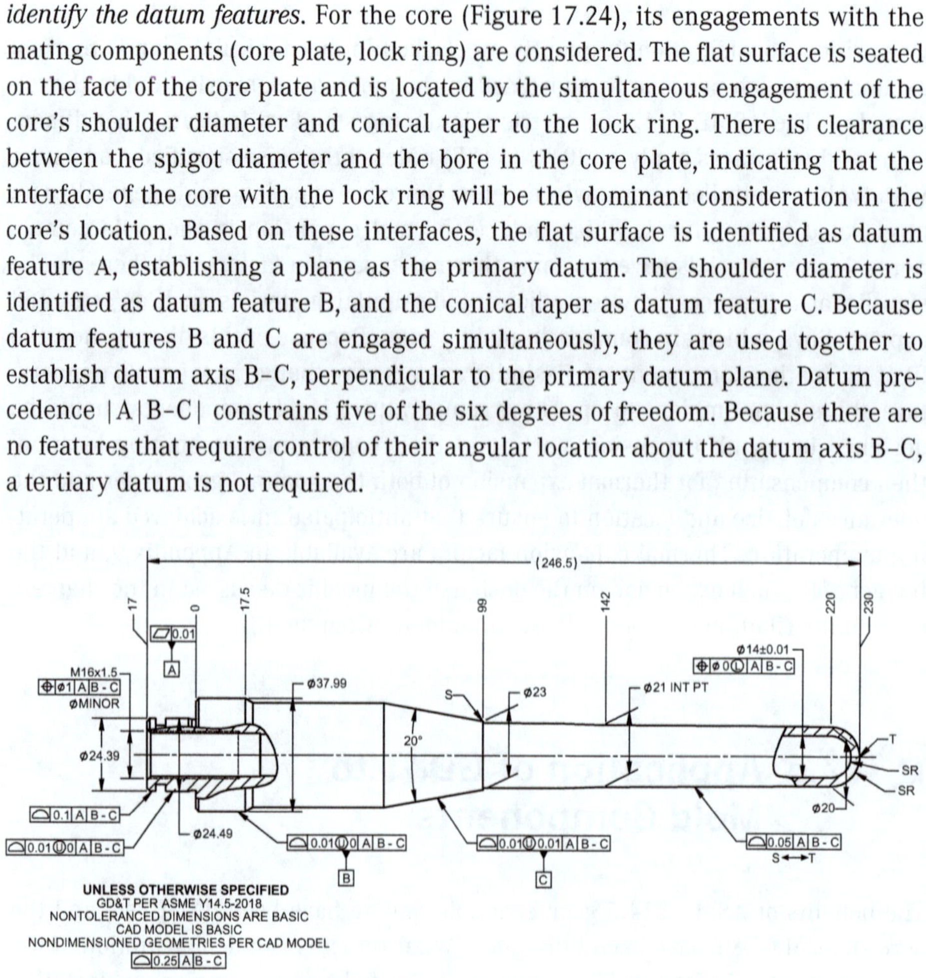

Figure 17.24 Application of GD&T to an injection mold core insert

The **second step** in GD&T application is *control of the datum features*. Datum feature A is controlled with flatness.

A brief explanation of profile of a surface and the circle-U modifier is appropriate here, but must be preceded by an explanation of a useful analogy for tolerances. Consider a hole with an equal-bilateral size tolerance; the positive tolerance results in a larger hole which reduces the mass of the workpiece, whereas the negative tolerance results in a smaller hole which increases the mass of the workpiece. Conversely, consider a shaft with an equal-bilateral size tolerance; the positive tolerance results in a larger shaft, which increases the mass of the workpiece, whereas the negative tolerance results in a smaller shaft, which decreases the mass of the workpiece. The tolerance portion of the profile control is always total and, by default, equally disposed about the nominal geometry. In simple terms, this means that by default the tolerance is split equally to both sides of the nominal geometry. The circle-U modifier allows an unequal disposition of the tolerance. When a circle-U modifier is included in the tolerance portion, the first number is the total profile of a surface tolerance, and the second number is the portion of the total tolerance that goes to the side that adds mass.

Datum feature B engages the shoulder of the lock ring with a flush-to-clearance fit; this is addressed by using a profile of a surface control with a circle-U modifier indicating that all of the tolerance decreases the diameter, or that it goes to the side that decreases mass. Datum feature C engages the conical taper in the lock ring with a flush-to-preload fit; this is addressed using a profile of a surface control with a circle-U modifier indicating that the entire tolerance goes to the side that adds mass. Both datum features B and C are related to primary datum A, and to datum axis B–C.

The **third step** in GD&T application is the *location of all non-datum features with respect to the datums*. ASME Y14.5 requires that the geometries and locations of all features are controlled. Rather than apply controls to every feature, a general profile of a surface control is specified, and applies to all features not otherwise directly controlled. Beyond that, individual features are controlled directly. For features where the primary concern is that they are contained within an inner and outer boundary, a profile of a surface control is applied to the surface directly. For the core, the spigot diameter, lead-in diameter, and the molding surface are controlled directly with profile of a surface control. The cooling hole and the thread at its end are separately controlled with position controls. Note the LMC (least material condition) modifier on the position tolerance of the gundrilled hole; this is an indication that the primary concern is that the combination of size and location of the hole do not result in an unacceptably thin wall.

17.6.2 GD&T Applied to an Injection Mold Lock Ring

For the lock ring (Figure 17.25), the three steps of GD&T application include the following considerations:

Step 1: Identify Datum Features

Primary datum feature: The base of the lock ring sits flush on the core plate. Labeled datum feature A.

Secondary datum feature: The lock ring is aligned by the core interfaces at the lock ring's lower bore diameter (labeled datum feature B) and at the bore's conical taper (labeled datum feature C). As the two datum features (B and C) engage simultaneously, they create datum B–C, an axis that is perpendicular to datum A.

Tertiary datum feature: Collectively, the primary and secondary datums constrain five of the six degrees of freedom, leaving only rotation about datum axis B-C unconstrained. Because there are features that require control of their angular location about the datum axis B–C (i.e., four mounting holes, two vent holes in the shallow conical taper pocket, two vent holes in the base), a tertiary datum is required. The lock rings are spaced closely to each other on the core plate. Therefore, the width of the base is selected as tertiary datum feature D.

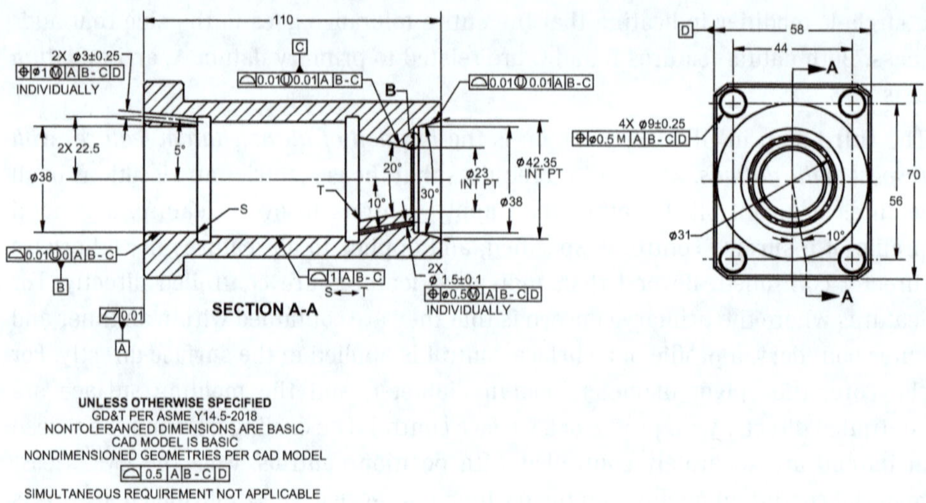

Figure 17.25 Application of GD&T to a lock ring insert

It is important to note that the placement of the datum feature identifier indicates whether the datum feature is the width of the base or one side-face of the base. The triangle of the datum feature identifier is aligned with the arrow of the width dimension, indicating that the width is the datum feature and the datum is the center plane of that width.

If the datum feature identifier had been shifted out of alignment with the arrow of the width dimension, it would have indicated the face to which the extension leads as the datum feature, establishing a datum plane perpendicular to primary datum A, parallel to the axis of secondary datum B-C, and coincident with the indicated surface. ■

Step 2: Control Datum Features

Primary datum feature: The base of the lock ring is nominally flat, therefore a flatness control is applied to datum feature A.

Secondary datum feature: Datum feature B engages the shoulder of the core with a flush-to-clearance fit; this is addressed by using a profile of a surface control with a circle-U modifier indicating that all of the tolerance increases the diameter; that is, to the side that reduces the mass of the part. Datum feature C engages the conical taper in the lock ring with a flush-to-preload fit; this is addressed using a profile of a surface control with a circle-U modifier indicating that the tolerance can only allow the cone to be smaller than nominal; that is, it all goes to the side that adds mass to the part. Both datum features B and C are related to primary datum A, and to datum axis B–C.

Tertiary datum feature: The general profile of a surface control locates each surface of the periphery of the base of the lock ring. This indirectly controls the width of datum feature D.

Step 3: Locate All Non-Datum Features with Respect to the Datum Reference Frame

Notice that the location of the top and bottom vent holes in the lock ring (Figure 17.25) are toleranced with a position control modified at MMC (maximum material condition). This indicates that the primary concern for the holes is that the combination of size and location of the holes does not result in an unacceptably small interface area with the airline in the core plate and with the vent groove in the shallow conical taper pocket of the lock ring.

To ensure adequate venting in the shallow conical taper pocket that is the neckring seat (Figure 17.26), the longitudinal location of the vent relief is controlled with a unilateral profile of a surface control that ensures the relief can only get deeper from its nominal location.

The molding surface features of the lock ring are controlled by a profile of a surface tolerance between points K and M.

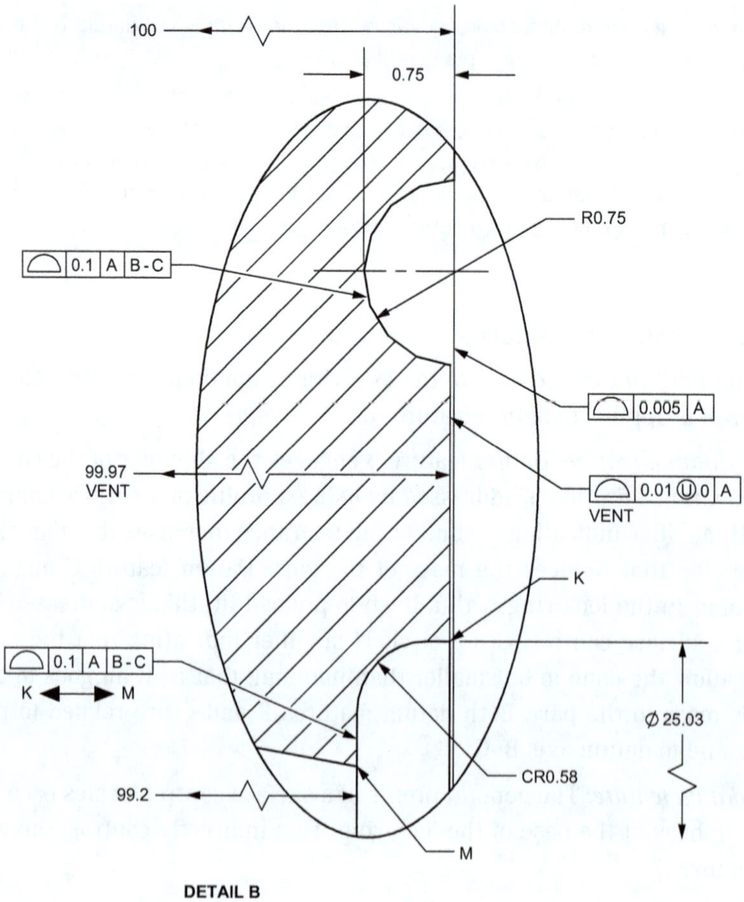

Figure 17.26 Application of GD&T to a lock ring insert – detail

17.6.3 GD&T Applied to an Injection Mold Core Plate

For the core plate (Figure 17.27), the three steps of GD&T application include the following considerations:

Step 1: Identify Datum Features

Primary datum feature: The core plate mounting face sits flush on the mold platen. Labeled datum feature A.

Secondary datum feature: Most features on the insert mounting face are visually symmetrical about the center of the plate. The datum feature identifier is aligned with the arrow of the width dimension of the plate, establishing datum B as the center datum plane.

Tertiary datum feature: The datum feature identifier is aligned with the arrow of the length dimension of the plate, establishing datum C as the center datum plane.

Auxiliary datum feature: There are evident repeated patterns on the insert mounting face of the core plate, each centered on the core spigot interface hole. Centered on each core spigot hole is a local pattern that includes four lock ring mounting holes and an air hole on the surface of the plate, and two holes coaxial to the core spigot interface hole (Figure 17.28, Detail A, Section B–B). Because the relationship of each of these features with respect to its local center hole is more important than their relationship to the center of the plate, auxiliary datum feature K is identified and will be used as the secondary datum reference for these features. Notice that the identifier for datum feature K includes a notation "32X INDIVIDUALLY". This means that there are 32 patterns, and for each there is a single local datum feature K.

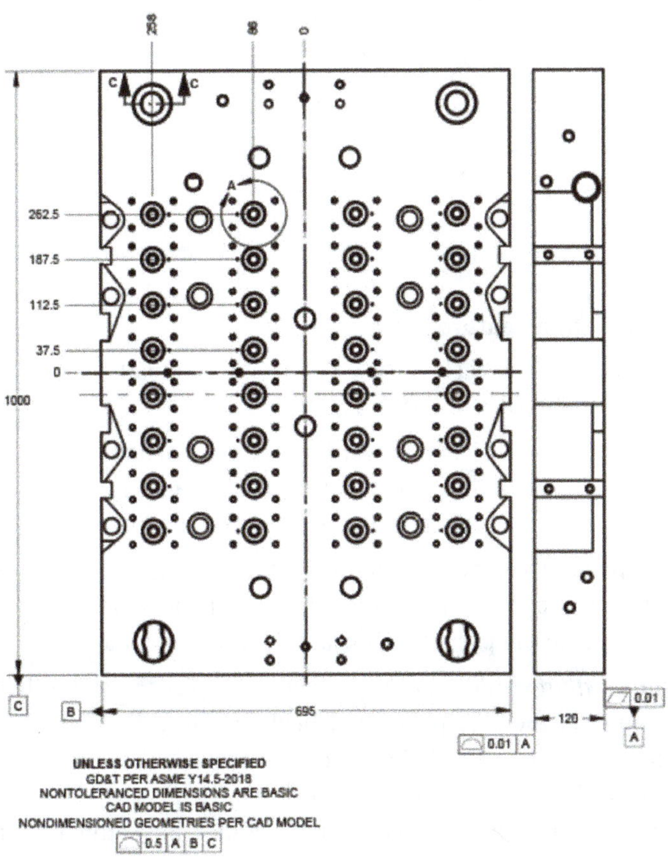

Figure 17.27 Application of GD&T to an injection mold core plate

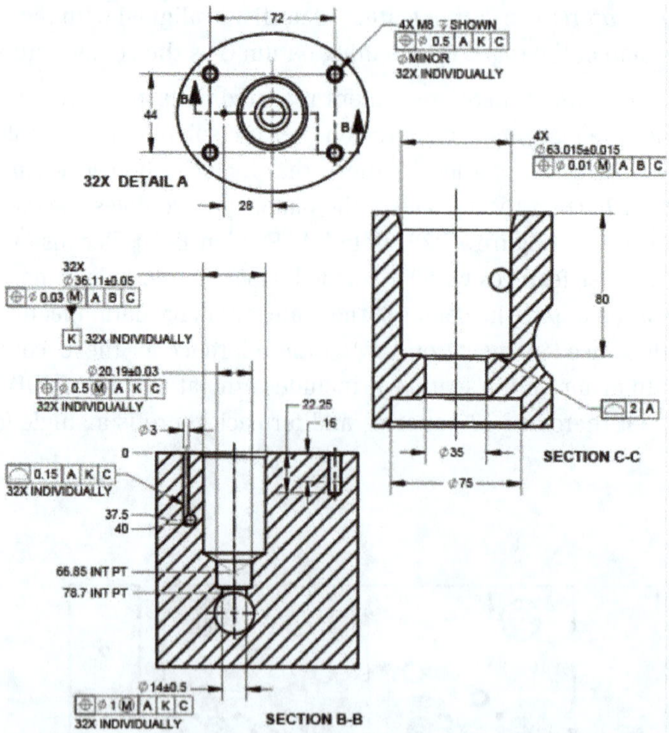

Figure 17.28 Application of GD&T to an injection mold core plate - detail

Step 2: Control Datum Features

Primary datum feature: The back of the core plate is nominally flat, therefore a flatness control is applied to datum feature A.

Secondary datum feature: The actual width of the core plate is not critical; rather, each surface needs to be within a boundary zone. The general profile of a surface control is applicable to the periphery surfaces of the core plate.

Tertiary datum feature: The actual length of the core plate is not critical; rather, each surface needs to be within a boundary zone. The general profile of a surface control is applicable to the periphery surfaces of the core plate.

Auxiliary datum feature: The location of each core spigot interface hole, with respect to the center of the core plate, is critical. Furthermore, the core spigot has some clearance in its interface hole, which means that the core is able to move around slightly with respect to its basic location. Therefore, auxiliary datum feature K is controlled for position with an MMC modifier, indicating that the primary concern is the combined effects of size and position for the smallest hole diameter.

Step 3: Locate All Non-Datum Features with Respect to the Datum Reference Frame

The general profile of a surface control locates all surfaces with respect to datum precedence |A|B|C| unless otherwise specified.

The top surface of the core plate is controlled for location using a profile of a surface tolerance with respect to datum A; note that this also limits the parallelism with respect to datum A and the flatness within the same tolerance value. Similarly, the location of the bottom of the leader pin mounting pocket is controlled using a profile of a surface tolerance with respect to datum A (Figure 17.28, Section C–C). In general, the profile of a surface should be used to control the location of the surface of a spotface, the bottom of a counterbore, planar seating surfaces, and other similar features where the orientation and flatness of the surface are also important; see the GD&T Hierarchy (Figure 17.22).

Each of the features within the previously mentioned pattern is controlled for position with respect to datum precedence |A|K|C|, where there is a different datum K for each pattern. Note that when there are multiples of the same feature within a pattern, the local quantity within each pattern is stated, then the quantity of individual patterns is also provided. When there are single features within a pattern, the local quantity (1X) is not included, and the quantity of individual patterns is provided.

17.6.4 GD&T and Mold Engineering

Molds are complex devices with close-fitting moving components, multiple alignment and locating features, preloads, vents, molding surfaces and other features that each need to be controlled. Conventional tolerancing does not indicate how to determine the actual size or location of individual features and has no way of controlling patterns of features. Conventional tolerancing does not communicate how to set up a part for inspection, and allows tolerance accumulation that is dependent on the layout of dimensions on the drawing. The result of these systemic deficiencies is a poorly defined product that requires mold fabricators to over-manufacture the components in a *best effort* production philosophy. This wastes valuable machine-center capacity and frequently results in *use as-is, subject to test and rework dispositions* on workpieces, all of which increase mold production costs. Even though metrologists have been measuring conventional dimensions and tolerances for over a century, their results are not repeatable with significant accuracy.

There is a common misperception that geometric tolerancing automatically increases production costs, therefore you should only use GD&T on critical features. The truth is that conventional tolerancing practices increase ambiguity, waste and cost, whereas geometric tolerancing communicates the design intent through control of location, orientation, size, and form, and indicates the relative criticality of a feature by the magnitude of the tolerance value stated for each control.

Geometric tolerancing in accordance with ASME Y14.5 is used to communicate the design intent for every feature on each workpiece. GD&T indicates to the metrologist how to set up the workpiece for inspection, and can guide the inspection methodologies used. This basis in design intent means that mold designers can focus on providing appropriately toleranced controls for each feature rather than over-specifying tolerances that drive production costs higher and may necessitate more stringent manufacturing methods.

18 Mold Design References

This chapter contains examples of many different types of molds to illustrate material discussed earlier in the book and to provide references of actual mold designs. Explanations of the design intent are provided to give color and commentary on how and why the mold was designed a certain way. These designs were graciously provided by mold-makers in order to help new mold designers learn from past designs.

18.1 Plastic Housing (Using Shut-offs)

Courtesy of Don Smith

This rectangular housing (Figure 18.1) along with the 2+2 power tool housing shown in Section 18.16 are used as part of an interactive online course offered by Don Smith on mold design. This illustrative part teaches a mold designer how to design a basic mold with straight ejection using ejector pins, blade ejectors and ejector sleeves, and how to deal with drafts, cooling, alignment and shut-offs. The part is nominally 0.100" thick but can be 0.115" at the base of the ribs to account for draft needed on the ribs and bosses. Note that the bosses are created using two pins, one from the cavity side and one from the core side to eliminate any localized thick areas in the part.

There are no undercuts on this part, and the drafts on the part are such that it will stay on the core side for ejection. Therefore, the only ejectors required are ejector pins, sleeve ejectors, and blade ejectors. The ejector pins are placed under all bosses and strategically on the ribs and sidewalls to ensure that the part will not hold up in any deep-rib section. A 1.5" ejection stroke is used to clear the parts from the core. The ejector pins are guided by four leader pins inside the ejector box.

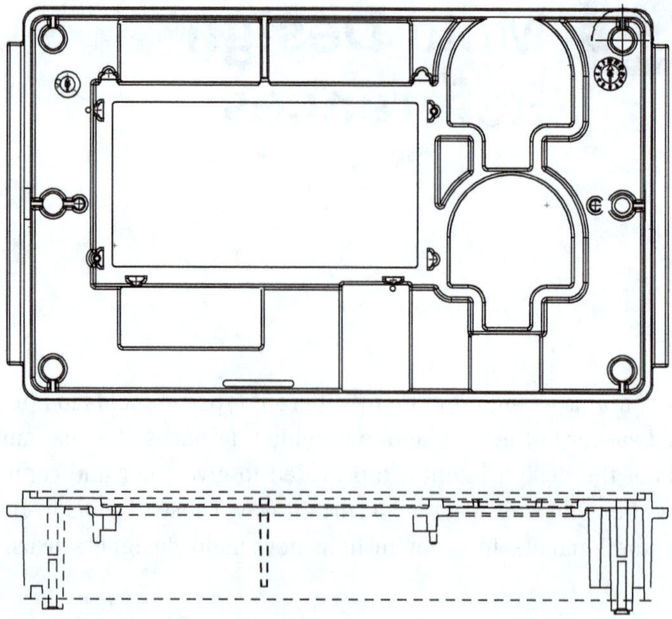

Figure 18.1 Plastic housing: Part drawing

The design of the core side (Figure 18.2) is such that the core insert creates the inside of the part and supporting core plate itself creates the outside vertical side of the housing. The cavity is more or less a flat surface.

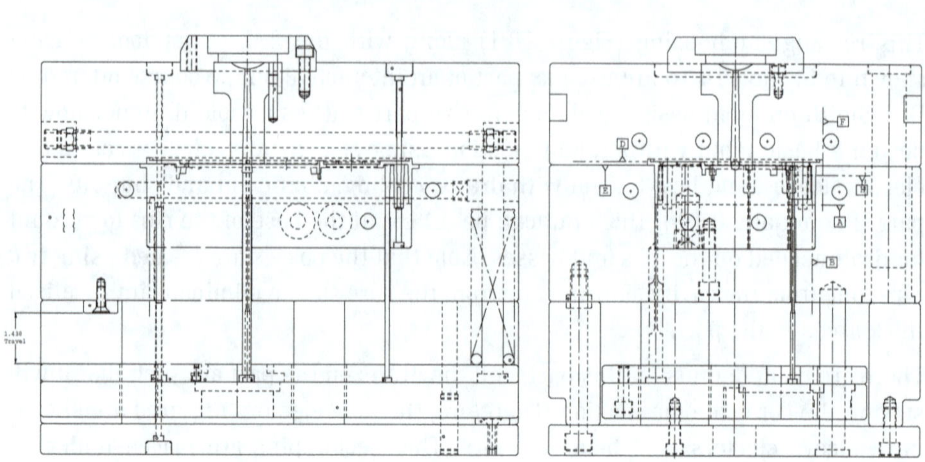

Figure 18.2 Plastic housing: Section assembly views of mold

Cooling of the cavity side is achieved by a series of four lines drilled into the cavity. The core-side cooling is a circuit with short baffles in the core insert, and an outer cooling circuit in the core plate. The housing is filled via a cold runner with two tunnel or submarine gates. The cold runner is ejected with the part, but is naturally degating due to the tunnel gate system. A sucker pin is used on the core side to ensure that the part transfers to the core side when the mold is opened.

18.2 Facemask

Courtesy of Haidlmair

This face mask mold was produced by Haidlmair in 6 weeks when the COVID-19 pandemic hit in the spring of 2020. The machine was also supplied by Engel in record time to allow the Germany-based molder, Providee, to respond to the need for masks. The 50 g mask is made with TPE and is 1.8 mm thick. It runs with 200 tonnes of clamp force at 14.5 s. The facemask mold can contain disposable filter cloth, while the mask can be reused (see Figure 18.3).

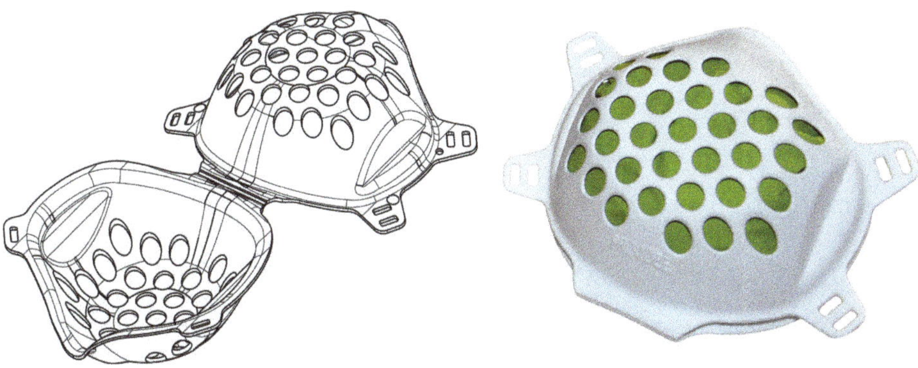

Figure 18.3 Facemask - part drawing

The four-cavity mold has the stacks recessed in plates with H7/G6 tolerancing for solid alignment. The two halves of the mold are aligned by four side interlocks. An ejector plate with 42 ejector pins per part removes the part from the core for ejection. There is a U-shaped ejector on the cavity side which is spring-loaded forward to free the part from the cavity side and ensure that it stays on the core side during mold opening. Return pins move the ejector pins back before mold closing. Once all the ejector pins were added to the mold, the remaining open space in the ejector plate was cut out, and pillars were added between the core backing plate and the core plate for support and to prevent deflection of the mold during injection and clamping (see section assembly).

Cooling of the part is accomplished by drilled lines and baffles in the core and cavity blocks.

Since this part is molded with a high-flow TPE and it would be used close to the face, there could be no flash on the parts. The venting was therefore cut shallow all around the part for the first test, and tonnage blocks were added to support the parting line. These blocks can be seen on the cavity side plan view assembly. Once the mold was tested, the cavity blocks were shimmed forward gradually to increase the venting to the point that good parts could be made without any flash.

A four-cavity hot runner system is used to direct gate the part in the hinge area of the part. A valve gate is used to create a cosmetic gate on the part. A front-mounted rectangular gate insert contains and cools the nozzle and gate area.

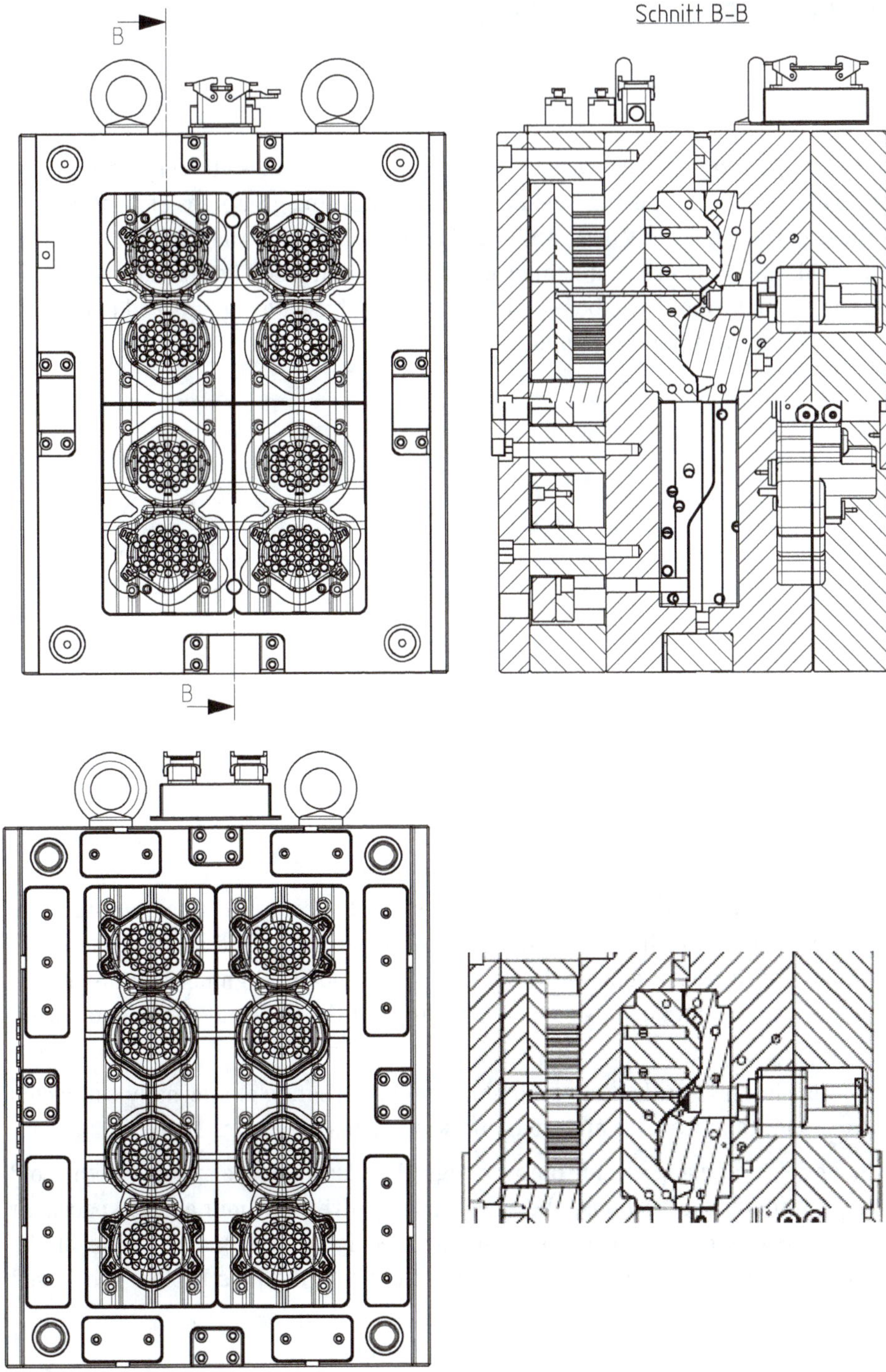

Figure 18.4 Facemask: Plan view and stack section view of four-cavity mold

18.3 Plastic Cutlery

Courtesy of iMFLUX

In the case of long slender parts like cutlery, razors, handles etc. (Figure 18.5) the molds are usually built using modular blocks containing groupings of parts. Gating of these parts is normally done at the end or near the end of the part. In the case of this mold, the parts are cold runner gated on the end (see Figure 18.8). A full hot runner solution would also be possible using hot edge gate bars instead of the cold runner bars shown here (see Figure 18.9).

Figure 18.5 Plastic cutlery

The mold shown in this reference contains eight blocks, each containing eight spoons which are machined into the blocks (Figure 18.6). The parting line is not flat in this design, as it follows the contour of the spoon. Alignment of the mold is with leader pins and bushings as the alignment of the core and cavity is not critical.

A cooling circuit is drilled into each block, and the cooling is internally connected between blocks in the core and cavity backing plate (see Figure 18.8).

Parts are ejected using ejector pins strategically placed on the spoon and the cold runner system's intersections. The ejector plate is driven by four evenly spaced air ejection pistons housed in the ejector box (see Figure 18.7). The stroke of the ejector plate is 0.438". Return pins are used to bring the pins back to the closed position after ejection.

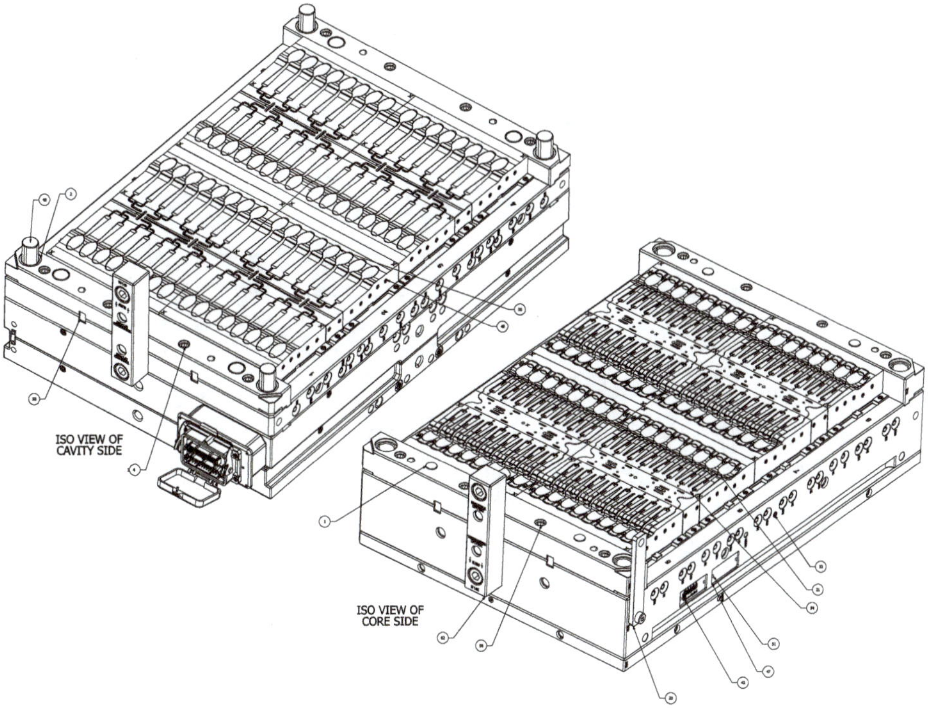

Figure 18.6 Plastic cutlery: Isometric view of 64-cavity spoon mold core and cavity (hot to cold runner)

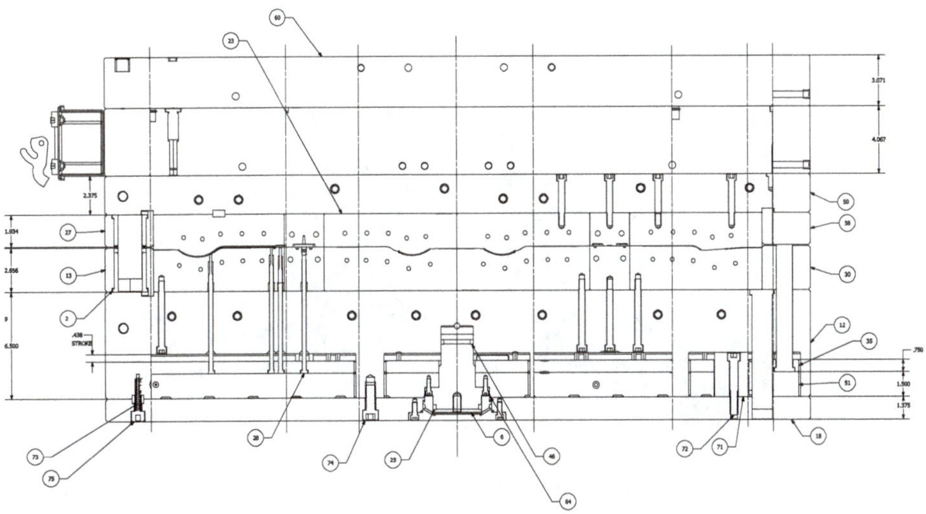

Figure 18.7 Plastic cutlery: Section assembly of 64-cavity spoon mold

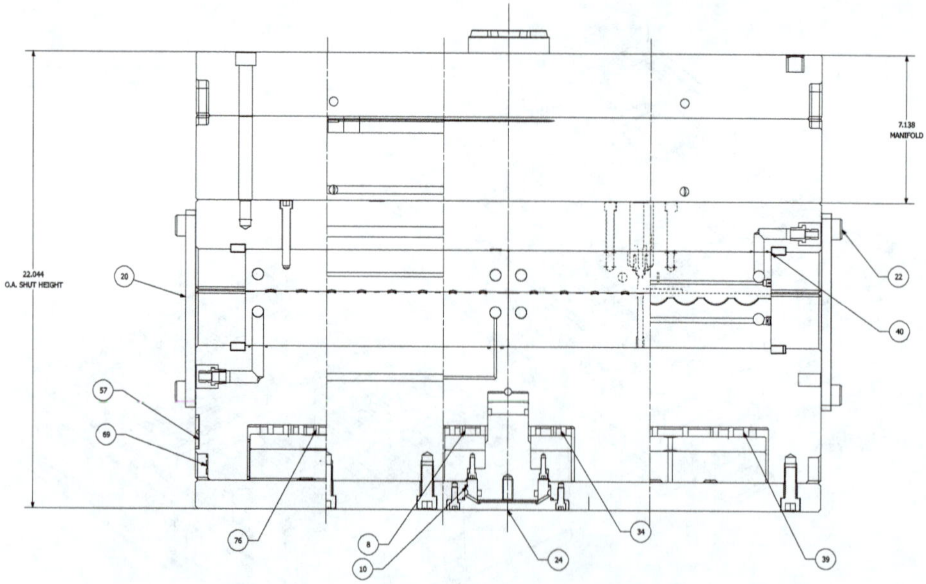

Figure 18.8 Plastic cutlery: Section assembly of 64-cavity spoon mold

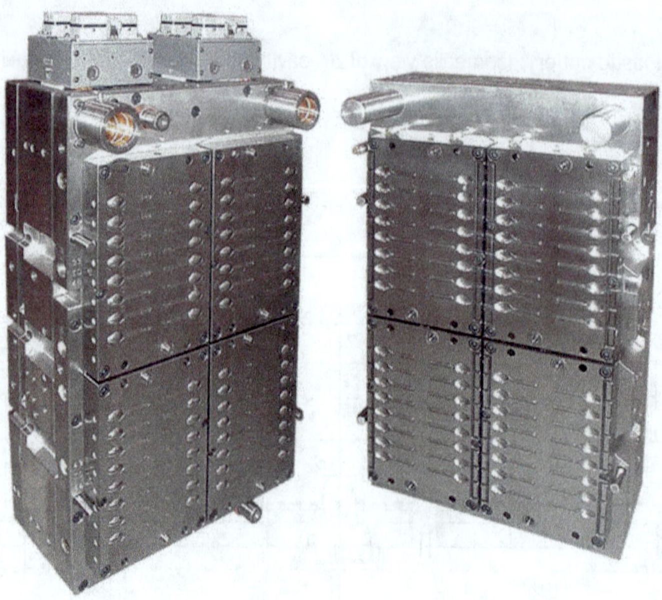

Figure 18.9 Plastic cutlery: Reference spoon mold (not the same mold as the drawings above and using a hot runner rather than a cold runner system) (Courtesy of StackTeck Systems)

■ 18.4 Vial

Courtesy of iMFLUX

This polypropylene random copolymer vial (Figure 18.10) is tunnel-gated near the rim, where the internal thread is on the part due to the customer requirement to keep the top of the part free from any marks. The gate of 0.020" in diameter and it is fed by a 0.094" trapezoidal runner. The offset position of the gate makes the part susceptible to core shift and ovality.

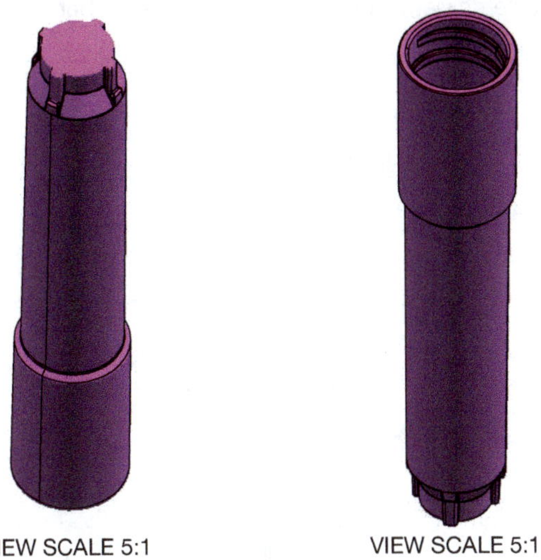

Figure 18.10 Vial: Part drawing

The eight-cavity mold (10" × 11" × 12") (Figure 18.11, Figure 18.12) is laid out in a circular pattern to feed each part with a runner. Since the part is so small, it only required a few tons to hold it closed.

The thread on the internal diameter of the part is strippable, so no rotation of the cores or retractable cores are required (nor would they be feasible in such a small mold). The 1.875" stripper plate stroke removes the parts and the eight-cavity runner. The stripper ring is one piece for all eight cavities. Note the shoulder bolt used to prevent the stripper plate from moving too far forward. This bolt is not stressed in each cycle, and acts only as a safety mechanism. The machine ejectors are double-acting so a return pin is not needed.

A taper between the stripper ring and the cavity side keeps alignment between the cores and cavities. Rigidity is added by adding a cavity plate around the cavity (which contains all eight parts machined into one piece of steel). The core is aligned to the stripper ring with a taper.

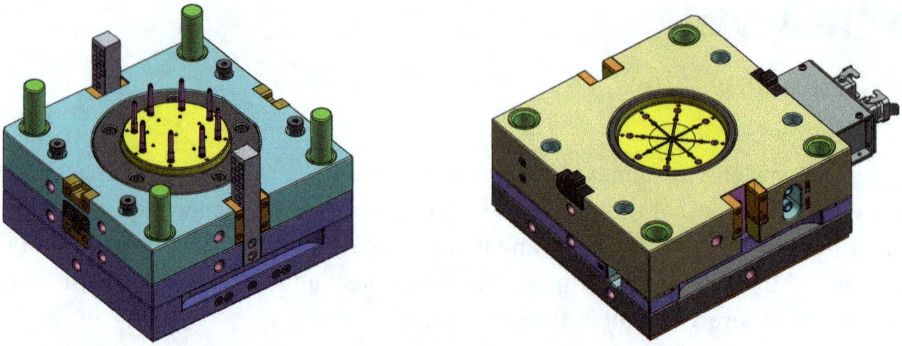

Figure 18.11 Vial: Isometric views of the core and cavity side. Note the side interlocks for alignment

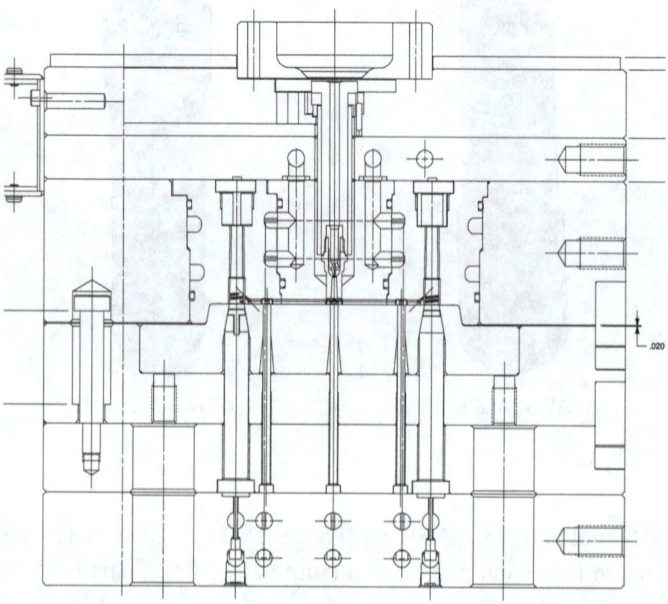

Figure 18.12 Vial: Section view of mold

18.5 Open-Ended Floss Spool Tube

Courtesy of iMFLUX

Designing a mold which has no closed end presents a number of challenges for a mold designer. The gate location in particular can affect molded part. There is no obvious location for the gate, and any gate(s) location could result in warpage, ovality, weld lines and other molding defects.

In the case of this 64-cavity floss spool mold the part is edge-gated by a hot runner edge gate system which feeds groups of four parts (16 nozzles on the hot runner manifold). Since the part weight of this PP part is only 0.67 g the single nozzle gating four parts helps to reduce the inventory of plastic in the hot runner system and increase the shot volume per nozzle.

The stack design consists of four parts in each module (Figure 18.13–Figure 18.16). Cooling is supplied by a water circuit for each grouping of four parts on the cavity and stripper ring. The cores are individually cooled with bubblers.

The parts are ejected with a stripper ring/stripper plate actuated form the machine ejection system. Although the stack has a flat parting line the mold is aligned using side interlocks on four sides of the mold.

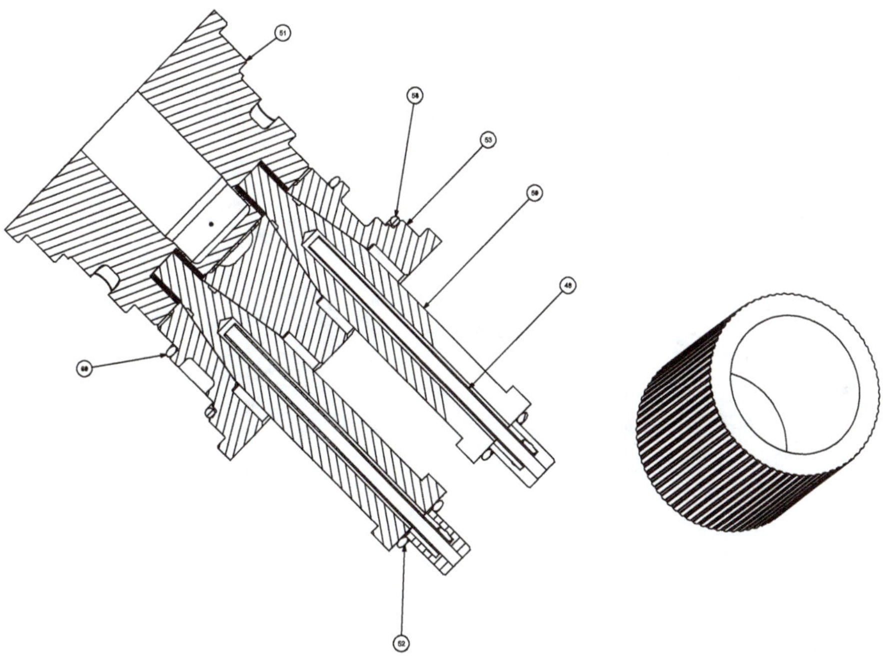

Figure 18.13 Open-ended floss spool tube: Stack section view and part drawing. A single stack consists of four molded parts

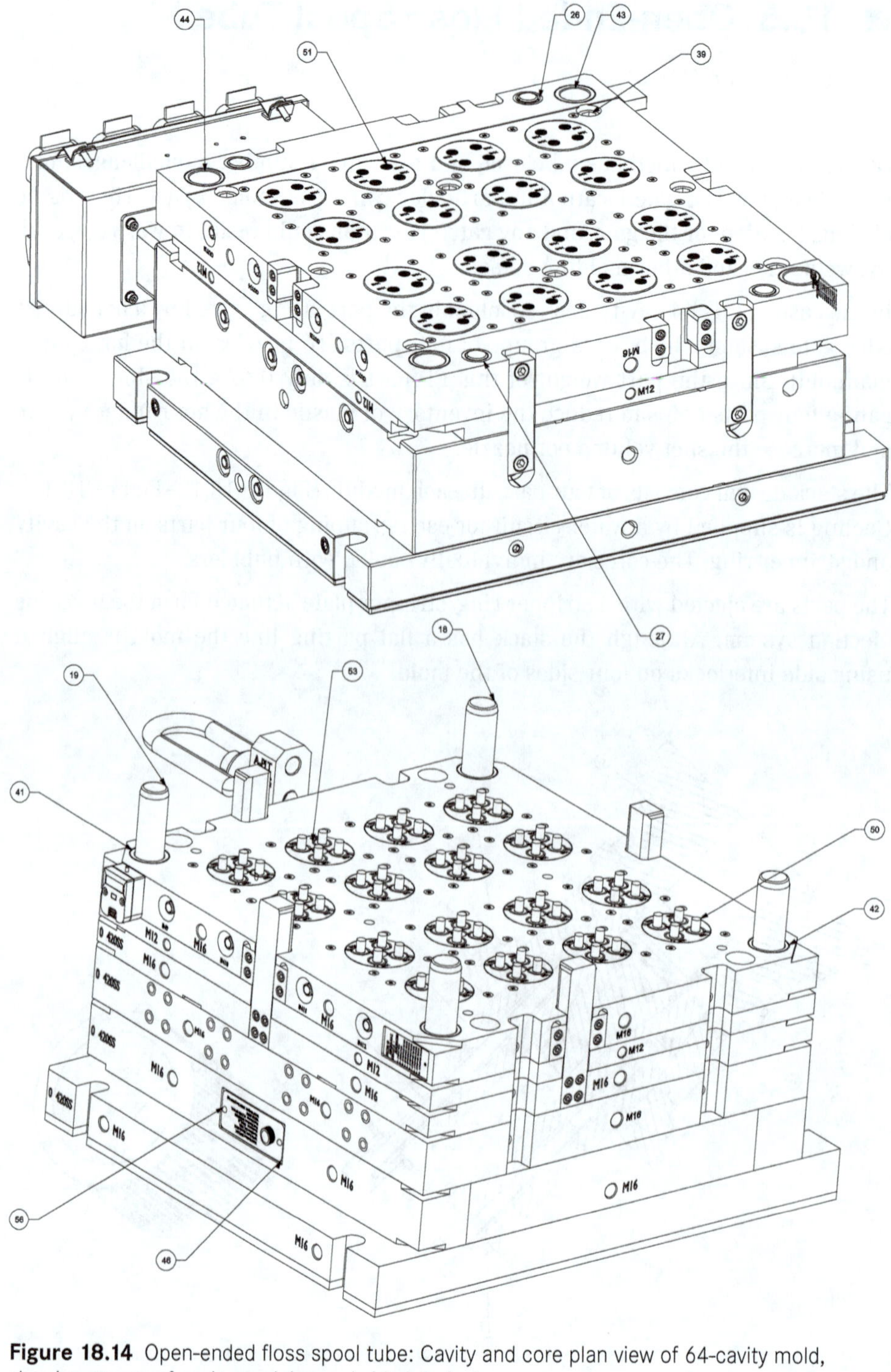

Figure 18.14 Open-ended floss spool tube: Cavity and core plan view of 64-cavity mold, showing groups of cavity modules and side interlocks

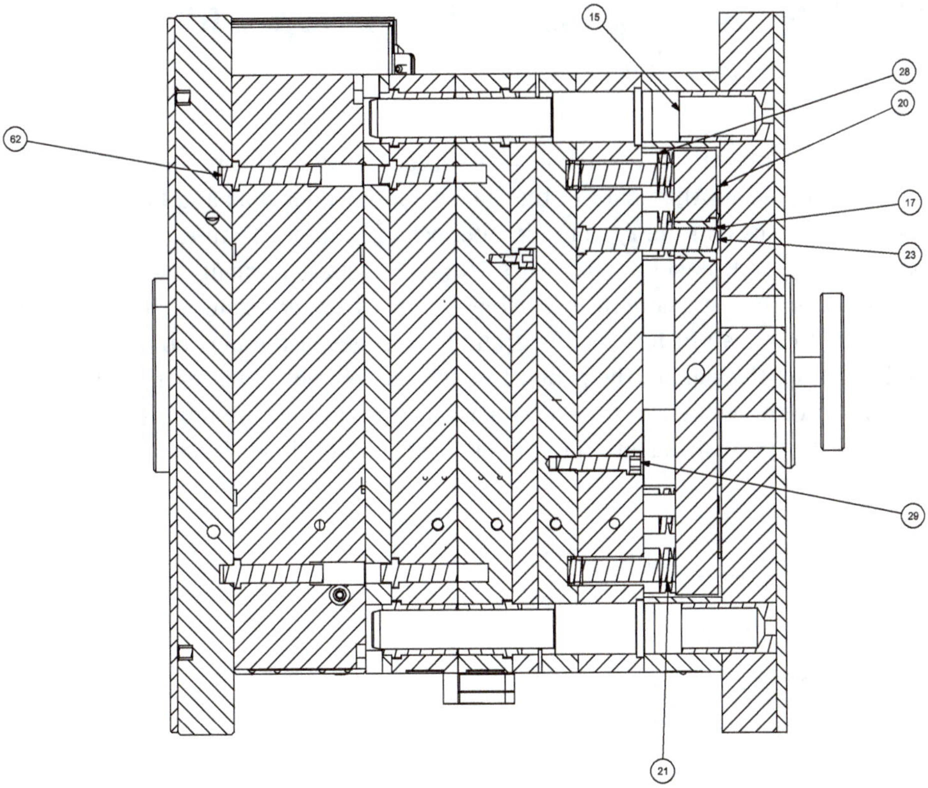

Figure 18.15 Open-ended floss spool tube: Section view assembly of 64-cavity mold, showing the ejector box and leader pins

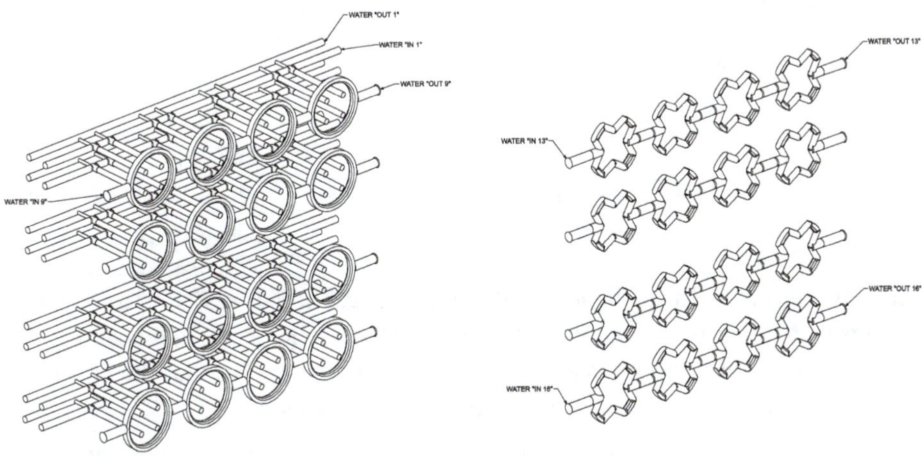

Figure 18.16 Open-ended floss spool tube: Water cooling circuits on the core and cavity side of 64-cavity mold

18.6 Overcap Lid (Using Stripper Ring Ejection)

For flat or shallow parts a stack design without individual core and cavity taper locks can be used. Alignment of the stacks using leader pins and bushings, and round taper locks on the P/L or side interlocks is sufficient. An example of a flat parting line stack is shown in Figure 18.17. The stack normally consists of five components: the core, core insert, stripper ring, cavity, and gate insert. This overcap lid with a snap fit has cooling around the edges of the part on the core and cavity but not throughout the panel. This is used to create a flat part, as the outer ring will freeze before the panel. The gate insert and the core top are both made from BeCu to promote faster cooling, as the cycle time will be 4–6 s.

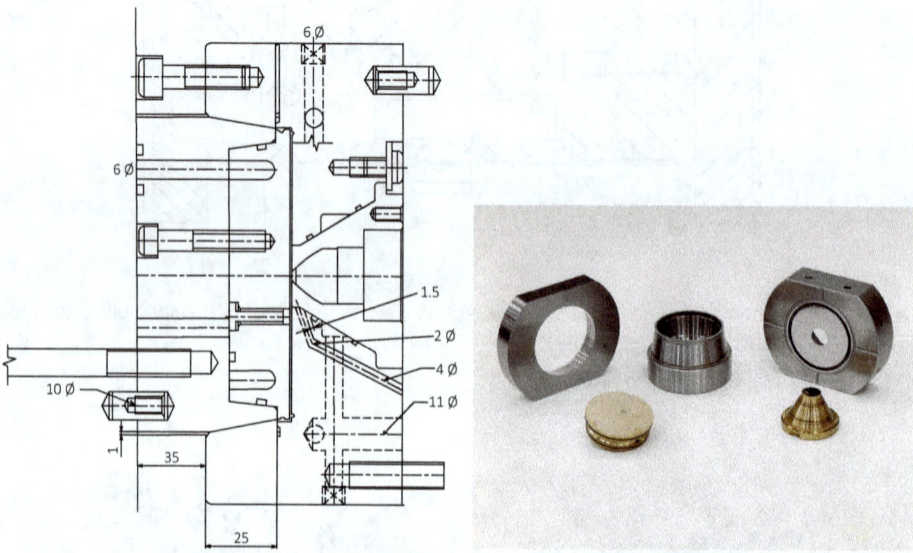

Figure 18.17 Overcap lid: Flat parting line stack design (Courtesy of Husky Injection Molding Systems Ltd.)

The part is ejected with a simple round stripper ring. There is cavity air around the gate insert to ensure that the part transfers to the core side on mold opening. An additional air poppet circuit is added to the core side to ensure that the part does not stick to the core side when the stripper ring is activated.

In the case of the stack design shown in Figure 18.17, the mold half alignment interlocks are between the cavity and the stripper plate. The stripper ring is fixed to the stripper plate and the core floats. In this case, the cavity is fixed in location and so is the stripper ring. The core floats to meet the stripper rings. The core plate is not aligned to the cavity plate, as this causes alignment problems and the interlocks become insufficient to hold the components in place. There is no preload between the stripper ring and the core base. The fit between the taper on the core and the stripper ring is size-to-size, with -0.013 mm clearance due to tolerances. This prevents locking of the stripper ring to the core base.

DU bushings (self-lubricating bushings) are used for the stripper ring tie rods. The stripper ring pushes on half of the wall section being ejected to avoid deformation on ejection. A 0.2 mm gap between the stripper ring and the core is maintained to prevent any interference and damage between these two components.

Figure 18.18 shows a six-cavity lid mold for a 6" diameter lid. BeCu is used on the core insert and gate insert to promote faster cooling for this 6 s cycling mold. Two leader pins are used at the top of the mold for assembly, and two taper interlocks are used for alignment between the stripper plate and cavity.

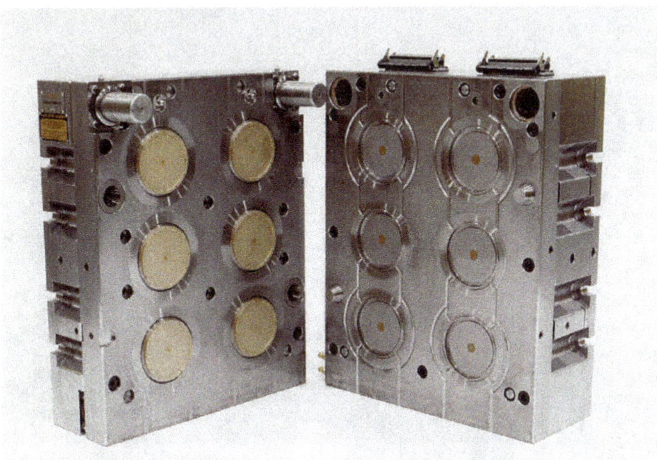

Figure 18.18 Overcap lid: Six-cavity mold with flat parting line and taper interlock alignment (×2) (Courtesy of Husky Injection Systems Ltd.)

■ 18.7 Plug Lid

Courtesy of StackTeck Systems

This lid mold (Figure 18.19) is designed for an in-mold labeled (IML) application. Of special note is the use of a double-sided label with a clarified PP resin. Double-sided IML lids offer twice the printing space of a traditional IML label. This is not only attractive for those looking to improve brand awareness, but it also allows space for additional information about a product. Due to the inserted label, the part required reverse (core side) gating and ejection. As noted in Chapter 14 on gating and Chapter 3 on part design, core-gating the part usually requires a much longer hot runner nozzle. A good example is shown here in the section assembly view of the mold.

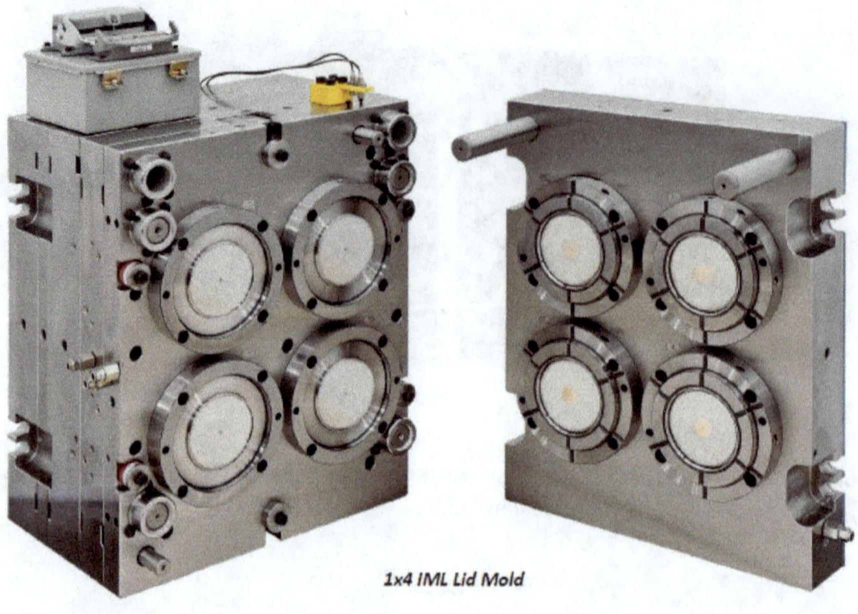

Figure 18.19 Plug lid: Reverse-gated four-cavity mold with core-to-cavity alignment

IML has become a very popular solution for creating vivid graphics on molded parts, as can be seen in the part photo (Figure 18.20). The process of IML is precise and economical when multi-color graphics are required for the product like those seen here. By using a robot to insert the labels and pick up the part, the process of labeling and handing is fully automated.

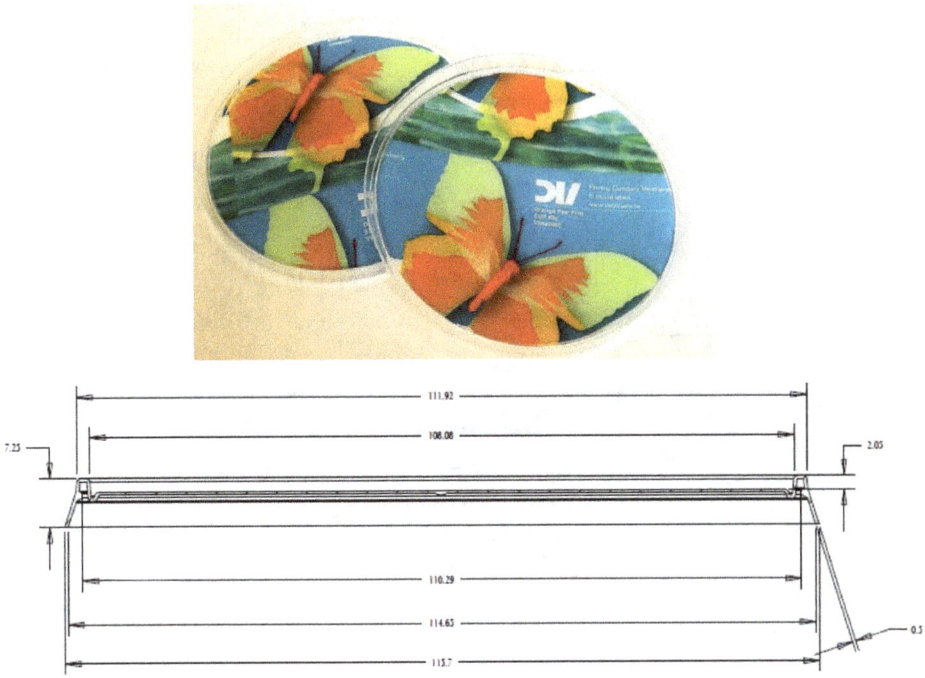

Figure 18.20 Plug lid: Typical plastic part

The skirt is large enough that the mold maker decided that core-to-cavity alignment was also necessary. What is unusual for this design is that the stripper ring acts as both the alignment member and ejection member of the stack (Figure 18.21, Figure 18.22). Two cylinders are used for ejection and four leader pins are used to align the stripper plate. Two additional leader pins are used for mold alignment.

The core, cavity and stripper ring are all front-mounted for ease of assembly, serviceability, and adjustment in the molding machine.

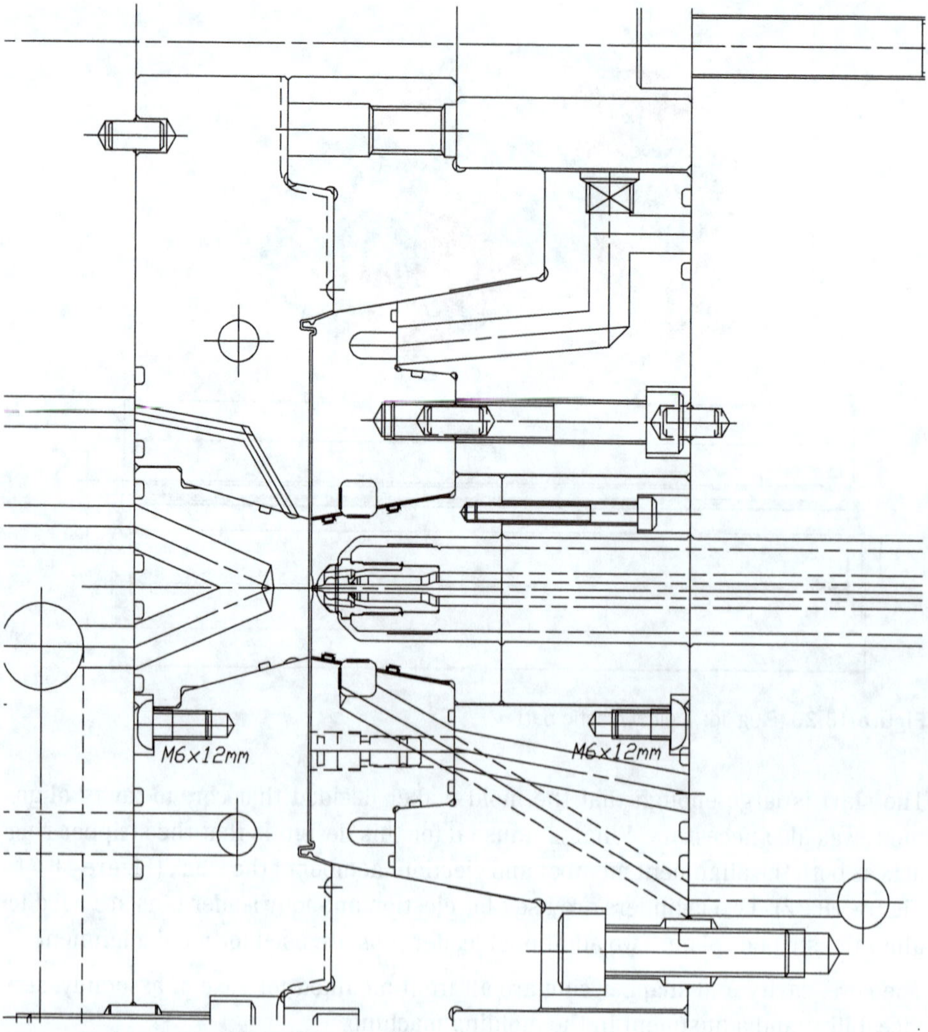

Figure 18.21 Plug lid: Mold stack section

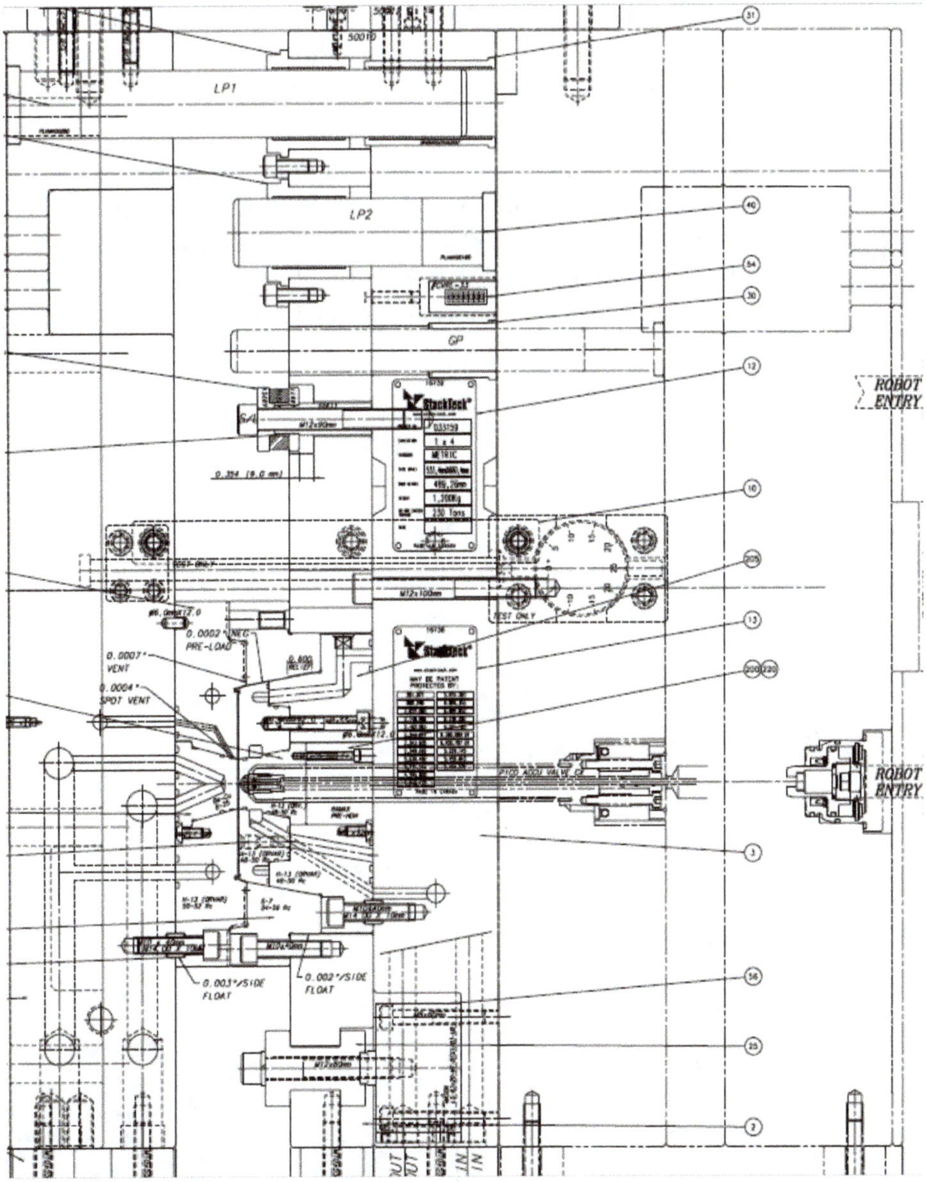

Figure 18.22 Plug lid: Section assembly of mold

On the plan view of the core assembly (Figure 18.23), note the typical cooling pattern for lid molds. The lid is cooled in the cavity opposite the gate and also around the rim area. Cavity and core cooling on the panel is limited to allow the part to cool flat.

Stroke limiters are used to prevent the stripper plate from over-stroking and for service. The water lines have been internally plumbed in the plates, and one IN and one OUT water line is used on each plate. Note the illustrations in the top right of the plan view (Figure 18.23) that show how the water lines are routed within the plates. This can very helpful to the process engineer and mold setup technician.

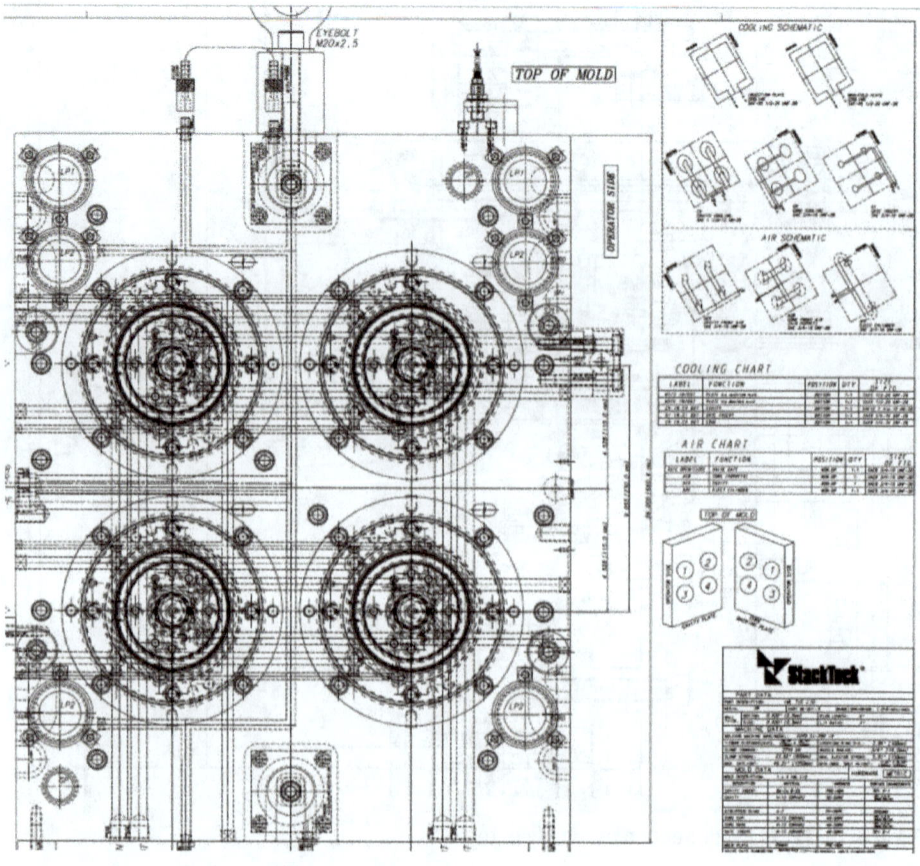

Figure 18.23 Plug lid: Plan view of core assembly

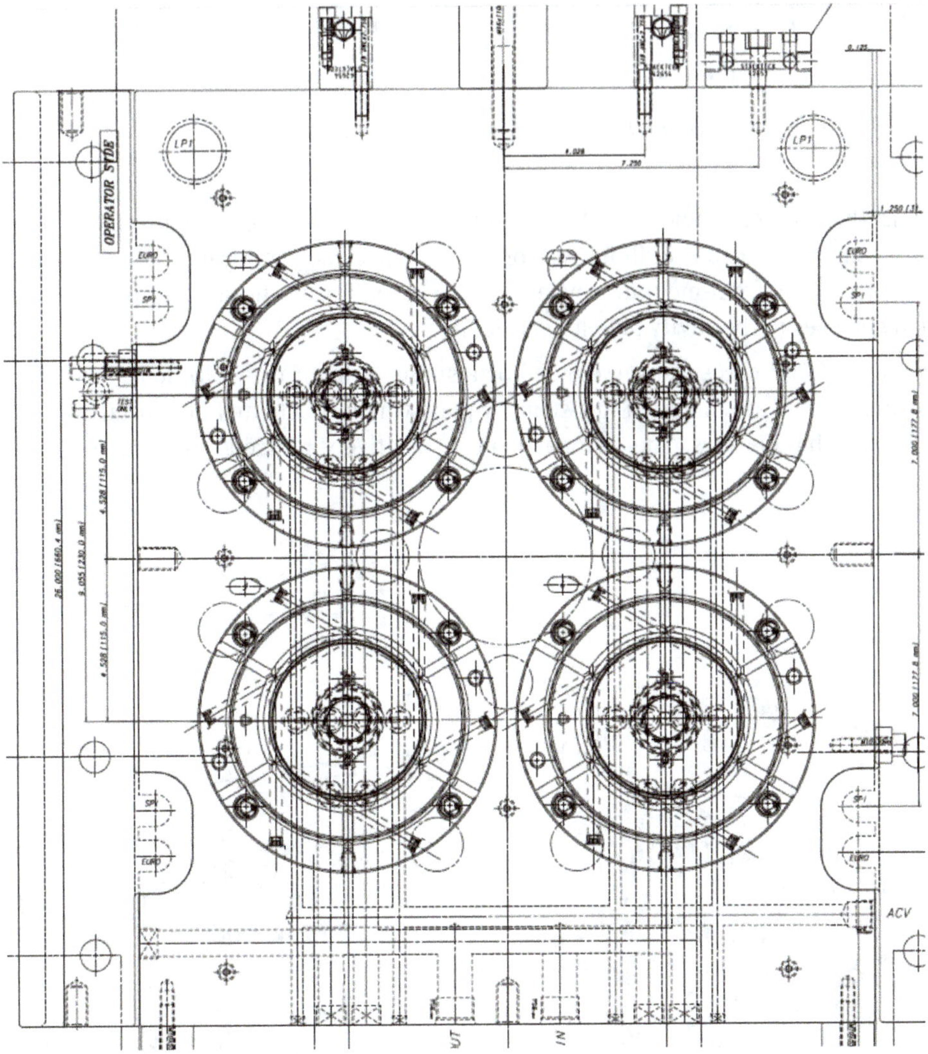

Figure 18.24 Plug lid: Plan view of cavity assembly

18.8 Urine Tub Cap (Using Stripper Ring Ejection)

Courtesy of Nypromold

This 48-cavity cap mold is built in modules of four, as shown in Figure 18.25–Figure 18.28. The mold is built for ease of maintenance and serviceability in the press. All inserts are front-mounted and accessible in the press. The stripper plate can be latched over to the cavity side for servicing.

Groups of four cores are surrounded by a single stripper ring with four parts. The stripper rings are mounted to a stripper plate which is actuated forward 0.345" by the machine ejectors. The stripper plate is positively returned by springs in the mold.

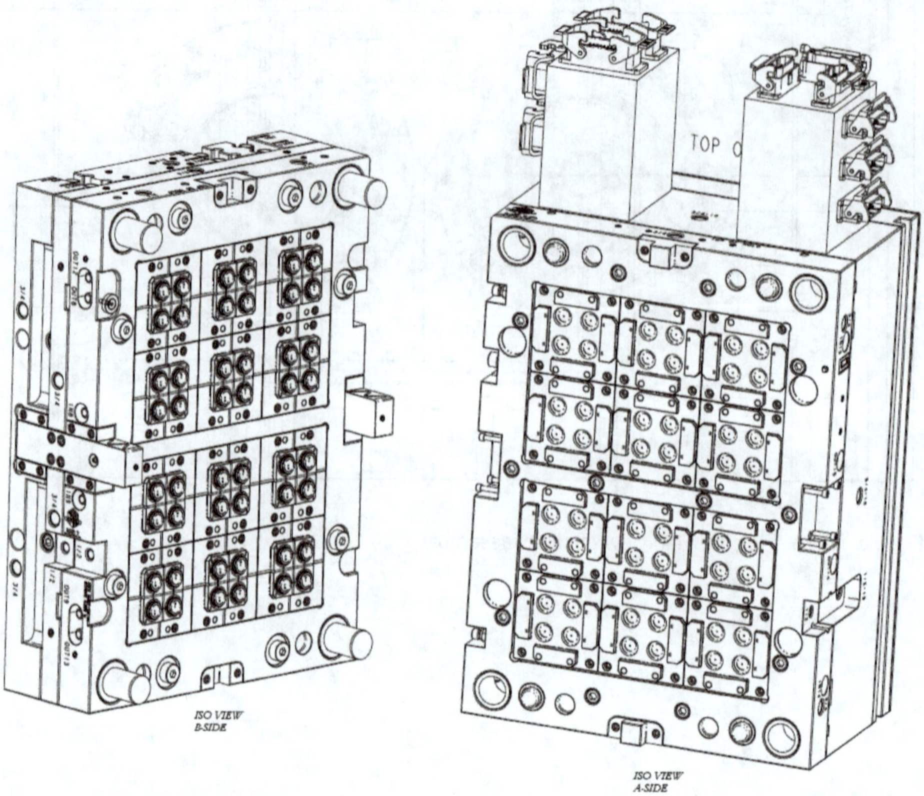

Figure 18.25 Urine tub cap: ISO assembly view of 48-cavity mold

Alignment of the mold is achieved by four side interlocks and tapers between the fixed cavities and the floating stripper rings and cores. The cores are held into a core retaining plate and they align to the stripper ring on a shallow taper. Two of the side interlocks also serve a dual purpose of keeping the stripper plate aligned during its stroke.

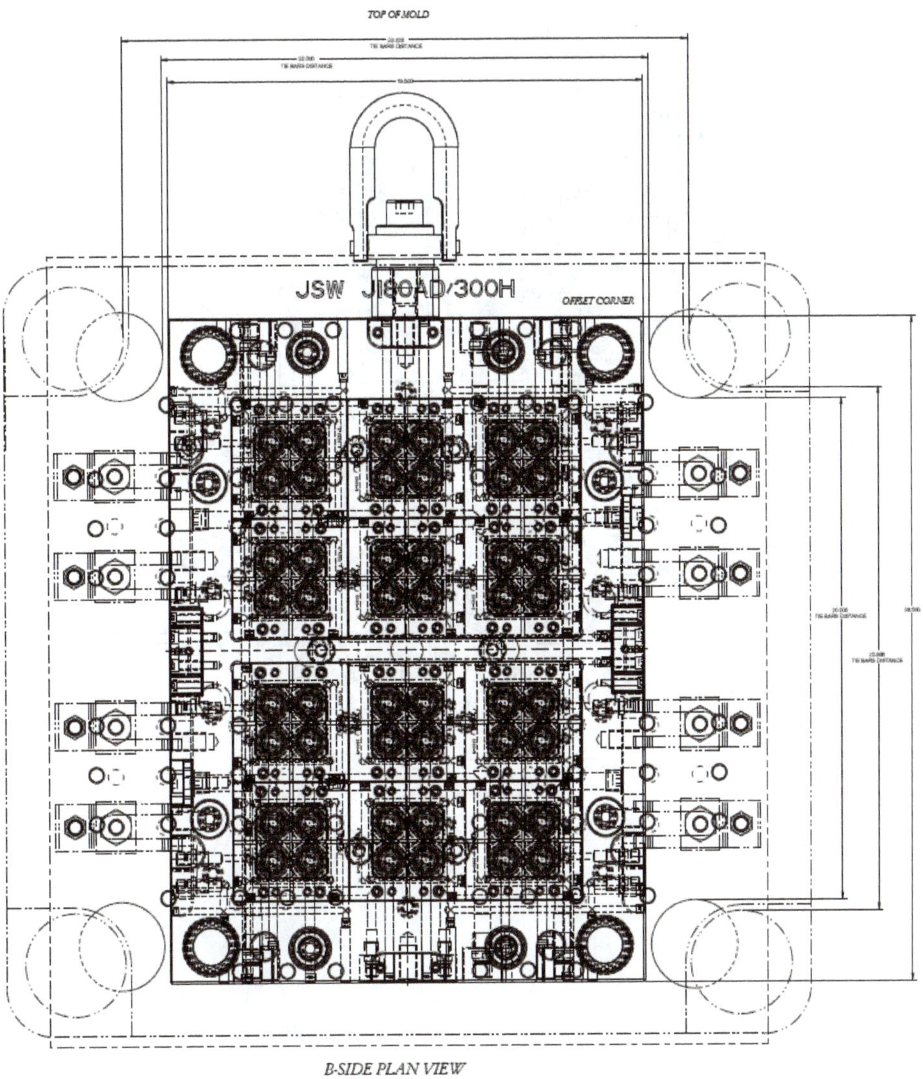

Figure 18.26 Urine tub cap: Plan view of mold core assembly

The stripper rings are cooled via the stripper plate and the cores are cooled with bubblers. A hot runner system is used to feed each of the 48 caps.

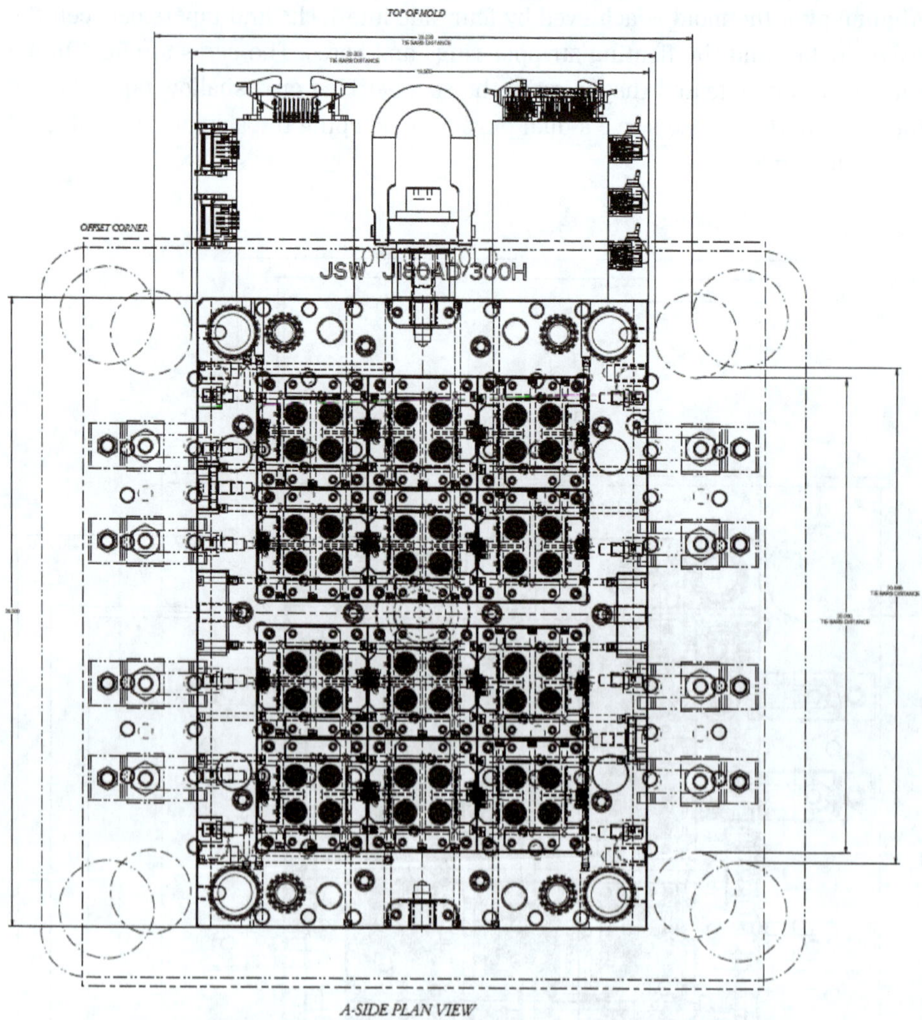

Figure 18.27 Urine tub cap: Plan view of mold cavity assembly

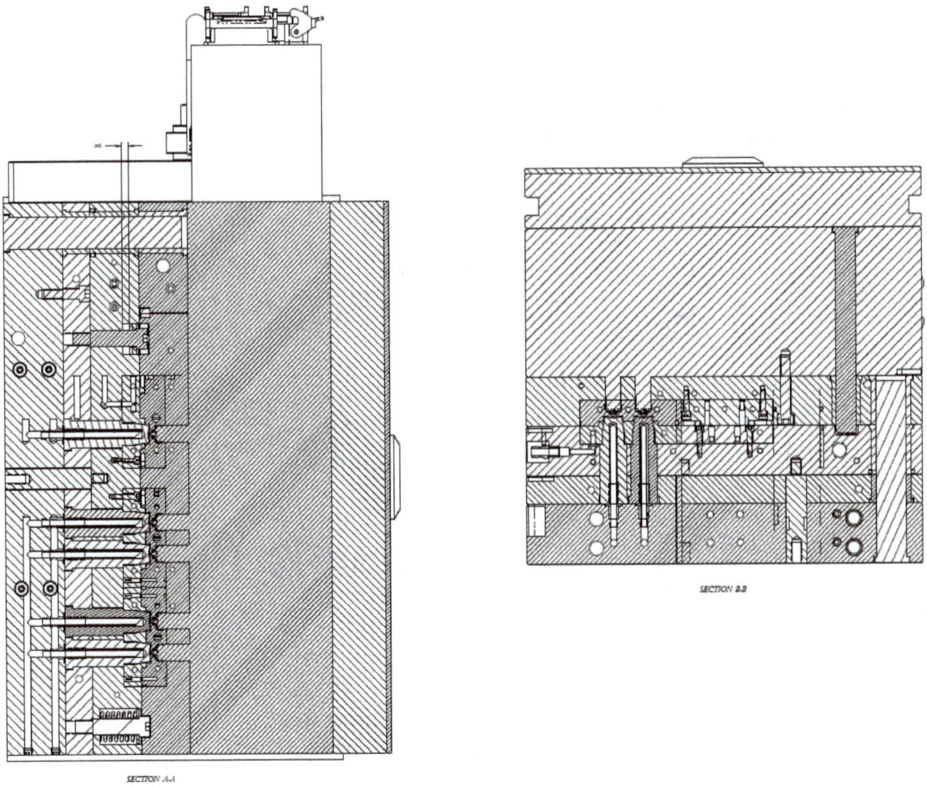

Figure 18.28 Urine tub cap: Section view of 48-cavity mold assembly

18.9 Deli or Yellow Fats Container (Using Air Ejection and with Cavity Lock Alignment)

Figure 18.29 shows a typical cavity lock stack design. Cavity lock refers to the fact that the cavity surrounds the core taper and locks it in place. This design is more cost-effective to produce than the core lock design described in Section 18.10. A cavity lock design works well to align the individual stacks on a multi-cavity tool with *L*/*t* ratios of less than 200–250. If core shift is a primary concern, then a more expensive but stronger core lock design is advised.

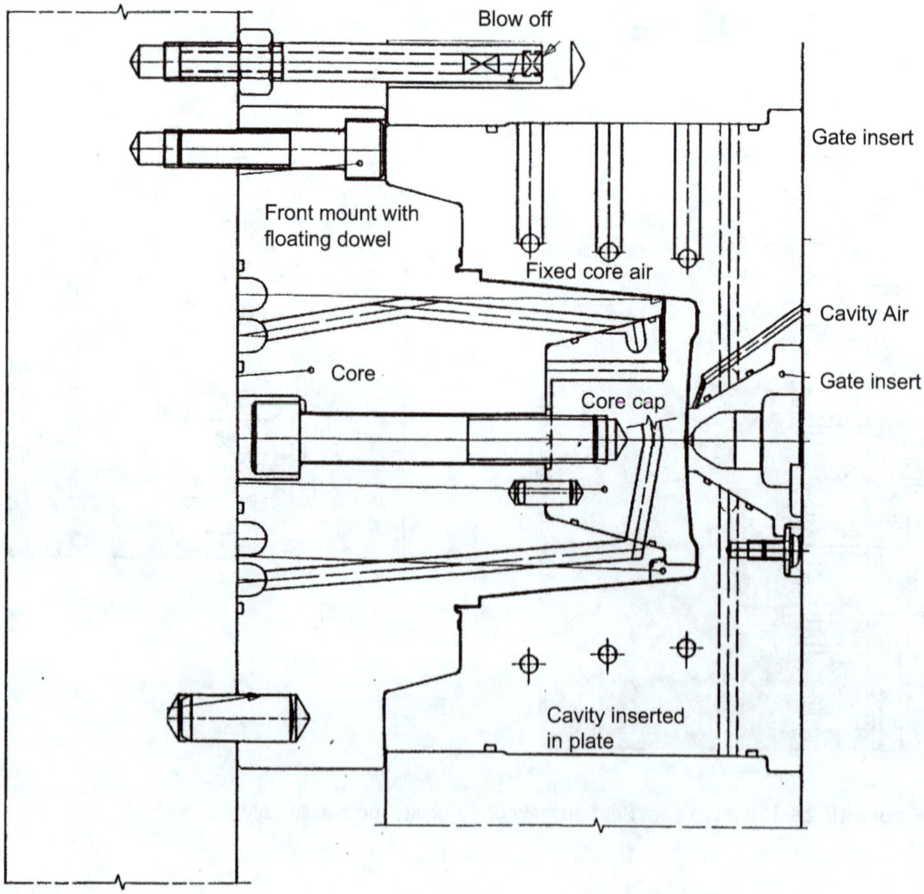

Figure 18.29 Deli or yellow fats container: Cavity lock with cavity inserted in a plate

The cavity lock design shown in Figure 18.29 shows a solid core construction. A modification of this design is to create a separate wear ring for the taper between the core and cavity that can be easily replaced. However, in general this should be avoided, as it is more costly and the vent on the lip is hard to control.

The vent grooves should be located on the core side on the primary shut-off and on the cavity side for the tapers and secondary shut-off. This allows easy chroming of the core taper when the mold requires rework. Note that chrome is strongest in compression, and it should be applied to the taper that is in compression (in this case the core).

For air ejection the blow off feature must not be overlooked in cavity lock molds, as it may result in many problems at test. Figure 18.29 shows a fixed adjustable blow-off pin mounted to the core plate with a clearance hole in the cavity plate. Consideration should be given to sharing the blow-off pin between two or four cavities to reduce the required number of pins. This can be accomplished by adding additional holes to the top of the pin. Note that with all air circuits the blow-off air must be an independent circuit.

In a cavity lock stack design there are commonly five components: the core, cavity, gate insert, gate pad, and core cap. The stack design shown in Figure 18.30 has only four components are the container is small (3 oz) and there is not enough space for a separate gate pad, as described in Chapter 12 on cooling and Chapter 13 on alignment. Since these molds run at cycle times of less than 6 s, high-conductivity materials such as BeCu are used for the core cap and gate insert, to conduct heat from the plastic faster. Note that these components are not subjected to any compressive loads from the clamp, as they are soft and would hob quickly.

Figure 18.30 Deli or yellow fats container: Stack components for a cavity lock mold (Courtesy of Husky Injection Molding Systems Ltd.)

Note the large size of the stack relative to the part. In many cases the stack will be > 2× the diameter of the part to provide the support needed for the high-pressure injection. Large vent grooves on the core taper and the cavity face are used to quickly evacuate air during injection. These are required to allow the fast fill times of less than 0.2 s needed in these thin-wall molds, which have wall thicknesses in the 0.5 mm range.

The cores are normally front-mounted to allow for adjustment of the core to cavity in the press. When the mold is installed, the cores are loosened off slightly. The mold is clamped up and the cores are allowed to move slightly on the loose dowel to align to the fixed cavity. Once the mold is opened, the core mounting bolts are torqued to the required setting to prevent any further movement during molding.

Due to the large tapers on each of the cores and cavities, only two leader pins are needed for mold assembly and alignment. Removing the bottom leader pins reduces the chances that the air-ejected parts will hit the greased leader pins (normally located below the stacks). Figure 18.31 also shows the blow-off pins (four per stack) and the blow-down nozzles at the tip of the mold that blow the part off the core and then down to the conveyer below the mold.

Figure 18.31 Deli or yellow fats container (2 lb): Four-cavity mold with surface-mounted cores and cavities (Courtesy of Husky Injection Molding Systems Ltd.)

■ 18.10 Stadium Cup (Using Core Lock Alignment)

For cases where the wall thickness variation is critical or where core shift is a high risk (due to a tall part, thin sidewall or low flow rate material), a core lock design should be considered. The term "core lock" refers to the fact that the core side wraps around the cavity taper, as shown in Figure 18.32. Figure 18.33 shows a picture of a typical eight-cavity 32 oz stadium cup mold. These molds can be run at sub-6 s cycles in high-performance injection-molding machines. The parts are normally made with high-flow (50+ MI) HDPE or PP. The parts are filled normally in 0.25–0.4 s, and the parting line vents are nominally 0.018 mm in HDPE and 0.010 mm in PP.

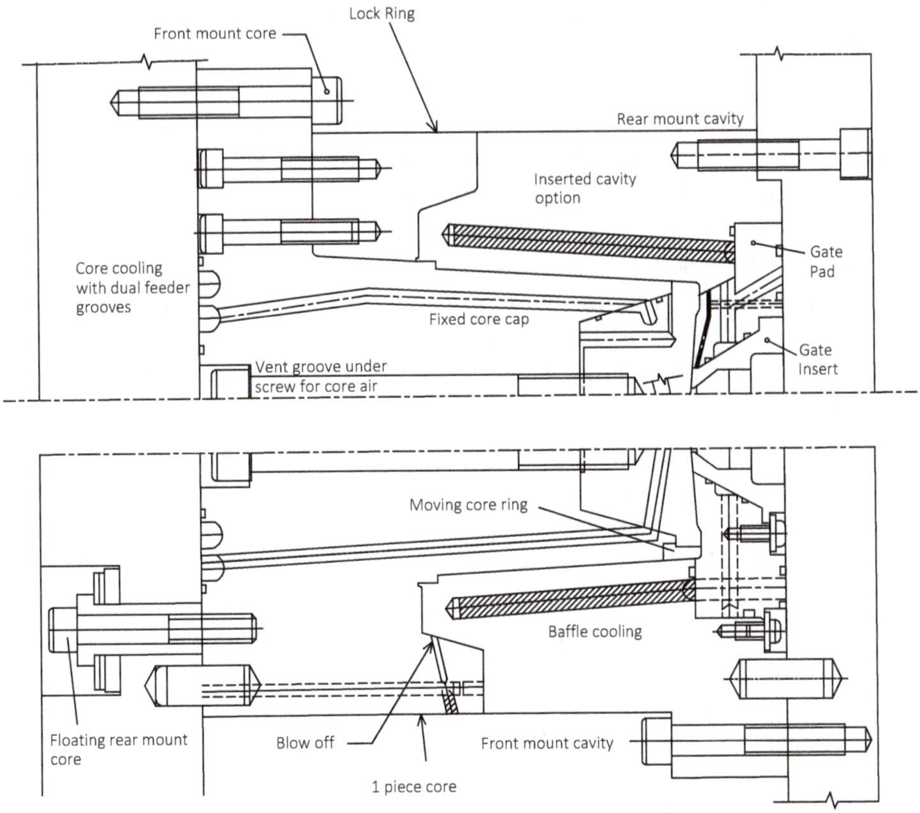

Figure 18.32 Stadium cup: Core lock stack design with options

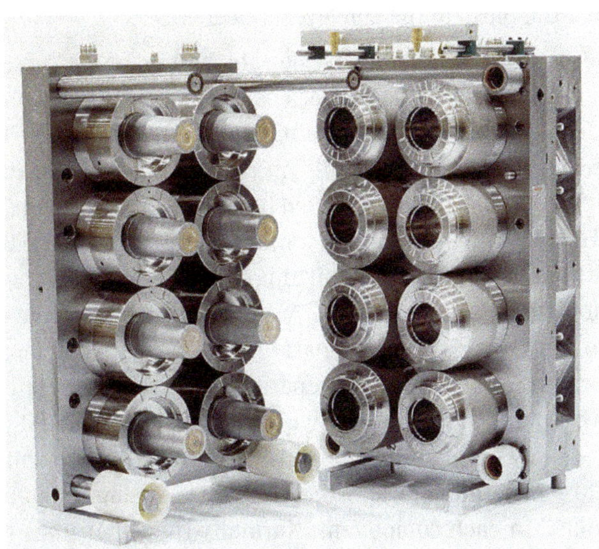

Figure 18.33 Stadium cup: Eight-cavity core lock mold (Courtesy of Husky Injection Molding Systems Ltd.)

In Figure 18.32 a few design options are illustrated. The module size used for these stacks is large due to the forces needed to fill the parts. A 300 mm to 350 mm module can be used typically. The cores can be either front-mounted (for in-press alignment with M16 bolts) with a floating dowel (typically 16 mm diameter), or rear-mounted with 4 × M16 bolts with float using bevel washers and a sleeve. The cavity side will be fixed solid and can be either front-mounted or rear-mounted in a spigot. Both options for core and cavity mounting are shown in the figure. Depending on the geometry of the part, the gate pad split can be on the bottom of the cup or on the sidewall after the corner radius. As discussed in Chapter 2, the diameter of the gate pad should be nominally smaller than the cavity by 0.03 mm (0.001"), to prevent any possibility of an undercut in this area, which would hold the part in the cavity.

The stadium cup is air-ejected from the core. Just before mold opening, cavity air is turned on to ensure the part transfers to the core on mold opening. There is an air circuit around the gate insert for this purpose. Once the mold is open, the core cap air is turned on. The core cap air is supplied from the M12 center bolt hole. A groove is cut on the face of the counter-bore and the air travels up the hold and into a groove on the bottom face of the core cap. It is then fed to the fixed core cap vent (usually 0.013–0.025 mm in size with a 1 mm land) or to the floating core ring (with a 0.13 mm vent in the open position). Note that both fixed and floating core cap air designs are shown in Figure 18.32. The split between the core and the core cap should be as close to the top of the core as possible, generally in the 5 mm range from the top of the core. Once the core cap air has freed the part from the molding surface, the blow-off air, which originated from the lock ring, is activated. This blows the part clear of the core and into the take-off robot. Blow-down air is then used to push the parts to the conveyer if needed.

Core cooling is designed with two grooves on the bottom of the core. One groove is the inlet feed and the other is the outlet channel. The BeCu core cap is used to connect the IN feeds to the OUT feeds using a star-shaped pattern. Chapter 12 on cooling of the core provides more details on core cap cooling. In this design, the cavity cooling is accomplished using angled holes and an inserted baffle. There are typically 12 or 16 holes (8.73 or 11 mm) containing the baffles. The baffles must be held in position to prevent them from turning. A groove cut into the bottom of the cavity is used to feed the baffled holes. A separate gate pad/gate insert cooling circuit is used to cool the bottom of the part and the gate area. A separate circuit is used, so that this sensitive area can be separately controlled for gate vestige. In all cases the cooling circuits are contained with O-rings. There should be a minimum of 1.5 mm between an O-ring and any feature to prevent the possibility of leakage. In order to ensure that the core and the cavity are strong enough, the stresses should be calculated in each component. Normally the water lines in the core are kept at a minimum of 12–15 mm from the molding surfaces, and the cavity water lines are kept 18–20 mm from the molding surfaces.

The leader pins on the mold shown in Figure 18.33 are longer on the top and shorter on the bottom, so that they do not interfere with the falling parts. The leader pins have been covered over with a HDPE sleeve to avoid contamination with grease.

A hot runner is normally used for these parts with large channel sizes in the manifold and a 1–1.5 mm hot tip gate or a slightly larger (2.5 mm) valve gate.

18.11 Polystyrene Tumbler (Using Stripper Ring Ejection)

Figure 18.37 shows an example of an eight-cavity tumbler mold. The mold is installed horizontally to allow for faster ejection (less distance to fall to the conveyer). The molding surfaces are high gloss and the material to be molded is clear polystyrene (see Figure 18.34), and because of this stripper ring ejection is used.

Figure 18.34 Polystyrene tumbler: Samples

Since the part is very thin (< 0.5 mm), each stack is individually aligned using a core lock design. The cavities are fixed in a plate for high rigidity, and the cores are fixed front-mounted to allow for alignment in the press. The stripper rings are attached to the stripper plate via tie rods. Note the BeCu25 core insert to facilitate good cooling opposite the gate and to allow for venting on ejection. Leader pins are used for assembly purposes only. The mold cooling is internally balanced, and there is only one IN and one OUT cooling line per plate for ease of setup and installation. This mold runs in a high-speed press at sub-3 s cycles. The parts free-fall from the press and are re-oriented using a cup stacker built by CBW Automation.

Figure 18.35 shows a cross-section of a typical PS cup. Note that the bottom is preferably domed to prevent a "potato chip" twisting effect on the bottom of the part. While shrinking, the curvature of the dome will change somewhat, but it will not pull inward and thereby deform the side wall of the container. It is always quite difficult to mold any straight surface, especially from high-shrinkage plastics, unless the cooling cycle is greatly extended to permit the product to reach the mold temperature before ejection.

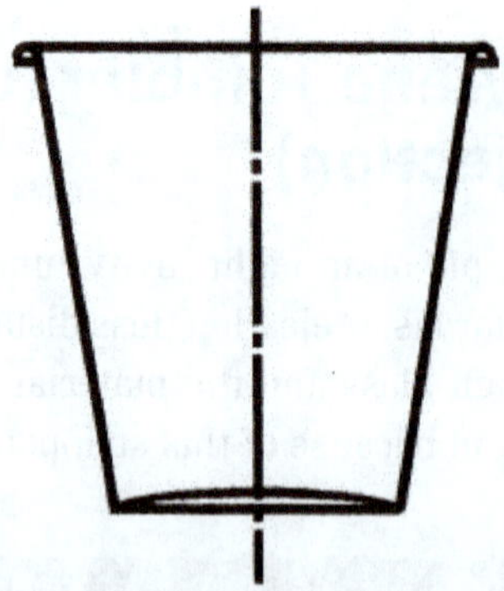

Figure 18.35 Polystyrene tumbler: Cross-section of a simple cup shape

A typical stack section for such a product is illustrated in Figure 18.36 and Figure 18.38. The gate pad, cavity, and core are normally made from H13 steel hardened to 49–51Rc. The lock ring is made from A2 steel hardened to 52–54Rc and the stripper ring is made from the same material but hardened to 54–56Rc. Sometimes S7 steel is used for the stripper rings. The core insert is made from BeCu25 hardened to 38–41Rc.

Alignment of the cores and cavities is best achieved using a core lock design due to the high forces involved in molding these parts. The module side is 170 mm. The lock ring is preloaded by 0.1 mm onto the core with eight M8 bolts, and jacking holes are added to separate these two components if needed as the taper between the two parts is typically 5°. The preload between the lock ring and the cavity is 0.013 mm at 15° taper angles. The stripper ring has no preload between it and the core. The stripper ring has 0.5 mm clearance to the lock ring. The cavity is held rigidly inside a plate and heeled into the plate from behind.

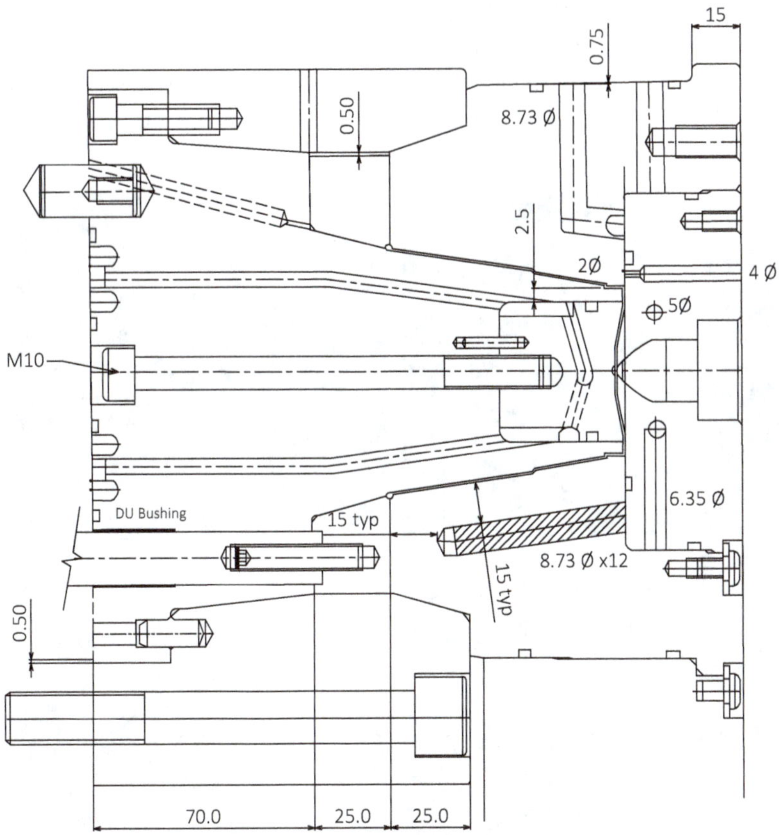

Figure 18.36 Polystyrene tumbler: Stack section with stripper ring eject and core lock construction

The parting line vents for these molds are typically 0.018 mm + 0.003 mm. A vent land of 2 mm is used, and the vent grooves are cut into the front of the stripper rings and then onto the cavity tapers. The air then vents down the lock ring and into four holes cut into the lock ring (see Figure 18.36). Natural venting between the gate pad and cavity is used but no vent is added there to avoid a sharp edge. An addition blow-off vent is added behind the stripper ring in the core to assist in the final blow-off of the part onto the conveyer.

Cooling in the cavity is achieved with a series of 12 baffles tied together with a groove between the gate pad and the cavity. Core cooling uses a series of 12 5 mm holes which intersect at the core insert. The core insert is drilled in a spoke pattern to connect the lines and provide cooling opposite the gate.

Figure 18.37 shows an example of an ejector box system for this type of mold. Four tie rods are attached to each stripper ring. The tie rods are guided by DU® bushings in the core base and the fastened to the stripper rings with studs. No screws are permitted on the face of the stripper ring to avoid flash getting into these holes. The ejector plate is actuated by the machine ejector system. Note the extensive support pillars in the core plate to support the cores and cavities. This is needed due to the high clamp tonnage and injection pressures needed for these molds.

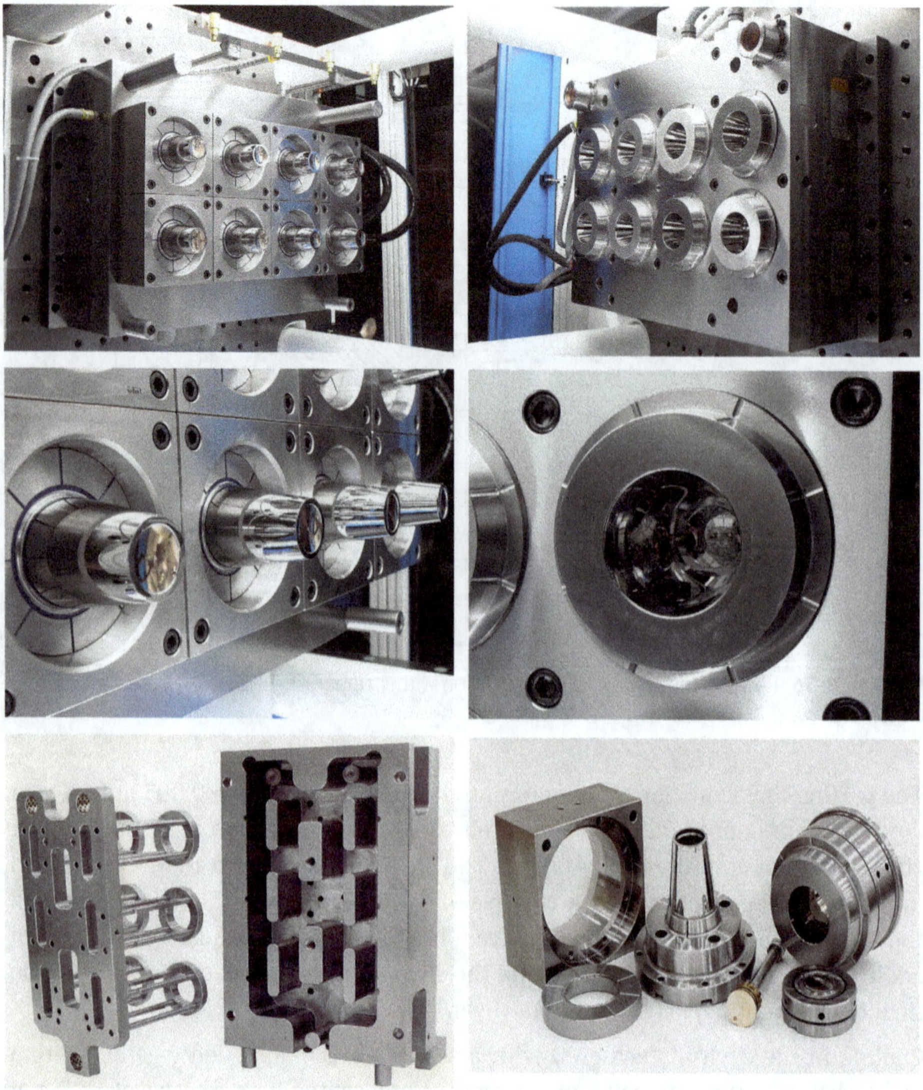

Figure 18.37 Polystyrene tumbler: Eight-cavity mold. Six-cavity ejector box shown for illustration purposes only (Courtesy of Husky Injection Molding Systems Ltd.)

Normally a hot-tip hot runner is used for these types of molds, with a gate diameter of 1–1.5 mm. Sometimes a valve gate system is used if the appearance of the gate area is critical. A dimple is cut into the core to allow the partially cooled slug from the gate to be injected into the part on the next shot.

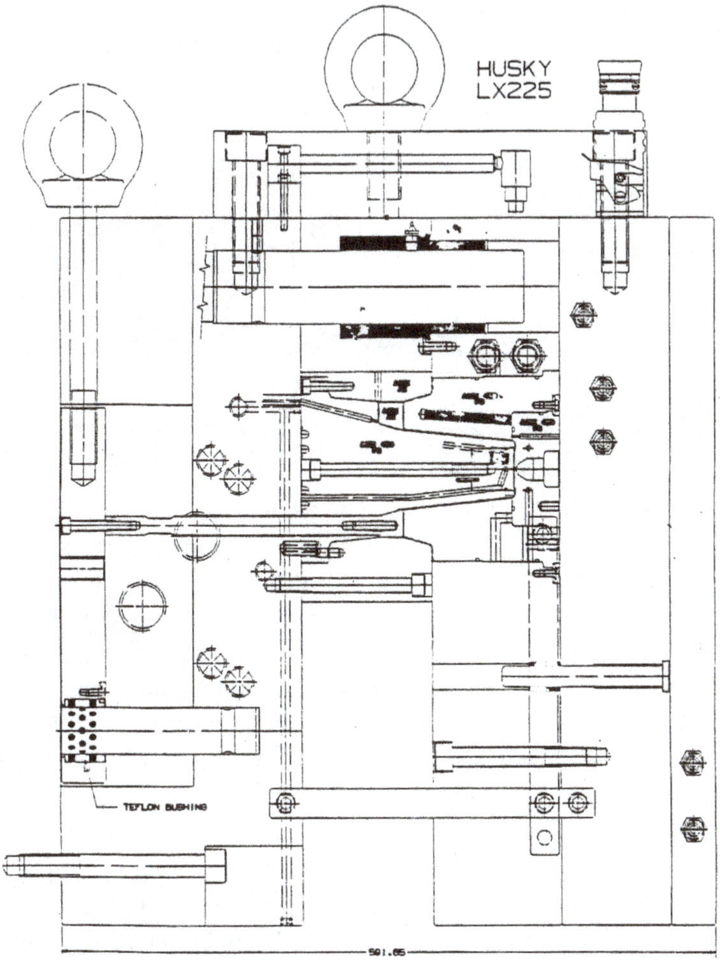

Figure 18.38 Polystyrene tumbler: Section assembly view of a mold with stripper ring ejection

18.12 False-Bottom Cup (Using Moving Cavity and Air Ejection)

Figure 18.39 depicts a cup with a false bottom with a (preferably) domed bottom. The false-bottom technique is used by some product designers to create the effect of a larger container. A false bottom where the dimension *f* is larger than 2× the wall thickness creates some challenges in ejection.

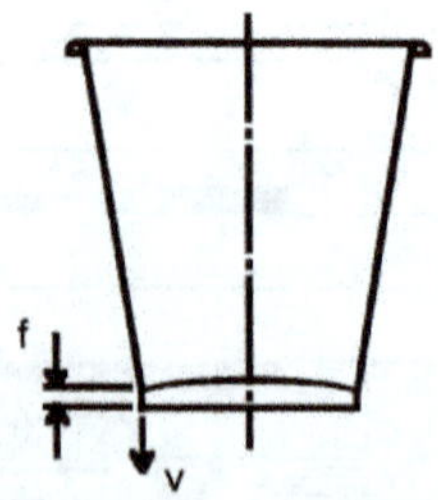

Figure 18.39 False-bottom cup: Cross-section. Such a design requires moving-cavity ejection

Because of the rim of the false bottom, it will be difficult or even impossible to fill this portion of the bottom and reliably eject the part from the core side. If a piece of plastic breaks off in that narrow section and remains there, it would be very difficult to remove it without dismantling the mold. Therefore, special measures must be provided in the mold. The cavity of the mold must follow the core as the mold opens, for a short distance (≈ *f*) until the mold part that forms the inside of the re-entrant, which usually also contains the gate, is completely withdrawn from the molded plastic piece. Only after this happens is the mold allowed to separate at the regular parting line. This method also facilitates good venting at the bottom and allows proper filling of the part in this area. If it was not vented, this thin rib section would be a "dead pocket" and not fill, as already discussed. Note that this method is called moving-cavity (see Figure 18.40). It is, in principle, a type of two-stage ejection system. Moving-cavity ejection is covered in more detail in Chapter 11 on ejection.

The cavity plate is guided on a separate set of guide pins to control its location relative to the gate retainer plate (or hot runner plate or cavity backing plate, as should be the case). Its stroke is limited to be only slightly larger than the dimension *f*. Air actuators (usually four) built right into the backing plate push the cavity plate so that it follows the mold opening motion until the set limit is reached. The product is now easily ejected from the core, and there is no danger that the rib on the bottom of the cavity gets trapped between the gate pad and the cavity. There must be ample venting provided where the alignment ring meets the gate pad.

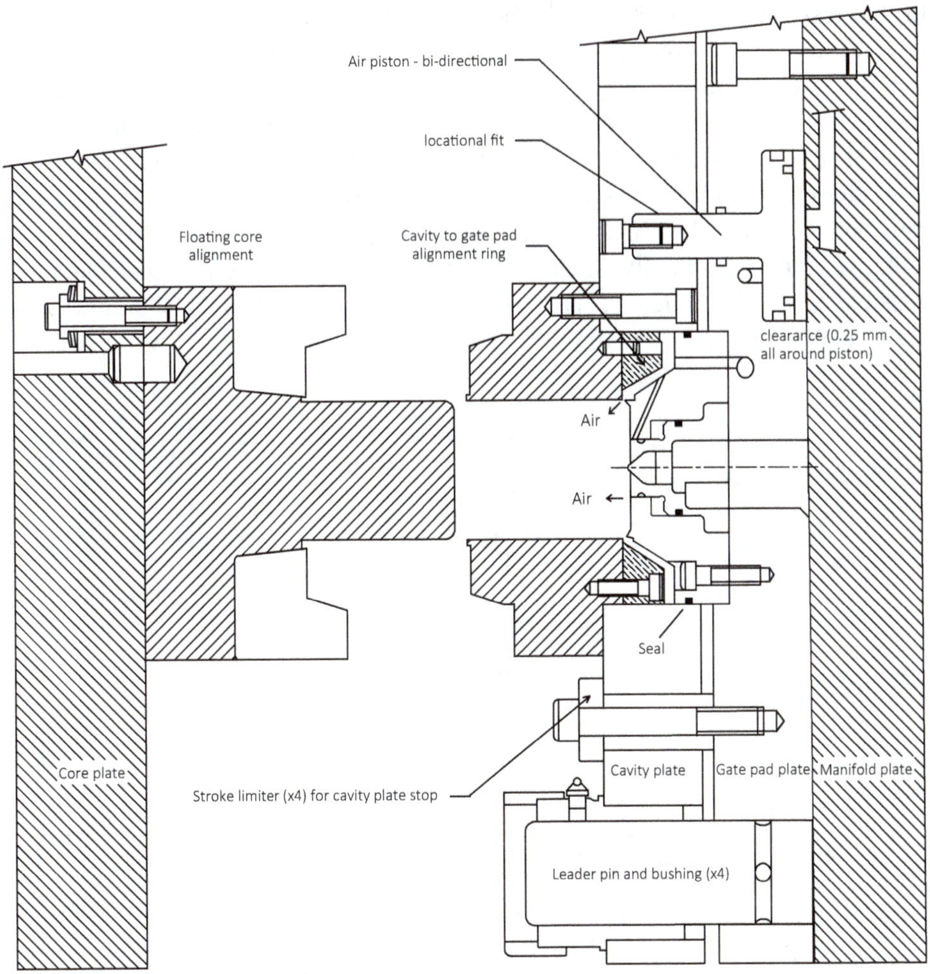

Figure 18.40 False-bottom cup: View of moving-cavity section assembly

18.13 Round Closure (Using Stripper Ring (Bump-off) Ejection)

Courtesy of StackTeck Systems

This four-cavity 110 mm round closure has a thread profile that allows for the undercuts (threads) to be stripped off the core without the need for any secondary motions. The HDPE lid has a continuous thread finish and runs at a 5 s cycle with 125 tonnes of clamp force.

In this case, because of the size of the part (110 mm diameter) and the thin-wall section, individual core-cavity alignment through the stripper ring is designed into the stack (Figure 18.41). The stripper ring has a taper to mate to the core as well as a small taper that mates with the cavity side. This provides added alignment that is not normally seen in closure molds. In many closure mold applications the part is relatively small (< 43 mm diameter), the wall section is in the range of 1 mm and the required cavitation is large (32, 48, or 72 cavities). Therefore, the added cost of individual core-cavity alignment is both not needed and not possible due to the limited space available for each stack. Instead, side interlocks are used to align the core to the cavity side.

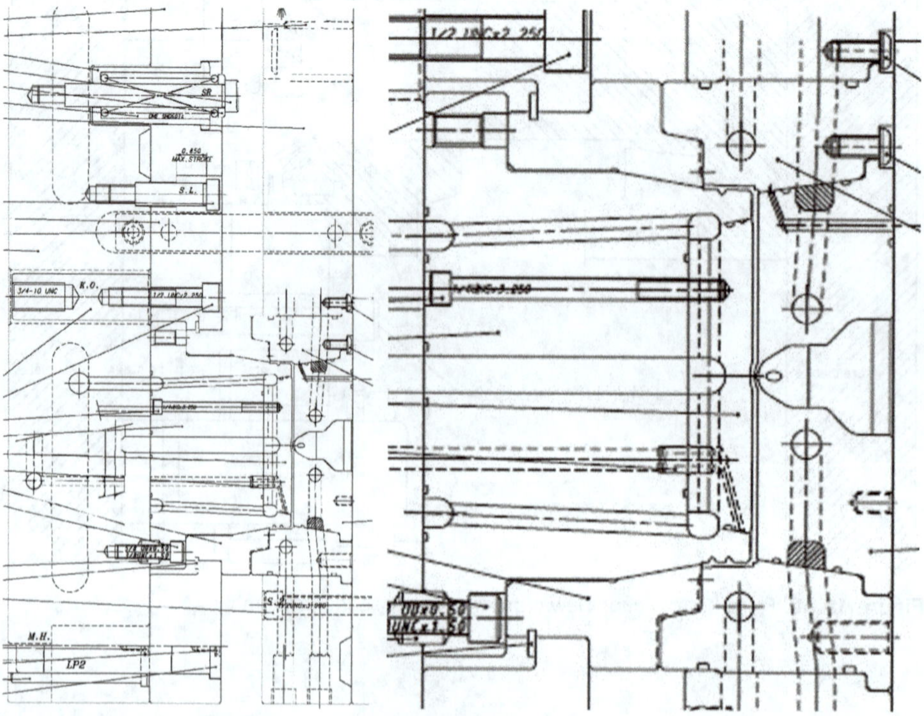

Figure 18.41 Round closure: Section view of stack

The cavity is rear-mounted into a cavity plate for rigidity and held in from behind. The cores are front-mounted onto the core plate with a loose hollow dowel (where the core is mounted), which allows for some float. The stripper rings are held in a stripper plate with snap rings that allow the stripper ring to float slightly.

The 110 mm closure is ejected in the following sequence. Before the mold opens, the cavity air circuit around the gate pad is activated to ensure that the part stays on the core side. The mold then opens. The core insert air, which is fed from the bottom of the core into an air groove and vent, is activated just slightly ahead of the stripper ring to eliminate any vacuum and to break the part from the top of the core. The stripper plate then moves forward enough to clear the core, and the parts fall onto a conveyer. The ejector plate is returned via springs. There is a stroke limiter to ensure that the stripper plate is not over-stroked, but the stripper plate should never hit the stroke limiter during normal operation. Note that the machine ejectors act directly on the stripper plate, eliminating the need for any ejector box.

The cooling in a stripped closure mold is fairly straightforward. Angled drilled lines in the cavity create a circuit around the thread, and a separate circuit is drilled to cool the bottom panel and the nozzle/gate area. The arrangement can be seen in Figure 18.42. On the core side, a drilled line provides water up the center of the core opposite the gate. This water then divides into six drilled lines that feed into a groove and then into drilled lines down the core to another collector groove. A high-conductivity core insert is used to provide higher cooling rates (see Figure 18.43). In the section view, note the location of the O-rings on the cavities and the steps to prevent the O-rings from pinching during assembly.

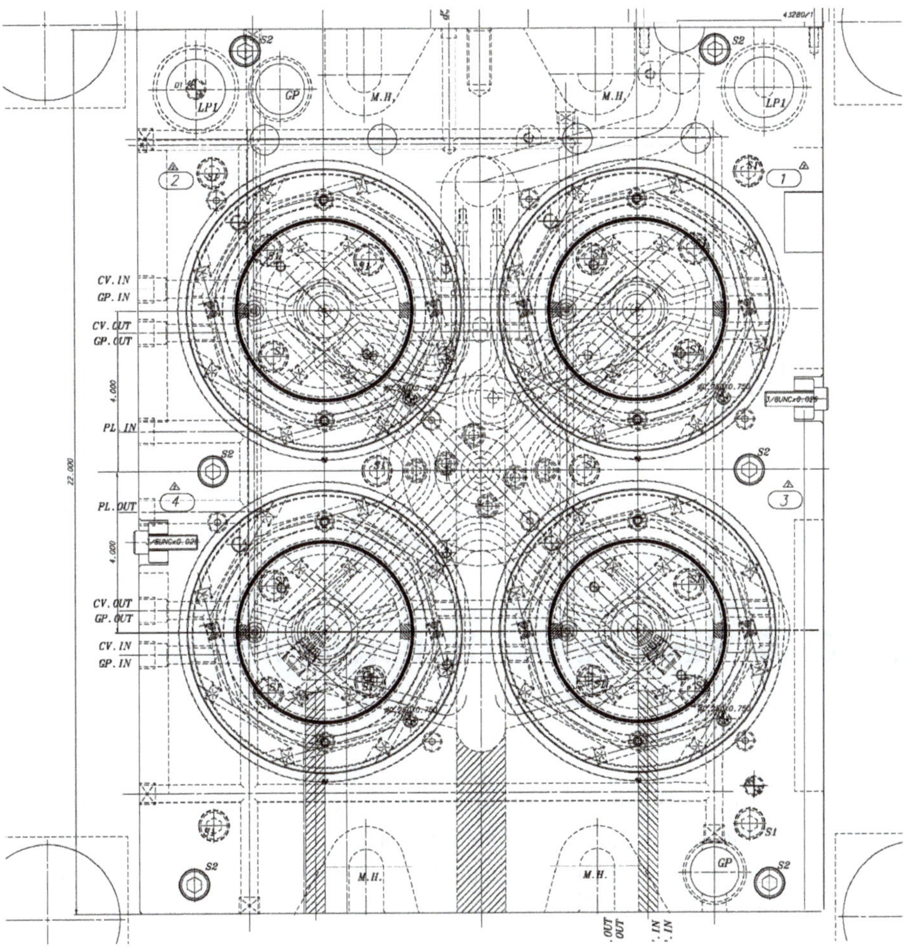

Figure 18.42 Round closure: Plan view of cavity

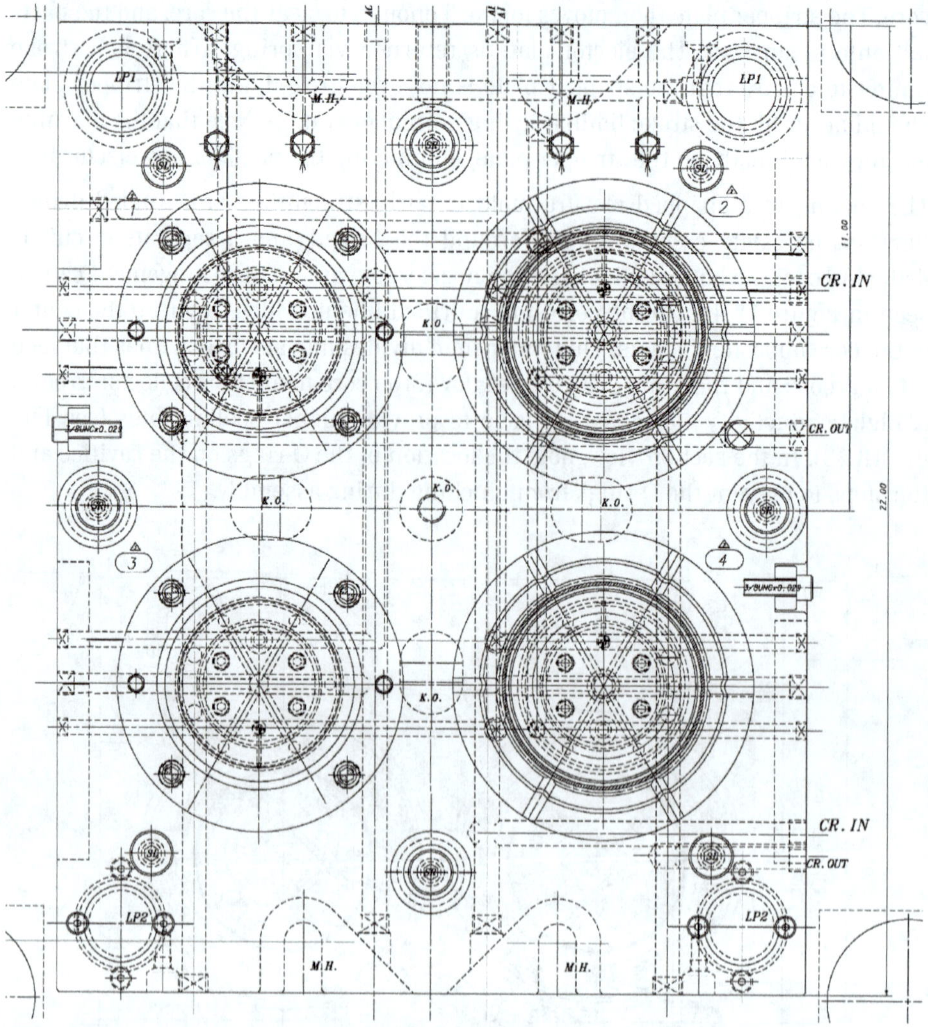

Figure 18.43 Round closure: Plan view of core

■ 18.14 Paint Pail with Handle Inserts

Courtesy of Plasdan

This 422 g single-cavity paint pail is made from PP and runs at 8–10 s cycles. It is gated with one hot sprue at the bottom center of the part. The sprue must be well-cooled so that it does not impact the cycle time. The average wall thickness of the part is 1.6 mm.

The core and cavity are aligned using wedges (shown in dark blue) between the core and cavity, with the ability to adjust the alignment using spacers, shim, and grinding (Figure 18.44, Figure 18.45). The wedges shown here are on the four sides on the module, but a preferred location would be on the four corners in diagonals. This allows for easier alignment to eliminate any variation in wall section.

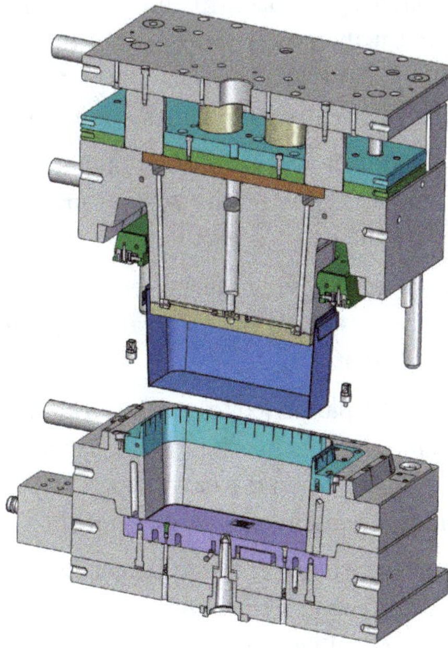

Figure 18.44 Paint pail with handle inserts: Single-cavity stack

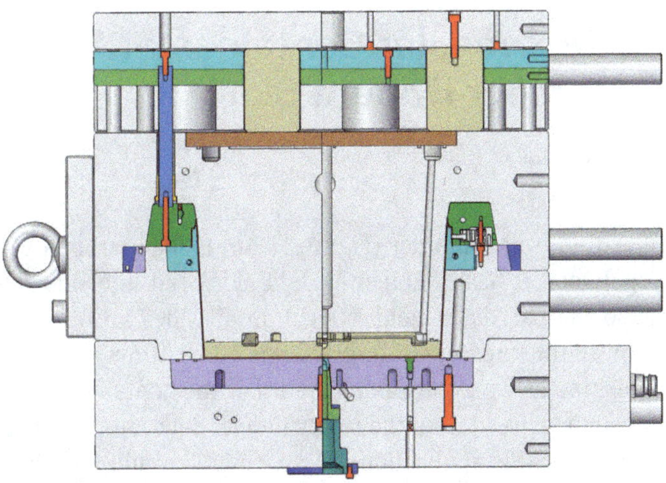

Figure 18.45 Paint pail with handle inserts: Single-cavity stack

To achieve the fast cycle times, the core and cavity are inserted with high conductivity Moldmax® inserts at the lip (in teal blue), at the gate pad (in purple) and at the core cap. The cooling circuit on the core has a central water channel opposite the gate that then divides into 12 radial lines that join with drilled lines that travel down the side of the core. This ensures equal cooling all around the circumference of the core. An additional cooling circuit is added to the Moldmax® insert at the lip on the core to keep this area well-cooled. At the bottom of the core, the water is gathered into a groove and it then exits the core. On the cavity side, the cooling consists of a separate cooling circuit of grooves in the gate pad, a cooling circuit for the hot sprue, and a series of circuits with baffles around the outside of the cavity.

The cavity is split at the gate pad area before the radius at the bottom of the part which allows the cavity to be EDM-wired for ease of manufacture and to achieve accurate tolerances.

Ejection is achieved with a stripper ring (dark green) tied to an ejector plate built into the mold. Machine ejectors drive the stripper plate, which has the stripper ring rod extensions fastened to it.

The two holes in the side of the part, used to attach the handle of the paint pail, are made with spring-loaded inserts. The insert is spring-loaded back so that when the mold opens the pin moves back. When the mold is closed, the spring is again compressed, moving the pin into location (on a taper - not shown) to mold the hole.

The large light-yellow supports shown around the ejector plate supply critical support to the mold during clamp-up and injection of plastic. Without strong support behind the core, there would be unacceptable wear and core shift on the part.

■ 18.15 Rectangular Container with Tamper-Evident Skirt

Courtesy of Fourmark

This rectangular thin-wall container (Figure 18.46) is made from a 50 melt index random PP copolymer. It has a tamper-evident skirt and is inside-gated through the core because the part is in-mold labeled on all sides and the bottom of the cavity. The part wall thickness is 0.6 mm and it is 106 × 106 × 80 mm high. It runs in a 180 ton machine at 9 s cycles. Both the lip of the container and the tamper-evident skirt need to be freed from the cavity for ejection. A unique two-stage ejection system is used for this application. An alternative would have been to use an internal or external latch lock system (see Chapter 11 on ejection for more details).

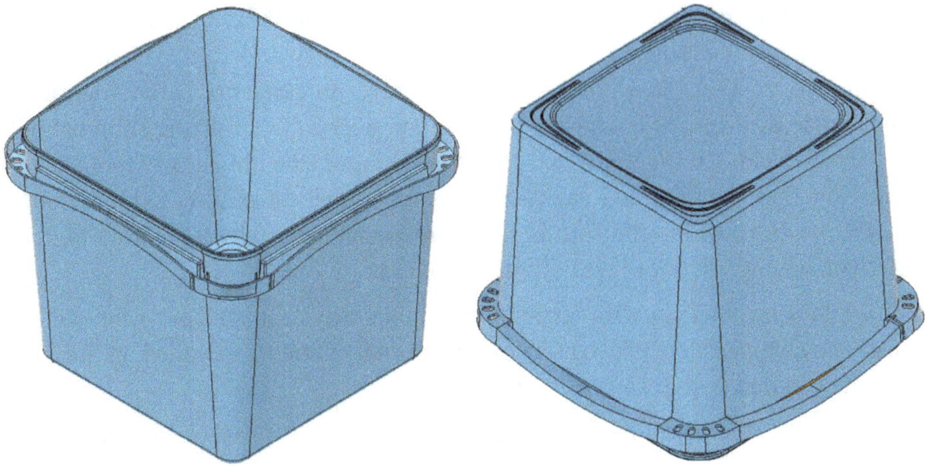

Figure 18.46 Rectangular container with tamper-evident skirt

The cavity is cooled with baffle cooling evenly spaced around the outside of the cavity (Figure 18.47). A separate cooling circuit opposite the gate provides cooling to the bottom of the part. On the core side, the cooling is restricted due to the hot runner nozzle and the two-stage stripper ring ejector. W drilling is used to cool the core in the areas where cooling lines can be drilled. The dual stripper rings are not cooled directly. A core cap of high-conductivity material (BeCu) is added to the core to allow for air-ejection assist and to allow the cooling to be routed through the core. A gate insert around the nozzle allows for additional cooling and air-ejection assist.

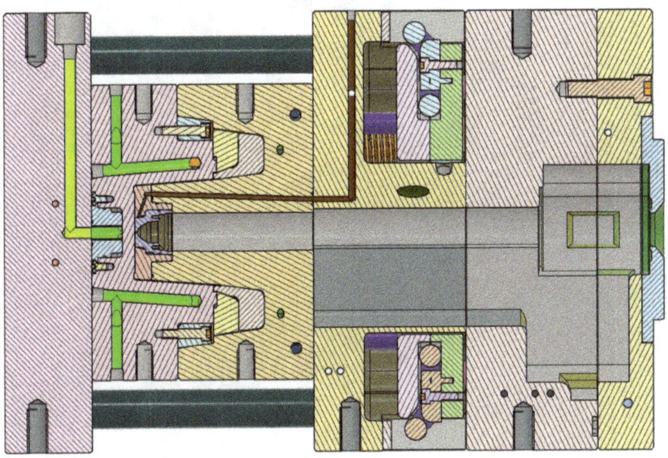

Figure 18.47 Rectangular container with tamper-evident skirt: Section view

Alignment of the core and cavity is accomplished with a wedge lock system. Since this part is rectangular, the wedge lock system is very effective in maintaining uniform wall sections on all four sides of the part. Adjustments to the alignment can be made by grinding the spacer behind the wedge or by inserting shim behind the spacers. Note that the wedge lock system is contained within a very robust core block, and that the core and cavity are front-mounted for ease of servicing and adjustment. In this case the core is rigidly held and the cavity is allowed to float slightly to ensure even wall sections on all sides.

Ejection is accomplished with a combination of a two-stage stripper ring and air assist. The stripper plate is driven by two hydraulic cylinders mounted on the core side. The ejection sequence is as follows:

1. Gate insert and core cap air are activated to break the part free from the core
2. The two stripper rings move together for the first part of the stroke
3. Once the roller hits the stop, the lever pivots and drives only the top stripper ring to its full stroke (see Figure 18.48)
4. After the stripper rings are in their forward positions, core blow-off air from behind the stripper rings is actuated to propel the part off the core
5. The stripper rings are returned with the internal spring system.

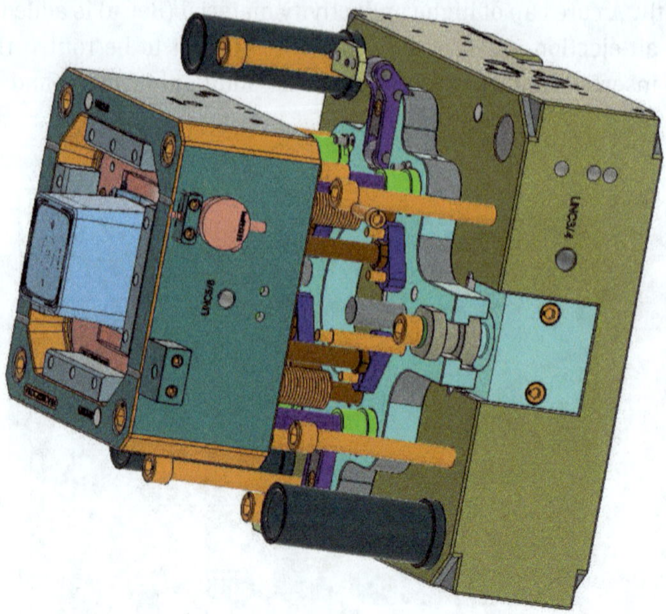

Figure 18.48 Rectangular container with tamper-evident skirt: Ejection mechanism (two-stage lever in purple, ejector plate in cyan)

18.16 Power Tool Housing

Courtesy of Don Smith

This power tool housing is made from Nylon (PA6) with 30% glass filler. The shrinkage used is 0.003"/1". The mold (Figure 18.49, Figure 18.50) consists of two left-hand and two right-hand sides of the power tool housing that when molded are assembled together to produce two full housings. The average part wall section is 0.098", and the mold runs at a cycle time of 26 s in a 280 ton injection-molding machine.

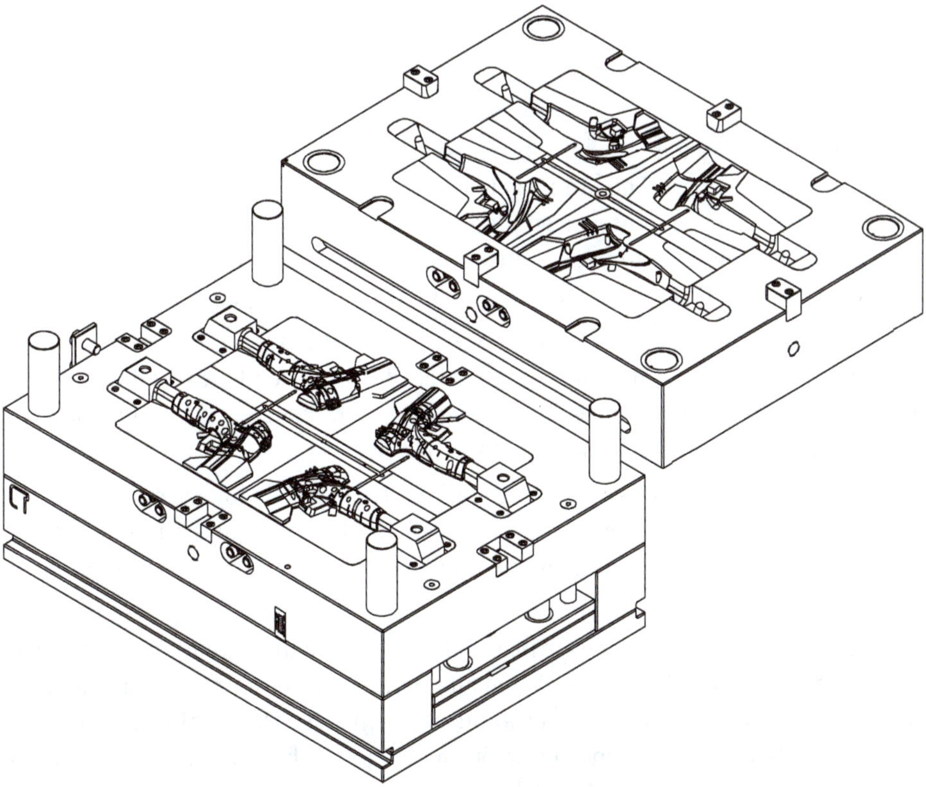

Figure 18.49 Power tool housing: Isometric assembly of 2+2 mold

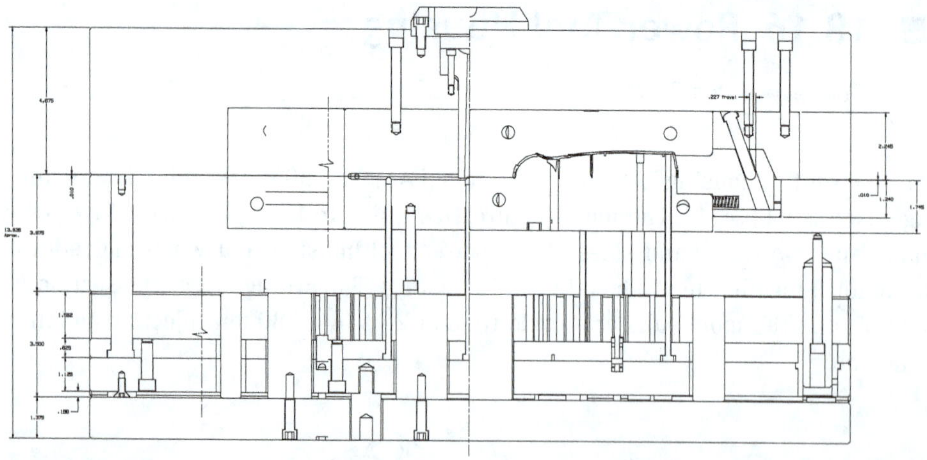

Figure 18.50 Power tool housing: Section view of mold

Each of the parts are fed by a cold runner system and then to a curved tunnel gate on the inside of the part. The tunnel gate is ejected from the core side as part of the ejection stroke. Ejector pins placed along the runner ensure that it is positively ejected with the parts. A runner sub-insert has been created to allow easy exchange of this part, or modification if necessary to adjust the filling pattern.

The core and cavity halves of the mold are aligned with four straight interlocks on the parting line face. Additionally four leader pins are used for assembly and initial alignment of the mold halves. The only other alignment mechanism is the core pull back-up taper located at the end of the handle.

Each of the core and cavity blocks is made from H13 steel and inserted into a plate for additional support and mounted from behind. The parts can be removed from the pockets using jacking holes. The core and cavity inserts are made from S7 steel hardened to 54–56Rc.

Each of the deep ribs and bosses is inserted to create a natural venting condition at the bottom of the rib. The part is vented all around and the support of the mold for clamp tonnage comes from the core and cavity plates. The parting line vent is created simply by the difference between the height of the insert and the depth of the pocket in the core and cavity plates.

The ejection sequence is as follows. As the mold opens, the slide at the end of the handle retracts on the angled horn pin. The insert is spring-loaded back so that it retracts on the horn pin as the mold opens. The insert retracts a total of 0.227" and then it contacts the pocket in the core plate where it is inserted. On closing, the horn pin compresses the spring and moves the insert forward. On mold closure, the tapered slide lock insert on the cavity side solidly backs-up the slide insert and prevents movement due to plastic injection pressure. The slide is held in position by gibs on both sides of the slide, and a lamina wear plate on the bottom ensures that there is no sliding metal pick-up during movement. Note that the horn pin angle is 20° and the taper for the backup is 25°, so that the slide insert firmly backs-up the insert without interfering with the horn pin.

Once the mold is in the fully open position, the ejector plate is activated forward by 1.562". The ejector plate is guided by four leader pins inside the ejector box. This moves the ejector pins (between 15 and 20 per part) forward and at the same time the two lifters start to move forward and inward on a 5° angle, so that in the fully open position the plastic undercuts on the part have cleared any metal interference from the core. The angled hole and lifter insert cut into the core can be seen in Figure 18.52 (section A–A). The lifter mechanism can be seen in Figure 18.51. The lifter is fastened to the ejector plate with a pivoting insert, so that as the lifter moves forward the 5° angle does not fight the ejector plate. Once the part is removed, the ejectors return to the back position on stop buttons. The machine ejector is tied to the ejector plate with stud spacers so that the machine ejector rods do not protrude into the platen area. The lifters are set back against a ledge between the core and the lifter shown in Figure 18.51.

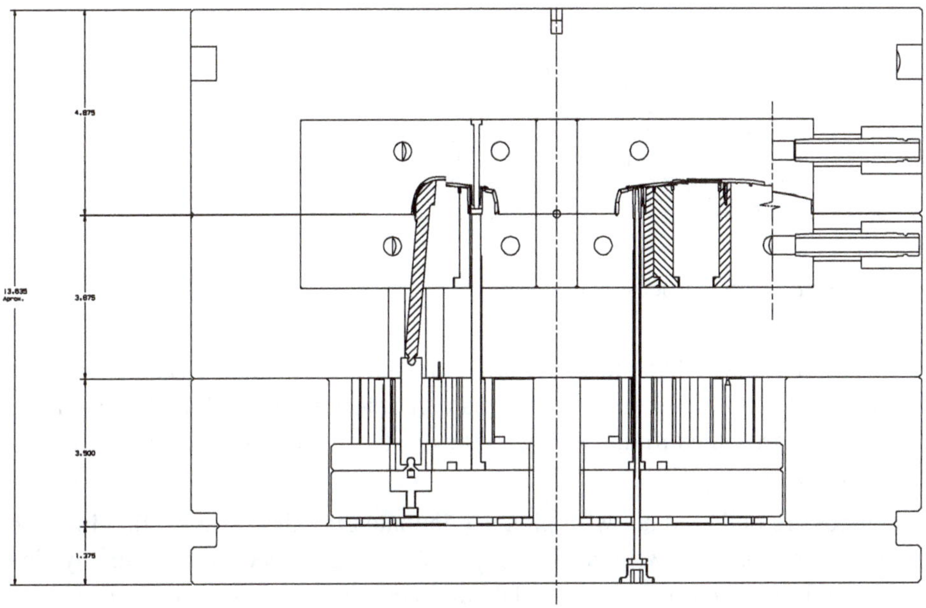

Figure 18.51 Power tool housing: Section view of mold

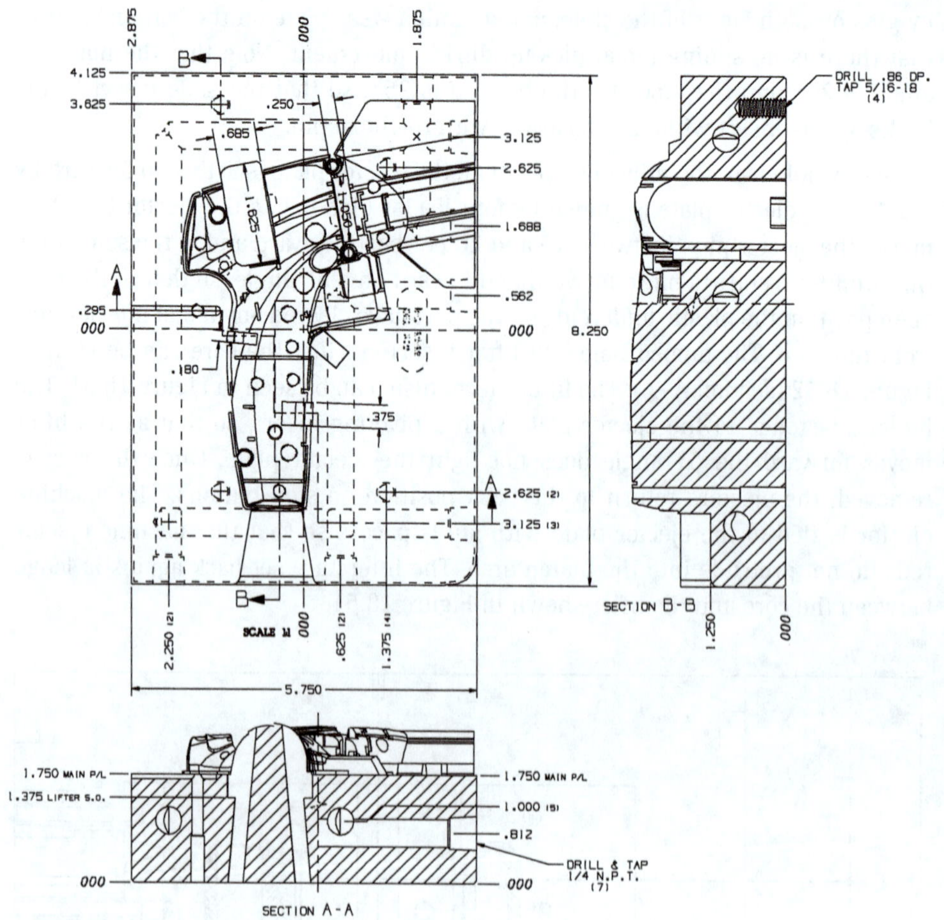

Figure 18.52 Power tool housing: Core drawing

To ensure that the mold is fully supported, there are several pillars added between the core plate and the ejector backing plate. All ejector pins with molding surface on them are prevented from rotating on the ejector retainer plate with a flat on the heads of the pins.

All bosses are hollowed out with pins to eliminate the thick cross-section in these areas. Figure 18.51 shows a good example of a pin-and-sleeve boss assembly. The pin is held in the ejector backing plate and backed-up by a threaded insert. The sleeve is held in the ejector plate and moves forward to eject the boss. Typically, these areas are naturally vented by the clearance between the two parts.

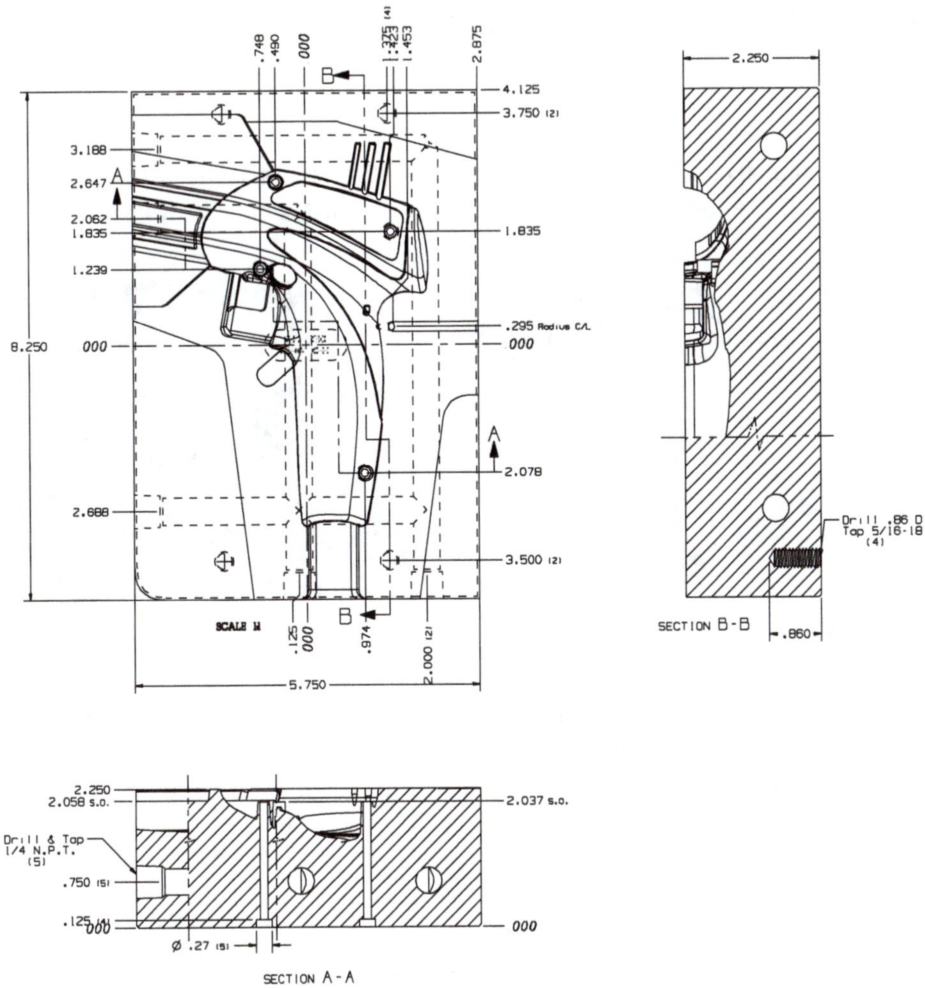

Figure 18.53 Power tool housing: Cavity drawing

18.17 Rear Hatch Door

Courtesy of Haidlmair

In this automotive application, the part is 1150 g and is made with ABS (Figure 18.54). It uses 630 tonnes of clamp force and a part is molded every 90 s.

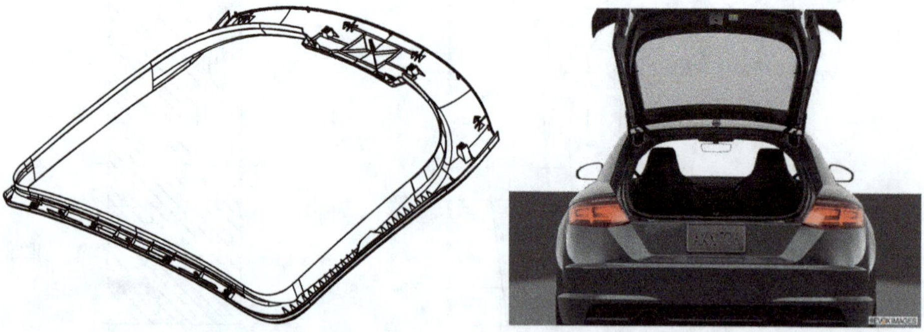

Figure 18.54 Rear hatch door: Part drawing

This single-cavity mold (Figure 18.55) uses four side interlocks on the outside of the mold. There are an additional four interlocks on the inside of the mold, but these were later removed as they were redundant. Cooling lines are drilled in the core and cavity to provide ample cooling for the 90 s cycle.

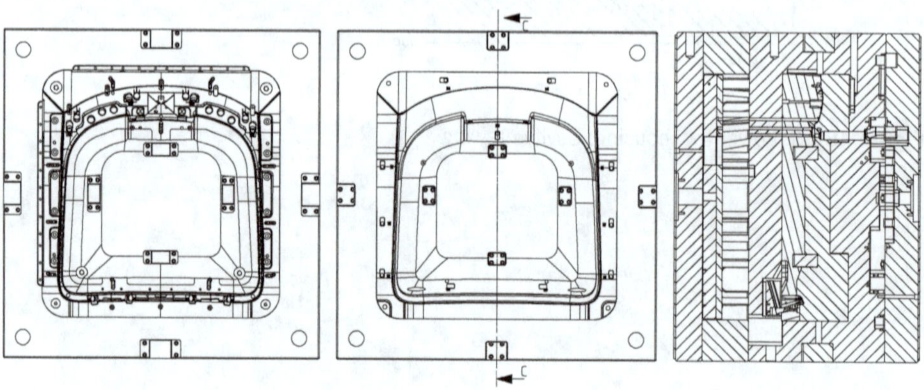

Figure 18.55 Rear hatch door: Mold assembly drawings

The molten plastic is fed to the cavity via an 11-drop hot-to-cold runner system. The small cold runners, which can be seen on the core side plan view around the part, are removed after the part is ejected. Each of the valve gates can be actuated individually, so that the flow fronts and weld lines can be adjusted after looking at the part samples. This gives the process engineers a higher degree of flexibility when processing the part for weld line locations.

Ejection is accomplished with six ejector bars on the side of the part, ejector pins, and several sliders on the top and bottom of the part. All these ejector functions are actuated by an ejector plate. The location of the sliders and stripper bars can be seen on the core plan view (see angled holes). A small core pull is used to free a window on the bottom side of the mold, as can be seen on the bottom of the section view. After clamp tonnage is released, the core pull is actuated and then the mold is opened. This action is interlocked so that the mold cannot open until the core pull has retracted.

■ 18.18 Crinkle Bag

Courtesy of Haidlmair

Due to the undercuts on this plastic bag (Figure 18.56), the cavity needs to split and open before the mold can open and eject the part from the core. The core side also has four sliders to free the undercuts on the core side before ejection from the stripper plate. The part is molded with PP and weighs 115 g. It is 2 mm thick and needs 250 tonnes of clamp force to mold a part. The mold runs at 18.5 s.

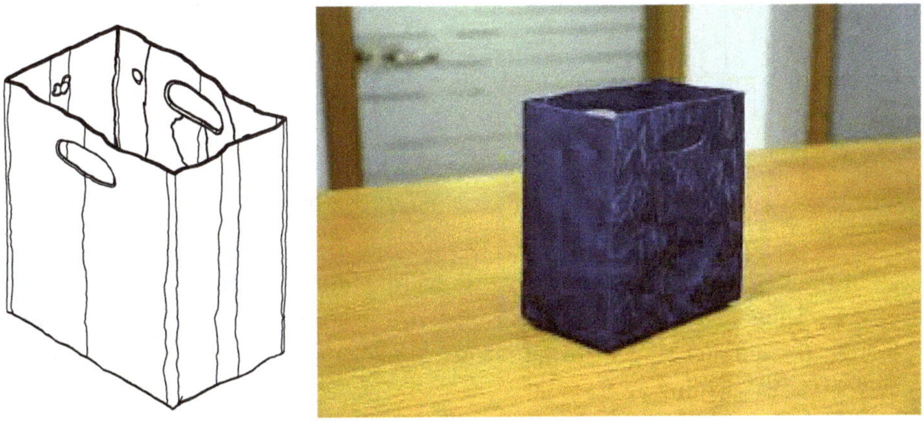

Figure 18.56 Crinkle bag: Part drawing and part photo

The alignment of the core and cavity is accomplished by large tapers between the split cavities and the cavity plate (Figure 18.57). Given the wall section of 2 mm, exact alignment is not as critical as with thin-wall molds.

The ejection sequence is as follows:

1. As the mold opens, gas springs push the split cavity sections onto the core side. They move outwards according to the angle of the cavity taper until the gas spring is extended and the undercuts are freed (approx. 10 mm).
2. The mold then opens fully and the core inserts are pushed forward and inward to both eject the part and free the undercuts on the core.
3. The center ejection mechanism using the machine ejectors pushes the part off the core the remaining distance using long ejector pins.

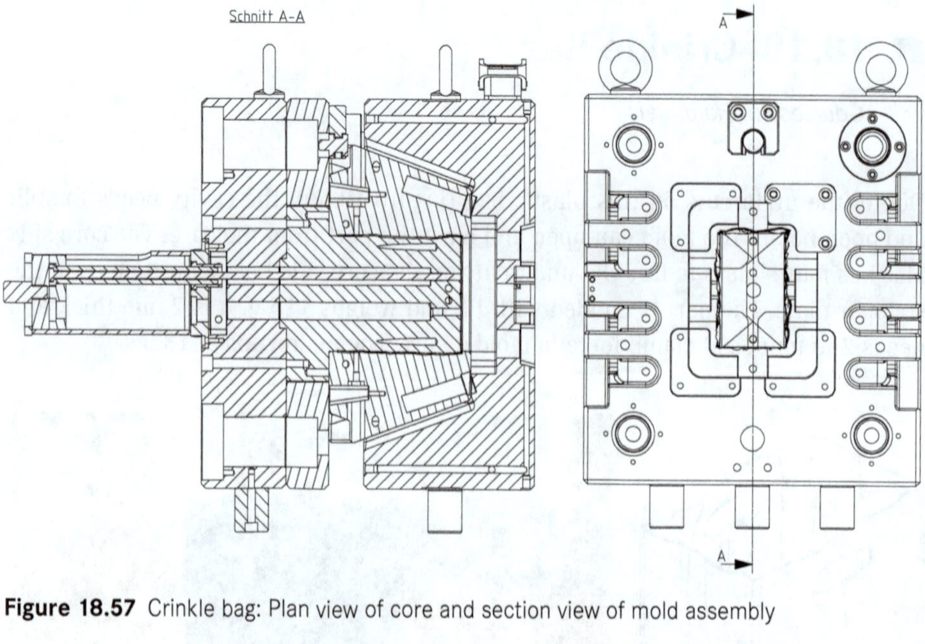

Figure 18.57 Crinkle bag: Plan view of core and section view of mold assembly

Each of the cavity sections is cooled with drilling and plugging, as shown in Figure 18.58. There are four lines for each section. The hoses for these cooling circuits exit from the cutouts shown in the isometric view of the mold. The core side is cooled with "W" drilling around the core.

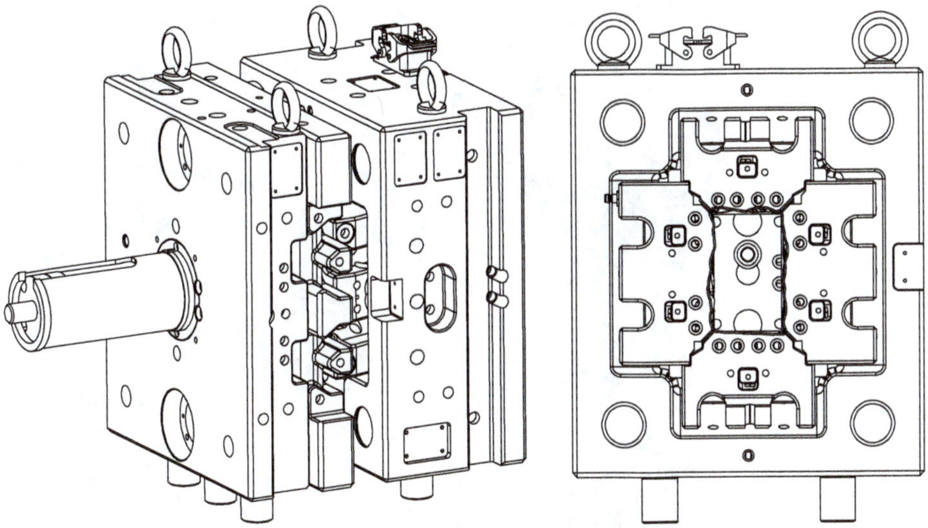

Figure 18.58 Crinkle bag: Plan view of cavity and isometric view of mold

The part is gated with a single gate (FDU style - see Chapter 14).

18.19 Preform (Using Neck Ring Ejection)

Courtesy of StackTeck Systems

A preform is a semi-finished part that looks like a tube with a thread on the top. After the preform is manufactured in an injection-molding process, it is re-heated and blown into a bottle such as shown in Figure 18.59. While bottles like this are also made in a one-step injection/blow process, the two-step process of manufacturing the preform has many advantages, such as lighter-weight parts, better thread definition and lower production costs at high volumes. Preforms can be made from PP or HDPE, but the most popular material is PET. PET strain-hardens biaxially when it is re-heated and blown, creating a glass-like clarity with high strength and toughness.

Figure 18.59 Preform: Typical preform and blown bottles

Preform mold cavitations can range from two up to 216 cavities (see Figure 18.64 for an example of a 72-cavity preform mold). These molds are highly standardized and are likely the single most common type of mold in the world given the popularity of PET bottles for holding liquids.

Preform parts can be as thick as 8 mm for returnable bottles to around 2 mm for thin-wall single-use water bottles. The average wall section for a PET bottle is around 3–4 mm. Normally, one would think that these molds would run in the 20–30 s cycle range. However, due to post-mold cooling and other innovative technologies developed by leading mold makers, these molds run between 7 and 15 s. The part is only cooled in the mold long enough so that is rigid for ejection. The rest of the cooling is done in innovative post-mold cooling devices developed initially by Husky Injection Molding Systems.

Regardless of the cavitation, the alignment, stack design, cooling, and ejection remain the same. A typical core lock preform stack is shown in Figure 18.60 and Figure 18.62), and the components are shown in Figure 18.61. The stack consists of five elements: the cavity and gate insert on the stationary side and the core, lock ring, and neck rings on the core side.

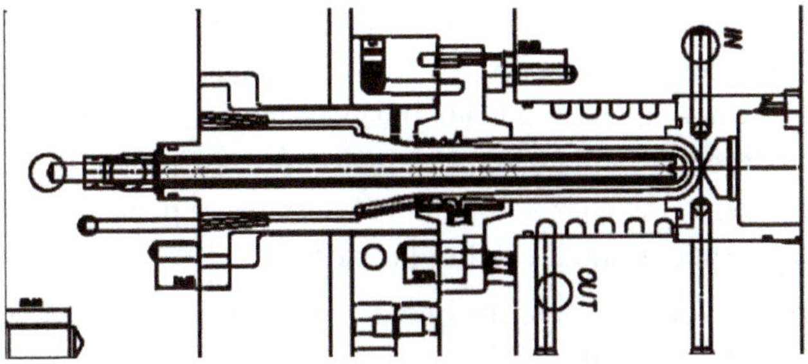

Figure 18.60 Preform: Stack section of mold (core lock design)

Figure 18.61 Preform: Mold stack parts

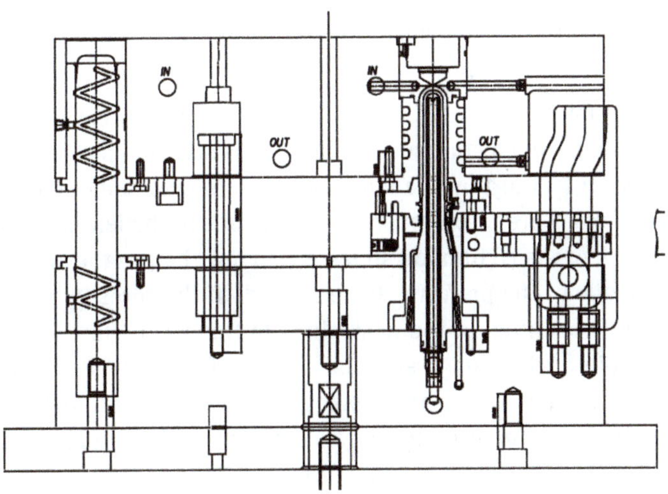

Figure 18.62 Preform: Section assembly of eight-cavity mold

Alignment between the core and cavity side is accomplished with the neck ring, which has both a taper to mate with the fixed cavity side and another taper to align to the very slightly floating core side via the lock ring and then the core. The cavity is held rigidly in the cavity plate and can be front- or rear-mounted. The lock ring is heavily preloaded to the core and mounted to the core plate. The neck rings are mounted on slides that are tied together for ejection.

The part is gated on the bottom of the dome of the part with a valve-gated hot runner system. Since PET is semi-crystalline, it will tend to crystallize if it is not cooled rapidly below its crystallization temperature. PET is also very shear-sensitive so the gate is relatively large at 4 mm. The flow of material in the hot runner must be very balanced so that each cavity fills the same. Machines built for molding PET preforms normally have a reduced injection pressure, as the filling pressure needed for these molds is less than 14,000 psi (100 bar). In terms of clamp tonnage, a 48-cavity mold normally runs in a 225–300 tonne injection-molding machine.

Venting is cut into the mold on the parting line between the lock ring and neck ring. There is also venting on the faces of the neck ring. A vent between the lock ring and the core provides additional venting between these two pieces, as the top of the part needs to be well-filled but not flashed. The top of the part is where people will drink the beverage from, so this area is highly quality-controlled.

Cooling of cavity is accomplished with a single- or double-helix cooling spiral. Several cavities in columns are tied into the same cooling circuit. The gate insert or gate pad normally has its own cooling circuit to control the gate area (see Figure 18.63). The core side of the mold is cooled with a bubbler that runs up inside the core. Several cores are connected in series. For large cavitation molds there is a series/parallel cooling arrangement. The neck rings are cooled by drilling in the neck ring. The cooling is supplied by the slides.

Ejection of the preforms is done with a combination of opposing slides and stripper plate (see Chapter 11 on ejection for details on this style of ejection). As the stripper plate is pushed forward by the machine ejectors (and sometimes boosters for more force), the slides are forced open by cams. Each of the left-hand slides is tied together, and each of the right-hand slides is tied together, so they move in unison. The cam profile allows the preforms to be ejected off the cores (which have a very slight taper) while opening up the neck rings to free the undercuts of the thread profile. The profile of the cam system is adjusted for the type of preform being molded. A ring at the top of the thread provides the last forward motion of the preform, which is inserted into cooling sleeves for post-mold cooling.

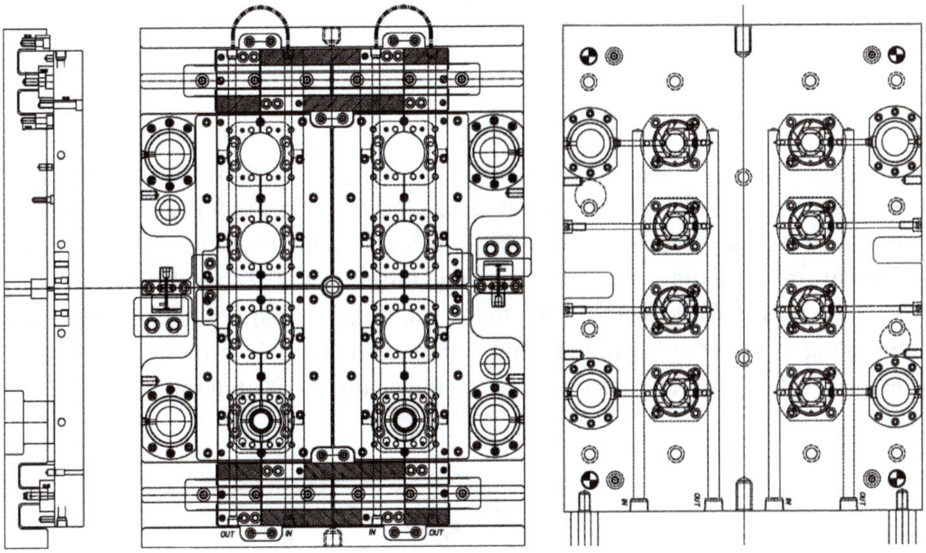

Figure 18.63 Preform: Cavity and core plan view assemblies for mold

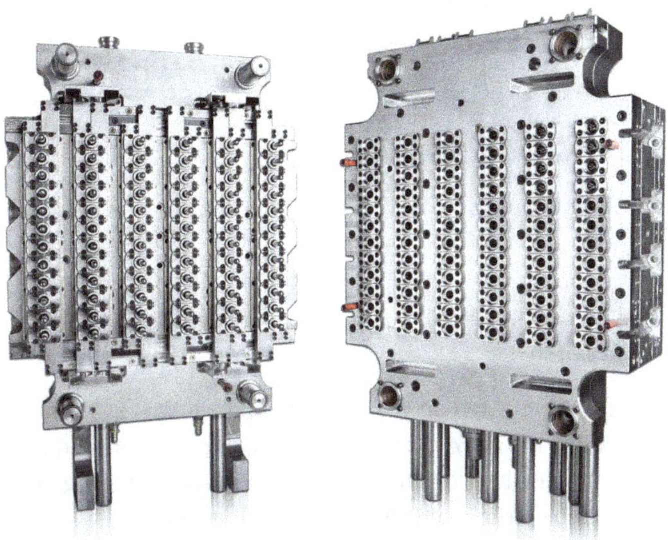

Figure 18.64 Preform: 72-Cavity mold

■ 18.20 Living Hinge Cap

Courtesy of iMFLUX

A polypropylene flip-top (6 g) with a living hinge part as shown in Figure 18.65 presents several challenges for the mold designer. Gate location must be carefully selected so that the living hinge does not break over time. The thread on the inside of the part must be removed, and is typically unscrewed. Then the flip top and the rest of the part must be ejected. Due to the rotating core requirement for ejection, the cooling of the core is limited. This four-cavity pilot mold runs in a 200 tonne press at 10–11 s. The 16-cavity production mold runs in a 600 tonne press.

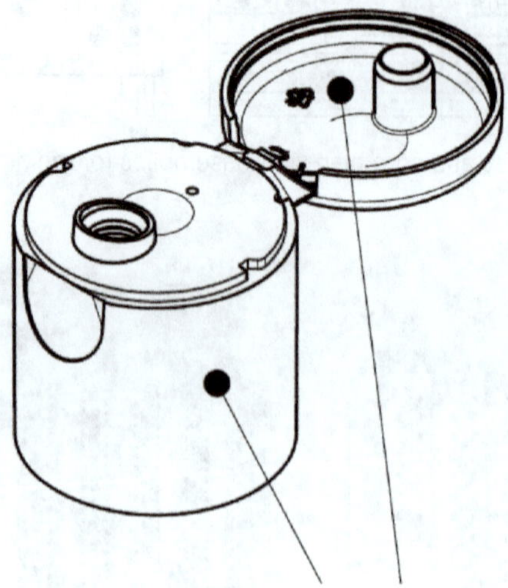

Figure 18.65 Living hinge cap

The part is directly valve-gated on the top of the part just behind the opening hole (can be seen in Figure 18.65). A very long and small nozzle is used to feed the part and produce a cosmetically acceptable part. The location of the gate ensures that the living hinge will be strong and the part will fill evenly.

The ejection sequence for a rotating-core design uses a rack-and-pinion and sine bar system. The sine bar can be seen in purple and pink on the superstructure containing the hydraulic cylinder and rack (Figure 18.66). The sine bar moves the stripper plate forward at the precise time that the rack unscrews the part from the core. The cap is kept from rotating by ratchet lugs at the bottom of the part. The core rotates on a pinion that is turned by the rack, as can be seen in Figure 18.67.

Figure 18.66 Living hinge cap: Isometric views of mold

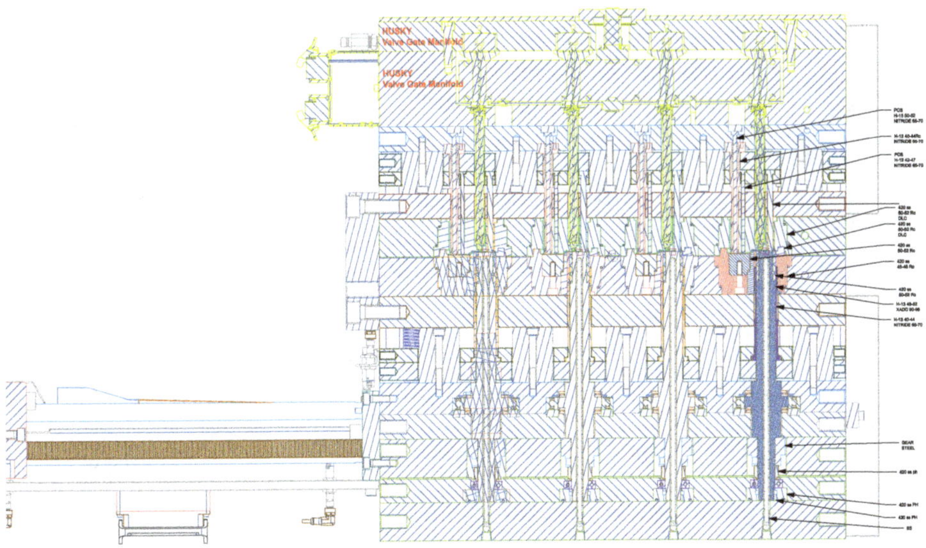

Figure 18.67 Living hinge cap: Section assembly view of mold

To keep the flip top tight on the core as the mold opens, air-activated (double-acting) pistons drive ejector pins from the cavity side to press on the part. A sleeve ejector removes the plug boss from the cavity side as the thread advances. After the rotation has completed and the thread is freed up, the stripper sleeve is advanced 30 mm to remove the rest of the part remaining on the core (see stack section view in Figure 18.68).

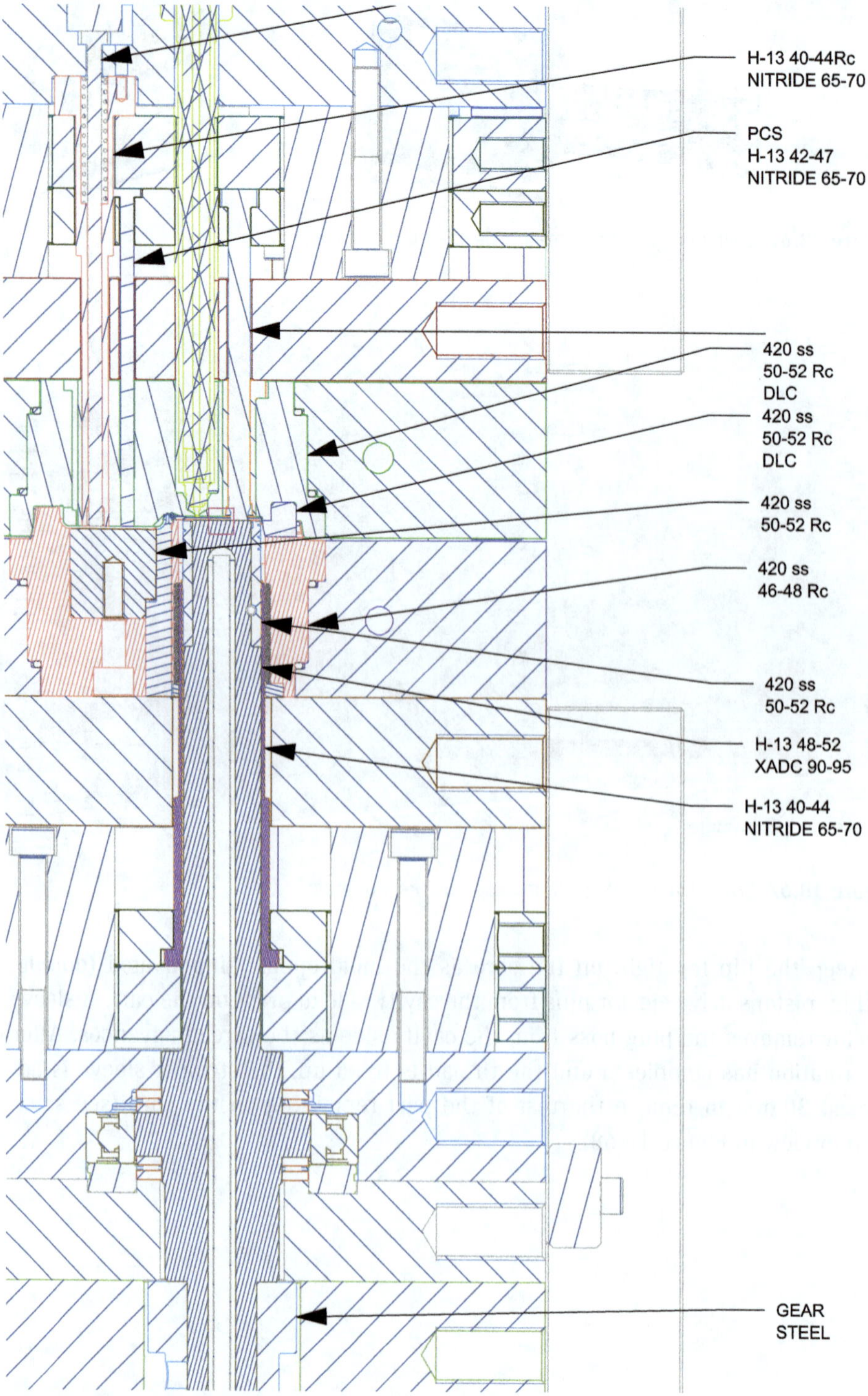

Figure 18.68 Living hinge cap: Stack section of mold

18.21 Rectangular Box with Undercuts (Using Split-Cavity Mold)

Courtesy of Manual do Projectista

Shown in Figure 18.69 is an example of a mold construction with undercuts on the sidewall of the part. In this case the cavity splits open as the mold opens. Specifically, during the first part of the opening stroke the split cavity moves with the core until the undercuts are free. At this point the core moves away from the split-cavity sections, and then a stripper ring pushes the parts off the core.

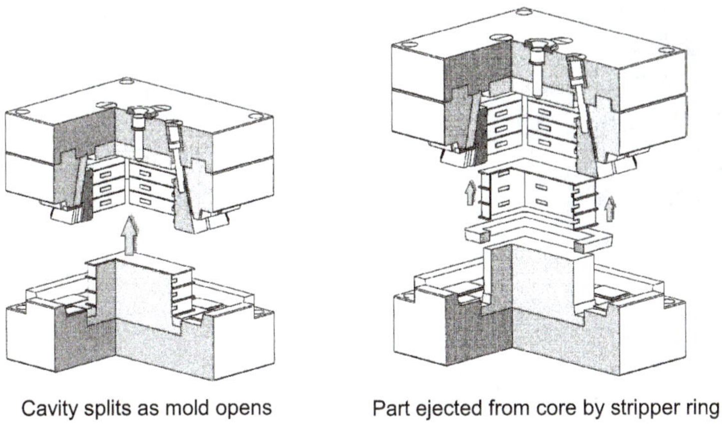

Figure 18.69 Rectangular box with undercuts: Split-cavity mold with stripper ejection (Courtesy of Manual Do Projectista)

The angle of the angled pin (Figure 18.70) will always be 3–5° less than the back-up taper on the split-cavity sections, to ensure that the mold locks up tightly on the tapers and not on the pins. The split-cavity sections are held in a dovetail to ensure that they stay aligned.

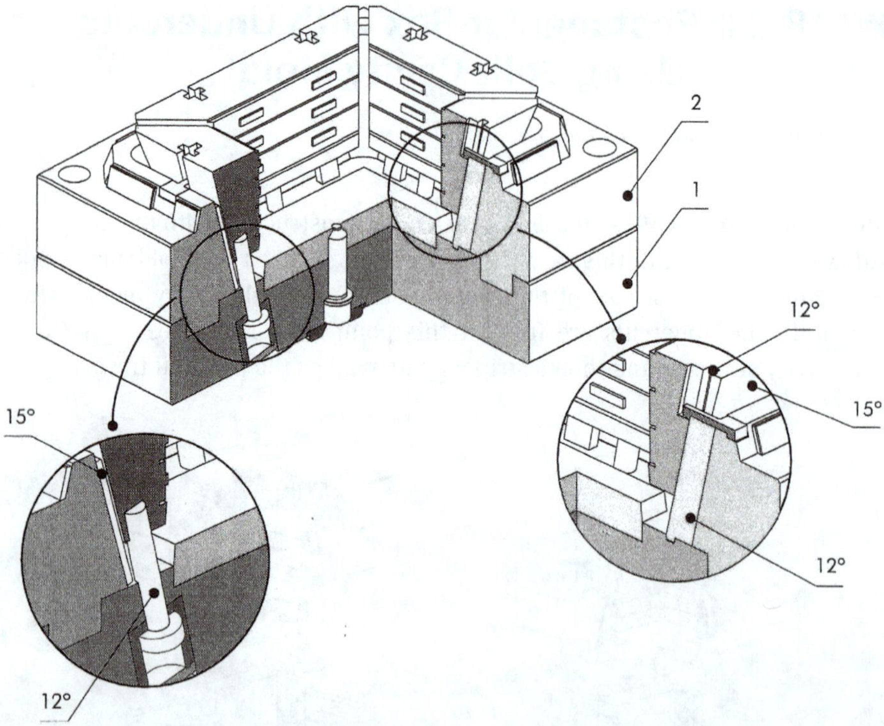

Figure 18.70 Rectangular box with undercuts: Ejection angles for split-cavity mold (Courtesy of Manual Do Projectista)

■ 18.22 Detergent Cap (Using Unscrewing Ejection)

Courtesy of StackTeck Systems

Unscrewing molds are normally used when there is a thread on the core that cannot be stripped off. An alternative would be to use a collapsible core, but this method is quite effective in removing deep-threaded parts and has been used successfully for many decades. The 32-cavity mold shown in Figure 18.71 is for a detergent cap which has deep threads under the skirt that cannot be stripped off.

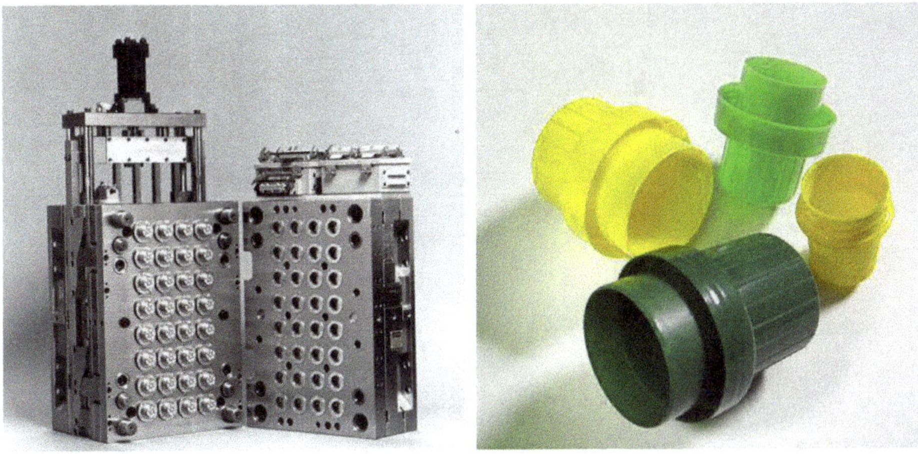

Figure 18.71 Detergent cap: 32-Cavity unscrewing mold - rack-and-pinion

Recently, servo motors have been used to replace the rack-and-pinion (see Figure 18.72). The servo motor provides much better speed and force control of the stroke, and it eliminates the superstructure needed for the rack above the mold.

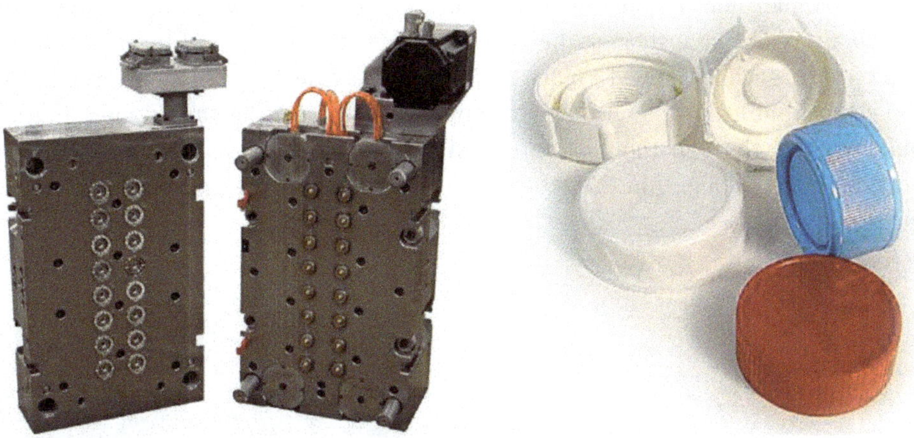

Figure 18.72 Detergent cap: Unscrewing mold for a threaded closure - servo unscrewing

The example mold shown in Figure 18.73 is a traditional four-cavity unscrewing mold, which uses a rack-and-pinion to eject the part. The part is made from 12–20 MI HDPE.

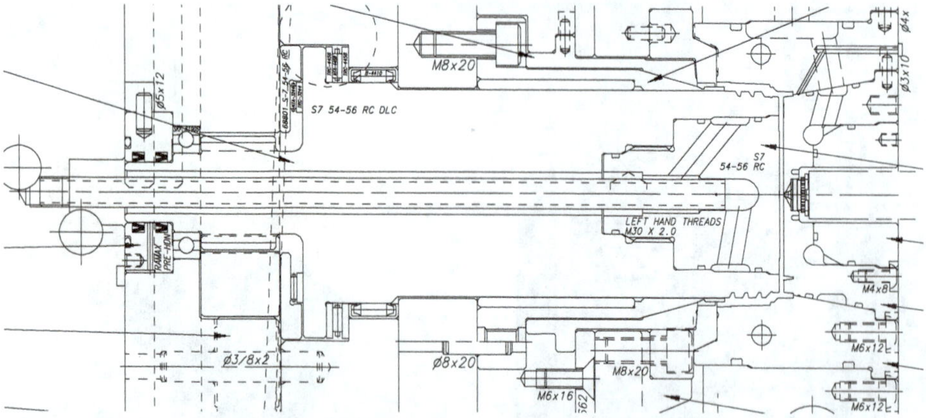

Figure 18.73 Detergent cap: Stack section of rotating core ejection

Cooling on the cavity side is conventional (Figure 18.73 and Figure 18.74), with two cooling circuits, one that is drilled around the cavity to cool the skirt of the closure, and the other that cools the bottom panel and the hot runner nozzle and gate area. Cooling on the core side is restricted due to the rotating motion of the core. In many cases, only a bubbler can be used on the core side. In this case, both a bubbler and a small cooling circuit in the core insert cool the part and thread. Since the core rotates, the water must be sealed from leaking in the core plate. In the section view of Figure 18.73 a double-V sealing can be seen to keep the water from leaking. The water in the core flows from the supply line, up the bubbler tube, through the cooling circuit at the top of the core, and down the outside of the bubble into the return line. The stripper ring and stripper sleeve have no cooling.

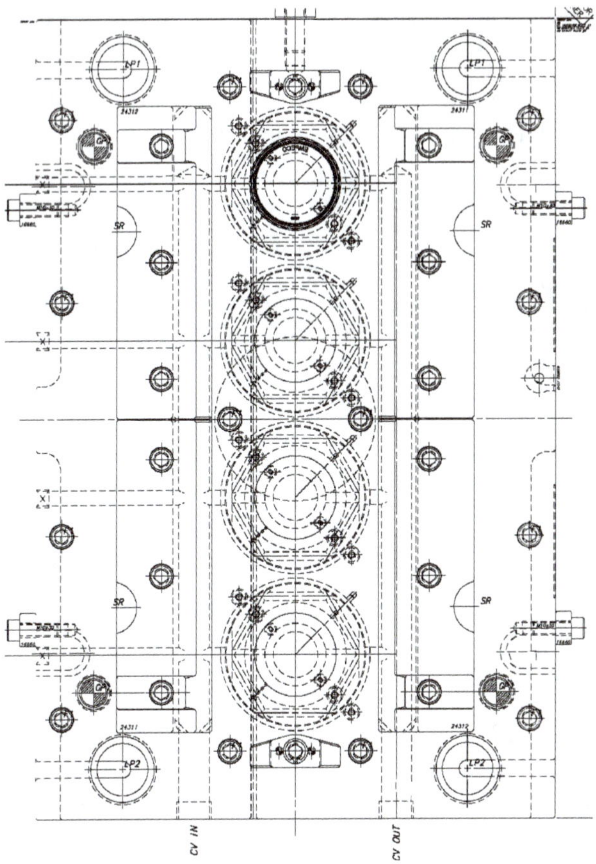

Figure 18.74 Detergent cap: Plan view of unscrewing mold cavity assembly

Ejection of the molded closures requires the coordination of several movements, which can be seen in Figure 18.75 and the stack section in Figure 18.73. The details of the design of an unscrewing mold can be found in Chapter 11. The ejection process starts with activating the hydraulic cylinder on the top of the superstructure containing the rack and cam bars. As the hydraulic piston extends, it drives the rack and cam bars downward. The rack engages with the pinions on the core and the core begins to turn. As the core turns, the stripper sleeve (which has teeth (or ratchets) in it that protrude into the part) prevents the molded part from turning. The stripper plate, stripper ring and stripper sleeve together advance on the cam bars (along rollers which can be seen in hidden lines), which have an angle on them that matches the lead angle of the thread profile. The stripper sleeve and stripper ring are held together for the first part of the stroke by spring force. After a given stroke, in this case 0.45", the stripper sleeve stops and the stripper plate advances (on its rollers) to eject the part off the teeth of the stripper sleeve.

Once the parts eject, the hydraulic cylinder must be allowed to reset to the top position before the mold is closed.

The core is supported in the core plate by two side bearings and a thrust bearing, which allow the core to turn but not move forward.

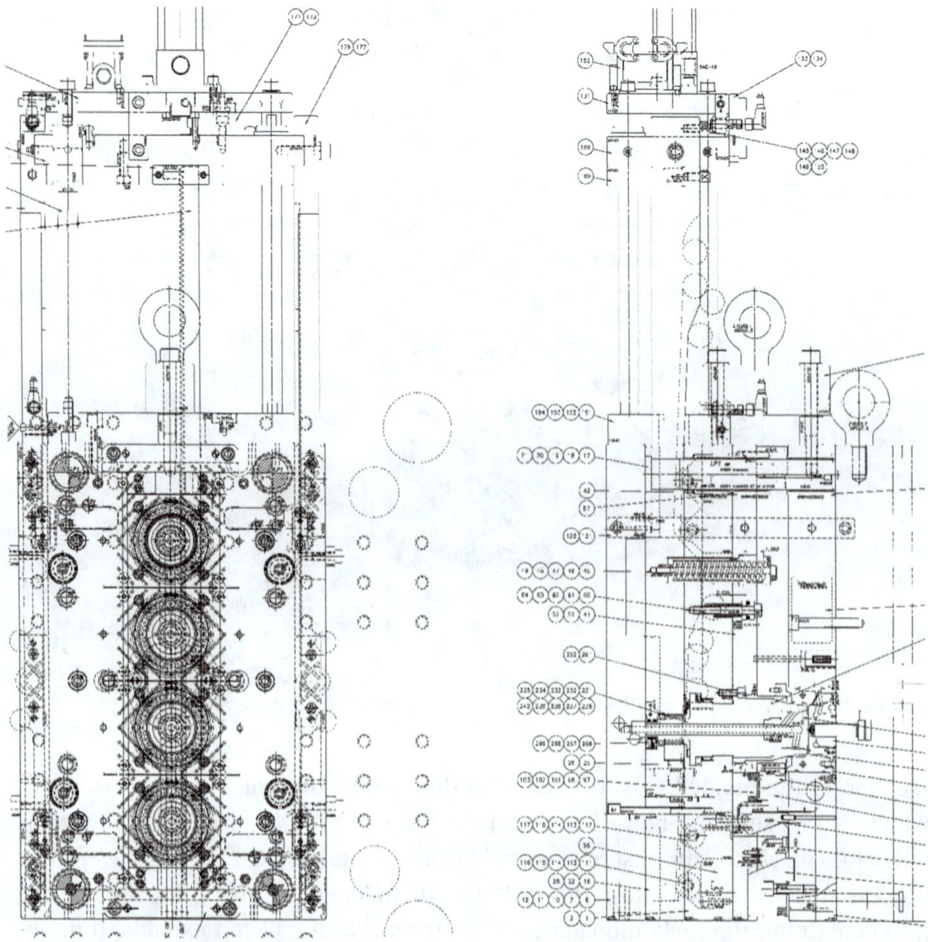

Figure 18.75 Detergent cap: Plan view of mold core and section assembly

18.23 Folding Crate

Courtesy of Haidlmair

This is essentially a five-piece family mold (Figure 18.76). All parts are molded in a single shot and then assembled by automation in a connected post-molding operation. The box is made from PPC and all the parts weigh 1650 g. The wall thickness is 2.4 mm, and it requires 900 tonnes of clamping force to mold a part. The cycle time is 40 s. There are many variations to folding boxes. This one is a simple design, as it has no need for any side movement in the mold.

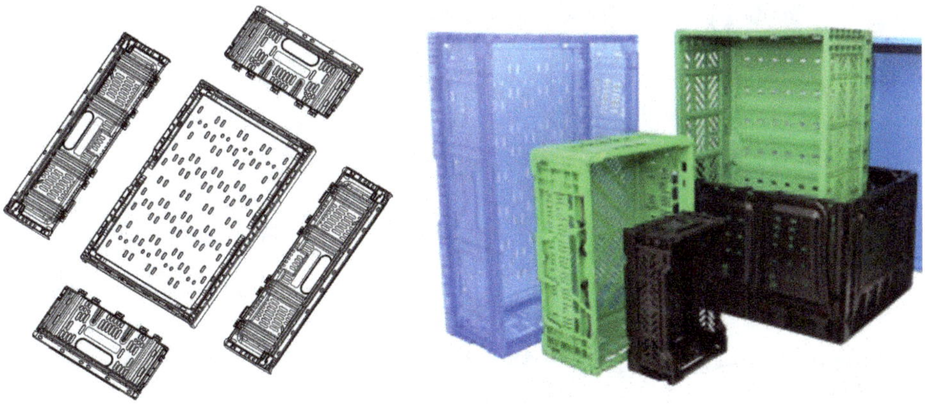

Figure 18.76 Folding crate: Part drawing

Figure 18.77 shows the core side of a folding crate mold with side core retractions in multiple locations on the parts. This is not needed in the case of the part design shown in Figure 18.76, but the principle of the mold is the same.

The two halves of the mold are aligned by sets of side interlocks on each side of the mold, as shown in Figure 18.78. Leader pins are used for assembly and additional alignment. Note that both the core side and the cavity side have a locating ring. This is due to the fact that the mold is loaded into the machine in two pieces, to allow each side to be properly located relative to the machine platens. Clamp slots and clamps are used to fix the mold to the platens.

The mold contains interchangeable rear-mounted inserts on the panels, to make the final product suitable for different types of fruits and vegetables. These inserts are made from high-conductivity materials to achieve better cooling of the five panels. The cooling on the core and cavity side must be controlled so that they are the same, to prevent warpage and twisting of the parts after molding. A CAE simulation analysis on the mold for cooling was essential. The cooling lines are drilled and plugged in the inserts in a pattern that creates uniform cooling.

Figure 18.77
Folding crate: Core side of mold

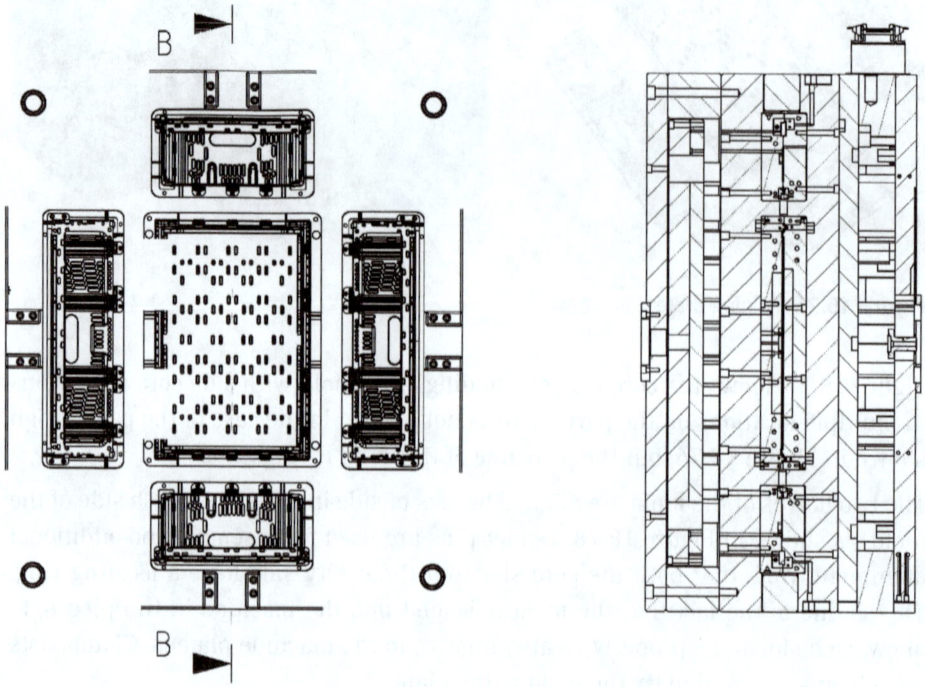

Figure 18.78 Folding crate: Core plan and section assembly views of mold

The part is ejected in a single-stage ejection motion using ejector pins. A robot enters the molding area to accept the part and lift it off the ejector pins. Return pins ensure that he ejector plate is back before the mold is closed. The ejector plate contains approximately 100 ejector pins. Since there are many holes in the part requiring shut-off between the cavity and core side, the mold must be very rigid and well-supported. The space in-between the ejector pin locations is used to provide support pillars between the core plate and the core backing plate. There are pillars in the hot runner plates between the hot runner manifold locations, to ensure that there is ample support for the mold to resist the injection pressure and tonnage. This prevents flashing into the many openings and windows in the mold parts.

The mold is filled with a 16-cavity valve-gated hot runner system. A 3D filling analysis was used to ensure that the parts fill evenly. Additional control of the filling is accomplished with a sequential valve gate filling capability.

The most critical part on the mold are the hinges and venting inserts. These areas are subject to high wear, and so separate replaceable inserts made with hardened steels are machined into these areas. This allows the molder to replace and repair the inserts without significant downtime.

Note the recessed areas on the side of the mold in Figure 18.79, where hose fittings are located. It is good practice to recess these areas so that the fittings are not damaged when the mold is inserted into the press or being maintained.

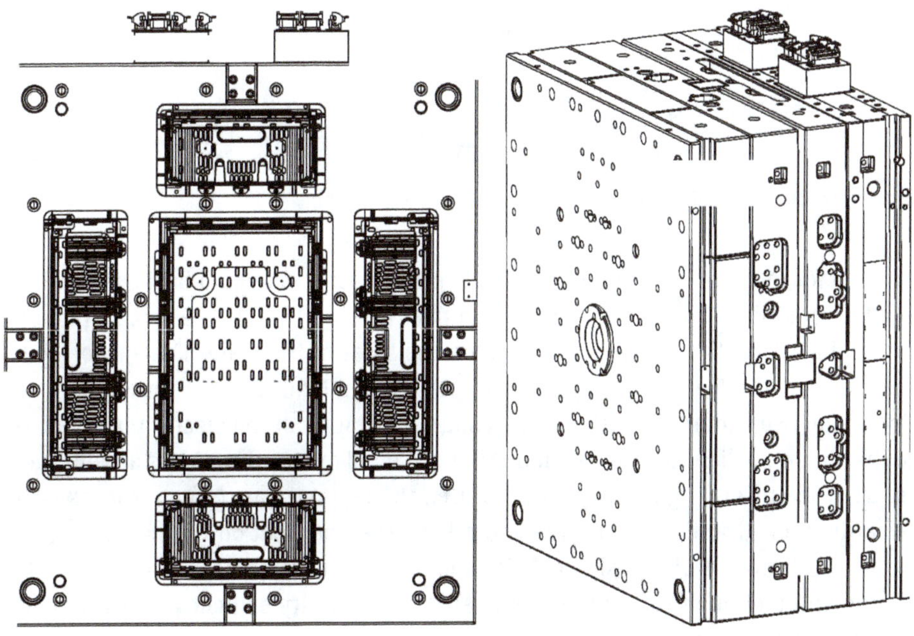

Figure 18.79 Folding crate: Cavity plan view and isometric view of mold

18.24 Automotive Panel Cover

Courtesy of Plasdan

Cooling of the part is achieved with a series of baffle cooling channels on the core side and a series of horizontal drilled circuits on the cavity side (Figure 18.80). The cooling circuits are separated by internal plugs on the cooling lines, which can be seen between the drilled baffle holes. The baffles are held in by threads at the back of the baffles (see Chapter 12 on cooling). Additional cooling is added to the light blue sliders to provide cooling in this area.

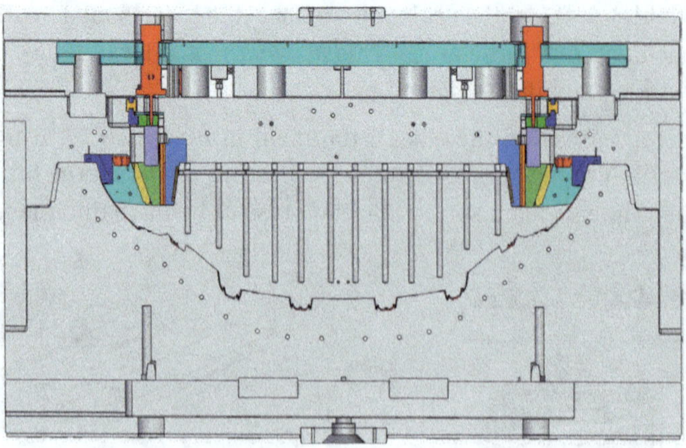

Figure 18.80 Automotive panel cover: Section view of assembly

The ejector pins are mounted into a single-piece ejector plate and backed-up by set screws that allow the pins to continue to float (Figure 18.81). The ejector plate is driven by hydraulic cylinders mounted on the sides of the plate. The ejector plate is guided by the leader pins (dark blue). A separate hydraulic circuit and pistons drives the two sliders. This allows for a slight hesitation of the ejector pins relative to the sliders, which can aid ejection. After the mold opens, the sliders are engaged to drive the insert (light blue) inwards (i.e., the tapered part (green) moves back towards the hydraulic cylinder (red) to relieve the undercut on the part). There is no forward motion of the slider. It is held by the insert (red) in a dovetail. The timing and motion of the sliders is signaled by proximity switches shown in dark blue between the yellow piece.

The two inserts were added to the top of the cavity to control venting and for ease of polishing in this area. It allows the molders to change or repair these inserts if needed. There are "tonnage blocks" added to the parting line and on the tapers, which allow the molder to fine-tune the mold venting and alignment if needed during operation. These are becoming more common, as they provide the molder some ability to easily modify the venting and shut-off if needed. These blocks also prevent hobbing of the parting line and of the vents, which can occur over time.

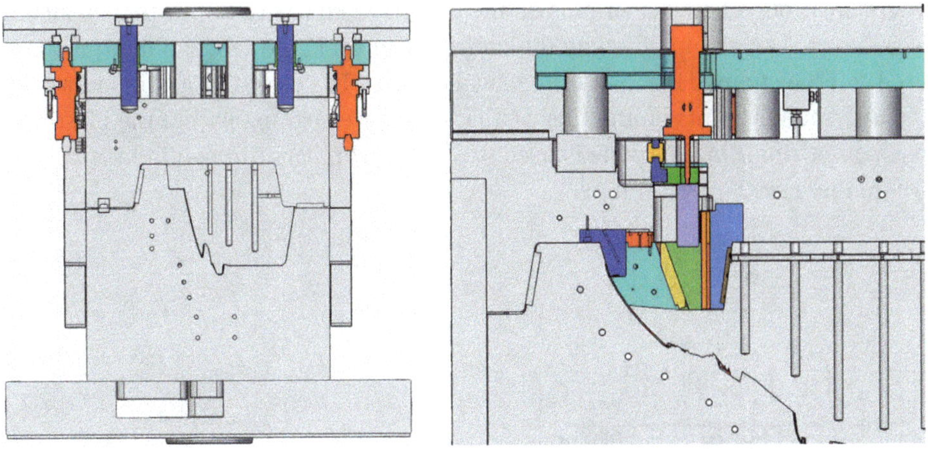

Figure 18.81 Automotive panel cover: Undercut removal detail

18.25 Pallet

Courtesy of Haidlmair

The pallet is made from HDPE and weights 7950 g (Figure 18.82). It is molded in a 1500 tonne injection molding machine at 48 s cycles. The mold weighs 15 tonnes and is loaded into the press in two parts. A locating ring on both sides of the mold helps to locate the two mold halves into the press separately.

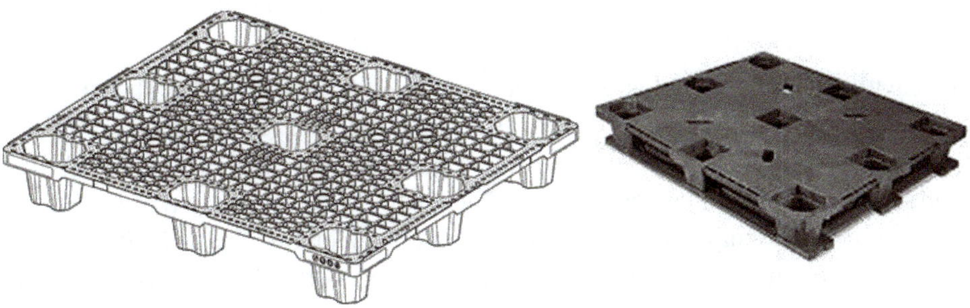

Figure 18.82 Pallet: Part drawing

Straight push-pull ejection from the machine ejection system is used to drive the ejector plate and ejector pins. Five locations are used on this mold. Optional locations for hydraulic cylinders (two on top and two on the bottom) are added in case the mold is mounted into other machines where the ejector positions don't line up. There are around 150 ejector pins on the mold (see Figure 18.83), with a diameter of 3–5 mm. Larger pins become difficult to keep cool. The larger 5 mm pins are used to protect any deflection of the smaller pins. The pins are mostly located at the intersection points on the ribs, so that they have a strong part of the plastic to push on. A single ejector plate is used with support pillars through it, in locations where there are no ejector pins.

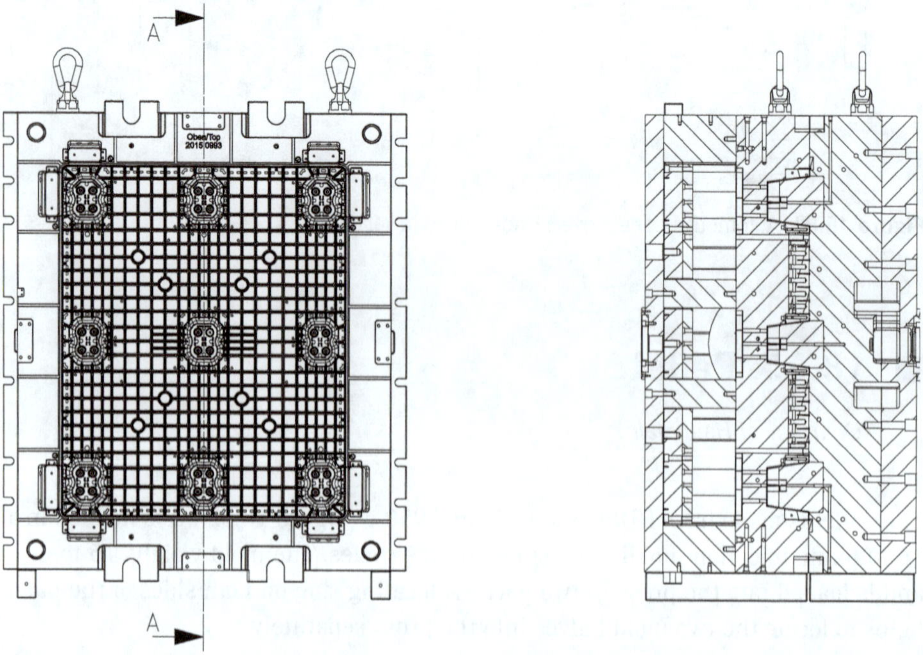

Figure 18.83 Pallet: Plan view of core and section view of mold assembly

The cavity side of the mold is mostly flat. Cooling lines are drilled in the cavity to create a uniform cooling pattern. The core side of the mold contains deep ribs for part strength. Baffle cooling is used liberally throughout the core side to bring cooling up between the ribbed sections. The cooling is designed so that each cooling circuit sees a temperature increase (ΔT) of less than 5 °C. Inserts with cooling are added around each of the nine feet on the pallet, as this is the thickest area of the part.

There are 16 gates on the part, fed by a hot runner system (see Figure 18.84). Venting is added to the bottom of the nine feet to evacuate air in these areas. Note that the hose fittings are recessed into the plates to protect them against damage during mold maintenance and installation into the machine. Mold feet are added to the bottom of the mold for ease of handling and storage. Alignment is accomplished with four large side interlocks on the mold.

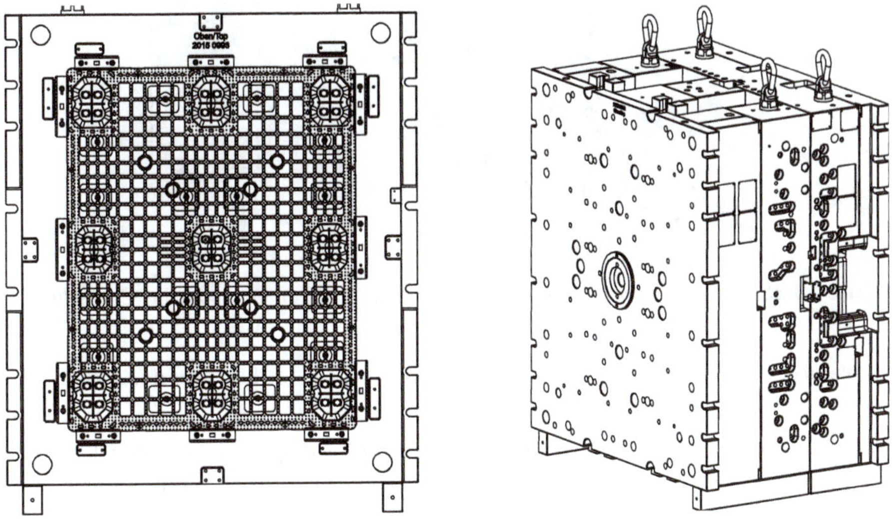

Figure 18.84 Pallet: Plan view of cavity and isometric view of mold

18.26 Automotive Covers

Courtesy of Plasdan

This two-cavity mold (Figure 18.85) produces a left-hand and right-hand side cover for an automotive application. Ejection of the part is accomplished with a side core retract of the pins (dark blue) and an ejector plate with ejector pins and lifters to free the undercuts (see Chapter 11 on ejection for more details). As the mold opens, the pins (dark blue) are retracted by a hydraulic cylinder (not shown). When the mold is fully open the ejector plate is engaged, moving the ejector pins and lifters forward. At the full stroke, the part is free from the core and from the undercut. It can now be removed by a robot. The ejector plate is returned by springs (shown) and return pins.

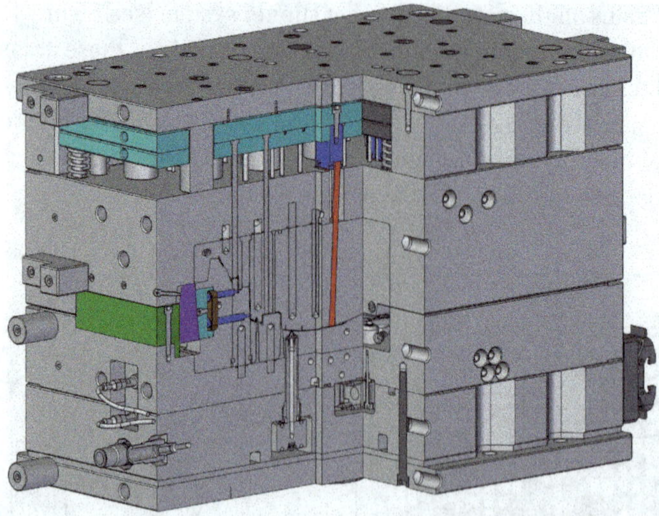

Figure 18.85 Automotive covers: Section view of mold for left-hand and right-hand side cover

There are several pins on both the core and cavity side that produce holes in the part and provide natural venting (Figure 18.86). Sometimes these inserts are also produced to ease manufacturing and polishing of the molding surfaces by providing more access to enable machining of fine tolerances and to polish molding surfaces. Cooling of the part is achieved by using a combination of drilled cooling lines (on the cavity side) and drilled lines with baffles (on the core side).

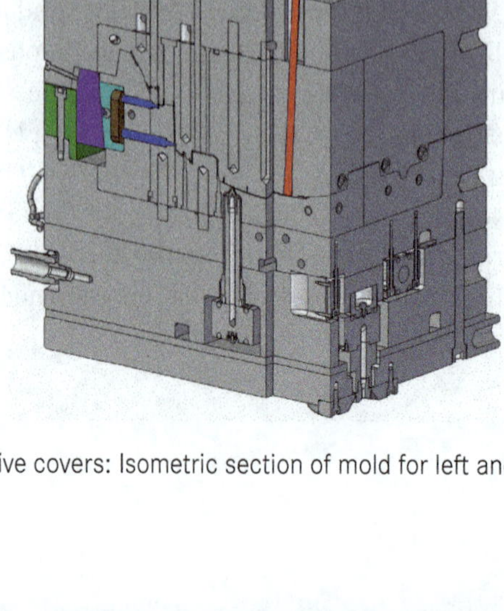

Figure 18.86 Automotive covers: Isometric section of mold for left and right covers

18.27 Double-Walled Threaded Cap (Using Collapsible-Core Mold)

Courtesy of iMFLUX

This double-walled PE cap mold runs at 10 s cycles with 25 tons of clamp force. When the undercuts on the core side require retraction, a collapsible core design can be used to effectively draw the steel out of the undercut before the part is ejected. In the case of this single-cavity double-wall threaded cap, the core is advanced by 50 mm in the first stage of ejection to allow the undercuts to move out of the molded part. A Rohr retractable core system is used to move the four collapsing sections inwards using a dovetail system (see plan view of core in Figure 18.87). The base of the collapsible core is held by two concentric rings. There is enough float between the rings that the collapsible sections can move inward with the angled dovetail. The part is then ejected off the core with a 15 mm stroke of the stripper plate. An external latch lock system controls the movement of the two ejection sequences. The section view of the assembly in Figure 18.88 shows the stripper and collapsible cores in the forward position. The four sections that collapse as the part is ejected can be seen in detail G on Figure 18.87. The Rohr collapsible core element can be seen in Figure 18.89. No ejector box is needed on this single-cavity version. The production mold required an ejector box to match the machine ejectors to the mold ejector locations.

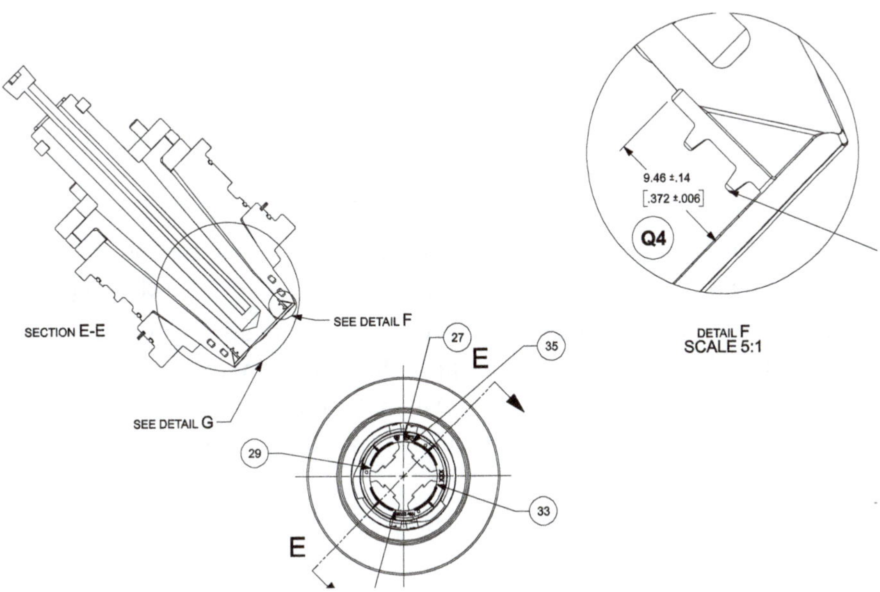

Figure 18.87 Double-walled threaded cap: The double-walled threaded design with undercuts requires a collapsible core to eject the plastic part

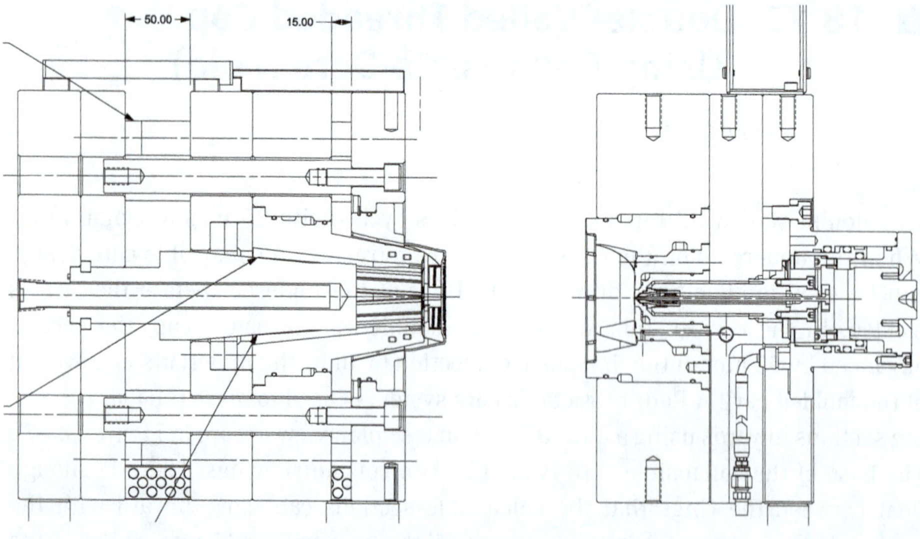

Figure 18.88 Double-walled threaded cap: Section view of single-cavity mold with collapsible cores

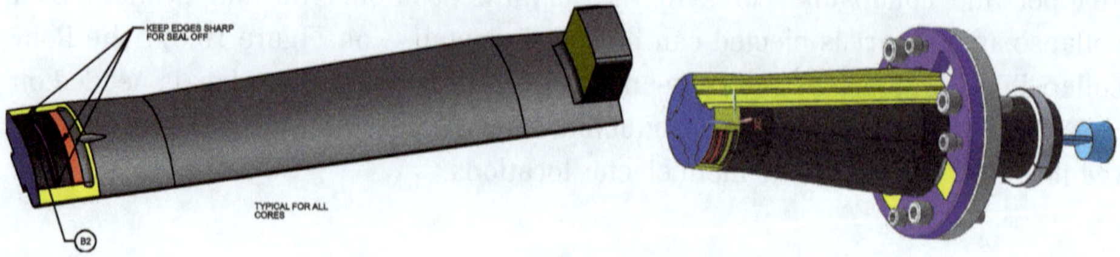

Figure 18.89 Double-walled threaded cap: Collapsible core element for the single-cavity mold

A single-cavity valve-gate hot runner is used to convey the melt into the cavity. The stroking air piston is directly over the center of the part, and the melt is diverted around the piston, and then into the nozzle and gate area.

Cooling to the core is accomplished with the use of conformal cooling channels in the core, as the retractable core inserts make cooling to the core a challenge.

Since this is a fairly thin-walled part, the core and cavity are aligned with a taper on the stripper ring and cavity. The stripper ring is held in the stripper plate with a retaining ring, and floats in the plate to align with the cavity and the core. Both the core and the cavity are heeled into their respective plates for more rigid support.

■ 18.28 Automotive Cover

Courtesy of Plasdan

This mold (Figure 18.90) is designed to mold an automotive cover. The part is ejected with ejector pins (in yellow) as well as lifters on the side of the part. Both the pins and the lifters are activated by the ejector plate. The inserts on the end of the lifters are cooled through the lifter mechanism to keep them from getting hot, and so avoid an extended cycle time. A bubbler inside the lifter provides a cooling circuit that returns the cooling to the fittings at the base of the lifter. A yellow key holds the insert to the lifter. Note the long ejector stroke needed to free the long horizontal undercut on the part. Angles of more than 15° on the lifters can cause them to break (see Chapter 11). The lifters are standard commercially available parts from mold part supply houses.

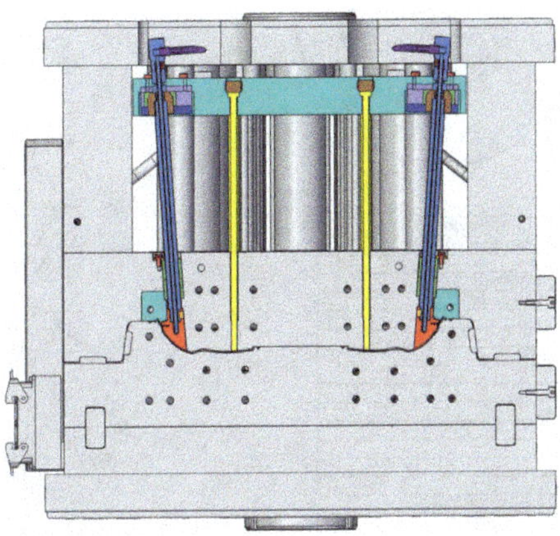

Figure 18.90 Automotive cover: Section view of mold

Removable inserts (light blue) were added onto the core to provide additional cooling and the ability to maintain and service these inserts.

On the parting line and the taper there are tonnage blocks that allow fine-tuning of the shut-off areas, adjustment of the vents, and help to prevent the mold from hobbing on the parting lines.

18.29 Externally Threaded Cap Mold

Courtesy of iMFLUX

This externally threaded cap (Figure 18.91 and Figure 18.92) has deep undercuts on the cavity side and a thread requiring a cavity that splits before ejection of the part. The two halves of the cavity are moved open by horn pins during mold opening, freeing up the threads (Figure 18.93). A stripper sleeve is then advanced to strip the part off the core. The part is cold runner gated with a 0.020" gate. The runner is ejected using the external latch lock system.

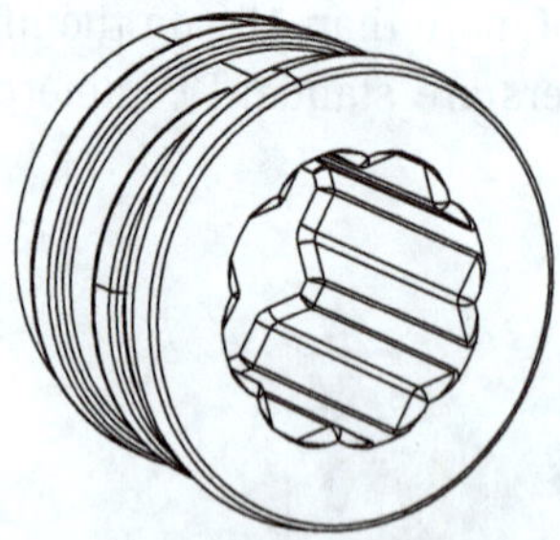

	SPECIFICATION
	8 CAVITY SECOND CAP MOLD
	STERIPACK
	GATES SECOND CAP V23
	2008COR-DD-second cap-01
	-
	-
	18.856 x 12 x 11.875
	DIRECT
	.020"
	.110"
	.56"
	A-SIDE - 550 lbs / B-SIDE - 240lbs
	8
	.016 in/in
	SPI-B2
	SPI-B2 / SPI-A2(3)
	.0005"
	0.14g / TOTAL - 3.24g

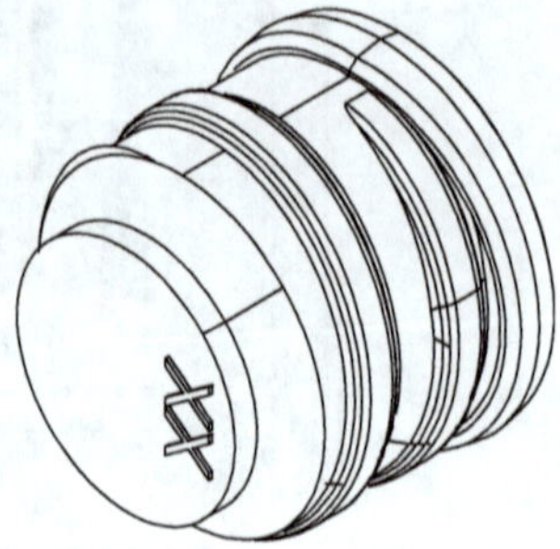

Figure 18.91 Externally threaded cap: Part drawing. Note the deep threads and cavity-side undercuts

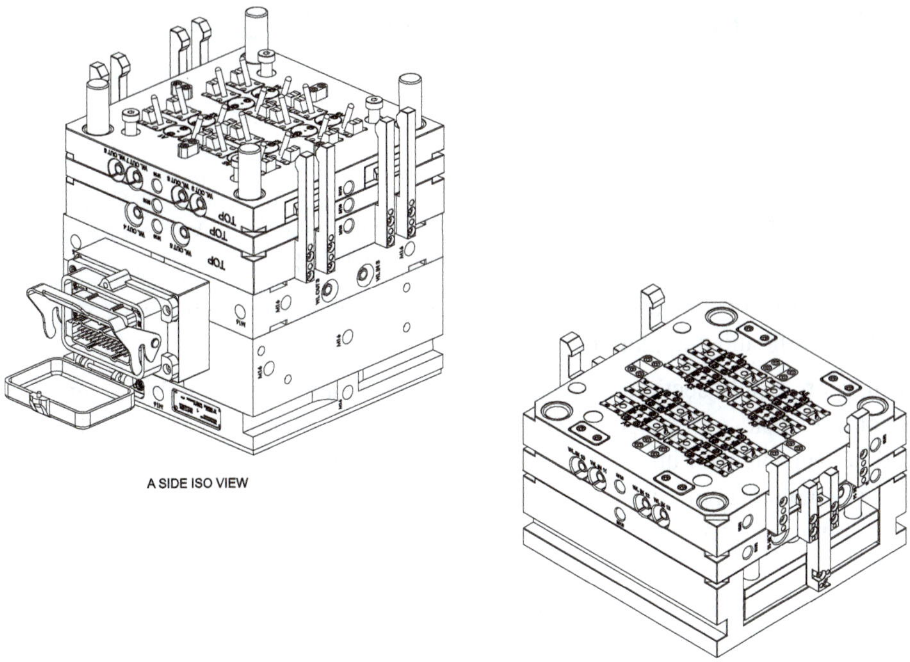

Figure 18.92 Externally threaded cap: Isometric views of mold assembly

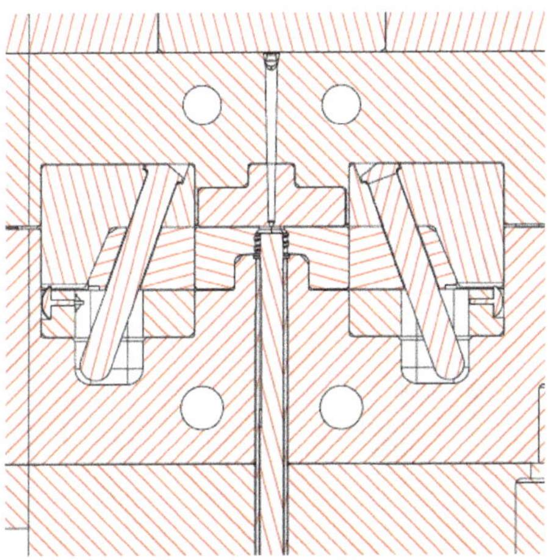

Figure 18.93 Externally threaded cap: Section view of mold stack

18.30 Oxygen Mask (with Side Core Retraction)

Courtesy of Plasdan

This four-cavity mold design is for an oxygen mask made from a very flexible PP and a TPE material similar to that shown in Figure 18.94.

The part is reverse-gated so that the cores are on the fixed injection side and the cavity is on the moving platen side.

The oxygen supply section in the mask requires a side core retraction before the rest of the mask can be ejected with the ejector plate. On the cavity side, pins are used to keep the part on the core. The ejector plate is internally spring-returned. Once the mold is open, the ejector mechanism on the core side is actuated and the part is removed by a robot.

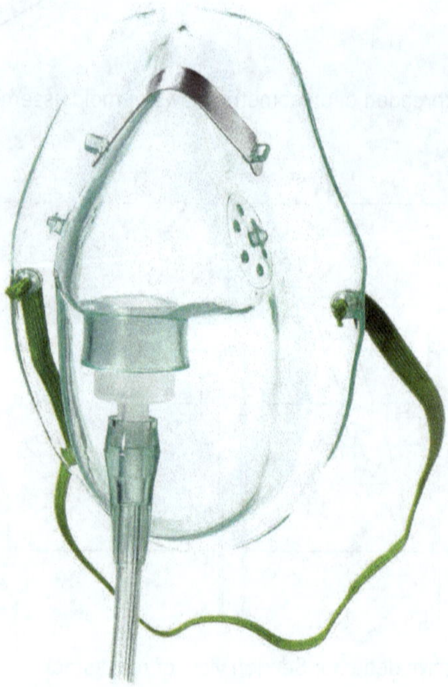

Figure 18.94 Oxygen mask: Typical product, similar (but not the same) to the one made by the mold described

The TPE is fed to the cavities with a hot-to-cold runner system. The tunnel-gated cold runner is ejected with the part. The PP part is fed with a hot runner system. An auxiliary secondary injection unit is mounted to the mold to inject the TPE material.

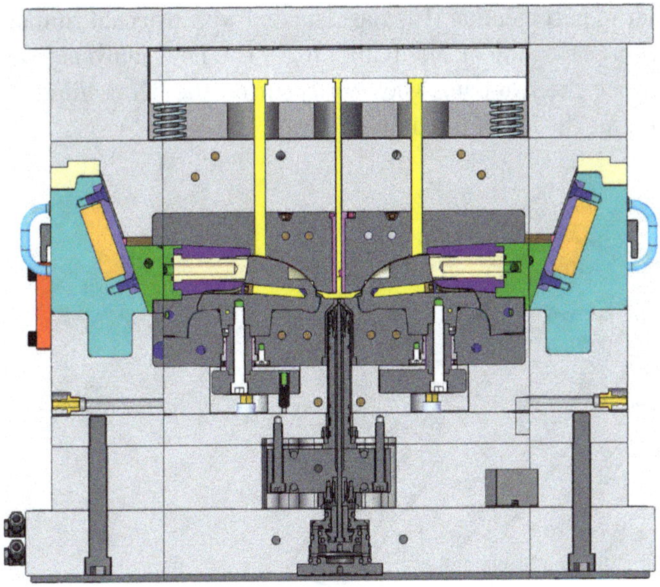

Figure 18.95 Oxygen mask: Top view section assembly

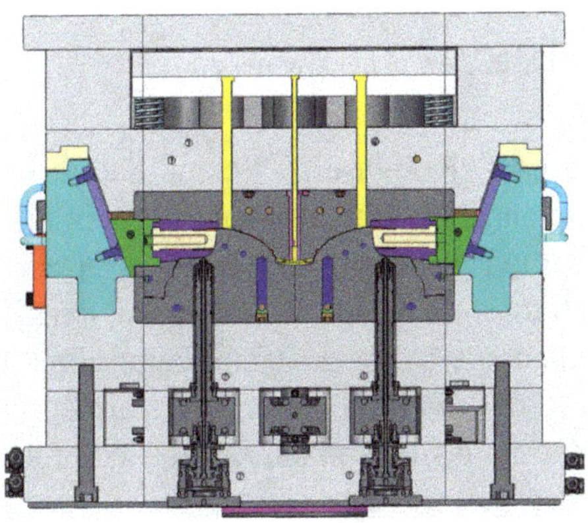

Figure 18.96 Oxygen mask: Side view section assembly

18.31 Living-Hinge Case

Courtesy of iMFLUX

The hinged case shown Figure 18.97 has many undercuts and holes that need core retraction prior to part ejection. The part is gated with two cold runner submarine gates per part on each side of the living hinge. The flow analysis reveals that the weld line between the two gates goes not occur on the living hinge. This is critically important for the life of the living hinge.

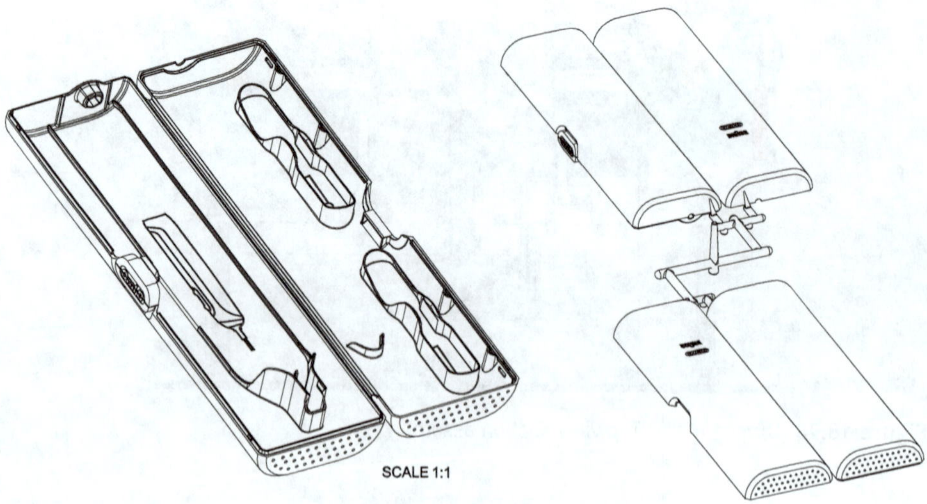

Figure 18.97 Living-hinge case: View of cold runner and part

As can be seen in Figure 18.98, Figure 18.99, and Figure 18.100, there are many core retractions in the mold. There are also some pins in the part that form holes. The horn pin shown in the right of the section view releases the closing snap on the case. These are retracted using horn pins on the cavity side. A lifter system from Progressive Components (www.procomps.com; Lifter Link™) was used for the lifter driver systems, but the lifters themselves are custom-manufactured. All of the lifters are driven from a single plate. A three-axis robot lifts the parts off of the lifters, as the part does not free-fall.

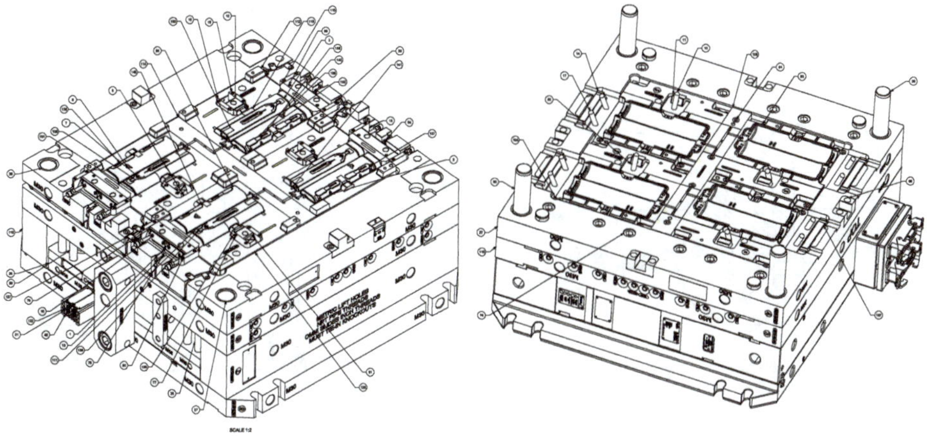

Figure 18.98 Living-hinge case: Isometric views of the core and cavity of the four-cavity mold

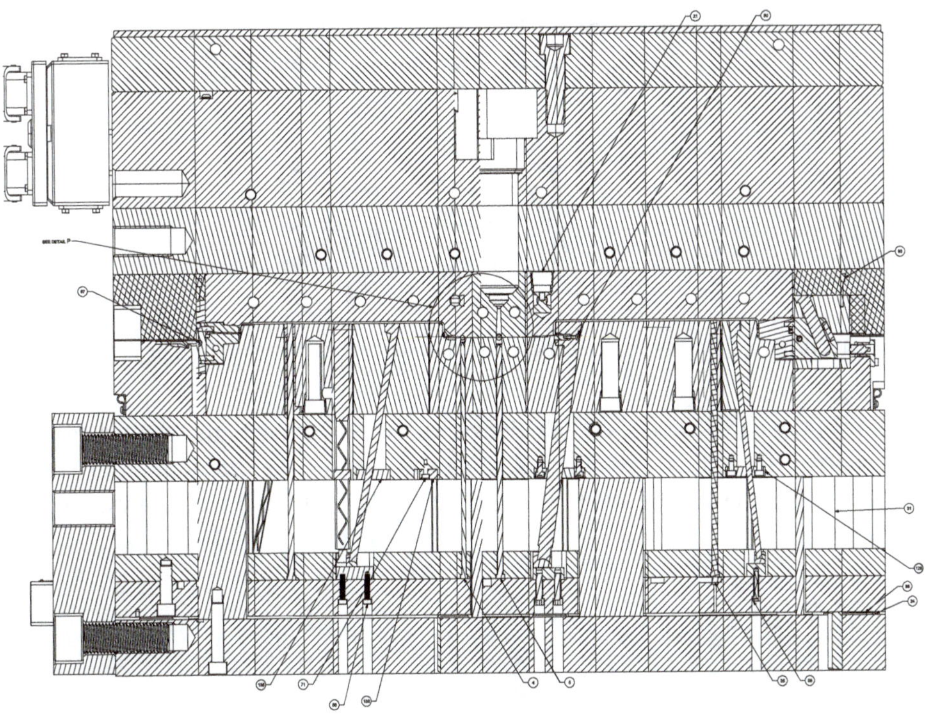

Figure 18.99 Living-hinge case: Section assembly showing ejection mechanism for four-cavity mold

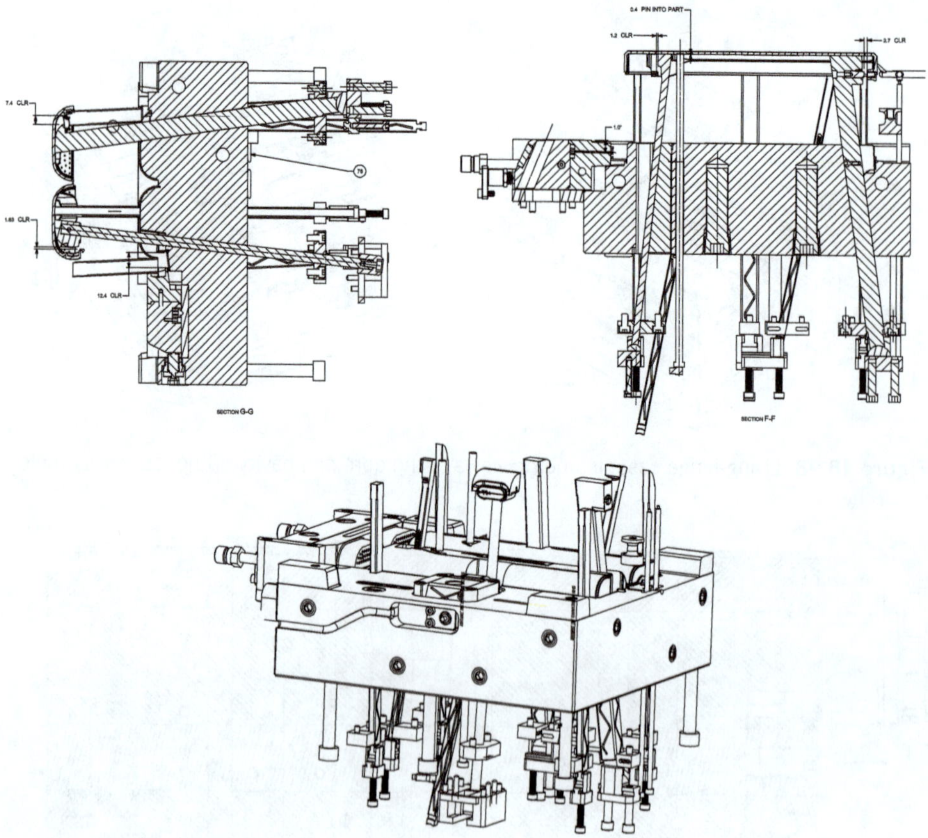

Figure 18.100 Living-hinge case: Section view showing undercut removal actions

Cooling of the core side is a real challenge with this many actions. Limited cooling is permitted (three baffle cooling holes) on the core, as the lifters themselves are not cooled. The result is a mold that runs much slower than if the core was cooled thoroughly.

18.32 Iron Housing (Using Split-Cavity Mold)

Courtesy of Plasdan

In order to eject this part (Figure 18.101) the cavity must be split into two sections. These sections are preloaded against the forces from the injection pressure to prevent flashing along the split lines.

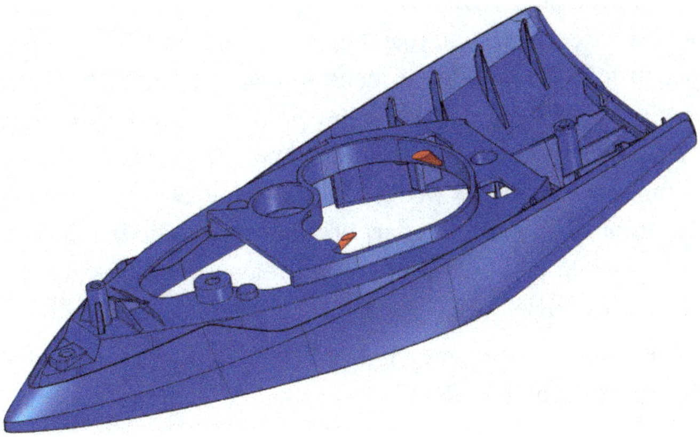

Figure 18.101 Iron housing: Part drawing

Figure 18.102 provides a simple reference of a cup mold with a handle to show the principle of how a split-cavity mold is built and functions.

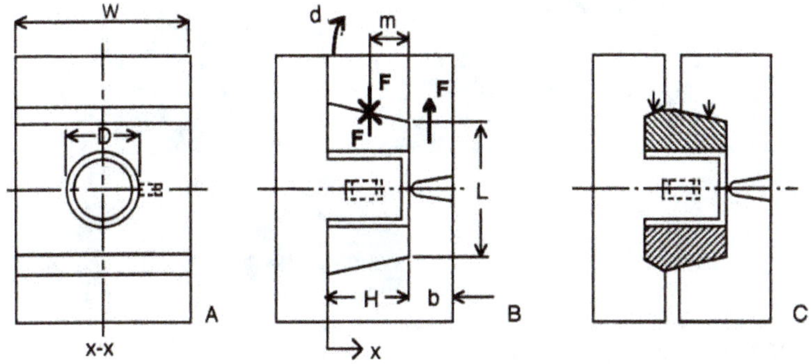

Figure 18.102 Example of a split-cavity mold (for a cup with a handle) to show how the iron housing mold functions: (A) Plan view into the cavity; (B) Section through a mold with wedges on the cavity half only; (C) A similar mold, but with wedges on both cavity and core sides. *W*, width of the plates; *L*, length of stretched cavity plate; *b*, thickness of cavity plate along *L*; *H*, height of cup; *D*, cup diameter; *F*, the forces to be contained

As the mold opens, the cavity "splits" move for a short distance with the core, while the splits open sideways. Only then can the housing be ejected. With the closed mold, during injection, the injection pressure p inside the cavity acts on the projected area of the sides of the housing, $F = p \times D \times H$. In mold B, the force F pushes against the wedge, which is part of the cavity plate and is counteracted by the steel of the cavity plate, with a cross-section of $b \times W$.

There are now two challenges to overcome: (1) The force F will stretch the portion of the plate with a length L, and create an undesired gap at the split line. The wedges must therefore be preloaded as explained in Chapter 11. (2) Because of the distance m between the forces and reaction forces, there will be a bending moment $m \times F$, which will force the wedge to bend outward as indicated (arrow d). This system is therefore only suitable for shallow products. For deep products, the side forces must be taken up on both the cavity and core sides of the mold. This is illustrated by mold C, which has wedges both in the cavity and the core side. The forces F trying to push the halves apart are thereby divided, and both cavity and core plates will provide reaction forces. The preload must be calculated and provided for each set of wedges.

Figure 18.103–Figure 18.105 show views of the iron housing mold that use the principles discussed in the boxed section above.

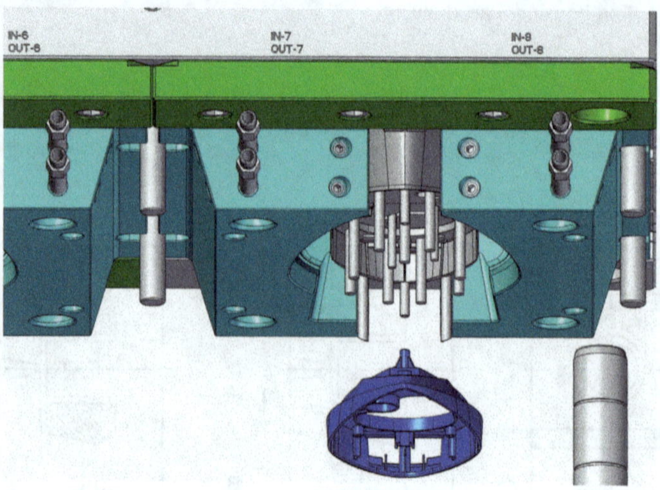

Figure 18.103 Iron housing: Mold ejection view

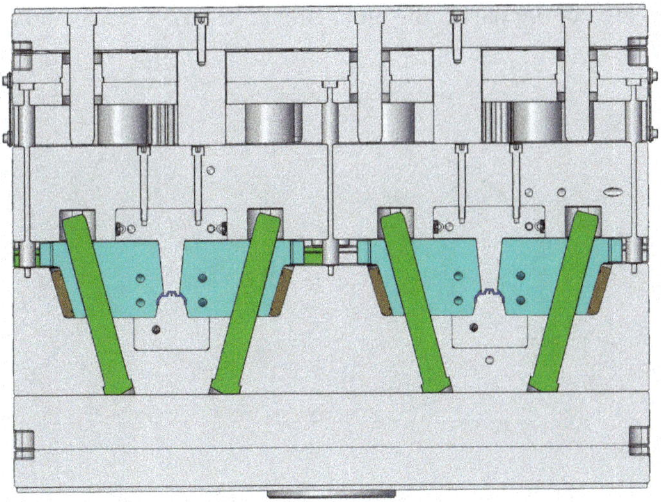

Figure 18.104 Iron housing: View of mold section assembly (1)

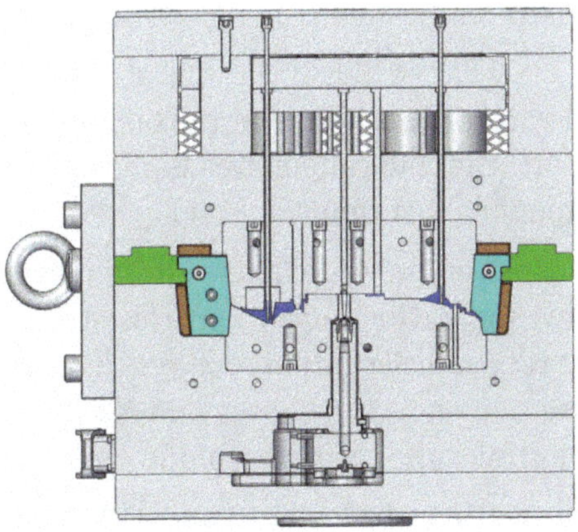

Figure 18.105 Iron housing: View of mold section assembly (2)

18.33 Oral Care Product (Using Split-Cavity Mold)

Courtesy of iMFLUX

This internal part (Figure 18.106) is made from Hostaform® copolymer acetal (POM). It runs in a 400 ton press due to the size of the tool, but only requires 30 tons of closing force due to the orientation of the part in the press and the limited projected area of the parts. The mold runs in the 20 s range and the parts are removed with a robot.

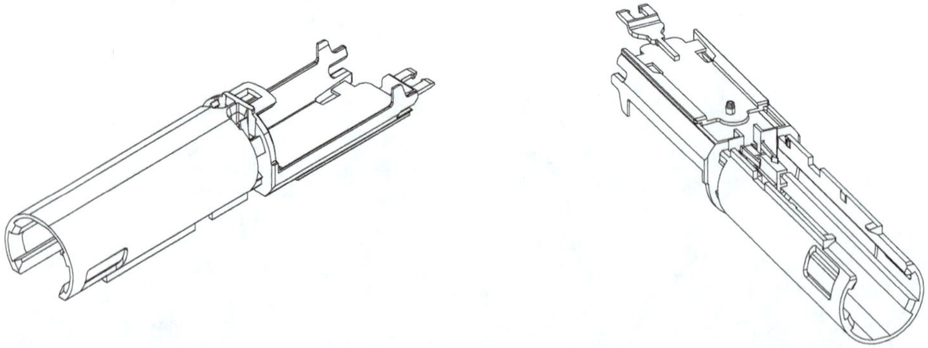

Figure 18.106 Oral care product: Part drawing

This mold contains many complex geometries, and it requires a split-cavity configuration to free up the part for ejection. The mold also has two core sides to make the internal geometry. The parting line for the tool is at the center of the part.

The part is cold runner gated. The cold runner runs up through the cavity side core into the center of the part. The cold runner area is cooled by a cooling sleeve. Both parts are fed by a conventional hot runner system.

The cavity-side action is driven by sets of angled horn pins that are mounted onto the core side and drive the cavities open and closed. A 60 mm ejector stroke connected to a stripper ring and ejector pin removes the part from the core for the final step in the ejection sequence. Latch locks on the side of the mold are used to control the limits of the stroke and the return of the ejector plates.

A pin (shown in Figure 18.107 by balloons 19 and 313) holds the part on the core during ejection until the part is pushed off the core by a long (59 mm) ejector pin in the final stage of ejection and into the robot.

Each of the two cavity blocks is cooled with its own circuit. Due to the movement in the mold, many of the parts of the mold are fed with hoses as opposed to internally with the plates. The two cores are cooled by spiral cooling circuits (see Figure 18.109 and Figure 18.110).

The mold is filled and packed using an innovative process called iMFLUX®. This constant low-pressure filling process allows the part to be filled and packed at the same time. There are tremendous benefits to the iMFLUX process for many injection-molding applications. The reader is encouraged to the visit *www.imflux.com* to learn more about how this process could help improve productivity, consistency, and part quality. The sensors shown on the mold are used by the iMFLUX process to monitor the pressures.

Alignment of the mold is through four plate interlocks on the face of the core and cavity plates. The latch locks shown on the sides of the isometric views are for the cold runner plate pullover.

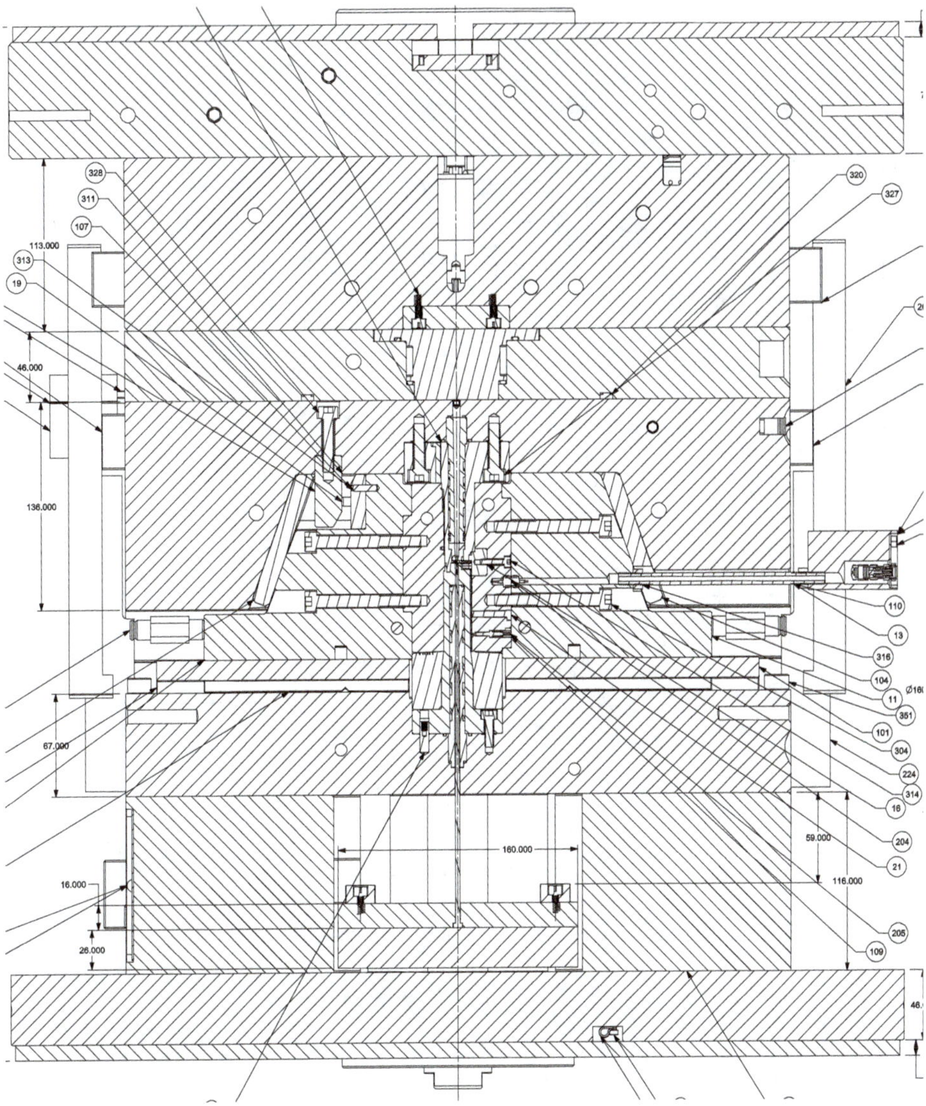

Figure 18.107 Oral care product: Section assembly for eight-cavity mold

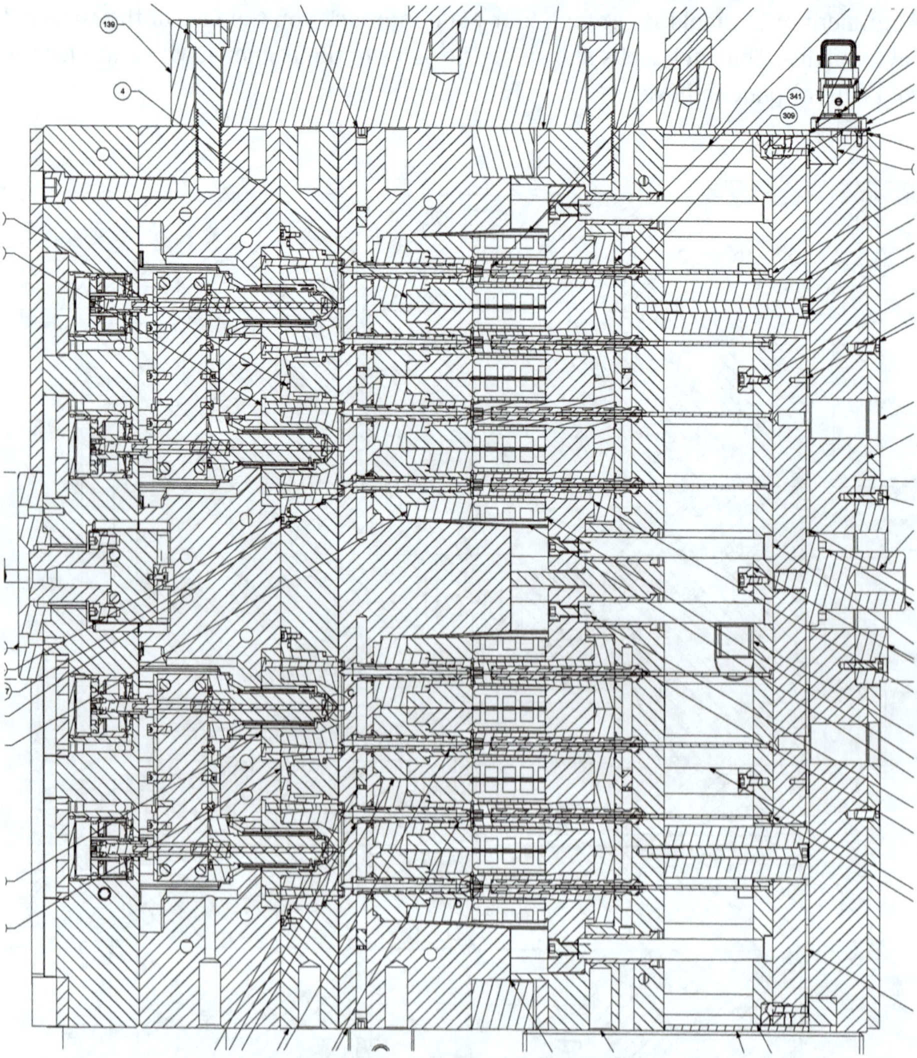

Figure 18.108 Oral care product: Section assembly for eight-cavity mold

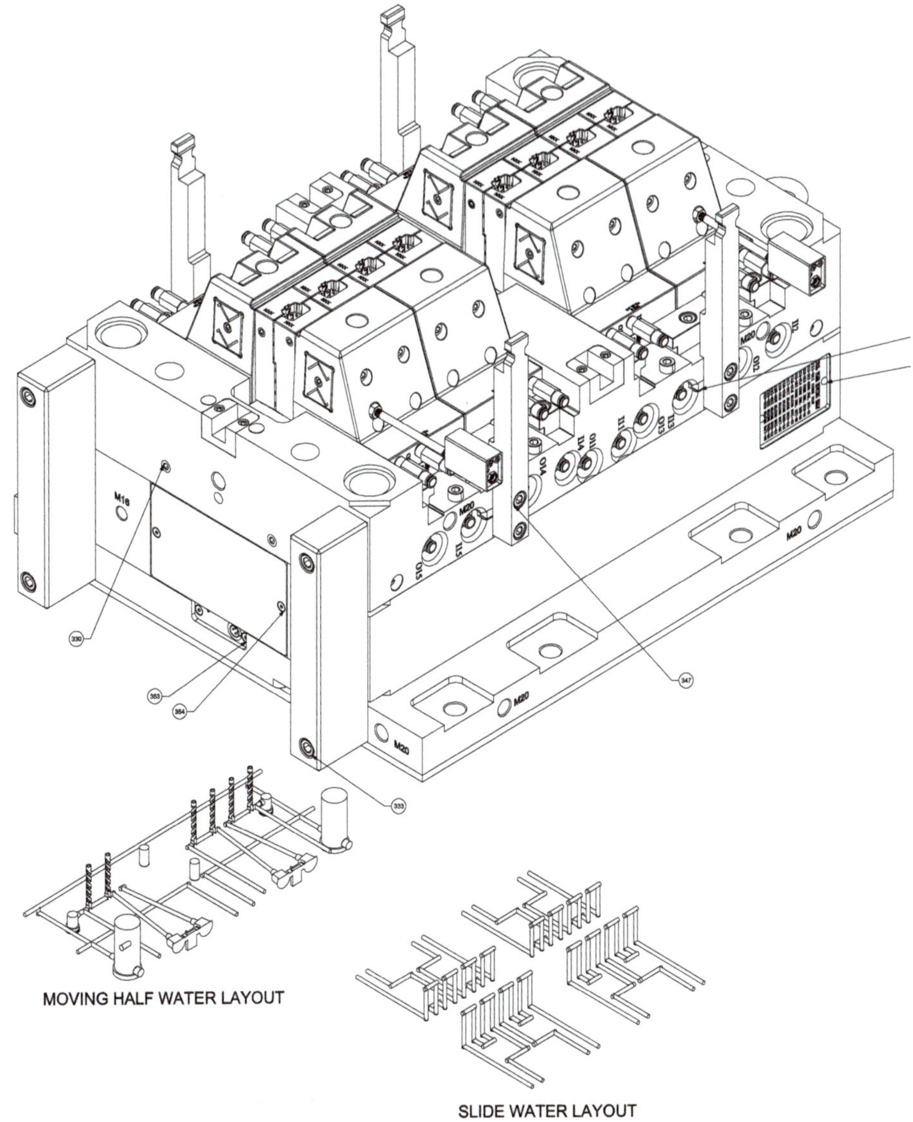

Figure 18.109 Oral care product: Isometric view of core side

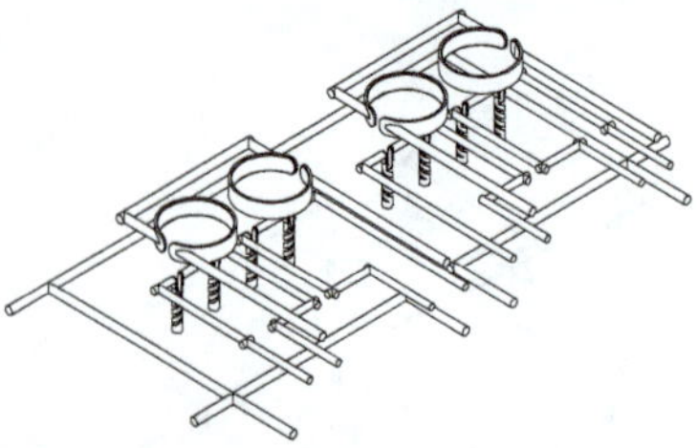

FIXED HALF WATER LAYOUT

Circuit #	Description	Smallest orifice size (mm/inch)	Fitting	Supply Temperature (°C/°F)	Flow rate (L/min / GPM)
I01/O01	Fixed clamp plate	9/.354	Xintech 09NA2100XS	90/194	11.5 / 3
I02/O02	Hot runner plate	9/.354	Xintech 09NA2100XS	90/194	11.5 / 3
I03/O03	Hot runner plate	9/.354	Xintech 09NA2100XS	90/194	11.5 / 3
I04/O04	Floating plate	9/.354	Xintech 09NA2100XS	90/194	11.5 / 3
I05/O05	Floating plate	9/.354	Xintech 09NA2100XS	90/194	11.5 / 3
I06/O06	Cavity plate	9/.354	Xintech 09NA2100XS	90/194	11.5 / 3
I07/O07	Cav. Ins. N41-42	1.53/.06	Xintech 09NA2100XS	90/194	1.5 / 0.4
I08/O08	Cav. Ins. N43-44	1.53/.06	Xintech 09NA2100XS	90/194	1.5 / 0.4
I09/O09	Cav. Ins. N45-46	1.53/.06	Xintech 09NA2100XS	90/194	1.5 / 0.4
I10/O10	Cav. Ins. N47-48	1.53/.06	Xintech 09NA2100XS	90/194	1.5 / 0.4
I11/O11	Core plate	9/.354	Xintech 09NA2100XS	90/194	11.5 / 3
I12/O12	Core Ins. N41-42	1.93/.076	Xintech 09NA2100XS	90/194	2.25 / 0.6
I13/O13	Core Ins. N43-44	1.93/.076	Xintech 09NA2100XS	90/194	2.25 / 0.6
I14/O14	Core Ins. N45-46	1.93/.076	Xintech 09NA2100XS	90/194	2.25 / 0.6
I15/O15	Core Ins. N47-48	1.93/.076	Xintech 09NA2100XS	90/194	2.25 / 0.6
I16/O16	Slide N41-42	8/.315	Xintech 09NA2100100XS	90/194	8.5 / 2.25
I17/O17	Slide N43-44	8/.315	Xintech 09NA2100100XS	90/194	8.5 / 2.25
I18/O18	Slide N45-46	8/.315	Xintech 09NA2100100XS	90/194	8.5 / 2.25
I19/O19	Slide N47-48	8/.315	Xintech 09NA2100100XS	90/194	8.5 / 2.25
I20/O20	Slide N41-42	8/.315	Xintech 09NA2100100XS	90/194	8.5 / 2.25
I21/O21	Slide N43-44	8/.315	Xintech 09NA2100100XS	90/194	8.5 / 2.25
I22/O22	Slide N45-46	8/.315	Xintech 09NA2100100XS	90/194	8.5 / 2.25
I23/O23	Slide N47-48	8/.315	Xintech 09NA2100100XS	90/194	8.5 / 2.25

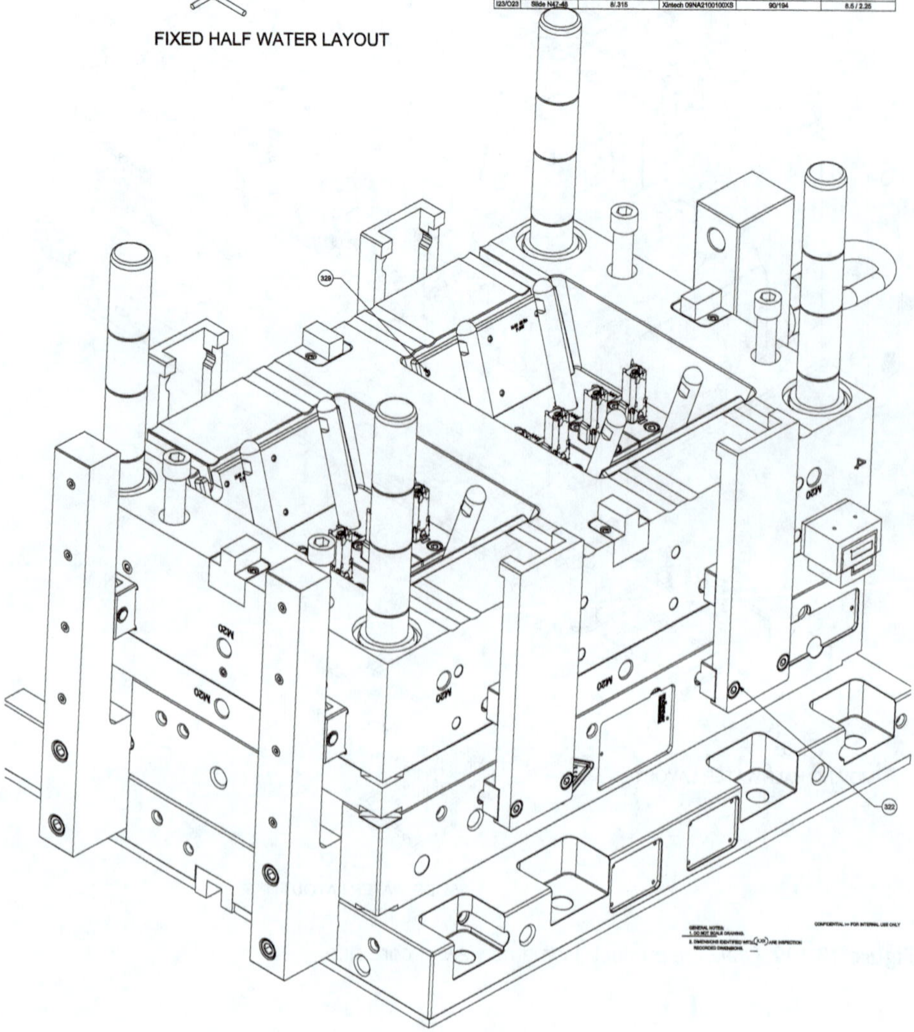

Figure 18.110 Oral care product: Isometric view of cavity side

18.34 Actuator (Using Split Cavity with Multi-Stage Ejection)

Courtesy of Nypromold

This 16-cavity medical inhaler actuator is a complex and challenging part that requires many shut-offs, multi-stage ejection, and a split-cavity mold design to mold and eject the parts (see Figure 18.111–Figure 18.113). Large tapers with generous wear plates are used to back-up the split cavities when the mold is closed to resist injection pressure. In addition to the large tapers, the mold is aligned using straight interlocks on the sides of the mold. Leader pins are used for assembly and maintenance of the mold.

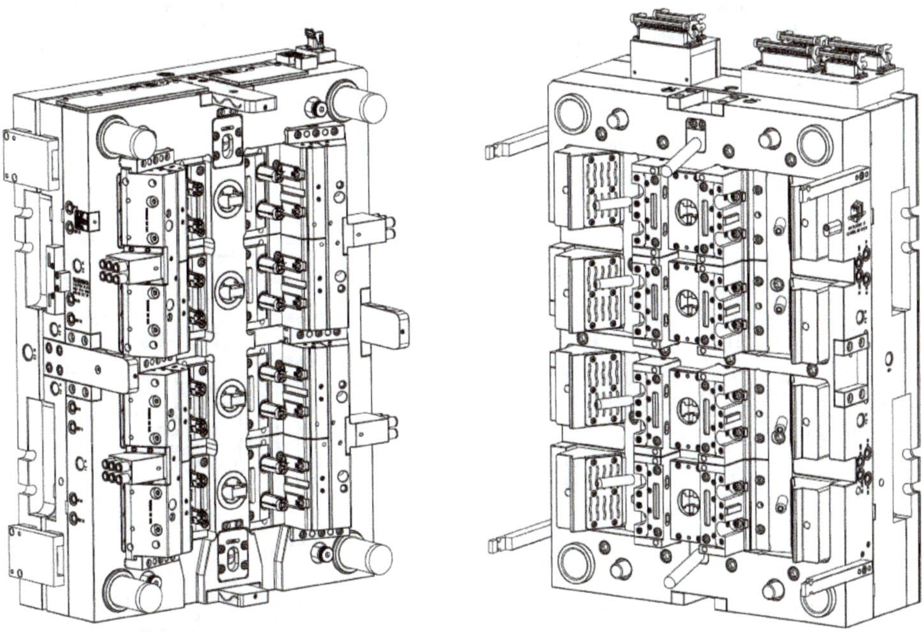

Figure 18.111 Actuator: ISO assembly view of 16-cavity mold

A latch lock system is used to actuate the multi-stage ejector system after the split cavities have opened and the horn pins have retracted. Cooling is supplied to the split cavities to ensure a reasonable cycle time is achieved. Other core and cavity cooling is manifolded through the plates of the mold to reduce the number of hose connections.

A hot runner system is used to feed the molten resin into the cavities.

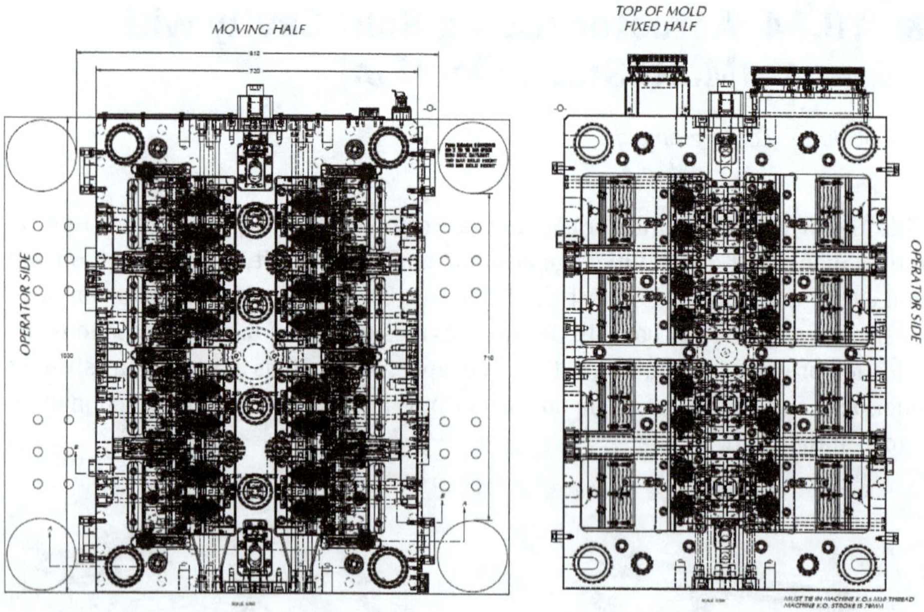

Figure 18.112 Actuator: Plan view of mold assembly

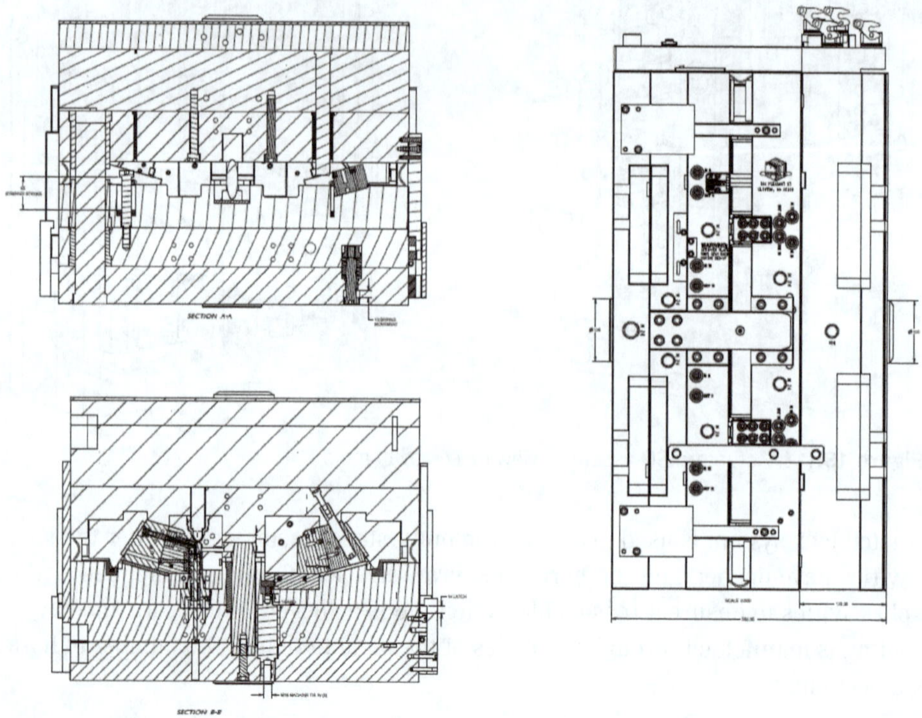

Figure 18.113 Actuator: Section view of mold assembly

19 Mold Testing, Approval, and Troubleshooting

Now that the mold has been designed, manufactured, and assembled, it is finally ready for testing. The following provides general guidelines on how to test and optimize the process for an injection mold. Testing of a mold must be done with great care and by those with skill and experience in starting up and testing new molds. While the designer is mostly an observer in the process, they are often called upon to be there at the start-up. Providing the tester with correct and accurate information for testing the mold is critical to ensure they know how the mold is to function. While the drawings and instructions in the mold manual provide much of the needed information, the designer can also provide additional support when needed. It is therefore important for designers to know what it takes to test a mold, and to have some understanding of how to resolve issues that come up in test. This chapter includes a general outline of how to test a mold, information on what should be tested and a troubleshooting guide to help with any issues that may occur during testing.

19.1 Information to Be Supplied to the Mold Tester

The following information should be supplied to the mold tester prior to the test to ensure that the mold test goes as planned:

- Date of test
- Project engineer
- Designer
- Test machine to be used
- Resin and colorant used
- Job number

- Similar job(s)
- Cavitation
- Expected wall thickness
- Expected cycle time
- Samples required and how they should be packaged
- Any specific instructions for the testing
- Assembly drawings
- Preliminary process setup guide
- Templates for any information required to be filled out (part weights, bluing, core shift, test log, etc.) (see Section 19.2 and Section 19.5 for more details).

19.2 Dry-Cycle Testing before Injecting Plastic

Before plastic is injected into the mold for the first time, the all of the items in Section 19.2 should be reviewed and completed to ensure mold function, prevent any damage, identify any mold design or manufacturing issues, verify cooling, and ensure correct hot runner connections.

Check the machine before installing the mold. Make sure it is square, parallel, and level. Make sure that the locating ring is flush to proud by 0.05 mm (0.002"). Ensure that the machine meets all the requirements for the mold in terms of space, stroke, processing conditions, mounting, and ejection. Check to see that all the required hardware to test the mold exists. Check ejector rod lengths and make sure you have stroke limiters.

Once the mold is safely installed in the machine, the following should be completed and checked:

- The mold is properly mounted in the press with the required bolts or clamps torqued to specification.
- The mold ejector mechanism has been correctly connected to the machine ejector plate.
- The machine orifice and radius and mold orifice are the same size.

- Ensure that all connections for water, air, and hydraulics are correctly installed and that they are tight, routed correctly, and have no pinch points. Turn on air, water, and hydraulics to ensure that no leaks exist. This should be done for the full stroke of the machine to ensure that the connections are free to move to whatever settings the operator gives the machine.
- All limit switches and safeties are connected and proper function is confirmed.
- Check and ensure that there is lubrication in all the required locations.
- Connect hot runner controller and cables to the mold. Turn on the heaters and check to see that the temperature increases to the entered set points. Ensure that all manifold heaters are coming up to the set point temperature equally and drawing the same power. Turn on the nozzle tip heaters and confirm the same. Check to see that all nozzle tips are consuming the same power.
- Set up the mold sequence as specified in the mold assembly drawings, mold manual, or supplied setup guide. Check to ensure that the entered machine setup parameters allow the mold to run through a complete dry-cycle without issue.
- Ensure that no cables and hoses are pinched, that they are long enough to allow the mold to stroke open fully for maintenance, and that they do not rub on one another or on a part of the machine or mold.
- Check the core and cavity mold alignment using a lead test (Section 19.2.3). Realign as necessary using the procedure in Section 19.2.2.
- Check for proper shut-off, venting, and taper fits using the procedure in Section 19.2.1.
- Check to ensure the machine nozzle and the sprue bushing are correctly aligned to ensure that no plastic will spray out in this area and cause potential injury during injection of plastic into the mold.
- Run the press in automatic without injecting plastic in to the mold (this is called a dry-cycle) to ensure proper function. Check to see that no hoses or cables rub or interfere with normal operation.
- Listen for any strange noises.
- Once the mold has been dry-cycled a few times, check to see that all guiding and sliding elements have no galling marks or damage.

- Record the following mold cooling data:

Circuit #	Mold half	Circuit	Flow (L/min)	Temp. IN (°C)	Temp. OUT (°C)	ΔT (°C)	Pressure IN (bar)	Pressure OUT (bar)
1	Stationary	Hot runner						
2	Stationary	Cavity plate						
3	Stationary	Cavity #1-#4						
4	Stationary	Cavity #5-#8						
5	Moving	Stripper plate						
6	Moving	Core #1-#4						

- Record the following hot runner control parameters:

Zone	Type		Temperature (°C)	Heating (%)	Current (A)
#1	Nozzle	Cavity 1			
#2	Nozzle	Cavity 2			
#3	Nozzle	Cavity 3			
#4	Nozzle	Cavity 4			
#5	Manifold	Top			
#6	Manifold	Bottom			
#7	Bushing	Inlet			

19.2.1 Bluing to Ensure Proper Shut-off, Venting, and Taper Fit

Before the mold is started up, the shut-off faces should be checked to ensure that there is good venting of the air from the cavity space, to ensure that the tapers are preloaded, and to verify that the alignment of the mold is good. Since these dimensions are difficult to measure exactly on a bench due to tolerance build-up, this process is often done visually using toolmaker's blue in the press.

The bluing compound is painted on the cavity side and then the mold is closed and minimal tonnage is built. The mold is then opened to observe where the bluing transferred to the mating core side. The results show the existence of the vents and the contact of mating tapers. The amount of transferred blue tells the mold maker how intense the contact is. The sheet shown in Figure 19.1, or a derivation of it that reflects the part geometry, can be used to record the results. Mark clamp/injection, operator/non-operator and cavity numbers on the drawing so that the orientation is easily understood.

Figure 19.1 Sample bluing mapping template for a round part. Instructions: Blue the mold on the primary tapers and secondary shut-off cavity side. Set the machine to its lowest clamp tonnage and clamp up the mold. The tapers should hit first, then the primary shut-off should hit. The bluing must transfer all around the shut-offs. Complete the diagram

19.2.2 Procedure for Aligning Mold Halves in the Injection Molding Machine

Procedure supplied by Vince Lomax.

This procedure should be followed when the mold is first started up and whenever the molder or mold maker suspects a mold alignment issue.

1. The mold must be clamped in the molding machine.
2. Set four indicators outboard on top and sides of the mold at a location about 3 mm further away than the guide pins are from the parting line on the moving side plate. Set all indicators to "0" (the indicators must stay on the moving side plate while the mold is opened slowly until interlocks are disengaged).
3. Open the mold until the interlocks just disengage each other but the leader pins are still engaged.
4. Stop the press, and note and record the indicator readings (deviation from zero).
5. Now open the mold until the guide pins disengage, and once again check the indicators and note the readings (deviation from zero).
6. Use a jacking (or lifting) system to lift the moving half of the mold by applying force to the ejector housing close to the platen. The jacking system must have very controlled movement. Loosen mold clamps on moving side just a little so that mold will move when applying pressure with jacks (do not loosen completely).
7. If out of alignment, the mold would have probably dropped when you opened it as described in the steps above. Use the jacking system to move the mold moving half, so that it gives the same indicator readings as when the mold was being clamped.
8. Close the mold slowly to check that no indicator movement takes place as the mold is closed and the interlocks and guide pins engage.
9. Now add clamp tonnage and once again check the indicators to ensure that the reading is correct.
10. Remove indicators from top and sides of mold. Re-check all bolts to ensure that the mold is good and tight, before removing the jacking system.

19.2.3 Lead Test Procedure to Check Wall Thicknesses and Alignment

Lead-test each stack in the mold to ensure that the mold will produce the correct wall thicknesses and that there is good alignment in the mold. The lead should be taped in place on the bottom and sidewalls of the core or cavity. Table 19.2 can be used to record the lead test results.

If the lead test shows poor alignment, align the cores to the cavities. The alignment procedure is as follows:

- Loosen the cores slightly from the core plate so they have 40 ft·lb (54 N·m) of torque left holding them to the core plate.
- Dry-cycle the mold using operating tonnage for approximately 10 cycles. This will allow the floating core to align to the fixed cavity.
- Torque all core-retaining screws to the torque specified on the assembly drawings (240 ft·lb for M16 screws, 100 ft·lb for M12 screws).

19.3 Initial Mold Start-up and First Shots

> **A qualified mold tester MUST ALWAYS conduct the mold installation and testing. The designer or project engineer is an important observer, but without proper education and training a designer should NOT take it upon themselves to install or test the mold. Serious injury or death could result from improper handling of the mold or the machine.**

An injection molding system is a powerful machine that could cause serious injury or death. These guidelines are NOT provided to allow any untrained person to conduct a mold test. It is a review of the process, so that the designer or project engineer knows what is happening and so they can point out anything that might be missing or incorrect. Prior to beginning any mold test, the following should be reviewed:

1. Ensure that a proper dry-cycle test has been completed (Section 19.2).
2. Ensure that the machine is set up specifically for the mold to be tested by the mold tester. A baseline setup sheet (similar to that shown in Appendix 6) should be provided to the tester by the designer or project engineer assigned to the job.
3. Ensure sufficient supply of the required resin and colorant is available and dried (if necessary).
4. Make sure that the tool is vented and that the air can escape easily.
5. Once these items have been completed, the mold is ready to receive its first shots and to begin the testing process.

19.3.1 Mold Start-up Procedure

1. Set the mold protection following the procedures in the machine manual.
2. Turn on the machine cooling water. Without cooling, the hopper throat could bridge and the machine could overheat.
3. Turn the mold and manifold plate cooling ON before the injection unit and hydraulic pump electrical power.
4. Depending on the size of the machine barrel, turn on the barrel heaters.
5. Approximately ½ to 1½ h prior to mold start-up, allow the barrel to come up to temperature. The temperature must be just below the operating temperature, low enough to prohibit the resin from degrading. If for any reason start-up is interrupted, the heat must be turned OFF to prevent overheating the plastic. For extended delays, the tester should follow the manufacturer's and resin supplier's procedures regarding machine shut-down.

> Remember that overheated plastics can form dangerous gasses within minutes, and that they can explode into the open when unblocking a gate in the mold. Retract the machine injection unit to allow the hot runner system to vent if the mold operation is interrupted.

6. Electrical power to the manifold and sprue bushing heaters should be turned ON approximately 30 min before the mold is to be run. Set the temperature controller supplying electrical power to the manifold(s) and the sprue bushing to the designated operating temperature. This temperature is specified on the assembly drawings, and is also shown on the nameplate mounted on the operator's side of each mold.
7. Once the power is applied, the controller read-out should be checked to ensure the temperature is rising. If the temperature indicator shows a declining temperature, the thermocouple leads may have been installed reversed. This must be corrected before continuing.

> Failure of the sprue bushing or sprue bar to come up to temperature at the same time or faster than the manifold may cause a build-up of internal pressure within the manifold. To prevent this, ensure that the heater machine nozzle is pressed against the sprue bushing/sprue bar. This will conduct additional heat into the sprue bushing/sprue bar to melt any cold material.

> Do not operate the system outside of the specified temperature difference (range). The temperature difference of the system is the difference between the manifold operating temperature and the cooling temperature in the manifold plate. Check the nameplate attached to your mold for specific design temperatures.

> A manifold operated below the specified temperature range will not allow sufficient thermal expansion of the components to properly seal the plastic within the manifold. This will result in leakage.
>
> A manifold operated above the specified temperature range may collapse the components due to excessive thermal expansion. A loss of sufficient seal will result and leakage will occur. Collapsed components must be replaced.

8. Enter the basic parameters into the machine to mimic the provided setup sheet, or use a similar setup sheet from another comparable mold. Set the mold cooling time to longer than expected so that the parts have time to cool properly before ejection.
9. Check all peripheral equipment to see that it is ready and operable (i. e., product handle, dryers, etc.).
10. Once the manifold has reached operating temperature, allow it to heat-soak for 10 min. For resins that have a very narrow thermal stability window, this time may have to be shortened.
11. With valve-gated molds, turn ON the air to the valve gates. The air must be activated to close the gates. If the air is turned ON before the manifold is heated, there may be air leakage. This air leakage will cool the manifold and delay start-up.
12. Turn ON the electrical power to the hot runner nozzle heaters. If the nozzle heaters are operating on manual control, they should be allowed to heat-soak for a minimum of 3–5 min. The controller should be set as recommended in the setup guide. Higher temperature resins may require a longer soak time, which must be determined through trial. Nozzle heaters controlled by thermocouples and automatic temperature controllers should only be allowed to heat-soak for 2–3 min after the controller indicates the operating temperature has been reached.
13. Move the material hopper slide to the open position.
14. Initiate extruder "screw start" so the screw starts plasticizing material.
15. Purge the machine extruder to bring clean, well-plasticized resin to the machine nozzle. Follow the machine manufacturer's instructions and safety procedures.

> Do not attempt to run the extruder screw or move the shut-off nozzle until operating temperatures are reached. Failure to do so could cause the screw to break.

16. Clamp up the mold and jog the injection unit carriage forward until the set point is adjusted correctly. Set decompression (suck back).

17. Open the mold and check with a mirror to see that the valve stems are forward and the gates are closed. If the valve stems are not in the proper position, the air lines may have been reversed during installation. This must be corrected.

> The injection unit carriage must be retracted while the mold is open for inspection. If this is not done, then the remaining pressure in the barrel and hot runner could cause plastic to explode uncontrollably from the gates of the mold.

18. With proper PPE (gloves, coat and face shield), check mold gates for drool using a hand-held mirror. DO NOT LOOK DIRECTLY AT THE GATES. Remove any material that has seeped from the gates. Never push on the valve gate stems. A hardwood or soft brass chisel and a vacuum cleaner are recommended to prevent damage to the polished molding surface.

> Do not look directly at a gate, use a mirror. Wear recommended safety clothing, consisting of a protective heat-resistant coat, heat-resistant gloves and a full face shield over safety glasses.

19. Where applicable, spray the mold with an appropriate mold release.
20. Close the mold. Cycle the mold open and closed a few times to ensure that the open and close positions and velocities are set correctly. For start-up, the clamp movement should be significantly slowed. The objective at this time is not to run the fastest cycle. Once the process is stable then the clamp can be sped up.
21. Check the ejectors. Ensure that they stroke freely and to the desired stroke. For now, leave the ejectors in manual mode so that they can be stroked manually. Initially the ejection should be run manually and slowly to ensure that the parts eject properly and so that no mold part breakage occurs.
22. Set the shot size. The first shot should be intentionally short (less material than required for the product) to avoid potential flashing. The exception is where damage or extensive cleaning would be required when a short shot gets stuck.
23. The machine should be set in the semi-automatic mode. That means it will cycle one complete shot and stop to wait for operator input.
24. Start the first cycle.
25. If the mold has a hot runner, then before plastic fills the cavity it must fill up the hot runner. In this case the hot runner must be first filled with plastic. This can be done manually with the mold open or in semi-automatic mode with the set shot size. Note that if it is done in semi-automatic mode then the first parts molded may be very short and it may not be possible to easily eject the parts.

They may have to be manually removed from the cores. If, however, there is no plastic in the cavity space after a few shots (usually two but no more than five), the mold should be stopped and checked for leakage of plastic in the hot runner. Once the hot runner is filled, then plastic should enter the cavities at the predetermined shot size each cycle.

> If no plastic is injected into the mold then there may be a hot runner leakage. With the manifold at operating temperature, check for smoke from the top of the hot runner plates. Also inspect the hot runner for leaks through holes or openings. If plastic is detected or leakage is suspected then the mold should be opened up to find out where and why the leak has occurred.

26. The machine nozzle to sprue bushing should also be checked for leakage or blockage if there is no plastic in the cavities.

> If the screw does not move forward to inject, check the machine temperature settings and ensure that the heaters or thermocouples are not faulty. Also check the manifold heaters, sprue heaters, nozzle heater band, their settings and thermocouples to ensure they are not faulty. Check the temperature of the mold gates with a pyrometer to ensure that they are warmer than the surrounding mold.
>
> Raise the tip temperature 28–56 °C (50–100 °F) above the operating temperature until all cavities fill, and then reduce the temperatures to normal operating values. If the problem persists, refer to the troubleshooting guide in Section 19.7.

27. After each shot, remove any short shots and/or drool from the gates.
28. Test to see that ejection works correctly and that the parts fall freely from the mold onto the conveyer.
29. Proceed cycling, gradually increasing the shot size until the cavities are all filled. Once the cavities are filled and the parts eject without hanging up, the tester can switch to the automatic mode.
30. Once in automatic mode, any automation can be started and tested. Start-up of the automation will not be covered in this guide. See the start-up guidelines from the automation manufacturer for details on this process.
31. Now that the mold is running in automatic mode, optimization of the process can begin. First a fully packed-out part must be produced by adjusting the injection and pack processes. Cycle time can then be reduced by speeding up the clamp and reducing process conditions. Clamp tonnage can be finally optimized to the minimum value.

32. Once an optimized process is reached (this may take some time), samples can be collected and inspected. A setup sheet like that shown in Appendix 6 should accompany all samples taken for inspection so that the process can be re-created if needed.

If a problem occurs while molding, refer to Section 19.7 for more information.

19.4 Changing Materials or Colors

Some molds require frequent color changes as part of their core functionality. In fact, some molds can require color changes many times a day. There are many factors affecting color change, and some of them are related to the mold design.

The hot runner must be specifically designed for color changes. If not, the color change procedure will take much longer, and it will be more difficult to remove all traces of the previous resin. In general, a hot runner designed for color change will have smaller melt channels to create more shear (wiping action) in the hot runner. This has the trade-off in that it will increase the pressure drop in the hot runner. The most important factor for improving color changes in the hot runner is to remove any and all places for material to hang up or stagnate. The key area where this occurs is in the gate region.

> For easier color change, the insulating bubble of plastic near the gate that separates the cold mold from the hot runner system must be reduced or replaced with a non-melting tip insulator material such as Vespel®.

The following is a list of suggestions for improving color change in a mold:

- In general, a liquid colorant will be more difficult to clean from the extruder than a dry (powder) colorant. A dry colorant will in turn be more difficult to clean up than a color concentrate pellet or fully colored resin.
- If color changes are required and an insulator made of Vespel® or similar material is not used, then once the extruder and hot runner melt channels have been purged of the previous color the cavity plate should be latched over, and the plastic bubble containing the previous color, which will reside around the nozzle tips, must be removed.
- The machine nozzle orifice bore and nozzle radius must match the hot runner sprue bushing orifice bore and radius. The machine nozzle orifice should not have any steps or tapers and should be the same through diameter as the sprue bushing it feeds.

- Production schedules should be arranged wherever possible to go from the lightest to darkest colors. This makes color changes faster than molding a light color after a dark color.

19.4.1 General Color Change Procedure

Color changes MUST always be conducted by personnel who have been properly trained in the operation of the machine, auxiliary equipment and hot runners. Serious injury or death could occur if this is conducted by an untrained person.

> Always wear the appropriate safety clothing (PPE) consisting of a protective, heat-resistant coat, heat-resistant gloves and a full face shield over safety glasses. Cover any exposed parts of the body (e. g. arms, chest, and head) and use adequate ventilation for fumes.

The following is a general procedure for changing colours in a mold designed with a hot runner system:

1. Shut OFF the resin supply to the hopper feed throat and continue processing the plastic until the machine extruder is almost empty.
2. Stop the extruder motor, stop the pump motor, and make sure the oil in the accumulators is dumped to tank and the extruder back-pressure reads zero (if the machine is hydraulic).
3. Before moving the hopper back to expose the feed throat, remember that overheated plastic may form dangerously hot gasses, which could explode into the open through the throat opening.
4. Remove any plastic or colorant from the hopper and its feed throat with a vacuum cleaner. Never use a compressed air blast for cleaning the feed throat, since particles of hot plastic may fly into the air and cause injury.

> Never look directly into the feed throat of an injection molding machine. Always use a mirror to inspect the inside of the feed throat.

5. Clean the hopper, magnets, and feed throat with a damp rag. Also, clean any color concentrate feeders and any dryers or feed lines.
6. Reposition the hopper and any other cleaned auxiliary equipment. Reload the hopper with the new resin and colorant.
7. Move the injection unit carriage forward so that the extruder nozzle is properly located inside the purge guards to help contain any hot plastic that may splatter.

8. Set the temperatures on the extruder heaters and hot runner heaters to the recommended settings for running the mold plus 20 °C (36 °F) to allow the runner walls to be scoured clean of resin.
9. With all extruder temperatures up to operating temperature, move the extruder to the purge position and start the screw turning to process and purge material. Repeat this process to purge material, until all resin coming from the machine nozzle is the correct color with no signs of the previous color.
10. Never handle purgings until they are cold.
11. Clean the machine nozzle and the sprue bushing or sprue bar nozzle seat of any previous colorant.
12. Begin cycling the mold at its normal operating parameters, until the previous color is eliminated and the molded product is acceptable. Run 40–50 shots (or 15 min) with new material in the hot runner, then return heat settings to normal.

> As a general guideline for hot runner systems, approximately 500 g of resin must pass through a nozzle tip (gate) to complete a color change. This means that if molding a 50 g product on a 1 min cycle, it will take about 10 min to change the color after the machine extruder has been cleaned and purged. A 5 g product running a 20 s cycle will take approximately 33 min to change. These are not fixed numbers but estimates gained through experience, and do not include the time to purge the machine.

19.5 Mold Test Information and Data to Be Collected

Once the mold has been started up and is making shots in automatic mode at or near the desired cycle, it is time to start collecting information on the parts and the process. The following information is generally required to help approve a mold for production:

Fill out a setup sheet with optimal processing parameters, and record the following parameters (see Section 19.2.1 and Section 19.5.1):

- Fastest cycle with breakdown: (s)
- Lowest clamp tonnage: (tonnes)
- Minimum cooling required: (gal/min or L/min) at (psi, kPa)

Wall thicknesses and core shift mapping should be similar to examples shown in Section 19.5 and the Appendices to ensure that the mold is made to print and that the tolerance stack-ups are acceptable.

Full and short-shot part weights (see Table 19.1) should be noted, to ensure that the mold is balanced and that the specified part weight is reached. A customer may request parts made with various weights.

Table 19.1 Sample Part Weight Chart

Cavity	Full weight (g)	Full weight (g)	70% full weight (g)	Cavity	Full weight (g)	Full weight (g)	70% full weight (g)
1				13			
2				14			
3				15			
4				16			
5				17			
6				18			
7				19			
8				20			
9				21			
10				22			
11				23			
12				24			
Tester's comments:							

19.5.1 Core Shift Mapping

The samples must be taken after the mold has run for at least 15 min continuously to ensure that equilibrium has been reached. Core shift should be measured at the most extreme point. In the case of a deep-draw part it would be at the tip of the core. The measurements should be marked on a diagram similar to Table 19.2 to show the maximum and minimum wall thickness and the direction of the core shift. The head of the arrow must point to the thinnest wall section.

Table 19.2 Core Shift Mapping Template

Cavity #: Max. wall: Min. wall: Direction of core shift thick to thin:	Cavity #: Max. wall: Min. wall: Direction of core shift thick to thin:
Cavity #: Max wall: Min wall: Direction of core shift thick to thin:	Cavity #: Max. wall: Min. wall: Direction of core shift thick to thin:

19.6 APQP and PPAP

It is not the intention of this book to teach APQP and PPAP, but to introduce their existence to mold designers. Designers who will need more knowledge on this subject are encouraged to look for sources or literature that provides the necessary detail.

APQP (Advanced Product Quality Planning) creates a roadmap for developing new products, complete with time-based milestones and decision points along the way. APQP was introduced in the mid-1990s by the AIAG (Automotive Industry Action Group) to eliminate confusion on the requirements from OEMs of vehicles. APQP provided a common platform for planning, developing, and producing a product or providing a service. A major objective of APQP is to provide a process and procedure for suppliers and customers to work together when developing a new product. A comprehensive APQP helps assure that the product is delivered on time and with the highest added-value. There are five phases to an APQP:

1. Planning and defining the program
2. Product design and development verification

3. Process design and development verification
4. Product and process validation and production feedback (PAPP is in this area)
5. Launch, assessment, and corrective action.

PPAP (Production Part Approval Process) was also developed by AIAG, and it is an important part of the comprehensive APQP process. PPAP is the process of reporting of activities, many of which are automatically done in the course of testing and validating a mold. The PPAP process assists by putting the reporting of activities and their results into a structured and consistent format. A major objective of PPAP is for molders to provide their customers with evidence that customer requirements are understood and to demonstrate that the molder's process has the capability to produce a product that consistently meets those requirements. A PPAP is normally required when production of a new product or part is planned, and when there is any significant change to a product or process. Although originally developed for the automotive industry, the process has gained acceptance in many other markets.

19.7 Mold Test Troubleshooting Guide

The troubleshooting guide in Table 19.3 is designed for commodity resins such as PP, PE, and PS but it can also be useful for engineering resins. There may appear to be conflicting solutions to problems, for example when it is suggested to increase and decrease the hold pressure as the solution to the same problem. The reason for this is that one mold or one resin may respond to an increase while another will respond to a decrease. I would also like to direct the reader to the troubleshooting guides provided by John Bozzelli in the trade publications and in Chapter 15.2 of the *Runner and Gating Design Handbook* [1].

Troubleshooting MUST be done in a systematic fashion to diagnose problems effectively. The basic rules for troubleshooting are:

- Define the problem. Make sure that you are not addressing a symptom of an underlying problem. Do not ignore potential machine problems
- Develop a method to isolate the problem
- Test one item at a time to verify results
- Monitor the solution for a time to verify the problem has been solved
- Document the solution to make sure the occurrence will not be repeated.

[1] J. P. Beaumont, *Runner and Gating Design Handbook* (3rd edn), Hanser Publishing, 2020.

If there is an issue with the mold, the following information should be obtained to determine where the problem may lie:

- Describe the problem in as much detail as possible
- Were there any other problems in the plant which could have caused this problem?
- Determine if the problem lies with a specific cavity and if the orientation is constant and repeatable
- Determine if the mold ever worked, or if it was always a problem. Does the problem come and go?
- Is the machine set up correctly? Is it level, parallel, and in 100% working order? Is the cooling or air hooked up correctly and clean?
- Is the correct grade of resin being used? Is there any pattern to the problem?
- Is there any damage or signs of wear on the mold?
- Obtain short shots and full shots for analysis
- Obtain samples of the problem for inspection
- Obtain a setup sheet.

Table 19.3 Mold Testing Troubleshooting Guide - Possible Causes and Solutions

Problem	Processing/machine	Mold design	Part design
Gate stringing	↓ Lower tip temp. ↑ Cycle ↑ Hold pressure or time ↓ Shot size ↑ Decompression time ▪ Damaged nozzle tip	▪ Wrong gate size ▪ Wrong tip/ position ↑ Gate cooling ▪ Different gate insert material ▪ Nozzle tip unscrewed	▪ Part too thick? ▪ Gate too large? ▪ Add a dimple
High gates Gates freezing	↓ Tip temp. ↓ Cycle ↓ Hold ↓ Decompression ▪ Obstruction in gate ▪ Heater touching gate ▪ Damaged nozzle tip ▪ Replace nozzle heater ▪ Replace nozzle thermo-couple ▪ Valve stem seizing ▪ Replace piston seal ▪ Check for plastic leakage ↑ Air pressure to valve gate	▪ Wrong tip/ position ↑ Gate size ↓ Gate cooling ▪ Nozzle tip unscrewed ▪ Gate insert collapsed	▪ Add a dimple

Problem	Processing/machine	Mold design	Part design
Sink marks near the gate	↑ Shot size ▪ Check valve faulty ↑ Hold time, pressure ↑ Cool time ↓ Melt/mold temp. ↓ Injection speed	↑ Cooling near gate	↓ Thickness variations ↓ Rib thickness ↓ Dimple size
Sink marks far from the gate	↑ Cushion, shot size ▪ Check valve faulty ↑ Hold time, pressure ↑ Cool time ↑ Injection speed ↑ Melt/mold temp. ▪ Check/clean vents	↑ Cooling	↓ Thickness variations ↓ Rib thickness ↓ Lip thickness
Burning/ streaks	▪ Check heat controls ▪ Clean, change screw ↓ Clamp tonnage ▪ Check color pigment ▪ Change resin ↓ Melt temp. ↓ Back-pressure, rpm ↓ Injection speed ↓ Residence time (cycle) ↓ Cavity air ▪ Check gate for debris ▪ Check that machine nozzle matches the sprue ▪ Check machine nozzle heater	▪ Hot runner plug loose, rotated, or leaking ↑ Vent size ▪ Sharp edges at gate ↑ Runner size ▪ Tip unscrewed ▪ Heater touching plate ▪ Check for thermo-couple fault	
Air streaks/ hooks	↓ Injection speed ↓ Decompression ▪ Use shut-off nozzle	↓ Vent at insert ▪ Mismatch of inserts	↓ Engraving depth ↓ Thick to thin areas
Flow lines/ streaks	▪ Clean molding surfaces ↑ Melt/mold temp. ↑ Injection speed/pressure ↑ Tip temp. ▪ Purge previous color	▪ Water leak ▪ Change nozzle tip ▪ Change vent size	
Dieseling Air entrap-ment	▪ Check vents for dirt ↓ Damp force ↓ Injection speed	▪ Is area vented? ▪ Core shift? ▪ Are the vents hobbed in?	

Table 19.3 Mold Testing Troubleshooting Guide – Possible Causes and Solutions *(continued)*

Problem	Processing/machine	Mold design	Part design
Dark spots/specks	▪ Clean or change screw & check valve ▪ Clean M/C nozzle ▪ Resin contaminated ↓ Melt temp. ↓ Back-pressure ↓ Screw speed ↓ Cycle, injection time ↓ Extruder size	▪ Hot runner plug loose, rotated, or leaking ↓ Melt channel diameter	▪ Check colorant used
Moisture streaks/splay	↓ Melt temp. ↓ Machine nozzle length ↓ Injection speed/pressure ↓ RPM/back-pressure ↓ Decompression	▪ Water leak into vents or insert split lines ↑ Gate size	
Weld lines	↑ Injection speed ↑ Injection pressure ↑ Hold time/hold pressure ↑ Cushion/feed ↑ Mold/melt temp. ▪ Check and clean vents ▪ Check heaters ▪ Contaminated resin	▪ Change gate location ↑ Venting at weld ▪ Realign cores ▪ Eliminate core shift	↑ Wall thickness ▪ Rougher surface finish ▪ Move gate location ↓ Color %
Jetting	↓ Injection rate ↓ Melt temp. ↓ Back-pressure	▪ Relocate gate to impact on core ↑ Gate diameter ▪ Round off transition from gate to molding surface	
Flow lines	↑ Melt temp. ↑ Injection speed ▪ Clean parting line	↑ Venting	
Record grooves	↑ Speed/pressure ↑ Melt temp. ↑ Mold temp.	↑ Runner size	
Stress whitening	↓ Melt/mold temp.	▪ Ejector pin marks?	
Dull spots @ gate	↓ Injection speed ↑ Melt temp. ↑ Mold temp. ↑ Screw back-pressure	↑ Gate diameter ↓ Gate cooling	

Problem	Processing/machine	Mold design	Part design
Delamination/ blisters	↓ Melt temp. ↑ Mold temp. ▪ Contamination ↓ Injection speed, pressure ↑ Back-pressure, RPM ↓ Decompression		▪ Eliminate regrind ▪ Change colorant
Grease on parts	↓ Grease frequency ▪ Oil leak in machine ▪ Change air ejector timing	▪ Oil leakage on mold ▪ Erratic air ejection	
Discoloration	↓ Melt temp. ↓ Manifold temp. ↓ Extruder size		
Brittleness	↓↑ Injection speed ↓ % Regrind ↓↑ Mold temp. ↓↑ Melt temp. ↓ Screw RPM ↓ Back-pressure	↑ Gate diameter	↑ Radii
Hobbing	↓ Clamp tonnage	↓ Shut-off area	
Taper wear	▪ Clean taper regularly ▪ Realign cores	▪ Taper too short? ▪ Correct materials?	
Dull spots	▪ Contamination on molding surface? ↓↑ Melt/mold temp. ↑ Back-pressure ↑ Screw RPM ↑ Injection speed		▪ Change surface finish
Voids	↑ Hold pressure/time ↓ Injection speed ↓ Injection pressure ↑ Mold temp. ↑ Shot size ▪ Leakage at sprue? ▪ Check valve sticking? ▪ Clean vents	↓ Sprue melt channel diameter to prevent drool ↑ Gate size	▪ Change resin

Table 19.3 Mold Testing Troubleshooting Guide – Possible Causes and Solutions *(continued)*

Problem	Processing/machine	Mold design	Part design
Inserts cracking	▪ Eliminate glycol ▪ Do not use Cl in water ▪ Machine slamming shut	▪ Water lines too close ▪ Plugs causing stress ↑ Radii ▪ Material defective ▪ Material hardness incorrect ▪ Thin sections in the steel	↑ Radii in part
O-Rings pinching		▪ Deburr water lines ▪ Cut step for O-ring	
Bubbles in part	↑ Melt temp. ↑ Back-pressure/RPM ↓ Suck back ↓ Decompression		
Difficult to fill	▪ Machine nozzle too small ▪ Faulty heaters/thermo-couples ▪ Gate too cold (see above) ↑ Melt temp. ↑ Injection speed	▪ Vents blocked/crushed ▪ Gate detail damaged ▪ Nozzle tip unscrewed ▪ Valve stem not opening	▪ Wall thickness too thin ▪ Eliminate regrind ▪ Use easier-filling resin
Hot cores or cavities	▪ Cooling incorrectly hooked up	▪ Cores/cavities mounted incorrectly ▪ Cooling lines too small ▪ Short circuit in cooling ▪ Network?	
Excessive wear or galling	▪ Poor maintenance ▪ Machine worn ▪ Machine platen too thin	↓ Preload ↑ Back-up on inserts ▪ Leader pins worn ▪ Stacks not symmetric about center-line of platen	
Premature hobbing	▪ Too much tonnage	↑ Shut-off area	

Problem	Processing/machine	Mold design	Part design
Ripples in the parts	↓ Melt temp. ↓ Injection speed ↑ Packing	▪ Core shift? ▪ Pinch point in the cavity space	↑ Wall thickness ▪ Pinch points in part? ▪ Change resin (higher melt index)
Core shift	▪ Taper wear ▪ Re-align cores to cavities ▪ Rotate lock ring ▪ Insufficient tonnage ↓ Injection time ↑ Melt/mold temp. ▪ Locating ring not shimmed 0.05 mm proud? ▪ Tie bar stretch uneven ▪ Machine not level	▪ Core bolts hitting cavity side? ↑ Float on locating dowels ▪ Are tapers preloaded? ▪ Mold support too small? ▪ Tapers too small	▪ Part wall section too thin ▪ Aspect ratio too high ▪ Resin melt index too low
Parts stick in cavity	• Change air sequence ↑ Cavity air ↑ Cycle time ↓ Injection speed ↓ Pressure ↓ Hold time/pressure ↓ Melt/mold temp. ▪ Re-align mold and machine ▪ Check water hook-up ↓ Mold open speed	↑ Cavity air ↑ Cavity/core cooling ↓ Undercut in cavity ↑ Draft	↑ Undercut on core ▪ Stack shoulder surface finish too fine, use draw-stone
Parts stick to the core	↑ Core air ↓ Mold temp. ↑ Cycle time ↓ Injection speed ↓ Injection pressure ↓ Melt temp. ▪ Change air sequence ▪ Check for plugged air ▪ Re-align mold and machine	↑ Core air ▪ Add side air ▪ Core damaged	▪ Change core surface finish ↓ Undercuts on core

Table 19.3 Mold Testing Troubleshooting Guide – Possible Causes and Solutions *(continued)*

Problem	Processing/machine	Mold design	Part design
Parts do not stack (they nest)			▪ Add undercut to core lip (0.15 mm max.) and/or stacking shoulder (0.12 mm max.) ↑ Air gap between parts ↓ Core undercut ▪ Check shrinkage ▪ Resin too sticky
Short shots	↑ Shot size ↑ Injection speed/pressure ↑ Melt/mold temp. ↑ Hold pressure/time ↑ Cycle time ↑ Screw speed/back-pressure ▪ Check and clean vents ▪ Check heaters ▪ Bridge in hopper throat ▪ Check for broken valve	↑ Melt channels ↑ Gates size ↑ Vent size	↑ Part thickness ▪ Pinch point in part ▪ Resin melt index too low
Flash	↑ Melt temp. ↓ Back-pressure, RPM ↓ Injection pressure/speed ▪ Poor melt quality ▪ Move transition back ↓ Packing, shot size ↑ Clamp tonnage ▪ Machine level/flat? ▪ Platen aligned? ▪ Thicker platens? ▪ Tie bar stretch uneven	▪ Ensure that the core is able to float by 0.05 mm ▪ Align cores ▪ Clean parting lines ▪ Eliminate wear ring ↓ Vent sizes ↑ Back-up ▪ Unbalanced runner ▪ Parting line not clean	↑ Part thickness ▪ Resin melt index too low ▪ Eliminate regrind

Problem	Processing/machine	Mold design	Part design
Part dimensions too large	↓ Cycle time to increase shrink ↓ Cooling ↑ Melt temp. ↓ Packing time or pressure	▪ Steel sizes correct?	▪ Shrinkages correct?
Part dimensions too small	↑ Cycle (cooler parts less shrink) ↑ Cooling ↓ Melt temp. ↑ Packing	▪ Steel sizes correct?	▪ Shrinkages correct? ↓ Lip thickness ▪ Part toeing in
Sink marks, voids	↑ Pack pressure/time	↑ Gate size	
Uneven fill	▪ Broken thermocouple, heater	▪ Steel the same? ▪ Gate sizes the same?	
Drool	↑ Suckback, decompression ↓ Melt temp. ↓ Cushion	↓ Melt channel size ↓ Sprue orifice	
Warpage	↓ Resin viscosity ↓ Injection/back-pressure ↑ Cool time ↑ Melt temp. ↑ Hold time/pressure ▪ Use nucleated resin ▪ Clean vents	▪ Uneven cooling ▪ Non-symmetrical ejection	↑ Wall thickness ▪ Add rib to part for rigidity ▪ Part non-symmetrical
Stripper sticks	▪ Clean stripper rings ▪ Leader pin/bushing worn? ▪ Worn rack and pinion ▪ Broken return springs? ▪ Air piston not pressurized?	▪ Stripper rings heating up? ▪ Ejector rods different lengths	
Burning in manifold	↓ Cavity air ↓ Hot runner temp.(s) ▪ Check fit between sprue and machine nozzle	▪ Surface plugs flat or incorrectly fit ▪ Hot spots in manifold ▪ Incorrect manifold heater pattern ▪ Melt channel drill lines create dead spot	

Table 19.3 Mold Testing Troubleshooting Guide - Possible Causes and Solutions *(continued)*

Problem	Processing/machine	Mold design	Part design
Long heat up time	▪ Check supply voltage ▪ Moisture in hot runner causing soft start option ▪ Faulty heaters, wires ▪ Plastic leakage		
Sprue bar won't heat up		▪ Check sprue bar heater length ↓ Melt channel size	
Air leak in hot runner	▪ Quick exhaust broken	▪ Check piston seal ▪ Check sealing surfaces ▪ Collapsed HR parts ▪ Insufficient cold clearance	
Plastic leakage	▪ System started cold ▪ Overheated ▪ Damaged components	▪ Nozzle sealing diameter damaged ▪ Cold clearance wrong	
Slow color change	▪ Gate damaged ▪ Tip unscrewed ▪ Screw/hopper not clean ▪ Damaged check valve	↓ Melt channel size ▪ Missing tip insulator	

20 Appendix

Appendix 1: Mold Material Properties

Mold Material Properties – Metric Units

Property	Temp (C)	Prehardened Steels			Tool steels									Stainless steels				Copper alloys		Alu-minum	Tita-nium
		P20	4140	NAK55	H13	H13	S7	A2	D2	A10	O6	CPM9V	CPM10V	DIN 1.2316 (prehrd)	420F (prehrd)	420 ESR	416T (prehrd)	BECU3 HT	BECU25	6061 T6	6A1-4V
Chemical comp.		0.33 C	0.40 C	0.15 C	0.38 C	0.38 C	0.50 C	1.0 C	1.55 C	1.35 C	1.45 C	1.78 C	2.45 C	0.38 C	0.33 C	0.38 C	0.15 C	0.4 Be	1.9 Be	97.6 Al	6.0 Al
		0.3 Si	0.3 Si	0.3 Si	1.0 Si	1.0 Si	0.3 Si	0.2 Si	0.3 Si	1.2 Si	0.9 Si	0.9 Si	0.9 Si	0.3 Si	0.4 Si	0.9 Si	1.0 Si	1.8 Ni	0.6 Co, Fe, Ni	0.6 Si	4.0 V
		1.4 Mn	0.8 Mn	1.5 Mn	0.3 Mn	0.3 Mn	0.7 Mn	0.6 Mn	0.3 Mn	1.8 Mn	1.0 Mn	0.5 Mn	0.5 Mn	1.1 Mn	1.4 Mn	0.5 Mn	1.2 Mn			0.28 Cu	0.3 Fe
		1.8 Cr	1.0 Cr	0.1 S	5.2 Cr	5.2 Cr	3.2 Cr	5.3 Cr	12.0 Cr	1.9 Ni	0.3 Mo	5.3 Cr	5.3 Cr	16.0 Cr	16.7 Cr	13.6 Cr	13.0 Cr			1.0 Mg	
		0.8 Ni	0.2 Mo	1.0 cu	1.3 Mo	1.3 Mo	1.4 Mo	1.1 Mo	0.8 Mo	1.5 Mo		1.3 Mo	1.3 Mo		0.1 S	0.3 V	0.2 S			0.2 Cr	
		0.2 Mo		3.0 Ni	1.0 V	1.0 V		0.2 V	0.8 V			9.0 V	9.7 V								
				1.0 Al									0.1 S								
Hardness (RC)	20	28–32 (270–300 HB)	28–32 (270–300 HB)	40 (375 HB)	42–44	49–51	54–56	54–56	56–58	58–60	58–60	53–55	60–62	28–36 (270–340 HB)	36 (340 HB)	49–51	26–36	18–26 (95–102 Rb)	38–41	(95 HB)	30–34
Ultimate strength (MPa)	20	1007	965	1262	1386	1696	(2448 comp)	(2454 comp)	1917 (2765 comp)	1917 (3599 comp)	(3392 comp)	1896		1076	1096	1765	945	924	1517	310	896
	200	951	910	1041	1303	1744									1000	1731		896	1310		689
	400	793	751	882	1158	1469												717	310		572
Yield strength (MPa)	20	800	758	1020	1117	1331	(2034 comp)	(2061 comp)	1475 (2123 comp)	1737 (2275 comp)	(2744 comp)	1620		924	910	1469	758	807	1344	275	827
	200	751	710	841	1048	1262									820	1172		786	1172		620
	400	627	586	683	931	1062												648	241		483
Elongation (%)	20	17	18	1–15	15	13								3–9	3–10	8	16	5–25	3–10	12	11
Imp. str. (J) (V-notch)	20	73	51	8	30	27	27	41 (C-notch)	47 (un-notch)	58 (un-notch)	30 (un-notch)	73 (C-notch)	35 (C-notch)	50	17	12	31				19
	200	81	60	12	37	30									30						
	400	98	76	11	41	34															
Elastic modulus (GPa)	20	207	207	207	200	200	200	193	193	193	207	221	221	221	200	200	200	138	131	70	114
	200	200	200	200	186	186	193	186	186					214	193	193		110	117		
	400	186	186	179	172	172	179	172	172					207		179		62	48		
Therm exp. (10E-6/C)	200	12.6	12.8	12.6	11.5	11.5	12.1	11.7	11.0	11.7	11.2	11.2	11.2	11.0	11.0	11.0	10.5	17.6	17.5	24	9.0
	400	13.5		13.5	12.2	12.2	12.4	11.3	10.8	11.3	12.6	11.7	11.7	12.1		11.5	11.5				9.4
Thermal cond. (W/M C)	20	29.1	42.7	24.7	25.9	25.9	29.1	26.1	20.1	25.9		21.4	21.4	16.9	23.0	23.0	24.9	249.1	103.9	167.0	6.9
	200	29.6	42.2	25.8	25.9	26.1	29.7	27.1	21.1	27.0				19.0	23.9	23.9	28.7				8.6
	400	31.1	37.7	26.8	26.5	26.5	31.0	28.7	23.0	28.5				24.7		24.9					11.4
Corrosion resistance		Poor	Poor	Poor	Poor	Poor	Poor	Poor	Poor	Poor	Poor	Poor	Poor	Good	Good	Good	Good	Good	Good	Fair	Good
Applications		Plates, misc. comp.	Plates, misc. comp.	Mflds	Nozs, H.R. comp.	Cores, cavs, gates	Gates, stripr rings	Lock rings, stripr rings	Wear parts	Sliding surf.	Sliding surf.	High wear gates	High wear parts	Mold plates, slides	Mold plates, misc. comp.	Cores, cavs, gates	Misc. comp.	Noz tips	Core caps, gate insers	Cooling manif., T.O. comp.	Insu-lators

(Note: Values shown are approximate depending on supplier and quality level of material.)

Mold Material Properties – Imperial Units

Property	Temp (F)	Prehardened Steels			Tool steels									Stainless steels				Copper alloys		Alu-minum	Tita-nium
		P20	4140	NAK55	H13	H13	S7	A2	D2	A10	O6	CPM9V	CPM10V	DIN. 1.2316 (prehrd)	420F (prehrd)	420 ESR	416T (prehrd)	BECU3 HT	BECU25	6061 T6	6A1-4V
Chemical comp.		0.33 C	0.40 C	0.15 C	0.38 C	0.38 C	0.50 C	1.0 C	1.55 C	1.35 C	1.45 C	1.78 C	2.45 C	0.38 C	0.33 C	0.38 C	0.15 C	0.4 Be	1.9 Be	97.6 Al	6.0 Al
		0.3 Si	0.3 Si	0.3 Si	1.0 Si	1.0 Si	0.3 Si	0.2 Si	0.3 Si	1.2 Si	0.9 Si	0.9 Si	0.9 Si	0.3 Si	0.4 Si	0.9 Si	1.0 Si	1.8 Ni	0.6 Co, Fe, Ni	0.6 Si	4.0 V
		1.4 Mn	0.8 Mn	1.5 Mn	0.3 Mn	0.3 Mn	0.7 Mn	0.6 Mn	0.3 Mn	1.8 Mn	1.0 Mn	0.5 Mn	0.5 Mn	1.1 Mn	1.4 Mn	0.5 Mn	1.2 Mn			0.28 Cu	0.3 Fe
		1.8 Cr	1.0 Cr	0.1 S	5.2 Cr	5.2 Cr	3.2 Cr	5.3 Cr	12.0 Cr	1.9 Ni	0.3 Mo	5.3 Cr	5.3 Cr	16.0 Cr	16.7 Cr	13.6 Cr	13.0 Cr			1.0 Mg	
		0.8 Ni	0.2 Mo	1.0 cu	1.3 Mo	1.3 Mo	1.4 Mo	1.1 Mo	0.8 Mo	1.5 Mo		1.3 Mo	1.3 Mo		0.1 S	0.3 V	0.2 S			0.2 Cr	
		0.2 Mo		3.0 Ni	1.0 V	1.0 V		0.2 V	0.8 V			9.0 V	9.7 V								
				1.0 Al									0.1 S								
Hardness (RC)	70	28–32 (270–300 HB)	28–32 (270–300 HB)	40 (375 HB)	42–44	49–51	54–56	54–56	56–58	58–60	58–60	53–55	60–62	28–36 (270–340 HB)	36 (340 HB)	49–51	26–36	18–26 (95–102 Rb)	38–41	(95 HB)	30–34
Ultimate strength (KSI)	70	146	140	183	201	246	(355 comp)	(356 comp)	278 (401 comp)	278 (522 comp)	(492 comp)	275		156	159	256	137	134	220	45	130
	400	138	132	151	189	253									145	251		130	190		100
	750	115	109	128	168	213												104	45		83
Yield strength (KSI)	70	116	110	148	162	193	(295 comp)	(299 comp)	214 (308 comp)	252 (330 comp)	(398 comp)	235		134	132	213	110	117	195	40	120
	400	109	103	122	152	183									119	170		114	170		90
	750	91	85	99	135	154												94	35		70
Elongation (%)	70	17	18	1–15	15	13								3–9	3–10	8	16	5–25	3–10	12	11
Imp. str. (Ft-lbs) (V-notch)	70	54	38	6	22	20	20	41 (C-notch)	35 (un-notch)	43 (un-notch)	22 (un-notch)	54 (C-notch)	26 (C-notch)	37	13	9	23				14
	400	60	44	9	27	22									22						
	750	72	56	8	30	25															
Elastic modulus (PSIE06)	70	30	30	30	29	29	29	28	28	28	30	32	32	32	29	29	29	20	19	10	16.5
	400	29	29	29	27	27	28	27	27					31	28	28		16	17		
	750	27	27	26	25	25	26	25	25					30		26		9	7		
Therm exp. (10E-6/F)	400	7.0	7.1	7.0	6.4	6.4	6.7	6.5	6.1	6.5	6.2	6.2	6.2	6.1	6.1	6.1	5.8	9.8	9.7	13.3	5.0
	750	7.5		7.5	6.8	6.8	6.9	6.3	6.0	6.3	7.0	6.5	6.5	6.7		6.4	6.2				5.2
Thermal cond. (BTU/FTHRF)	70	16.8	24.7	14.3	15.0	15.0	16.8	15.1	11.6	15.0		12.4	12.4	9.8	13.3	13.3	14.4	144.0	60.1	97.0	4.0
	400	17.1	24.4	14.9	15.0	15.1	17.2	15.7	12.2	15.6				11.0	13.8	13.8	16.6				5.0
	750	18.0	21.8	15.5	15.3	15.3	17.9	16.6	13.3	16.5				14.3		14.4					6.6
Corrosion resistance		Poor	Poor	Poor	Poor	Poor	Poor	Poor	Poor	Poor	Poor	Poor	Poor	Good	Good	Good	Good	Good	Good	Fair	Good
Applications		Plates, misc. comp.	Plates, misc. comp.	Mflds	Nozs, H.R. comp.	Cores, cavs, gates	Gates, stripr rings	Lock rings, stripr rings	Wear parts	Sliding surf.	Sliding surf.	High wear gates	High wear parts	Mold plates, slides	Mold plates, misc. comp.	Cores, cavs, gates	Misc. comp.	Noz tips	Core caps, gate insers	Cooling manif., T.O. comp.	Insu-lators

(Note: Values shown are approximate depending on supplier and quality level of material.)

Appendix 2: Frequently Used Conversion Factors

Conversion Factors for Units of Measure (Imperial and Metric)

Measure	Conversion factor	Measure
Density (mass density)		
lb/in^3	27.6799	g/cm^3
g/cm^3	0.03613	lb/in^3
Energy		
Joule (J)	1	N·m
British thermal unit (BTU)	1.055	kJ
British thermal unit (BTU)	0.252	cal
British thermal unit (BTU)	778.0	ft·lb
International calorie (cal)	4.1868	J
Kilowatt hour (kWh)	3.6	MJ (megajoule)
Watt second (W·s)	1	J
Force		
Newton (N)	0.1020	kgf
Newton (N)	0.2248	lbf
Kilogram-force (kgf)	9.8066	N
Tonne-force (tf)	9.8066	kN
Pound-force (lbf)	4.4482	N
Kilonewton (kN)	101.972	kgf
Heat (Temperature difference (ΔT))		
Degree Celsius (°C)	1	Kelvin (K)
Degree Celsius (°C)	$9/5$	°F
Degree Fahrenheit (°F)	$5/9$	°C
Heat flow (conductivity)		
BTU/ft^2·h·°F/inch thickness	0.1442	W/m·°C/cm
Length		
Meter (m)	39.3700	in
Meter (m)	3.2808	ft
Millimeter (mm)	0.03937	in
Inch (in or ")	0.0254	m
Inch (in or ")	25.4	mm

Measure	Conversion factor	Measure
Mass		
Short ton (2000 lb)	0.9072	t (tonne)
Long ton (2240 lb)	1.0160	t (tonne)
Metric tonne (t)	1000	kg
Kilogram (kg)	2.2046	lb
Pound (lb)	0.4536	kg
Pound (lb)	16	oz
Ounce (oz avoirdupois)	28.3495	g
Gram (g)	0.03527	oz (avoirdupois)
Power		
Watt (W)	1	J/s
BTU/h	0.2931	W
Horsepower (hp)	746.0	W
Pressure		
Pound force/in^2 (lbf/in^2, psi)	0.006895	MPa
Pound force/in^2 (lbf/in^2, psi)	6.8948	kPa
Pound force/in^2 (lbf/in^2, psi)	0.0703	kgf/cm^2
Kilogram force/cm^2 (kgf/cm^2)	14.2233	psi
Pascal (Pa)	1.	N/m^2
Kilopascal (kPa)	0.1450	psi
Pound force/in^2 (lbf/in^2, psi)	1	kN/m^2
Pound force/in^2 (lbf/in^2, psi)	101.972	kgf/m^2
Pound force/in^2 (lbf/in^2, psi)	0.01020	kgf/cm^2
Pound force/in^2 (lbf/in^2, psi)	0.1	N/cm^2
Column mercury inch (at 20 °C)	3.3741	kPa
Column water inch (at 20 °C)	0.2486	kPa
Atmospheric pressure (bar)	0.9869	kg/cm^2
Atmospheric pressure (bar)	100.	kPa
Temperature		
Degrees Fahrenheit (°F)	(°F – 32) × $^5/_9$	°C
Degrees Celsius (°C)	(°C × $^9/_5$) + 32	°F
Kelvin (K)	°C + 273.15	K
Torque (moment of force)		
Pound force (lbf)	1.3558	N·m
Pound force (lbf)	0.1383	kgf·m
Newton meter (N·m)	0.7376	lbf·ft
Newton meter (N·m)	0.1020	kgf·m
Kilogram force meter	7.2330	lbf·ft
Kilogram force meter	9.8066	N·m

Measure	Conversion factor	Measure
Velocity		
Foot/second (ft/s)	0.3048	m/s
Dynamic viscosity		
Poise (P)	1	$N \cdot s/m^2$
Poise (P)	1	$dyn \cdot s/cm^2$
Kinematic viscosity		
Stokes (St)	0.1	cm^2/s
ft^2/s	0.0929	m^2/s
Volume		
Cubic centimeter (cm^3)	0.0610	in^3
Cubic inch (in^3)	16.3871	cm^3
Cubic foot (ft^3)	0.0283	m^3
Cubic foot (ft^3)	28.3168	dm^3 (L or l)
Imperial (liquid) ounce (oz)	28.4131	cm^3
U.S. (liquid) ounce (oz)	29.5735	cm^3
Imperial gallon (gal)	4.5461	dm^3 (L or l)
U.S. gallon (gal)	3.7854	dm^3 (L or l)
Liter (L or l)	1000	cm^3

Appendix 3: Plastics and their Abbreviations

Plastic Materials and Their Abbreviations

Acronym	Full name
ABS	Acrylonitrile-butadiene-styrene
CA	Cellulose acetate
CAB	Cellulose acetate butyrate
CP	Cellulose propionate
EVA	Ethylene vinyl acetate
HDPE	High-density polyethylene
HIPS	High-impact polystyrene
IB	Polyisobutylene
LCP	Liquid-crystal polymer
LLDPE	Linear low-density polyethylene
LDPE	Low-density polyethylene
MDPE	Medium-density polyethylene

Acronym	Full name
PA66	Polyamide (Nylon 6/6)
PA6	Polyamide (Nylon 6)
PAEK	Polyaryl ether ketone
PAI	Polyamide imide
PAR	Polyarylate
PAS	Polyaryl sulfone
PBI	Polybenzimidazole
PBT	Polybutylene terephthalate
PC	Polycarbonate
PEI	Polyether imide
PEK	Polyether ketone
PEEK	Polyether ether ketone
PES	Polyether sulfone
PET	Polyethylene terephthalate
PETG	Polyethylene terephthalate (co-polymer)
PMMA	Polymethyl methacrylate (Acrylic)
PMP	Polymethylpentene
Polyester	Thermoplastic polyester
POM	Polyoxymethylene (Acetal)
PP	Polypropylene
PPO	Polyphenylene oxide
PPS	Polyphenylene sulfide
PS	Polystyrene
PSU	Polysulfone
PTFE	Polytetrafluoroethylene
PVC	Polyvinyl chloride
SAN	Styrene-acrylonitrile
TPUR	Thermoplastic polyurethane
UHMWPE	Ultra-high molecular weight polyethylene

Appendix 4: Mold Design Review Checklist

Additional checklist information supplied by Vince Lomax.

A process for reviewing the order such as a kick-off meeting should be held. Then a concept review of the design should be conducted. Finally, a full engineering design review should take place before the mold is released to manufacturing. This document can be used and modified to supply the necessary information for all three of these meetings.

Order # ________________ Cavitation ________________ Quote # ________________
Customer: ______________________ Description: ________________________________
Telephone # ____________ Contact: ____________________ E-Mail: ________________

PART INFORMATION

Weight _______ Wall thickness _______ Total shot weight ______ Est. cycle time ________
Print supplied _____ Y/N Samples provided _____ Y/N
Part finish inside ______________________ Part finish outside ________________________
Engraving: Logo ______________ Cavity ID _______________ ID Location ______________

PLASTIC MATERIAL (First shot – material #1)

Material type __________ Supplier __________ Material # _________ Fillers ___________
Melt flow _______ Process temp. ______ Release characteristics ______ Flash factor _____
Vent depth _______ Material shrinkages: Wall, x-y-z, specific geometry etc. _____________
____

PLASTIC MATERIAL (Second shot – material #2)

Material type __________ Supplier __________ Material # _________ Fillers ___________
Melt flow _______ Process temp. ______ Release characteristics ______ Flash factor _____
Vent depth _______ Material shrinkages: Wall, x-y-z, specific geometry etc. _____________
____

MOLD INFORMATION

Plate material _____________ Mold threads: Inch ______________ Metric______________
Type of ejection ______ Spare stacks (qty) _____ PKO Pattern _____ PKO Screw size _____

HOT RUNNER SYSTEM INFORMATION

Make ________________ Type ________________ Number of drops ________________

Gate diameter ____________ Center to center ______________ Mold temp. ____________

Hot runner spares ____

Notes: ____

CONTROLLER INFORMATION

Make ___________________ Volts ___________________ Phase ________________

Cable information ____

Notes: ____

MOLDING MACHINE INFORMATION

Make/tonnage _________________________ Spec. sheet encl. ________________

Locating ring diameter _____________________ Sprue diameter ________________

Nozzle radius ____________________ Distance between tie rods ________________

Max./min. press tonnage __________________ Max. shot size of press ______________

Maximum daylight ____

Minimum daylight ____

Opening stroke ____

Auxiliary equipment required, core pull etc. _____________________________________

OTHER

Read all notes, for any unusual requirements ______________________________________

Identify gate location ___________________ Gate vestige requirement ______________

Identify cavity ID location _______________

Steel safe recut needed?____________ Customer design standers needed____________

Spares required ____

Sample Concept or Mold Design Review Checklist

Question	Yes/No	Risk? Define and mitigation
PART DRAWING REVIEW		
Does part require CAE flow simulation: Fill? Cool? Warpage?		
Will part fill?		
Any weld lines or other potential molding defects?		
Are all pinch points eliminated?		
Possible sink marks eliminated?		
Are horizontal sections (bottom/stack shoulder) 0.05 mm thicker to account for stack compression and filling?		
Negative drafts on part eliminated?		
All geometry defined? (radii, angles, etc.)		
Show parting & split line locations on drawing. Customer approved?		
Is gate vestige an issue or not, dimple required?		
Stacking: Do parts stack and de-nest? Show on drawing		
Shrinkage (must be specified by the customer)		
Any geometry to be left off until after first test? (Pull rings, engraving, etc.)		
Leave stock for fit and sizing?		
Note deviations to standard tolerancing of molding surfaces & fits		
Show engraving and cavity numbering		
Has the part drawing been approved by the customer?		
GENERAL MOLD CONCEPT		
Stack design: Materials, core versus cavity lock, moving ring, gate pad, etc.		
Show ejection: Type, stroke, pattern (review air eject sequence). Will the part eject?		
Mold cooling: In and out locations, type of cooling in insert. Is the cycle time achievable?		
Do inserts need to shut off on plates? (Calculate compressive stress)		
Leader pin locations and size (top only preferred)		
Latching (provide parking position)		
Review machine mounting, electrical, ejector, size, stroke, shut height, tie bar spacing		

Question	Yes/No	Risk? Define and mitigation
Hot runner: Gate size, tip type, dimple, melt channel sizes		
Alignment: How are the core and cavity aligned? Is the core allowed to float?		
Are the dowels and/or screws offset to "dummy proof" installation?		
Is the size of the stack module sufficient?		
Cooling: What is the minimum distance to molding surface? Is it adequate?		
Are all chamfers and matching radii sized properly?		
Are screws less than 6 mm being used? Can they be increased?		
Verify adequate support and structural integrity of all plates. Review support and locking areas to prevent mold base flexing		
Venting		
Where is the part vented and what size?		
Are all cavity and runner vents detailed? Show depth and width, show escape to atmosphere		
Ring vent 360° where possible		
If mold vacuum required, show details		
Location of vents and vent depths. Do vent grooves allow removal of air?		
Stacks		
Split lines and parting lines: Are the split line locations acceptable for inserts and ejection?		
Preloads: Are they acceptable?		
Manufacturing tolerances acceptable on angles? Preloads on tapers?		
Servicing: Are the cores and cavities removable in the press? Other serviceability concerns?		
Are jacking holes required for removal of any parts?		
Lift bar required for stack components? (Lift holes on inserts)		
Gate pad/cavity mismatch to prevent hang-up in cavity? Gate pads must be 0.03 mm smaller than the cavity		
Cores and cavities fixed from rotation?		
Is orientation needed between inserts?		

Question	Yes/No	Risk? Define and mitigation
Are all alignment tapers 15°? Are there any tapers less than 7° (locking)?		
Determine surface finish for all molding surfaces and specify on detail. Note any special polishing techniques (radial or draw)		
Any special plating or surface treatment needed?		
All tie rod bushings should be located in the core		
Locate any possible stress risers (i.e., small radii <0.8 mm). If order is for stack only, release stack mounting hardware. Provide lead-ins on all parts		
Has the stack compression been calculated? Acceptable?		
Are the compressive stresses within limits to ensure no hobbing?		
Are the cavities sufficiently backed-up with steel? Are the axial stresses within limits?		
Have all spares been accounted for and numbered?		
Cooling review		
Trace all circuits to ensure even flow and no dead spots. Show separate schematic if necessary		
Have the cooling calculations been completed for Re number and flow requirements?		
Does the header have enough volume to feed entire circuit?		
Are the any brass baffles used on the cavity? If so, replace with machined-up stand to prevent cracking		
Does the machine(s) have enough ins and outs for mold?		
O-Rings: Will O-ring hang up on any sharp corners?		
Are O-rings clear of wire grooves, pipe plugs, etc.?		
Air circuits		
Are all air circuits for pistons balanced?		
Is a blow-down bar required for ejection?		
Is the piston designed to bottom-out? (Stroke limiter recommended)		
Part ejection		

Question	Yes/No	Risk? Define and mitigation
Do parts fall free without hitting leader pins, taper pins, automation, etc.?		
Are the customer's machine and test machine ejector positions available on the mold?		
Are air lines and supply lines large enough to permit ejection?		
Are there lift holes on all sides of the plates? (Never use M20 lift holes)		
Is the stripper plate fully supported at all times?		
Do mold feet or "toes" need to be put on plates?		
Are pry slots accessible in and out of machine?		
Review latching procedure: Is it acceptable?		
Is ceiling height or crane capacity an issue in customer's plant?		
Leader pins: Are length and diameter adequate to protect the mold?		
Do leader pins engage before molding surfaces?		
Is the mold stroke acceptable for the test and customer's machines?		
Can the machine stroke fully without any issues?		
Are necessary screws accessible when mold is closed?		
Do cutouts allow for 65° turn of screws minimum?		
Is there a nameplate on the mold?		
Melt distribution		
Is the gate in the best location?		
Gate size agreed to?		
Cold runner system		
Sprue diameter and length?		
Is the runner balanced?		
Runner ejection? Any conflict with parts?		
Hot runner system		
Is there proper cooling around manifold and nozzles?		
Check for good support for injection pressure		
Can the hot runner be latched away from the cavity plate in the press to service and clean tips and heaters?		
Other		

Question	Yes/No	Risk? Define and mitigation
Other potential issues or risks?		

Appendix 5: Contents of a Mold Manual

Contents of a Mold Manual

Description (possibly a photo) of the mold
Date of building Serial number, size, weight List of patents applying to the mold. List of third-party trade marks of components used.
Table of contents
1. Safety of operation 2. Specifications: What is required to operate the mold: Molding machine type, space, plant floor, hoist capacity, air supply, cooling water, lubricants, power, etc. 3. Safe handling of the mold 4. Installation, startup, and operation 5. Troubleshooting 6. Mold storage 7. Mold maintenance (frequency and procedures) 8. Setup guide from the original testing of the mold 9. Mold drawings and bill of materials Assembly drawing(s) Detail drawings of stack components Detailed drawings of all other mold parts, optional Drawings or specifications of special, commercial products used in the mold (valves, heaters, seals, etc.)

■ Appendix 6: Mold Setup Guide Template

The following information should be collected during the testing of the mold and added to the mold manual. The following table can be used as a template for a setup guide for the mold in a machine.

JOB INFORMATION

Process parameter: ______________________

Date: ______________________

Project: ______________________

Article: ______________________

Mold supplier: ______________________

Job number: ______________________

No. of cavities: ______________________

Carried out by: ______________________

RESIN, COLORANT, AND ADDITIVE

Resin:

Type: ______________________

Name: ______________________

Supplier: ______________________

MFI (g/10 min): ______________________

Lot no.: ______________________

Colorant:

Name: ______________________

Supplier: ______________________

Code: ______________________

% Addition: ______________________

Additive:

Name: ______________________

Supplier: ______________________

Code: ______________________

% Addition: ______________________

EQUIPMENT

Press: ______
Serial no.: ______
Fabrication date: ______
Screw type: ______
Screw diameter (mm): ______
Hot runner: ______
Job no.: ______
Fabrication date: ______
Hot runner controller: ______
Job no.: ______
Fabrication date: ______

KEY PROCESS PARAMETERS

Cylinder temperature (°C):

Nozzle: ______ Zone 1: ______ Zone 2: ______
Zone 3: ______ Zone 4: ______ Zone 5: ______

Hot runner temperature (°C):

Nozzle 1: ______ Nozzle 2: ______ Nozzle 3: ______
Nozzle 4: ______ Nozzle 5: ______ Nozzle 6: ______
Manifold 1: ______ Manifold 2: ______ Inlet: ______

Cooling water:

Water IN temp. (°C) ______ Water OUT temp. (°C): ______
Water pressure (bar): ______ Water flow total (L/min): ______

Injection:

Injection speed (cm^3/s): ______ ______ ______ ______
Screw position (cm^3): ______ ______ ______

Transition:

Position (cm^3): ______ Hydraulic (bar): ______ Time (s): ______
Cavity pressure (bar): ______

Press settings:

Clamp force (kN): ______ Opening stroke (mm): ______
Injection high set (bar): ______ Injection pressure peak (bar): ______
Holding pressure (bar): ______ Dosing volume (cm^3): ______
Dosing speed (mm/s): ______ Dosing pre-decomp. (cm^3): ______
Dosing post-decomp. (cm^3): ______ Back pressure (bar): ______
Cavity pressure max. (bar): ______

Timer and counter (s):

Mold closing: __________

Clamp force build-up: __________

Actual fill time: __________

Holding time: __________

Cooling time: __________

Mold opening: __________

Demolding time: __________

Delay time: __________

Cycle time: __________

MOLD SEQUENCE

Short description of the mold sequence:

- Mold closing
- Injection, hold, and cooling
- Mold opening
- Ejector forward
- Ejector backward, delay 0.5 s, signal ejector back
- Delay
- Mold closing.

MOLD SETUP GUIDE TEMPLATE

Mold Setup Guide Template

Mold Wizard Inc.
SET UP GUIDE

Customer: ____ Mold No: ____ Job No.: ____
Project Engineer: ____ Tested by: ____ Date: ____

MOLDED PART

Part Description: ____ Part Wt (g): ____ Shot Wt (g): ____
Overall Cycle: ___ (sec) ___ (shots/min) Avg Wall Thick.: ____ Flow Length: ____

MATERIAL

Type: ____ Manufactured by: ____ Pre-Dry Temp & Time: ____
Number: ____ Batch No: ____ Color: ____ % Concentrate: ____

MOLD

No. of Cavities: ____ Shut Height (mm): ____ Type: Single Face ☐ Stack ☐
Nozzle Bushing: ____ Locating Ring (mm): ____ Mounting Bolts: Metric ☐ Inch ☐
Ejection System: ____ Ejector Stroke (mm): ____ Runner System: ____

MOLD PROCESS TEMPERATURES AND PRESSURES

Core: ____ (°C) ____ (°F) Cavity: ____ (°C) ____ (°F) Gate Pad: ____ (°C) ____ (°F)
Pressure in (psi): ____ Pressure out (psi): ____ Air Pressure (psi): ____
Flow Rate: ____ (US gal/min or Litres/min) Cooling Tonnage Required: ____

MACHINE

Type: ____ Extruder Dia. (mm): ____ Clamp Stroke (mm): ____
Serial No.: ____ Inj. Pot Dia (mm): ____ No. of Accumulators: ____
Auxiliary Equipment: ____

HEAT PROFILE

Machine	NOZ	DIS1	DIS2	S/P2	S/P1	B/HD	EXT6	EXT5	EXT4	EXT3	EXT2	EXT1
	___°C	___°C	___°C	___°C	___°C	___°C	___°C	___°C	___°C	___°C	___°C	___°C
S/Point	___°F	___°F	___°F	___°F	___°F	___°F	___°F	___°F	___°F	___°F	___°F	___°F
Mold	N/T1	N/T2	N/T3	N/T4	S/BAR	X/MA	H/R1	H/R2	H/R3	H/R4		
Start	___°C	___°C	___°C	___°C	___°C	___°C	___°C	___°C	___°C	___°C	___°C	___°C
Auto	___°F	___°F	___°F	___°F	___°F	___°F	___°F	___°F	___°F	___°F	___°F	___°F

NOTES (For different zone temperatures and/or special start-up procedures see notes below)

CLAMP

CLAMP CLOSE

Timer(s) [ ]

Set Points (mm) | Close: [ ] [ ] [ ] [ ]

Velocity (%): [ ] [ ] [ ] [ ]

Clamp Tonnage: (%) (t) [ ] (%) (t) [ ]

Mold Break (mm) [ ] Clamp Pos. (mm) [ ]

CLAMP OPEN

Timer(s) [ ]

Open | Set Points (mm): [ ] [ ] [ ] [ ]

Velocity (%): [ ] [ ] [ ] [ ]

Mold Pos. (mm) [ ]

Mold Protection: Pressure (%) [ ] Set Point (mm) [ ]

EJECTOR & AIR

HYD. EJECTOR

	Set
Hyd. Ejec. Start S/P (mm)	
Hyd. Ejec. Back (mm)	
Hyd. Ejec. Stroke (mm)	
Hyd. Ejec. Fwd. Timer	
Hyd. Ejec. Counter	

AIR FUNCTION

		Set Point (mm)	Set Timer (t)
1			
2			
3			
4			
5			
6			
7			
8			
9			
10			

INJECTION

FILL

Pull Back ______ Shot Size ______

Set Points (mm)

Velocity (%)

Cycle Time(s) ______

Inj. Time(s) ______

Trans. Time(s) ______

Injection Pressure

______ (%) ______ (psi)

Piston Pos. ______ (mm)

TRANSITION

Time ______ (s) Position ______ (mm) Cavity Press ______ (psi) Inj. Pres. ______ (psi)

HOLD

Hold Time (t)

Hold Press. ______ (psi)

		(%)

SCREW

Back Press. (%)	Back Press. (psi)	Speed (%)	Speed (rpm)	Current (amp)

CARRIAGE

Fwd. Set P.	Stroke

EXTRUDER

Back Pos.

TIMERS & COUNTERS

CLAMP

	Set
Mold Close Timer	
Mold Open Timer	
Hyd. Ejec. Fwd Timer	
Hyd. Ejec. Stroke Ctr.	

PRODUCT HANDLING

Index Conveyor Counter	
Parts Drop Off Delay	
Tilt Forward Time	

INJECTION

	Set
Exc. Inject. Timer	
Exc. Cycle Timer	
Hold Zone #3 Timer	
Hold Zone #2 Timer	
Hold Zone #1 Timer	
Shooting Pot Pack Time	
Extended Decompression	
Shut-off Noz. Del. Tmr.	

VALVE GATE

Delay Valve Gate O/Tmr.	
Ext. Valve Gate O/Tmr.	

Appendix 7: Order Confirmation Template

Order Confirmation Template

Confirmation of Order

February 9, 2002

ABC Plastics
123 North Street
Jones, NY 90120

Attention: Mr. John Smith, Purchasing Manager

Reference: **Job Number 123456**
Purchase Order Number AB3214

Dear John:
We are pleased to acknowledge your Purchase Order Number PO#B5678, covering the following:

1 only **6 Cavity Widget**

MOLD MATERIALS			
Item	**Material**	**Hardness**	**Finish**
Cavity	420SS/H13	49–51 Rc	
Core Base	420SS/H13	49–51 Rc	
Core Insert	BeCu25	36–41 Rc	
Gate Insert	BeCu25	36–41 Rc	
Stripper Ring	S7/A2	54–56 Rc	
Wear Ring	A2	52–54 Rc	
Lock Ring	A2	52–54 Rc	
Gate Pad	420 SS/H13		
Wear Ring	A2	52–54 Rc	Ground
Shoe	SS	28–32 Rc	Ground

STANDARD SHOE FEATURES

- Tested mold complete with set-up guide
- Mold Technical Data Manual
- Stainless Steel Mold Base
- Mounting Bolts
- Heater Cables and Plugs
- Lifting Bar and Mold Feet
- Nozzle Tip Wrench
- Machine Stroke Limiter
- Multi-pin Connectors
- Safety Blocks
- Rack and Pinion Alignment
- Bearing blocks for Center Section Support

INCLUDED OPTIONAL SHOE FEATURES
• Horizontal Center Section Support • Water and Air Manifolds • Water and Air Connection Kit • Product Blow down system with Venturi Nozzles • Cable Holder • Leader Pin Shields • Customer Specified air and water location
INCLUDED OPTIONAL CONTAINER STACK FEATURES (CAVITY LOCK)
• Solid, Adjustable, Front Mounted Cavity Lock Cores • Fixed Core Cap • Blow-off Tube • Flush rear-mounted cavity inserted in a cavity plate • BeCu replaceable gate insert • Replaceable core wear ring • Moving Cavity
GATING
Outside Center, Hot Tip
EJECTION
Stripper rings
TOTAL PRICE
$60,000
DELIVERY SCHEDULE
Please note that we are not able to formally accept this order or place it in our production schedule until we receive a fully dimensioned, finalized product drawing along with resin shrinkage to be applied to the part. On the basis of your order, we have scheduled the mold into engineering. However, we must have any missing information before March 31, 2002 or your order may lose its place in the engineering schedule, moving it back to the next available opening. Providing that all outstanding information is received by the above date, we can schedule your mold to be ready for delivery to our test room on or before July 31,2002, 16 weeks from receipt of final part drawing. You should allow approximately two weeks for testing, acceptance of samples, and preparation of the mold for shipment. A formal confirmation of the delivery will follow at a later date, once all the information is received. If this order is placed on hold or product design changes are requested after commencement of engineering and/or manufacturing, please understand that we may be faced with considerable expenses; including production rescheduling, additional set-ups, handling, storage, inventory costs, etc. Accordingly, an interruption charge may be required on all or part of the order. Changes to the product design after test, such as stacking, lid fit or wall thickness, will be chargeable to ABC Plastics.

MOLD TESTING
The mold will be tested in our test room on a xxxx machine. Please send 1000 pounds of PP 2345 by July 20, 2002.
TERMS
As agreed, the terms applying to this order are as follows: 40% with order 40% 30 days prior to test 20% net 30 days from date of shipment F.O.B. our plant Payable in U.S. Funds Any local taxes payable by the customer. Since the initial deposit is now due, we have forwarded our Deposit Invoice Number AX2345, in the amount of $24,000 to your accounts payable department and would appreciate receiving your formal Purchase Order and deposit cheque at your earliest convenience. John, we thank you for your valued order and we look forward to working with you on this program. Should you have any questions or comments, please do not hesitate to contact either me or David Jones. Best regards, *Jack Brown* Project Engineer cc: David Jones Accounts Payable

Appendix 8: General Properties of Plastics

General Properties of Plastics – Typical Values

Name	Density (g/cm^3) [a]	Specific volume (cm^3/g)	Specific volume (in^3/lb)	Water absorption [b]
AB S	1.08	0.926	25.7	0.25
ABS 20% GR	1.2	0.833	23.1	0.12
Acetal copolymer	1.41	0.709	19.6	0.22
Acetal 20% GR	1.55	0.645	17.9	0.45
Acetal homopolymer	1.42	0.704	19.5	0.25
Acetal 20% GR	1.55	0.645	17.9	0.25
Acrylic	1.2	0.833	23.9	0.2
Modified acrylic MMA	1.16	0.862	24	0.3
Cellulose acetate	1.22	0.82	22.7	2.6
Cellulose acetate butyrate	1.15	0.87	24.1	0.9-2.2
Cellulose propionate	1.17	0.855	23.7	1.2-3
Ethyl cellulose	1.13	0.885	24.5	1.3
Chlorinated polyether	1.4	0.714	19.8	0.01

Name	Density (g/cm³) [a]	Specific volume (cm³/g)	Specific volume (in³/lb)	Water absorption [b]
CTFE	2.1	0.476	13.2	0
Polyvinylidene fluoride	1.75	0.571	15.8	0.04
FEP	2.12	0.472	13.1	0.01
Ionomer	0.93	1.07	29.8	0.1–1.2
Nylon 6	1.16	0.862	23.9	1.6
Nylon 6 30% GR	1.35	0.741	20.5	1.3
Nylon 6/6	1.14	0.877	24.3	1.5
Nylon 6/6 30% GR	1.35	0.741	20.5	0.7
Nylon 1/1	1.4	0.962	19.8	0.4
Nylon 6/10	1.09	0.917	25.4	0.4
Nylon 6/10 30% GR	1.35	0.741	20.5	0.2
Polysulfone	1.24	0.806	22.3	0.22
Modified phenylene oxide PPO	1.08	0.926	25.6	0.066
PPO 30% GR	1.29	0.775	21.5	0.06
Polyaryl ether	1.14	0.877	24.3	0.25
Polyaryl ether 30% GR	1.37	0.73	20.2	1.8
Polyarylsulfone	1.36	0.735	20.4	0.16
Polycarbonate	1.2	0.833	23.1	0.07
Polycarbonate 30% GR	1.43	0.699	19.4	< 0.015
Polyethylene LD	0.910–0.925	1.10–1.08	30.4–30.0	< 0.01
Polyethylene MD	0.926–0.940	1.08–1.06	30.0–29.4	–
Polyethylene HD	0.941–0.97	1.06–1.03	29.4–28.6	< 0.01
Polyethylene 30% GR	1.18	0.847	23.5	0.04
Polyethylene high-MW	0.94	1.06	29.4	0.09
Ethylene vinyl acetate (EVA)	0.935	1.07	29.6	< 0.01
Polypropylene	0.9	1.11	30.8	< 0.01
Polypropylene 30% GR	1.12	0.893	24.7	< 0.01
Polycopolymer	0.9	1.11	30.8	< 0.01
Polyallomer	0.9	1.11	30.8	0.01
Poly(4-methyl-1-pentene)	0.83	1.2	33.3	0.03–0.1
Polystyrene GP	1.07	0.935	25.9	0.05–0.1
Polystyrene GP 30% GR	1.33	0.752	20.8	0.05–0.6
Polystyrene HI	1.07	0.935	25.9	0.08
PTMT	1.31	0.763	21.1	0.25
Styrene-acrylonitrile (SAN)	1.07	0.935	25.9	–
Styrene-acrylonitrile 20% GR	1.22	0.82	22.7	0.08
Styrene-butadiene	0.93–1.1	1.07–0.909	29.6–25.2	0.2–0.4

Name	Density (g/cm³) [a]	Specific volume (cm³/g)	Specific volume (in³/lb)	Water absorption [b]
Rigid polyvinyl chloride (PVC)	1.4	0.714	19.8	0.07–0.4
Chlorinated PVC	1.5	0.666	18.4	0.02–0.15
PVC-PP	1.4	0.714	19.8	0.07–0.4
Vinylidene chloride	1.7	0.588	16.3	0.1
Urethane elastomer	1.2	0.833	23.1	1.6
Phenol-formaldehyde (no filler)	1.3	0.769	21.3	–

[a] In accordance with ASTM D792. To convert g/cm³ to oz/in³, multiply by 0.578. To convert g/cm³ to lb/in³, multiply by 0.03613.
[b] In accordance with ASTM D570: Percentage increase in weight over 24 h of a sample ⅛" thick.

Appendix 9: Thermal Properties of Plastics

Thermal Properties of Plastics [a]

Name	Deflection temp. (°F)		Deflection temp. (°C)		Thermal conductivity (10^{-4} cal·cm/s·cm²·°C)	Thermal expansion (10^{-5}/°C)
	At 264 psi [b]	At 66 psi	At 18.6 kg/cm²	At 4.64 kg/cm²		
ABS	200	210	93	99	7	8
ABS 20% GR	230	240	110	116	5.5	3
Acetal copolymer	230	316	310	158	6	8.5
Acetal 20% GR	320	325	160	163	–	–
Acetal homopolymer	255	330	124	115	5.5	8.1
Acrylic	180	190	82	88	7	7
Modified acrylic MMA	221	239	105	116	5	5.4
Cellulose acetate	140	150	60	66	6	13
Cellulose acetate butyrate	160	170	71	77	6	14
Cellulose propionate	170	200	77	93	6	14
Ethyl cellulose	140	150	60	66	5.5	13
Chlorinated polyether	210	285	99	141	3.1	8
CTFE	–	258	–	126	5	6
Polyvinylidene fluoride	195	300	91	149	3	8.5
FEP	–	160	–	71	6	9
Ionomer	100	110	38	43	5.8	12
Nylon 6	150	330	66	166	6	8.3

Name	Deflection temp. (°F)		Deflection temp. (°C)		Thermal conductivity (10^{-4} cal·cm/s·cm²·°C)	Thermal expansion (10^{-5}/°C)
	At 264 psi [b]	At 66 psi	At 18.6 kg/cm²	At 4.64 kg/cm²		
Nylon 6 30% GR	390	420	199	216	6	4
Nylon 6/6	185	395	85	202	6	8
Nylon 6/6 30% GR	480	490	249	254	5	2
Nylon 1/1	180	300	82	149	7	15
Nylon 6/10	180	300	82	149	5	9
Modified phenylene oxide PPO	240	260	116	127	5	5.2
Modified phenylene oxide 30% GR	300	310	149	154	4	2.2
Polyaryl ether	300	320	149	160	7	6.5
Polyarylsulfone	525	–	274	–	4.6	4.7
Polycarbonate	275	280	135	138	4.6	6.6
Polyethylene LD	100	112	38	44	8	17
Polyethylene MD	115	142	46	61	9	15
Polyethylene HD	120	165	49	74	11.5	12
Ethylene vinyl acetate (EVA)	93	146	34	63	–	18
Polypropylene	132	215	57	102	2.8	8
Polypropylene 30% GR	280	310	138	154	–	–
Polyropylene copolymer	127	206	53	96	3	8.5
Polyallomer	130	175	54	80	3	9
Poly(4-methyl-1-pentene)	–	239	–	115	4	12
Polystyrene GP	175	195	80	91	3	3
Polytetramethylene terephthalate (PTMT)	122	302	50	150	–	9
Polystyrene GP 30% GR	223	231	106	111	–	4.5
Polystyrene HI	200	207	93	97	2	4-20
Styrene-acrylonitrile (SAN)	205	215	96	102	2.9	3.7
SAN 20% GR	230	240	110	116	–	2.7
Styrene-butadiene	160	165	71	74	3.6	13.5
Rigid polyvinyl chloride (PVC)	218	232	102	111	4	7
Chlorinated PVC	160	–	71	–	3.3	7.2

Name	Deflection temp. (°F)		Deflection temp. (°C)		Thermal conductivity (10^{-4} cal·cm/ s·cm^2·°C)	Thermal expansion (10^{-5}/°C)
	At 264 psi [b]	At 66 psi	At 18.6 kg/cm^2	At 4.64 kg/cm^2		
PVC-PP	140	–	60	–	–	–
Vinylidene chloride	250	260	121	127	3	19
Phenol formaldehyde (no filler)	–	–	–	–	4.5	5

[a] These are typical values and vary with the different grades of material.
[b] In accordance with ASTM D648.

Appendix 10: Mechanical Properties of Plastics

Mechanical Properties of Plastics [a]

Name	Tensile strength (10^3 psi) [b]	Tensile strength (kg/cm^2)	Tensile modulus (10^5 psi) [b]	Elongation at yield (%) [b]	Flexural strength (10^3 psi) [c]	Flexural strength (kg/cm^2)	Flexural modulus (10^5 psi) [c]	Izod impact (ft·lb/in^2 of notch (½ × ½ bar)) [d]	Rockwell hardness (LRM scale, Shore-D scale)
ABS	6-8	422-562	3-4.2	5-60	6-13	422-914	2-4	3-8	R, 75-115
ABS 20% GR	8-19	562-1336	5.9-10	3	16-27	1125-1898	9-13	1-2.4	M, 65-100
Acetal copolymer	10	703	4.1	70	13	914	3.8	1.5	R, 120
Acetal homopolymer	8.8	619	5.2	25	14	984	4.1	1.4	M, 80
Acetal 20% GR	11	773	10	3	15	1055	8.8	0.8	M, 80
Acrylic	7-11	492-773	3.5-4.5	2-7	13-14	914-984	4.3	0.4	M, 100
Modified acrylic MMA	10	703	4.6	3	12	844	5	0.3	M, 105
Cellulose acetate	2-9	141-633	0.7-4	6-10	2-16	141-1125	–	0.4-5	R, 35-125

Name	Tensile strength (10^3 psi) [b]	Tensile strength (kg/cm^2)	Tensile modulus (10^5 psi) [b]	Elongation at yield (%) [b]	Flexural strength (10^3 psi) [c]	Flexural strength (kg/cm^2)	Flexural modulus (10^5 psi) [c]	Izod impact (ft·lb/in^2 of notch (½ × ½ bar)) [d]	Rockwell hardness (LRM scale, Shore-D scale)
Cellulose acetate butyrate	2.5–7	176–492	0.5–2	40–90	2–9	141–633	–	0.5–12	R, 30–116
Cellulose proprionate	2–8	146–562	0.6–2	30–100	3–11	211–773	1.5–3.4	1–6	R, 10–122
Ethyl cellulose	2–8	141–562	1–3	5–40	4–12	281–844	–	2–8	R, 50–116
Chlorinated potyether	6	422	1.6	60–160	5	352	1.3	0.4	R, 100
CTFE	4.5–6	316–422	1.5–3	80–230	9.3	654	2.2	2.6	R, 85
Polyvinyl-idene fluoride	5.5–7.5	387–527	1.2	100–300	7	492	2	4	R, 80
FEP	3	211	0.5	250–330	–	–	1	–	R, 25
Ionomer	2–4	141–281	–	150–450	–	–	–	6–11	R, 55
Nylon 6	7–12	492–844	1.1–3	100–400	7–15	492–1055	1	2	R, 103–120
Nylon 6 30% GR	21	1476	10	2.2	26	1828	10	3	–
Nylon 6/6	9–12	633–844	1.8–4	60–300	–	–	1.8	1–2	R, 108–120
Nylon 6/6 30% GR	20–28	1406–1968	10–12.8	5–10	30	2109	10–12	2	R, 110–120
Nylon 1/1	8	562	1.9	300	–	–	1.4	1.8	R, 110–120
Nylon 6/10	8.5	597	3.8	85–300	–	–	1.6	1.2	R, 110–120
Nylon 6/10 30% GR	18	1125	9	2.5	23	1617	8	1.6	R, 120
Polysulfone	10.2	717	3.6	50–100	15.4	1083	8–18	1.3	M, 100
Modified phenylene oxide (PPO)	8–10	562–703	3.7	20–30	13	914	16	1.8	R, 118
PPO 30% GR	17	1195	9–12	6	20	1406	20	1.5	L, 108

Name	Tensile strength (10^3 psi) [b]	Tensile strength (kg/cm^2)	Tensile modulus (10^5 psi) [b]	Elongation at yield (%) [b]	Flexural strength (10^3 psi) [c]	Flexural strength (kg/cm^2)	Flexural modulus (10^5 psi) [c]	Izod impact (ft·lb/in^2 of notch (½ × ½ bar)) [d]	Rockwell hardness (LRM scale, Shore-D scale)
Polyaryl ether	7.5	527	3.2	25-90	11	773	–	8	R, 117
Polyaryl sulfone	13	914	3.7	13	17	1195	–	5	M, 110
Polycarbonate	8-9.5	562-668	3.3	100-130	12.5	879	3.4	12-18	M, 70
Polycarbonate 20% GR	9	633	8	110	13.5	949	3.4	16	M, 91
Polyethylene LD	0.6-2.3	42-162	0.14-0.38	90-800	–	–	0.08-0.6	–	D, 45
Polyethylene MD	1.2-3.5	84-246	0.25-0.55	50-600	4.8-7	337-492	0.6-1.2	0.5-15	D, 55
Polyethylene HD	3.1-5.5	218-387	0.6-1.8	30-1000	–	–	1-2.6	0.5-20	D, 65
Polyethylene 30% GR	8	562	8	3	9.5	668	7	2	R, 70
Polyethylene high-MW	2.5-3.5	176-246	0.2-1.1	300-500	–	–	1.3	–	–
Ethylene vinyl acetate (EVA)	1.5-2.8	105-197	0.02-1	750-900	–	–	0.01-0.2	–	D, 20
Polypropylene	4.3-5.5	302-387	1.6-2.3	200-700	6-8	422-562	–	0.5-2	R, 85-110
PTMT	8.2	576		250	12	844	3.1	–	R, 117
Polypropylene 30% GR	7.3	513	0.8	3	10	703	7.6	2	R, 110
Polypropylene copolymer	2.5-4.5	176-316	1-1.7	200-700	5-7	352-492	–	1-20	R, 50-95
Polyallomer	34	211-281	–	400-500	–	–	–	–	R, 50-85

Name	Tensile strength (10^3 psi) [b]	Tensile strength (kg/cm^2)	Tensile modulus (10^5 psi) [b]	Elongation at yield (%) [b]	Flexural strength (10^3 psi) [c]	Flexural strength (kg/cm^2)	Flexural modulus (10^5 psi) [c]	Izod impact (ft·lb/in^2 of notch (½ × ½ bar)) [d]	Rockwell hardness (LRM scale, Shore-D scale)
Poly(4-methyl-1-pentene)	4	281	2.1	15	–	–	–	0.8	L, 70
Polystyrene GP	5–12	352-844	4–6	1–2.5	9–14	633-984	4.5	0.3	M, 72
Polystyrene GP 30% GR	9–15	633-1055	12	1.3	11–20	773-1406	8–10	4	M, 80
Polystyrene HI	1.5–7	105-492	1.5–5	2–80	5–12	352-844	3.8	0.5–11	R, 50–100
Styrene-acrylo-nitrile (SAN)	9–12	633-844	4–6	3	14–19	984-1336	6	0.4	M, 85
SAN 20% GR	9–20	633-1406	4–14	3	22–26	1547-1969	8–18	4	M, 100
Styrene-butadiene	0.6–3	42–211	0.01-0.5	300-1000	–	–	0.4–1.5	–	–
Rigid poly-vinyl chlo-ride (PVC)	5–9	352-633	3.5–6	100-1000	10–16	703-1125	–	0.4–20	D, 70
Chlorinated PVC	7.5–9	527-633	3.6–5	5–65	15–17	1055-1195	3.4–6	1–6	R, 120
PVC-PP	5–8	352-562	4	100-140	11–15	773-1055	3.5–5	0.4–32	R, 110
Vinylidene chloride	3–5	211-352	0.6	10–250	4–6	281-422	–	1	M, 60
Urethane elastomer	6.5–8	457-562	–	600	–	–	–	–	D, 52
Phenol-formal-dehyde (no filler)	7	492	7–10	1.5	9	633	10	0.25	M, 125

[a] These are typical values and vary with the different grades of material.
[b] In accordance with ASTM D638.
[c] In accordance with ASTM D790.
[d] In accordance with ASTM D256.

Appendix 11: Thermal and Processing Properties of Plastics

Properties of Plastics (Injection Molding Grade) – Part 1

Name		ASTM D792		ASTM D1003		ASTM D648	
		Solid density (g/cm³)	Melt density (g/cm³)	Transparent (TP) Translucent (TL) Opaque (O)	Crystallinity	Shrinkage (%)	Drying required (time (h) @ temp. (°C))
ASA	Acrylonitrile-styrene-acrylate	1.07	–	O, TP	Amorphous	0.4–0.6	–
ABS	Acrylonitrile-butadine-styrene	1–1.2	0.9	O, TP	Amorphous	0.5–0.8	2–3 @ 88–77
CA	Cellulose acetate	1.3	–	TP	–	0.3–0.7	2–3 @ 60–70
CAB	Cellulose acetate butyrate	1.2	–	TP	–	0.3–0.7	2–3 @ 60–70
CAP	Cellulose propionate	1.2	–	TP	–	0.3–0.7	2–3 @ 60–70
EVA	Ethylene-vinyl acetate	0.94 (avg.)	0.8	TL, TP	–	–	–
HDPE	High-density polyethylene	0.93–0.97	0.81	O, TP	Crystalline	1.2–2.2	None
HIPS	High-impact polystyrene	1–1.1	0.9	O, TP	Amorphous	0.4–0.7	None
Ionomer	Ionomer	0.94–0.96	–	TP	Semi-crystalline	–	None (8 @ 60)
LCP	Liquid-crystal polymer	1.5–1.7	–	TP, TL	Liquid-crystal	0.2–0.8	4–8 @ 150
LDPE	Low-density polyethylene	0.91–0.93	–	TP, TL	Crystalline	1.5–3	None
LLDPE	Linear low-density polyethylene	0.90–0.92	–	TP,TL	Crystalline	1.5–5	None
MDPE	Medium-density poly-ethylene	0.93–0.95	–	O, TP	Crystalline	1.2–2.2	None
PA66	Polyamide (Nylon 6/6)	1.1–1.4[a]	1.0–1.2	TL	Crystalline	1–2.2	3–4 @ 71–60

Name		ASTM D792		ASTM D1003		ASTM D648	
		Solid density (g/cm³)	Melt density (g/cm³)	Transparent (TP) Translucent (TL) Opaque (O)	Crystallinity	Shrinkage (%)	Drying required (time (h) @ temp. (°C))
PA6	Polyamide (Nylon 6)	1.1–1.4	1.2–1.3	TL	Crystalline	0.8–2.1	3–4 @ 71–60
PAEK	Polyaryl ether ketone	1.3–1.5[a]	–	O	Semi-crystalline	0.1–0.6	2–4 @ 175–150
PAI	Polyamide-imide	1.4–1.6	–	O	Amorphous	0.1–0.2	5 @ 175
PAR	Polyarylate	1.2	0.77	O	Amorphous/crystalline	0.6–0.9	3–4 @ 175–150
PAS	Polyaryl sulfone	1.36 (avg.)	–	O	Amorphous	0.6	3–4 @ 175–150
PBT	Polybutylene terephthalate	1.3–1.6[a]	1.1–1.2	TL	Semi-crystalline	1.5 –2 (0.5[a])	3 @ 150
PC	Polycarbonate	1.2–1.5[a]	1.1	TP	Amorphous	0.4[a]–0.7	3–4 @ 120
PEI	Polyether imide	1.3–1.5[a]	–	O	Amorphous	0.5–0.7 (0.2[a])	4–6 @ 150–130
PEK	Polyether ketone	1.3–1.5[a]	–	O	Semi-crystalline	0.1–0.6	2–4 @ 175–150
PEEK	Polyether ether ketone	1.3–1.4[a]	–	O	30% Crystal-line	0.1–1.4	2–4 @ 175–150
PES	Polyether sulfones	1.2–1.6[a]	–	TP	Amorphous	0.3[a]–0.6	3–4 @ 150–135
PET	Polyethylene terephthalate	1.4–1.7[a]	1.2	O, TP	Crystalline/amorphous	0.2–2	2–4 @ 180–160

Name		ASTM D792		ASTM D1003		ASTM D648	
		Solid density (g/cm^3)	Melt density (g/cm^3)	Transparent (TP) Translucent (TL) Opaque (O)	Crystallinity	Shrinkage (%)	Drying required (time (h) @ temp. (°C))
PETG	PET (co-polymer)	1.2–1.3	–	O, TP	Amorphous	–	4 @ 66
PMMA	Polymethyl methacrylate (Acrylic)	1.1–1.2	1.0–1.1	O	Amorphous	0.4–0.8	2–3 @ 77–93
Polyester	Thermoplastic polyester	1.3	–	O	Amorphous	1.5–1.8	2–4 @ 120–77
POM	Polyoxymethylene (Acetal)	1.4–1.6[a]	1.2	O	80% Crystalline	0.8[a]–2	None
PP	Polypropylene	0.9–0.92	0.77	TP, TL	Semi-crystalline	1–2.5	None
PPO	Polyphenylene oxide	1.1–1.2[a]	0.96–1.0	O	Amorphous	0.2[a]–0.7	2–4 @ 115–107
PPS	Polyphenylene sulfide	1.3–1.9[a]	1.5 (avg.)	O	65% Crystalline	0.1–0.5	3–4 @ 150
PS	Polystyrene	1–1.1	0.9	O, TP	Amorphous	0.4–0.7	None
PSU	Polysulfone	1.2–1.6[a]	–	O, TP	Amorphous	0.7	5 @ 120
PVC	Polyvinyl chloride	1.2–1.4	1.2–1.3	TP	Amorphous	0.2–0.5	None (2 @ 60)
SAN	Styrene-acrylonitrile	1.1–1.3	0.9	O, TP	Amorphous	0.3–0.7	2–3 @ 80–70
TPUR	Thermoplastic polyurethane	1.2–1.3	–	O,TL	Amorphous	0.8–2	2–3 @ 110–104
UHMWPE	Ultra-high molecular weight PE	0.93–0.94	0.9	O, TP	Crystalline	1.2–2.2	None

[a] Filled property.

Properties of Plastics (Injection Molding Grade) – Part 2

Name	Melt temp. (°C)	Mold temp. (°C)	Injection speed	Sensitive to shear rate or temp.?	Sensitive to residence time?	Max. L/t ratio (1 mm thick)	No-flow temp. (°C)	Tonnage required (tons/in^2)	Typical injection pressure (MPa)
ASA	230–260	40–90	–	–	–	–	–	–	–
ABS	195–240	38–93	Slow, even	No	Yes	30–150:1	135–150	4–6	120–140
CA	180–230	50–80	Various	–	Yes	–	130–170	–	69–180
CAB	180–230	50–80	Various	–	Yes	–	130–170	–	69–180
CAP	180–230	50–80	Various	–	Yes	–	130–170	–	69–180
EVA	200–210	20–60	–	–	–	–	100	–	–
HDPE	200–280	10–70	Fast	No	No	250:1	120–130	2–2.5	100–200
HIPS	180–280	10–85	Fast	No	No	200–250:1	130	2–4	100–200
Ionomer	210–260	5–50	Slow-moderate	Yes	No	–	100	2–3	70–110
LCP	400–430	240–280	Medium	Yes	No	200–300:1	370	2–3	80–120
LDPE	170–240	10–50	Fast	No	No	275:1	100–110	1.5–2	80–120
LLDPE	170–200	10–50	Fast	No	No	250–300:1	90	1.5–2	80–120
MDPE	190–260	10–70	Fast	No	No	250:1	110	2–2.5	100–150
PA66	270–320	20–100	Fast	Yes	Yes	140–340:1	240–260	4–5	100–150
PA6	260–310	20–100	Fast	Yes	Yes	140–340:1	190–200	3–5	90–150
PAEK	370–400	160–220	Slow	No	No	170–200:1	370	3–4 (6[a])	160–200
PAI	305–370	205–220	Fast	Yes	Mild	140–340:1	–	5–6	160–200
PAR	260–380	65 to150	Moderate	No	Yes	30–100:1	–	3–5 (4–6[a])	138–200
PAS	340–370	120–155	Moderate	No	No	140–170:1	–	–	138–200
PBT	240–270	50–100	Moderate-fast	Yes	Yes	160–200:1	220–250	3–5	80–120

Name	Melt temp. (°C)	Mold temp. (°C)	Injection speed	Sensitive to shear rate or temp.?	Sensitive to residence time?	Max. L/t ratio (1 mm thick)	No-flow temp. (°C)	Tonnage required (tons/in^2)	Typical injection pressure (MPa)
PC	270–325	80–110	Fast	No	Yes	30–100:1	195	3–5 (4–6[a])	138–200
PEI	340–425	65–175	Medium-fast	No	Yes	–	230	–	100–160
PEK	370–400	160–220	Slow	No	No	170–200:1	370	3–4 (6[a])	160–200
PEEK	370–400	160–220	Fast	No	No	200:1	370	2–4 (6[a])	160–200
PES	340–380	140–160	Fast	Yes	Yes	60–120:1	300	5–10	160–200
PET	260–300	7–80	Slow, even	Yes	Yes	80–200:1	240–250	2–6	70–160
PETG	190–275	20–30	Slow-fast	Yes	Yes	80–200:1	–	–	80–100
PMMA	200–260	38–60	Various	Yes	No	130–150:1	160–170	2.5–3	100–200
Poly-ester	230–260	40–100	Slow, even	Yes	Yes	80–200:1	70–80	3–5	80–100
POM	180–230	80–100	Medium-fast	Yes	Yes	100–200:1	160–170	3.5–5	100–170
PP	230–275	15–65	Fast	No	No	200–300:1	170–180	2–3	100–130
PPO	250–315	82–110	Fast	Yes	Yes	–	150–200	2.5–5[a]	120–180
PPS	300–360	40–150	Slow	Yes	No	150:1	260–280	2–3	50–140
PS	180–280	10–85	Fast	No	No	200–250:1	130–160	2–4	100–200
PSU	310–400	100–170	Slow	No	No	–	–	–	–
PVC	180–204	20–40	Slow-moderate	Yes	Yes	100:1	120	–	70–140
SAN	220–270	5–60	–	Yes	Yes	–	130–170	–	35–140
TPUR	190–220	30–65	–	Yes	–	–	120	0.5–2	70–140
UHM-WPE	200–280	10–70	Fast	No	No	150:1	120	3–4	130–200

[a] Filled property.

Appendix 12: Mold Preventative Maintenance Checklist

Mold maintenance is critical to maintaining the life of the tool. Although each tool will require slightly different maintenance based on its complexity, the following guide can be used to guide the customer on basic maintenance requirements.

Daily: (5 minutes downtime)

- Clean tapers and molding surface.
- Make sure the leader pins and other greasing points are greased.
- Remove any jammed plastic parts.
- Check for damage or unusual wear.
- Check that all air functions are working correctly.
- Check to make sure all actions on the mold run smoothly.
- Check for water leaks (replace O-rings if necessary).

Weekly: (5 minutes downtime, 30 minutes total)

- Grease leader pins.
- Measure samples for core shift and part weight, re-align cores if necessary.

6 months to 1 year: (4 hours downtime)

- Check nozzle tips and band heaters, replace if necessary.
- Check the size of the gate holes. Replace gate inserts if eroded.
- Blue-in tapers and shut-offs to make sure preload and vent exists. Re-dress tapers and re-cut vents if necessary.
- Check for hot runner leakage.
- Check for wear on the leader pins or leader pin bushings.
- Check the sprue bushing and machine nozzle interface for leakage.
- Re-torque all bolts; replace any worn O-rings.

Appendix 13: Surface Finishes

The following table can be used to determine some typical molding surface finishes.

Micron	Micro inch	SPI #	Symbol	Manufacturing method	Applications
0-3	1	A-1	0.025 ∇	Lapping, 8000 diamond	Petri dishes
5	2		0.05 ∇ and buff	900 stone, 8000 diamond polish	Test tubes
8	3		0.08 ∇ and buff	900 stone, 3000 diamond polish	Styrene tumblers
10	4	A-2	0.1 ∇ and buff	600 drawstone, 3000 diamond polish	Opaque Shiny surface
10-20	4-8	C-3	0.1-0.2 ∇	900 drawstone	Opaque surface (stack shoulders)
		D-2	0.1-0.2 ∇ and v. hone	900 drawstone, vapor hone	Matte surface (cores, gate pad)
15-20	4-6		0.15-0.2 ∇ v. hone, lt. buff	900 drawstone, vapor hone, light buff	Semi-opaque ("clear" cores)
10-15	4-6		0.1-0.15 ∇ v. hone and buff	900 drawstone, vapor hone, buff	Semi-opaque surface (cavities)
20	8		0.2 ∇ and buff	600 drawstone, 3000 diamond buff	Preform inserts
20-30	8-12	4	0.2-0.3 ∇	400-600 stone	Technical parts
		D-3	0.2-0.3 ∇ and sand blast	400-600 stone, sandblast	Texture finish
40	16		0.4 ∇	220-300 stone	

Notes:

- A tool steel finish kit is available through the D-M-E Company as an official SPE/SPI kit.
- The newest SPE/SPI codes were achieved using the specified media (diamond paste) as a guide and hand operations.
- Frosting (e.g., on tumbler sidewalls) should never be specified with any depth since it is a surface finish only (i.e., 21-24 Charmille).

Index

B

C

D

E

I

N

O

P

Q

R

S

U

V

W

Y

Z

www.ingramcontent.com/pod-product-compliance
Lightning Source LLC
LaVergne TN
LVHW082053011026
849162LV00020B/1630